Periodic Table of the Elements

Atomic number → 1
Symbol of element → H
Atomic weight → 1.0080
Name of element → Hydrogen

Inert Gas
Gas
Liquid

Light Metals

Transitional Elements

Heavy Metals

Nonmetals

IA	IIA
1 **H** 1.0080 Hydrogen	
3 **Li** 6.939 Lithium	4 **Be** 9.012 Beryllium
11 **Na** 22.990 Sodium	12 **Mg** 24.31 Magnesium
19 **K** 39.102 Potassium	20 **Ca** 40.08 Calcium
37 **Rb** 85.47 Rubidium	38 **Sr** 87.62 Strontium
55 **Cs** 132.91 Cesium	56 **Ba** 137.34 Barium
87 **Fr** (223) Francium	88 **Ra** 226.05 Radium

III B	IV B	V B	VI B	VII B		VIII B		I B	II B
21 **Sc** 44.96 Scandium	22 **Ti** 47.90 Titanium	23 **V** 50.94 Vanadium	24 **Cr** 52.00 Chromium	25 **Mn** 54.94 Manganese	26 **Fe** 55.85 Iron	27 **Co** 58.93 Cobalt	28 **Ni** 58.71 Nickel	29 **Cu** 63.54 Copper	30 **Zn** 65.37 Zinc
39 **Y** 88.91 Yttrium	40 **Zr** 91.22 Zirconium	41 **Nb** 92.91 Niobium	42 **Mo** 95.94 Molybdenum	43 **Tc** (99) Technetium	44 **Ru** 101.1 Ruthenium	45 **Rh** 102.90 Rhodium	46 **Pd** 106.4 Palladium	47 **Ag** 107.870 Silver	48 **Cd** 112.40 Cadmium
57 TO 71	72 **Hf** 178.49 Hafnium	73 **Ta** 180.95 Tantalum	74 **W** 183.85 Tungsten	75 **Re** 186.2 Rhenium	76 **Os** 190.2 Osmium	77 **Ir** 192.2 Iridium	78 **Pt** 195.09 Platinum	79 **Au** 197.0 Gold	80 **Hg** 200.59 Mercury
89 TO 103									

III A	IV A	V A	VI A	VII A
5 **B** 10.81 Boron	6 **C** 12.011 Carbon	7 **N** 14.007 Nitrogen	8 **O** 15.994 Oxygen	9 **F** 18.998 Fluorine
13 **Al** 26.98 Aluminum	14 **Si** 28.09 Silicon	15 **P** 30.974 Phosphorus	16 **S** 32.064 Sulphur	17 **Cl** 35.453 Chlorine
31 **Ga** 69.72 Gallium	32 **Ge** 72.59 Germanium	33 **As** 74.92 Arsenic	34 **Se** 78.96 Selenium	35 **Br** 79.909 Bromine
49 **In** 114.82 Indium	50 **Sn** 118.69 Tin	51 **Sb** 121.75 Antimony	52 **Te** 127.60 Tellurium	53 **I** 126.90 Iodine
81 **Tl** 204.37 Thallium	82 **Pb** 207.19 Lead	83 **Bi** 208.98 Bismuth	84 **Po** (210) Polonium	85 **At** (210) Astatine

Lanthanide series

57 **LA** 138.91 Lanthanum	58 **Ce** 140.12 Cerium	59 **Pr** 140.91 Praseodymium	60 **Nd** 144.24 Neodymium	61 **Pm** (147) Promethium	62 **Sm** 150.35 Samarium	63 **Eu** 157.25 Europium	64 **Gd** 158.92 Gadolinium	65 **Tb** 158.92 Terbium	66 **Dy** 162.50 Dysprosium	67 **Ho** 164.93 Holmium	68 **Er** 167.26 Erbium	69 **Tm** 168.93 Thullium	70 **Yb** 173.04 Ytterbium	71 **Lu** 174.97 Lutetium

Actinide series

89 **Ac** (227) Actinium	90 **Th** 232.04 Thorium	91 **Pa** (231) Protactinium	92 **U** 238.03 Uranium	93 **Np** (237) Neptunium	94 **Pu** (242) Plutonium	95 **Am** (243) Americium	96 **Cm** (247) Curium	97 **Bk** (249) Berkelium	98 **Cf** (251) Californium	99 **Es** (254) Einsteinium	100 **Fm** (253) Fermium	101 **Md** (256) Mendelevium	102 **No** (256) Nobelium	103 **Lw** (257) Lawrencium

1
6
7

INTRODUCTORY
OCEANOGRAPHY

INTRODUCTORY OCEANOGRAPHY

Ninth Edition

Harold V. Thurman
Mt. San Antonio College, Emeritus

Elizabeth A. Burton

Environmental Issues in Oceanography by:

Daniel C. Abel, *Coastal Carolina University*
Robert L. McConnell, *Mary Washington College*
Eric Koepfler, *Coastal Carolina University*

Prentice Hall
Upper Saddle River, New Jersey 07458

Library of Congress Cataloging-in-Publication Data

Thurman, Harold V.
  Introductory oceanography.—9th ed. / Harold V. Thurman, Elizabeth A. Burton.
    p. cm.
  Includes bibliographical references and index.
  ISBN 0-13-857061-2
  1. Oceanography.   I. Burton, Elizabeth A.   II. Title.

GC16.T45 2000
551.46—dc21   00–040056

Acquisitions Editor: *Patrick Lynch*
Editorial Director: *Paul F. Corey*
A. V. P. of Production and Manufacturing: *David W. Riccardi*
Executive Managing Editor: *Kathleen Schiaparelli*
Creative Director: *Paul Belfanti*
Art Director: *Joseph Sengotta*
Cover Design: *Joseph Sengotta*
Manufacturing Manager: *Trudy Pisciotti*
Buyer: *Michael Bell*
Art Manager: *Gus Vibal*
Art Editor: *Karen Branson*
Photo Editor: *Beth Boyd*
Photo Research: *Tobi Zausner*
Marketing Manager: *Christine Henry*
Assistant Editor: *Amanda Griffith*
Editorial Assitant: *Sean Hale*
Cover Photo: *Chad Ehlers/Tony Stone Worldwide*

 © 2001 by Prentice-Hall, Inc.
Pearson Education
Upper Saddle River, New Jersey 07458

Printed in the United States of America.

10   9   8   7   6   5   4   3   2   1

ISBN 0-13-857061-2

Prentice-Hall International (UK) Limited, *London*
Prentice-Hall of Australia Pty. Limited, *Sydney*
Prentice-Hall Canada, Inc., *Toronto*
Prentice-Hall Hispanoamericana, S.A., *Mexico*
Prentice-Hall of India Private Limited, *New Delhi*
Prentice-Hall of Japan, Inc., *Tokyo*
Simon & Schuster Asia Pte. Limited, *Singapore*
Editora Prentice-Hall do Brasil, Ltda., *Rio de Janeiro*

*To Annie with all my love for
50 years of loving support*

*To Greg, Clara, and Christopher,
and my parents*

Brief Contents

Contents

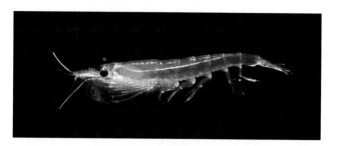

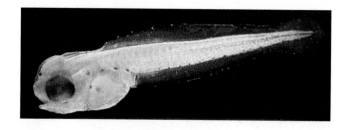

 Animals of the Benthic Environment   446

 Exploitation and Pollution of Marine Resources   479

Preface

To the Professor

Welcome to the Ninth Edition The first thing you need to know about this edition is that it includes the efforts of a new co-author with disparate views and expertise than that of the original author. As a result, material has been added, condense, and refined throughout the text. In particular, it is the only major introductory oceanography book that gives students exposure to something many of you reviewing the text have asked for: coverage of chemical oceanography as opposed to seawater chemistry. This coverage, in Chapter 7, includes discussion of gases in the sea, acidity and alkalinity (moved from Chapter 2), organic chemistry of seawater, the sources and cycling of salts in the ocean, geochemical cycling (moved from Chapter 15), and the distributions of tracer ions that oceanographers use to decipher the critical geologic, biologic, and physical processes that occur within the ocean. For students who need a little refresher, Appendix I Logarithms and Scientific Notation, and Appendix V A Chemical Background, should be helpful.

Features of the book that you have told us you like have been retained—rigorous coverage of tides and coverage of biological processes. Note that unlike many new texts and continuing in the tradition of previous editions, we have chosen to use text space to emphasize basic fundamentals of the science, presented in what we think is a most understandable way. Although they make interesting reading, we have rejected the current trend of including peripheral "sea stories" which do not help the student learn the science.

This edition, like its predecessors, emphasizes fundamental marine processes, the basic issues of interactions among humans, the ocean, and the atmosphere. A chapter on marine pollution has been included since the second edition, but the emphasis was on chronic and acute coastal water problems related to point sources of petroleum, sewage, dredge spoils, persistent chemicals, and heavy metals. The coverage was broadened in the sixth edition to address the human activities that are clearly causing a broad range of changes in the Earth environment—ozone depletion and greenhouse gas buildup. This edition emphasizes the link between ocean resource exploitation and marine pollution by combining the two into Chapter 18.

We have added twelve "Issues in Oceanography" critical thinking projects to increase the focus on environmental issues. These are written by three writers with a special interest in critical thinking regarding environmental issues. They are Dan Abel and Eric Koepfler of Coastal Carolina University and Robert McConnell of Mary Washington College. These features will help students better understand the ecological significance of problems that are picked up by the news media because of their negative biological impact. They also make clear the need to better understand the physical marine conditions, modifications of which are often related to the biological manifestations. Titles of some of the features are "Oil in the Marine Environment," "Global Warming and Sea Level Rise," "Illegal Imigration: Ballast Water and Exotic Species," "Greenhouse Gases, Global CO_2, Emissions, and Global Warming," and "Sharks." The authors discuss the Issues later in this Preface.

It is always a high priority to update the topics covered in the text and improve the illustrations. One of the characteristics of this text over the years, that has been appreciated by users and commented on frequently, is the fact the material is up to date. Every effort has been made to include the latest information available on topics covered throughout the text. Numerous new and interesting photographs, clear and easy to read graphs and line drawings have been added. Line drawings that were not as clear as they should have been in the 8th edition have been corrected.

The sequential organizational has not changed except for gathering together the discussions of seawater chemistry from chapters 2, 5 and 14 from the last edition and combining them with the new material presented in Chapter 7, "The Chemistry of Seawater." The present organization is as follows:

A. Two **introduction** chapters on the (1) "History of Ocean Exploration and Ocean Science" and (2) "Beginnings of the Universe, Earth and Life."

B. Three chapters on **marine geology**: (3) "Global Plate Tectonics," (4) "Marine Provinces," and (5) "Marine Sediments."

C. Two chapters on the **nature of seawater**: (6) "The Physical Properties of Water and Seawater." and (7) "The Chemistry of Seawater."

D. Four chapters on **physical oceanography**: (8) "Air-Sea Interaction," (9) "Ocean Circulation," (10) "Waves," and (11) "Tides."

E. Two chapters on the **margins of the ocean**: (12) "The Shore," and (13) Coastal Waters and Marginal Seas."

F. Four chapters on **marine biology and ecology**: (14) "The Marine Environment," (15) "Biological Production and Energy Transfer," (16) "Animals of the Pelagic Environment," and (17) "Animals of the Benthic Environment."

G. One chapter on **human interaction with the oceans**: (18) "Exploitation and Pollution of Marine Resources."

We hope you will find the changes have improved the book by making it more interesting to your students and providing coverage of topics that were missing from previous editions.

The New Instructional Package

For the Instructor:

- A **Transparency Set**, with 125 full-color acetates of select illustrations from the text. (Note: All figures are available on the Digital Image Gallery CD-ROM.)
- A **Slide Set**, with 100 full-color images of photographs primarily from the text. (Note: All photographs for which we could get permission are available on the Digital Image Gallery CD-ROM.)
- The **Digital Image Gallery CD-ROM**, which includes all illustrations, tables, and photographs (for which we could get permission) from the text in high-resolution, 16-bit JPEG files. The JPEG files are organized by chapter and can be easily imported into your lecture presentation software (e.g., PowerPoint). The CD-ROM comes with a comprehensive reference list of all figures and their captions.
- A new **Instructor's Manual**, authored by Dr. Darlene Richardson, of Indiana University of Pennsylvania, which includes a new **Test Item File** authored by Dr. Marsha Bollinger, of Winthrop University. Also included are instructor resources and test questions for the "Issues in Oceanography," written by Dan Abel, Robert McConnell and Eric Koepfler, the Issues authors.
- **PH Custom Test** (available in both Windows and Macintosh formats), with which you can easily create and tailor exams to your own needs: produce multiple versions of your exam, with answer keys; select questions by number, type, level of difficulty, or random distribution; export your exam to a rich text file for import into your word processing program; and more. The software comes with a comprehensive reference guide, which includes the toll-free technical support line.

For the Student:

- The **Companion Web Site** at www.prenhall.com/thurman, written by Al Trujillo, of Palomar College and co-author of *Essentials of Oceanography 6/e,* and Molly Trecker. Organized by chapter for easy integration into your course, the Companion Web Site offers numerous review exercises (from which students can get detailed, automatic feedback which they can then transmit to you via e-mail) and continuously updated, annotated Internet links for further exploration. It is here that the students will find the complete "Issues in Oceanography" projects.
- The **New York Times Themes of the Times— Oceanography** is a unique newspaper-format supplement featuring recent articles about oceanography culled from the pages of the New York Times. This supplement, available for wrapping with your text at no extra charge, encourages students to make connections between the classroom and the world around them.

From the Authors of "Issues in Oceanography"

Included in this text are a series of twelve "Issues in Oceanography," brief projects which use pressing environmental issues as a means to develop your students' critical thinking skills in a deliberate and structured way. You'll find these Issues modules *between* various chapters because, by their nature, they require students to integrate the topics from across chapters to measure, analyze, and evaluate the issue using the discipline and method of a scientist. (Contained in the text are brief, two-page introductions. The complete projects are found on the Companion Web Site at www.prenhall.com/thurman.)

These Issues, like our previous *Environmental Issues* text, grew out of our desire to encourage in the students a more "active" style of learning and discourage what we see as a passive and unhealthy dependency on the faculty person as the "expert." One of our major objectives is to foster math literacy (*numeracy*) among today's students; not necessarily arcane math, but the kind of math needed to properly quantify environmental issues. This lack of math skills often leaves students unprepared to deal with the complexity of today's environmental issues. We provide detailed introductions for each of these topics, as well as a detailed answer key that shows the step-by-step calculations used to determine the answers. The Issues projects are designed to take from one to three hours to complete. The Instructor's Manual contains suggestions for employing the Issues in your course, as well as test questions based on their content.

The Issues and critical thinking questions have been designed to be provocative. Although we have made the content as factual as possible, we admit to having strong convictions about these issues. Convictions are not, however, biases. Our views as scientists are subject to change as evidence supporting our convictions changes. Indeed this aspect can be turned to a major advantage. Ask your students to look for examples of bias in the questions, and then discuss with them the difference in science between "bias" and "conviction." No doubt it will prove a fruitful activity, and may lead students into research (perhaps to "prove us wrong"), which is the essence of progress in the search for scientific truth.

To the Student

Welcome to the study of the oceans, which we hope you will approach with a sense of adventure. With the rapid expansion of knowledge concerning Earth, life on Earth, and the entire cosmic realm, it is becoming difficult to find aspects of our existence that present much mystery. However, the oceans are full of mystery. The maps you see in Figures 4-7 and 4-8 can mislead you. Although they depict much detail, only a fraction of one percent of the ocean floor has been investigated first-hand – observed by the human eye. We also have only a rudimentary knowledge of the inhabitants of the seas, especially the deep sea. However, enough is known to give us enticing clues to encourage us to further investigate all of the unknowns of the marine world.

It is our aim, and that of your professor, to give you an interesting introduction to the marine processes and phenomena that we do understand as a foundation for further learning about the oceans. We have attempted to organize the introduction of important concepts and processes in a logical and interesting way. We begin with a history of human interaction with the oceans and marine science followed by coverage of the origins of the universe, Earth, and life on Earth. We then examine the tectonic processes that have created the ocean basins and the sediments that have accumulated on the ocean floor. The physical and chemical properties of the water filling the ocean basins is then covered, leading us into the study of physical oceanography that deals with the motions of currents, waves and tides - and then consider the combined effects of geology and physical processes on the shore and coastal waters. Our attention is then focused on marine ecology and coverage of life forms inhabiting the waters of the ocean and the ocean floor. Finally, we consider the changes human activities have rendered on the oceans.

Occasionally, "boxes" that focus on recent ecological events of significance or the lives of pioneering marine scientists have been inserted, and a dozen issue-related features bring to your attention specific conditions of interest or concern that have resulted from human interaction with the marine environment.

Some of the issues you read will make a case for the fact that we should be concerned about the health of the marine ecosystem—especially with regard to the shore and coastal waters. Rapidly expanding technology gives us the power to do more harm—or if used properly—greatly reduce the harm done by human interactions with the oceans. It is hoped that your experience in this course will give you a great appreciation for the uniqueness of our oceans as an aesthetic and practical necessity for human well-being. If so, the oceans shall have a healthy future.

Harold V. Thurman
Elizabeth A. Burton

From the Authors of "Issues in Oceanography"

As environmental scientists, we care deeply about environmental issues, and we feel that you, as a responsible citizen who will have to make increasingly difficult choices in the years ahead, need to be concerned about them as well. We hope you will find the "Issues in Oceanography" contained in this text to be a provocative introduction to a number of these issues, including many that you may have never even thought about. (Contained in the text are brief, two-page introductions. The complete projects are found on the Companion Web Site at www.prenhall.com/thurman.) These are real-life issues, not hypothetical ones, and you need certain basic skills to fully understand them:

- You must be familiar with the units of the metric system.
- You must be able to use a few mathematical formulas to quantify the issues you will be debating, and you must be able to carry out the calculations accurately.
- You must rigorously assess your thinking and apply certain critical thinking skills and techniques when discussing the implications of your calculations.

We understand that many students have some "math anxiety," so we use a step-by-step method to take you through many of the calculations in the Issues. Math proficiency is one of the important skills necessary for fully understanding environmental issues, and without these skills, your only option is to make choices on the basis of which "expert" you believe. But becoming educated is much more than simply acquiring skills. Therefore, we have another two fundamental objectives: to provide you with the knowledge and intellectual standards necessary to apply critical thinking to environmental studies, and to foster your ability to critically evaluate issues.

As our national and global population grows and changes and our relationships with other nations and peoples evolve, environmental issues will become increasingly complicated. We hope that you will be challenged by the issues discussed in this text and that you will research them and become an "expert" on the topics yourself. In fact, if we may be allowed a hidden agenda, this is it.

Daniel C. Abel
Robert L. McConnell
Eric Koepfler

About the Authors

Harold V. Thurman

Hal Thurman retired in May 1994, after 24 years of teaching in the Earth Sciences Department of Mt. San Antonio College in Walnut, California. Interest in geology led to a B.S. degree from Oklahoma A&M University, followed by seven years working as a petroleum geologist, mainly in the Gulf of Mexico. Here his interest in oceans developed, and he earned a M.A. degree from California State University at Los Angeles, then joined the Earth Sciences faculty at Mt. San Antonio College. Hal Thurman has also co-authored a marine biology textbook, written articles on the Pacific, Atlantic, Indian, and Arctic oceans for the 1994 edition of *World Book Encyclopedia*, and served as a consultant on the National Geographic publication, *Realms of the Sea*. He still enjoys going to sea on vacations with his wife Iantha.

Elizabeth A. Burton

Elizabeth A. Burton received a bachelor's degree in Geology from Bryn Mawr College, a master's in Marine Geology from the University of Miami, and a Ph.D. in Earth and Planetary Sciences from Washington University in St. Louis. Growing up in California, she developed an early affinity for the oceans, but her interest in scientific study of the oceans did not take root until a college field trip to study the reefs and associated marine environments of the Bahamas and Florida Keys. She pursued research in this area for both her graduate degrees and since 1988, while serving as a professor of Geology at Northern Illinois University.

Research and a quest for fun have taken her on diving trips throughout the Caribbean but have also made her eyewitness to the devastation that anthropogenic changes in ocean chemistry, including sewage outfall, soil runoff, and perhaps the rise in atmospheric CO_2, have caused to shallow marine environments. Dr. Burton's teaching and research evolved away from an emphasis on natural geologic phenomena toward integration and understanding of these anthropogenic changes. She and her students have since investigated a diversity of environmental issues, including landfill gas emissions, fertilizer runoff, and the integrity of radioactive waste disposal sites. Bringing these real-world applications into the classroom has made teaching science both more fun and more relevant.

In a quest to bridge the gap further, Dr. Burton took a hiatus from teaching in 1998 in order to pursue environmental and geochemical research at Chevron. She has published over 30 scientific papers, serves as an Associate Editor for the Geological Society of America Bulletin, is a licensed professional geologist, and a member of the American Association of Petroleum Geologists and the American Chemical Society.

Coastal Population Growth: A Global Ecosystem at Risk

- How does human population growth threaten coastal areas?
- How can we measure these threats?
- What can we do about them?
- Who is responsible for solving the problem?

Introduction

Chances are, you live within 100 km (62 mi) of the beach. According to the World Resources Institute, at least 60 % of the planet's human population live that close to the coastline. Coastal areas have the fastest growing populations as well. Not surprisingly, over half the world's coastlines are at significant risk from development-related activities. Some of these activities are:

- Conversion of tropical mangrove communities to fish and shrimp farms
- Expansion of many coastal cities which are in the direct path of hurricanes, monsoons, and tropical storms
- Pollution from Western-style industrial agriculture, including industrial meat production
- Releases of untreated or partially treated human sewage (between one-third and two-thirds of the human waste generated in developing countries is not even collected).

Sewage and Animal Waste Impacts

Sewage and animal waste may contain a number of harmful substances, including pathogens such as bacteria and viruses.

In addition to pathogens, untreated or partially treated sewage contains high concentrations of nitrogen and phosphorus compounds, which are essential nutrients for growth of plants and algae. Excess nitrogen and phosphorus can stimulate rapid growth of algae, which can form a thick layer at the water surface. This surface layer of algae may thrive in the short term, but it can block sunlight from lower layers, whose plants eventually die. Moreover, rapidly-growing algae can quickly deplete nutrients, resulting in rapid die-offs. In both cases, oxygen-using bacteria begin to decompose the dead algae. This can cause hypoxia (dangerously low oxygen levels) or anoxia (absence of oxygen).

Sewage can also contain heavy metals such as copper and zinc, which are toxic to plants and algae (these metals are often fed to pigs as supplements in amounts that exceed the animals' ability to metabolize them, so they end up in the animals' waste).

Coastal cities are reaching sizes unprecedented in human history. Here are but a few examples: Sao Paolo, Brazil, 17 million; Bombay, India, 16.5 million; Lagos, Nigeria, 11 million (and growing at a rate of 5.7% per year); Dhaka, Bangladesh, 8 million (growth rate nearly 6% per year). Most of these cities make little attempt to treat the sewage produced by their residents. On our web site, we evaluate the impact of the growth of some of these cities.

Case Study: Bangladesh

Bangladesh (Figure 1), a country about the size of Illinois, Wisconsin, or Florida, lies on the northern shore of the Indian Ocean. Dr. Selina Begum of the U.K.'s Bradford University wrote[1]:

"Bangladesh is a delta of the Ganges, Brahmaputra and Meghna Rivers. Tributaries and distributaries [branches] of the river system cover all of the country. The rivers rise in the Himalayas and drain a catchment area of about 1.5 million km^2 [577,000 mi^2], only 7.5 percent of which lies in Bangladesh [Our emphasis. Stop and explain why the responsibility for Bangladesh's destiny may lie with others far away.]."

"The country is prone to meteorological and geologic natural disasters, due to its geographic location, climate, variable topography, dynamic river system and exposure to the sea. The steady increase in population continuously increases the potential for natural disaster."

Bangladesh thus makes an ideal case illustrating the impacts of coastal growth.

Background

Bangladesh was originally part of Pakistan after the Indian subcontinent gained independence from Britain in 1947. In 1970, Bangladesh seceded from the rest of Pakistan. It had a 1999 population of 127 million on a land area of 144,000 km^2 (55,000 mi^2).

Bangladesh is vulnerable to catastrophic flooding from river discharges as well as from tropical storms. In addition, it is defenseless against sea level rise from global climate change since more than half the country lies at an elevation of less than 8 meters (26 ft) above sea level. Moreover, 17 million people live on land that is less than 1 m (3.3 ft) above sea level. As a result, about 25–30% of the country is flooded each year, which can increase to 60–80% during major floods.

Flooding in Bangladesh can result from some combination of the following: (1) excess rainfall and snowmelt in the catchment basin, especially in the foothills of the Himalayas (an area where many of the Earth's rainfall records were set); (2) simultaneous peak flooding in all three main rivers; and (3) high tides in the Bay of Bengal, which can dam runoff from rivers and cause the water to "pond up" and overtop its banks.

In addition, changes in land use in the far reaches of the catchment area can result in profound changes in the flood potential of Bangladesh's rivers. For example, deforestation of the Himalayan

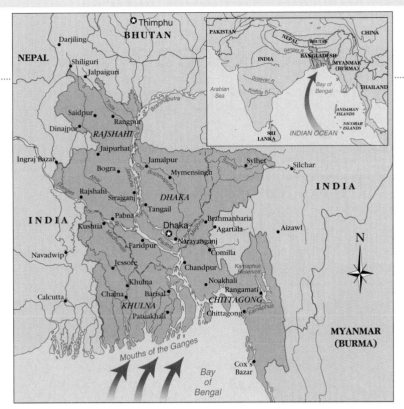

Figure 2

Bangladesh countryside following the 1991 cyclone that killed approximately 145,000 people. *(Gamma Photograph/Bartholomew)*

Figure 1

Map of Bangladesh Showing Generalized Tracks of Twentieth Century Cyclones.

foothills for agriculture, fuelwood, or human habitation can make the hills more prone to flash flooding. Then, during the torrential monsoon rains, mudslides can dump huge volumes of sediment into the rivers. This sediment can literally pile up in the river channels, reducing channel depth and thereby the channel's capacity to transport water. Result: more water sloshing over the rivers' banks during floods, and more disastrous floods in Bangladesh. Another result: the silt can smother offshore reefs when it finally reaches the ocean. Geologists estimate that the drainage system dumps at least 635 million metric tons (1.4×10^{12} lb) of sediment a year into the Indian Ocean, which is four times the amount of the present Mississippi River.

Some researchers feel that Western-style development is contributing to Bangladesh's increased vulnerability to natural disasters. Farhad Mazhar, director of the Bangladeshi research and development group Ubinin, concludes that the basic problem is Western-style development. He observes that "to be modern" the Bangladeshis built roads, which blocked floodwaters. These and other "modern improvements" increased flooding from cyclones and have also increased the damage and loss of life. Mazhar concludes,"… if you compare the history of the old cyclones and flood waters, you'll see… the misery has increased with this kind of 'engineering' solution to water[2]."

Truly catastrophic floods can result when high tides coincide with tropical storms. In the early 1990's, such a combination took over 125,000 lives (Figure 2). About one tenth of all Earth's tropical cyclones occur in the Bay of Bengal (Figure 1), and about 40% of the global deaths from storm surges occur in Bangladesh.

Storm surges result when cyclones with winds that can exceed 250 km/hour (155 mi/hr), move onshore, and push a massive wall of seawater onshore with them. Surges in excess of

5 meters (16.4 ft) above high tide are not uncommon, so you can imagine the impact such a storm can have on a country in which half the land is within 8 meters (26.2 ft) of sea level!

Computer models forecast an average rise of sea level of 66 cm (26 in) by 2100 due to global warming, but sea level could rise anywhere from 20–140 cm (8–55 in). There is, however, a complicating factor: The deltas on which Bangladesh is built are subsiding, meaning the area will be more susceptible to the flooding that will accompany sea level rise.

Subsidence is a natural feature of deltas. The water in a large river may reach the sea in one or more branches, or distributaries. As the distributaries evolve, they may meander more and more, thus lengthening the river's route to the sea. During floods, the distributary can overtop its banks and cut a new and shorter route to the sea. This can result in a rapid shift in where sedimentation occurs at the delta's mouth. An existing delta can sometimes be quickly abandoned, and without the constant addition of new sediment, it subsides.

Now go to our web site to continue the analysis.

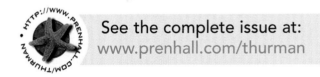

See the complete issue at:
www.prenhall.com/thurman

[1] Begum, S. 1996. Climate Change and Sea-Level Rise: Its Implications in the Coastal Zone of Bangladesh. *In* Global Change, Local Challenge: HDP (Human Dimensions of Global Environmental Change Programme) Third Scientific Symposium 20–22 September 1995. Geneva.

[2] Mazhar, Farhad."What We Want From Kyoto". OTN explores Global Warming. Available: http://www.megastories.com/ warming/bangla/kyoto.htm. 6 April 1998.

1 The History of Ocean Exploration and Ocean Science

The DSRV (deep sea research vehicle) Alvin returns to its home base on Lulu during Project Famous in 1974. Lulu was towed to a location near the Azores above the Mid-Atlantic Ridge by the R/V (research vessel) Knorr.   (Courtesy of Harold V. Thurman)

With each passing year, human interaction with the oceans intensifies. Fishing efforts, the search for offshore petroleum and minerals, and the introduction of pollutants all increase. The importance of marine resources to the economy and the impact of human activities on marine ecosystems make it imperative for us to increase our knowledge of the oceans. About 70 percent of Earth's surface is covered by oceans, but humans are basically limited to life on land. If this page represented the area of the ocean bottom, the period at the end of this sentence would be all that has been seen by human eyes.

Early Evidence of Ocean Travel

Primitive peoples living near the coast undoubtedly discovered the bountiful food resources that could be harvested from the ocean. They may have experimented with building primitive boats or rafts to improve their fishing capabilities, but there is no evidence remaining of such vessels. Archaeological evidence for long-distance travel by sea is comparatively recent. The first seafarers to cross the open ocean were most likely the immigrants who settled the scattered islands of the southern Pacific Ocean a few thousand years ago, but where they came from has been the subject of great controversy.

The Pacific Islands

The habitable islands in the southwestern Pacific have been divided into three groups based on cultural differences among the peoples living there: Micronesia (small islands), Melanesia (black islands) and Polynesia (many islands) (Figure 1-1). The first human inhabitants somehow had to navigate over hundreds of miles of open ocean accurately and safely enough to make landfall, perhaps by traveling in outrigger canoes, seagoing double canoes, or balsa rafts.

Most anthropologists believe the migration came out of Asia, but alternative hypotheses of the origins of the Polynesians began to surface in the 1920s when it was shown that peoples from North and South America may have had vessels capable of making the journey. Similarities in the cultures and physical appearance of Polynesians and the coastal peoples of southwestern Canada led some researchers to suggest that these people had traveled via Hawaii into Polynesia and New Zealand. In support of this hypothesis, in 1928, Captain J. C. Voss sailed a 38-foot canoe, the *Tilikum,* from Vancouver Island, Canada, to Tongareva in central Polynesia and on to New Zealand.

In the 1940s, Thor Heyerdahl, suggested that there was compelling evidence for a seafaring pre-Incan culture in Peru, including sixteenth century records of Spanish explorers who wrote of very seaworthy balsa rafts used by coastal Peruvians and of tales the Incas told of having visited islands in the Pacific. Also, equatorial currents flow from South America to Polynesia. In 1947, Heyerdahl successfully sailed the *Kon Tiki,* a balsa raft constructed like those described by the Spanish, from Callao, Peru, to the island of Raroia in Polynesia.

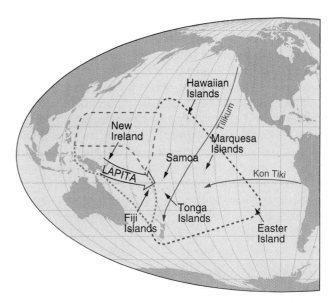

Figure 1-1   Peopling of the Pacific Islands
Micronesia (brown), Melanesia (red), and Polynesia (green) were settled progressively by the Lapita people. They were living in New Ireland about 5000–4000 B.C.; by 1500 B.C. they immigrated to Fiji, Tonga, and Samoa. By 500 B.C., settlement of the rest of Polynesia had begun. Voyages of *Tilikum* and *Kon Tiki* also are shown.

The controversy created by these voyages prompted archaeologists to do more research in the area. Pottery shard evidence indicated that the Lapita people of New Ireland (see Figure 1-1) had migrated through Melanesia via Fiji, settling Tonga in western Polynesia by 1140 B.C., the Marquesas by about 30 B.C., and Easter Island by 400 A.D. They also had migrated to the Hawaiian Islands by 500 A.D. and to New Zealand by 900 A.D. Comparative DNA studies also showed a strong genetic relationship between the peoples of Easter Island and Polynesia, but none between these groups and the natives of coastal North America.

The Atlantic and Mediterranean

Over time, the focus of power in the Mediterranean region shifted from the Phoenicians on the northern coast of Africa to the Greeks, and then to the Romans. By 2000 B.C., the Phoenicians had explored the Mediterranean Sea and the Red Sea, and had ventured into the Indian Ocean. They circumnavigated Africa in 590 B.C. and may also have sailed as far north along the European coast as the British Isles. By 250 B.C., the Greeks had reached Great Britain and Ireland. A few centuries later, the Romans had conquered the indigenous peoples of the British Isles and built extensive settlements and fortifications there.

The Scientific Discoveries of the Greeks and Romans

Herodotus mapped what the ancient Greeks knew of the world in 450 B.C. (Figure 1-2). His map shows the Mediterranean Sea surrounded by three continents that, in turn, are surrounded by a continuous ocean. By 325 B.C., however, the Greek astronomer-geographer **Pytheas** sailed northward to Iceland or Norway. He had worked out a simple method for determining latitude by finding the angle between two lines of sight—one on the northern horizon and the other on the North Star (see Appendix II). **Eratosthenes** (276–192 B.C.), was the first to accurately determine the Earth's circumference. His value for a meridional (pole-to-pole) circumference was 40,000 km (24,840 mi), which compares well with the figure of 40,032 km (24,860 mi) accepted today (Figure 1-3).

The Roman **Strabo** (63 B.C.–A.D. 24), by observing volcanic activity in what is now Italy, concluded that the land periodically sank and rose, causing the sea to invade and recede from the continents. He also recognized the role of streams in eroding the continents and depositing sediment in the seas. In approximately A.D. 150, **Ptolemy** produced a map of the world that represented Roman knowledge at the time. He introduced lines of latitude and longitude on his map and included the continents of Europe, Asia, and Africa (Figure 1-4 and Appendix II). This map was later a great incentive to

Figure 1-2   The World According to Herodotus (450 B.C.)

(From the Challenger Report, Great Britain, 1895).

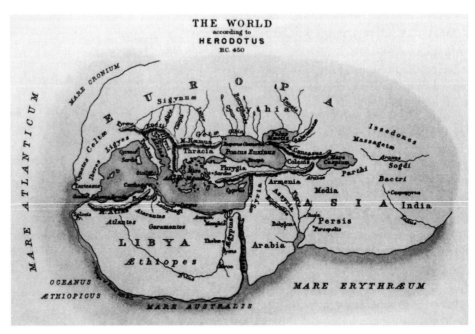

Figure 1-3   Eratosthenes' Determination of the Circumference of Earth (circa 200 B.C.)

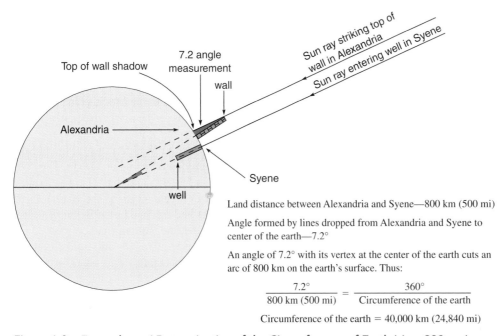

Land distance between Alexandria and Syene—800 km (500 mi)

Angle formed by lines dropped from Alexandria and Syene to center of the earth—7.2°

An angle of 7.2° with its vertex at the center of the earth cuts an arc of 800 km on the earth's surface. Thus:

$$\frac{7.2°}{800 \text{ km (500 mi)}} = \frac{360°}{\text{Circumference of the earth}}$$

Circumference of the earth = 40,000 km (24,840 mi)

Eratosthenes determined quite precisely the meridional circumference of Earth by using the simple geometric principle that when two parallel lines are cut by a third straight line, corresponding equal angles are formed.

Figure 1-4   Ptolemy's Atlas of the Known World (circa 150 A.D.)

Ptolemy's atlas depicts a reasonable map of the Mediterranean region, Europe including the British Isles, northern Africa, and western and central Asia. The lack of accurate longitudinal measurements gives the map a stretched look in the east–west direction. Also, Ptolemy used a value for Earth's circumference of 29,000 km, an error that later caused Columbus to believe he had encountered Asia rather than the Americas.

explorers keen to seek out the unknown lands beyond the known seas.

The Middle Ages and the Ming Dynasty

After the fall of the Roman Empire and the rise of Christianity, the achievements of the Phoenicians, Greeks, and Romans were either suppressed or lost by the peoples who then dominated Europe. The Arabs who controlled northern Africa and Spain retained this knowledge, enabling them to become the dominant navigators in the Mediterranean region and to trade across the Indian Ocean.

In the rest of southern and eastern Europe, the concept of world geography degenerated considerably. **Cosmas**, a sixth-century navigator, drew a map of the Earth as a flat rectangle 10,000 by 20,000 km (6200 by 12,400 mi) (Figure 1-5). Scientific inquiry counter to religious teachings was actively suppressed.

The Vikings

The Vikings of Scandinavia were active explorers. Late in the ninth century, a warming global climate freed the North Atlantic of ice, allowing the Vikings to sail westward. They conquered Iceland, looting the Christian settlements and driving out the population. From there, Eric the Red, who was banished from Iceland in 981, sailed west and discovered Greenland. He may also have traveled to Baffin Island. He returned to Iceland to collect colonists and supplies for a new settlement in Greenland. In 995, **Leif Eriksson**, the son of Eric the Red, set off in search of timber for the Greenland colony and made landfall in North America. He called this land **Vinland**, now a part of Newfoundland, Canada, and established a

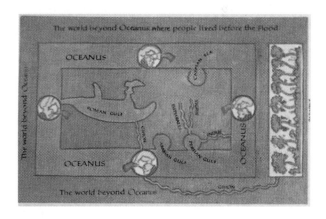

Figure 1-5   Cosmas' Map of the Known World (6th century A.D.)

temporary settlement. Figure 1-6 shows the paths of Viking exploration. By the beginning of the thirteenth century, a cooling trend in climate covered the North Atlantic once again with ice throughout most of the year, isolating the Greenland colonies and ending the Viking expansion.

Exploration by the Chinese

From 1405 to 1433, the Chinese Ming Dynasty sent large convoys of ships out on what were apparently diplomatic missions. Seven voyages were made, involving a total of 317 ships and 37,000 men. These ships were larger and more technologically advanced than anything in Europe. They had up to five masts, were divided into separate watertight compartments in the hull, and were equipped with magnetic compasses and navigational charts. The Ming voyages reached as far as Africa.

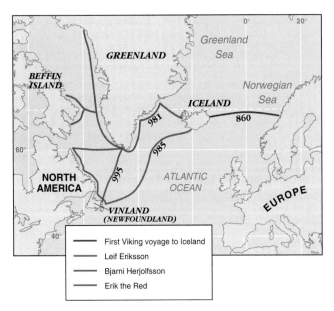

Figure 1-6   The Vikings Reach the New World

Viking influence spread from Europe, reaching Iceland by 860 (gray). Erik the Red sailed to Greenland from Iceland in 981 (red). Bjarni Herjolfsson's trip from Iceland, along the Newfoundland coast, and back to Greenland occurred in 985 (blue). In 995, Leif Eriksson voyaged from Greenland to Newfoundland, set up a camp at Tickle Cove in Trinity Bay, and named the land Vinland after the grapes he found there (green).

European Exploration and the Renaissance

In 1410, Ptolemy's map was republished in Europe. Other records of Greek and Roman accomplishments probably also became accessible to the Christian Europeans about this time because Christians gained control of Spain from the Arabs, and the knowledge stored in Spanish libraries.

The Portuguese, under the leadership of Prince Henry the Navigator (1392–1460), led a renewed drive to explore outside Europe. Portuguese ships, captained by **Bartholomew Diaz**, reached the southern tip of Africa in 1488. **Vasco da Gama** sailed completely around Africa to reach India in 1498, thus opening up a new trade route to Asia. Trade between Europe and Asia had been carried out by hazardous overland caravans.

In an attempt to find a way to reach India from the west, **Christopher Columbus** set sail from Spain with 88 men and three ships on August 3, 1492 (Figure 1-7). He first glimpsed the "New World" on October 12, 1492. Columbus had reached the Bahama Islands—probably landing at San Salvador Island. In 1513, Balboa, from a mountaintop in the Isthmus of Panama, was the first European to see the ocean that lay west of the new land that Columbus had encountered, the "Pacific".

The culmination of this period of discovery is marked by the first known circumnavigation of the globe. This Spanish expedition of 5 ships and 230 sailors, captained by **Ferdinand Magellan**, sailed south through the Atlantic

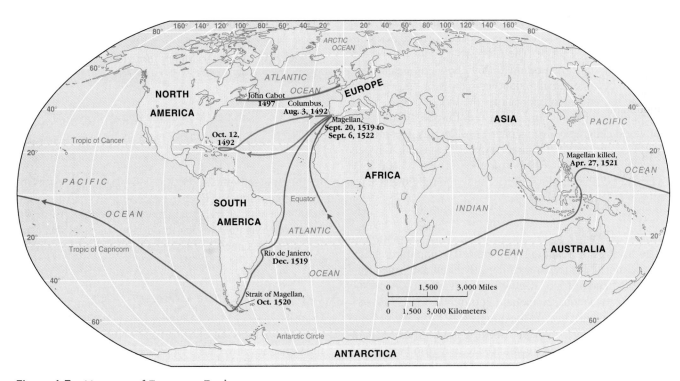

Figure 1-7   Voyages of European Explorers

Christopher Columbus' voyage in 1492 opened the New World to Europe (red), and Ferdinand Magellan circumnavigated the globe in 1519 (blue). John Cabot's exploration took him from England to the New World in 1497 (green).

Ocean, cut through the southern islands of South America, and then westward across the Pacific Ocean to the Philippines. In addition to natural hazards, Magellan had to evade marauding Portuguese and English ships vying for control of the seas and lucrative trade routes. Although Magellan was killed by natives in the Philippines, his servant, **Sebastian del Caño**, led the one remaining ship, *Victoria,* and the 18 surviving men back to Spain.

England also was exploring the Western Hemisphere. **John Cabot** landed somewhere on the northeast coast of North America in 1497 (Figure 1-7). Sir Francis Drake, knighted by Queen Elizabeth but viewed as a notorious pirate by Spain and Portugal, circumnavigated the world from 1577–1580, looting settlements and treasure-laden Spanish and Portuguese galleons as he went.

Spain maintained dominance of the seas until 1588, when it attempted to punish England for openly supporting piracy. The Spanish Armada, consisting of over 100 fighting and support vessels, set sail to invade England. As the fleet lay anchored in Dunkirk harbor the night before the invasion, the English sent frigates they had set on fire into the formation. Most of the remaining Spanish ships, fleeing into the treacherous waters of the North Sea and Atlantic, were wrecked by storms along the Scottish and Irish coasts. From that time until early in the twentieth century, the British dominated the seas.

Technological breakthroughs during this period helped to make long ocean voyages less risky. The magnetic compass, imported from China, showed up in Europe in the thirteenth century. The three-masted ship was invented and became the first ship in Europe large enough to carry the men and supplies needed for a long voyage.

Scientific understanding also was improving. In the last part of the fifteenth century, Leonardo da Vinci recorded his observations of currents and waves in the Mediterranean, and noted that seashells found in rocks in the mountains indicated that the land must once have been under the sea. In 1569, Gerhardus Mercator constructed the Mercator projection for mapping the world in a way that aided navigation. Robert Boyle published his experiments and observations on seawater chemistry in 1674. Copernicus and Galileo did their work on the motions of planetary bodies and the structure of the solar system. Isaac Newton and Leibnitz invented calculus. Newton established the fundamental laws of physics and laid the foundation for modern tidal theory in 1687. Linnaeus created the biological classification system that we still use today. Coriolis developed the equations to describe pathways of motion on the surface of a rotating sphere. The Coriolis effect (see Chapter 8) bears his name.

The Beginning of Ocean Science

The Voyages of Captain James Cook

To help maintain their marine superiority, the English navy undertook many voyages of discovery with scientific objectives. **Captain James Cook** (1728–1779), between 1768 and 1779, undertook three voyages and made the first accurate maps of many regions in the oceans with the aid of a new invention, the marine chronometer (Figure 1-8). This chronometer, invented by **John Harrison**, was the first timepiece capable of keeping accurate time aboard a ship at sea, a necessity for accurately determining longitude. Cook mapped the South Georgia Islands, South Sandwich Islands, and Hawaiian Islands. He pioneered sampling of subsurface temperatures, measuring winds and currents, taking depth soundings, and collecting data on coral reefs.

Benjamin Franklin and the Gulf Stream

Benjamin Franklin, deputy postmaster general for the American colonies, determined that mail ships coming from Europe by a northerly route took two weeks longer to reach the colonies than ships that came by a longer, more southerly route. His nephew Timothy Folger, a Nantucket whaling captain, had intimate knowledge of the current patterns of the northern Atlantic. He was particularly familiar with the warm Gulf Stream because whales were commonly found near it. He drew a chart of the Gulf Stream for his uncle, who had it printed in 1777 and distributed to the captains of the mail ships. They shortened their outbound and inbound voyages by avoiding this eastward flowing current on the trip to the colonies, and by sailing within the Gulf Stream on the trip back to England (Figure 1-9).

The "Father of Oceanography," Matthew Fontaine Maury

Matthew Fontaine Maury (1806–1873) (Figure 1-10), a career officer in the U.S. Navy, was placed in charge of the Depot of Naval Charts and Instruments after suffering an injury early in his career. Maury turned this job into a great opportunity for contributing to safe navigation and the international standardization of methods for marine data collection and reporting. Maury organized the first International Meteorological Conference held in Brussels in 1853 to establish uniform methods. In 1855, Maury published a summarized version of these data in the first textbook on oceanography, *The Physical Geography of the Sea*. Maury is often referred to as the "father of oceanography."

Darwin, Coral Reefs, and Biological Evolution

Charles Darwin (1809–1882), a naturalist for the H.M.S. Beagle, circumnavigated the globe from 1831 to 1836, collecting massive amounts of data from the southern oceans and oceanic islands (Figure 1-11). From his observations of the various forms of coral reefs and volcanic islands in the Pacific, Darwin was able to correctly deduce how the process of subsidence of oceanic islands produces a progressive change in reef form, from fringing reef to atoll. Darwin's observations of birds and other organisms on isolated oceanic islands, primarily the Galápagos Islands in the eastern Pacific, formed the foundation of his evolutionary theories, published in 1859 as the book, *On the Origin of Species.*

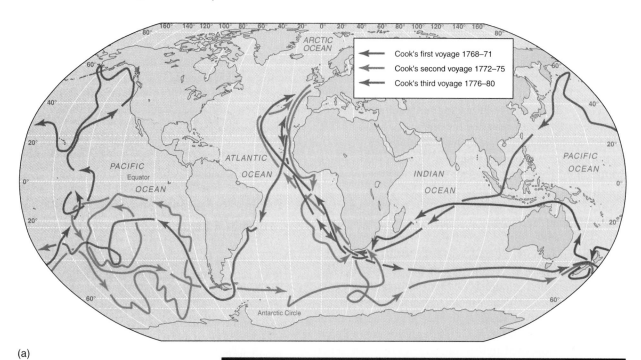

(a)

Figure 1-8 **(a) Voyages of Captain James Cook and (b) John Harrison's chronometer**

Routes taken by Cook on his three scientific voyages. John Harrison's chronometer allowed Cook to determine longitude accurately during his voyages. *(National Maritime Museum Picture Library, London, England)*

(b)

The Rosses, Edward Forbes, and Life in the Deep Sea

The first major controversy in marine science involved studies done by **Sir John Ross** (1777–1856), his nephew **Sir James Clark Ross** (1800–1862), and Edward Forbes (1815–1854), all eminent British scientists. John Ross took bottom measurements and collected animals from the ocean floor at depths of 1.8 km (1.1 mi) in Baffin Bay, Canada, during 1817 and 1818. James Clark Ross later undertook a similar mission in the Antarctic, recovering bottom dwellers at depths up to 7 km (4.3 mi) (Figure 1-12). Sir James observed that the Antarctic animals were the

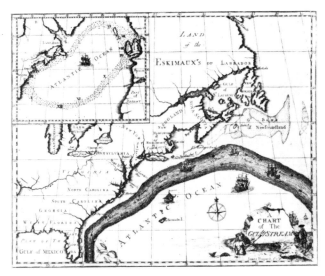

Figure 1-9   Chart of the Gulf Stream Compiled by Benjamin Franklin in 1777

Franklin's charts of the Gulf Stream originated from a drawing made by his nephew, a whaling captain.   *(Courtesy U.S. Navy.)*

Figure 1-10   Lieutenant Matthew F. Maury (1806–1873)

Photograph of an engraving by Lemuel S. Punderson after a daguerreotype, autographed and inscribed by Lt. Maury. *(Released Naval Historical Center photograph)*

same species that his uncle had found in the North Atlantic. Because these animals were intolerant of warmer waters, he concluded that there must be a cold-water corridor keeping them in genetic communication. This discovery led him to correctly conclude that the waters throughout the deep Atlantic must be uniformly cold.

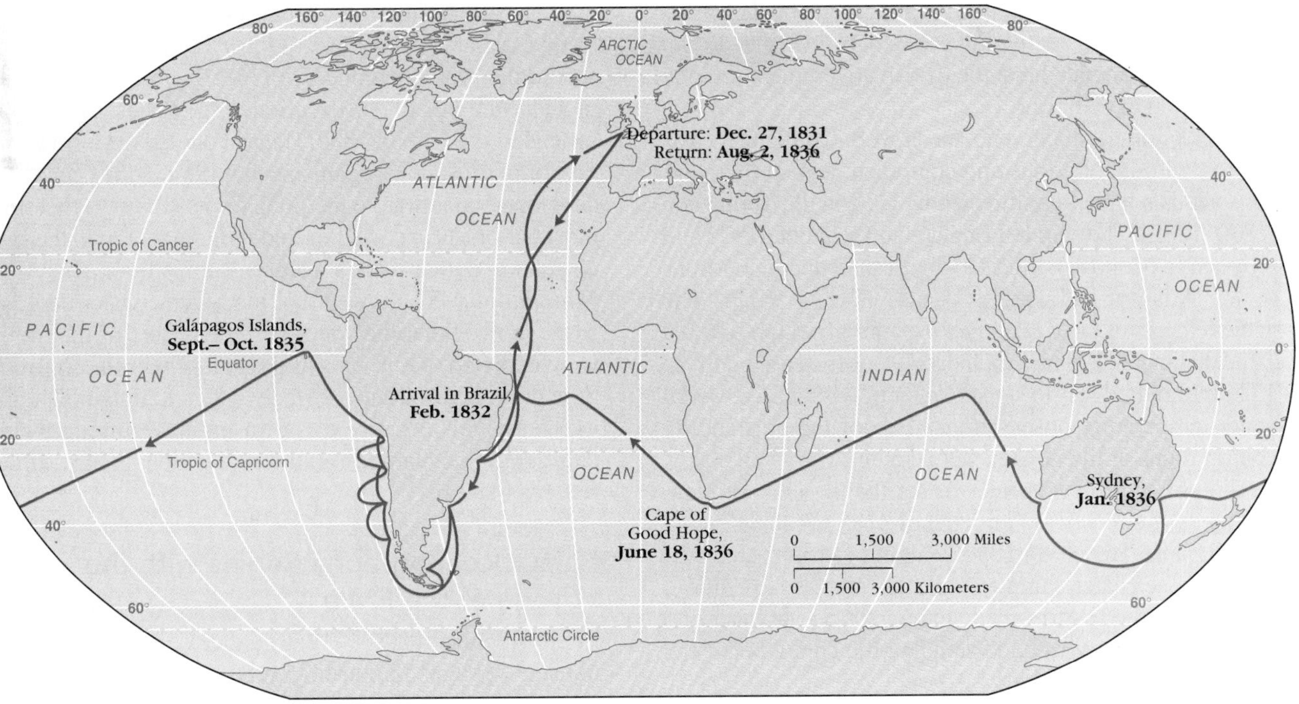

Figure 1-11   Darwin's journey aboard the *H.M.S. Beagle*

Sailing as naturalist aboard *H.M.S. Beagle*, Charles Darwin gathered the evidence that enabled him to develop his theory of biological evolution through natural selection.

Figure 1-12   The Ross Expedition to Antarctica

Members of the Sir James Clark Ross expedition to Antarctica make landfall on Possession Island. The penguins seem to be at least as curious about the visitors as the members of the expedition are about them.   *(Reprinted by permission of The Bettmann Archive)*

About the same time, **Edward Forbes** was studying the vertical distribution of life in the oceans. He concluded that plant life is limited to the sunlit zone near the surface and that consequently, animal concentrations are also greater near the surface, decreasing with increasing depth. Forbes thought that only a small trace of life, if any, could exist in the deepest waters. This view contrasted with the findings of the Rosses and caused a split in the British scientific community for decades.

The Ocean as a Laboratory: The Challenger Expedition

The Ross-Forbes controversy over the distribution of life in the ocean helped to stimulate public and political support in Britain for the first large-scale voyage made specifically for scientific study of the oceans. In 1871, the Royal Society recommended that the British government provide funds for an expedition.

A 2306 ton naval corvette, the **H.M.S. Challenger**, was refitted with laboratories and equipment and staffed with six scientists who were directed by C. Wyville Thompson (1830–1882). The ship left England in December 1872 and returned in May 1876 after having traversed large portions of the Atlantic and Pacific oceans (Figure 1-13). The achievements of the *Challenger* expedition, which covered 127,500 km (79,178 mi), included 492 deep-sea soundings, 133 bottom dredges, 151 open-water trawls, and 263 serial water-temperature observations. Two of the outstanding contributions of the voyage were the discovery and classification of 4717 new species of marine life and the measurement of a record water depth of 8185 m (26,847 ft) in the Marianas Trench. In 1884, **William Dittmar** analyzed 77 of the seawater samples collected by the expedition. These were the first detailed analyses of ocean water and contributed greatly to understanding the components and constancy of ocean salinity.

Alexander Agassiz

Alexander Agassiz (1835–1910), the son of the great Swiss scientist Louis Agassiz, was both a respected scientist and a multimillionaire benefactor to oceanography (Figure 1-14). Agassiz was a friend of **C. Wyville Thompson** and had a prominent role in the processing of the *Challenger* collection in 1876. He also contributed substantial money and effort to the development of oceanographic studies in the United States. He was the first to use steel cables for deep-sea dredging on his 1877 voyage aboard the *R.V. Blake* and developed other ingenious devices to improve the quantitative value of biological samples recovered.

Victor Hensen and Marine Ecology

In the late nineteenth century, there was great international concern about fluctuations in the abundance of commercial fish. Although scientific committees were formed to investigate the problem, **Victor Hensen** (1835–1910) solved the problem alone by pioneering an ecological approach, studying the distribution of the fishes' food supply instead of the fish. In 1887, Hensen coined the term *plankton* for microscopic organisms that float in the oceans and developed quantitative methods for its study. At that time, most biologists expected to find more plankton in the tropics than in colder waters. Hensen's expedition aboard the *National* in 1889 showed that the reverse was true. Later work showed that the warm surface waters of the tropics were depleted of nutrients as a result of thermal stratification; cold, high-latitude surface waters maintained higher nutrient levels because of storm-induced mixing. Nutrients determine plankton abundance which, in turn, affects fish populations.

Polar Oceanography Begins with the Voyage of the *Fram*

Fridtjof Nansen (1861–1930) undertook one of the most unusual voyages in oceanography. He set out aboard the Norwegian ship, *Fram,* on June 24, 1893, with 13 men provisioned for 5 years. The ship embarked from Oslo, Norway, bound for the New Siberian Islands. Nansen thought that the moving polar ice pack would carry the ship across the North Pole. Before it could reach the New

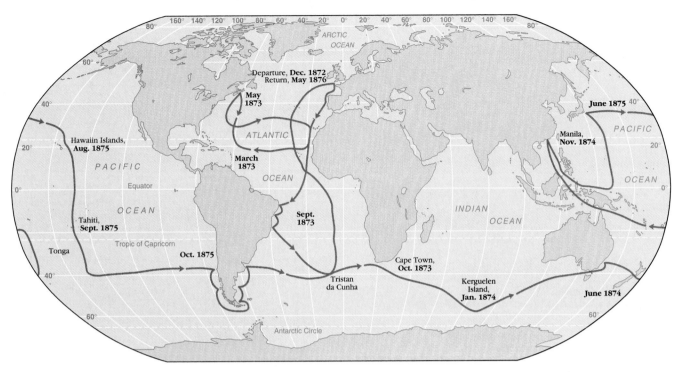

Figure 1-13   *Voyage of H.M.S. Challenger*

Above, the route traveled by *H.M.S. Challenger* during the first major oceanographic voyage (December 1872 to May 1876). Left, *H.M.S. Challenger*. *(From the Challenger Report, Great Britain, 1895).*

Figure 1-14   Alexander Agassiz (1835–1910)

Contemporary illustration of Professor Agassiz with friends, gathering specimens at night near Newport, R.I. *(Photo courtesy Corbis–The Bettmann Archive)*

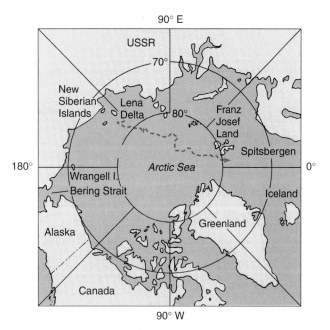

Figure 1-15   Voyage of the *Fram*

The course followed by the *Fram* after becoming frozen in ice near the New Siberian Islands, September 21, 1893. On August 13, 1896, the little ship was released by the ice off of Spitsbergen.

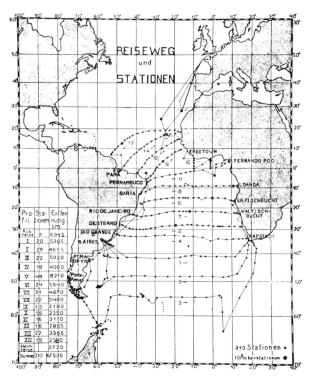

Figure 1-16   Voyage of the *Meteor*

This 1925 cruise crisscrossed the South Atlantic Ocean.   *(From G. Wüst, 1935. The stratosphere of the Atlantic Ocean. Scientific results of the German American Expedition of the Research Vessel Meteor 1925–1927, Vol. VI, sec. 1. W. J. Emery (trans. and ed.). Amerind Pub. Co., New Delhi, 1978)*

Siberian Islands, the ship became frozen in the ice and drifted with it for 3 years (Figure 1-15). The direction of drift, however, caused the ship to miss the North Pole by 394 km (245 mi). A frustrated Nansen left his ship on March 14, 1895 and set off on dogsled. After 14 months he had only reached 86°14' latitude and he abandoned the quest. On August 13, 1896, the *Fram* broke free of the ice near Spitzbergen, Norway, after drifting a total of 1658 km (1030 mi).

The drift of the *Fram* proved that no continent existed in the Arctic Sea and that the freely moving polar ice pack must have formed directly on the ocean surface. The crew had measured ocean depths of more than 3000 m (9840 ft) and discovered relatively warm, high salinity water at intermediate depths 150–900 m (500–3000 ft). Nansen correctly described this water as a mass of high salinity Atlantic water that had sunk below the less saline Arctic surface water.

Nansen developed the Nansen sampling bottle, a design that permitted accurate depth sampling and that remained in use for over 70 years. **V. Walfried Ekman** (1874–1954) later used Nansen's observations of the direction of ice drift, relative to the wind direction, to develop a mathematical explanation for the dependence of ocean current directions on wind directions, a relationship we now call the Ekman spiral (see Chapter 9).

Twentieth Century Oceanography ...

Voyage of the *Meteor*

In 1925, the German ship *Meteor* set the twentieth century standard for detailed multidisciplinary studies of the ocean. For 25 months, scientists studied the topography, currents, and chemistry of the South Atlantic, setting up a series of 310 sampling stations (Figure 1-16). The ship also carried two acoustic depth recorders which gave new details of the configuration of the ocean floor, making it easier to understand the patterns of water circulation observed. The chief scientist on the expedition, **George Wüst**, introduced the four-layer structure of circulation that is still accepted as correct (see Chapter 9).

Oceanography Gets Institutionalized

In the early twentieth century, major universities founded institutions exclusively for marine studies (Figure 1-17). In Europe, Prince Albert I of Monaco established the Musée Océanographique in 1903. Jacques Cousteau was among its alumni. The first oceanographic institution in the United States, founded in 1912 in La Jolla, California, is now called the Scripps Institution of Oceanography and is part of the University of California at San Diego. In the 1930s, Woods Hole Oceanographic Institution grew out of the summer seaside laboratory that Ellen Swallow Richards had established in the late nineteenth century to teach science to women; it is now affiliated with both Harvard University and the Massachusetts Institute of Technology. The Lamont-Doherty Earth Observatory, associated with Columbia University, was founded at Torrey Cliffs, Palisades, New York, in 1949 as the Lamont Geological Observatory. The University of Miami's Rosenstiel School of Marine and

(a)

(b)

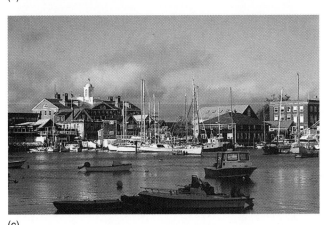

(c)

(d)

Figure 1-17   Oceanographic Institutions
....

(a) Musee Oceanographique, Monaco; (b) Scripps Institution of Oceanography; (c) Woods Hole Oceanographic Institution; (d) Rosenstiel School of Marine and Atmospheric Science.   (Part (a) courtesy of Patrick Aventurier, Liason Agency, Inc; (b) courtesy of Scripps Institution of Oceanography; (c) courtesy of Clyde H. Smith, Peter Arnold, Inc.; (d) (Courtesy of E. Hunt Augustus)

Atmospheric Sciences in Miami, Florida was founded in the 1930s. Texas A&M University's program, created in 1949, was chosen in 1984 to administer the Ocean Drilling Program (ODP).

The Expansion of Oceanography

Ironically, both World War I and World War II were of great benefit to oceanography. For example, the German U-boat in World War I led to the invention of the echo sounder to detect submarines. After the war, the technology was adapted to make depth surveys of the ocean floor much more quickly and cheaply than the old method of measuring cable dropped over the side of the ship. The *Meteor* used echo sounders extensively in its South Atlantic expedition.

In World War II, because so many of the battles were fought at sea, knowledge of currents and other physical properties of the ocean gave navies a significant advantage. The military performed and supported many studies on the transmission of sound in the ocean, waves, currents, and ocean-floor topography.

After World War II, the United States government saw the need to maintain its lead in oceanographic research. It established the Sea Grant program to fund oceanographic research at a greater number of universities.

Large-Scale and International Oceanographic Research

The Deep Sea Drilling Project (DSDP) was the first U.S. large-scale, cooperative effort in academic ocean research. In 1963, Scripps, the Rosenstiel School, Lamont-Doherty, and Woods Hole united to form the Joint Oceanographic Institutions for Deep Earth Sampling (JOIDES). They were later joined by the departments of oceanography of the University of Washington, Texas A&M, the University of Hawaii, Oregon State University, and the University of Rhode Island.

A ship had to be designed that was capable of drilling the ocean bottom at depths of 6000 m (3.7 mi). The *Glomar Challenger* (Figure 1-18a), constructed and launched March 23, 1968, began the first leg of the Deep Sea Drilling Project on August 11, 1968. The *Glomar Challenger,* at 122 m (400 ft) long, 20 m (66 ft) in beam, and with a displacement of 10,500 tons loaded, was capable of remaining at sea for 90 days. In 1983, the Deep Sea Drilling Project became the **Ocean Drilling Program**

(a)

(b)

Figure 1-18   Deep-Sea Drilling Ships

(a) The *Glomar Challenger* could produce 8800 continuous or 10,000 intermittent hp for propulsion and for operating drilling equipment. To remain over the drill site, the ship used dynamic positioning that could move the vessel in any direction.   *(Photo courtesy of Victor S. Soleto, Deep Sea Drilling Project)*
(b) *JOIDES Resolution*, replaced the *Glomar Challenger* as the new drilling ship for the Ocean Drilling Program.   *(Photo courtesy of the Ocean Drilling Program)*

(ODP), which is supported by France, the United Kingdom, Russia, Japan, Germany, and the United States. The *Glomar Challenger* was decommissioned after a highly successful 15 years of drilling. During this time, the information developed by the DSDP program helped to revolutionize earth science (see Figure 1-18). The ODP drill ship, *JOIDES Resolution,* which is much larger than the *Glomar Challenger* (Figure 1-18b), conducted its first scientific cruise in January 1985 and the JOIDES program continues today.

There are many more major national and international programs that have contributed to ocean science. The International Geophysical Year (IGY, 1957–1958), the International Decade of Ocean Exploration (IDOE, 1970s) and the International Indian Ocean Expedition (IIOE, 1959–1965), Geochemical Ocean Sections (GEOSECS, 1972–1978), the Joint Global Ocean Flux Study (JGOFS, 1987–present) and the World Ocean Circulation Experiment (WOCE, 1990–1997) are some examples.

The History Behind Plate Tectonic Theory

The history of studies related to the origin of the ocean basins deserves special mention. Although any small child can see how the shape of Africa's western coastline allows Africa to fit neatly between North and South America, for many years scientists refused to believe that these conti-

nents could once have been joined together because they had no explanation for how they could have subsequently moved apart.

In 1915, **Alfred Wegener** (1880–1930) developed the theory of continental drift. To explain the climate patterns indicated by fossil evidence in rocks of South America and Africa, he conceived of a single ancient landmass, which he called *Pangaea* that began to break apart about 180 million years ago along a system of global fractures (Figure 1-19). He suggested that the continents plowed through the ocean basins to reach their present positions and that the leading western edges of the continents deformed into mountain ridges because of the drag imposed by the ocean rocks. The driving mechanism he proposed was a combination of the gravitational attraction of Earth's equatorial bulge and tidal forces from the sun and moon.

Geophysicists rejected the idea as too fantastic and against the laws of physics. Material strength calculations showed that ocean rock was too strong for continental rock to plow through it; gravitational and tidal forces were too small to move the great continental land masses. However, even without an acceptable mechanism, many geologists who studied rocks in South America and Africa accepted continental drift. Only in the 1960s did geologists working in the Northern Hemisphere begin to seriously consider the idea because of the availability of new data from the ocean floor.

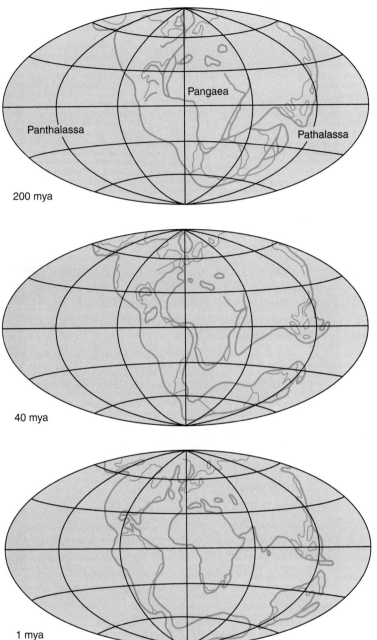

200 mya

40 mya

1 mya

Figure 1-19   Alfred Wegener's concept of continental drift

After Pangaea broke apart, the pieces plowed through the Panthalassa ocean basin to reach their present positions.

Harry Hess (1906–1969), when he was a U.S. Navy captain in World War II, had developed the habit of leaving his depth recorder on at all times. After the war, compilation of these depth records showed lines of flat-topped mountains oriented transversely to mountain ridges near the centers of ocean basins and extremely deep, trench-like valleys at ocean basin edges. In 1960, Hess published the idea of sea floor spreading with mantle (a layer deeper within the earth) convection as the driving mechanism (Figure 1-20). He suggested that the flat-topped mountains were volcanoes that formed on new ocean crust created at the ridges. The crust later disappeared back into the deep earth at the trenches. Mindful of the resistance of North American scientists to the idea of continental drift, Hess referred to his own work as "geopoetry."

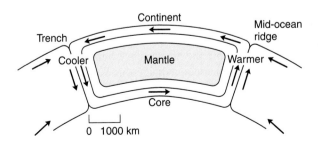

Figure 1-20   Harry Hess' concept of sea floor spreading

Mantle convection drives sea floor spreading at the mid-ocean ridges, carrying continents along in between. Sea floor material returns to the mantle at the trenches.

Figure 1-21

Magnetic anomalies on the sea floor provided Vine and Matthews with the definitive proof of sea floor spreading. Times of normal and reverse magnetization are shown symmetrically about the ridge axis.

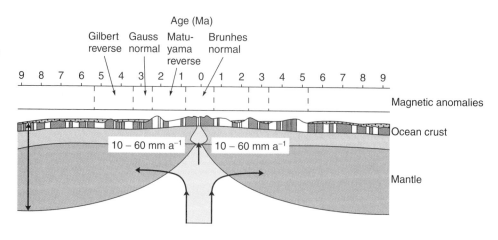

Frederick Vine and Drummond Matthews of Cambridge University provided conclusive evidence for sea floor spreading in 1963. In the process of using magnetic detectors to locate submarines during WWII, the magnetic patterns of the ocean floor had also been mapped. These maps showed parallel bands of similarly magnetized rocks on either side of oceanic mountain ranges, but no one could effectively explain them until Vine and Matthews realized that these were records of changes in Earth's magnetic field over time. Thus, the ocean floor was literally a conveyor belt that was being continuously formed at the mountain ridges and destroyed at the trenches (Figure 1-21). The continents were just riding along on the conveyor!

In 1965, Sir Edward Bullard reworked Wegener's fit of the continents, without any overlaps, by considering only the portions of coastlines that were continental crust and were 180 million years old or older. Dating of ocean floor samples from the Deep Sea Drilling Project showed that the ocean crust was progressively older with increasing distance from the ridges. These pieces of evidence were conclusive proof that, indeed, Pangaea had existed and continental drift was real. Thus, the theory of plate tectonics was born. We will discuss this theory in much more detail in Chapter 3.

Humans Invade the Deep Ocean

Table 1-1 lists some of the significant events in the history of human progress toward deep-sea descent. Humans have tried entering the ocean depths in submarines, in diving chambers, and with air supplies carried on their backs or piped down from the surface. Whatever the method, putting humans into the marine environment to observe it directly has always been a dangerous business.

The effects of increased pressure on body cavities and gases dissolved in body tissues limit the depth and duration of dives. The greater the depth, the longer the time that the diver must spend decompressing on the return trip to the surface. Decompressing involves waiting for specified lengths of time at specified depths to allow the dissolved gases in body tissues to equilibrate gently. Rapid ascent causes the dissolved gases to form bubbles in blood

vessels. These bubbles can block blood flow, causing an extremely painful illness known as the bends, or possibly death if a bubble blocks blood flow to the brain. At great depths or for long dives, divers must breathe a mixture of oxygen, helium, and nitrogen, and follow a strict and lengthy decompression schedule.

The need for divers to work in deep water to construct and repair offshore platforms and bridge foundations is pushing diving technology to make deeper and longer dives possible. One breakthrough is *Jim*, a diving suit which allows a diver to walk about freely on the ocean floor and maintains the inside of the suit at surface pressure, thus eliminating the need for decompression (see Table 1-1).

Submersibles

Manned submersibles have provided a very successful means for directly observing the deep-sea environment. Humans can travel housed in chambers in which the pressure remains the same as at the surface. In 1934, zoologist William Beebe descended to a depth of 923 m (3027 ft) off Bermuda in a tethered bathysphere to observe deep-sea life. Jacques Piccard designed an untethered vessel, *Trieste,* which was bought by the U.S. Navy and submerged to a record depth of 10,915 m (35,801 ft) in the Marianas Trench off Guam on January 23, 1960. Two submersibles that are now widely used in deep-sea research are untethered vessels. *Alvin* can descend to a depth of 4000 m (13,120 ft), and *Sea Cliff II* to a depth of 6000 m (19,680 ft) (Figure 1-22). Japan's *Shinkai 6500* is being used to study microbes in the deep-sea environment.

Some of the negative factors associated with manned submersibles are the risk to human life, the high cost of the systems required to accommodate humans, and on deep dives, the relatively short time that can be spent making observations. Of 12 hours of launch time, only 4 hours may be spent on bottom. Consequently, unmanned, remotely operated submersibles armed with cameras and a variety of measurement and manipulative devices have become popular for research and industry applications.

Remotely operated vehicles (ROVs) can explore hundreds of square kilometers per month. They can make

TABLE 1-1	Some Important Achievements in the History of Human Descent Into the Ocean.
360 B.C.	In his *Problematum*, Aristotle records the use of air trapped in kettles lowered into the sea by Greek divers.
330 B.C.	Alexander the Great descends in a glass, berrylike bell during the siege of Tyre.
1620	Dutch inventor Cornelius van Drebbel tests the first submarine, descending to a depth of 5 meters (16.5 feet) in the Thames River with King James I of England on board.
1690	Sir Edmond Halley descends in a lead-weighted diving bell. Air was replenished from barrels lowered into the Thames River.
1715	Englishman John Lethbridge tests a leather covered barrel of air with viewing ports and waterproof armholes to a depth of 18 m (60 ft).
1776	David Bushnell's *Turtle* attempts the first military submarine attack on the British ship *H.M.S. Eagle*. The attempt failed, but the British moved their fleet to less accessible waters.
1800	Robert Fulton builds the submarine *Nautilus*. It was the first of several to bear this name and was powered by a hand-driven screw propeller when submerged.
1837	Augustus Sieve invents a prototype of the modern helmeted diving suit.
1913	The German company Neufeldt and Kuhnke manufactures an armored diving suit with articulated arms and legs.
1930	William Beebe and Otis Barton descend in a bathysphere to a depth of 923 m (3027 ft) off Bermuda.
1943	Jacques-Yves Cousteau and Emile Gagnan invent the fully automatic, compressed air aqualung (Self-contained Underwater Breathing Apparatus: SCUBA).
1954	George Bond, Ed Link and Cousteau begin testing the concept of underwater living chambers initially using mice in simulated saturation dives to greater than 1000 ft.
1960	Jacques Piccard and Donald Walsh descend to a depth of 10,915 m (35,801 ft), the bottom of the Marianas Trench, in the bathyscaphe *Trieste*.
1962	Hannes Keller and Peter Small make an open-sea dive from a diving bell to a depth of 304 m (1000 ft) using a special gas mixture. Small dies during decompression.
1962	Ed Link stays for 14 hours in the first undersea living chamber at 60 ft. A month later, Robert Stenuit stays in the same chamber for 24 hours at 200 ft. Both use a mixture of helium and oxygen. Cousteau completes construction of *Starfish House*, an undersea living structure for five people and uses it in the Red Sea at 400 ft.
1963	Cousteau's second expedition, Conshelf II, uses *Starfish House* and a second chamber, *Deep Cabin* in 90 ft of water in the Red Sea.
1964	In *Sealab I*, a four man U.S. Navy team led by George F. Bond stays 11 days at 59 m (193 ft) off Bermuda.
1965	Albert Falco and Claude Wesly live for 1 week under conditions of saturation in Cousteau's *Conshelf III* in 328 ft of water off Marseilles, France.
1966	Submersibles *Alvin*, *Aluminaut*, and *Cubmarine*, with the aid of unmanned remote-controlled vehicle, CURV, locate and recover an atomic bomb from water over 800 m (2624 ft) deep off Palomares, Spain.
1969	Pressurized diving suit, *Jim*, is built to operate at depths below 304 m (1000 ft).
1969–1970	*Tektite I* and *II* accommodate 12 scientific teams for 14 to 60 days at a depth of 15 m (50 ft) off the U.S. Virgin Islands, including an all-woman team headed by Sylvia Earle.
1975	U.S. Navy divers use a pressurized, tethered sphere as transport to and from an open water dive at 1148 ft.
1976	Jacques Mayol free-dives to 100 m (328 ft) while holding his breath for 3 minutes, 40 seconds.
1979	Sylvia Earle makes a dive in *Jim* to 1250 ft off Hawaii, the first dive made in the suit without a tether to the surface.
1981	Diving scientists at Duke University simulate a dive to 686 m (2250 ft) in a pressure chamber.

computer-assisted maps based on sonar data they gather, record what they see with video and still cameras, and collect specimens with remotely operated arms. They can stay on the bottom as long as the ship to which they are tethered can stay on location, but at a much lower cost than for manned submersibles. Working teams often combine a manned submersible and a ROV deployed by the submersible. In 1985–1986, a prototype team, the ROV *Argo-* *Jason* and *Alvin*, provided the first images of the *Titanic* since she sank in 1912 (Figure 1-23).

Autonomous underwater vehicles (AUVs) are being developed for underwater surveys. These vehicles are preprogrammed to carry out specific data gathering missions of long duration without human intervention. AUVs have not received the wide usage that ROVs have because of technical limitations. For example, power supplies are inadequate

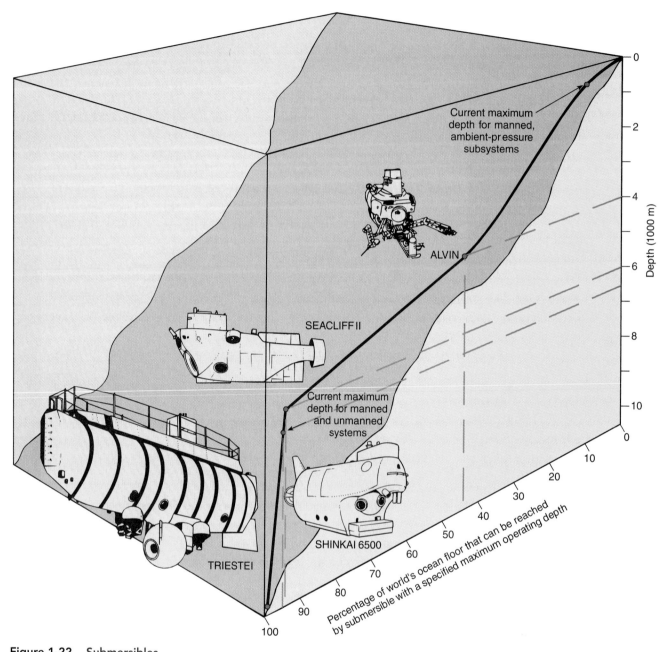

Figure 1-22 Submersibles

Percentage of ocean floor reachable by four manned submersibles is shown by the graph on the side of the cube. *Alvin* can explore about 44 percent of the ocean floor with its maximum operating depth of 4000 m (13,120 ft); *Sea Cliff II* has an operations limit of 6000 m (19,680 ft). The *Shinkai 6500*, the world's deepest-diving manned research submarine that is presently operational, can explore 97 percent of the ocean floor. It's maximum operational depth is 6500 m (20,865 ft.). The 1960 dive of *Trieste* to a depth of 10,915 m (35,801 ft) demonstrated its capability to explore any part of the deep ocean. However, its immobility made it impractical for deep-sea exploration and it is no longer in operation.

for propulsion and computing power needs, and better smart systems and/or artificial intelligence are needed for data collection, vehicle control, and navigation. ABE (Autonomous Benthic Explorer), an AUV developed at Woods Hole Oceanographic Institution, is shown in Figure 1-24. Many researchers hold high hopes for AUVs as the submarine information gathering system of the future.

Living Under the Sea

Researchers were motivated to create underwater living chambers when it was shown that divers, once their bodies have equalized with the increased pressure at any depth, can stay down indefinitely without any increase in decompression time required to return to the surface. Living chambers are held at the equivalent depth pressure

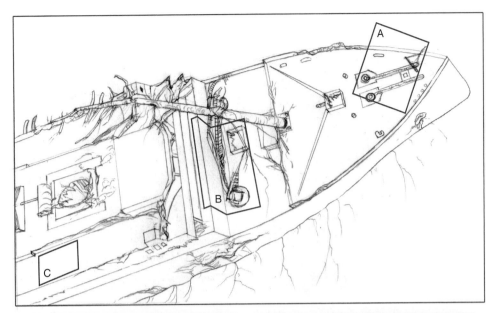

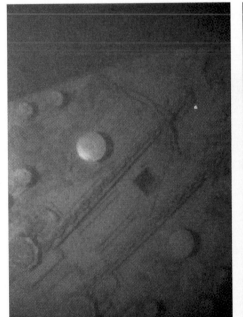

(a)

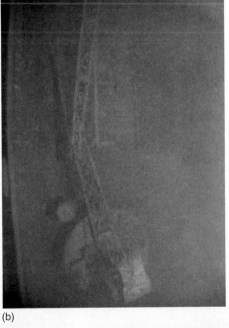

(b)

Figure 1-23   Using ROV and *Alvin* to photograph the *Titanic*

The *R.M.S. Titanic,* unseen by human eyes since 1912 when it sank 600 km (373 mi) south of Newfoundland, is at last visible in video and photographic images. The ship was discovered during sea trials of the deep-sea investigative instrument *Argo* on September 1, 1985. Photos (a), showing anchor chain, capstans and windlasses, and (b), showing loading cranes on the bow, were taken by a towed photographic unit, ANGUS (Acoustically Navigated Geological Underwater Survey System). When *Alvin* returned for the sea trials of *Jason Jr., Jason Jr.* took photo (c) of an electric winch on deck, and *Alvin* photographed *Jason Jr.* and its platform on *Alvin* photo (d).   *(Courtesy of Woods Hole Oceanographic Institution)*

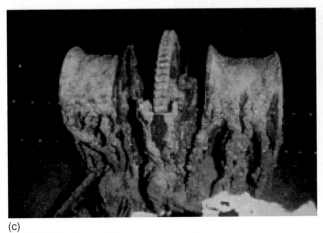

(c)

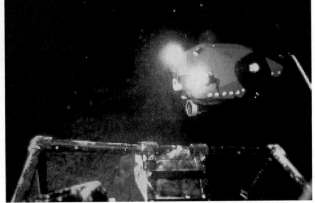

(d)

(a)

(b)

Figure 1-24 ROVs and AUVs

(a) *Jason* is the remotely operated vehicle (ROV) used in conjunction with the manned submersible *Alvin*. It is operated on a tether and can also be operated from a surface ship.
(b) ABE (Autonomous Benthic Explorer) represents a new class of underwater vehicles that is untethered and can remain beneath the surface for months, gathering data continuously. *(Courtesy of Woods Hole Oceanographic Institute).*

Figure 1-25 Undersea Living Chambers

using mixtures of oxygen and helium and may be placed on the sea bottom or suspended in the water. One of the counterintuitive things that happens with these chambers is that when the door is opened into the sea, the water doesn't come rushing in as we would expect. Because the air inside and the water outside are at equal pressure, there is a flat interface between the two through which the diver steps just as if he were entering the water at the surface.

Jacques Cousteau, George Bond, and Ed Link began designing and testing undersea living chambers in the 1950s. By the 1970s, teams were living in undersea chambers for up to 60 days (see Table 1-1). Several undersea chambers are in use today (Figure 1-25).

Remote Sensing

Sometimes, the best way to see the ocean is to go out into space. Instruments aboard satellites can measure the temperature, ice cover, color, and topography of the ocean surface as well as collect data that allows scientists to map the topography of the ocean floor. *Seasat A,* launched in 1978, was the first dedicated oceanographic satellite. Unfortu-

(a)

Figure 1-26   Seasat A

(a) Launched July 7, 1978, this satellite pioneered the remote sensing of the oceans. It shorted out on October 10, 1978, but provided sufficient data to prove its value to oceanography. The ghostly image below *Seasat* is the *H.M.S. Challenger.*

(b–d) continued on following page

nately, it shorted out after 3 months; but it still provided enough data to prove the great benefit of remote sensing. The change in sea surface temperature throughout the ocean during the lifetime of **Seasat A** is shown in Figure 1-26. **Nimbus 7**, from 1978–1981, collected chlorophyll spectral response data from which phytoplankton populations were mapped (Figure 1-27). A U.S.-French collaboration produced *TOPEX/Poseidon,* which began in 1993 to map global sea level and collect data on ocean-atmosphere interaction.

NASA has a program involving a series of satellites, collectively called the Earth Observing System (EOS), that is designed to look at the Earth in an integrated way, using some of the same instrumentation that NASA has employed to observe other planets. Modern navigation also uses a network of satellites known as Global Positioning System (GPS) that allows ships to determine their positions to within one meter.

(b)

(c)

(d)

Figure 1-26 (continued)

(b) Sea surface temperature (SST) averaged over a 14-day period just before *Seasat* shorted out. (c) SST averaged over a 12-day period just after *Seasat* was launched. (d) Change in SST that occurred between the two images *(Courtesy of Jet Propulsion Laboratory, NASA).*

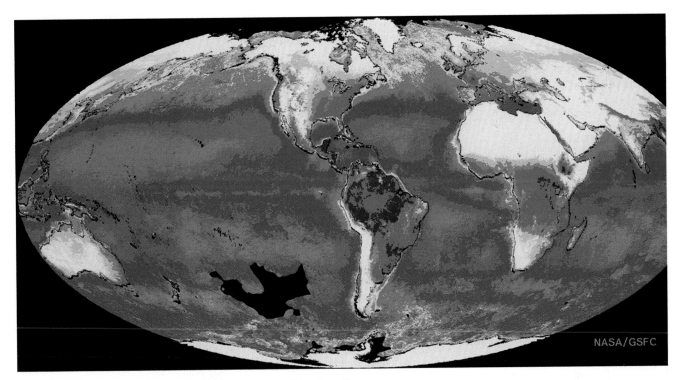

Figure 1-27   Photosynthetic Production Measured by Satellite

Data from the Coastal Zone Color Scanner (CZCS) aboard the satellite *Nimbus 7*, collected between November 1978 and June 1981. Colors correspond to chlorophyll content of surface waters. Low values are magenta, covering most of the open ocean; highest values are red, visible along some continental margins. Black indicates insufficient data.   *(Courtesy of Jane A. Elrod and Gene Feldman, NASA/Goddard Space Flight Center).*

Summary

War, trade, colonization and, probably as a distant fourth, curiosity have driven ocean exploration throughout human history. Early exploration was usually limited by the ability of ship builders to produce sufficiently large and seaworthy ships for long, open-ocean voyages. Advances in science and technology have been deciding factors in the history of naval and trade domination of the open ocean, as well as in our progress toward exploring the ocean's depths.

The Lapita people were probably the first to undertake long ocean voyages when they settled the islands of the Pacific. The Phoenicians were the first civilization in the Mediterranean region to undertake journeys as far as the Indian Ocean for trading purposes. The Greeks and Romans who succeeded them also were successful traders, explorers, and conquerors of regions around the Mediterranean and extending to the British Isles. Their scientific accomplishments included latitude determination and calculation of Earth's circumference.

After the fall of the Roman Empire, the scientific knowledge of the Greeks and Romans was lost to Europe until the fifteenth century. While the rest of Europe languished with respect to ocean exploration, the Vikings settled Iceland, Greenland, and established a colony in Newfoundland in 995. The Chinese Ming Dynasty sent diplomatic ships as far as Africa in the early fifteenth century. When Europe emerged from the Dark Ages, Spain and Portugal led the way, intent on finding new trading routes to India and China.

The beginnings of ocean science are marked by the voyages of Captain James Cook in the late eighteenth century. Matthew Fontaine Maury developed international standards for reporting oceanographic data and wrote the first oceanography textbook. By the mid-nineteenth century, the English biologists Charles Darwin, Sir John Ross, Sir James Ross, and Edward Forbes had made significant contributions to our knowledge of life in the oceans.

The voyage of *H.M.S. Challenger* from 1872 to 1876 was the first major full-scale expedition solely to gather scientific data. Alexander Agassiz was a major figure in the development of oceanographic studies in the United States. Victor Hensen developed the foundations for the study of marine ecology. In 1893, the Norwegian explorer Fridtjof Nansen, entrapped his ship in Arctic ice for three years.

Twentieth century oceanography is characterized by the use of new technologies, many of which were developed in World Wars I and II. Institutions dedicated to oceanography also were established in this century. A major focus of this century has been drilling the sea floor, begun by the Deep Sea Drilling Project and continued by the Ocean Drilling Program.

The use of submersibles and less expensive, safer, unmanned remote systems has made it possible for oceanographers to see the deep ocean first hand. Satellites have greatly improved our ability to study the large-scale phenomena of the ocean and to map ocean features, and the Global Positioning System (GPS) has greatly improved navigation accuracy.

Key Terms

People

Agassiz, Alexander
Beebe, William
Cabot, John
Columbus, Christopher
Cook, Captain James
Cosmas
Cousteau, Jacques
da Gama, Vasco
del Cãno, Sebastian
Diaz, Batholomew
Dittmar, William
Ekman, V. Walfrid

Eratosthenes
Eriksson, Leif
Forbes, Edward
Franklin, Benjamin
Harrison, John
Hensen, Victor
Herodotus
Hess, Harry
Magellan, Ferdinand
Matthews, D.
Maury, Matthew Fontaine
Nansen, Fridtjof
Piccard, Jacques
Ptolemy

Pytheas
Ross, Sir James Clark
Ross, Sir John
Strabo
Thompson, C. Wyville
Vine, Frederick
Wegener, Alfred
Wüst, George

Places, things, and acronyms

Autonomous Underwater Vehicle (AUV)

Bathysphere
Deep Sea Drilling Project (DSDP)
H.M.S. Challenger
Jim
Nimbus 7
Ocean Drilling Program (ODP)
Remotely Operated Vehicle (ROV)
Seasat A
Vinland

Questions and Exercises

1. Construct a time line that includes the major events of human history that have resulted in a greater understanding of our planet in general and the oceans in particular.

2. State the plane geometry relationship used by Eratosthenes to calculate the circumference of Earth. Determine the circumference of Earth from the proportion stated in Figure 1-3.

3. Why is accurate measurement of time at sea important to navigation?

4. If you were to make a map of the world, including only those places to which you have actually traveled, what would it look like?

5. What was Matthew Fontaine Maury's major contribution to an increased knowledge of the oceans?

6. List some major achievements of the voyage of *H.M.S. Challenger.*

7. What important oceanographic inventions and data came out of World Wars I and II?

8. What were some of the objections to Wegener's original concept of continental drift and how were these satisfied by later concepts and data?

9. Describe your own ideas of the barriers to human exploration of the deep ocean.

10. List the features of the oceans that might be studied remotely by use of satellite-borne sensors.

References

Allmendinger, E., 1982. *Submersibles: Past-Present-Future.* Oceanus 25:1, 18–35.

Bailey, H. S., Jr., 1953. *The Voyage of the Challenger. Scientific American* 188:5, 88–94.

Borgese, E. M., N., Ginsberg, and J.R. Morgan, 1994. *Ocean Yearbook 11*. Chicago: University of Chicago Press.

Brewer, P., ed., 1983. *Oceanography: The Present and Future.* New York: Springer-Verlag.

Carson, Rachel, 1951. *The Sea Around Us.* Oxford: Oxford University Press.

Earle, Sylvia, 1995. *Sea Change, A Message of the Oceans.* New York: Fawcett Columbine.

Heyerdahl, T., 1979. *Early Man and the Ocean.* New York: Doubleday & Company.

Irion, R., 1998. Instruments Cast Fresh Eyes on the Sea. *Science* 281: 194–196.

Mowat, F., 1965. *Westviking*. Boston: Atlantic-Little, Brown.

Oceanography from Space. 1981. *Oceanus* 24:3, 1–75.

Pirie, R.G., 1996. *Oceanography: Contemporary Readings in Ocean Sciences.* 3rd ed. Oxford: Oxford University Press.

Ryan, P. R., 1986. The Titanic Revisited. *Oceanus* 29:3, 2–17.

Sears, M., and D., Merriman, eds., 1980. *Oceanography: The past.* New York: Springer-Verlag.

The Titanic: Lost and Found. 1985. *Oceanus* 28:4, 1–112.

Suggested Reading

Sea Frontiers

Baker, S., 1981. *The Continent that Wasn't There.* 27:2, 108–14. A history of the Search for Terra Australis Incognita, the Unfound Southern Land.

Charlier, R. H., and Charlier, P. A., 1970. *Matthew Fontaine Maury, Cyrus Field, and the Physical Geography of the Sea.* 16:5, 272–81. A biography of Maury emphasizing his accomplishments as superintendent of the Department of Charts and Instruments and his book, *The Physical Geography of the Sea.*

Denzel, J. F., 1976. *Edward Forbes and the Birth of Marine Ecology.* 22:1, 16–32. Includes the studies by Edward Forbes in the Mediterranean and in the seas around the British Isles.

Gaunt, A., 1975. *Marking Time for Three Hundred Years.* 21:6, 322–30. An information rich history of Greenwich Observatory.

Maranto, G., 1991. *Way above Sea Level.* 37:4, 16–23. Describes Global Positioning System navigational satellites used to study changes in sea level due to climate change.

Rice, A. L., 1972. *H.M.S. Challenger—Midwife to Oceanography.* 18:5, 291–305. Summarizes the 1872 voyage of *H.M.S. Challenger,* which laid the foundation for most branches of marine science.

Schuessler, R., 1984. *Ferdinand Magellan: The Greatest Voyage of Them All.* 30:5, 299–307. A brief history of the voyage initiated by Magellan to circumnavigate the globe.

Van Dover, C., 1987. *Argo Rise: Outline of an Oceanographic Expedition.* 33:3, 186–194. Describes the first scientific use of the Argo system to survey hydrothermal vent activity on the East Pacific Rise.

Scientific American

Edmunds, P.J., 1996. Ten Days under the Sea. 275:4, 88–95. *Aquarius,* the only underwater habitat devoted to scientific research is located near the reefs off of Key Largo, Florida.

Giles, D.L., 1997. Faster Ships for the Future. 277: 4, 126–131. New designs for ocean freighters could double their speeds.

Hawkes, G.S., 1997. Microsubs Go to Sea. 277:4, 132–135. Tin one-man subs may someday take us to the deepest reaches of the ocean.

Herbert, S. 1986. Darwin As a Geologist. 254:5, 116–123. Darwin's early interest in the developing science of geology aided him in the development of his theory of biological evolution.

Revelle, R. 1969. The Ocean. 221:3, 54–65. The author discusses human use of the ocean and the geological creation of its present basins.

Oceanography on the Web

Visit the Introductory Oceanography home page for on-line resources for this chapter. There you will find an on-line study guide with review exercises and links to ocean-ography sites to further your exploration of the topics in this chapter. Introductory Oceanography is at http://prenhall.com/thurman (click on the Table of Contents menu and select this chapter).

2 Beginnings Of The Universe, Earth, And Life

Earth from space.   (Photo courtesy of NASA.)

Why is Earth the only planet in the solar system that has an ocean? How has the composition of our planet changed over time? What distribution of energy and mass were needed to create the right conditions for life to originate? Is an ocean essential to the origin of life? These are questions that we will explore in this chapter. Unfortunately, we don't have very good answers. We also will discuss how humans are gradually altering the composition of the atmosphere and perturbing the cycling of elements through the various layers of the Earth system. As yet, we have no real idea what the lasting impact of these actions will be.

The Universe and the Solar System.

Humans have always been fascinated by the stars, and some who took their stargazing seriously, gradually figured out that the points of light visible in the night sky were not all the same. In fact, eight of the closest ones are not even stars. Earth and its eight neighbors are planets that orbit around an average sized star, the Sun. We call this group the solar system.

The solar system is but a tiny, insignificant part of the universe. It lies about two-thirds of the way from the center of a huge spiral of stars called the **Milky Way** galaxy (Figure 2-1). All the stars you can see with your naked eye belong to the Milky Way. Proxima Centauri, the star nearest to the Sun, is more than 4×10^{13} (40 trillion) km away. See Appendix I for a discussion of scientific notation.

The Milky Way is only one of countless **galaxies** in the universe, and it is smaller than most others. On clear nights, if you look very carefully from a vantage point in the Northern Hemisphere, you might see a hazy patch of light within the constellation Andromeda. This patch is one of the galaxies closest to us. It is named Andromeda after the constellation through which we can see it; it is about 3×10^{19} km away. Using telescopes, we can observe many more galaxies. Within 10^{22} km of our galaxy, there are at least 100 million others. Astronomers estimate that there are on the order of 100 billion billion (10^{20}) stars in the known universe.

To deal with the long distances involved in astronomy, scientists developed a unit for measuring distance, the **light-year**. As a unit of distance, a light-year is perfectly incomprehensible within the realm of human experience. A light-year is equal to the distance light travels in one year. Measuring distance with a unit of time seems illogical—we all know that we can vary the time it takes to get some place by changing our speed. Light, however, always travels at exactly the same speed of 299,792 km/s (186,282 mi/s). In one year, light travels almost 10 trillion km (6.21 trillion miles). Using light-years makes the expression of

Figure 2-1   The Milky Way

The solar system is located about two-thirds the distance from the center of the Milky Way to the edge. The Milky Way completes a rotation every 200 million years. To maintain its position in the Milky Way, the solar system travels around the center of the galaxy at a velocity of about 280 km/s (174 mi/s).
(*Reprinted with the permission of Macmillan Publishing Company from Earth Science, 6th Edition., Figure 20.17, by Edward J. Tarbuck and Frederick K. Lutgens. © 1991 by Macmillan Publishing Company.*)

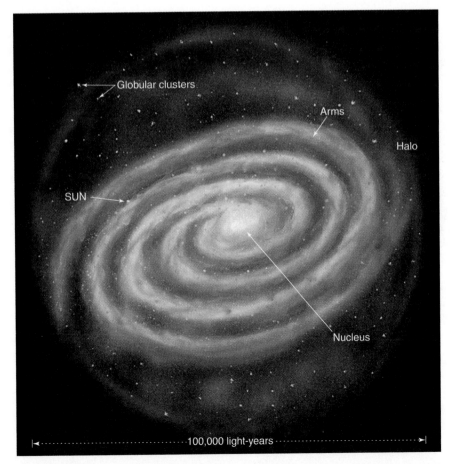

astronomical distances much simpler. For example, we can say that Proxima Centauri is about 4 light-years from the Sun. For the rest of this chapter, we will use light-years for astronomical distances.

The Milky Way galaxy is 100,000 light-years in diameter. Stars are much closer together near the center of the galaxy and much farther apart toward the edges. The central nucleus of the galaxy is only about 10,000 to 15,000 light-years across. Within the Milky Way, there are approximately 100 billion stars. From statistical considerations, astronomers estimate that tens of millions of these stars have families of planets, and millions of these planets could be inhabited by intelligent creatures.

By observing changes in the light radiating from these distant galaxies, astronomers have determined that most of them are moving away from us. Figure 2-2 shows how patterns of light emission vary with the velocity of the galaxy and its distance from the sun. The most distant galaxies are traveling more rapidly than the closest ones. Velocities of more than 250,000 km/s (155,250 mi/s), which is 80 percent of the speed of light, have been calculated for these galaxies.

It is important to remember that we see the motion of objects in the universe as passengers. Earth orbits the Sun at about 30 km/s, and the solar system spins around at 220 km/s as the Milky Way rotates in space. The Milky Way is also moving out from the center of the universe at high speed. We can remain happily ignorant of all this motion because everything that we perceive is moving at the same speed we are.

To understand this better, think about your experiences traveling in a car. If another car is traveling at the same speed as your car, you cannot perceive any change in distance between the vehicles. You are unable to determine that you are moving by observing the other car. However, if you see a car pass, you know it is traveling faster than you are. If the ride is perfectly smooth, you can't determine whether you are going 60 mph and being passed by

a car going 80 mph, or whether you are standing still and being passed by a car going 20 mph. The change in the relative positions of the two vehicles would be the same for a fixed time interval.

From our point of view, looking out at the universe from Earth, it appears that most of the galaxies in the universe are moving away from us. But, if we remember that we are moving too, a more reasonable explanation is that we are riding along within an expanding universe. The galaxies are moving away from a center and from one another as if they were fragments from some ancient explosion. The **Big Bang** is the name used for this idea of an exploding universe. Unlike a bomb explosion, however, where pieces lose velocity due to friction and gravity as they move away from the center, the pieces of the universe are gaining velocity as they move away from the center.

If all of these galaxies are moving away from a central point, we might suppose that they all originated from a single large mass. If so, we can calculate from their speeds how long it has taken for them to reach their present positions. In this way, astronomers estimate the age of the universe to be 12–15 billion years.

According to the Big Bang theory, all matter probably originated at the center of the universe at extremely high temperature and pressure. Elementary particles formed within the first one-billionth of a second; after one-hundredth of a second, neutrons, protons, and electrons formed. In about twenty-five minutes, the temperature dropped to one billion degrees, allowing neutrons and protons to combine to form atomic nuclei. At about 3000 K, nuclei attracted electrons to form the first atoms. All matter began as atoms of hydrogen and helium, the two lightest elements.

In the Big Bang, this matter exploded out from the center. Stars were born from the clouds of helium and hydrogen that moved outward. When these clouds became large enough, they started to contract and increase in tempera-

Constellation in which galaxy may be seen	Velocity km/s (mi/s)	Distance from sun (light-years)	Shift of absorption lines Violet Red
Virgo	1,200 (745)	43,000,000	
Corona Borealis	21,500 (13,351)	728,000,000	
Hydra	61,000 (37,881)	1,960,000,000	

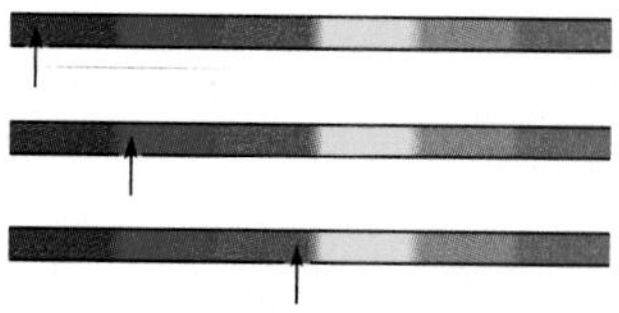

Figure 2-2 Galaxies and Spectral Shift

Three galaxies moving away from the Milky Way show characteristic relationships among velocity, distance from us, and the apparent shift of absorption lines toward the red end of the light spectrum. The absorption line shown here is a wavelength of violet light. In the closest galaxy, it still appears to represent violet light. However, in the more distant galaxies, it appears to represent longer-wavelength radiation that falls within the blue and green portions of the visible spectrum—a shift toward the red end of the spectrum.

ture. If temperatures within the clouds were high enough, nuclear fusion reactions began within the star, consuming hydrogen as fuel and producing helium. At some point in its history, every star undergoes another contraction, and the helium is burned producing carbon. With successive contractions, the elements oxygen, silicon, and iron are produced. When a star has developed an iron core and contracts, it explodes, producing a supernova. The material from this explosion contains many of the heavier elements. When this material is ejected into space, it forms interstellar clouds that can coalesce to form new stars.

Origin of the Solar System and of Earth

The Sun formed about 5 billion years ago from an interstellar cloud consisting of hydrogen, helium, and many of the heavier elements (Figure 2-3) . Most of the helium and hydrogen in the cloud collected to form the proto-Sun, which was probably as large in diameter as the present solar system. As the gas contracted under the force of gravity to form the Sun, a small amount of material was left behind in smaller eddies that flattened into a disk that became increasingly denser and more unstable. The disk broke into separate, small clouds that eventually, through collisions over about 100 million years, became protoplanets (Figure 2-3a).

Proto-earth was a huge mass, perhaps 1000 times greater in diameter than Earth today and 500 times more massive. Gravity pulled the heavier elements to the center of the forming planet, creating a dense core surrounded by lighter elements. Throughout this process, meteorites from space continued to bombard the planet. Late in the stage of planetary formation, a large planetesimal about the size of Mars struck Earth. Planetary scientists believe that the object's rocky outer layer was propelled into orbit around Earth as the Moon, while its metallic interior remained with Earth.

In addition to light, the Sun also emits intense streams of electrically charged particles that make up the **solar wind**. In the early stages of the solar system, the solar wind was strong enough to blow away any hydrogen and helium that escaped capture by the four inner planets. Eventually, these light gases were literally boiled away from the four inner planets as the planetary atmospheres were heated up by the Sun and by heat created from gravitational contraction and radioactivity within the planet interiors. Much of this lost gas was captured by Jupiter, Saturn, Uranus, and Neptune, the huge planets beyond Mars (Figure 2-3b and c).

(a)

Figure 2-3   The Nebular Hypothesis for Solar System Formation

(a) (A) A huge rotating cloud of dust and gases begins to contract; (B) most material is swept to the center, producing the Sun; (C) the remaining material orbits as a flattened disk and gradually coalesces to form the planets; (D) remaining debris is swept up by the planets or swept in to space by the solar wind.   *(From Thurman and Trujillo, Essentials of Oceanography, 6th ed.)*

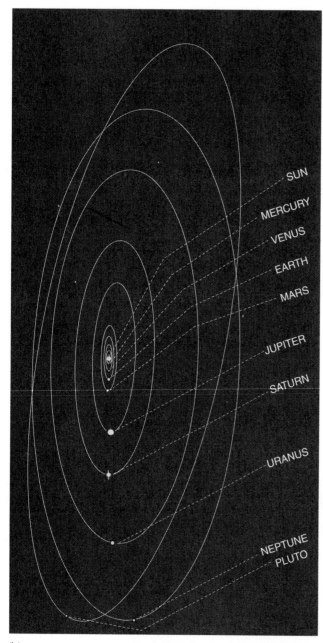

(b)

Figure 2-3 (Continued)

(b) Orbits of the planets of the solar system, drawn to scale;
(c) relative sizes of the Sun and the planets. Distance scale is not
maintained (right). *(Reprinted with the permission of
Macmillan Publishing Company from Earth Science, 6th Edition,
Figures, 19.1 and 19.2, by Edward J. Tarbuck and Frederick K.
Lutgens. Copyright © 1991 by Macmillan Publishing Company.)*

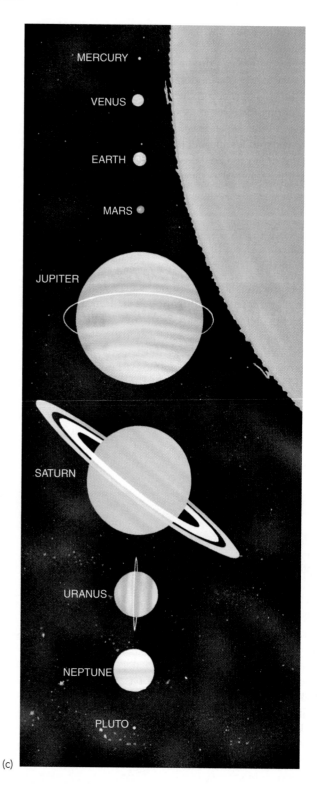

(c)

The energy released by meteoric bombardment, contraction, and internal radioactivity was enough to keep Earth's surface from forming a rock crust for probably the first hundred million years and allowed further separation of the elements within the planet according to their densities. The heaviest elements became concentrated in the core, intermediate combinations of elements formed a layer around the core (this layer is called the **mantle**), and the lightest elements remained in the outermost layers,

forming the crust and atmosphere. The solid rocks that crystallized from the molten crust material are called primary crystalline rocks.

The Composition of Earth

The thicknesses and compositions of the core, mantle, and crust (Figure 2-4a) are estimated from the action of seismic waves that pass through Earth's interior. The center of

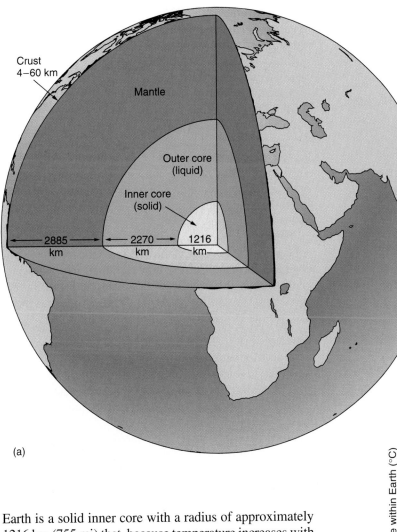

(a)

Figure 2-4   Earth's Internal Structure
....
(a) The cross section shows Earth's three major subdivisions: core, mantle, and crust. The inner core, outer core, and mantle are drawn to scale, but the thickness of the crust is enlarged about five times. *(Reprinted with the permission of Macmillan Publishing Company from The Earth, 4th Edition, Fig. 1.11, by Edward J. Tarbuck and Frederick K. Lutgens. Copyright © 1993 by Macmillan Publishing Company.)* (b) Temperature changes with depth (the geotherm) within the Earth control the density and phase changes of minerals of the mantle, inner core, and outer core. The transition from the liquid outer core to the solid inner core occurs when the geotherm crosses the melting curve.

(b)

Earth is a solid inner core with a radius of approximately 1216 km (755 mi) that, because temperature increases with depth, grades to a liquid outer core approximately 2270 km (1410 mi) in thickness (Figure 2-4b). The **core** contains abundant iron and nickel. The **mantle** surrounding the core is 2885 km (1790 mi) thick and is composed of very dense minerals formed from iron and magnesium combined with silica. The outermost, solid **crust** of Earth ranges in thickness from 4 to 60 km (2.5 to 37 mi). The thinner crust under the oceans is composed of basalt (Figure 2-6); the thicker crust forms the continents and is granitic in composition. The division between the crust and the mantle is marked by a distinct density change. This division, named after the Yugoslavian seismologist who discovered it, is called the **Mohorovičić discontinuity**, or Moho. We will learn more about these layers in the next chapter. Now, we will focus briefly on the nature of the rocks that compose Earth before continuing on to discuss the formation of the oceans and atmosphere.

Common Rock-Forming Minerals

Geologists recognize three main categories of rocks on Earth: igneous, sedimentary, and metamorphic. Just after Earth solidified, the only types of rocks present were igneous. Igneous rock form from the crystallization of molten material. Sedimentary rocks are the products of erosion and weathering of other rock types. High pressures and temperatures in Earth convert both igneous and sedimentary rock to metamorphic rock. The assemblage of minerals composing the rock is used to name it and to deduce the conditions under which it formed.

Major Minerals   A **mineral** is a chemical compound that has a general chemical formula and specific crystal structure, but may also contain varying amounts of minor constituents. For example, $CaCO_3$ is the chemical formula for the compound calcium carbonate. The mineral calcite has the same general chemical formula, but the natural mineral usually contains minor amounts of magnesium, iron, or other elements. The major minerals on Earth are either silicates or

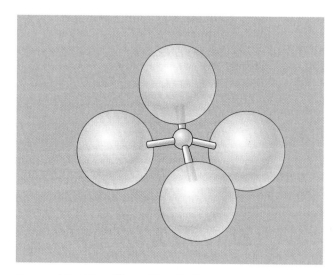

Figure 2-5   The Silicate Tetrahedron

Small silicon ion (Si^{4+}) bonded to and surrounded by four larger oxygen ions (O^{2-}), creating a $(SiO_4)^{4-}$ tetrahedron.

carbonates. (If you are unfamiliar with chemical terms such as ion, cation, bonds, or van der Waals forces, refer to Appendix V and the glossary before continuing).

Silicates are minerals built from arrangements of silicon ions (Si^{4+}) surrounded by four oxygen ions (O^{2-}), creating a **silicate tetrahedron** $(SiO_4)^{4-}$ (Figure 2-5). These tetrahedra may be isolated from one another by other cations or connected by sharing oxygen atoms. Tetrahedra may be connected to form chains, double chains, sheets or three-dimensional frameworks. Chains, double chains, and sheets are held together by cations.

Olivine is the most common mineral formed from isolated silicate tetrahedra bound together by iron (Fe^{2+}) and magnesium (Mg^{2+}). *Pyroxene* minerals consist of single chains of silica tetrahedra and incorporate Fe^{2+}, Ca^{2+} and Mg^{2+}. *Amphiboles* are silicate minerals that consist of double chains of silicate tetrahedra with hydroxyl ions (OH^-) within the double chains. In addition, aluminum ions (Al^{3+}) substitute for Si^{4+} in some of the tetrahedra. Because Al^{3+} and Si^{4+} do not have equal electrical charges, the substitution creates a charge imbalance in the structure that is compensated by 3+ cations replacing 2+ cations. Amphiboles contain Mg^{2+}, Fe^{2+}, Ca^{2+}, Na^+, K^+, Al^{3+} and Fe^{3+} between the chains.

Sheet silicates come in two-layer and three-layer varieties. The two-layer type consists of a sheet of silica tetrahedra connected to a sheet of aluminum or magnesium and hydroxide ions. Stacks of two-layer sheets are held together by hydrogen bonds. *Kaolinite* and *serpentine* are common two-layer sheet silicate minerals. The three-layer type consists of two silicate tetrahedral layers with a hydroxyl sheet in between. Cations, including Mg^{2+}, Fe^{2+}, Ca^{2+}, Na^+, K^+, Al^{3+} and Fe^{3+}, occupy spaces within the three-layer structure. Stacks of three-layer sheets are held together by van der Waals forces. Three-layer sheet silicates

include several large groups of minerals, including *talc, mica, montmorillanites, chlorites,* and *vermiculites. Clay minerals* are included in many of these groups.

In *framework silicates,* the oxygen atoms at every tetrahedral corner are shared with other tetrahedra. In *feldspar* minerals, aluminum may substitute for some silicon atoms, and Ca^{2+}, Na^+, and K^+ cations occupy spaces within the framework. Quartz is a mineral made only of silicate tetrahedra. If the silicate tetrahedra retain some attached water molecules, the crystal framework fails to form completely, creating amorphous minerals, such as *opal. Zeolites* are another important group of hydrated framework silicates that also have aluminum and other cation substitutions.

The most common *carbonate minerals* are *aragonite, calcite,* and *dolomite.* Calcite and aragonite have the same chemical formula, $CaCO_3$; however, their ions fit together in slightly different crystal structures. Dolomite has about equal proportions of Mg and Ca, giving it the chemical formula, $CaMg(CO_3)_2$.

Igneous rocks   Igneous rocks form from the resolidification of **magma,** the molten rock found at the high temperatures and pressures deep within Earth. The crystal size, or texture, of igneous rocks is determined by how rapidly the molten rock cools. Magma extruded at Earth's surface during volcanic activity cools rapidly and forms small crystals. If cooling is very rapid, crystals do not have a chance to form at all, and the rocks that form resemble glass. These fine-grained and glassy rocks are called **extrusive rocks. Basalt** is a typical example of a fine-grained igneous rock. It is the dominant rock type making up the ocean crust and is the most common igneous rock in volcanic oceanic islands, such as Hawaii.

Coarse-grained igneous rocks form at great depths in the Earth's crust as a result of slow cooling. Great pressure moves molten material up from the mantle, intruding it into crustal rocks, but the material crystallizes before it reaches the surface. Because the heat dissipates much more slowly within the rock than in the atmosphere, these **intrusive rocks** cool more slowly and develop larger crystals. The most common intrusive rock is **granite.** Continental crust is composed mostly of granite.

Further divisions of igneous rocks are made on the basis of the component silicate minerals. Figure 2-6 shows the most common varieties of igneous rocks of both intrusive and extrusive origin arranged in order of mineralogy. On the left, granite and *rhyolite* are intrusive and extrusive equivalents. They are light-colored rocks composed primarily of minerals, with no iron or magnesium. Granite and rhyolite are less dense than other igneous rock types. Such rocks are sometimes referred to as sialic because they characteristically contain high percentages of the elements silicon and aluminum.

Basalt and its intrusive equivalent, *gabbro,* are on the right side of Figure 2-6. They are dark in color and have high contents of magnesium and iron-bearing minerals. They also have higher densities. These rocks are some-

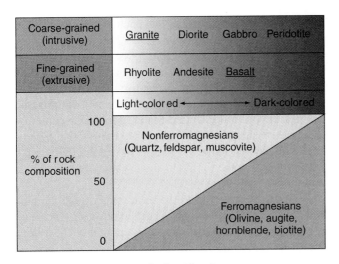

Figure 2-6   Igneous rock classification

The most common type of igneous rocks. Coarse-grained rocks cool slowly deep beneath the Earth's surface; fine-grained rocks cool rapidly near or at the surface. Minerals become richer iron and magnesium toward the right.

times referred to as *mafic,* a contraction of magnesium and ferric (iron). *Peridotite* is composed almost entirely of ferromagnesian minerals; this composition is extremely rare at the surface, thus there is no significant extrusive equivalent. Earth's mantle is made of peridotite.

Sedimentary Rocks   Sedimentary rocks are divided into clastic and carbonate types. *Clastic rocks* are formed from sediments composed of particles weathered from other rocks. *Carbonate rocks* are made up of fragments of the hard parts of marine organisms. The major minerals of clastic sedimentary rocks include only those that are hard enough to withstand weathering by wind and rain. Many of the minerals in igneous rocks dissolve and disintegrate rapidly when exposed to water. As the rock breaks down, some components dissolve and some remain as loose solid particles to be carried away by wind or water and deposited as sediment, sometimes thousands of miles away from the original rock.

Quartz is the hardest mineral in igneous rocks and survives weathering as solid particles. It is the most common mineral in sediments. Feldspar minerals weather slowly, eventually forming clay minerals and contributing dissolved Ca, Na, and K ions to natural waters. The minerals present in mafic igneous rocks, such as basalt, weather rapidly and are therefore rarely found in sediments.

Clastic rocks are divided into four groups on the basis of dominant particle size (Figure 2-7). *Conglomerates* are composed of boulder to gravel-sized particles; *sandstones* contain particles from 2 mm to 0.062 mm; *siltstones* contain particles from 0.062 mm to 0.004 mm; and *shales* or *mudstones,* are composed of particles that are less than 0.004 mm. It is difficult to see particles less than 0.062 mm in diameter with the naked eye. The classic trick of old-time field geologists for determining whether a rock is siltstone or shale is by using their teeth. Siltstones give a gritty sensation; shales are smooth.

Carbonate rocks are composed of two minerals, calcite and dolomite. Many marine invertebrate organisms secrete shells of calcite or aragonite. Accumulations of their shell fragments form carbonate sediments, which eventually harden into the rock we call limestone. (Limestones are

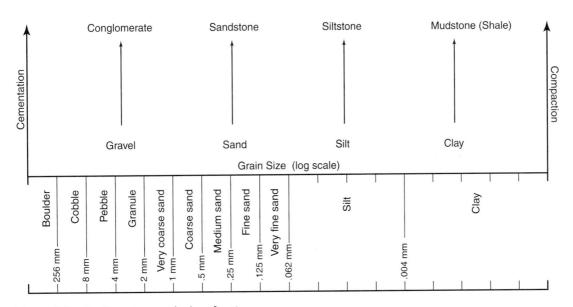

Figure 2-7   Sedimentary rock classification

Clastic rocks are divided into four groups on the basis of dominant particle size. *Conglomerates* are composed of boulder to gravel-sized particles; *sandstones* contain particles between 2 mm and 0.062 mm; *siltstones* contain particles from 0.062 mm to 0.004 mm; and *shales* or *mudstones,* are composed of particles that are less than 0.004 mm.

not green, in spite of their name; they are mined to make lime, CaO, a component of mortar and cement.) During the process of becoming limestone, the aragonite changes to calcite, a more stable form of calcium carbonate. Dolomite is formed when magnesium-rich waters react with limestone.

Metamorphic rocks   **Metamorphic rocks** are formed by the transformation of other rocks at temperatures and pressures that are much higher than at Earth's surface, but not high enough to melt the rock. They form in areas of the Earth's crust where mountain building is occurring and at the margins of magmas where the heat literally cooks the surrounding rock. Metamorphic transformations of rocks can occur with or without water.

The type of metamorphic rock formed depends on the original rock and the temperature and pressure (Figure 2-8). Shales can be cooked and squeezed to produce *slate* or *phyllite*. Coarser grained sedimentary or igneous rocks produce *schist* or *gneiss*. Quartz sandstones form *quartzite;* limestones become *marble*. At low pressures and temperatures, basalts form zeolites, and at intermediate pressures, they become *greenschists, amphibolites,* or *granulites* as temperature increases. Basalts and gabbros become *blueschists* or *eclogites* at very high pressures.

Relative and Absolute Dating

One of the fundamental principles of geology is that, in undisturbed horizontal layers of sedimentary rocks, the underlying layers are older than the overlying layers. **Relative dating** of rocks began in the early nineteenth century when geologists realized that they could correlate layers of sedimentary rocks in different areas using the fossils they contained. Assemblages of certain fossils were unique in time; rocks laid down earlier or later had distinctly different collections of plant and animal fossils. In places where great piles of sedimentary rocks existed, for example, the Grand Canyon in the United States, geologists could assign relative ages to the fossils. Once these fossil assemblages were recognized, geologists could date rocks containing the same fossils in areas where the sedimentary sequence was incomplete (Figure 2-9a).

Note that we have used the term relative dating. This method tells us only whether a particular rock is the same age as, younger than, or older than another rock; it does not tell us the actual age in years. It also does not help us with rocks that do not contain fossils.

The discovery of radioactivity by Marie and Pierre Curie led to huge gains in understanding the geology of our planet. **Radioactive dating** allows us to determine rock ages in years, giving absolute rather than relative ages. Most rocks contain small amounts of radioactive elements such as uranium, thorium, and potassium. The nuclei of some atoms of these elements are unstable, having too much mass. The nucleus eventually breaks down, giving off radiation and/or subatomic particles, until it is gradually transformed into a stable atom. This process, called **radioactivity** or radioactive decay, can be used for dating because it occurs at a constant rate (Figure 2-9b). If we know the rate of decay for a radioactive element, the age of a sample can be determined by using the ratios of radioactive atoms to atoms of the stable decay product. Of course, this method works only if we have a closed system, that is, no product atoms are lost or

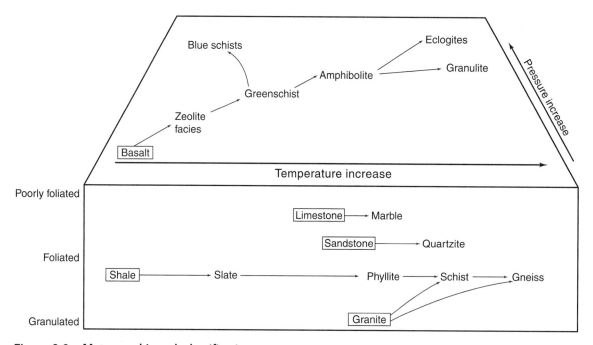

Figure 2-8   Metamorphic rock classification

Metamorphic rocks are classified by mineral assemblages that indicate the temperature and pressure under which they formed, and the composition of the original rock that was deformed. The original rocks are boxed.

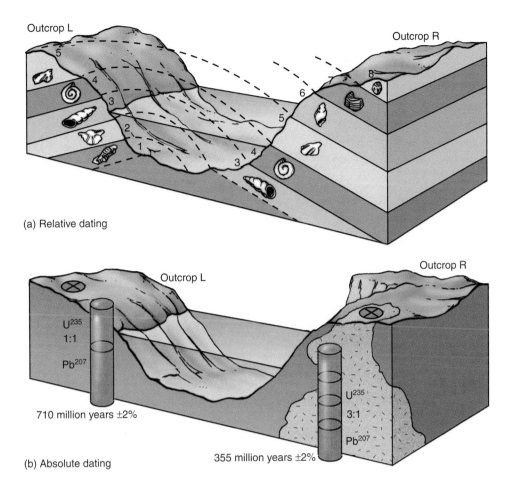

(a) Relative dating

(b) Absolute dating

Figure 2-9   Relative and Absolute Dating

(a) Fossil assemblages used in relative dating can be found in outcrops L and R. Assemblages 3, 4, and 5 may be matched, telling us the rocks in segments of the outcrops formed at the same time. We could not, however, tell how many years ago this formation occurred when fossil evidence was the only dating tool. The concept of superposition states that younger sediments are laid down on older deposits. (b) Outcrops L and R contain the radioisotope uranium 235. It has a half-life of 710 million years and decays to Pb^{207}. Dating with radioisotopes is possible for many types of rocks but is more broadly possible with igneous rocks like the basalt and granite that make up most of the oceanic and continental components, respectively, of Earth's crust. At outcrop L the ratio of U^{235} to Pb^{207} of 1:1 means that half of the U^{235} atoms have decayed to Pb^{207}, so the rocks are one half-life of U^{235} old, or 710 million years. In outcrop R, the ratio of U^{235} to Pb^{207} is 3:1. This means that one-fourth of the original U^{235} atoms have decayed to Pb^{207}, so the rock is one-half of a half-life old, or 355 million years.

gained by the rock, and only if we can identify product atoms as distinct from other atoms of the same element. Because atoms are more easily exchanged when fluids are present than when rock is solid, radioactive dating measures the age at which a crystalline rock solidified or at which a sedimentary rock lithified. Scientists can reproduce measured ages of rocks to within 2–3 percent.

Origin of the Atmosphere and Oceans

Earth's geologic history began about 4.5 billion years ago, when the planet had cooled sufficiently for the crust to become solid. At that time, only a small fraction of Earth's original gas envelope (atmosphere) remained, but there was still a lot of volcanic activity resulting from outgassing of the interior. Intense **meteorite** bombardment of the surface also continued until about 3.9 billion years ago.

The composition of the atmosphere at that time was different from the original hydrogen and helium gases that surrounded proto-earth and from the atmosphere of the present day. There was probably little free oxygen and nitrogen, but large amounts of carbon dioxide, water vapor, sulfur dioxide, and methane were present. By 4 billion years ago, most of the water had condensed to form the first permanent accumulations of water on Earth's surface, the oceans (Figure 2-10).

Neighboring planets do not have oceans. What is unique about Earth given that the histories of formation for all the terrestrial planets are similar? We all know that

Figure 2-10   Formation of Earth's Oceans and Early Atmosphere

Widespread volcanic activity released water vapor and smaller quantities of carbon dioxide, chlorine gas, and hydrogen. This produced an atmosphere containing water vapor, carbon dioxide , methane, and ammonia. As Earth cooled, the water vapor (1) condensed and (2) fell to Earth's surface. There it accumulated to (3) form the oceans.

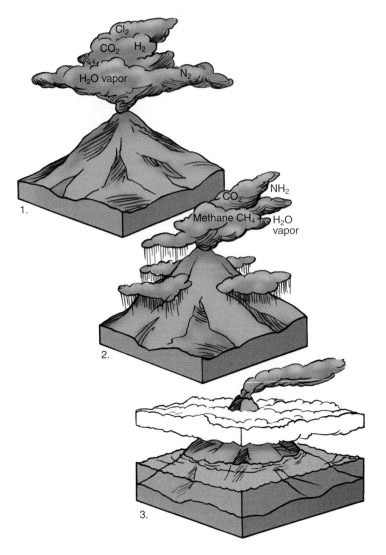

water must remain at temperatures between 0 and 100°C (32 to 212°F) to be liquid. Once Earth had cooled enough to have a solid crust, the Sun became the major temperature regulator at Earth's surface. Thus, average distance from the Sun and rotational period become critical factors in Earth's surface temperature. Earth's average distance from the Sun is about 150,000,000 km (93,000,000 mi); this does not vary much throughout the year because Earth's orbital path is nearly circular. Even though Mars and Venus also have nearly circular orbits, Mars orbits much further from the Sun and so is much colder; Venus orbits closer to the Sun and so is much warmer. Earth's rotational period is also relatively rapid, once every 24 hours, thus, the hemispheres do not warm up or cool down very much while facing toward or away from the Sun. Even with Earth's orbit and rotational period, the surface would average only −21°C if not for the atmosphere. An atmosphere acts like an insulating blanket around a planet, blocking both incoming solar energy and escaping re-radiated energy. The effectiveness of the atmosphere depends on the gases it contains and the wavelengths of the solar and re-radiated energy. Earth's atmosphere raises the average surface tem-

perature from −21°C to 14°C, virtually guaranteeing that the oceans will neither freeze nor boil away.

The oceans have presumably existed continuously since their formation, but their chemical compositions have changed. The high carbon dioxide and sulfur dioxide content in the early atmosphere would have created a very acidic rain, capable of dissolving the minerals in the crust to a much greater extent than occurs today. In addition, volcanic gases such as chlorine dissolved in the rain. These dissolved compounds accumulated in the newly forming oceans, however some were removed by other chemical reactions between ocean water and rocks on the ocean bottom. Eventually, these processes produced an ocean with a chemical composition similar to today's oceans and a balance between inputs and outputs has maintained that composition relatively constant through time.

Cycling and Mass Balance

If input from outer space is discounted, the elements that make up the atmosphere and the ocean must have come from within Earth's interior and been brought to the surface initially by volcanism. Can you get a whole ocean full of water from volcanoes? Is there enough chloride in volcanic emissions to account for all the chloride in the sea? We can answer these questions using simple mass balance calculations.

Sources of Water

Let's look at water. The material brought to Earth's surface by volcanoes comes from the lower crust or the upper mantle. The mantle of Earth has a volume of 10^{27} cm^3 and an average density of 4.5 g/cm^3. To get the mass of material in the mantle, we multiply these figures together.

$$10^{27} \text{ cm}^3 \times 4.5 \text{ g/cm}^3 = 4.5 \times 10^{27}\text{g} \qquad (2.1)$$
$$\text{volume} \ \times \text{density} \quad = \text{mass}$$

By the same method, we obtain the mass of water in the present-day oceans as 1.4×10^{24} g. If all of this water came from the mantle, how much mass has been lost from the mantle? To answer this, we need to compare the mass of the ocean to the mass of the mantle before water loss (the sum of the present-day mantle mass and the mass of ocean water). We calculate:

$$1.4 \times 10^{24}/(4500 \times 10^{24} + 1.4 \times 10^{24}) = 0.00031 \ (2.2)$$
$$\genfrac{}{}{0pt}{}{\text{ocean}}{\text{mass}} \Big/ \genfrac{(}{)}{0pt}{}{\text{mantle mass}}{+\text{ocean mass}} = \genfrac{}{}{0pt}{}{\text{fraction of mantle}}{\text{mass lost to ocean}}$$

Therefore, the mantle must have lost 0.031 percent of its mass as water to have produced Earth's oceans.

The next question to ask is how much water, by weight percent , the mantle could have contained. To figure this out, we need to find some analogous material that is similar in composition to the original mantle material. Scien-

tists think that silicate-containing stony meteorites are good analogs. The average water content of these meteorites is about 0.5 percent by weight. That is about 16 times more than the 0.031 percent we calculated was necessary to account for the present oceans. Therefore, the mantle could very adequately have served as a source for the water of Earth's oceans if there was a sufficient rate of escape from the mantle to the surface. There have been oceans on Earth for about 4 billion years. If we consider the amount of water in volcanic emissions and take an average rate of discharge over the last 4 billion years, we find volcanoes have produced enough water vapor to fill the oceans more than 100 times. Even if 99 percent of this water was recycled, there would still be enough to account for the present-day oceans.

An astronomer, Louis Frank, wondered whether the water on Earth could have come from objects from space that have collided with Earth. He thinks that water has been coming to Earth throughout its history in the form of cometlike balls of ice, estimating that such comets enter Earth's atmosphere at a rate of 20 per minute. He got the idea from examining images of Earth's atmosphere produced from ultraviolet (UV) photometric measurements made by *Dynamic Explorer I*, a polar-orbiting satellite. The UV image captures the "**dayglow**" created by absorption and re-radiation of solar energy by atomic oxygen at an altitude of 290 km (180 mi) above sea level. The dayglow image, in which Earth looks like a yellow ball, had a number of short-lived dark spots, some up to 48 km (30 mi) across and lasting for up to 3 minutes (Figure 2-11). Frank thinks that these spots are the result of ice balls from space breaking up and vaporizing in Earth's atmosphere at an altitude of 1600–3200 km (1000–2000 mi). At the observed rate of occurrence, Earth would receive 0.0025 mm (0.0001 in.) of water per year. Four billion years of such

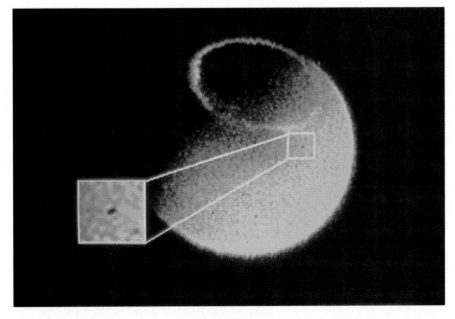

Figure 2-11   Image of Earth's Dayglow

Recorded by ultraviolet photometer aboard the satellite Dynamic Explorer I. The dark spot, or hole, thought to be caused by the vaporization of cometlike balls of ice, is shown in the inset. The circular emission at the top is the northern auroral oval.   *(Courtesy Dr. L. A. Frank, University of Iowa.)*

bombardment would give us enough water to fill the oceans to their present volume. More recent satellite data have produced clearer images that support Frank's claims, but other researchers have not yet been able to substantiate his findings; thus, the concept remains controversial.

Sources of Salts in the Sea

To understand the sources and processes that control the amount of salt in the ocean, we need to look more closely at the chemical and physical weathering processes that break down rock. Chemical weathering releases elements contained in rock by dissolving them. Physical weathering breaks down rocks by various natural processes that crack, split, smash, pulverize, and grind rocks into smaller pieces. Water carries both dissolved and particulate materials to the oceans. In an oversimplified view, water flowing from the mountains toward the oceans encounters increasingly more gentle slopes thereby losing velocity along the way. Because the size of the particles that water can carry depends on its velocity, the larger particles are deposited after relatively short distances, and the finest particles are carried all the way to the ocean where they finally settle out in deep, quiet water. Dissolved materials are also carried to the ocean.

Volcanic gases emitted into the atmosphere may also end up in the ocean. All gases dissolve to some extent in water. Chlorine, sulfur gases, and carbon dioxide in the atmosphere all dissolve in rain and thereby enter the ocean.

Element Mass Balances

In the preceding discussion, we stated that all the material that is now sediment, all the dissolved salt in the ocean, and all the gases in the atmosphere are thought to have come from the **primary crystalline rocks** of the solid Earth. If this is true, then we should be able to add up the masses for each of the elements in sediments, the oceans, and the atmosphere to see whether the total balances with what could have been weathered from primary crystalline rocks (Figure 2-12). In this context, the primary rocks, sediments, oceans and atmosphere are referred to as *reservoirs*.

Estimates for the total element masses in each reservoir are obtained by multiplying the volume of the reservoir by the average element concentration in the reservoir. The mass estimates are most accurate for those reservoirs for which the volumes are known and in which the element concentrations are not highly variable, such that the average value is representative. Ocean and atmosphere element mass estimates are much more accurate than the estimates for primary crystalline rocks and sediments. Rocks and sediments are compositionally heterogeneous and their volumes are difficult to estimate. Another difficulty in making these estimates is separating recycled material from primary material.

For the most common elements in crystalline rocks (Na, Ca, K, Si, Mg, and Fe), the balance sheets agree well, within the error of the estimates. For the volatile elements (Cl, S, C (as CO_2), and N) it appears that large quantities of these elements have come from somewhere other than primary crystalline rock. These volatile elements are much more abundant in the atmosphere, ocean, and sediments than they should be if they were all derived from the volume of crystalline rocks estimated to have weathered out during Earth's history (Table 2-1). This imbalance supports the idea that volcanic activity rather than surface weathering is the source for these elements and for the water in the oceans and atmosphere.

What Happens Over Time?

We have made estimates showing that the present-day mass of the oceans could have accumulated during Earth's history through two separate processes: mantle outgassing through volcanic emissions, and ice comet impact. For these calculations, we used present-day rates, assuming that water accumulated gradually and that rates were relatively constant over 4.5 billion years. Does this mean that the oceans will continue to grow at similar rates?

The area that a given volume of ocean water covers is determined by the depth of the ocean basins; the deeper the ocean, the less area covered by the same ocean volume, and vice versa. The depth of ocean basins depends on

Figure 2-12 Geochemical Balances

All the material lost from primary crystalline rocks is accounted for by material in the atmosphere, oceans, and sediments.

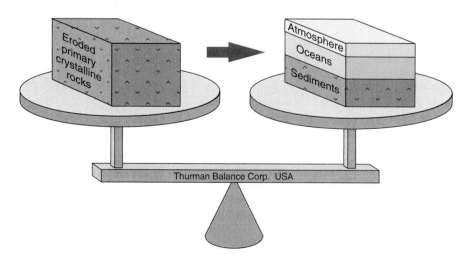

TABLE 2-1   Excess Volatiles

The volatile elements and compounds found in the atmosphere, ocean, and sedimentary rocks in far greater amounts than would have been made available by chemical weathering of primary crystalline rocks are called excess volatiles. Instead, they are believed to have been released from Earth's interior by volcanic activity. Note the absence of free oxygen among these substances. Mass of Elements or Compounds in 10^{20} g

Element	H_2O	C as CO_2	Cl	N	S	H, B, Br, Ar, F, etc
Found in:						
Atmosphere and ocean	14,600	1.5	276	39	13	1.7
Buried in sedimentary rocks	2,100	920	30	4	15	15
Total	16,700	921.5	306	43	28	16.7
Supplied by:						
Weathering of crystalline rocks	130	11	5	0.6	6	3.5
Excess volatiles (unaccounted for by rock weathering)	16,570	910.5	301	42.4	22	13.2

Source: After Rubey, 1951 Courtesy of the Geological Society of America.

the difference in elevation between the continents and the ocean floor. Recall that the continental and oceanic crusts have different compositions and thicknesses. Continental crust is much thicker and less dense, and therefore, sits higher than ocean crust.

The continental crust may not have been present when Earth's crust first solidified. The primary crust of Earth probably consisted of *anorthosite,* a rock containing primarily calcium-rich feldspars. The highlands of the moon are composed of similar material. Very little of this primary crust remains on Earth today, having been recycled by plate tectonic processes into secondary and tertiary crust composed of basalt and granite respectively. Basalt is the product of heat, caused by radioactivity, melting small amounts of Earth's mantle; the melted rock reaches the surface as lava. A basaltic **secondary crust** covers Earth's ocean floors, the lunar maria (flat, low-lying surfaces on the Moon), and the surfaces of Mars and Venus. **Tertiary crustal material** (granite) is produced when oceanic crust and sediments are carried down into the mantle by tectonic processes and remelted. The hot molten rock rises back to the surface, but resolidifies before it reaches the surface. Earth is the only planet in the solar system on which plate tectonic processes are known to occur, and thus, as far as we know, the only planet with granitic tertiary crust development.

The formation of continents by plate tectonic recycling has occurred for at least 3 billion years, while the oceans were forming. As continental crust gradually differentiated from oceanic crust, differences in their elevations increased. In this scenario, the capacity of the ocean basins increased, along with the volume of water produced, simply because the basins were getting deeper. In fact, the surface area covered by the oceans has probably decreased as the amount of continental crust has grown. These processes will be discussed further in Chapter 3.

We have also estimated the source and rates of addition of elements to seawater, assuming present-day rates are representative of average rates since the oceans first formed. Does this mean that the salinity of the oceans has been increasing with time and will continue to increase at similar rates?

We will focus on chlorine because it is by far the most important component of salinity. If salinity changed, we could assume that its largest component would also have changed. Recall that the chlorine in seawater comes from dissolved chlorine gas emitted by volcanoes and that water is also a component of these emissions. If we assume that this process accounts for most of the water and chlorine in the oceans, then for seawater salinity to be constant, the ratio of chlorine to water in volcanic gas must also be constant. Although it is impossible to know, geologists find nothing in Earth's rock record indicating any major fluctuation in this ratio. Thus, on the basis of present evidence, ocean salinity appears to have been relatively constant, but ocean volume and the volume of continental crust have increased throughout Earth's history, and these trends will likely continue.

The Origin of Life

Recent research has led to a number of competing hypotheses for the origin of life on Earth. All hypotheses must identify an environment that contains the right combination of elements to make organic matter, and a source of energy to combine that organic matter into complex molecules. Some scientists have studied the possibility that the organic building blocks for life came to Earth embedded in meteors, comets, and cosmic dust. Some are convinced that life originated at hydrothermal vents on the deep ocean floor, using the heat energy escaping from the Earth. Others favor hypotheses involving lightning as the energy source for forming complex organic molecules in tide pools. Whatever the origin of life on Earth may have been, we are well aware of the importance of oxygen to the more complex life forms.

The Importance of Oxygen

Today, oxygen makes up about 21 percent by volume of Earth's lower atmosphere. In the upper atmosphere, oxygen in the form of ozone protects Earth's inhabitants from solar ultraviolet radiation. Much of the energy represented by this ultraviolet radiation is exhausted in our upper atmosphere, converting diatomic oxygen (O_2) into ozone (O_3). Ozone and oxygen both absorb ultraviolet light. When oxygen molecules absorb this ultraviolet light, the energy taken up exceeds the molecular bond strength. The bond breaks to give two charged oxygen atoms:

$$O_2 + \text{Ultraviolet light} \rightarrow O + O$$

The oxygen atoms produced are very reactive and quickly combine with another oxygen molecule to produce a molecule of ozone:

$$O_2 + O \rightarrow O_3$$

These reactions occur so rapidly in the stratosphere that diatomic oxygen is present in undetectable concentrations.

Note that we did not include free oxygen in the excess volatiles that were produced at Earth's surface by volcanic activity. Although most of Earth's iron is found in the core, there was probably sufficient iron in early volcanic rocks to chemically bind most of the oxygen in volcanic gases. Iron occurs in two forms in volcanic rock, ferrous iron (Fe^{2+}) and ferric iron (Fe^{3+}). Most of the iron in volcanic rocks is ferrous. Any oxygen released by volcanic activity was used up in the conversion of ferrous iron to the ferric state so that no free oxygen was found in the early atmosphere. Without oxygen in the atmosphere, ultraviolet radiation would have reached Earth's surface, damaging or destroying any life forms. Therefore, life must have originated in water deep enough to block the ultraviolet light.

The First Organic Substances

The main elements in organic compounds and in living things are hydrogen, carbon, and to a lesser extent, nitrogen. **Amino acids** and **nucleotides** are the two types of organic compounds that are the building blocks of living tissue on Earth. Only twenty different amino acids exist; there are only five nucleotides. From these, all of the more complex organic molecules are formed, **proteins** form chains of amino acids in different combinations, and the nucleic acids DNA and RNA form chains of the five nucleotides.

One advantage of the absence of oxygen in the early atmosphere was the chemical stability of gases containing reduced C and N, such as methane and ammonia, as well as the more complex molecules formed from them. These gases formed by combination of the hydrogen, nitrogen, and carbon outgassed from the mantle. Stanley Miller's experiments in Harold Urey's laboratory at the University of Chicago, done in the early 1950s, showed that exposing a mixture of hydrogen, methane, ammonia, and water to electrical spark produces a large assortment of organic molecules, including amino acids (Figure 2-13). Miller suggested that lightning in the early atmosphere could have set the stage for life. Throughout the 1960s, scientists continued to work with the Miller-Urey apparatus and, using either electricity or ultraviolet radiation, succeeded in synthesizing all 20 types of amino acids and the five nucleotides. All of these compounds must have been present in the "primordial soup," the ocean water within which life arose.

The production of these small organic molecules in natural environments or laboratory experiments is still a huge step away from making the large, complex molecules found in living organisms, and from those producing a self-organizing, self-replicating organism. In water, many of these small organic molecules are dissolved compounds, separated from other molecules by layers of attached water molecules (hydration spheres). In this setting, it seems impossible for the small molecules to form complex chains. However, experiments have shown that evaporating or heating the organic soup to remove the water, or allowing the organic compounds to settle out onto clay mineral surfaces are ways in which these individual molecules could become linked into larger molecules.

The First Organisms

The step from large organic molecules to living organism is enormous. In organisms, the amino acids and nucleotides that make up proteins and nucleic acids, respectively, are linked in very specific order. Chains produced

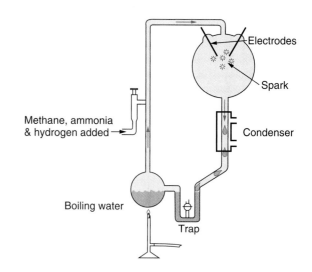

Figure 2-13   Synthesis of Organic Molecules

The apparatus used by Stanley L. Miller and Harold Urey in the 1952 experiment that resulted in the synthesis of the basic components of life, amino acids. A mixture of water vapor, methane, ammonia, and hydrogen was subjected to an electrical spark that provided the energy for synthesis. This mixture is thought to resemble the composition of the atmosphere and oceans that existed when the first organic molecules were formed on Earth.

experimentally contain randomly ordered molecules. Also, organisms are self-organized, able to control and regulate their internal environment, and able to self-replicate. Self-organization requires that an organism be physically separated from the external environment so that its internal chemistry can remain within certain limits, regardless of changes in outside conditions. The first living organisms were probably little more than simple membranes surrounding internal fluids that were very similar in composition to the primordial soup. The membrane had to allow needed molecules to enter the organism, waste products to leave the organism, and regulatory molecules (proteins) to be retained. The membrane also had to be easily mended after being torn in two when the organism divided to form two new individuals.

The earliest life forms were **heterotrophs,** probably obtaining energy by taking in and breaking down organic molecules from the primordial soup. Modern fermenting bacteria have a similar lifestyle; they obtain energy by breaking down complex organic molecules, such as sugars and carbohydrates, which they take in from their environments. The wastes of this process are simpler molecules such as ethanol and acetic acid, which are excreted. Humans utilize fermenting bacteria to create foods, including wine, beer, cheese, and vinegar. These bacteria are heterotrophic, that is, dependent on external organic matter as their food supply. The process of turning food into energy is termed cellular respiration.

The First Autotrophs

Autotrophs, organisms that can synthesize organic molecules from simple inorganic molecules, have a distinct advantage over heterotrophs, organisms that must obtain their organic nutrients from the environment. The first heterotrophs could only survive and increase their numbers if organic molecules were created rapidly enough by ultraviolet radiation and lightning strikes to keep pace with their expanding populations. They were also limited to living only within the nutrient-rich primordial soup. Autotrophs can prosper simply by making what they need wherever they can find the raw materials, but they need to be able to harness an energy source. Possible energy sources include chemical reactions that release energy, electricity (as was used in the Miller-Urey experiments), and light. Organisms that use chemical energy to synthesize organic matter are called **chemosynthesizers**; organisms that use light energy are **photosynthesizers**. We don't know of any organisms that have managed to harness electricity to make organic matter! The first autotrophs may have been chemosynthesizers. Bacteria living today at deep-sea hydrothermal vents use this process, which is one reason that some scientists think life may have begun in these strange environments.

The most abundant energy source around was, and still is, sunlight. Although ultraviolet radiation (light of very short wavelengths) provided the energy that synthesized the first organic molecules out of the primordial soup,

organisms could not harness that energy without sustaining damage to their existing molecules. In that water blocks transmission of ultraviolet radiation, the first organisms must have lived far enough below the water's surface to escape damage. Photosynthesizing organisms evolved to use the non-damaging, longer wavelength, visible light that penetrates water to greater depths.

The first photosynthesizers were probably similar to modern sulfur bacteria that use H_2S as a source of hydrogen and CO_2 as a carbon source. Both of these compounds were abundant in Earth's early atmosphere and in the primordial soup. The waste product of these photosynthesizing reactions is pure solid sulfur, which is excreted harmlessly by the organism. At some point, however, autotrophs evolved that could use water as a source of hydrogen. These organisms, ancestors of modern cyanobacteria, had a great advantage in that water was abundant everywhere. Just as the splitting of H_2S produced sulfur as a waste product, the waste product of those reactions was oxygen, which was toxic to many organisms.

Some of the oldest rocks on Earth, dated at 3.8 billion years old, contain equivocal chemical evidence for bacterial photosynthesizers. Any fossils of these organisms are missing because the rocks have been extensively metamorphosed. The oldest unequivocal evidence, preserved fossil remnants, exists in rocks that are 3.465 million years old (Figure 2-14). Some resemble cyanobacteria; others resemble coccoidal bacteria, and some look as if they were preserved in the act of dividing. From about 2 billion years ago, there is ample evidence that oxygen created by bacterial photosynthesis was abundant in the atmosphere. Over just a few hundred million years, atmospheric concentrations of oxygen increased from less than 1 ppm to about 21 percent, the level at which it remains today (Figure 2-15).

For bacteria that had grown successfully in an oxygen-free world, all this oxygen was a catastrophe! The ozone concentration in the upper atmosphere built up, shielding Earth's surface from ultraviolet radiation and ending their food supply of abiogenically synthesized organic molecules. Also, recall that organic matter is reduced carbon, which is highly reactive with and destroyed by oxygen. This reaction is particularly rapid in the presence of light. Buildup of waste oxygen gas created an environment that was highly toxic to bacteria that lived at the surface. Descendants of such bacteria survive on Earth today in isolated microenvironments that are dark and free of oxygen (anaerobic), deep in soil or rocks, in garbage, and inside other organisms. Death is instantaneous for these anaerobic bacteria when they are exposed to oxygen and light.

Cyanobacteria, however, evolved to exploit this new situation. A metabolic pathway evolved that enabled these organisms to use oxygen to release energy from organic matter. They were the first organisms with the capacity for aerobic respiration! Up to that time, respiration was strictly anaerobic (without oxygen). Because oxygen is so reactive with organic matter compared to the compounds metabolized in anaerobic respiration, it also yields much

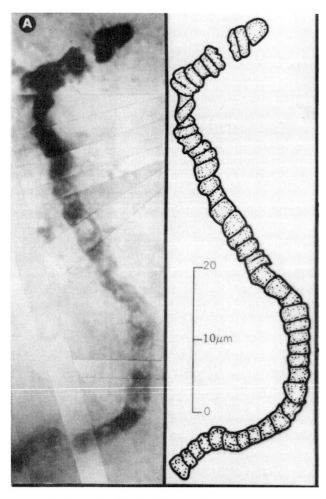

Figure 2-14   Microfossil of the Early Archean Apex Chert

This carbonaceous prokaryotic fossil microrganism (with interpretive drawing) is shown in petrographic thin section of the Early Archean (~3,465 million year-old) Apex chert of northwestern Western Australia.   *(Reprinted with permission from J. William Schopf, from "Microfossils of the Early Acchean Apex Chert: New Evidence of the Antiquity of Life,"* Science *260: 640–46. © 1993 AAAS.)*

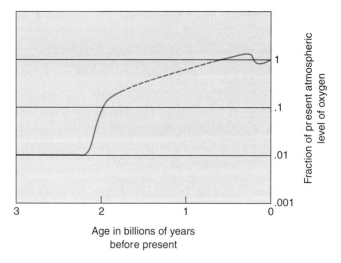

Figure 2-15   Oxygen Concentration in Earth's Atmosphere

The estimated increase in atmospheric oxygen over the past 3 billion years based on data from geochemical and fossil evidence. The steep rise about 2 billion years ago is believed to have resulted from photosynthesis by photosynthetic bacteria. The most uncertain part of the curve is dashed.

Symbiosis and multicellular life

The earliest living organisms were **prokaryotic cells**, they consisted of a single cell with no central nucleus and no internal membranous structures. The genetic material was loose in the cell and not surrounded by a membrane. **Eukaryotic cells**, including modern plant and animal cells, contain membrane-bound nuclei, have complex membrane systems and other intracellular bodies such as mitochondria and plastids. The central nucleus contains the genetic material of the cell, tightly coiled into structures called chromosomes. The oldest preserved eukaryotic cells, about 1.4 to 1.6 billion years in age, are single-celled organisms called **acritarchs**.

There seem to be no transitional forms between prokaryotic and eukaryotic cells. This fact and other chemical and genetic evidence have led biologists to suggest that eukaryotic cells began as cooperative interactions between groups of bacteria. Mitochondria may have begun as predatory, oxygen-respiring bacteria that ate their hosts from the inside out. Eventually these respiring bacteria survived within the host without killing it and at the same time, the host had the advantage of utilizing their metabolic byproducts as food. Similarly, plastids may have been photosynthesizing bacteria that were ingested by larger organisms, ended up surviving within the host and providing it with new organic matter as food. These types of advantageous arrangements between organisms are called **symbiotic relationships**, and are still employed by many modern organisms (see Chapter 15). Prokaryotic cells undergo simple division in order to reproduce. Sometimes, the division is incomplete and two cells remain attached to each other.

more energy. Compared to anaerobic respiration, aerobic respiration yields nearly twenty times more energy! Needless to say, this lifestyle really caught on.

In early cyanobacteria, the same structures were used in photosynthesis and respiration, with photosynthesis occurring during the day and respiration occuring at night. Modern algae and plants can carry on both processes simultaneously because their cells have separate structures for the two processes, mitochondria for respiration and plastids for photosynthesis. Mitochondria are also the site of respiration in animal cells, but those cells lack plastids and cannot synthesize organic matter. Modern cells are much more complex than the cells of the cyanobacteria and their early contemporaries, but the first critical steps in their design likely occurred at that time in Earth's history as a result of the oxygen crisis.

(Life forms in which a number of similar cells live attached together are called colonies.) Scientists think that "mistakes" of this sort led to prokaryotic and eukaryotic multicellular organisms. With these colonial arrangements, member cells could become specialized for specific tasks, such as locomotion, sensing, photosynthesis, respiration. or reproduction. In particular, eukaryotic cell colonies proved to be such an advantageous lifestyle that by about 1 billion years ago, all sorts of eukaryotic organisms appeared based on this design. By 700 million years ago, numerous types of complex, soft-bodied animals existed. Today, the descendants of some types survive nearly unchanged from their ancestors. By 580 million years ago, organisms were making hard shells or skeletons and the record of their history is left much more clearly in the rock record. Animals finally evolved forms that allowed them to live on land just 425 million years ago. Land plants evolved about 460 million years ago (Figure 2-16).

The living world on Earth is a powerful example of symbiosis on a grand scale. Whether in the ocean or on land, photosynthetic organisms and animals live in a beautifully balanced environment where the waste products of one fill the vital needs of the other. This balance maintains Earth's oxygen level at 21 percent and, in the absence of human interference from burning fossil fuel, regulates atmospheric carbon dioxide levels. Fossil fuels, in fact, have linked the inanimate machinery of our civilizations with the biotic symbiosis of Earth. Petroleum and coal deposits are the remnants of plants that were buried more quickly than bacteria could break them down. When we burn these fossil fuels, we oxidize the reduced carbon they contain by combustion with oxygen, a process chemically equivalent to respiration. These deposits provide us with more than 80 percent of the energy we use today. Thus, not only are we dependent upon the present productivity of plants to supply energy required by our life processes, but our modern civilization relies on energy stored by ancient plants from the geologic past. It remains to be seen whether the Earth can tolerate this perversion of the symbiotic contract.

Figure 2-16   Major Events in the Evolution of Earth

About 4.5 billion years ago, Earth's solid crust formed. Oceans are believed to have existed by 4 billion years ago. The earliest non-fossil evidence of photosythetic organisms dates back 3.8 billion years. Although fossils that resemble present-day photosynthetic bacteria exist in rocks 3.5 billion years old, an oxygen-rich atmosphere is not indicated until 2 billion years ago.

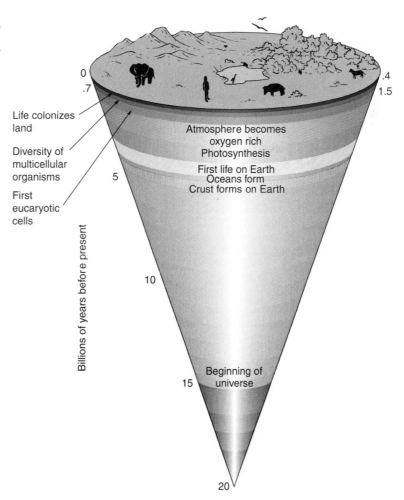

Life on Mars?

Some of the meteoritic material that has fallen to Earth is thought to have originated on the planet Mars. One of these meteorites, called the Allan Hills meteorite, caused a great scientific controversy in 1996. Researchers from NASA reported that they had found evidence of life within the meteorite. One group of skeptics immediately countered with the fact that the meteorite has been on Earth for thousands of years and has had plenty of opportunities to become contaminated by terrestrial life. Another group was skeptical as to the biologic origin of the evidence; they claimed it could have been created by inorganic processes. What tools can scientists use to tell the difference?

The evidence the NASA scientists relied on consisted of mineral shapes resembling those that are precipitated by terrestrial bacteria (Figure 2A). They claimed that these morphologies and mineral associations could only occur through the oxidation-reduction reactions performed by bacteria. The meteorite material was treated to a host of tests designed to test these claims.

Figure 2A   Martian Meteorite

This one half μm tubelike structure is one of the controversial structures found in the meteorite from Mars. *(Courtesy of NASA.)*

One of the signatures of terrestrial life is its preference for light isotopes of elements. If an organism needs to use carbon, it will preferentially use ^{12}C instead ^{13}C; if it needs sulfur, it will use ^{32}S instead ^{34}S, and so on. Thus, organic matter and other compounds generated by metabolic processes will be enriched more in the light isotopes than the heavy isotopes relative to the source material. Isotopic analyses of a test sample do not show enrichment in light isotopes. But what if the life forms that evolved on Mars didn't preferentially use light isotopes? Other researchers are analyzing the samples for the presence of amino acids. Are they present in chains? If so, are the sequences in the chains random or present in definite patterns, a hallmark of life? It is possible that we might mistake a Martian pattern for randomness? Can we distinguish between amino acids from Earth and those from Mars if they both originate from living material?

Whatever the final outcome of the Allan Hills meteorite controversy, we will continue our fascination with the question of life on other planets. The evidence that Mars once had quantities of liquid water at its surface suggests that one of the most important conditions for the origin of life was present on Mars. Did whatever life form may have begun on Mars die out because the water became trapped at the poles? Or, were other conditions never just right for life to form? Are Martian and Earth life forms both seeds from space—Earth's seed flourished, but Mars' seed died?

At this time, our attitude toward life in space seems to be predominantly that life forms must presently exist wherever conditions were right for life to begin—in other words life can adapt and sustain itself, and will likely evolve into intelligent species with whom we can communicate. Thus, we have "listening" telescopes set up, and have sent radio messages and probes containing messages out into space in hopes that someone else might be looking and listening, too.

But what if we find that many of the planets only contain fossil evidence of a past living world? This may be a more likely scenario. Life has existed on Earth very nearly since its beginnings, but we know that it can only survive here temporarily. The life cycle of the Sun will cause that star to increase in size someday far into the future. All of Earth's atmosphere and oceans will boil away and surface temperatures will be far more extreme. Eventually, the Earth itself will be consumed in the solar expansion that precedes solar collapse.

Within the history of life on the planet, there have been precarious times. The buildup of atmospheric oxygen was one of the first. There have been several major subsequent extinctions. The Tertiary-Cretaceous extinction that wiped out the dinosaurs is the most famous, but not nearly the most severe. In every case so far, some life forms have managed to escape demise and to carry on.

Finding lifeless worlds only seems to convince us of our own uniqueness, but finding dead worlds that once lived might give us a different perspective. Is this the kind of evidence we need to appreciate the delicate balance required for coexistence between the living and nonliving world?

Summary

Our solar system, consisting of the sun and nine planets, belongs to the galaxy of stars we call the Milky Way. The galaxies are thought to be accumulations of the debris from the Big Bang, the explosion that formed the Universe. Stars result from accumulation of gaseous masses that collapsed in on themselves, resulting in internal temperatures and pressures extreme enough to set off fusion reactions, burning hydrogen to produce helium. Stars give rise to heavier elements by successive expansions and contractions. The Solar System contains a star, the Sun, and nine planets that formed from gases that were not swept up by the Sun.

Proto-earth was composed mostly of hydrogen and helium, but as it condensed and heated up, these elements were driven off into space. There are indications that Earth became molten and then developed an atmosphere rich in water vapor and carbon dioxide produced by volcanic activity. Methane and ammonia also may have been present in significant quantities. As Earth's surface cooled sufficiently, the water vapor condensed and accumulated in depressions on the surface to give Earth its first oceans.

Life is thought to have begun in the oceans. Ultraviolet radiation combined dissolved methane and ammonia to produce simple organic compounds that are now formed naturally on Earth only by living organisms. As these chance combinations became more complex, the first life forms evolved; these organisms were probably similar to fermenting bacteria. Photosynthesizing bacteria evolved around 3.4 to 3.8 billion years ago. The release of free oxygen as a by-product of photosynthesis began the buildup of oxygen in the atmosphere. Oxygen was lethal to many existing organisms, but eventually organisms evolved that could use oxygen for respiration. Symbiotic relationships led to the evolution of eukaryotic cells, and from these, the great diversity of plants and animals that inhabit the Earth today has evolved over the last 1 billion years. About 460 to 425 million years ago plants and animals, respectively, evolved into forms that could survive on land. In a balanced environment, autotrophic and heterotrophic activities support all kinds of organisms. Fossil fuel burning has perturbed this balance and we have yet to learn the magnitude of the consequences.

Key Terms

Acritarchs
Amino acids
Autotrophs
Basalt
Big Bang
Chemosynthesizers
Core
Crust
Cyanobacteria
Dayglow
Eukaryotic cell
Extrusive rock
Galaxy
Granite
Heterotrophs
Igneous rock
Intrusive rock
Isotope
Light-year
Magma
Mantle
Metamorphic rock
Meteorites
Milky Way
Mineral
Mohorovičić discontinuity (Moho)
Nucleotides
Photosynthesizers
Primary, secondary, and tertiary crystalline rock
Prokaryotic cell
Proteins
Proto-earth
Radioactive or absolute dating
Radioactivity
Relative dating
Sedimentary rock
Silicate
Silicate tetrahedron
Solar wind
Symbiotic relationship

Questions and Exercises

1. Construct a graph with velocity in thousands of km/s plotted along the vertical axis from 0 to 100,000 km/s and light-years plotted along the horizontal axis from 0 to 2 billion. On this base, plot the values given for the constellations Virgo, Corona borealis, and Hydra in Figure 2-2. Establish a line by connecting the three points. How much farther from us is a galaxy for each increase in velocity of 10,000 km/s?

2. Why is it theorized that the observed motions of galaxies were initiated by an explosion?

3. During its history, Earth may have possessed three distinctly different envelopes of gases, or atmospheres. Describe the atmosphere of proto-earth, of Earth when the oceans first formed, and of Earth at present.

4. Why does the fact that new water is continually being released to the atmosphere by volcanic activity not necessarily mean the oceans will progressively cover an increasing percentage of Earth's surface?

5. How does the presence of oxygen (O_2) in our atmosphere help reduce the amount of ultraviolet radiation that reaches Earth's surface?

6. Describe some basic characteristics of living things and list the order in which they evolved.

7. Discuss photosynthesis and respiration, and explain their relationship to the chemical processes of storing and releasing energy by organisms.

8. Describe how geologists date rocks and fossils.

9. As plants evolved on Earth, great changes in Earth's environment were produced. Describe some of the major changes caused by plants.

10. What events must have occurred for life to evolve?

References

Coles, P., 1998. The end of the old model universe. *Nature* 393: 741–744.

Frank, L. A., J. B. Sigwarth, and J. D. Craven, 1986. On the influx of small comets into the Earth's upper atmosphere. *Observations and Interpretations Geophysical Research Letters* 13:4, 303–310.

Glaessner, M. F., 1984. *The Dawn of Animal Life: A Biohistorical Study*. Cambridge: Cambridge University Press.

Gregor, B. C. R. M. Garrels, F. T. Mackenzie, and J. B. Maynard, eds., 1988. *Chemical Cycles in the Evolution of the Earth.* New York: Wiley.

Hagene, B. and C. Lenay, 1986. *The Origin of Life.* New York: Barrons.

Holland, H. D., 1984. *The Chemical Evolution of the Atmosphere and Oceans.* Princeton, NJ: Princeton University Press.

Lineweaver, C.H., 1999, A younger age for the universe. *Science* 284: 5419, 1503–1507.

Margulis, L., 1982. *Early Life.* Boston: Science Books International.

Rawls, R.L., Earth's first organics. *Chemical and Engineering News,* December 1, 1997: 20–22.

Richer, H.B., 1998, White dwarfs sing the blues. *Nature* 394: 825–826.

Rubey, W. W., 1951. Geologic history of sea water—An attempt to state the problem. *Geological Society of America Bulletin* 62:1110–19.

Schopf, J. W., 1993. Microfossils of the early archean apex chert: New evidence of the antiquity of life. *Science* 260:5108, 640–646.

Strom, K. M., and S. E. Strom, 1982. Galactic evolution: A survey of recent progress. *Science* 216:4546, 571–780.

Suggested Reading

Sea Frontiers

Stephenson, F., 1992. *Evolution at Sea.* 38:2, 46–51. Study of marine fossils supports the gradualism of natural selection predicted by Charles Darwin.

Scientific American

Angel, J. R. P., and N. J. Woolf, 1996. *Searching for Life on Other Planets.* 274:4, 60–77. Describes the telltale elemental signatures astronomers must look for to identify conditions similar to our own on other planets.

Barrow, J. D., and J. Silk, 1980. *The Structure of the Early Universe.* 242:4, 118–28. Although the universe is not homogeneous on the small scale of the solar system or even a galaxy, it is homogeneous as a whole.

Binzel, R. P., M. A. Barucci, and M., Fulchignonni, 1991. *The Origin of Asteroids.* 265:4, 88–95. Research reveals that asteroids originate in the belt of the missing planet between Mars and Jupiter. Insight is also gained into the process by which the solar system formed.

Bothun, G.D., 1997. *The Ghostliest Galaxies.* 276: 2, 56–61. Huge galaxies too diffuse to be seen until the 1980s contain mass equal to that of all the previously known visible galaxies, providing new ideas as to how mass is distributed in the universe.

Clarke, B., 1975. *Causes of Biological Diversity.* 223:2, 50–61. Diversity within species and its relationship to natural selection are pursued. The physiological and social implications of the data are discussed.

de Duve, C., 1996. *The Birth of Complex Cells.* 274:4, 50–59. A Nobel Prize recipient explains how natural selection made possible the development of more complex eukaryotic cells from the symbiotic relationships between simpler prokaryotic cells.

Frieden, E., 1972. *Chemical Elements of Life.* 227:1, 52–64. The roles of the 24 elements known to be essential to life are discussed, including some background on how they may have been selected from the physical environment in which life evolved.

Gehrels, T. T., 1996. *Collisions with Comets and Asteroids.* 274:3, 54–61. The nature of asteroids and comets is reviewed. The age and size of some objects involved in major Earth-jarring impacts by extraterrestrial bodies, and the odds of impacts by different sized objects are discussed.

Krauss, L. M., and G. D. Starkman, 1999. *The Fate of Life in the Universe.* 281:5, 58–65. Thermodynamics holds the key to whether life can survive in an ever expanding universe.

Luu, J. X. and D. C. Jewitt, 1996. *The Kuiper Belt.* 274:5, 46–53. Beyond Pluto lies a belt of objects a few tens to hundreds of kilometers across which may be the source of short period comets.

Miley, G. K., and K. C. Chambers, 1993. *The Most Distant Radio Galaxies.* 268:6, 54–61. Radio signals from remote galaxies allow the galaxies to be seen as they were when the universe was one billion years old. This information offers clues to the formation of galaxies and the origin of the universe.

Special Issue. 1994. *Life in the Universe.* 271:4, 44–122. This special issue covers topics from the evolution of the universe through the origin and evolution of life on Earth. It also raises questions about how we will be able to sustain life on Earth.

Stebbins, G. L., and F. J. Ayala, 1985. *The Evolution of Darwinism.* 253:1, 72–85. Advances in molecular biology and new interpretations of the fossil record add to the knowledge of evolution.

Wilson, A. C., 1985. *The Molecular Basis of Evolution.* 253:4, 164–75. Genetic mutations play an important role in evolution at the organismal level.

Oceanography on the Web

Visit the Introductory Oceanography home page for on-line resources for this chapter. There you will find an on-line study guide with review exercises and links to ocean- ography sites to further your exploration of the topics in this chapter. Introductory Oceanography is at http://prenhall.com/thurman (click on the Table of Contents menu and select this chapter).

3 Global Plate Tectonics

The Torres del Paine mountains near Puerto Natales in southern Chile are visible from the Straights of Magellan. They are at the southern end of the Andes mountains, produced by the subduction of the Nazca and Antartic plates beneath the South American plate. (Photo by Michael Giannechini, Courtesy of Photo Researchs, Inc.)

We live on a dynamic planet where movement is the rule rather than the exception—and this refers to more than just our penchant for traveling. Massive amounts of energy released by volcanic eruptions and earthquakes are part of the tectonic process that constantly changes the face of our planet. People who live near the coasts surrounding the Pacific Ocean are particularly aware of this process. Plate tectonic theory explains the motion of the continents and the changing shapes of the ocean basins. As we discussed in Chapter 1, this theory caused a revolution in the field of geology comparable to that caused by Darwin's evolutionary theory in the field of biology. The word Tectonics derives from the Greek word, tektonikos, which means to construct and refers to the building of Earth's crust. As you will discover, however, half of the process is destruction, not construction.

Earth's Crust is not uniformly old nor is it a hard rigid shell as you might expect from our discussions about crustal formation. We do not live on the surface of a hard-boiled egg. A pot of boiling chocolate pudding (or your flavor of choice) is a better analogy. As the pudding surface cools, it develops a hard skin like the crust, but convection cells produced by the heat below carry this crust back down into the liquid where, with luck, it remelts so as not to make lumps in the pudding! The crust often cracks into separate pieces and liquid pudding wells up in the cracks. The separate pieces of the crust are the tectonic plates. The liquid pudding is lava extruded at plate margins. If you throw some marshmallows in, you'll find that the convection cells have trouble pulling them down. These marshmallows are the continents!

Isostasy

Isostasy describes the relative elevations that materials of different densities and thicknesses reach at equilibrium with gravity. Isostatic adjustment of Earth's crust accounts for the difference in heights of the continental and oceanic crusts and contributes to the mechanism of global plate tectonics.

The average density of the basaltic oceanic crust is about 3.0 g/cm^3 and averages 6 km in thickness. The granitic continental crust ranges in density from 2.67 g/cm^3 near the surface to 2.8 g/cm^3 deep beneath the continental mountain ranges. The continental crust averages about 35 km (22 mi) in thickness, but may reach a maximum thickness of 70 km (43 mi) beneath the highest mountain ranges. The granitic continental blocks and basaltic oceanic crust float on the denser mantle beneath (Figure 3-1). This concept of crustal flotation may be compared with the flotation of ice on water. If we take a regularly shaped cube of ice with a density of 0.91 and float it on water with a density of 1.0, the ice will sink into the water until it displaces water equal in mass to the ice. When this displacement, called buoyancy, occurs, 91 percent of the ice mass will be submerged. Similarly, about 55 percent of the mass of the continents is submerged in the mantle. Actually, we would expect about 91 percent of the continents to be submerged because they are, on the average, about 91 percent of the density of the upper mantle. However, where heat rising up from the mantle heats and reduces the density of the upper mantle beneath the continents, this low-density mantle provides isostatic compensation. In situations such as those that exist beneath the southwestern United States and eastern Africa, the hot, low-density mantle material provides the buoyancy that would normally be provided by the deep root of continental crust and causes them to rise to higher elevations than would be expected.

Evidence from the Continents for Plate Tectonics

The supporting evidence for plate tectonic theory comes from the ocean basins and the continents. Collection of much of the supporting data from the sea floor did not begin until the Deep Sea Drilling Project began in 1968; data from the continents has been collected and analyzed for several hundred years, but it took the evidence from the oceans to provide the pivotal clues for construction and

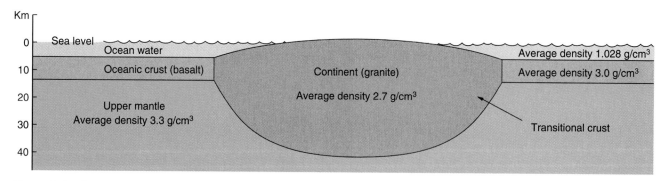

Figure 3-1   Isostasy

Isostasy is a state of equilibrium reached by different components of Earth's crust and the mantle. The lithosphere thereby maintains a uniform average density. Continental crustal units that extend up to 14 km (8.7 mi) above the floor of ocean basins must send deep roots into the mantle.

confirmation of plate tectonic theory. We will begin by discussing the evidence from the continents and examine sea floor evidence in the next section.

Continental Jigsaw

Attempts were made by many early investigators, most notably Alfred Wegener, to arrange the continents together in a manner that would achieve a reasonable fit and support data on ancient climates and correlative rock formations from the continents. Most of these attempts used the existing shorelines as the margins of the continents. Sir Edward Bullard, an English geophysicist, constructed a computer fit of all the continents in 1965. The arrangement that produced minimum overlaps and gaps was formed by continents outlined by the 2000m (6560 ft) depth contour. This contour represents a depth that is approximately halfway down the continental slope. We now know that this position is a better marker for the edge of the continental crust than the coastline (Figure 3-2).

Continental Geology

To test the fit of the continents, we may compare the rocks along the margins of two continents that were at one time united. We need to identify rocks of the same type and age on the continents along their common margin. Identification is not easy in some areas, because during the millions of years since the continents separated, younger rocks have been deposited and cover those rocks that might hold the key to the past history of the continents. However, there are many areas where the older rocks are available for observation, and in these areas we can compare their ages by the use of (1) fossils, the remains of ancient organisms preserved in rocks, and (2) radiometric dating of the rocks.

Ancient Life and Climates   The fossil record of plants and animals in sedimentary rocks can reveal much about the environments of the past. For instance, it is relatively simple to determine whether an organism lived in the ocean or on land. This distinction can normally be made by the characteristics of the sedimentary rock itself. However, the fossils reveal even finer differences in environmental conditions.

Upon examining the location of some fossil assemblages and characteristics of the rocks that give clues to the climate under which they were formed, we find some fossil organisms that could not have lived under present climatic conditions. These anomalous locations of fossil assemblages and rock types may be explained by assuming that either (1) the factors that control climatic belts today are different than those that controlled climates in the past, or (2) the factors controlling climate throughout geologic time have not changed. The second assumption, which seems more logical, lets us conclude that the puzzling assemblages and rock types representing sediments of the past must have been carried to their present positions by the movement of the continental masses.

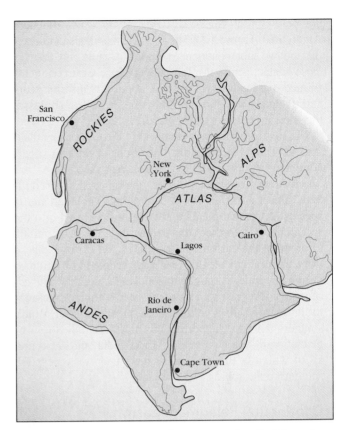

Figure 3-2   Computer Fit of Continents by Sir Edward Bullard

Sir Edward Bullard's fit of the continents was attempted in 1965 after a convincing fit pattern had been achieved in 1958 by the Australian geologist S. Warren Carey without the aid of computers. Bullard's fit was in complete agreement with that of Carey.   (From Continental Drift by Don and Maureen Tarling, (c) 1971, G. Bell & Sons, Ltd. Reprinted by permission of Doubleday & Co., Inc.)

The dominant factor controlling climatic distribution is latitudinal position on the rotating Earth. Assuming that Earth's axis of rotation has not changed significantly throughout history, we may conclude that latitudinal belts have possessed climatic characteristics that have not changed greatly during the evolution of life on Earth.

Laurasia and Gondwanaland   Modern reef-building corals exist only in environments where the water is clear, shallow, and always above 18°C (64°F). Although we cannot be sure, we can probably assume that similar conditions were required for ancient corals. The same species of coral are found in 350 million-year-old rocks in western Europe and eastern North America, as well as throughout the Alps and Himalayas, implying that these areas were in proximity at that time. This fact, as well as other evidence, indicates that from 350 to 250 million years ago a major ocean, the **Tethys Sea**, separated the supercontinent **Laurasia** (composed of what are now North America, Europe, and Asia) from the supercontinent **Gondwanaland** (composed of what are now South America, Africa, India, Australia, and Antarctica) (Figure 3-3). The geologic record also indicates that the two supercontinents periodically came into direct or close contact across the Tethys Sea. These supercontinents formed one landmass, **Pangaea**, about 200 million years ago. Pangaea was surrounded by the ocean **Panthalassa**, the ancient precursor of the Pacific Ocean.

Fossils from sediments that were laid down on land aid in determining the latitudinal positions of the two supercontinents. From 350 to 285 million years ago, there were two distinct floral assemblages, one on each supercontinent. The Laurasian floral assemblage included many species of tropical plants that were incorporated into the sediment that formed the extensive coal beds mined throughout the eastern United States and Europe. These tropical flora indicate that Laurasia occupied low latitudes during that time. Throughout Gondwanaland, the floral assemblage is represented by a few species of plants thought to have grown in a cold climate, presumably at a high southern latitude. Supporting this belief are indications of glaciation in South America, Africa, India, and Australia, all of which at that time must have been very near the southern polar region.

Continental Paleomagnetism

Although we cannot reconstruct the relative positions of continents prior to 200 million years ago with much accuracy, there is enough evidence from that time to the present to give us a very good idea of the paths followed by the continents after the breakup of Pangaea. The first clues to such movements came from studies of continental **paleomagnetism**, that is, studies of the orientation and intensity of magnetism in rocks over time.

Most igneous rocks contain some particles of magnetite (an iron oxide, Fe_3O_4), which become magnetized by and align themselves with Earth's magnetic field at the time that the molten igneous material solidifies into rock. Volcanic lavas such as basalt are high in magnetite and solidify from molten material at temperatures in excess of 1000°C (1832°F); however, the magnetic signatures are not set until the rock cools below 600°C (1112°F), the temperature called the **Curie point**. At the Curie point, the

magnetite particles become fixed in the direction of Earth's magnetic field, permanently recording that field relative to the rock location. If Earth's magnetic field changes subsequent to the formation of the igneous rock, the alignment of these particles is not affected.

Magnetite is also deposited in sediments. While the deposit is in the form of sediment, magnetite particles have an opportunity to align themselves with Earth's magnetic field. This alignment is preserved when the sediment is buried and becomes rock.

Although a number of rock types may be used for the study of Earth's paleomagnetism, the basaltic lavas and other igneous rocks rich in magnetite are best for such studies. The magnetite particles act as small compass needles, as shown in Figure 3-4. They not only point in a north-south direction but also point into Earth at an angle relative to Earth's surface called the **magnetic dip**, or *inclination*, which is related to latitude. At the equator, the "needle" does not dip at all but lies horizontally. It points straight into Earth at the magnetic north pole and straight out of Earth at the magnetic south pole. At points between the equator and the pole, the angle of dip increases with increased latitude. This dip is retained in magnetically polarized rocks. By measuring the dip angle, we can estimate the latitude at which the rock formed. Therefore, when rocks near the equator, such as those in southern India, are found to have a high dip angle, we know they were not formed at that location, but at a higher latitude.

Apparent Polar Wandering Because the present magnetic poles do not coincide exactly with the geographic poles which are the reference points for latitude, we might ask if determining latitude by magnetic dip is misleading. However, if we average the positions of the magnetic poles over the last few thousand years, their positions coincide much more closely with those of the geographic poles. If we assume that this is true for the past, we can determine average positions of the magnetic poles during a time interval and consider them to represent the geographic poles of that time.

As determined from rocks on the continents, the average positions of the magnetic poles have changed significantly through time. The **apparent polar wandering**

Figure 3-3   Fossil and Glacial Evidence Supporting Continental Drift

(a) The distribution of continents 250–350 million years ago showing the three fossil assemblages that help in plate reconstruction. A tropical assemblage (green) is found in North America, Europe, and Asia, supporting the idea that these continents formed Laurasia between 280 and 350 million years ago. A cold-climate flora (gray stippled) is found in association with evidence of glaciation. This assemblage indicates that a large Southern Hemisphere continent, Gondwanaland, which contained present-day South America, Antarctica, Africa, India, and Australia, was in existence between 280 and 350 million years ago. Fossil corals (orange stripes) are believed to have grown in the shallow Tethys Sea that separated the two large continents. (b) By 200 million years ago, Laurasia and Gondwanaland combined to produce the single large continent Pangaea. The ocean that covered the rest of Earth, Panthalassa, may be considered the ancestral Pacific Ocean, which has been decreasing in size since the breakup of Pangaea about 180 million years ago. (c) The present-day configuration of the continents. Arrows show direction of ancient glacial flow.

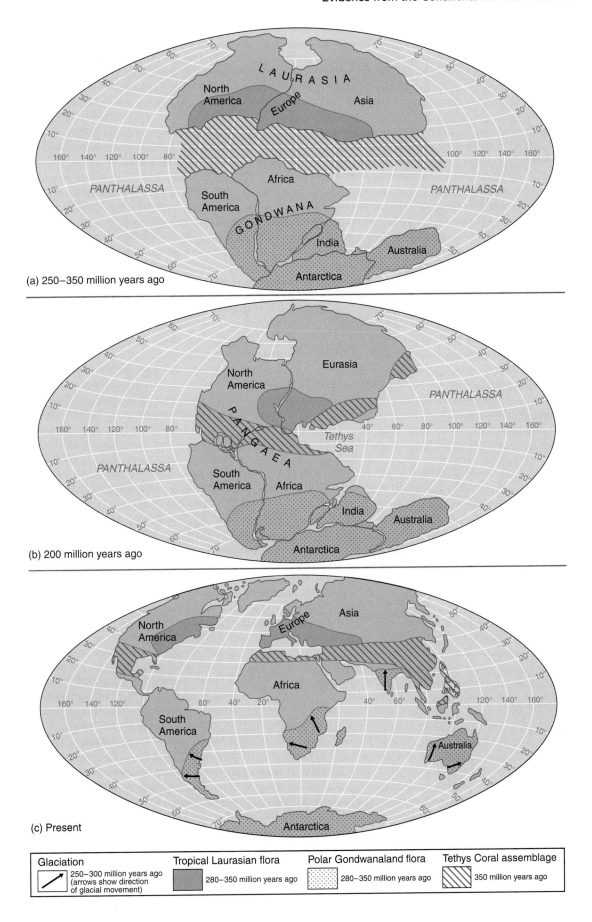

(a) 250–350 million years ago

(b) 200 million years ago

(c) Present

Glaciation	Tropical Laurasian flora	Polar Gondwanaland flora	Tethys Coral assemblage
250–300 million years ago (arrows show direction of glacial movement)	280–350 million years ago	280–350 million years ago	350 million years ago

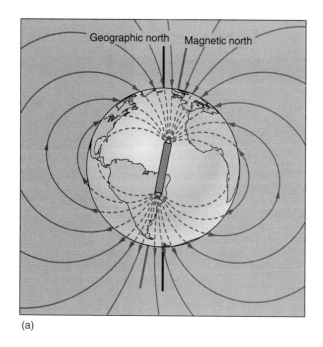

(a)

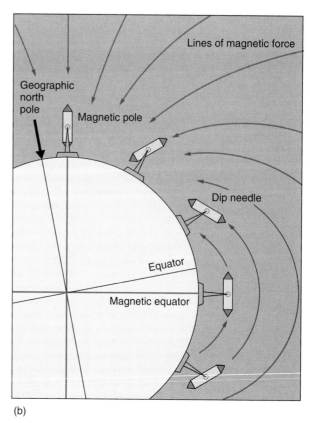

(b)

Figure 3-4 Nature of Earth's Magnetic Field

(a) Earth's magnetic field consists of lines of force much like those a giant bar magnet would produce if placed inside Earth. Although the magnetic poles seldom coincide with the geographic poles, over long periods of time, the average position of the magnetic poles can be considered to represent the position of the geographic poles. (b) Magnetite particles in newly formed rocks are aligned with the lines of force of Earth's magnetic field. This causes them to dip into Earth. The angle of dip increases uniformly from 0° at the magnetic equator to 90° at the magnetic poles. When the rocks solidify, the magnetite particles are frozen into position and become fossil compass needles that tell today's investigators about the strength and alignment of Earth's magnetic field when the rocks formed. The angle at which they dip into Earth also tells at what latitude they formed. (*Reprinted with permission from* The Earth, *4th Edition, Figures 18.8 and 18.9, by Edward J. Tarbuck and Frederick K. Lutgens. Macmillan Publishing Company, 1993*).

curve of the pole determined from North America shows an interesting relationship with that determined from Europe. Both curves have a similar shape, but for all rocks older than 70 million years, the pole determined from North American rocks lies to the west of that determined from European rocks. This difference implies that North America and Europe have changed position relative to the pole and relative to each other (Figure 3-5).

If there can only be one north magnetic pole at a time and its position must be at or near the north geographic pole, a problem exists that can be solved only by moving the continents. The two wandering curves can be made to coincide by moving the two continents together as we go back in time. Now we have a single wandering curve, which shows the magnetic north pole is too far south during the interval of time from 200 million to 300 million years ago. Although the poles have wandered some over geologic time, this wandering is small compared to the dis-

tances shown in Figure 3-5. Rotating the merged continents can bring the poles into the proper position. The fact that moving the continents is the only solution to this problem is very strong evidence in support of the movement of continents throughout geologic time.

Estimating the latitude of formation for rocks from many areas studied throughout the continents on the basis of paleoclimatic and paleomagnetic evidence produces similar results. The most logical explanation for changes over time in both climate and magnetic dip of the rocks at a given location is that the continents have moved.

Magnetic Polarity Reversals Not only have the magnetic poles as determined from continental rocks seemed to wander throughout geologic time, but also the polarity, the direction of the magnetic field, seems to have reversed itself periodically. Reversals in polarity can be described in the following way. A compass needle that points to the

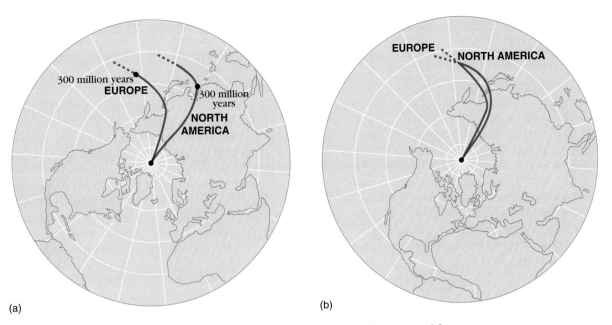

(a)

(b)

Figure 3-5   Apparent Wandering Curves of North Magnetic Pole Determined from Rocks of North America and Europe

(a) Apparent wandering curves of the north magnetic pole, as determined from North America and Europe from the present to 300 million years ago, follow divergent paths. (b) If the apparent position of the pole as determined from North America is moved to a position that makes it coincide with the pole as determined from Europe, the direction and distance equal that required to close the Atlantic Ocean between the two continents. This still leaves the apparent north magnetic pole out in the North Pacific Ocean about 40° south of the geographic north pole. Again, movement of the continent is required to solve this problem. Rotation of Laurasia, which is formed by closing the Atlantic Ocean and joining Europe and North America, can bring the wayward apparent north magnetic pole to a position coincident with the north geographic pole.   *(Reprinted with permission from* The Earth, *4th Edition, Figure 18.10, by Edward J. Tarbuck and Frederick K. Lutgens. Macmillan Publishing Company, 1993.)*

magnetic north today, would point south during a period of reversal. We can not fully explain why these reversals occur, but for the last 76 million years they have happened once or twice each million years. A change in polarity occurs over a few thousand years. It is identified (on the basis of magnetic properties of rocks) by a gradual decrease in the intensity of the magnetic field of one polarity, followed by a gradual increase in the intensity of the magnetic field with opposite polarity. The time during which a particular paleomagnetic condition existed can be determined by the radiometric dating of the igneous rock from which the paleomagnetic measurements were taken. Earth's present magnetic field has been weakening for the last 150 years, and some investigators believe that the present polarity will be reversed in another 2000 years.

Evidence for Plate Tectonics from the Oceans

So far, we have considered only data obtained from the continents. This continental evidence alone did not convince many geologists that the continents had moved. For many years, no other data were available because exten-sive sampling of the deep-ocean bottom was not technologically feasible until the 1960s.

The Deep Sea Drilling Project, which began in 1968, and the Ocean Drilling Program, which followed it in 1983, have added greatly to our knowledge of the ocean floor. The impetus for such programs was largely provided by the need for observations of oceanic sediment and crust to check the theory of sea-floor spreading (see Chapter 1). Extensive geophysical studies, including those related to the paleomagnetic characteristics of Earth's crust, were first carried out in the 1950s. The results of these studies gave the planners of the Deep Sea Drilling Project clues as to where to concentrate the drilling to gain the greatest amount of new knowledge.

Seafloor Paleomagnetism

A detailed Deep Sea Drilling Project study of seafloor paleomagnetism in the Pacific Ocean by scientists at Scripps Institution of Oceanography identified narrow strips of sea floor where the magnetic properties of the oceanic crust differed from those in crusts currently forming. These strips represent **magnetic anomalies**. The magnetic anomalies run parallel to the Juan de Fuca Ridge. Each anomaly represents a period when the polarization of Earth's magnetic field was reversed compared with today's. The anomalies are separated by bands of ocean crust that display present-day polarity.

Frederick Vine and Drummond Matthews observed that the sequence of polarity changes on one side of an **oceanic ridge** (an oceanic mountain range) is identical to the sequence of reversals on the opposite side of the ridge. Dating of the reversal points on both flanks of the ridge showed that the rocks became older with increased distance from the ridge axis. This evidence indicated that new oceanic crust was being formed at the oceanic ridges and moving away on each side of the ridge (Figure 3-6). Having determined from continental studies the dates when many of the more recent magnetic reversals occurred, it

was possible to determine the age of the ocean floor at each strip boundary. Dividing the width by the number of years that the polarity lasted produced the rate at which the ocean floor appeared to be moving away from the ridge (Figure 3-7).

Confirmation of the spreading of the oceanic crust away from the oceanic ridges required radiometric dating of the crust from various locations. In addition, because sediment could not be laid down on crustal material until it formed at the oceanic ridges, it would be expected that the fossil assemblages observed in sediments immediately overlying the oceanic crust would be representative of organisms that existed at the time of crustal formation. It would also be expected that the sediment thickness would be greater on older sea floor than on younger sea floor. A

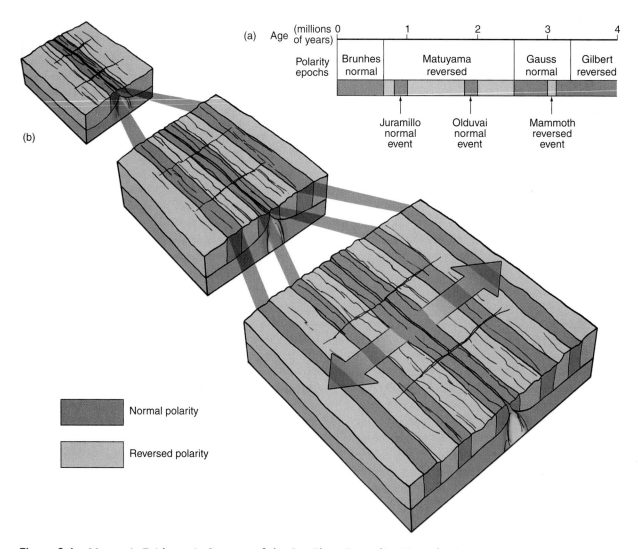

Figure 3-6   Magnetic Evidence in Support of the Sea-Floor Spreading Hypothesis

(a) This time scale of Earth's magnetic field in the recent past was developed by establishing the magnetic polarity for volcanic lavas of known ages. (b) New sea floor records the polarity of the magnetic field at the time it was formed. Hence it behaves much like a tape recorder, as it records each reversal of Earth's magnetic field.   *(Data from Allan Cox and G. B. Dalrymple. Reprinted with permission from* The Earth, *4th Edition, Figure 18.13, by Edward J. Tarbuck and Frederick K. Lutgens. Macmillan Publishing Company, 1993.)*

Figure 3-7 Magnetic Time Scale, Heat Flow, Ocean Depth, and Lithosphere Thickness

Black bands represent periods when the magnetic polarity of Earth was the same as at present; and white bands represent periods when the polarity was reversed. The curve shows that the ocean gets deeper and the lithosphere thicker with increasing age, while the rate of heat flow to Earth's surface through the lithosphere decreases as the lithosphere ages. Although this curve cannot be accurately applied to any particular region of the ocean basins, it indicates that, in general, heat flow decreases while ocean depth, thickness of the lithosphere, and age of the ocean floor increase with increasing distance from the spreading centers. Time is the independent variable for all of these changes. The geologic time scale is shown along the left margin (see inside front cover).

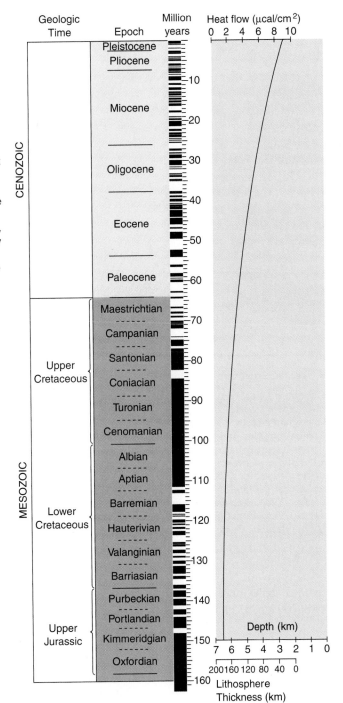

significant task of the Deep Sea Drilling Project was to check the age of the ocean bottom by drilling through sedimentary sections into the oceanic crust. Although not all attempts to do this met with success, fossil assemblages immediately overlying the crust and radiometric age determinations of the crust confirmed that the oceanic floor is moving in the manner indicated by the paleomagnetic data. The youngest oceanic crust was found at the axes of the oceanic ridges, and the age of the crust increased with increasing distance from the ridge crests (Figure 3-8).

Sea-Floor Spreading

New ocean floor appears to be forming at the oceanic ridges and rises. It then moves away from the ridge axes, and volcanic processes create yet a younger strip of ocean floor. Thus, the axes of oceanic ridges and rises are referred to as **spreading centers**. Harry Hess (see Chapter 1) was right, and we now use his term, **sea-floor spreading**, to describe the process. It is, however, more than the sea floor that is moving. Although the details are far from clear, the mechanism of spreading involves two fundamental units

Figure 3-8   Relative Age of Oceanic Crust beneath Deep-Sea Deposits

Notice that the youngest rocks (bright red areas) are found along the oceanic ridge crests and the oldest oceanic crust (brown areas) is located adjacent to the continents. When you observe the Atlantic basin, a symmetrical pattern centered on the Mid-Atlantic ridge crest becomes apparent. This pattern verifies the fact that sea-floor spreading generates new oceanic crust equally on both sides of a spreading center. Further, compare the widths of the yellow stripes in the Pacific basin with those in the South Atlantic. Because these stripes were produced during the same time period, this comparison verifies that the rate of sea-floor spreading was faster in the Pacific than in the South Atlantic.   *(Reprinted with permission from* The Earth, *4th Edition, Figure 19.14, by Edward J. Tarbuck and Frederick K. Lutgens. Macmillan Publishing Company, 1993. After* The Bedrock Geology of the World, *by R. L. Larson et al. W. H. Freeman, 1985.)*

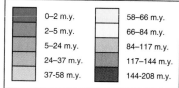

0–2 m.y.		58–66 m.y.	
2–5 m.y.		66–84 m.y.	
5–24 m.y.		84–117 m.y.	
24–37 m.y.		117–144 m.y.	
37–58 m.y.		144–208 m.y.	

of Earth's structure that are confined to the lithosphere and the asthenosphere–the uppermost 700 km (435 mi) of Earth (Figure 3-9).

The **lithosphere** is a relatively cool, rigid shell that includes the crust and upper mantle. It is broken into a dozen or so lithospheric plates (Figure 3-10). Our first clue as to the locations of lithospheric plate boundaries was that most tectonic events such as earthquakes and volcanic activity are confined to linear belts. Underlying the lithosphere is a high-temperature, plastic layer within the mantle. This layer, called the **asthenosphere** (weak sphere), flows slowly, allowing the rigid lithospheric plates resting on its upper surface to move (Figure 3-11). The ultimate fate of a lithospheric plate, if mass is to be conserved, is **subduction**—a process by which the plate descends beneath another plate and is ultimately resorbed into the mantle.

A Possible Mechanism for Sea-Floor Spreading   The mechanism of lithosphere formation at the oceanic ridges and its movement away from those ridges is not yet fully understood. However, the lithosphere thickens, from a few kilometers near the ridge to more than 200 km (124 mi)

beneath some continental regions. The increasing thickness of the lithosphere correlates well with increasing depth of the ocean and decreasing **heat flow** as the lithosphere ages (Figure 3-7), as we will discuss below.

Within Earth, radioactive atoms are breaking down and releasing energy that must find its way to the surface as heat. It is probable that this heat moves to the surface through **convection cells**, circulation cells that carry the heat upward in regions of oceanic spreading centers. If this is the case, there must be regions on Earth where the cooler portions of the mantle descend to complete the convection cell. Measurements of heat flow taken throughout Earth's crust show that the quantity of heat flowing to the surface along the oceanic ridges is as much as eight times greater than the average. In addition, areas with known **ocean trenches** (linear seafloor depressions) have heat flow that is as little as one-tenth the average.

The lateral movement that occurs between the regions of creation (ocean ridges and rises) and the regions of destruction (trenches) of the lithosphere may originate within the asthenosphere. Movements in the asthenosphere may carry the more rigid lithosphere, which contains the

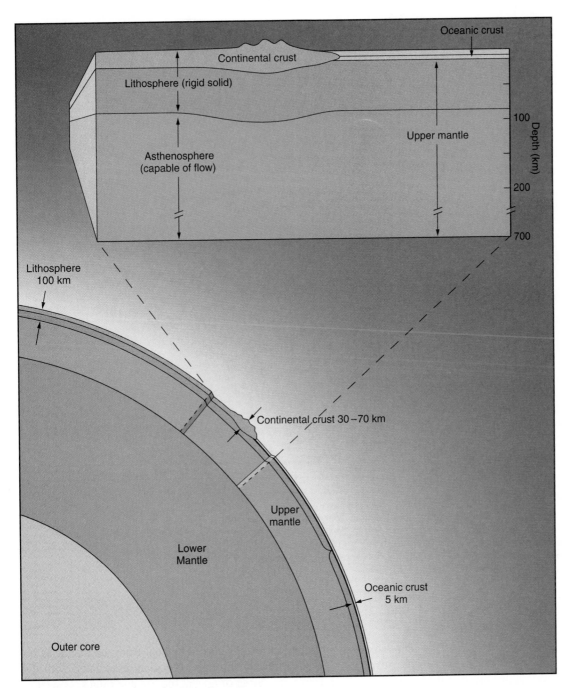

Figure 3-9   Lithosphere and Asthenosphere

The lithosphere is the rigid outer shell of Earth's structure that includes the crust and cooler
upper mantle. It is supported and allowed to move across Earth's surface by flow within the
warmer, plastic, underlying portion of the mantle called the asthenosphere. This weaker, possibly
partially molten, layer is thought to extend to a depth of approximately 700 km (435 mi). The
three major subdivisions of Earth—core, mantle, and crust—are shown to clarify the position of
the lithosphere and asthenosphere.   *(Adapted from* Earth Science, *6th Edition, Figure 5.19, by
Edward J. Tarbuck and Frederick K. Lutgens. Macmillan Publishing Company, 1991.)*

crust and upper mantle. It has been observed that the
trenches do not fit the isostasy patterns characteristic of
the continents, and it appears that some force other than
gravity may be pulling the lithosphere down in these regions. The descending limbs of convection cells might
well provide that force.

The observed relationship of increasing lithospheric plate
thickness, age of ocean floor, and ocean depth, along with
the decrease in rate of heat flow through the ocean floor
with increasing distance from spreading centers can be explained by the convection process. The lithosphere at the
spreading centers is thin because of high temperatures

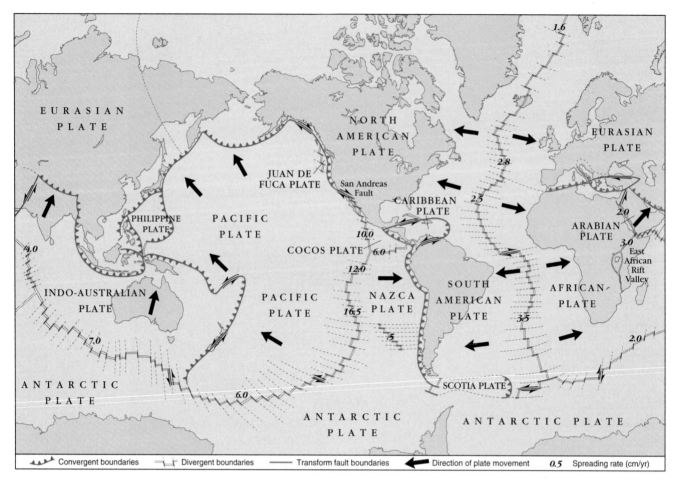

Figure 3-10   Lithospheric Plates

Lithospheric plates are defined by the axes of submarine mountain ranges (gray double line) offset by transform boundaries (green) and by deep-ocean trenches (red toothed curves). The plates move from 1.5 centimeters (0.8 inches) per year (North and South American plates) to more than 10 centimeters (5.4 inches) per year (Pacific Plate).   *(Reprinted with permission from E. J. Tarbuck and F. K. Lutgens,* The Earth: An Introduction to Physical Geology, *3ʳᵈ Edition, Figure 5.11. Macmillan Publishing Company, 1991.)*

immediately beneath the ocean floor. This heat causes expansion of the upper mantle rocks, raising the oceanic ridges and rises to elevations well above those of the ocean floor on either side of the upwelling heat. As the lithosphere moves away from the spreading centers toward zones of downward convection (downwelling) beneath the trenches, it encounters decreasing rates of heat flow from the underlying mantle through the ocean floor. This results in the lithosphere increasing in thickness as the cooling asthenosphere is converted to rigid lithosphere. The thermal contraction of hot, fluffy asthenosphere into cold brittle lithosphere accounts for the increasing depth of the ocean away from the spreading centers.

Hot Spots and Mantle Convection

Evidence indicates that about 94 percent of the energy driving the plate tectonic process is derived from radioactivity within the mantle. The other 6 percent appears to originate in Earth's core. The heat from the mantle is released primarily through volcanism associated with the passive upwelling of lava at the oceanic ridges, as well as volcanism associated with subduction. The heat from the core is released by plumes of hot material that may rise from the **core-mantle boundary (CMB)**, through the mantle, to create what are called **hot spots**. Some 41 hot spots are presently active on Earth's surface (see Chapter 4). Most hot spots are closely associated with oceanic spreading centers, and the large amount of lava they provide has created islands such as Iceland. Others are at midplate locations and produce island chains such as the Hawaiian Islands. Still others rise beneath continents and are responsible for hydrothermal phenomena like those in Yellowstone National Park.

The basalt typically produced at spreading centers contains reduced concentrations of potassium, rubidium, cesium, uranium, and thorium, relative to basalt typical of hot spots. This observation has led to the conclusion that

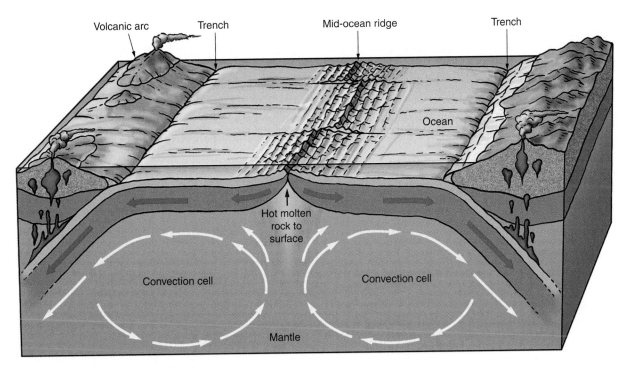

Figure 3-11 Movement in the Asthenosphere

Lateral flow within the asthenosphere is a possible mechanism for the lateral movement of the lithosphere, including the crust. The asthenosphere is a low-strength, plastic zone within the mantle where pressure is low enough and temperature sufficiently near the melting point of the material to allow it to flow slowly. Some form of thermal convection within the asthenosphere appears to create new lithosphere at the oceanic ridges and rises and may carry old lithosphere back into the mantle to be subducted beneath oceanic trench systems. *(Reprinted with permission from* The Earth, *4th Edition, Figure 18.11, by Edward J. Tarbuck and Frederick K. Lutgens. Macmillan Publishing Company, 1993.)*

the spreading-center basalt, or **midocean ridge basalt (MORB)**, is derived from the upper mantle, whereas the hot-spot basalts are from the lower mantle near the core-mantle boundary.

Evidence suggests that the mantle is layered, the bottom being 2 to 3 percent denser than the top. The most distinct density interface, occurring at a depth of 670 to 700 km (416–435 mi), divides the mantle into upper and lower layers. This interface may serve as a barrier to mixing between these layers. However, it appears that this barrier can be penetrated by the lower mantle plumes carrying heat up from the lower mantle, and by what may be subducted lithospheric plates detected more than 2000 km (1240 mi) beneath trenches (Figure 3-12).

If heat from the core is a significant source of energy driving mantle convection, a three-dimensional spherical model of convection in Earth's mantle should show that upwelling occurs as cylindrical plumes arising from the base of the mantle (Figure 3-13). In this model, the downwelling occurs in sheets that form a ringlike pattern around the upwelling plumes. The overall pattern is similar to what we now see occurring in the Pacific Ocean. A discussion of some of the surface phenomena resulting from hotspot activity is found in Chapter 4.

Plate Boundaries

The process of sea-floor spreading sets up stresses within the lithosphere. Some plates move at relative velocities up to 16 cm (6 in)/yr. There are seven major and a half-dozen or more minor plates. There are three fundamental types of plate boundaries. A boundary where new lithosphere is added along an oceanic ridge is called a **divergent (constructive) boundary**. A boundary where plates collide and one subducts the other is called a **convergent (destructive) boundary**. Lithospheric plates may also move past one another along a **transform (shear) boundary** (Figure 3-14). All of these plate motions cause earthquakes. Records of this activity collected at seismic (seismos = earthquake) stations throughout the world allow seismologists to identify the locations and nature of the motion along plate boundaries (Figure 3-15).

Seismic Waves

The forces that cause earthquakes generate low-frequency **seismic waves.** The speed of these waves as they travel through Earth depends on the density of the rocks or sediments they pass through. As a wave travels outward

Does the Wind Blow in Earth's Mantle?

Seismic techniques are becoming so sophisticated that it is now possible to detect small variations in shear wave velocities that can be interpreted as temperature differences with in the Earth's mantle from the surface to the core-mantle boundary (CMB). It is now possible to detect the location of hot plumes rising from the CMB to the surface and to locate the position of cold slabs descending in the opposite direction as is shown in Figure 3-12.

A similar investigation of mantle shear wave velocities has revealed an interesting phenomenon beneath Africa. A low shear wave anomaly (increased temperature anomaly) is continuously rising from the CMB beneath the southeastern Atlantic Ocean to a position beneath the East African rift system. This plume can be detected in the cross-section of figure 3A by the red shading on which are marked the location X, Y, and Z (they are at depths of 2500, 1500 and 700 kilo-meters, respectively). The position of X, Y, and Z beneath Africa are shown on the maps below the cross-section. The horizontal distance from the origin of the plume at the CMB to the position of the East African rift zone is 4000 km (2480 mi). As most models of plumes rising from the CMB show them rising vertically, the explanation for the 4000 km deviation from vertical by this plume must be explained.

The dashed line in Figure 3A represents the seismic discontinuity at 670 km depth that separates the upper mantle from the lower mantle. It is clear that the plume's tilt to the northeast through the lower mantle continues into the upper mantle where it surfaces at the East African rift zone. The explanation for the slanting rise of this plume is not known, but it has been speculated that there is an advection (horizontal flow) or "mantle wind" present in the lower mantle beneath Africa.

Figure 3A   Mantle Plume Beneath Africa

The red line on the map shows the alignment of the 140° cross-section of shear wave velocity anomalies in the Earth's mantle. The plots to the right of the maps show a dimensionless plot of the deviation of the shear wave velocity from reference value along a vertical profile. Movement to the right indicates the amount of amount of velocity decrease due to high temperature relative to the reference value. The plume is clearly identified as present at the depths shown on each map.   *(Reprinted with permission from Jeron Ritsema, et al. Complex shear wave velocity structure imaged beneath Africa and Iceland. Science 286: 1925–28. Figure 4, p 1927. © 1999, American Association for the Advancement of Science.)*

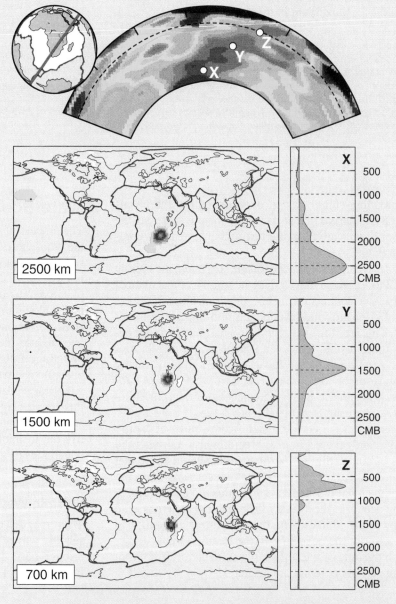

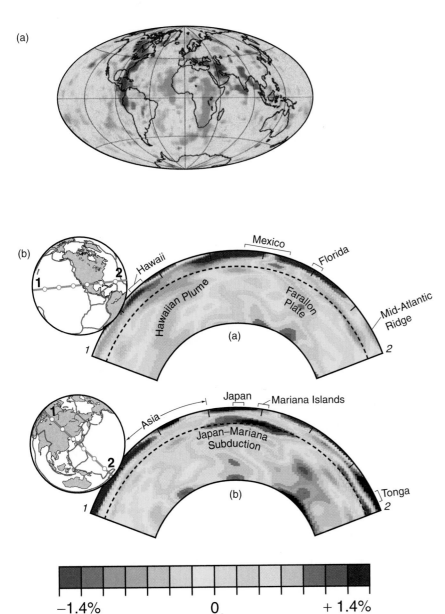

Figure 3-12 **Shear-Wave Tomography. A Model at a Depth of 1500 Km in the Mantle**

(a) This image shows hot, low-velocity (red) and cold, high-velocity (blue) regions at the 1500-km (937-mi) depth in Earth's mantle. The high-velocity anomaly beneath North America and South America is the Farallon Plate that subducted along the eastern margin of the Pacific Ocean more than 100 Ma. It can be detected to a depth of 2800 km (1740 mi). (Courtesy of Hans-Peter Bunge. Reprinted with permission from Nature 397:571b, 203 © 1999. Macmillan Magazines Lt. (www.nature.com.)) (b)Two 140° cross-sections of mantle beneath the eastern (a) and western (b) Pacific Ocean. The dots on the location maps are separated by 20° of arc. The Hawaiian plume is shown in (a) by the red low velocity anomaly and the Farallon plate is shown by a blue high velocity anomaly. (b) The plates being subducted at the Mariana and Japan trenches are identified by the blue high velocity region of cross-section. Reprinted with permission from Jeroen Ritsema et al, Complex shear wave velocity structure imaged beneath Africa and Iceland. (Science 286:5446, 1925–28. Supplementary Figure 1. © 1999, American Association for the Advancement of Science.)

from the source (the earthquake's epicenter), it divides into component waves that travel at different speeds. The most important of these in earthquake studies are the **P and S waves**.

The P wave (for primary or compressional) compresses and extends the material it travels through. Springs exhibit this type of phenomenon when you pull them, causing extension, or push them, causing compression (Figure 3-16a). The S wave (secondary or shear) creates transverse motions like those created by moving the end of a spring up and down or back and forth (Figure 3-16b). There are two S waves, S1 and S2, that move at right angles to each other. Together, the P and S waves create motion in three dimensions (Figure 3-16c).

Solid rocks transmit P and S waves, but liquids transmit only P waves because liquids cannot be sheared. The pat-

tern of S wave arrivals at the Earth's surface are a major line of evidence that the outer core of the Earth is liquid (see Chapter 2).

First Motion Studies

Patterns of seismic waves recorded at seismograph stations distributed over a large area give information about fault orientation and direction of fault motion (Figure 3-17a). The first earthquake waves to arrive at recording stations are P waves, which compress the rocks in the direction they travel through them. This compression causes the first motion recorded at the seismographs to be an upward displacement. In contrast, opposite to the direction of movement, rocks are stretched or dilated, and the first motion recorded at stations in these areas is downward. Thus, the area around an

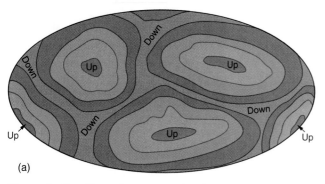

(a)

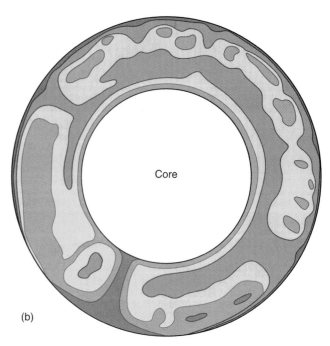

Core

(b)

Figure 3-13 Three-Dimensional Spherical Model of Convection in Earth's Mantle

According to some models, the circulation in the mantle appears as shown here, with half the heat coming from within the mantle and half from the core/mantle boundary. (a) At a depth midway between the surface and the core, mantle circulation shows plumelike upwelling surrounded by rings of downwelling. Upwelling is shown as high (red) or moderate (orange). Green shows little vertical motion. Blue and violet show increasing rates of downwelling. (b) This longitudinal cross section of the mantle shows relative temperature. Blue represents the coldest near-surface mantle material. Green, yellow, and orange indicate increasing temperature, with red representing the hottest material associated with the upwelling plume and the core/mantle boundary.   *(After D. Bercovici et al., Three-Dimensional Spherical Models of Convection in the Earth's Mantle, Science 244, pp. 950-55, 26 May, 1989. Used with permission of the author and publisher. Copyright 1989 A.A.A.S.)*

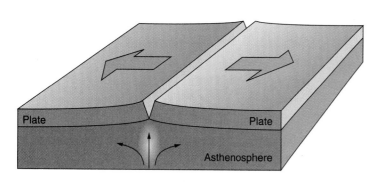

(a)

Figure 3-14 Lithospheric Plate Boundaries

(a) Divergent (constructive) plate boundary. New material is added to plates at the spreading centers. The plates then diverge from one another as they move away from the spreading center. (b) Convergent (destructive) plate boundary. Lithospheric plates are destroyed when they converge at trenches and one subducts to be melted back into the mantle. (c) Transform (shear) plate boundary. Spreading centers are offset by transform faults where plates slide past one another without the creation or destruction of lithosphere.   *(Reprinted with permission from Earth Science, 6th Edition, Figure 6.7, by Edward J. Tarbuck and Frederick K. Lutgens. Macmillan Publishing Company, 1991.)*

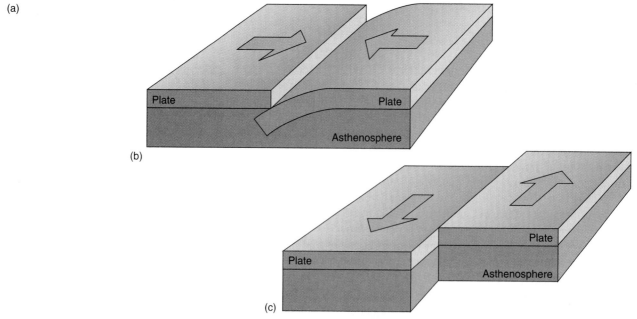

(b)

(c)

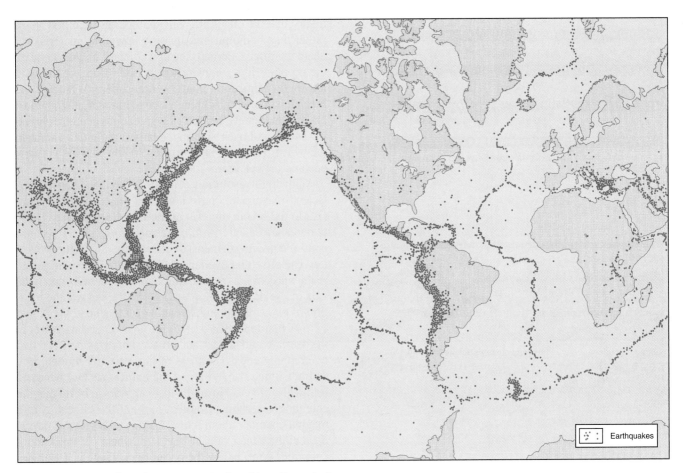

Figure 3-15   Earthquake Patterns Define Plate Boundaries

Orange dots show locations of major earthquakes worldwide over a nine-year period. Patterns of earthquake locations correlate well with plate boundaries.   *(Reprinted with permission from* The Earth: An Introduction to Physical Geology, *3ʳᵈ Edition, Figure 5.11, by E. J. Tarbuck and F. K. Lutgens. Macmillan Publishing Company, 1991.)*

earthquake is divided into quadrants on the basis of these first motions, and this information indicates the orientation and motion of the fault. The surface orientation of a fault is the **fault strike**; the angle of the fault plane relative to the surface is the **fault dip**. Movement of faults along the fault dip are either *normal fault motion* (where the ground on either side moves apart or extends) or *reverse fault motion* (where the ground moves together or compresses). Motion also may be *strike-slip* (the ground moves horizontally parallel to fault strike), or, most commonly, some combination of dip and strike-slip (Figure 3-17b).

Divergent Boundaries

The term *spreading rate* applies to the widening rate of an ocean basin resulting from the motion of both plates away from the spreading center where they are created. We call the mountain ranges associated with oceanic spreading centers oceanic ridges or oceanic rises. Because deepening

of the ocean floor correlates directly with the amount of time the lithosphere has cooled, the rate of sea-floor spreading significantly effects the steepness of the flanks of associated mountain ranges. The faster the spreading rate, the broader the mountain range. These gently sloping features are the **oceanic rises.** The best example is the East Pacific Rise between the Pacific Plate and the Nazca Plate, where spreading rates are as high as 16.5 cm/yr (6.5 in./yr). The Mid-Atlantic Ridge, with spreading rates of 2 to 3 cm/yr (0.8-1.2 in./yr), displays the steeper slopes characteristic of slow spreading (Figure 3-18).

Project FAMOUS, the French-American Mid-Ocean Undersea Study conducted in 1974, made observations in the Mid-Atlantic Ridge's **rift valley**, the depression that runs along the axis of the oceanic ridge. Those observations provide a clue to the spreading process. Fissures oriented parallel to the ridge axis range from hairline size near the center of the rift valley to more than 10 m (32.8 ft) in width near the margins of the valley, supporting the

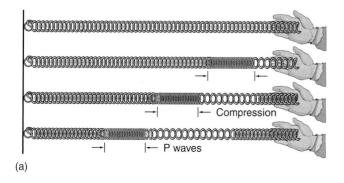

(a)

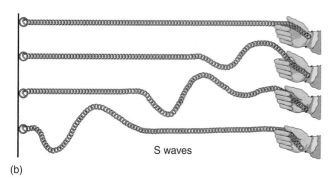

(b)

P wave

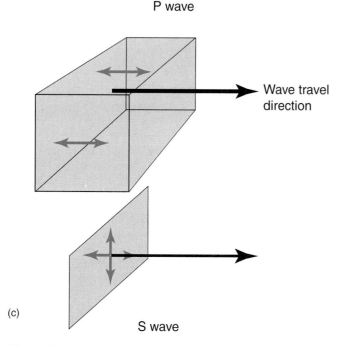

(c)

S wave

Figure 3-16   P and S Waves

(a) P waves created when you compress or stretch a spring cause compression and extension along the length of the spring. (b) S waves created when you shake the spring up and down (S1) or back and forth (S2) cause shearing of materials. (c) Taken together, P, S1, and S2 waves record motion in three-dimensions.

theory that the plates are being continuously pulled apart, rather than being pushed apart by upwelling material beneath the ridges. Perhaps the upwelling of magma beneath the oceanic ridges is merely a passive filling of the void left by the separating plates of lithosphere. Whatever the mechanism of emplacement, a large mass of magma must exist immediately beneath the spreading centers, elevating ridges and rises to form the largest mountain chains in the world. This process produces 20 km^3 (4.8 mi^3) of new ocean crust per year.

With high-resolution acoustical devices such as Sea Beam, investigators have identified offsets of 2 to 10 km (1.2 to 6.2 mi) in the axis of the East Pacific Rise. These features are called *overlapping spreading centers* because ends of adjacent segments of offset spreading centers overlap. The offsetting overlaps have an interesting relationship to the mass of hot, partially molten rock found beneath the axis of the East Pacific Rise. Figure 3-19 shows the location of two overlapping spreading centers at 12°37'N and 12°54'N latitude. The axial ridge is lower and narrower at the overlaps and highest and broadest midway between two overlapping ends. The mass of hot, partially molten rock extends uninterrupted along the entire axis and bends to follow the offsets. However, there appears to be no magma chamber (reservoir of molten rock) at the offsets. A thin magma chamber about 4 km (2.5 mi) wide is most developed at the center of each segment, where its top is about 1.4 km (0.9 mi) below the ocean floor. There is a well developed rift, or summit graben, at midsegment, where the magma chamber is present; the graben weakens and disappears toward the ends of the segments, where the offsets occur. These data clearly show that segmentation on a relatively small scale (some segments are only 20 km long) is a fundamental property of oceanic ridges and rises, and is associated with segmentation of magma chambers beneath the axis of the spreading center (Figure 3-20).

The magnitude of energy released by earthquakes along the divergent plate boundaries is closely related to spreading rate. As measured by the *moment magnitude scale,* earthquakes occurring in the rift valley of the slow-spreading Mid-Atlantic Ridge reach a maximum magnitude of about 6, while those occurring along the axis of the fast-spreading East Pacific Rise seldom exceed 4.5. On the moment magnitude scale, each unit increase in magnitude represents an increase in energy release of about 30 times. All spreading-center earthquakes are confined to the lithosphere, so the difference in observed magnitude may result from the fact the slow-spreading centers have thicker and colder lithosphere at the plate boundaries.

Transform Boundaries

Segmentation of the oceanic ridges and rises is associated with the emplacement of separate magma chambers beneath their axes. Each segment may range from 10 to 80 km (6 to 50 mi) in length and is centered over a magma chamber. The lithosphere is typically thicker at the boundary between adjacent segments, and the segments may be offset

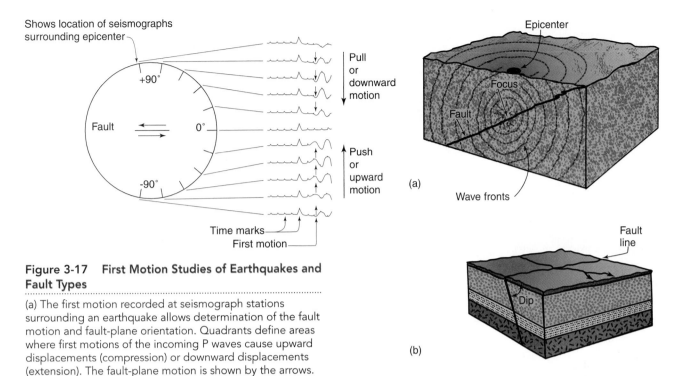

Shows location of seismographs surrounding epicenter

+90°

Fault   0°

-90°

Pull or downward motion

Push or upward motion

Time marks
First motion

(a)

Figure 3-17   First Motion Studies of Earthquakes and Fault Types

(a) The first motion recorded at seismograph stations surrounding an earthquake allows determination of the fault motion and fault-plane orientation. Quadrants define areas where first motions of the incoming P waves cause upward displacements (compression) or downward displacements (extension). The fault-plane motion is shown by the arrows.

Epicenter

Focus

Fault

Wave fronts

(a)

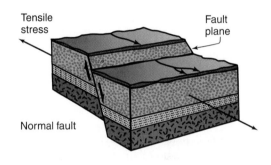

Fault line

Dip

(b)

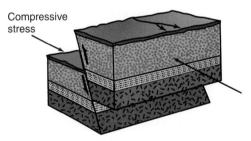

Tensile stress   Fault plane

Normal fault

Compressive stress

Thrust reverse fault

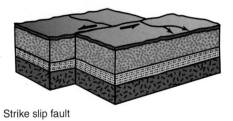

Strike slip fault

at this boundary. The offset is called a transform boundary along the plate edge. The amount of ridge offset may be very small or it may extend for hundreds of kilometers.

Why do these offsets occur? As newly formed lithosphere moves away from the axis of an oceanic ridge, it travels perpendicular to the axis. On a spherical surface like that of Earth, all of this new lithosphere will converge toward two points. For instance, if the equator were an oceanic ridge, the points of convergence would be the North and South poles. Obviously, all this new matter cannot be accommodated at two points, and stresses are set up within the lithosphere. The stresses help cause the lithosphere to break along lines perpendicular to the ridge axis. Movement along the breaks offsets the ridge axis, producing **transform faults** (Figure 3-21). Along such faults, one plate slides past the other, producing shallow earthquakes in the lithosphere. Since the lithosphere is thicker at transform boundaries than along the spreading axes, these earthquakes may be larger. Magnitudes of 7 have been recorded along oceanic transform boundaries.

An example of such a fault that has come ashore is the San Andreas Fault, which runs across California from the head of the Gulf of California to the San Francisco area. Where the San Andreas Fault runs through continental crust, the lithosphere is much thicker than at oceanic sites. Earthquakes along this fault have had magnitudes up to 8.5.

Fracture Zones   Continuing along the line of the transform fault, beyond the offset axes, are fracture zones. The transform faults represent active displacements of the axes; the fracture zones are evidence of past transform fault activity. On opposite sides of the transform fault, the

Figure 3-17   (continued)

(b) Types of fault motion: dip-slip in normal or reverse directions, and strike-slip

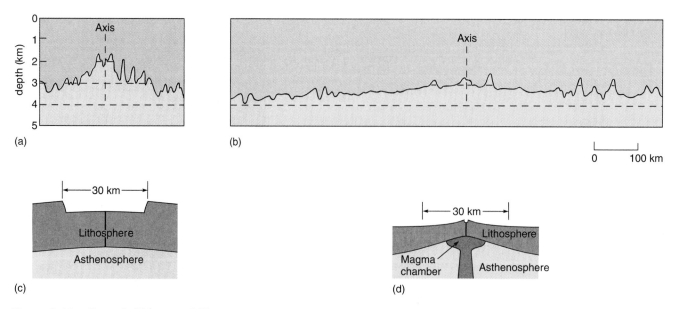

Figure 3-18 Oceanic Ridges and Rises

(a) The Mid-Atlantic Ridge has a steep and irregular topography as a result of a low spreading rate. (b) The high spreading rate of the East Pacific Rise has produced a gently sloping and smooth bottom topography. (Vertical exaggeration 50x.) (c) The axial valley of a slow-spreading ridge is typically 30 km wide and from 100 to 3000 m deep. The lithosphere may be from 3 to 10 km thick at the plate boundary near the center of the axial valley. (d) At fast-spreading rises, an axial ridge a few hundred meters high and from 2 to 10 km in width typically overlies the plate boundary above a magma chamber that may be about 4 km wide and 1 km thick. An axial valley less than 1 km wide and up to 100 m deep may run along the axis of the axial ridge. The thickness of the lithosphere at the rise axis may be less than 2 km. *(Part (b) from* The Ocean Basins: Their Structure and Evolution, *The Open University and Pergamon Press, 1989.)*

lithospheric plates are moving in opposite directions, whereas there is no relative plate motion occurring along a fracture zone. In other words, the transform faults serve as plate boundaries, while the **fracture zones** are ancient fault scars embedded in a single plate (see Figure 3-21). Earthquakes above a depth of 10 km are common along the transform faults, whereas the fracture zones are aseismic (without earthquakes).

The age of the ocean floor on one side of a fracture zone is greater than that on the opposite side, and the older ocean floor will have subsided more from thermal contraction. This results in a vertical escarpment that faces the older ocean floor; the escarpment increases in height with increased difference in the ages of the ocean floor on opposite sides of the fracture zone. The Mendicino Fracture Zone is a good example of a fracture zone with a large vertical escarpment.

Convergent Boundaries

Along with oceanic ridges and oceanic rises, trench systems represent significant discontinuities in Earth's crustal structure and are characterized by earthquakes and volcanic activity. Spreading centers create new ocean floor at the rate of 3.5 km²/yr, and trenches destroy an equal amount at an average subduction rate of 8.7 cm/yr.

However, there is a significant difference in the earthquake activity of these regions. Shallow quakes, usually less than 10 km (6 mi) in depth, are associated with the constructive ridge boundaries where the lithosphere is thin. Earthquakes in the trench regions may occur at depths to 670 km (416 mi). The largest magnitude earthquakes are associated with these convergent (destructive) plate boundaries.

The earthquake focus, the point at which the movement that causes the quake occurs, is shallow in the immediate area of the trench. The earthquake focus in the subduction zone is deeper, reaching a maximum depth of 670 km. The *Wadati-Benioff seismic zone* is a band 20 km (12.4 mi) thick that dips from the trench region under an overriding plate, within which all of the earthquake foci are located. The angle of dip is approximately 45°, becoming steeper at greater depths. Because the descending lithosphere would become too ductile to produce earthquakes below depths of 70 to 100 km, some other mechanism must be found to explain them to depths of 670 km. It is possible that these deep-focus earthquakes result from the changing of a mineral, olivine, to a denser form, spinel. Olivine makes up a large percentage of the descending lithospheric mantle. This process is considered a likely cause of deep-focus earthquakes because it should occur between the depths of 400 and 670 km.

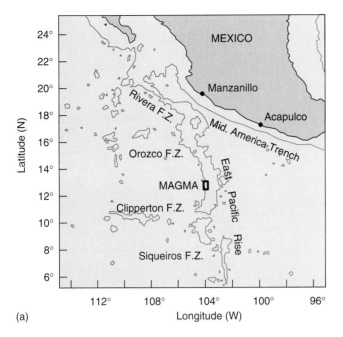

(a)

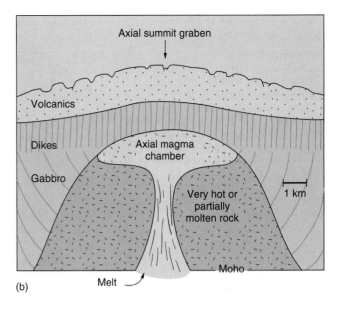

(b)

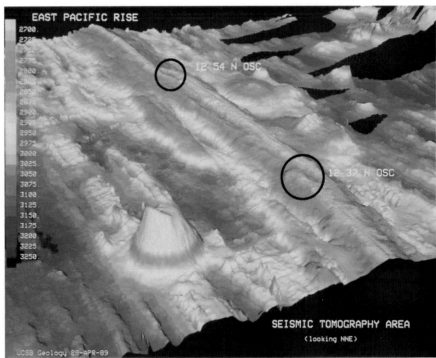

(c)

Figure 3-19   Magma Chambers and Their Relationship to Offset Spreading Centers

(a) Location of MAGMA Expedition that investigated the relationship between offset spreading centers and the distribution of magma chambers beneath them. The axis of the East Pacific Rise is outlined by the 3000-m depth contour. (b) Schematic cross section of the crust beneath the East Pacific Rise, based on gravity and seismic data. A mass of very hot rock that is, for the most part, not molten is found beneath the rise axis. The mushroom-shaped magma chamber that caps this mass of hot rock is fed by molten material rising from the upper mantle and is best developed where the ridge axis stands highest at the center of segments of the offset spreading centers. The magma chamber, which reaches a maximum width of 4 km (2.5 mi), gradually disappears toward the ends of the segments and is absent beneath the overlapping offsets. (c) False-color map of the MAGMA Expedition area of study with depths indicated by the color scale to the left. This map was prepared using data from Sea Beam and SeaMARC II side-directed sonar systems. It has a vertical exaggeration of 6. The axial ridge is broken into three segments by two overlaps, which are circled and identified by the latitude values 12°37′ and 12°54′N. The axial magma chamber is present beneath the rest of the rise but disappears at the two overlapping spreading centers.   *(Part (a) Map courtesy of Mark S. Burnett, Scripps Institution of Oceanography, University of California, San Diego. (b) Courtesy of K. C. Macdonald, University of California, Santa Barbara. (c) Map produced by S. P. Miller using NECOR software and unpublished data of K. C. Macdonald and P. J. Fox and provided courtesy of K. C. Macdonald, University of California, Santa Barbara.)*

Figure 3-20 Segmentation of Oceanic Ridges and Rises

Beneath the ridge axis, the heat-bearing asthenosphere (yellow) rises between the separating lithospheric plates (green). A zone of partially molten asthenosphere (red) is less dense than the cold lithosphere above, and diapirs of magma rise from it to form magma chambers at intervals along the ridge axis. The ridge axis rises over the magma chambers and is deepest at the segment boundaries midway between magma chambers.

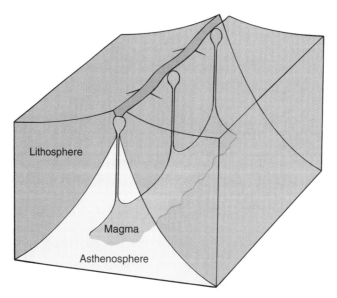

Figure 3-21 Transform Faults-Fracture Zones

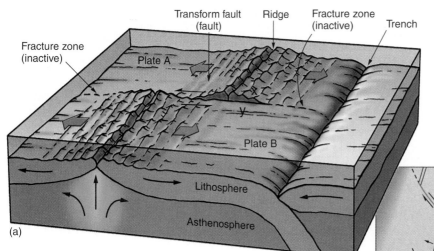

(a) The axis of an oceanic ridge is offset by a transform fault due to the stresses from the motion of rigid plates. Transform faults are so-called because they cut across the larger physiographic features of the ocean floor, oceanic ridges and rises. Earthquakes are common and of relatively great magnitude along transform faults because of their active movement. Extending beyond the offset ridges are fracture zones, where earthquakes are rare because there is no relative movement on the opposite sides of the fracture zone. Fracture zones, scars of old transform-fault activity, remain significant topographic features because the age of the ocean floor on one side of the fracture zone is older than the ocean floor immediately opposite it. For example, point X is younger than point Y. Therefore point Y has subsided more because of cooling and thermal contraction than point X. Thus there is an escarpment (cliff) along the fracture zone in Plate B that faces the older ocean floor on the side where point Y is located. There is also an escarpment on Plate A, but it faces the opposite direction.

(b) The San Andreas Fault is a classic example of a transform fault that offsets an oceanic ridge or rise and forms a boundary between plates that are sliding past each other. Along this contact, the Pacific Plate is moving north relative to the North American Plate. *(Reprinted with permission from* The Earth: An Introduction to Physical Geology, *3rd Edition, Figures 18.22 and 18.23, by E. J. Tarbuck and F. K. Lutgens. Macmillan Publishing Company, 1990.)*

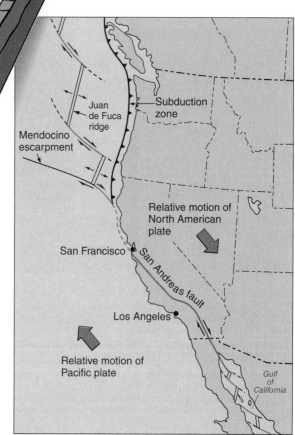

Ophiolites

It is a well accepted tenet of the global plate tectonic theory that new oceanic lithosphere created at spreading centers is ultimately subducted and returned to the underlying mantle. However, about one one-thousandth (0.001) percent of the oceanic lithosphere does not return. Instead, it is obducted onto the edge of a continental mass (Figure 3B). These fragments of oceanic lithosphere embedded into the continents are called **ophiolites** (serpent stone) because of their long, slender and curvy shapes. They characteristically consist of a thin layer of sedimentary rock that formed in the ocean, underlain by a sequence of crystalline rocks typical of oceanic crust and the underlying mantle. Many of the ophiolites contain rich metallic ores with patterns of enrichment similar to those occurring in the sediments and oceanic crust near hydrothermal vents on oceanic spreading centers.

With our present understanding of the sea-floor spreading process and its role in global plate tectonics, we have developed a reasonably clear view of how the units of an ophiolite assemblage may have formed. How they become obducted onto the continents is not so well understood. In the idealized ophiolite sequence of rock types, there is a layer of marine sediments and an underlying series of crystalline rocks. Immediately beneath the sediments are pillow basalts that flowed out on the ocean floor and show the characteristic pattern of submarine cooling. Descending through the lava flow, there is an increasing concentration of basaltic dikes that replace the flows completely about 1 km (0.62 mi) below the pillow lava surface (Figure 3C). It is believed that these sheet dikes form as vertical sheets of lava that have filled narrow gaps formed as lithospheric plates are pulled away from the axis of spreading. At a distance of approximately 3 km (1.9 mi) beneath the pillow lava surface, the dikes give way to massive gabbro, a rock similar to basalt in composition but of a coarser texture due to slower cooling. The gabbro may represent the cooling of magma near the roof of the magma chamber that was the source of the overlying basalts. The massive gabbro may be up to 300 m (984 ft) thick. It is underlain by layered gabbro that may have crystallized within the upper region of the magma chamber. Layered peridotite, a mantle rock composed mostly of the iron and magnesium-rich minerals pyroxene and olivine, is found beneath the layered gabbro. The boundary between the gabbro and peridotite is thought to represent the Moho, or the contact between the ocean crust and the underlying mantle. The Moho is usually found about 4.5 km (2.8 mi) below the top of the pillow lavas.

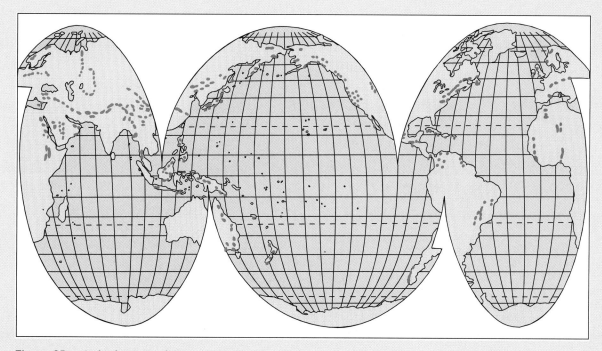

Figure 3B   Ophiolite Distribution

Distribution of ophiolites throughout the world. Ophiolites near the margins of the present continents where subduction is occurring are generally younger (less than 200 million years old) than those embedded deep within the continents. Some ophiolites are thought to be as much as 1.2 billion years old.

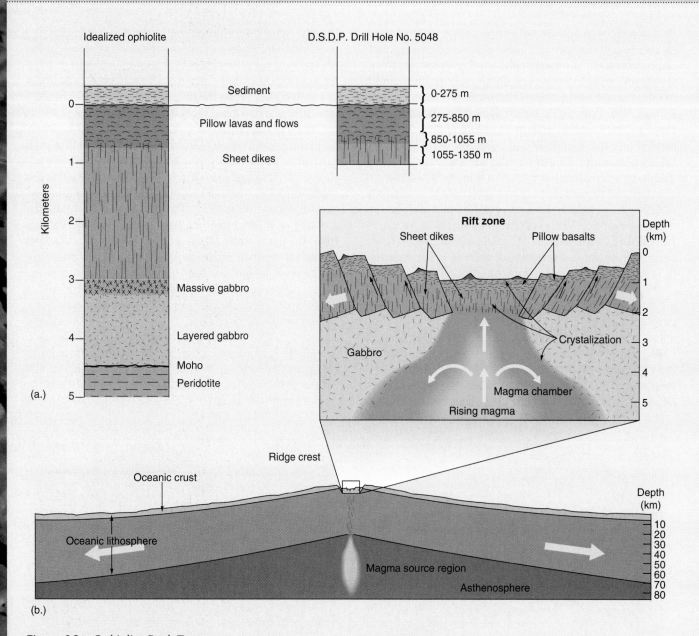

Idealized ophiolite

D.S.D.P. Drill Hole No. 5048

- Sediment
- Pillow lavas and flows
- Sheet dikes
- 0-275 m
- 275-850 m
- 850-1055 m
- 1055-1350 m

Kilometers

Massive gabbro

Layered gabbro

Moho

Peridotite

(a.)

Rift zone

Sheet dikes

Pillow basalts

Depth (km)

Crystalization

Gabbro

Magma chamber

Rising magma

Ridge crest

Oceanic crust

Depth (km)

Oceanic lithosphere

Magma source region

Asthenosphere

(b.)

Figure 3C Ophiolite Rock Types

(a) The column on the left shows the sequence of rock types found in ophiolite complexes such as the Troodos Massif of Cyprus. On the right is the sequence of rock types encountered in the Deep Sea Drilling Project drill hole No. 504B, drilled to a total depth of 1350 m (4428 ft) beneath the ocean floor. It closely resembles that of the upper ophiolite complex and adds support to the belief that ophiolites are pieces of oceanic lithosphere that have been obducted onto the margins of continents. (b) Diagram showing how the pillow basalts, sheet dikes, and gabbros of an ophiolite might form.   *(Reprinted with permission from* The Earth. *4th Edition, Figure 19.15, by Edward J. Tarbuck and Frederick K. Lutgens. Macmillan Publishing Company, 1993.)*

The main features of the idealized ophiolite sequence observed on the margins of continents were partially encountered in place beneath the ocean floor. The Deep Sea Drilling Project drill hole No. 504B, located on the flank of the eastern Galapagos Spreading Center, was drilled to a depth of 1.35 km (0.84 mi) beneath the ocean floor in 3.46 km (2.15 mi) of water. The crustal rocks penetrated are thought to be just over 6 million years old, and the sequence of rocks penetrated in the hole is as follows:

1. Sediment   0-275 m (0-902 ft)
2. Pillow lavas   275-850 m (902-2788 ft)
   and flows
3. Pillow lavas, flows,   850-1055 m (2788-3460 ft)
   and sheet dikes
4. Sheet dikes   1055-1350 m (3460-4428 ft)

Figure 3B shows how well this sequence resembles that observed in the ophiolite complexes. One of the oldest known ore deposits is the Troodos Massif on the island of Cyprus (Figure 3D). It has been mined for copper since the time of the Phoenicians. The main sulfide ore bodies are found in the volcanics near the top of the ophiolite sequence, and the sediment section above the volcanics is rich in iron and manganese. Such a sequence of metal enrichment occurs at oceanic spreading centers as a result of hydrothermal activity.

The spreading centers that produced most ophiolites appear not to have been oceanic ridges or rises, because most ophiolite rocks are too rich in silica (SiO_2) and too poor in iron. They more closely resemble the spreading centers such as the Palaus or Mariana Back-Arc Spreading Centers discussed in Chapter 4.

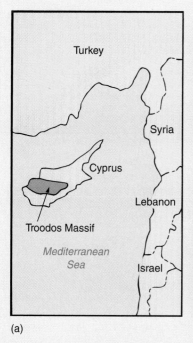

(a)

(b)

Figure 3D   Ophiolite Mining

(a) Location of the Troodos Massif ophiolite on the island of Cyprus in the eastern Mediterranean Sea. (b) Mining the Troodos Massif. The sulfide ore is the black material in the bottom of the pit.   *(Photo courtesy of R. Koski, U.S.G.S.)*

The difficult question of how the ophiolites become obducted onto the continent has numerous possible answers. One of the more straightforward explanations (Figure 3E) is as follows:

1. A subduction zone with a volcanic island arc, back-arc spreading center, and adjacent continent provides the initial setting.

2. The subducting oceanic plate contains water-bearing minerals such as zeolites and amphiboles. Subduction heats these minerals and releases water. The heated water converts the mantle rock, peridotite, to serpentine, which is much less dense than the surrounding peridotite. The serpentine expands and rises, pushing up the overlying back-arc mantle and crust.

3. Continued compressional stress causes the back-arc lithosphere to break into narrow strips, some of which are observable above sea level along the margin of the continent.

Continued study of ophiolites is needed to answer many questions that remain about the nature of all the processes involved in forming and obducting these ore-rich bodies along the margins of continents.

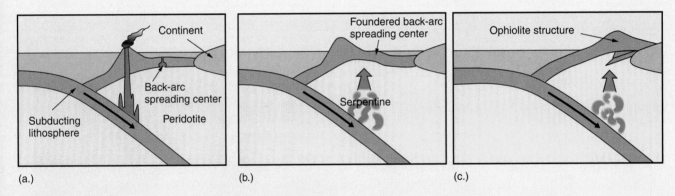

(a.)   (b.)   (c.)

Figure 3E   The Obduction of an Ophiolite onto a Continent

(a) An island arc system in which seaward migration of the trench has produced a back-arc spreading center has lithosphere that contains water-bearing minerals subducting beneath it. (b) Water-bearing minerals are heated and the hot water is released into the peridotite, which is altered to less dense serpentine. Seaward migration of the trench stops, and compressive forces move the subduction system toward the continent. The expansion of the serpentine lifts the overlying mantle and oceanic crust. (c) Continued compressional stress breaks up the uplifted back-arc lithosphere and pushes it onto the continent where it remains as an ophiolite structure.

The band within which the earthquakes originate must lie within the descending slab of lithosphere. The fact that the lithosphere generated at the oceanic rises and ridges is subducted at the trenches is supported by the fact that no oceanic crustal material more than about 200 million years old has been recovered.

Island Arc Trench Systems   In many subduction zones, particularly in the Pacific, the leading edges of both converging plates contain oceanic crust. The trench may be located hundreds of miles seaward of the edge of a continent. In these cases, the plate with no continent on it generally subducts beneath a plate with a continent at or within 1000 km (630 mi) of its leading edge. As the subducting oceanic lithosphere reaches sufficient depth, it is heated. Water and other volatiles are freed, forming a low-density mixture that rises to the surface and produces volcanoes on the overriding plate about 100 km (62 mi) landward of the trench. If the continent on the overriding plate is more than 100 km from the leading edge of the plate, the volcanoes are usually composed of basaltic rock. The basaltic composition of the volcanoes may occur because the rising lava only passes through mantle rock and basaltic ocean crust on its rise to the surface.

The volcanoes in the chain running parallel to and on the continental side of the trench grow to become islands—producing an **island arc system** separated from the continent by a marginal sea (Figure 3-22a). Examples are the Bering Sea separating the Aleutian Islands from Siberia and Alaska, and the Sea of Japan separating the Japanese Islands from the Asian mainland.

Continental Arc Trench Systems   Should an oceanic plate subduct a plate with a continent at its leading edge, the melting of the subducting oceanic plate occurs beneath the continent. Consequently, the rising basalt melt passes through and mixes with the granite of the continental crust. This results in a **continental arc** of volcanoes along the edge of the continent (Figure 3-22b). The volcanoes are composed of andesite, a rock with a composition between that of basalt and granite, which is thought to form from the mixing of the two basic crustal rock types.

If the spreading center producing the subducting plate is far enough from the subduction zone, an oceanic trench is well developed along the margin of the continent. The Peru-Chile Trench is an example. It is associated with the Andes Mountains continental arc of volcanoes from which andesite takes its name. No trench is visible where the Juan de Fuca Plate subducts beneath the North American Plate (NAP) off the coasts of Washington and Oregon to produce the Cascade Mountains continental arc. There, the Juan de Fuca Ridge is so near the NAP that the subducting lithosphere is less than 10 million years old and has not cooled enough to produce deep-ocean depths. Also, the large amount of sediment carried to the ocean by the Columbia River has filled any trench that may have developed. Many of the Cascade volcanoes have been active within the last 100 years. Most recently, Mount St. Helens erupted in May of 1980 and killed 62 people (Figure 3-23).

Continental Mountain Systems   If two lithospheric plates collide and both contain continental crust near their leading edges, the surface expression is a mountain range composed of the folded sedimentary rocks; these rocks are derived from sediments deposited in the sea that previously separated the continental blocks (Figure 3-22c). For the most part these sedimentary rocks do not subside in a zone of subduction because of their relatively low density derived from the continents. The oceanic crust however, may subside beneath such mountains. The exact nature of the subsidence of oceanic crustal material in these regions has not been determined. The Himalaya mountains were formed by this process.

The Growth of Ocean Basins

So far in this chapter we have discussed some specific features associated with construction, transport, and destruction of tectonic plates. In this section we concentrate on the broader changes in plate configurations that have occurred since the breakup of Pangaea 180 million years ago. The study of the historical changes of shapes and positions of the continents and oceans is called **paleogeography**. **Paleoceanography** applies to the study of the changes in physical and biological character of the oceans brought about by paleogeographical changes.

Figure 3-24 shows the changing positions of the continents over the last 150 million years. Observing the changes in the Northern Hemisphere (Figure 3-24a), it is clear that the major change was the separation of North America and South America from Europe and Africa that produced the Atlantic Ocean. During this process, Europe, Africa, and North America all moved a bit northward. Many investigators of plate motion conclude that the evidence indicates rates of spreading have historically been higher at lower latitudes. This is apparent in the pattern of formation of the North Atlantic Ocean. Europe and North America have separated at a much lower rate than have Africa and North America. It was not until 60 Ma (million years ago) that a noticeable separation began, and the island of Greenland appears to have been moving on the same plate as Europe. By 40 Ma, the north end of the Mid-Atlantic Ridge shifted to the east of Greenland, putting it back on the North American Plate. North America and South America were not fully connected by the Isthmus of Panama until about 5 Ma. This event had a marked effect on patterns of ocean circulation and the distribution of biological species.

The early stages of continental breakup in the Southern Hemisphere, depicted in Figure 3-24b, show South America and a continent composed of India, Australia, and Antarctica separating from Africa. By 90 Ma, there was a clear separation between South America and Africa, and India had moved north, away from the Australia-

Figure 3-22 Effects of Plate Collisions

(a) Classic oceanic trench systems develop where oceanic lithosphere subducts beneath a plate with oceanic crust at its leading edge and a continent trailing some distance behind. A basaltic island arc trench system develops. The sloping belt in which earthquake foci are located is the Wadati-Benioff seismic zone. (b) When an oceanic lithospheric plate subducts beneath a plate with continental crust at its leading edge, an andesitic continental arc develops.
(c) As two plates carrying continents near their leading edges converge, trenches do not develop because sediments that accumulated in the sea between the continents are too light to be subducted. The shortening of Earth's crust in such regions is accommodated by the folding of the sediments into mountain ranges such as the Appalachians, Alps, Himalayas, and Urals. The lower lithosphere and asthenosphere may be involved in low rates of subduction beneath the mountains. (*Reprinted with permission from* Earth Science, *6th Edition, Figure 6.11, by Edward J. Tarbuck and Frederick K. Lutgens. Macmillan Publishing Company, 1991.*)

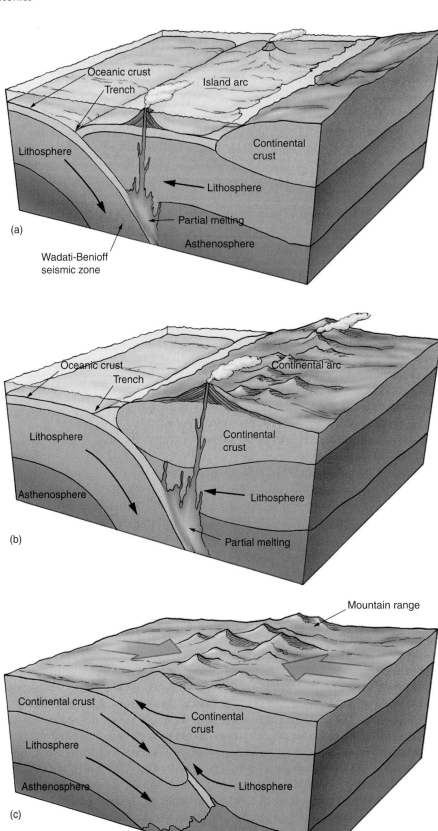

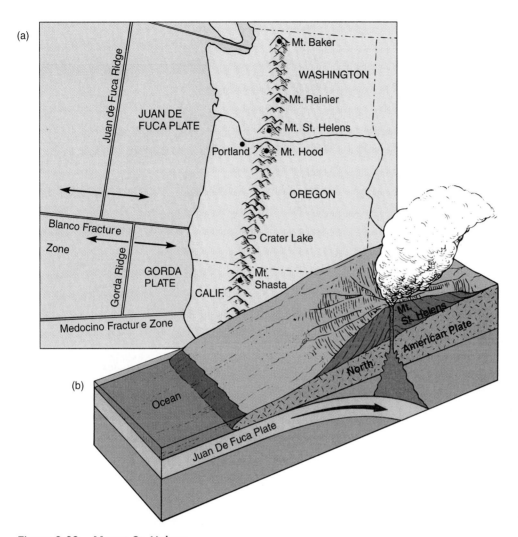

Figure 3-23   Mount St. Helens

(a) Tectonic features of the Mount St. Helens region. The mountains shown are volcanoes within the Cascade Mountain Range. (b) The eruption of Mount St. Helens resulted from the subduction of the Juan de Fuca Plate beneath the North American Plate.

Antarctica mass, which began moving poleward. As the Atlantic Ocean continued to open, India moved rapidly north and collided with Asia about 35 Ma. Subsequent to this collision, the spreading center that ran between India and Australia foundered and was replaced by one that began to separate Australia from Antarctica. The scars representing shear-plate boundaries during the northward movement of the Indian Plate are the Chagos-Laccadive Ridge to the west and the Ninety East Ridge to the east. These features are clearly visible in the map of the ocean floor in Chapter 4.

The overall effect of global plate tectonic events over the past 180 million years is the creation of the new Atlantic Ocean, which continues to grow, and the division of Panthalassa into the Pacific Ocean and Indian Ocean. The fate of the Indian Ocean in terms of its changing size is questionable, but it is clear that the Pacific Ocean continues to shrink as oceanic plates produced within it subduct into the many trenches that surround it and continent-bearing plates bear in from east and west.

Much also can be learned of the history of ocean basins from maps of age distribution of the ocean crust. In Figure 3-8, it can be seen that the Atlantic Ocean has the simplest and most symmetrical pattern of age distribution. As Pangaea was rifted by the incipient Mid-Atlantic Ridge, the first separation involved North America and Africa more than 180 Ma. The Pacific Ocean has the least symmetrical pattern because of the large amount of subduction that surrounds it. The ocean floor that is older than 40 Ma east of the East Pacific Rise has been subducted, whereas ocean floor in the northwestern Pacific is more than 170 Ma. The Rise itself has disappeared under North America. The broader age bands on the Pacific Ocean floor are clear evidence that the spreading rate has been greater in the Pacific Ocean than in the Atlantic or Indian Oceans throughout history.

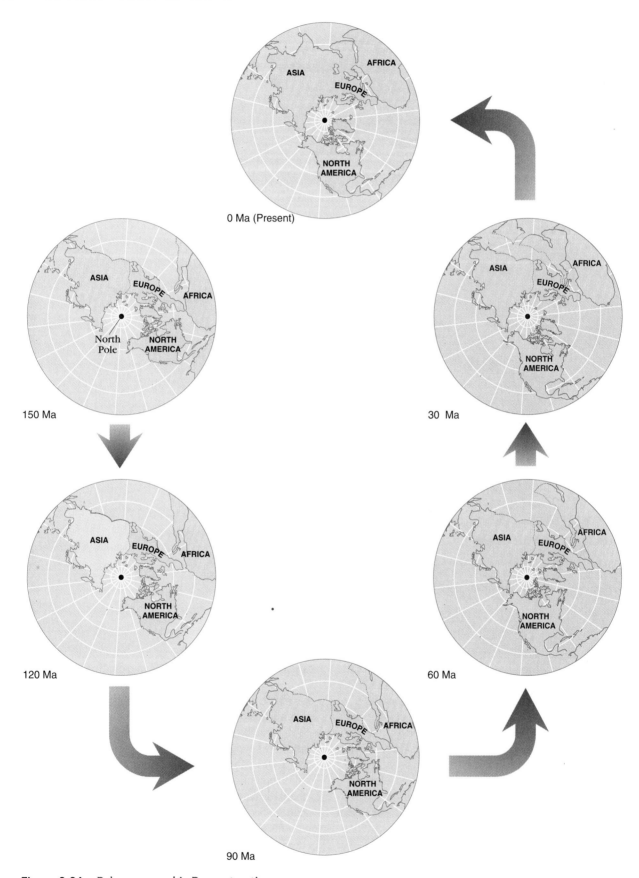

0 Ma (Present)

150 Ma

North
Pole

120 Ma

90 Ma

60 Ma

30 Ma

Figure 3-24 Paleogeographic Reconstructions

(a) These North Polar views show the positions of the continents over the last 150 million years.
Ages are in Ma (millions of years ago) at intervals of 30 million years.

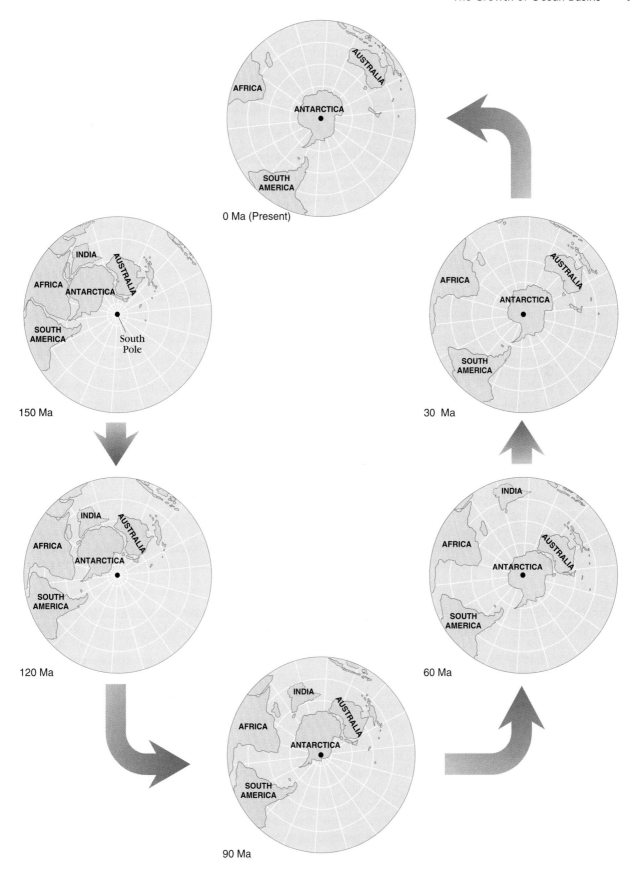

Figure 3-24   (continued)

(b) These South Polar views are based on data from the first 60 legs of the Deep Sea Drilling Project, conducted during a decade of deep-ocean drilling, 1968 to 1978.   *(Courtesy of P. L. Firstbrook, B. M. Funnell, A. M. Hurley, and A. G. Smith, in association with Scripps Institution of Oceanography, University of California, San Diego.)*

How Fast are the Continents Growing?

After we consider all of the volcanic activity and subduction associated with global plate tectonics, the erosion of continents, and the deposition of marine sediment, we may ask how fast the continents are growing. A study conducted by the U.S. Geological Survey led to a possible budget for the growth and denudation of continents. By calculating the yearly flux of various components of marine sediment, seamounts, and island arc volcanoes, a somewhat tentative rate of growth for continents since the breakup of Pangaea was determined.

It was determined that ocean crust is recycled every 110 million years. Its average age is 55 million years. Lithogenous particles accumulate in the deep ocean at an average rate of 1.27 km^3/yr. Biogenous particles, the chemical components of which come ultimately from continents, accumulate at an average rate of 0.38 km^3/yr. This sums to a rate of continental denudation of 1.65 km^3/yr. Adding this to the 0.05 km^3/yr of accumulated volcanic sediment, gives us a total yearly rate of sediment accumulation of 1.7 km^3/yr. Seamounts add 0.2 km^3/yr of mass to the ocean crust, and volcanic island arcs add 1.1 km^3/yr. The total material available to accrete to continents during subduction is the sum of all these rates: $1.7 + 0.2 + 1.1 = 3$ km^3/yr. Subtracting the rate of continental denudation (1.65 km^3/yr) from this total leaves 1.35 km^3/yr of possible continental growth. Whatever amount of deep-sea sediment or seamounts is subducted must be subtracted from this growth rate, but the numbers for these variables are not yet known.

Summary

By the late 1960s, data gathered from the geology of the continents and the oceans provided the evidence necessary for plate tectonic theory to be formulated. The theory states that all the present continents move within a "sea" of ocean crust that is continually formed and destroyed. The continents once formed one large continent, Pangaea, about 200 million years ago. Just before the formation and possibly after the breakup of Pangaea, two large continents appear to have existed, Laurasia in the north and Gondwanaland to the south.

The discovery that Earth's magnetic field has changed polarity throughout Earth's history and that the record of these changes was permanently recorded in the rocks of the oceanic crust made it possible to develop the hypothesis that sea-floor spreading was the process that moved the continents. According to this hypothesis, new rock was added to rigid lithospheric plates by intrusion and extrusion of molten material along submarine mountain ranges. The lithospheric plates are composed primarily of dense mantle rocks with a thin crust of basaltic lava beneath the ocean basins. The continents float isostatically on these plates because they are composed of a less dense igneous rock, granite. As material is added to the plates near the axis of the oceanic ridges and rises, it moves away to make room for new material surfacing at the axis. Thus the crustal rocks beneath the oceans increase in age with increasing distance from the axis of the oceanic ridges and rises down their slopes into the deep-ocean basins. Continents located on a lithospheric plate are carried passively along as the plate moves away from the oceanic ridge or rise.

The movement of the lithosphere appears to be made possible by the existence of a plastic region of mantle material—the asthenosphere—immediately beneath the lithosphere. The temperatures of this region are near the melting point of the mantle. As the lithospheric plates move away from high-heat-flow oceanic ridges and rises, toward low-heat-flow trenches, they cool and thicken. The age and depth of the ocean floor increase with increasing distance from the spreading centers. Evidence suggests that about 94 percent of the heat released at Earth's surface is derived from the mantle and is vented through volcanism at spreading centers and subduction zones. About 6 percent comes from the core and is released at hot spots. Mid-ocean ridge basalt (MORB), which forms at spreading centers, can be traced to the upper mantle because it has lower concentrations of certain elements than does the basalt of hot spots. Most hot spot basalt is believed to come from near the base of the mantle.

As new mass is added to the lithosphere at the oceanic rises and ridges (divergent, or constructive, plate boundaries), the leading edges of the plates are subducted into the mantle at ocean trench systems or at continental mountain ranges such as the Himalayas (convergent, or destructive, plate boundaries). In addition, there are transform boundaries, along which oceanic ridges and rises are offset, and plates slide past one another. Although Earth is more than 4.6 billion years old, no rocks more than 175 million years old have been found on the ocean floor. It is becoming clearer that sinking at the leading edges of the plates, rather than upwelling of new material at the ridges, drives the convection-like movement and that the process recycles material between the lithosphere and the deeper mantle.

The paleogeography, or changing shape and location of ocean basins and continents, has produced changes in the physical and biological character of the oceans. The study of these changes is called paleoceanography. Since the breakup of Pangaea, which began 180 Ma, the Atlantic Ocean has come into existence and Panthalassa has been reshaped into the Pacific and Indian oceans.

Key Terms

Apparent polar wandering	Fault strike	Mid-Ocean Ridge Basalt	Rift valley
Asthenosphere	Fracture zones	(MORB)	S waves
Continental arc	Gondwanaland	Oceanic ridge	Sea-floor spreading
Convection cells	Heat flow	Oceanic rise	Seismic waves
Convergent (destructive)	Hot spots	Ophiolites	Spreading centers
boundary	Island arc systems	P waves	Subduction
Core-mantle boundary	Isostasy	Paleoceanography	Tethys Sea
Curie point	Laurasia	Paleogeography	Transform fault
Divergent (constructive)	Lithosphere	Paleomagnetism	Transform (shear) boundary
boundary	Magnetic anomalies	Pangaea	Trenches
Fault dip	Magnetic dip	Panthalassa	

Questions and Exercises

1. Describe the evidence found on the present continents that suggests Pangaea broke into two continents, Laurasia and Gondwanaland, between 250 and 350 million years ago.

2. How does the magnetic dip of magnetic particles found in igneous rocks tell us at what latitude they were formed?

3. How does continental movement account for the two apparent wandering curves of the north magnetic pole determined from Europe and North America?

4. What property of the magnetic record found in oceanic crustal rocks gave rise to the idea of sea-floor spreading?

5. Describe a possible mechanism responsible for moving the continents over Earth's surface. Include a discussion of the lithosphere and asthenosphere.

6. Describe the general relationships that exist among distance from the spreading centers of the oceanic ridges, heat flow, age of the oceanic crustal rock, and ocean depth.

7. Discuss the roles of divergent and convergent plate boundaries, as compared to those of hot spots, in the release of heat from Earth's interior.

8. Discuss the three types of plate boundaries, spreading centers of the oceanic ridges, trenches, and transform faults. Explain why earthquake activity is usually confined to depths less than 10 km (6 mi) along the spreading centers and transform faults, whereas it may occur as deep as 670 km (415 mi) near the trenches. Construct a plane view and cross section showing each boundary type and direction of plate movement.

9. Describe the relationship between fracture zones and transform faults, including a comparison of the earthquake activity along each.

10. When does evidence for the following events first show up on the paleogeographic reconstructions of Pangaea?

    a. Greenland separates from Europe

    b. South America begins to move away from Africa

    c. India separates from Antarctica

    d. Australia separates from Antarctica

References

Anderson, D. L., T. Tanimoto, and Y. Zhang, 1992. Plate tectonics and hotspots: The third dimension. *Science* 256:1645–51.

Bercovici, D., G. Schubert, and G. A. Glatzmaier, 1989. Three-dimensional spherical models of convection in the Earth's mantle. *Science* 244:4907, 950–54.

Bullard, Sir Edward, 1969. The origin of the oceans. *Scientific American* 221:66–75.

Burnett, M. S., D. W. Caress, and J. A. Orcutt, 1989. Tomographic image of the magma chamber at 12°50' on the East Pacific Rise. *Nature* 339:206–8.

Carrigan, C. R., and D. Gubbins, 1979. The source of the Earth's magnetic field. *Scientific American* 240–2:118–33.

Detrick, R. S., et al., 1993. Seismic structure of the southern east pacific rise. *Science* 259:499–503.

Dietz, R. S., and J. C. Holden, 1970. The breakup of Pangaea. *Scientific American* 223:30–41.

Elthon, D., 1991. Geochemical evidence for formation of the bay of islands ophiolite above a subduction zone. *Nature* 354:6349, 140–42.

Hamilton W. B., 1988. Plate tectonics and island arcs. Bulletin Geological Society of America 100:1503–26.

Hess, H. H., 1962. History of ocean basins. *Petrologic Studies: A Volume to Honor A. F. Buddington.* Engel, A. E. J., H. L. Lames, and B. F. Leonard, eds., 559–620. New York: Geological Society of America.

Jordan, T. H., 1979. The deep structure of the continents. *Scientific American* 240–1:92–107.

Kearey, P. and F. J. Vine, 1996, *Global Tectonics.* 2nd ed. Blackwell Science, Oxford.

Menard, H. W., 1986. *The Ocean of Truth: A Personal History of Global Tectonics.* Princeton, NJ : Princeton University Press.

Olson, P., P. G. Silver, and R. W. Carlson, 1990. The large-scale structure of convection in the Earth's mantle. *Nature* 344:209–14.

Renne, P. R., et al., 1992. The age of parana· flood volcanism, rifting of Gondwanaland, and the Jurassic-Cretaceous boundary. *Science* 258:975–79.

Ritsema, J. H. J. van Heijst, and J. H. Woodhouse, 1999. Complex shear wave velocity structure imaged beneath Africa and Iceland. *Science* 286:5446, 1925–28.

Richards, M. A., 1999. Prospecting for jurassic slabs. *Nature* 397:6716, 203–04.

Sclater, J. G., B. Parsons, and C. Jaupart, 1981. Oceans and continents: similarities and differences in the mechanisms of heat loss. *Journal of Geophysical Research* 86:11, 535–52.

Shepard, F. P., 1977. *Geological Oceanography.* New York: Crane, Russak.

Smith, D. K., and J. R. Cann, 1993. Building the crust at the mid-atlantic ridge. *Nature* 365:6448, 707–15.

Tarbuck, E. J., and F. K. Lutgens, 1993. *The Earth: An Introduction to Physical Geology.* 4th. ed. New York: Macmillan.

———. 1991. *Earth Science,* 6th ed. New York: Macmillan.

Tarling, D., and M. Tarling, 1971. *Continental Drift: A Study of the Earth's Moving Surface.* Garden City, N.Y.: Doubleday.

Vine, F. J., and D. H. Matthews, 1963. Magnetic anomalies over oceanic ridges. *Nature* 199:947–49.

Suggested Reading

Sea Frontiers

Burton, R., 1974. *Instant Islands.* 19:3, 130–36. The 1973 eruption on the island of Heimaey south of Iceland and its relationship to plate tectonic processes are described.

Dietz, R. S., 1976. *Iceland—Where the Mid-Ocean Ridge Bares Its Back.* 22:1, 9–15. Description of the rift zone of Iceland and its relationship to the Mid-Atlantic Ridge and sea-floor spreading.

———. 1977. *San Andreas: An Oceanic Fault that Came Ashore.* 23:5, 258–66. Discusses the San Andreas Fault and its relationship to global plate tectonics.

———. 1971. *Those Shifty Continents.* 17:4, 204–12. A very readable and informative presentation of crustal features and a possible mechanism of plate tectonics.

Emiliani, C., 1972. *A Magnificent Revolution.* 18:6, 357–72. Advances in studying earth science from the sea, including climate cycles, plate tectonics, and the economic potential of resources lying within marine sediments.

Mark, K., 1974. *Earthquakes in Alaska.* 20:5, 274–83. The origin, nature, and effects of earthquakes in Alaska.

———. 1972. *Ocean Fossils on Land.* 18:2, 95–106. Significance of marine fossils found in rocks now many miles inland is centered on the work of an early American paleontologist, James Hall.

Rona, P., 1984. *Perpetual Seafloor Metal Factory.* 30:3, 132–41. How metallic mineral deposits form in association with hydrothermal vents located on oceanic ridges and rises.

Taylor, S. R., and S. M. McLennan, 1996. *The Evolution of Continental Crust.* 274:1, 76–81. Review of the knowledge gained by planetary study and Earth-based research into the nature of planetary crusts reveals much about the evolution of Earth's crust.

Scientific American

Bloxham, J., and D. Gubbins, 1989. The evolution of the earth's magnetic field. 261:6, 68–75. The past record of the flow of molten iron in the outer core yields insight into the process by which Earth's magnetic field is generated and how it may behave in the future.

Bonatti, E., 1994. The Earth's mantle below the oceans. 270:3, 44–51. The study of rocks on the floor of the Atlantic Ocean shed new light on the nature of hot spots. There is some indication that the large volume of basaltic lava released at some hot spots results from the upper mantle rocks being unusually wet rather than unusually hot.

Dalziel, W. D., 1995. Earth before Pangaea. 272:1, 58–63. Discusses the location of the continents 250 million years ago and earlier. Also covers the effects of tectonic shifting of land masses on the environment and on the evolution of early lifeforms.

Dvorak, J., C. Johnson, and R. Tilling, 1992. Dynamics of Kilauea Volcano. 267:2, 46–53. This article looks at the behavior of Kilauea volcano since it began erupting in 1983.

Francis, P., and S. Self, 1983. The Eruption of Krakatau. 249:5, 172–87. Analysis of the 1883 event based on the volcanic deposits and the timing of air and sea waves created by the eruption.

Gass, I. G., 1982. Ophiolites. 247:2, 122–31. The relationship between ophiolite emplacements and global plate tectonic processes is discussed.

Green, H. W. II, 1994. Solving the paradox of deep earthquakes. 271:3, 64–71. Study of simulated deep earthquakes shows that dehydration and increased pressure changes the crystal structure of minerals in the subducting lithosphere and causes earthquakes.

Hallam, A., 1975. Alfred Wegener and continental drift. 232:2, 88–97. A history of Wegener's development of his theory of continental drift presented in 1912 and the reasons for its general rejection until the 1960s.

Pollack, H. N., and D. S. Chapman, 1977. The flow of heat from the Earth's interior. 237:2, 60–76. Based on data determined from thousands of heat-flow measurements on continents and the ocean floor, this article describes the relationship between lithospheric thickness and the heat flow through it.

Taylor, S. R., and S. M. McLennan, 1996: Vol 1 76–81.
Continental Crust within the Solar System is Unique to Earth. The reasons for this and the processes responsible are discussed.

Oceanography on the Web

Visit the Introductory Oceanography home page for on-line resources for this chapter. There you will find an on-line study guide with review exercises and links to ocean-ography sites to further your exploration of the topics in this chapter. Introductory Oceanography is at http://prenhall.com/thurman (click on the Table of Contents menu and select this chapter).

Coral Rocks!
The Value of the World's Coral Reefs

- What are coral reefs, where are they located, and what organisms contribute to reef construction and function?
- What is the importance of coral reefs as reservoirs of biodiversity?
- How important are coral reefs to marine fisheries?
- What is their significance in the carbon cycle?
- What are the global threats to the health of coral reefs?

Introduction

Coral reefs (Figure 1) are tropical, shallow-water, limestone mounds formed mainly by coral animals and plants which remove calcium carbonate ($CaCO_3$) from seawater and deposit it as skeletal material, a process known as "carbonate-fixing."

Coral refers to marine invertebrate organisms belonging to the Phylum Cnidaria, Class Anthozoa, or to the hard, calcareous structures made by these organisms. An individual coral animal, known as a polyp, resembles a minute anemone. It is the ability of these polyps to remove dissolved calcium carbonate from seawater and to deposit it as part of their rocky skeleton that allows the formation of coral reefs. How coral animals do this is not precisely known. What *is* known is that corals need algae actually living in their tissues in order to precipitate the $CaCO_3$. (More about this below.) This type of relationship, termed symbiosis, or more specifically, mutualism, occurs among a wide variety of marine organisms.

Coral reefs represent some of the most important real estate on the planet. They cover approximately 600,000 km² (230,000 mi²) in the tropics, an area roughly comparable to the state of Texas. They are a major oceanic storehouse of carbon and contain thousands of species of organisms, only a fraction of which have been identified.

Coral reefs are increasingly under threat from human actions. For example, nearly half a billion people live within 100 km (62 mi) of coral reefs. This issue focuses on the importance of coral reefs, their role in global carbon balance, and what can be done to protect them.

Background

While reef environments have been fairly common in Earth's history, and reefs have been built by organisms on and off for the past several billion years, not all reef structures have been built by corals. Corals have been important contributors to reefs only since about 300 million years ago, and only within the past 50 million years have modern corals assumed their reef-building roles. Noncoral reefs built by cyanobacteria were accumulating more than 3 billion years before present, and thus are among the most ancient structures built by organisms.

Along with tropical rainforests, coral reefs, particularly those in the tropical Indo-Western Pacific Ocean, have the highest

Figure 1

Satellite photograph of the Great Barrier Reef. The Great Barrier Reef is the largest structure built by living organisms. It stretches along Australia's east coast for 2,000 km (1,200 miles). Like most organisms, corals live near the maximum temperature they can tolerate. Will rising sea temperatures result in the disappearance of corals? *(Paul Chelsey, National Geographic Society)*

known biodiversities of any ecosystem on earth. But globally, coral reefs are not prospering; indeed, their very existence is being threatened by nutrient pollution, sedimentation, overfishing, global warming, and even ecotourism. The severity of this problem can be appreciated if you realize that tourists spend upwards of $100 billion each year to visit locations near reefs. Florida reefs, for example, bring in at least $1.66 billion annually to that state's economy.

Coral Reefs and Fishing

Although reefs cover less than 0.2% of the ocean surface, they harbor a quarter of all marine fish species. Fish in coral communities are of two basic types - herbivores that feed on algae and carnivores that eat other animals.

Fishers have been plying reefs for millennia, and today reefs provide employment for millions of fishers. During the past few

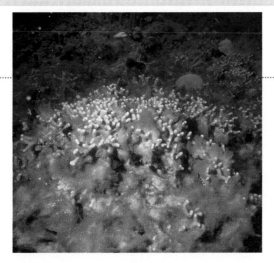

Figure 2

Algae overgrowing staghorn coral in the Caribbean.
(© Stephen Frink/CORBIS)

decades however, the intensity of fishing has begun to degrade reef communities. Damage to Philippine reefs has resulted in the loss of 125,000 jobs. You may be able to anticipate some of the reasons for the degradation of coral reefs: if too many herbivores are removed, the marine algae that they eat may grow out-of-control and smother the coral (Figure 2). Removing carnivorous fish can also upset the reef's ecological balance. Can you see why?

While removal of the fish may harm reefs ecologically, some methods of collection, such as dynamiting, kill the hard coral colonies directly. Another method, called muro-ami, involves bouncing rocks tethered to lines off the coral to herd the fish. This method, employed in the Philippines, typically destroys about 17 m^2 (183 ft^2) of coral cover per hectare (10,000 m^2 = 108,000 ft^2) per operation. Typically thirty muro-ami boats repeat the process about ten times a day. It may take forty or more years for reefs that are destroyed by fishing practices to recover (if allowed to).

Macroscopic marine algae grow amid and on the coral. Some of these algae are themselves carbonate producers and may be second only to corals in their carbonate production. These algae, as well as noncalcareous algal species, may be eaten by parrotfish and other herbivores, who find safety in the numerous nooks and crannies of the reef itself. These fish in turn become food for predators.

Reef Health

That coral reefs are widely distributed in the tropics does not imply that the coral animals themselves are very hardy. In fact, quite the opposite is true; corals are sensitive to such a variety of environmental factors that it is a wonder they exist at all. These factors are light, temperature, salinity, sedimentation, and nutrient levels.

Light Reef-building (also called *hermatypic*) corals are restricted to shallow waters because they require a certain quality and quantity of light. Why would simple, eyeless, invertebrate animals such as corals have a need for light? This requirement is due to the presence of simple, one-celled algae known as zooxanthellae which live within coral cells.

Zooxanthellae are dinoflagellates, a group that also includes the organism responsible for red tides. These organisms, thousands of which live within the cells of a polyp, require light to carry on photosynthesis. The high-energy end-products of photosynthesis are used to nourish not only the zooxanthellae but the coral as well. Some of the organic products are transferred to the coral polyp as a nutritional supplement to the food obtained when the coral feeds using its tentacles. Zooxanthellae also contribute oxygen and remove some waste materials, and they are involved in calcification.

Temperature In addition to light, corals require a relatively constant, moderately high (but not too high) temperature. The global distribution of coral, in fact, correlates best with surface temperature: corals are not generally found where winter surface water temperatures fall below 20°C, and they generally expel their algae and may die if water temperatures exceed 30°C. Thus, reefs are generally confined to waters between the Tropics of Cancer and Capricorn. This location protects them from low water temperatures, but means that most corals live near their upper thermal limit of tolerance.

Salinity cannot vary much from the 35 part per thousand average for seawater for corals to survive (except for Red Sea corals, which are adapted to higher salinities).

Sedimentation rates should be low and grain size relatively coarse or corals can be easily smothered. First, the corals' filtering apparatus can be clogged, making it difficult for them to feed on the tiny creatures that comprise their diet, and second, the sediment can blanket the colony and keep the zooxanthellae from photosynthesizing. Sedimentation due to runoff from construction sites onshore or mining activities is one of the reefs' most lethal enemies.

Nutrient levels (that is, the concentrations of phosphorus and nitrogen compounds) must also be low. In fact, most coral reefs flourish in nutrient "deserts." The reason again is fairly easy to understand. High nutrient concentrations may stimulate the growth of algae, which can smother the corals. Expansion of intensive, Western-style agriculture on land (with its massive doses of water-soluble, nutrient-rich fertilizer) may lead to die-offs in offshore reefs. And in the United States, treated, nutrient-rich sewage from Florida's West Coast may be among sources of degradation affecting the once-healthy coral reefs in the Florida Keys.

Now go to our web site to complete the analysis.

See the complete issue at:
www.prenhall.com/thurman

4 Marine Provinces

- Bathymetry
- Continental Margin Marine Provinces
- Deep Ocean Basin Marine Provinces
- Summary

Eruption of Mount St. Helens in the Cascade Range of Washington, 1980. (U.S.G S. photo.)

The moving lithospheric plates that continually re-shape the surface of our planet, and the boundaries between plates usually are not obvious features when viewed at the surface, because they are often buried under sediment. In this chapter, we will look at the characteristic of surface features originating from plate tectonic forces, including features within plates and at plate margins, both on land and under the sea. Satellite data and photos taken from orbiting satellites and astronauts on the space shuttle have recently given us a better view and understanding of these large-scale features. Sometimes, a distant vantage point is much better than a close-up view!

Bathymetry

Bathymetry is the measurement of depths from the water's surface to underwater features in the same way that topography is the measurement of elevations of land features. The general bathymetric features of the ocean are related to global plate tectonic processes, but are modified by Earth surface processes, including physical, geological, chemical, and biological phenomena.

The Hypsographic Curve

The importance of the oceans on Earth is illustrated by the fact that about 70 percent of the Earth's surface is covered by ocean and less than 30 percent by land. This can be shown graphically by a **hypsographic curve**, a cumulative plot of area vs. depth and elevation (Figure 4-1). Present-day sea level is the zero point; ocean areas plot below this point (negative values), and land areas plot above it (positive values). There are 510 million square kilometers (about 200 million square miles) of total surface area shown on the horizontal axis; elevations from 11 km (7 mi) below sea level to 9 km (6 mi) above sea level are shown on the vertical axis. You can see that the curve is composed of five differently sloped segments. On land, mountains are represented by the first steep segment, and the coastal plain by the gentle slope. Below sea level, the gentle slope continues as the continental shelf. The curve gets steeper again for the continental slope (and midocean ridges), followed by the abyssal plains (the longest, flattest part of the whole curve), and the trenches (the last steep part).

Looking at the plot, note that there is a very uneven distribution of area at different depths and elevations. A distribution in which equal amounts of land occurred for each depth or elevation interval would plot as a straight line. Instead, we see two almost flat areas, and three steeply sloped areas. Most of the area of Earth is the flat intraplate expanses of the abyssal plains, occurring at depths between 4 and 6 km, and the low elevations on land from 0 to 1 km, including flat plate interiors and coastal plains along passive continental margins (see below). Mountains, continental slopes, midocean ridges, and deep ocean trenches are all features related to plate boundaries and represent relatively small amounts of the total surface area.

Thus, from a simple hypsographic curve, a knowledgeable person can deduce the existence of plate tectonic activity. In fact, planetary scientists construct hypsographic curves using topographic data from interplanetary probes as one line of evidence for or against plate tectonics on other planets and moons.

A second point to note is that sea level is not a constant reference point in time! In the geologic past, sea level has been both higher and lower than at present. If all present-day glacial and polar ice melted, sea level would be 60 meters higher; during the glacial stages in the Pleistocene, glacier expansion left sea level as much as 120 meters lower than it is today. Note that small changes in sea level make big differences in the amount of land area because the hypsographic curve has such a low slope near sea level. A 60-meter rise in sea level would turn a lot of our coastal cities into underwater amusement parks! We will look at this topic more in Chapter 12.

Finally, note that because a hypsographic curve is a plot of area vs. height, an area on the graph actually represents a volume of crust. We can also use the curve to determine the mean elevation of the continents above the sea, 840 m (2755 ft.), and mean depth of the oceans, 3800 m (12,464 ft.). The difference, recalling from our study of isostasy in Chapter 3, results from the greater density and lesser thickness of the oceanic crust compared with the continental crust. If we change our frame of reference from one based on the surface area above or below sea level to one based on the area occupied by oceanic vs. continental crust, we have to modify our claims about the importance of the oceans. The continental crust makes up about 47 percent of Earth's surface area. Of this, 28.8 percent is exposed as continents rising above sea level and 18.2 percent is submerged continental margin area. Ocean basins underlain by oceanic crust make up only 53 percent of Earth's surface. Because of the current sea level, some continental area has been lost to the ocean.

The area covered by oceans is traditionally considered as three separate ocean basins, but as they are all connected, the Earth really has only one ocean. Figure 4-2 shows a map of the continents, continental margins, and oceans, as well as depth and area data for each ocean. The Pacific is the largest and deepest ocean.

Slopes on the Ocean Floor

The hypsographic curve shows the area of a specific depth or elevation interval, but does not give any information about the rate of change for depth or elevation with distance, the *slope*. If we were walking from one elevation to another, would we go down a gentle slope or would we fall off a cliff? Study of the slopes on the continents and ocean floor reveals some interesting features (Figure 4-3). In this figure, slopes have been averaged for all points within a 250 meter depth-altitude increment; slopes that occupy small areas (i.e., vertical cliffs) become insignificant.

In the oceans, steep slopes are associated with trenches, volcanoes, and midocean ridges. The steepest average

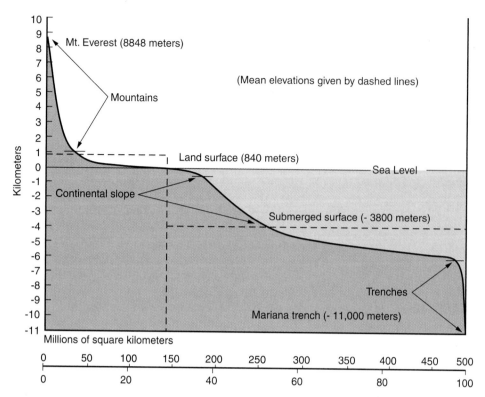

Figure 4-1   The Hypsographic Curve

Plot showing amount of Earth's surface area at various elevations and depths. Horizontal dashed lines indicate average height of the continents at 840 m (2750 ft) above sea level and average depth of the oceans at 3800 m (12,460 ft) below sea level. The vertical dashed line marks the division between land and sea at present-day sea level. *(Reprinted with the permission of Macmillan Publishing Company from* Earth Science, *6th Edition, Figure 10.2, by Edward J. Tarbuck and Frederick K. Lutgens. Copyright © 1991 by Macmillan Publishing Company.)*

slopes found on Earth are the inner trench slopes approaching averages of 5° in the Java, Philippine, Ryukyu, Japan, Aleutian, and Kurile trenches of the Pacific Ocean. The average slope for all trenches exceeds 3.4°. Chains of volcanic islands and seamounts also show steep slopes. The steepest such average slopes, about 4°, are found along the Hawaiian Ridge-Emperor Seamount chain. The slopes on the 72-million-year-old northwest end of the chain are just as steep as the younger slopes. Steepness is associated also with the rough slopes of slow-spreading centers, such as the Mid-Atlantic Ridge. By contrast, the smooth, fast-spreading East Pacific Rise has average slopes of less than 0.5°. In fact, the rise is undetectable on the basis of slope steepness. The average oceanic ridge slope decreases exponentially from a high of about 1° for spreading rates of 2 to 3 cm/yr (0.8–1.2 in/yr) to about 0.1° for a spreading rate of 10 cm/yr (3.9 in/yr). The fracture zones in the northeastern Pacific contain the most pronounced transform fault scarps, or steep linear slopes, on the ocean floor. On the slow-spreading Mid-Atlantic Ridge and Carlsberg Ridge, transform faults contribute significantly to general roughness.

The steepest portion of the continental slope occurs between the depths of 1 and 3 km (0.6 and 1.9 mi); it aver-

ages about 2°. The peaks of steep slopes at 1 and 2 km depths are separated by gentler slopes caused by the relatively flat surfaces found at a depth of 1.5 km (0.9 mi) on the tops of seamounts and large plateaus. The gentlest slopes in the deep-ocean basin are those of abyssal plains. These features have an average slope of 0.3° and an average depth of 5.8 km (3.6 mi).

The gentlest slopes on Earth are on the shores. This narrow strip at sea level (0 km) represents base level for *subaerial* erosion and the upper limit of marine sediment deposition. The average slope there is an imperceptible 0.2°. Moderate slopes are associated with the surfaces of the continental ice sheets covering Greenland and Antarctica from 3 to 4 km in elevation. Steep slopes of about 1°, comparable to those found for oceanic ridges, are found on mountains above 4 km in elevation. The largest mountainous area on Earth is the Himalayas and the neighboring Tibetan plateau.

Bathymetric Techniques

Seismic surveys of various types are used to determine the shapes and positions of boundaries within rocks and sediments. Sound waves are influenced by differences in

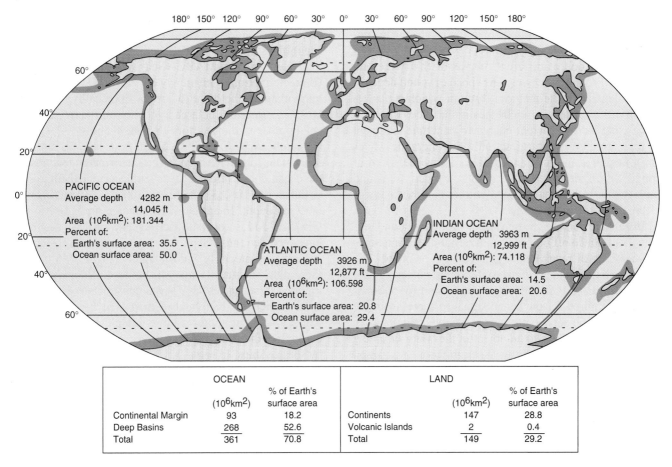

Figure 4-2   Earth's Oceans

The three ocean basins, showing data for average depths, percentage of total surface and of total ocean surface they cover. Box shows data on amounts of area on ocean and land occupied by continental margins and deep basins.

Figure 4-3   Slopes on Earth's Surface

Average slope is calculated for all surface area in 250 m (820 ft) intervals, with polar data eliminated beyond 75° latitude. Slopes of 1°, 2°, and 3° are shown to help visualize steepness.   *(After Moore and Mark, Eos 67:48, 000,1986.)*

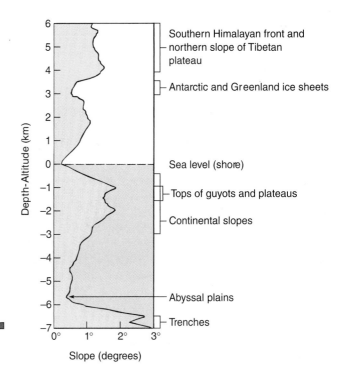

density between materials. At the interface between ocean water and the ocean bottom there is a large change in density; there are also density differences between layers of sediment or rock of different origins or that were laid down by different processes. The speed of sound depends on the density of the material through which it passes. Density differences cause changes in the speed of sound waves that have sufficient energy to penetrate sediment or rock, creating wave refraction. Less energetic waves are reflected at the boundaries. If we know the speed of sound within these various materials, we can use sound travel times to image the ocean bottom and the layers of sediment or the rocks below.

In seismic surveys, a transmitter produces sound signals that bounce off object boundaries and return to a receiver or set of receivers, where they are recorded. The distance to the object is equal to the sound velocity in the material it has traveled through, multiplied by one-half the time required for the sound signal to travel from the surface to the object and back to the receiver (Figure 4-4). To obtain the distance to an object in the water or on the ocean floor, the relatively weak, high-frequency sound signals of an echo sounder can be used. However, to determine the thicknesses of various rock or sediment layers beneath the ocean floor requires more penetrative, low-frequency signals that are generated by setting off small explosions.

One commonly used type of survey is **side-scan sonar**, a technique that uses transmitters directed away from both sides of a ship to map strips of ocean floor up to 60 km (37 mi) wide. Figure 4-5a shows a schematic of

the GLORIA [Geological Long-Range Inclined ASDIC (Acoustical Side Direction)] system for deep-sea side-scanning sonar surveys. Other early systems include SeaBeam and SeaMARC (Sea Mapping and Remote Characterization). One newly developed Deep-Tow system enhances the detail of ocean floor bathymetry, but covers a more limited area by towing the source and receiver close to the sea floor (Figure 4-5b).

Another way to map the ocean floor is to go out into space! Due to gravitational effects, bathymetric features on the ocean floor cause the ocean surface above them to rise or fall about 1 m (3.3 ft) for every 1000 m (3300 ft) change in ocean depth. Radar altimeters aboard the U.S. Navy's GEOSAT and European Space Agency's ERS-1 satellites have recorded these changes in the elevation of the ocean surface very precisely and accurately (Figure 4-6). From these data, the gravity map of the ocean floor shown in Figure 4-7 (see pages 92–93) was constructed. Although it is not strictly a bathymetric map (see Figure 4-8 on pages 94–95), it shows a more complete picture of the ocean floor than one constructed from measurements made by surface ships. In some parts of the Southern Ocean, for example, there are areas as large as the state of Kansas where no ship soundings have been made. The radar altimeter tracks have a maximum distance between them of 8 km (5 mi) and can locate any bathymetric feature that is at least 1 km (0.6 mi) tall and 10 km (6 mi) across. Half of the seamounts on the ocean floor were known only to the military until the U.S. Department of Defense released the altimetry data. Scientists from NASA and Scripps Institution of Oceanography produced the map in Figure 4-7 in 1996.

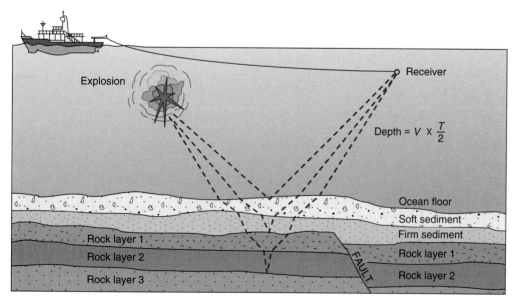

Figure 4-4 Seismic Profiling

An explosion emits low-frequency sound waves that can penetrate bottom sediments. They reflect off the ocean bottom and the boundaries between rock layers and return to the receiver where travel times are used to create a bathymetric map. Because of differences in the densities of water and different rock layers, the sound also refracts—the path bends.

(a)

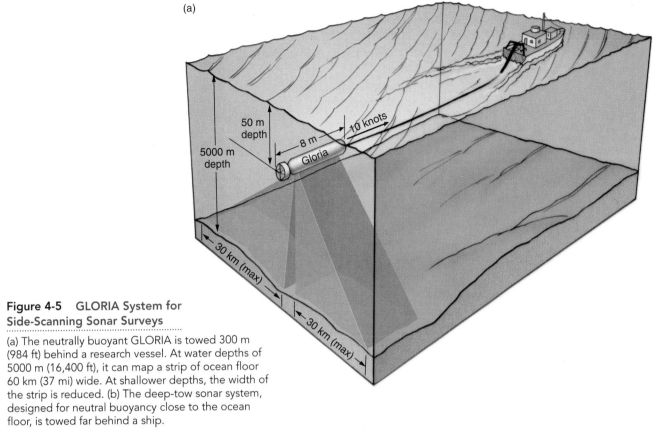

Figure 4-5   GLORIA System for Side-Scanning Sonar Surveys

(a) The neutrally buoyant GLORIA is towed 300 m (984 ft) behind a research vessel. At water depths of 5000 m (16,400 ft), it can map a strip of ocean floor 60 km (37 mi) wide. At shallower depths, the width of the strip is reduced. (b) The deep-tow sonar system, designed for neutral buoyancy close to the ocean floor, is towed far behind a ship.

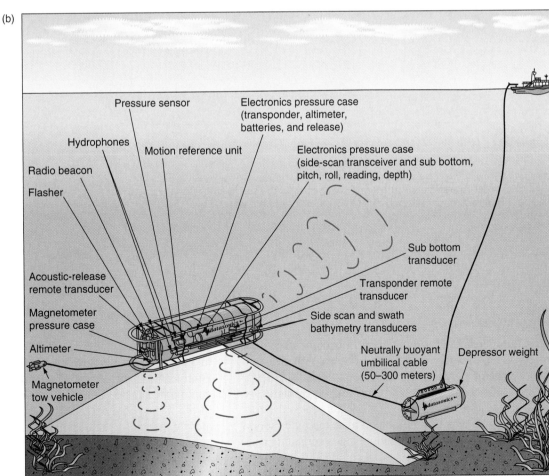

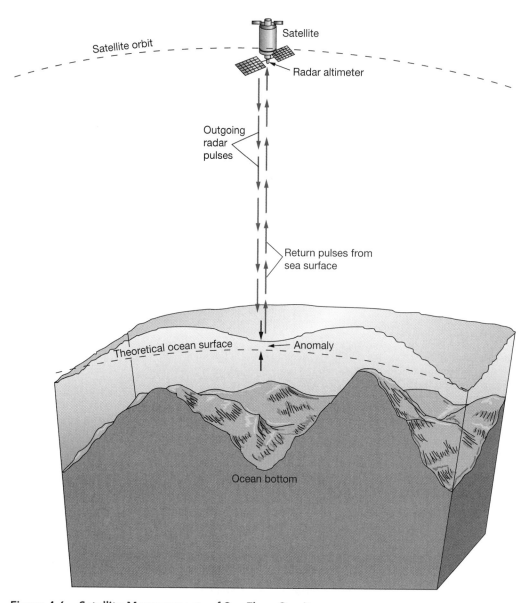

Figure 4-6   Satellite Measurements of Sea Floor Gravity

The shape of the ocean surface can be measured very accurately by radar signals sent and received by a satellite. The shape of the surface, when corrected for waves, tides, and currents, reflects gravitational attraction from shapes on the sea floor. Large objects such as seamounts attract more seawater, increasing the height of the water above them in proportion to their masses.

Marine Provinces

As bathymetric data accumulated, it became clear that most of the area covered by the oceans fell into two major provinces, the continental margins and the deep-ocean basins, and that these features were related to plate tectonic processes. Figure 4-9 illustrates the formation of an ocean basin based on plate tectonic theory. A **deep-ocean basin** forms gradually at a spreading center, starting with initial rifting of a continental mass over a magma chamber and eventually forming a mid-ocean

ridge. The boundary between a continent and the ocean, the **continental margin**, is composed of rift blocks of continental crust that gradually become covered by thick masses of sediment carried by rivers to the edges of the continents. Over time, the ocean basin deepens and the continents subside as a result of the thermal contraction of the lithosphere as it cools. The weight of sediments also contributes to subsidence of the continental margin.

Figure 4-10 shows a composite of marine physiographic provinces where subduction and spreading are taking place. The flat expanse of ocean plate that gen-

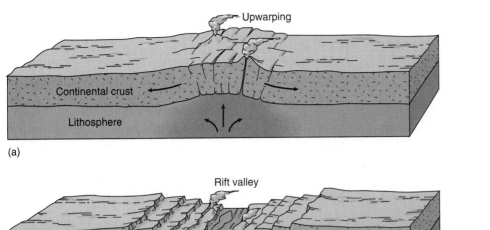

(a)

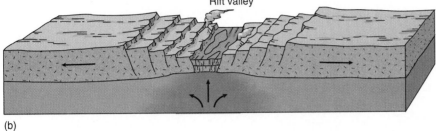

(b)

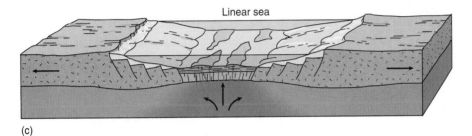

(c)

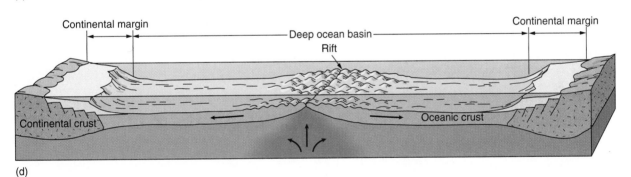

(d)

Figure 4-9 The Evolution of Deep-Ocean Basins and Continental Margins

(a) A continent begins to rift apart above a region of magma development in the lithosphere.
(b) Rifted blocks drop down on either side of a rift valley above the spreading center.
(c) Further spreading produces a narrow ocean, and sediment derived from the continents begins to accumulate on the down-dropped rift blocks. (d) A deep-ocean basin forms with continental margins produced through continental sediment deposition. *(Adapted with the permission of Macmillan Publishing Company from* Earth Science, *6th Edition, Figure 6.9, by Edward J. Tarbuck and Frederick K. Lutgens. Copyright © 1991 by Macmillan Publishing Company.)*

tly slopes away from the ridge is called the **abyssal plain.** In some parts of the ocean basins, notably in the Pacific, the abyssal plain surface is interrupted by small (less than 1 km in relief) hills that appear to be remnants of faulting and volcanoes that were active when the crust was younger and closer to the ridge. Fracture zones also are shown cross-cutting the ridge. The *trench* and *volcanic island arc* provinces of an active margin are shown on the right side; on the left, a passive continental margin with a broad continental shelf is shown. We begin our discussion of marine provinces with the continental margin.

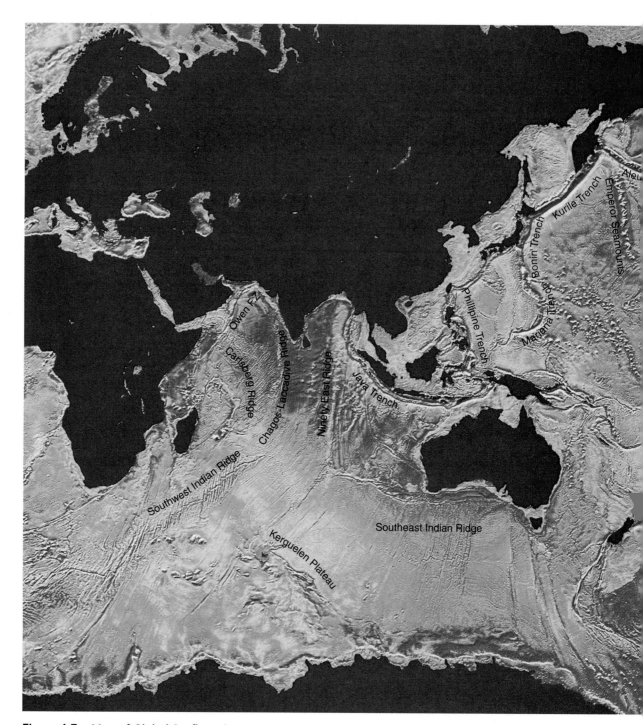

Figure 4-7 Map of Global Seafloor Gravity

This gravity map of the ocean floor appears at first glance to be a map of ocean bathymetry. Green areas are relatively flat surfaces on the deep-ocean floor where normal gravitational pull results in a flat ocean surface. Higher parts of the ocean floor are reflected by shadings from yellow to orange to red. Deeper parts of the ocean range from light blue, to dark blue, to violet in the deepest trenches. Violet colors seen around India, East Africa, California, and Baja California, much of the western Atlantic Ocean, and Hudson Bay reflect low gravity readings resulting from thick layers of continental sediment or the presence of low density continental crustal rock, not great ocean depths.   *(Created by Dr. David T. Sandwell, Scripps Institution of Oceanography, and Dr. Walter H. F. Smith, National Oceanic and Atmospheric Administration. Courtesy Scripps Institution of Oceanography.)*

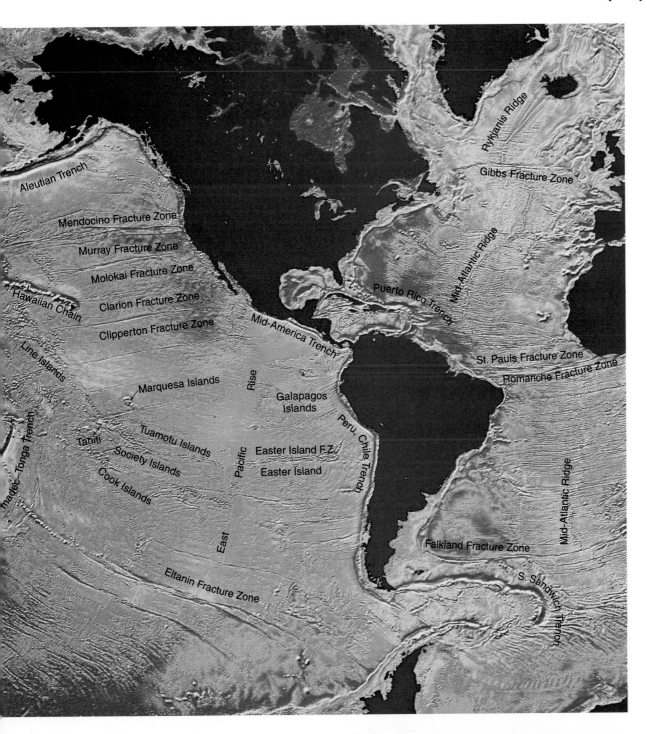

Aleutian Trench

Mendocino Fracture Zone

Murray Fracture Zone

Molokai Fracture Zone

Hawaiian Chain

Clarion Fracture Zone

Clipperton Fracture Zone

Mid-America Trench

Line Islands

Marquesa Islands

Rise

Galapagos Islands

Kermadec–Tonga Trench

Tahiti

Tuamotu Islands

Society Islands

Cook Islands

Pacific

Easter Island F.Z.

Easter Island

East

Peru, Chile Trench

Eltanin Fracture Zone

Rykjanis Ridge

Gibbs Fracture Zone

Mid-Atlantic Ridge

Puerto Rico Trench

St. Pauls Fracture Zone

Romanche Fracture Zone

Mid-Atlantic Ridge

Falkland Fracture Zone

S. Sandwich Trench

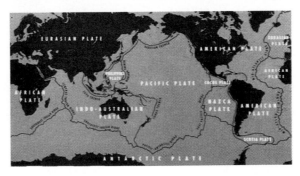

Figure 4-8 The Floor of the Oceans

Map made by compiling ship-track sonar data. *(© 1980 by Marie Tharp.)*

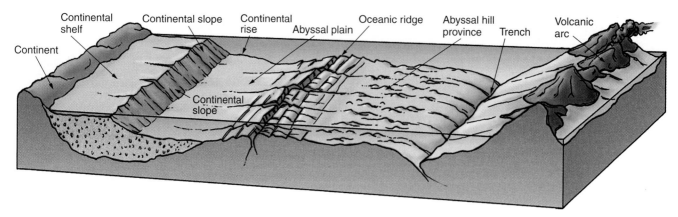

Figure 4-10   Marine Physiographic Provinces

Profile across a hypothetical ocean basin showing marine physiographic provinces. To the left of the oceanic ridge is a passive continental margin typical of the Atlantic Ocean with a wide continental shelf, slope, and rise created by a thick wedge of continental sediments. Features to the right are more typical of active margins like those of the Pacific where trenches and associated volcanic island arcs are common. In the deep-ocean basin, abyssal plain development is shown to the left of the ridge, and abyssal hills interrupt the plain to the right.

Continental Margin Marine Provinces

Continental margins of the type shown in Figure 4-9 are referred to as **Atlantic-type margins**, or **passive margins**, because there is little associated seismic activity. Sediments may accumulate to a thickness of 10 to 20 km. Along other continental margins, including most of the margins surrounding the Pacific Ocean, lithospheric plates converge and are destroyed through subduction. Here, volcanism creates arc-shaped chains of volcanic islands adjacent to trenches, and strong earthquakes are common. Along these **Pacific-type margins**, or **active margins**, continental margins are usually narrower and sediments are much thinner than those found along Atlantic-type margins. As shown in Figure 4-10, the continental margin is divided into the continental shelf, continental slope, and continental rise on the basis of depth and slope changes from the shoreline toward the open ocean.

The Continental Shelf, Slope, and Rise

The **continental shelf** begins at the shoreline and extends with a very gentle seaward slope to the **continental slope**, where the slope increases markedly. The line where this increase in slope occurs is referred to as the **shelf break**. The **continental rise** is at the bottom of the continental slope and has a gentler slope. Continental rises are generally not found on active margins because the converging plates do not allow sediment accumulation. Beyond the rise is the abyssal plain.

With few exceptions, coastal features visible on land extend beyond the shore and onto the continental shelf. They may be modified to some extent by marine processes. We will examine these features in more detail in Chapter 12. Widths of continental shelves vary from a few tens of meters to over 1000 km. The broadest shelves occur off the northern coasts of Siberia and North America in the Arctic Ocean,

off the coast of Alaska in the North Pacific, and off the coast of Australia in the West Pacific. The average width of the continental shelf is about 70 km (43 mi), and the average depth at which the shelf break occurs is about 135 m (443 ft). Around the continent of Antarctica, however, the shelf break occurs at a depth of 350 m (1150 ft). The average slope of the continental shelf is 0°07′, or 1.9 m/km (10 ft/mi). Sediment is transported across the continental shelf by storm waves, tidal currents, and submarine gravity or density flows (like avalanches on land) often triggered by earthquakes.

The continental slopes beyond the shelf break are features similar in relative relief to the fronts of mountain ranges on the continents. The break at the top of the slope may be from 1 to 5 km (0.62 to 3 mi) above the deep-ocean basin floor. In areas where slopes descend into submarine trenches, even greater vertical reliefs are measured. Continental slopes vary in steepness from 1° to 25° and average about 4°. Around the margins of the Pacific Ocean, compressive tectonic processes forming coastal mountain ranges and submarine trenches create steeper continental slopes than those found on passive margins. Slopes in the Pacific Ocean average more than 5°, while those in the Atlantic and Indian oceans average about 3°.

Submarine Canyons and Turbidity Currents   The continental slope and, less commonly, the continental shelf, are cut by large **submarine canyons** that resemble the largest canyons cut on land by rivers. Like river-cut canyons, submarine canyons have tributaries and steep, V-shaped walls that expose a wide range of rock types of various geological ages.

Side-scan sonar work done along the Atlantic continental margin shows a continental slope dominated by submarine canyons from Hudson Canyon near New York City to Baltimore Canyon off the Maryland coast (Figure 4-11a). Canyons occur more frequently where the continental slope is steeper. No canyons are present where slopes are less than

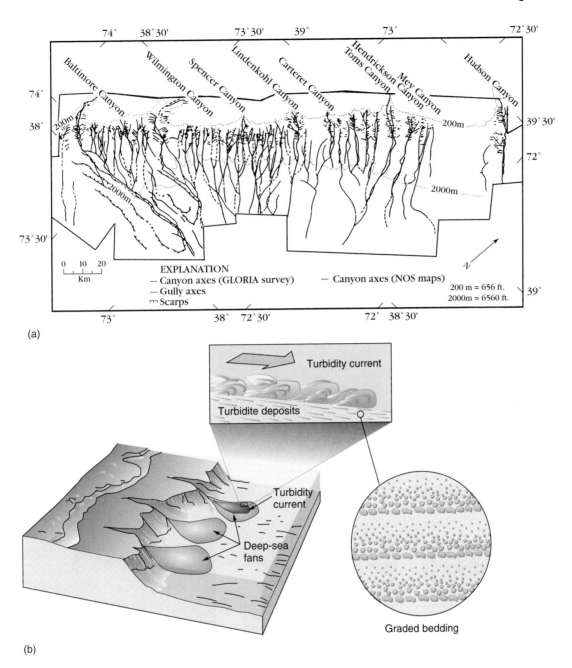

Figure 4-11   Submarine Canyons

(a) Submarine canyons along the Atlantic coast from New York to Maryland. There are no submarine canyons between Hudson Canyon and Mey Canyon where the slope gradient is less than 3°. Distance between adjacent submarine canyons decreases as the slope gradient increases. Between Lindenkohl Canyon and Baltimore Canyon, the gradient exceeds 6° for most of the continental slope. Where the slope is steepest, the submarine canyons are spaced as little as 1.5 km (0.9 mi) apart. (b) Erosional and depositional features associated with the formation of submarine canyons on the continental shelf, slope, and rise.

3°. Canyons are separated by 2 to 10 km (1.2–6 mi) where slopes are between 3° and 5°, and by only 1.5 to 4 km (0.9–2.5 mi) where slopes exceed 6°. Canyons are also more frequent on the middle and upper sections of the slope than on the lower sections. Canyons confined to the continental slope are straighter and have steeper canyon floors than those extending into the continental shelf.

Probably the most widely supported theory of the origin of submarine canyons is erosion by turbidity currents.

Turbidity currents are flows of sediment-laden water that periodically move down the continental slope. Some turbidity flows result from river input near the heads of submarine canyons that extend well into the continental shelf. However, turbidity currents in many canyons on the continental slope apparently begin after sediment moves across the continental shelf and into the head of the canyon. An earthquake may initiate the current, causing the sediment accumulation to collapse and move like an

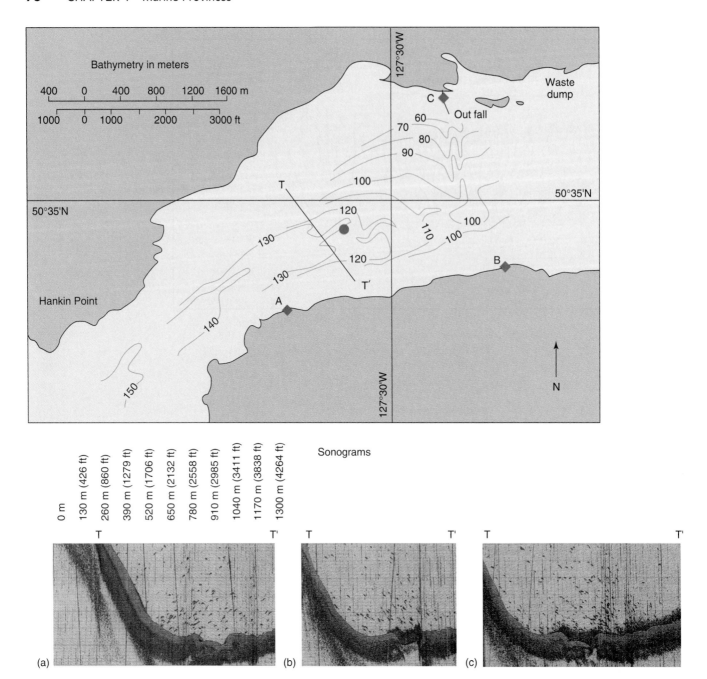

Figure 4-12   Turbidity Current in Rupert Inlet, British Columbia

Map shows bathymetry of submarine channel and line T-T′ followed by the Canadian Survey ship, *Vector*, on August 26, 1976. The red circle shows the approximate location of a bottom-moored current meter, and the red diamonds A, B, and C are locations of positioning transponders. The steep wall in the sonograms is on the north side of the channel, and the vertical lines result from transmitted pulses of another sounder. The spots in the water column above the bottom probably are fish. In sonogram (a), recorded at about 12:53 P.M., the ship moved across the channel and no turbidity current was flowing. Note levees on either side of the main channel. In (b), the ship was over the channel at 1:03 P.M., and the turbidity current had begun to flow in the main channel. By 1:28 P.M. in (c), the turbidity current had spilled over the levees. By 2:30 p.m., the turbidity current had passed through completely and was undetectable at that point in the channel.   *(Reprinted by permission from A. E. Hay, et al., Remote Acoustic Detection of Turbidity Current Surge. Science 217:4562, 833–835, August 27, 1982. Copyright 1982 by the AAAS. Photos courtesy of Alex E. Hay.)*

avalanche down the canyon, propelled by gravity. As the current loses energy, the sediment begins to settle out. The coarser grains fall out first, high on the slope; finer particles are deposited at the base of the slope or on the abyssal plain. Sediments form deep-sea fans at the bases of submarine canyons. Submarine fans are similar to the deltas of rivers and streams. As the sediment-laden current moves out across the fan, patterns of distributary channels bordered by levees develop (Figure 4-11b and 4-13).

Sonar recordings of a turbidity current generated on August 26, 1976, by the discharge of mine tailings into Rupert Inlet, British Columbia, may be the only real-time observation of a marine turbidity current moving through its channel (Figure 4-12). The sonograms show acoustic sounding profiles across the submarine channel in Rupert Inlet before and during the event, which lasted for less than 1.5 hours. Transoceanic telegraph cables have been snapped by turbidity currents. On the continental shelf and slope south of Newfoundland, several cables were laid at different distances down the slope. An earthquake on November 18, 1929, with an epicenter near the slope

triggered a turbidity current that snapped the cables sequentially. Because scientists knew the time of the earthquake and the times that the cables broke, they were able to calculate the speed of the current, a serendipitous outcome of an otherwise disastrous event. The calculated velocity of the current was 80 km/h (50 mi/h) on the steepest part of the slope, decelerating to about 24 km/h (15mi/h) on the gentle lower slope. Sediment masses moving at these velocities have enough energy to erode the submarine canyons observed on continental slopes and shelves.

Deep Sea Fans and the Continental Rise As previously noted, deep-sea fans accumulate at the mouths of submarine canyons when current velocities decrease. The merging of these deep-sea fans along the base of the continental slope results in the development of the continental rise.

The Amazon Cone, one of the largest deep-sea fans, extends 700 km (435 mi) northeast off the coast of Brazil. It is a typical passive-margin fan. Figure 4-13 shows a schematic map of the fan made from GLORIA sonar data. From Amazon Canyon (shown in the southeast corner of

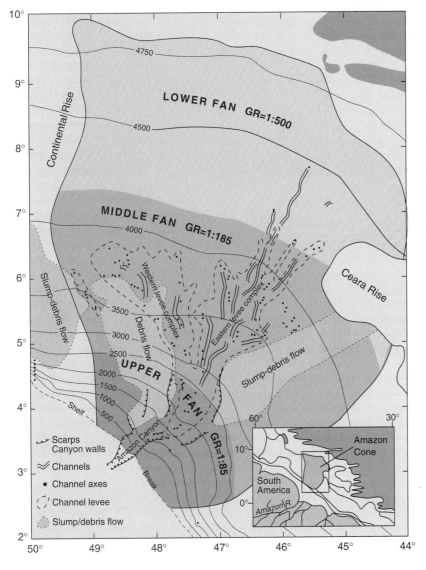

Figure 4-13   The Amazon Cone

The Amazon Canyon branches into two major distributary systems composed of meandering channels bounded by levees. Note the large debris flows originating on the steep upper slope of the fan cone. Note that the steepness gradient decreases from 1:85 on the upper fan to 1:500 on the lower fan. A gradient of 1:85 means the slope drops a vertical distance of 1 m over a horizontal distance of 85 m.
(Map courtesy of John E. Damuth and Roger D. Flood. Permission granted by the Geological Society of America.)

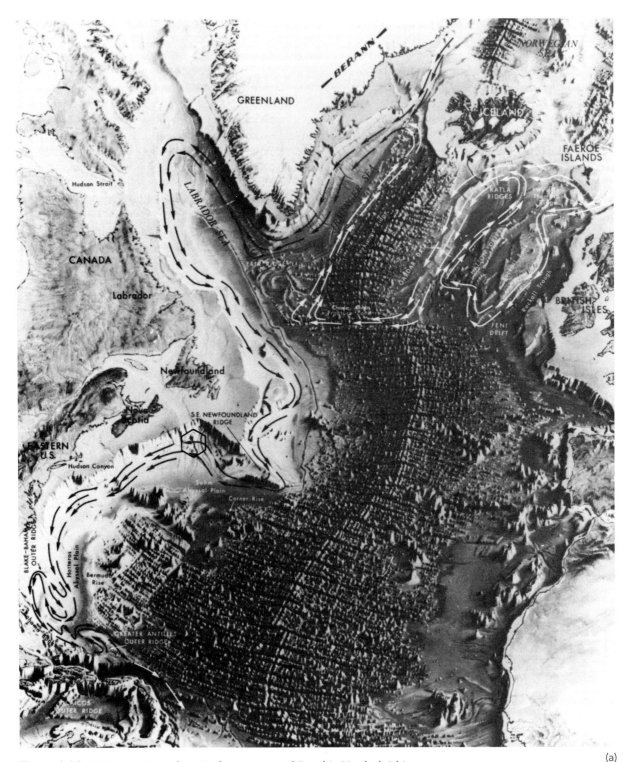

(a)

Figure 4-14   Western Boundary Undercurrent and Benthic Nepheloid Layer

(a) Black arrows indicate the path of the western boundary undercurrent along the North Atlantic. The hexagon indicates cable breaks associated with the 1929 Grand Banks earthquake.   (*Excerpted from The World Ocean Floor by Bruce C. Heezen and Marie Tharp, published by the Lamont-Doherty Geological Observatory of Columbia University and sponsored by the U.S. Navy through the Office of Naval Research. Copyright © 1977 by Marie Tharp.*) (b) Map of the Atlantic Ocean showing distribution of suspended material in the benthic nepheloid layer. Note that the highest concentrations are found in western portions of the ocean overlying the continental rise and coincident with the location of the western boundary undercurrent shown in (a).   (*After Biscaye and Eittreim, Marine Geology 23:155-72, 1977.*) (c) Schematics of light-scattering curves indicate the high turbidity in the nepheloid layer (yellow) that lies about 200 meters above the ocean floor.

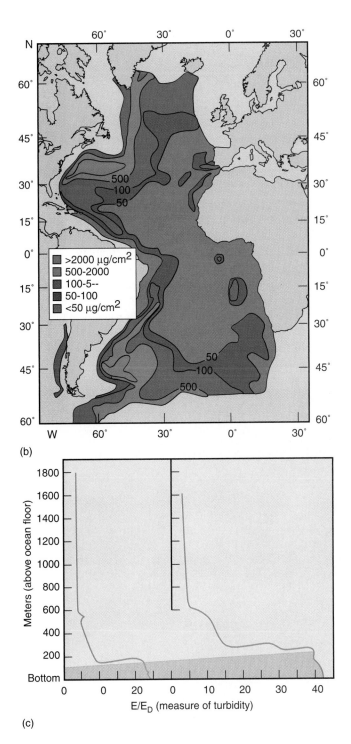

(b)

(c)

features of mature river systems such as the lower Mississippi River. On the lower fan, the surface has a very low slope, and flow which is no longer confined to channels can spread sheets of fine sediment across the fan surface.

The continental rise also is shaped by undersea contour currents that flow along the base of the continental slope. The strongest contour currents occur on the western boundaries of ocean basins. In the Atlantic, the **Western Boundary Undercurrent (WBUC)**, originates in the Norwegian Sea, flows southward through deep troughs around Iceland, and continues on toward the equator along the base of the North American continental slope (Figure 4-14a). This current, and the continental rise over which it flows, have been extensively studied along the east coast of North America using submersibles, surface ships, and side-scan sonar. The Coriolis effect (see Chapter 8) forces the WBUC to veer to the right and flow snugly against the base of the continental slope. Flowing at velocities up to 40 cm/s (about 1 mi/h), the WBUC carries significant quantities of sediment in suspension, including volcanic debris from Iceland, and sediment entrained from portions of the continental slope and rise. This suspended sediment forms a permanently turbid zone just above the ocean floor that is called the **benthic nepheloid layer (BNL)** (Figure 4-14b and c).

Mud waves also are found along much of the lower continental rise. These features are like giant ripples or sand dunes, with an average distance of 2 to 3 km (1.2–1.9 mi) between crests. They are composed of very fine, mud-sized grains instead of the sand particles that are more typical of these types of sediment structures. The internal structure of mud waves (Figure 4-15) shows that these features probably formed as a result of sediment falling rapidly out of suspension after high energy events.

Deep-ocean Basin Marine Provinces

Beyond the continental slopes lie the **deep-ocean basins**, where depths exceed 4000 m (13,000 ft). Bathymetric features result from the interplay of tectonic processes creating and destroying crust, fine-grained sediments brought in by slow gravitational settling from the surface, and the dying gasps of turbidity currents from the continental slopes. Ridges, trenches, fracture zones, and abyssal plains are the major provinces of the deep ocean.

Oceanic Ridges and Rises

Oceanic ridges and rises are segments of a continuous mountain chain extending some 65,000 km (43,000 mi) across the deep-ocean floor (see Figure 4-8). The chain somewhat resembles the raised seam of a baseball. The central axis of the mountain chain averages about 1.5 km in relief above the abyssal plain. With their associated fracture zones, ridges and rises influence bathymetry across about one-third of the ocean basins.

the map), the main channel extends seaward onto the fan, but quickly divides into several distributary channel systems as the current velocity ebbs at the bottom of the slope. At a depth contour of about 2200 m (7200 ft), within the middle fan, the channel has split into western and eastern systems that show extensive branching and development of meanders. Channel meandering starts where the slope of the fan surface decreases (note the wider spacing of the contours on the middle fan relative to the upper fan). Meander patterns and other channel features, including abandoned meander loops and cutoffs, closely resemble

Marie Tharp—Oceanographic Cartographer

Marie Tharp, working with Dr. Bruce Heezen in 1952, began mapping the ocean floor. Using primarily sounding data taken aboard Woods Hole Oceanographic Institution's first oceanographic vessel, *Atlantis*, in the late 1940's, Tharp made the first detailed transoceanic seafloor profiles of the North Atlantic Ocean. These profiles illustrated the existence of a rift valley running the length of the Mid-Atlantic Ridge. When enough data were plotted, Tharp, Bruce Heezen and Maurice Ewing, the director of Lamont Geological Observatory (now Lamont-Doherty Earth Observatory), discovered that the Mid-Atlantic Ridge was part of a 64,000 km (40,000 mi) long mountain range that ran as a continuous feature through the world's ocean basins—and that the crest of this mountain range was where many major earthquakes occurred. Finally, in 1977 Bruce Heezen and Marie Tharp completed and published the now famous map of the ocean floor, *The Floor of the Oceans* (see Figure 4-8).

In 1998 Marie Tharp was recognized as one of four individuals that have made major contributions to the field of cartography, and in 1999 she was honored for her achievements as the fourth recipient of the *Women Pioneers in Oceanography Award* awarded by the Women's Committee of Woods Hole Oceanographic Institution.

With publication of this map, it was as if the ocean had been drained. For the first time, the spectacular features of the ocean floor that are hidden by the oceans became visible. It helped broaden the acceptance of global plate tectonics as the features that formed the boundaries of the plates, the mid-ocean ridges with their transform fault offsets and ocean trenches, came into view—and revolutionized our image of the Earth.

Figure 4A   Floor of the Oceans

Marie Tharp and Bruce Heezen examine their map, *Floor of the Oceans*, shortly after its publication in 1977. (*Photo courtesy of Marie Tharp.*)

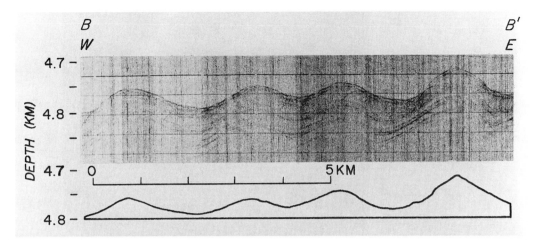

Figure 4-15   Mud Waves

Echo sounding of mud waves taken along an east-west line across the Blake-Bahama Outer Ridge, a large depositional feature on the continental rise north of the Bahamas. The near-bottom bathymetric profile beneath the echo sounding gives a more accurate shape of the waves.   *(Courtesy of Roger Flood, Lamont-Doherty Geological Observatory, Columbia University.)*

Hydrothermal Circulation at Spreading Centers   Near the axes of oceanic ridges and rises, where magma chambers may be less than 1 km (0.6 mi) beneath the ocean floor, seawater seeps down into fractures and other permeable zones in the ocean crust and is heated by magma. This hydrothermal water rises back to the surface, reentering the ocean at vents of various temperatures (Figure 4-16a). The temperature of the water at the vents controls geochemical reactions that occur as the water travels through the crust; it also controls the precipitation of minerals when the hydrothermal water remixes with seawater at the ocean floor. Warm-water vents, ranging from 10 to 20°C (50–68°F), precipitate few if any minerals; **white smokers**, ranging from 30 to 330°C (86–626°F), are named for the white barium sulfate that precipitates; and **black smokers**, at temperatures above 350°C (>660°F), precipitate black metal sulfides (Figure 4-16b). The entire volume of ocean water is cycled through this hydrothermal circulation system in about 3 million years, therefore, the chemical exchange between ocean water and the basaltic crust has a significant influence on the chemical composition of seawater (see Chapter 7).

Metallic sulfide deposits around black smokers contain many economically important metals, including lead, iron, nickel, copper, zinc, chromium, and silver. Mining these modern seafloor deposits is neither economically nor politically feasible at present, but many of the ancient deposits of these metals that we exploit on land probably originated as hydrothermal deposits at spreading centers active in the geologic past. These deposits are associated with *ophiolites,* pieces of ocean crust that have been shoved up onto the continent (see Ophiolites, Chapter 3).

Oceanic **hydrothermal vents** support unusual biological communities that first were discovered in 1977 when

geologists were exploring the Galápagos Rift between South America and the Galápagos Islands (Figure 4-16c). Giant clams, mussels, tube worms, and a variety of other animals make up a community that depends on chemosynthetic bacteria as the base of their food chain because no light penetrates to these depths. The bacteria use the oxidation of hydrogen sulfide to sulfate as a source of energy to turn inorganic compounds into organic matter.

Fracture Zones

Transform faults are parts of plate boundaries between spreading ocean plates that connect offsets of the ridge axis. Recall from Chapter 3 that the motion along transform faults is opposite to the direction of ridge offset. Fracture zones are the extensions of the transform fault beyond the ocean ridge offset. Relief associated with transform fault zones and fracture zones that offset ridge and rise axes is similar to that of cliffs on land, but these features may extend laterally for thousands of kilometers and be more than 200 km wide (Figure 4-17a). The ocean floor on either side of a fracture zone may differ in depth by as much as 1500 m (4900 ft). The Mendocino Fracture Zone in the North Pacific is a good example (see Figure 4-8). The ocean bottom on the south side is more than 1000 m (3280 ft) deeper than the floor on the north side, creating a feature sometimes referred to as the Mendocino Escarpment. Much of this difference in elevation is attributable to the 40-million-year age difference between the ocean floor on the south side and the north side. The older south side has undergone much more thermal contraction and consequent deepening than the younger ocean floor to the north.

Although fracture zones are not known to be sites of much earthquake activity, there is structural evidence in

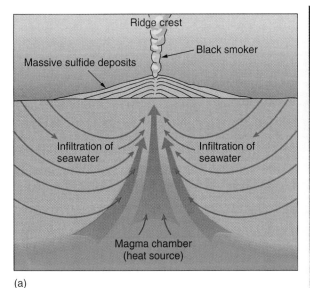

(a)

Figure 4-16 Hydrothermal Vents along Spreading Center Axes

(a) Cross section of the axis of a spreading center to illustrate the hydrothermal circulation in oceanic crust. Seawater percolates down along a broad zone of fractured crustal rock. It is heated as it approaches the underlying magma chamber, it then rises in a narrow zone near the axis of the spreading center. The resulting vents usually lie within 200 m (656 ft) of the ridge or rise axis. *(Reprinted with the permission of Macmillan Publishing Company from* Earth Science, *6ᵗʰ Edition, Figure 10.15, by Edward J. Tarbuck and Frederick K. Lutgens. Copyright © 1991 by Macmillan Publishing Company.)* (b) Black smoker photographed during a 1979 expedition to 21°N on the East Pacific Rise. Water was measured at 315°C (599°F). In the foreground is Alvin's sample basket. (c) Tube worms observed at an East Pacific Rise warm-water vent. The tubes are about 1 m (3.3 ft) long. These vestimentiferan worms have no mouth or digestive tract. Symbiotic chemosynthetic bacteria oxidize hydrogen sulfide gas dissolved in vent water and use the energy to produce food for themselves and the worms. *(Parts (b) and (c) courtesy of Scripps Institution of Oceanography, University of California, San Diego; © by Dr. Fred N. Spiess.)*

(b)

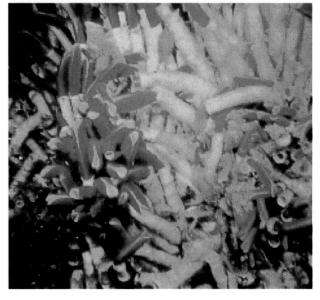

(c)

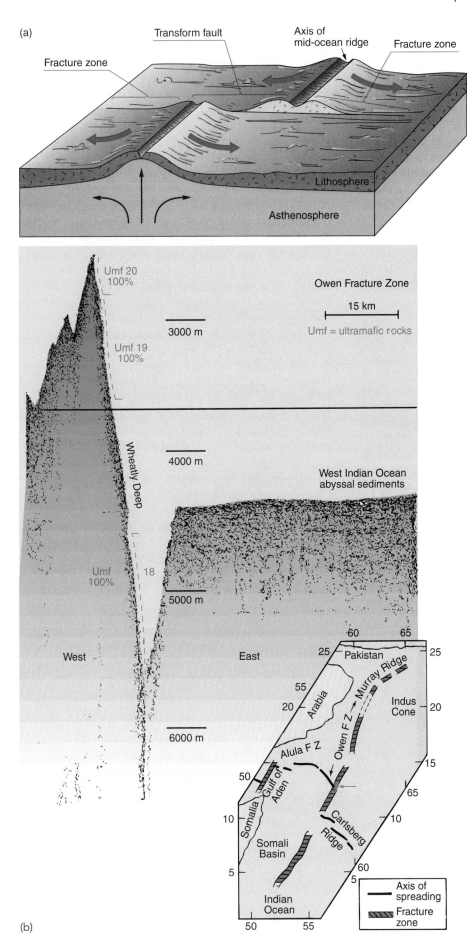

(a)

Fracture zone

Transform fault

Axis of mid-ocean ridge

Fracture zone

Fracture zone

Lithosphere

Asthenosphere

(b)

Umf 20 100%

Owen Fracture Zone

3000 m

15 km

Umf 19 100%

Umf = ultramafic rocks

Wheatly Deep

4000 m

West Indian Ocean abyssal sediments

Umf 100%   18

5000 m

West

East

6000 m

60   65

25   Pakistan   25

55   Arabia   Murray Ridge   Indus Cone

20   Owen F Z   20

Alula F Z   15

50   Gulf of Aden   65

Somalia   Carlsberg Ridge   10

10   Somali Basin

5   60

50   55   Indian Ocean   5

Axis of spreading

Fracture zone

Figure 4-17   Fracture Zones

(a) Schematic of a transform fault offsetting the axis of a midocean ridge and the fracture zones extending on either side of the transform fault beyond the ridges. (b) Cross-section of the Owen Fracture Zone (black arrows on map inset) showing the transverse ridge to the west of the Wheatly Deep. Dredge samples of the ridge at the location of the blue arrow recovered 100 percent ultramafic mantle-type rocks (Umf on cross-section), indicating the ridge probably resulted from uplift of mantle material rather than volcanism. *(Courtesy of Enrico Bonatti, Lamont-Doherty Geological Observatory of Columbia University.)*

fracture zones of the equatorial Atlantic and Indian oceans of tectonic activity. We will focus on these three fracture zones: the St. Paul Fracture Zone and Romanche Fracture Zone in the equatorial Atlantic Ocean, and the Owen Fracture Zone in the western Indian Ocean (see Figure 4-8).

The St. Paul islets are peaks on a transverse ridge that runs parallel to the St. Paul Fracture Zone. Charles Darwin collected samples of peridotite, an ultramafic rock typical of Earth's mantle, from these islets during his famous voyage on the H.M.S. Beagle. Mantle rocks also have been recovered from the two Romanche Fracture Zone transverse ridges, and from the transverse ridge of the Owen Fracture Zone. A cross-section of the Owen Fracture Zone shows the bathymetry of the transverse ridge and the locations from which ultramafic samples have been recovered (Figure 4-17b). Geologists can find no way to explain the presence of these rocks except by tectonic uplift of mantle material into these fracture zones.

Abyssal Plains and Intraplate Features

Abyssal plains begin at the toe of the continental rise, generally at depths of 4000–6000 m (13,000–20,000 ft), and are entirely underlain by basaltic oceanic crust. Because of their sediment blankets they are close to flat and feature-less, having slopes of less than 0.5 to 1°. **Abyssal hills** are volcanic and faulting remnants less than 0.6 km high. There are at least several hundred thousand of these features worldwide, but they are only visible at the sea floor in the Pacific Ocean. Seismic surveys reveal them buried by sediment in the Atlantic and Indian oceans. Sediment from the continents can more easily reach the abyssal plains in the Atlantic and Indian oceans, partly due to the much greater sediment inputs from continental rivers along passive margins, but also due to the smaller widths of those basins. The sediments bury the small hills. In the Pacific Ocean basin, abyssal hills are evident on the ocean floor because sediment from the continents is trapped behind or within trenches along the active margins. Also, the center of the Pacific Ocean is comparatively far from land, out of reach of turbidity flows.

Volcanic features are particularly abundant in the Pacific Ocean—more than 20,000 have been found. Volcanic peaks in the abyssal plain of more than 1 km (0.6 mi) relief are called **seamounts**. Seamounts with flattened tops are called **tablemounts**, or **guyots**. Tablemounts are less common than seamounts and are found in linear trends in the Pacific Ocean. The tops of most tablemounts in the Pacific are from 1800 to 3000 m (5900 and 9800 ft) below the present ocean surface.

The formation of seamounts and tablemounts probably occurs during development of oceanic ridge systems (Figure 4-18). Active oceanic volcanoes are characteristic of the ridge axes where cracks in Earth's crust serve as conduits for lava rising from the magma chamber below. As crust moves away from the ridge axis and cools, the cracks are sealed off and volcanic activity gradually decreases. Active volcanoes more than 30 million years old are rare in the oceans. If we consider the life history of a volcano that began at a crack near a ridge axis and progressively moved away from the ridge, carried by the lithospheric plate, we see how a seamount might form. As long as a volcano is near the magma source, it will grow in size from numerous lava eruptions that contribute to its relief above the sea floor. By the time it is 10 million years old, if it has been active enough to grow faster than the rate at which it is sinking due to crustal subsidence, it may develop into an island. The island may last for another 10 to 15 million years if the volcano remains active. Within 30 million years, however, the volcano will probably become dormant. As it continues to subside by plate motion away from the oceanic ridge, the top will be eroded by wave action, and it will finally become submerged. The flat-topped structure sinks deeper and deeper beneath the ocean surface as it descends the slope of the ridge. At this point it becomes a guyot, or tablemount, most of which are at least 30 million years old.

Volcanoes and Reefs   The first theory of coral atoll development was proposed by Charles Darwin. Although the development of a comprehensive theory of global plate tectonics was not developed until more than 100 years

Figure 4-18   Formation of Seamounts and Tablemounts

Volcanic activity develops near the crest of a ridge because of tensional fractures that develop perpendicular to the ridge. These fractures serve as conduits for lava but usually seal within 30 million years.

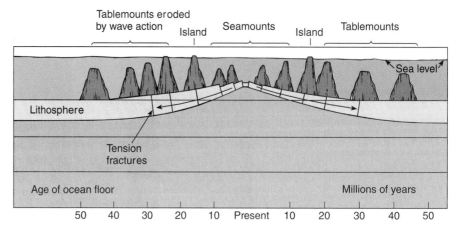

after his work, Darwin's atoll theory depended on the progressive sinking of volcanic islands, which we now know occurs because of lithospheric plate subsidence. Darwin published his theory in 1842 but conclusive proof was not obtained until the middle of the twentieth century when core material from atolls was radiometrically dated. The theory also may apply to tablemounts in warm tropical waters. Coring along the flanks of the tablemounts has revealed remnants of coral reefs that have colonized the sides of the volcanoes sometime during their history, forming flat-topped crowns around them. Eventually, as the reef failed to keep up with subsidence, the flat-crowned volcano became a tablemount on the sea floor.

There are three basic types of reefs—fringing, barrier, and atoll (Figure 4-19). **Fringing reefs** develop at the margins of any landmass where the temperature, salinity, turbidity, and depth of water are suitable for reef-building corals. The greatest concentration of living reef is on the seaward margin, which is more protected from sediment and salinity changes due to runoff. Without some change in sea level relative to the landmass during the existence of the fringing reef, it is not likely to transform into one of the other two types of reefs: the barrier reef or the atoll.

Barrier reefs are linear or circular structures that are separated by a lagoon from the adjacent continent or island. Barrier reefs form when the substratum on which a fringing reef is built begins to subside. With subsidence, the reef maintains its position at the optimum water depth by growing upward. In other words, subsidence gives the reef more room to grow. In the recent geologic past, reefs have grown typically at rates of 3 to 5 m/1000 yr (10–16 ft/1000 yr). Some Caribbean reefs, however, have grown at more than twice this rate. The most outstanding example of a barrier reef is the Great Barrier Reef along the northeast coast of Australia. It is 150 km (90 mi) wide and more than 2000 km (1250 mi) long. Smaller barrier reefs are found along the Caribbean coast of Belize and Honduras, around the island of Tahiti, and surrounding many other islands in the western Atlantic and Pacific oceans.

Atolls form either on the subsiding continental shelves around islands or around sinking volcanic islands in the open ocean. The greatest number of atolls occurs in the

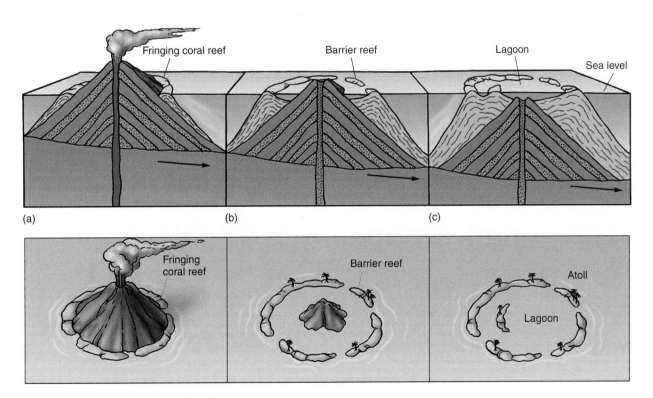

Figure 4-19   Development of Reef Types

(a) A volcanic island forming in warm-water latitudes becomes the substrate for a fringing coral reef in the shallow, sunlit waters adjacent to the island shore. (b) After the volcano becomes dormant and gradually subsides due to thermal contraction of the lithosphere, the reef maintains its position near the surface by growing upward. This produces a barrier reef where the actively growing reef is separated from the volcanic island by a well-developed lagoon. (c) Continued subsidence submerges the volcano, but the reef continues to grow upward to produce an atoll with active reef growth confined mostly to the outer edge of the reef. Deposition of reef debris may produce low-lying islands inside the protective ring of coral reef. Within the atoll ring lies a broad shallow lagoon.   *(Reprinted with the permission of Macmillan Publishing Company from* Earth Science, *6th Edition, Figure 10.20, by Edward J. Tarbuck and Frederick K. Lutgens. Copyright © 1991 by Macmillan Publishing Company).*

equatorial Pacific. Atolls may be circular or irregular in shape, but all surround a lagoon that is usually 30 to 50 m (98–164 ft) deep. The volcanic peak that initially occupied the lagoon is usually nearly gone or absent. The reef continues to grow upward as long as it keeps up with the rate of subsidence. In this way, the reef records vertical plate motion. The atoll reef usually has several tidal channels cut through it that allow circulation between the lagoon and the open ocean. The reef bottom is commonly broader on the windward side of the atoll, and channels are more numerous on the leeward side (Figure 4-20).

In addition to vertical plate motion, horizontal motion can also be recorded by reef development. The Great Barrier Reef clearly records the northward horizontal movement of the Australian continent. Figure 4-21 shows that the reef began forming about 30 million years ago when the land under the northern end of the reef reached a latitude near the Tropic of Capricorn and temperate climatic conditions were replaced by conditions sufficiently tropical to allow for coral reef development. The reef is progressively younger toward the south.

Hot Spot Traces   Some bathymetric features may result from the movement of lithospheric plates across hot spots, places within plates that have anomalously high heat flow and magma chambers close to the crust. Hot spots may remain active for about 120 million years. Forty-one hot spots are presently active (Figure 4-22). Most are found in the ocean basins, primarily in two large clusters, one running along the equatorial Pacific and the other in the southern Atlantic Ocean.

Many northwest-to-southeast trending island chains are located on the Pacific Plate. The most intensely studied of these is the chain of islands between Midway Island and Hawaii, and the Emperor Seamounts that extend beyond Midway to the north. The Emperor Seamounts are the oldest features of this chain and the islands decrease in age from Midway to Hawaii. Volcanic activity on Midway Island appears to have ceased at least 20 million years ago. The Hawaiian Islands are the only islands in the chain with active volcanoes; Kilauea on Hawaii is less than 1 million years old. An active submarine volcano, Loihi, exists southeast of the island of Hawaii and may

Figure 4-20   Atolls

View from space of a group of atolls in the Pacific Ocean. Note the windward margins on the right sides of atolls where reefs are wider.   *(Photo courtesy of NASA.)*

Figure 4-21 Australia's Great Barrier Reef and Global Plate Tectonics

The Great Barrier Reef extends for more than 2000 km (1242 mi) along the northeastern coast of Australia. It extends from 9° to 24°S latitude. The reef first developed about 30 million years ago when northern Australia moved into tropical waters from colder Antarctic waters to the south. The Great Barrier Reef's southern tip has developed very recently.

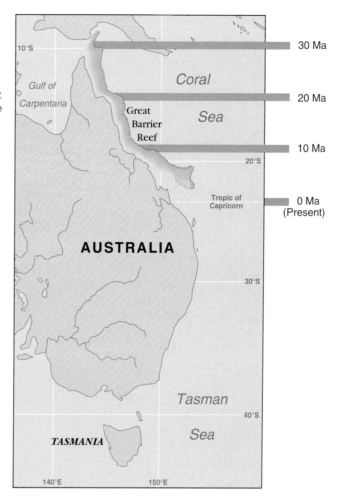

Figure 4-22 Hot Spots and Associated Features on Earth's Surface

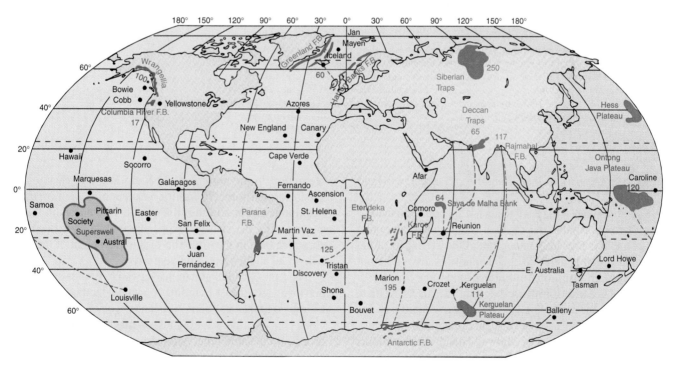

The map shows positions of forty-one known hot spots (black dots) and continental flood basalts and oceanic lava plateaus (red patches). The ages of some of the flood basalts and oceanic plateaus are given in Ma (millions of years ago). The Superswell (blue patch) that underlies the islands of French Polynesia is shown between the Samoa and Easter Island hot spots in the South Pacific.

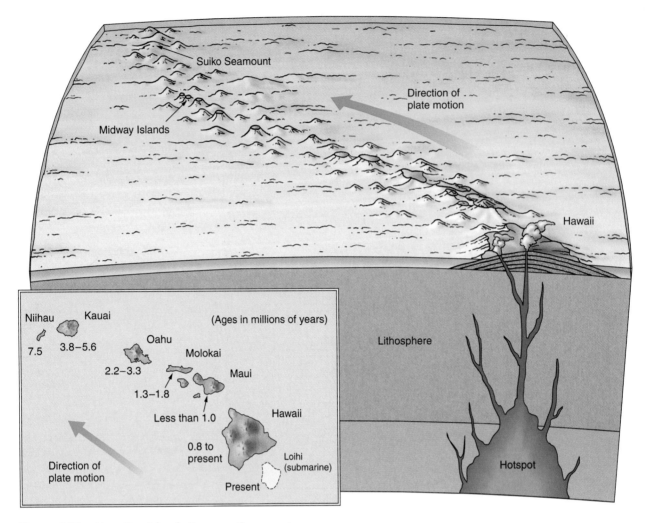

Figure 4-23 Hawaiian Islands-Emperor Seamounts

The chain of islands and seamounts that extends from Hawaii to the Aleutian trench results from the movement of the Pacific Plate over an apparently stationary hot spot. Radiometric dating of the Hawaiian Islands shows that the volcanic activity decreases in age toward the island of Hawaii. *(Reprinted with permission of Macmillan Publishing Company from* The Earth, *4th Edition, Figure 18.28, by Edward J. Tarbuck and Frederick K. Lutgens. © 1993 by Macmillan Publishing Company.)*

either become part of that island or create a new island (Figure 4-23).

Scientists have suggested that a hot spot beneath the Pacific Plate has created this chain of islands and seamounts; this hot spot is presently under the southeastern part of the main Hawaiian Islands. As the plate moves across the hot spot, volcanic eruptions produce islands, the volcanoes eventually become dormant, erode, and subside as the plate moves off to the northwest. The island and seamount chain, however, does not record a constant direction of plate motion. Directly northwest of Midway Island, the seamounts have the same alignment as the island chain from Midway to Hawaii, but at about 188°W longitude and 30°N latitude, an abrupt change in the alignment of the seamounts occurs. From that point on, the alignment is slightly west of true north. The age of the seamounts at this juncture of the two trends is about 43 million years, whereas the oldest

seamounts in the northward-trending Emperor Seamount Chain formed some 70 million years ago. If the Emperor Seamount Chain and the Midway-to-Hawaii chain were generated by the same hot spot, there must have been an abrupt change in the direction of the movement of the Pacific Plate about 43 million years ago.

The Line Islands south of the Hawaiian Islands and other northwest–southeast trending volcanic island chains—the Marquesas Islands, Tuamotu Islands, Society Islands, and Cook Islands of French Polynesia—may have formed more or less at the same time. Although there is evidence these chains of islands also may have a hot spot origin, the evidence is not as clearly developed as for the Hawaiian Islands-Emperor Seamount chain.

The origin of hot spots is still controversial. The Emperor Seamount–Hawaiian Islands hotspot trace supports the concept of a stationary lower-mantle source that always heats the same position below the crust. Some investigators believe that hot spots are associated with rifts that form in the lithosphere as the lithosphere contracts from cooling and is subjected to tensional forces from plate interactions. Whatever their origin, there is now clear evidence that at least some hot spots are not stationary, but move at rates of 1 to 2 cm (0.4–0.8 in.) per year, comparable to rates of lithospheric plate motion.

The Superswell is an enigmatic feature in the area of French Polynesia and may or may not be related to hot spots (Figure 4-22). This feature occupies ocean floor that is between 60 and 80 million years old, but the ocean floor is up to 1000 m (3300 ft) too shallow and its lithosphere is too thin for it to be this age. Although it occupies only 3 percent of Earth's surface, the Superswell accounts for 30 percent of all hotspot heat flow. Volcanoes within the Superswell generate lava with high concentrations of radioactive isotopes. A mantle reservoir with higher than normal radioactive heat generation may explain the high heat flow of the area.

Flood Basalts   Some marine volcanic activity produces widespread, gently sloping surfaces where large volumes of lava flowed out in broad sheets, solidified, and became *archipelagic aprons*. Such features commonly surround volcanic islands, but sometimes are large enough to form elevated plateaus called large igneous provinces.

When found on the continents, such deposits are called **continental flood basalts** (Figure 4-22). Where hot spot mantle plumes reach the bottom of the lithosphere, they form broad heads that may range from 1000 to 2000 km (620 to 1240 mi) in diameter. The temperature of the mantle within the plume head is 100–200°C (210–390°F) hotter than the surrounding mantle and produces a broad dome at Earth's surface from thermal expansion of the rock below. If the continental crust above thins from erosion of the uplifted dome or from incipient rifting, hot magma floods out at the surface. Continental flood basalts range from 1 million to 2 million km^3 (240,000 to 480,000 mi^3) and erupt at rates of about 1 km^3 (0.24 mi^3) per year over a relatively short interval of about 1 or 2 million years. However, the 120 million year old Ontong Java Plateau in the equatorial western Pacific contains about 65 million km^3 of lava—more than 30 times the volume of most continental flood basalts. This large volume was produced over a period of no more than 3 million years, requiring an eruption rate of about 22 km^3/yr.

Subduction Zones

Around the margin of the Pacific Ocean, *deep-sea trenches* are the dominant bathymetric features. In trenches, older lithospheric material is being subducted, maintaining a global balance with the amount of new lithosphere produced along the ridges and rises. The Pacific trenches range in depth from just under 7 km (4 mi) to just over 11 km (6.8 mi). The Mariana Trench, at 11.03 km, is the deepest place in the oceans (Table 4-1).

The positions of trenches are not constant over time. For example, the Wadati-Benioff seismic zone mentioned in

Chapter 3 marks the position of a subducting slab of lithosphere, but the slab sinks more steeply than the angle of slope of the Wadati-Benioff seismic zone (see Figure 3-21). This indicates that the bend in the subducting plate is moving backward with time as the overriding plate advances over it. This process is progressively decreasing the size of the Pacific Ocean because its oceanic plates are subducting beneath the North American, South American, Eurasian, and Indo-Australian plates. Trenches are notably absent from the Atlantic which is increasing in size, therefore, if Earth's surface is to remain a constant area, the Pacific subduction must balance new crust created in the Pacific and in the Atlantic. The Atlantic is growing at the expense of the Pacific.

When we think of plates colliding at subduction zones, it is easy to envision them as zones of compression. However, due to the seaward migration of hingelines, tensional stresses are much more commonly observed. This may well result from the fact that the thick, massive plate that is already subducted pulls the subducting plate over the hingeline with a force that greatly exceeds any other force that may contribute to plate motion. Figure 4-24a illustrates the process as it has developed over the past 27 million years of subduction of the Pacific Plate beneath the Philippine Plate.

Figure 4-24b shows an example of tensional stress within the Mariana Island Arc subduction system. The subducting plate is presently 100 km beneath the active volcanic island arc containing Pagan Island. About 200 km to the west is the original island arc of the Mariana Trench, now a remnant arc, the West Mariana Ridge. The volcanic peaks on this ridge are old and dormant, and those that once were islands have been submerged by subsidence. It appears that the 100 km depth to the top of the subducting plate has migrated 200 km eastward relative to the West Mariana Ridge.

TABLE 4-1 Dimensions of Trenches.

Trench	Depth (m)	Width (km)	Length (km)
Middle America	6,700	40	2,800
Aleutian	7,700	50	3,700
Java	7,450	65	2,200
Peru-Chile	8,100	100	5,900
South Sandwich	8,264	70	1,610
Japan	8,400	100	800
Puerto Rico	8,414	75	1,750
Kurile	10,500	120	2,200
Kermadec	10,000	40	1,500
Tonga	10,800	55	1,400
Philippine	10,500	60	1,400
Mariana	11,033	70	2,550

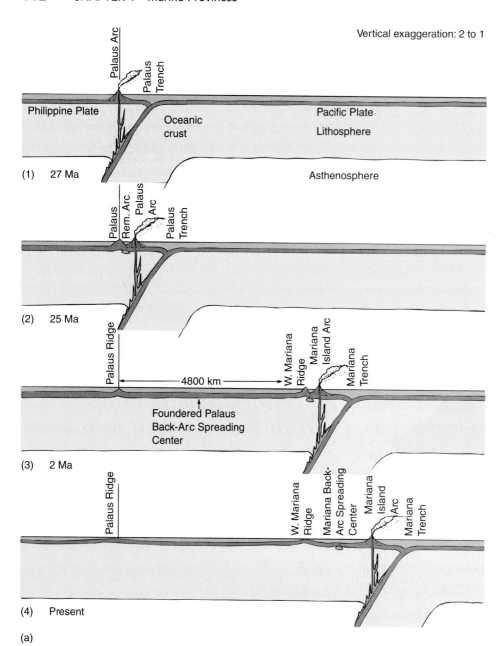

Vertical exaggeration: 2 to 1

Figure 4-24 Mariana Island Arc Subduction System

(a) Formation of simple remnant arcs as a result of seaward migration of a trench system.
(1) Pacific Plate was subducted beneath the Philippine Plate about 27 million years ago
(27 Ma), producing the Palaus island arc and the Palaus Trench. (2) By 25 Ma, the seaward migration of the Palaus Trench created tensional stresses that split the Palaus arc and created a back-arc sea-floor spreading system. The active half of the original Palaus arc continued eastward, keeping pace with the migrating trench while the western half remained more or less stationary and became volcanically inactive (Palaus remnant arc).
(3) By 2 Ma, the Palaus remnant arc (now called the Palaus Ridge) and the active Palaus arc were separated by 4800 km (3000 mi). The spreading center foundered and was replaced by a new rift of the active Palaus arc. The remnant arc created by this new rifting is what we now call the West Mariana Ridge. The active half created the Mariana island arc and the Mariana Trench. (4) The present relationships of the remnant arcs, the Palaus Ridge, and West Mariana Ridge, relative to the present Mariana back-arc spreading center, the Mariana island arc, and the Mariana Trench. (b) A location map, detail strip map, and cross section of the Mariana subduction system. The remnant arc (West Mariana Ridge) represents an ancient dormant volcanic arc separated from the present active arc (Mariana Islands) by a back-arc spreading center. All of these features are products of subduction initiated at the Mariana Trench. As the trench migrates seaward, the region of active volcanism that produces island arcs above the 100 km depth of the descending slab of lithosphere also migrates seaward. Tensional stresses associated with the seaward migration of the trench fracture the thin oceanic crust of the overriding plate and initiate the back-arc spreading. Continued migration maintains the back-arc

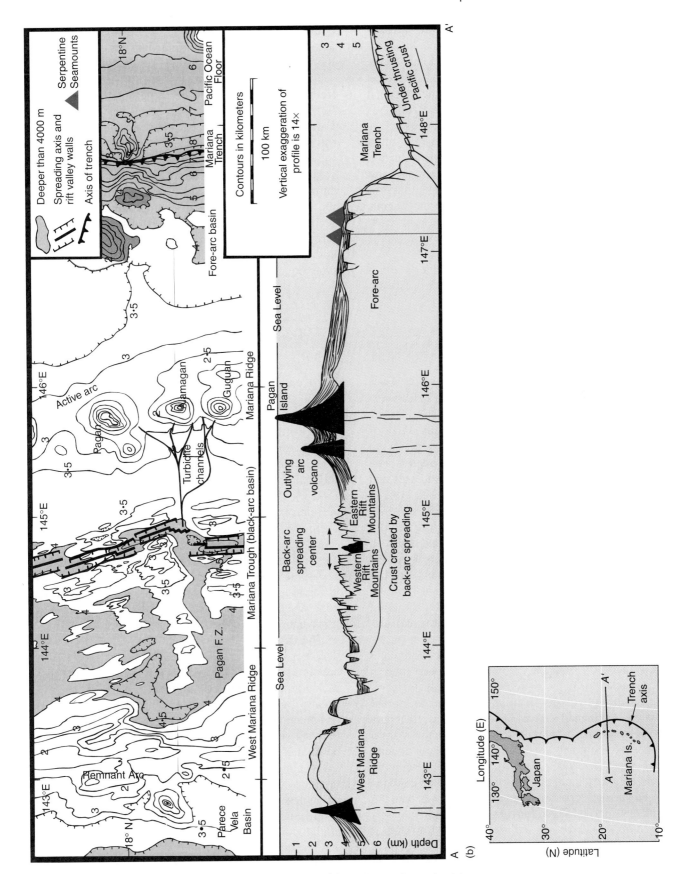

spreading, increasing the distance between the remnant arc and the active arc, but maintaining a position above the 100 km depth of the descending Pacific Plate. The seamount-like structures on the leading edge of the fore-arc are composed of serpentine produced by water combining with peridotite (mantle rock) in the overriding plate at temperatures above 100°C. *(Courtesy of Peter Lonsdale, Scripps Institution of Oceanography, University of California, San Diego.)*

Between the two volcanic arcs lies a **back-arc spreading center**, a new ocean basin forming behind the active arc. Here we have a small-scale sea-floor spreading system with its rift valley, hydrothermal circulation, and even fracture zones. This is clear evidence of the tensional forces at work within the island arc–subduction zone system. Although the origin of back-arc spreading centers is controversial, it is widely believed they form as a result of the rapid seaward migration of the trench and subduction "hinge" that lies seaward of the trench. This pulling apart of the arc system produces linear fractures in a thin volcanic crust, allowing the upwelling of underlying magma. As long as the seaward migration continues at a sufficient rate, new ocean floor is created within the rift valley of the spreading center.

A previously unknown feature of subduction zones was observed in 1989 during Ocean Drilling Program op-erations in the Mariana fore-arc (Figure 4-24b). The fore-arc is that part of the overriding plate that lies seaward of the volcanic arc. What were thought to be volcanic seamounts in a belt 50 to 150 km (31–93 mi) from the trench axis turned out to be serpentine, a mineral that forms when water reacts with peridotite, the rock from Earth's mantle. Apparently, seawater entered the mantle through fractures in the fore-arc crust and from dewatering of subducted sediments. Because of its low density, serpentine flowed up to the surface of the fore-arc block, carrying blocks of peridotite and basalt (possibly from the subducting ocean crust) to produce volcano-like seamounts near the seaward front of the fore-arc structure. See "Ophiolites" at the end of Chapter 3 for a discussion of special features associated with destructive plate boundaries.

Summary

The hypsographic curve shows the amount of Earth's surface area at different elevations and depths. The distribution of area is uneven with respect to height above or below sea level; most of the area is between −4 to −6 km (abyssal plains) and 0 to +1 km (coastal plains). The shape of the curve reflects plate tectonic processes and shows that small changes in sea level can result in large differences in the amount of exposed land area.

Most of our knowledge of the ocean floor has been obtained using echo sounders, seismic surveys, and side-scan sonar. Seismic surveys can reveal the bathymetry of the ocean floor and the location of rock unit boundaries beneath it. Side-scan sonar is used to make strip maps of the ocean floor along the course of the survey ship. Satellite data also are used to make gravity maps that reveal sea-floor bathymetric features.

The continental margin is composed of the continental shelf, slope, and rise. Gently sloping continental shelves extend seaward from the shoreline of continents to an average distance of 70 km (43 mi). Sediment is transported across the continental shelves by storm waves, tidal currents, and submarine gravity flows. Where the margin steepens, the continental shelf ends and the continental slope begins. The slope ranges from 1° to 25°, averaging 4°, and ends at the continental rise. Submarine canyons cut deep into the slopes and sometimes reach well onto the continental shelves. These canyons may be caused by turbidity currents which are sediment gravity flows commonly triggered by earthquakes. Turbidity currents deposit their sediment loads at the base of the continental slope to produce deep-sea fans. These fans merge to produce the continental rise. Boundary currents, particularly on the western edges of ocean basins, are major factors in redistributing sediment along the continental slope and in creating the benthic nepheloid layer.

Spreading centers create mid-ocean ridges and rises that make a continuous mountain chain stretching 65,000 km (40,000 mi) across the deep-ocean floor. Along the axes of oceanic ridges and rises, seawater percolates down through fractures to be heated and returned to the surface as hydrothermal vents with water temperatures exceeding 350°C (660°F). This process has produced most of Earth's metal sulfide ore deposits and serves as an energy source for chemosynthetic biological communities.

Fracture zones exhibit great amounts of relief. Examples include the 1000 m (3280 ft) cliff of the Mendocino Escarpment, which results from the differential cooling of different-aged lithosphere on opposite sides of the fracture zone, and the deep, narrow valley formed between the two transverse ridges of the Romanche Fracture Zone.

The extensive, flat, sediment-covered features of the deep-ocean basin are called abyssal plains. These plains are often interrupted by volcanic features protruding above the sediment. Scattered throughout much of the ocean floor are volcanic peaks called abyssal hills, seamounts, or tablemounts, depending on their heights and whether they are cone shaped or flat-topped. These features may be related to ridge volcanism or created by mantle hot spots. Other intraplate features include large regions of thermally elevated ocean crust and flood basalts. The development of coral reefs around volcanic islands or along continental margins, from fringing reef, to barrier reef, to atoll, records the subsidence of the ocean floor over time and, in some cases, horizontal plate motion.

Subduction zones include compressional and tensional features. Trenches created by subduction change position over time, migrating away from the overriding plate. This backward migration of the trench produces remnant arcs and active volcanic island arcs, often separated by back-arc spreading centers.

Key Terms

Abyssal hills	Benthic nepheloid layer (BNL)	Fringing reefs	Side-scan sonar
Abyssal plains	Black smokers	Guyots	Submarine canyons
Active margin	Continental flood basalts	Hydrothermal vents	Tablemounts
Atlantic-type margin	Continental margin	Hypsographic curve	Turbidity currents
Atolls	Continental rise	Pacific-type margin	Western boundary undercurrent
Back-arc spreading center	Continental shelf	Passive margin	(WBUC)
Barrier reefs	Continental slope	Seamounts	White smokers
Bathymetry	Deep-ocean basins	Shelf break	

Questions and Exercises

1. Describe what is shown by a hypsographic curve and explain why its shape reflects the presence of active tectonic processes.

2. Contrast the detail and area covered by different types of bathymetric surveys.

3. Describe the most widely accepted explanation for the development of submarine canyons on a continental slope. Discuss the supporting physical evidence.

4. What are the sources of the sediment that becomes suspended in the benthic nepheloid layer? Why is the benthic nepheloid layer most developed along the western margins of ocean basins?

5. Describe the general process of hydrothermal circulation associated with spreading centers and their relationships to ore deposits and chemosynthetic biocommunities.

6. Explain why seamounts and guyots increase in age and depth with increased distance from the oceanic ridges and rises. Discuss the length of time they probably were active volcanoes.

7. Discuss the possible relationship of coral reef development to sea-floor spreading. Include the effects of both vertical and horizontal motions.

8. How are the alignments and age distribution patterns of the Emperor Seamount and Hawaiian Island chains explained by hot spots?

9. Describe the process thought to be responsible for the development of the remnant arc, active arc, and back-arc spreading center of the Mariana arc subduction system.

10. Make a list of major oceanic bathymetric features and the plate tectonic processes associated with each.

References

Bevia, M., et al., 1995. Geodetic observations of very rapid convergence and back-arc extension at the Tonga Arc. *Nature* 374:6519, 249–51.

Biscaye, P. E., and S. L. Eittreim, 1977. Suspended particulate loads and transports in the nepheloid layers of the abyssal Atlantic Ocean. *Marine Geology* 23:155–72.

Bonatti, E., 1978. Vertical tectonism in oceanic fracture +zones. *Earth and Planetary Science Letters* 37, 369–79.

Damuth, J. E., V. Kolla, R. D. Flood, R. O. Kowsmann, M. C. Monteiro, M. A. Gorini, J. J. Palma, and R. H. Belderson, 1983. Distributary channel meandering and bifurcation patterns on the Amazon deep-sea fan as revealed by long-range side-scan sonar (GLORIA). *Geology* 11:2, 94–98.

Darwin, C., 1859. *On the Origin of the Species by Means of Natural Selection, or the Preservation of Favoured Races in the Struggle for Life.* London: John Murray.

Davies, G. F., 1992. Plates and plumes: Dynamos of the Earth's mantle. *Science* 257:5069, 493–94.

Davies, P. J., P. A. Symonds, D. A. Feary, and C. J. Pigram, 1987. Horizontal plate motion: A key allocyclic factor in the evolution of the Great Barrier Reef. *Science* 238:4834, 1697–99.

Davis, R. A., 1983. *Depositional Systems.* Englewood Cliffs, NJ: PrenticeHall.

Duncan, R. A., 1991. Ocean drilling and the volcanic record of hotspots. *GSA Today* 1:10, 213–19.

Green, H. W., and P. C. Burnley, 1989. A new self-organizing mechanism for deep-focus earthquakes. *Nature* 341:6224, 733–37.

Hamilton W. B., 1988. Plate tectonics and island arcs. *Geological Society of America Bulletin* 100:10, 1503–27.

Hay, A. E., R. W. Burling, and J. W. Murray, 1982. Remote acoustic detection of a turbidity current surge. *Science* 217:4562, 833–35.

Kennish, M.J. 1989. *Practical Handbook of Marine Science* Boca Raton, FL: CRC Press.

Machetel, P., and P. Weber, 1991. Intermittent layered convection in a model mantle with an endothermic phase change at 670 km. *Nature* 350:6113, 55–57.

Molnar, P., and J. Stock, 1987. Relative motions of hotspots in the Pacific, Atlantic and Indian oceans since late cretaceous time. *Nature* 327:6123, 587–91.

Moore, J. G., and R. K. Mark, 1986. World slope map. *Eos* 67:48, 1353, 1360–62.

Olsen, P.E., 1999. Giant lava flows, mass extinctions, and mantle plumes. *Science* 284: 5414, 604–605.

Silver, P. G., et al., 1995. Rupture characteristics of the deep Bolivian earthquake of 9 June, 1994 and the mechanism of deep-focus earthquakes. *Science* 268:5207, 69–73.

Sverdrup, H. U., M. W. Johnson, and R. H. Fleming, 1942. Renewal 1970. *The Oceans: Their Physics, Chemistry, and General Biology.* Englewood Cliffs, NJ: PrenticeHall.

Tarbuck, E. J., and F. K. Lutgens, 1993. *The Earth: An Introduction to Physical Geology,* 4th ed. New York: Macmillan.

———. 1991. *Earth Science,* 6th ed. New York: Macmillan.

Twichell, D. C., and D. G. Roberts, 1982. Morphology, distribution and development of submarine canyons on the United States Atlantic continental slope between Hudson and Baltimore canyons. *Geology* 10:8, 408–12.

Suggested Reading

Sea Frontiers

Golden, F., 1991. *Birth of the Caribbean.* 37:5, 20–29. Presents the history of the Caribbean Plate and its relationship to the existing islands, volcanism, and earthquakes in the region.

Mark, K., 1976. *Coral reefs, Seamounts, and Guyots.* 22:3, 143–49. This article discusses the role of global plate tectonics in explaining the distribution of seamounts, guyots, and the evolution of coral reefs.

Schafer, C., and L. Carter, 1986. *Ocean-Bottom Mapping in the 1980s.* 32:2, 122–30. Describes the use of SeaMARC (Seafloor Mapping and Remote Characterization), a side-scan sonar device, in mapping the continental margin off the coast of Labrador.

Shepard, F. P., 1975. *Submarine Canyons of the Pacific.* 21:1, 2–13. Describes canyons around the Pacific margin and discusses their possible origins.

White, R. S., and D. P. McKenzie, 1989. *Vulcanism at Rifts.* 261:1, 62–77. Covers new information about the size, depth, and length of magma chambers beneath spreading center rifts.

Scientific American

Emery, K. O., 1969. The Continental Shelves. 221:3, 106–25. Discusses nature of the continental shelves and the effects of the advance and retreat of the shoreline across them as a result of glaciation.

Fryer, P., 1992. Mud Volcanoes of the Marianas. 266:2, 46–53. Discusses how the subduction of upper mantle peridotite and its subsequent combination with water produces low-density serpentine that rises and produces mudlike volcanoes on the fore-arc of the Mariana island arc system.

Heezen, B. C., 1956. The Origin of Submarine Canyons. 195:2, 36–41. Presents theories on the origin of submarine canyons along with data on the location and nature of such features.

Herbert, S., 1986. Darwin as a Geologist. 254:5, 116–23. Before publication of *On the Origin of the Species,* Charles Darwin considered himself to be primarily a geologist. This article outlines his contributions in this field, with special attention to his theory of coral reef formation.

Hyndman, R. D., 1995. Giant Earthquakes of the Pacific Northwest. 273:6, 68–75. Although they have not yet been experienced by the modern populations of Seattle and Vancouver, there is evidence of numerous past earthquakes of magnitudes greater than 8 in the geologic record.

Larson, R. I., 1995. The Mid-Cretaceous Superplume Episode. 272:2, 82–89. From 125 million to 80 million years ago, sea level was 250 meters higher than it is today, and global temperatures were about 10°C warmer. Huge populations of marine plankton died and were buried in the sediment to produce half of the world oil supply, and most of today's diamond deposits were created. The article explains how all of this may have happened.

Malahoff, A., 1998. An Island is Born. *Scientific American Presents the Oceans* 9:3, 98–99. A visit to the undersea summit of Loihi, the newest creation of the Hawaiian hot spot.

Murphy, J. B., and R. D., Nance, 1992. Mountain Belts and the Supercontinent Cycle. 266:4, 84–91. Building on the ideas of Alfred Wegener, the authors identify a 500-million-year cycle of continental collision and breakup.

Oceanography on the Web

Visit the Introductory Oceanography home page for on-line resources for this chapter. There you will find an on-line study guide with review exercises and links to ocean-ography sites to further your exploration of the topics in this chapter. Introductory Oceanography is at http://prenhall.com/thurman (click on the Table of Contents menu and select this chapter).

5 Marine Sediments

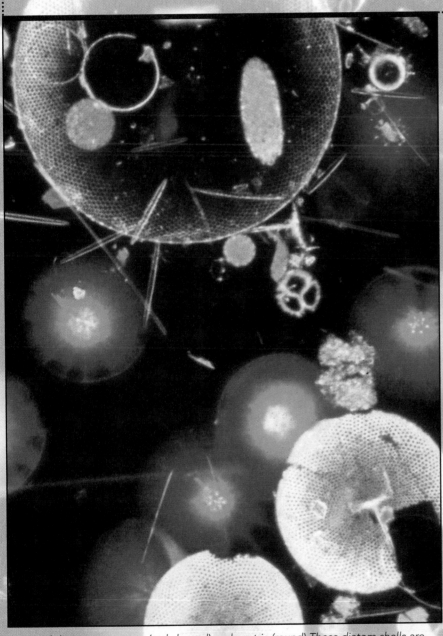

- Sediment Sources

- Classification of Marine Sediments

- Controls on Oceanic Sediment Accumulation

- Distribution of Oceanic Sediments

- Summary

Assorted diatoms—pennate (rod-shaped) and centric (round) These diatom shells are preserved in deep-sea sediments and give investigators clues as to the oceanic and climatic conditions at the time they were alive. (Photo by Jan Hinsch/Science Photo Library courtesy Photo Researchers, Inc.)

Geologists once believed that the sediments in the deep-ocean basins were undisturbed records of all of Earth's history since the oceans formed 4 billion years ago. They were greatly surprised to find that the oldest ocean crust was little more than 200 million years old. The dynamic nature of plate tectonic processes has removed the sedimentary record along with the subducting ocean crust, but in some places on the continents, ancient marine sediments are preserved. In fact, more than half of the rocks exposed on the continents above the present shoreline were formed from sediments laid down in past ocean environments. Most of what we know of Earth's past geology, climate, and biology has been learned through study of these ancient marine sediments—they are our "information highway into the past." Although plate tectonics has destroyed the record on the sea floor, it also has preserved marine sediments on the continents.

Sediment Sources

Lithogenous Particles From Continental Rocks

Lithogenous particles are derived from preexisting rock. In Chapter 2, we described the major minerals of silicate and carbonate rocks and reviewed the major rock types—igneous, sedimentary, and metamorphic. Many common minerals form at temperatures and pressures well above those present at Earth's surface and break down when they are exposed at the surface. Individual mineral grains, freed from their parent rock masses by the physical and chemical processes of weathering, are transported by wind or water to the ocean basins. The oceans' major sources of river-borne and wind-borne sediments are shown in Figure 5-1. Rivers dominate at latitudes with high rainfall; windblown sediments derive from the great deserts at arid latitudes.

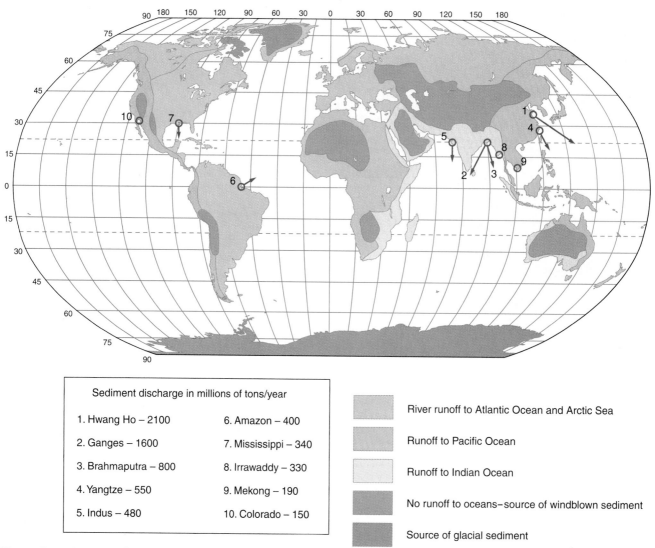

Sediment discharge in millions of tons/year

1. Hwang Ho – 2100
2. Ganges – 1600
3. Brahmaputra – 800
4. Yangtze – 550
5. Indus – 480
6. Amazon – 400
7. Mississippi – 340
8. Irrawaddy – 330
9. Mekong – 190
10. Colorado – 150

River runoff to Atlantic Ocean and Arctic Sea

Runoff to Pacific Ocean

Runoff to Indian Ocean

No runoff to oceans–source of windblown sediment

Source of glacial sediment

Figure 5-1 Sources of Windblown, Glacial, and River-borne Sediments to the Major Ocean Basins

The intensity of weathering is determined by the amount of rainfall and the temperature. When granitic rocks weather, grains of quartz, feldspar, and some minor mafic minerals are released. With very light or short periods of weathering, all of these primary minerals are present in the sediment, but the mafic minerals show signs of dissolution. With increased weathering, feldspars start to dissolve and mafic minerals dissolve completely. With intensive weathering, only the quartz remains. Much of the weathering process takes place in soils where organic acids from plant decay act as dissolving agents. When the sediment minerals dissolve into soil solutions, other minerals may re-precipitate. For example, dissolution of feldspars results in precipitation of clay minerals of various compositions, depending on the intensity of weathering. At temperate latitudes, with moderate rainfall and temperature variations, illite clays are found in soils; however, at tropical latitudes, where temperatures are warmer and rainfall is higher, kaolinite clays and aluminum oxides are the typical soil constituents. At high latitudes, much of the continental crust in the Northern Hemisphere consists of metamorphic rocks. The metamorphic minerals that break down yield chloritic clays. This latitudinal variation in weathering and weathering products is reflected in the continentally derived sediment distributions on the sea floor.

Sediment Texture and Transport   Sediment texture, the way a sediment appears according to the sizes of the particles and the way the grains are packed together, is indicative of the energy conditions under which the sediment was deposited. A grain size scale, called the **Wentworth scale** (Table 5-1) provides a standard naming scheme for particles with diameters from fractions of a meter to fractions of a micron, μ (1/1000 of a millimeter or one-millionth of a meter). The size categories use common names, but apply them to specific size intervals of sediment particles. The divisions are set up using a geometric progression. For example, the division between medium sand and coarse sand is 0.5 mm, the next division, between coarse sand and very coarse sand, is twice that size, 1 mm. The next division is another doubling to 2 mm. Between the extremes of boulders and colloids are cobbles, pebbles, granules, sand, silt, and clay-sized particles.

Moving water separates sediments by size. Deposits laid down by strong currents or waves (high-energy environments) are composed of larger particles; clay-sized particles are only deposited when current velocity and wave action are negligible (low-energy environments). Lots of repetitive transport by currents over time or long distances produces sediments composed of particles that are all close to the same size. Such sediments are classified as well sorted in contrast to poorly sorted deposits that consist of particles of various sizes, resulting from rapid deposition and/or short transport distances. Beach sand is typically a well sorted sediment because of repetitive wave action. Glacial deposits dumped on continental shelves are poorly sorted sediments. As the glaciers melt, particles carried

TABLE 5-1   Wentworth Scale of Grain Size for Sediments.

Size Range (mm)	Particle	
256	Boulder	
64	Cobble	
4	Pebble	
2	Granule	
1	Very coarse sand	
	Coarse sand	
$\frac{1}{2}$	Medium sand	SAND
$\frac{1}{4}$	Fine sand	
$\frac{1}{8}$	Very fine sand	
$\frac{1}{16}$	Course silt	
$\frac{1}{32}$	Medium silt	
$\frac{1}{64}$	Fine silt	SILT
$\frac{1}{128}$	Very fine silt	
$\frac{1}{256}$	Coarse clay	
$\frac{1}{640}$	Medium clay	
$\frac{1}{1024}$	Fine clay	CLAY
$\frac{1}{2360}$	Very fine clay	
$\frac{1}{4096}$	Colloid	

Source: Wentworth, 1922, after Udden, 1898

within the glacial ice, ranging in size from clay to boulders, are dropped out together.

Sediment mineralogy reflects transport history as well as source rocks. As sediment is carried from the source, the less resistant minerals break down faster. Quartz is the most common resistant mineral. Sediment created initially from the weathering of granite consists of quartz, feldspars, and perhaps some mafic minerals. Over time and increasing distance from the source, the proportions of all minerals except quartz will decrease in the sediment. The final sediment will be nothing but quartz grains that have become progressively more rounded by the continuous chemical and physical abrasion they have encountered during their journey. Taken together, sorting, mineralogy, and particle shape are indicators of **sediment maturity**. A sediment containing nearly pure, well-rounded quartz grains that are uniform in size is mature; a sediment consisting of feldspars and clays as angular, variously sized particles is immature (Figure 5-2).

Figure 5-3a shows how sediment particle size relates to horizontal current velocities needed for erosion, transportation, and deposition. Note the log-log scale. The settling curve between the tan and blue fields quantifies what happens when a current carrying sediment decreases in velocity. Larger particles settle out first where current velocities are high, but smaller particles stay in suspension until velocities are very low. The relationship between

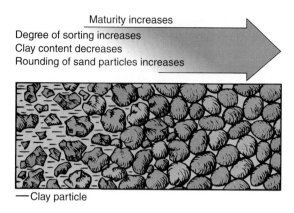

Maturity increases

Degree of sorting increases
Clay content decreases
Rounding of sand particles increases

—Clay particle

Figure 5-2   Sediment Maturity

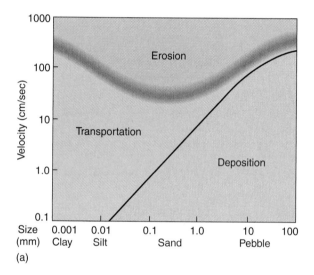

(a)

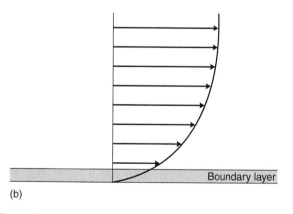

(b)

Figure 5-3   Hjulstrom's Curve and the Boundary Layer

(a) Horizontal current velocity versus erosion, transport and deposition of sediment particles by size.   *(Reprinted with the permission of Macmillan Publishing Company from* The Earth, *4th Edition, Figure 10.15, by Edward J. Tarbuck and Frederick K. Lutgens. Copyright © 1993 by Macmillan Publishing Company.)* (b) Velocity profile of a current passing over the bottom. Frictional forces cause the current velocity to decrease with proximity to the bottom. In the boundary layer, there is no flow.

grain size and velocity is constant (linear on a log-log scale) for grain sizes below about 10 mm. However, the curve defining the velocities needed to erode sediments shows that once deposited, the energy required to erode sediment is much higher than for deposition. Surprisingly, clay-sized particles require higher velocities than sand-sized particles. Can you guess why this might occur?

Water flowing over a substrate encounters resistance when it gets very close to the surface. A plot of velocities in a current near the bottom shows that velocities decline as the substrate is approached (Figure 5-3b). At some very small distance from the bottom, the forces of resistance to flow (friction) overcome the current force, and the water is stagnant. This stagnant film is called the *boundary layer*. The thickness of the boundary layer decreases with increasing current velocity. Small sediment grains can "hide" in this boundary layer, safe from being picked up by the flowing current—at least until the current velocity is high enough to reduce the boundary layer thickness sufficiently to expose them. Many clay-sized particles also happen to be clay minerals (sheet silicates described in Chapter 2). Clay minerals are flat and tend to have charged surfaces, both features that create a cohesive attraction among the particles. Because they stick together, a high velocity current is required to erode them.

Wind currents also carry sediment particles, but because the density contrast between air and sediment is much greater than between water and sediment, and because of the lower viscosity of air, wind is not as efficient as water. This is good news—otherwise, we would be sandblasted every time the wind blew. It takes very high wind speeds to pick up sand. Wind commonly carries only particles less than 20 μm (0.008 in) in diameter; the particles are often carried long distances high in the atmosphere, and much of this windblown sediment eventually ends up in the oceans. The particles settle out as the velocity of the wind decreases or they find their way into the ocean carried by rain. Dust from China is carried by the prevailing westerly winds across the Pacific and has been found at monitoring stations in the Bering Strait and as far south as Antarctica. Equatorial trade winds transport Saharan dust across the Atlantic Ocean to the Caribbean Sea and into the Pacific. The amount of wind-borne sediment worldwide has increased since data collection began in the 1960s. This may be the result of the increasing aridity of the major source regions for these particles, the deserts of China and Africa.

Volcanogenic Particles

Volcanogenic particles are ejected by volcanic eruptions and range in size from boulders to fine dust. They are a special class of lithogenous particles because, although they are derived from rock, they are not produced by weathering processes. The larger sized particles become sediments on the sea floor near their volcanic sources, but the fine dust may be carried in the upper atmosphere for years before it settles out or is washed out by rain. The dust from major volcanic eruptions has been implicated as a cause of short-term cooling trends in global climate because the particles act to reflect the sun's energy back into space.

Glacially Derived Particles

As snow falls and is added to the top of a glacier, the toe of the glacier moves downslope. By this process, glaciers act like very slow rivers transporting sediments to lower elevations. Glaciers passing over rock literally scrape away the rock's surface. The debris created ranges in size from boulders plucked from the surface to fine rock flour created by ice grinding over rock. All of this material gets incorporated into the ice as the glacier moves along. At high latitudes, glaciers may flow directly into the ocean, melt, and release their sediment loads. Many of the glacial deposits found today on continental shelves were laid down by glaciers that extended over these areas during the Pleistocene Epoch. Glacial deposits also are still forming around the continent of Antarctica and around the island of Greenland.

Biogenous Particles: Organism Hard Parts

Biogenous sediments are derived from the remains of marine organisms. The chemical composition of mineral particles produced by marine organisms is usually *calcareous*, consisting of **calcium carbonate** ($CaCO_3$), or *siliceous*, consisting of hydrated **silica** ($SiO_2 \cdot nH_2O$), called **opal**. These minerals are sculpted by invertebrate marine organisms into truly beautiful protective outer coverings (Figure 5-4). In the marine environment, protozoa, algae, plants, and animals make these hard parts.

The largest populations of these organisms are found in the surface waters of the ocean. They have a planktonic lifestyle, meaning they more or less float with the currents and have little control over their own mobility. When they die, their hard parts sink to the ocean floor to become part of the sediment. Microscopic algae called **diatoms** and protozoans called **radiolaria** contribute most of the siliceous biogenous particles, which rarely exceed 100 μm in diameter. **Foraminifera** are the protozoans and **coccolithophorids** are the algae responsible for most of the calcium carbonate in deep ocean sediments. Both of these organisms precipitate calcium carbonate in the form of calcite. Another significant contributor is the pteropod snail. These animals make their shells from calcium carbonate precipitated as aragonite. Also, up to 10 percent of the carbonate in deep sea sediments may be the remains of benthic (bottom-dwelling) foraminifera.

In shallow marine environments on the continental margins, there are many types of invertebrate organisms, including clams, corals, sponges, snails, algae, foraminifera, sea urchins, starfish, and fire corals, that make their hard parts from calcium carbonate, and also a few siliceous organisms, primarily diatoms and sponges. Shallow-water organisms live at all latitudes, but carbonate-secreting organisms flourish in certain low-latitude settings where they sometimes form massive reefs. Modern reefs are colonies of aragonite coral. In the geologic past, ancient corals as well as other kinds of organisms also built reefs that are sometimes preserved in the rock record as impressive and picturesque mountain peaks. Reefs truly are the cities of the marine invertebrate world. We will return to coral reef communities in Chapter 17. See Coral Rocks! The Value of the World's Coral Reefs on page 82–83.

Stromatolites **Stromatolites** are another type of massive biogenous deposit. Cyanobacteria, descendants of the first photosynthetic organisms, produce these deposits by trapping fine sediment in mucous mats. Other types of algae produce long filaments that bind the particles together. As layer upon layer of these algae colonize the surface, a structure rises off the sea floor. These soft organic structures become cemented by calcium carbonate that apparently precipitates chemically as the algae raise pH by removing CO_2 for photosynthesis and release calcium as a waste product of metabolic processes.

Today, stromatolites are found in extreme environments where corals cannot grow and where they are safe from grazing invertebrate populations. Some modern settings where stromatolites occur include hypersaline tidal pools such as Hamelin Pool in Shark Bay, western Australia, and the shifting carbonate sand shoals on Eleuthera Bank in the Bahamas (Figure 5-5). The geologic past, particularly from about 1 to 3 billion years ago, was the *golden age of stromatolites*. Cyanobacteria were the major reef-building organisms and had their pick of the best sites. Their ability to photosynthesize and use oxygen for respiration allowed them to easily beat the competition at that time. Stromatolitic structures hundreds of meters high are common in rocks from those ages.

Hydrogenous Deposits

Hydrogenous deposits form as the result of chemical reactions that take place in ocean water or from chemical interactions between seawater and the sediments on the ocean floor. Manganese nodules, metal sulfides, phosphorites, and glauconite pellet sands are common hydrogenous sediment types. For all of these sediment types, there is some debate as to whether the formation process is purely chemical or the sediments are formed as by-products of chemical reactions carried out by bacteria.

Manganese Nodules and Crusts Manganese nodules, looking like misshapen metallic snowballs, litter the deep-ocean floor (Figure 5-6). Since the voyage of the *Challenger*, we have known that they are relatively abundant in all the major ocean basins. **Manganese nodules** consist mainly of manganese dioxide (MnO_2) and iron oxide (Fe_2O_3), with an average manganese dioxide content of about 30 percent and an average iron oxide content of about 20 percent by weight. Copper, cobalt, and nickel also occur in concentrations of up to 2 percent in the nodules. Needless to say, these deposits are of great economic interest as sources of metal, but the price of metals and political climate prevent their exploitation at this time.

(a)

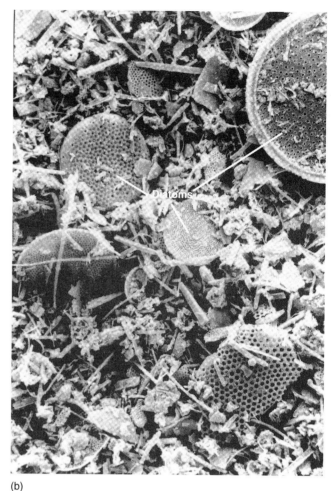

(b)

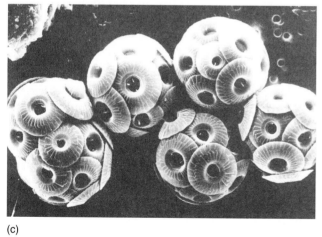

(c)

(d)

Figure 5-4   Microscopic Skeletons from the Deep Sea

(a and b) Scanning electron micrographs (SEM) of pelagic ooze containing skeletons of foraminifera, diatoms and radiolaria (photo a: 160×, photo b: 250×). (c) Coccolithophores (diameter of each individual is about 20 μm). (d) Foraminifera (diameter of most individuals is about 400 μm).   *(Part (a) courtesy of the Deep Sea Drilling Project, Scripps Institution of Oceanography, University of California, San Diego. (b) Photo courtesy of World Minerals, Inc., Lompoc, California (Sample from Celite Corporation Diatomite Mine in Lompoc California). (c) Reprinted by permission of Hallegraeff, G. M. Plankton: A Microscopic World, 1988, page 8. Courtesy of E. J. Brill, Inc. (d) Courtesy of Memorie Yasuda, Scripps Institution of Oceanography, University of California, San Diego.)*

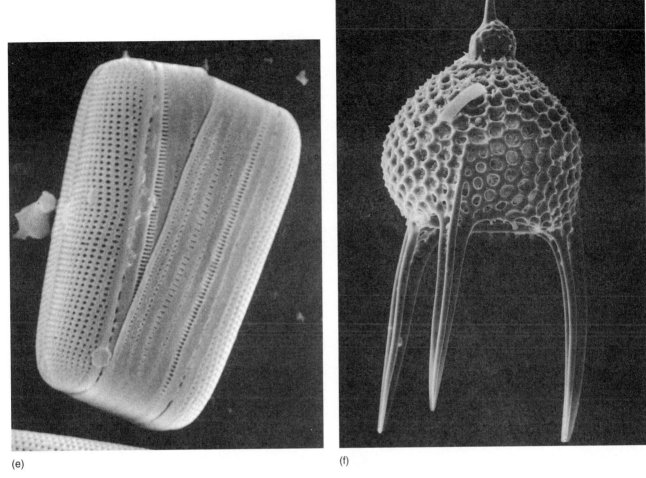

(e)   (f)

Figure 5-4   Microscopic Skeletons from the Deep Sea (continued)

(e) Diatom (length is about 30 μm). (f) Radiolarian (length is about 100 μm).   *(Part (e) reprinted by permission of Hallegraeff, G. M. Plankton: A Microscopic World, 1988, page 46. Courtesy of E. J. Brill, Inc. (f) Courtesy of Warren Smith, Scripps Institution of Oceanography, University of California, San Diego.)*

Because iron and manganese both occur in these nodules at concentrations greater than is found almost anywhere else, it is difficult to explain how these deposits formed. The chemical composition of deep-ocean water is essentially constant throughout the oceans and does not contain significant quantities of either of these metals. Researchers have identified three possible sources of the manganese and iron: (1) weathering of volcanic material produced by volcanic activity on the ocean floor (2) concentration in hydrothermal waters at spreading centers and (3) runoff that carries the minerals as soluble compounds from the continents. However, there still is no wide agreement on the mechanism that concentrates these metals to the point that oxide minerals would precipitate out. One idea that is gaining acceptance is that bacteria using iron and manganese oxidation in their metabolic processes may be involved. These bacteria operate in a thin zone just below the sediment surface, perhaps only coincidentally the same place that nodules often are found.

Manganese nodules are found only in deep-ocean environments where other sediments accumulate at rates of less than 7 mm/1000 yr (0.3 in/1000 yr). The nodules grow very slowly, adding layers from 1 to 200 mm (0.04 to 8 in) thick every million years. There also appears to be a correlation between the biological productivity of overlying waters and the compositions and growth rates of manganese nodules. Beneath unproductive areas near the center of the subtropical oceans, nodules are rich in iron and cobalt and grow slowly—less than 5 mm/million yr (0.2 in/million yr). Beneath waters of moderate biological productivity, such as those north and south of the equatorial Pacific Ocean, nodules have higher concentrations of copper, nickel, and manganese and grow at rates of 5 to 10 mm/million yr (0.2–0.4 in/million yr). Beneath very productive waters, nodules are rich in manganese but poor in the economically more important copper, nickel, and cobalt. These nodules grow very rapidly, from 10 to 200 mm/million yr (0.4–8 in/million yr). Scientists have no good explanation for these observations.

Figure 5-5 Stromatolites in the Bahamas and Shark Bay, Australia

(a) Subtidal oolitic stromatolites on the crest of an oolitic tidal bar on Eleuthera Bank, Bahamas. Most are 0.5–1.0 m (1.6–3.3 ft) high. (b) map of the Bahamas showing location of Eleuthera Bank. (c) Shark Bay stromatolites in hypersaline lagoons of Hamelin Pool. (d) Map of western Australia showing Shark Bay. Arrow denotes approximate location of Hamelin Pool. *(Maps and photos courtesy of Jeff Dravis. Permission granted by Exxon Production Research Company and American Association for the Advancement of Science.)*

(a)

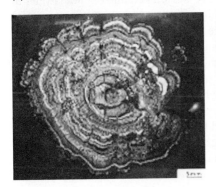

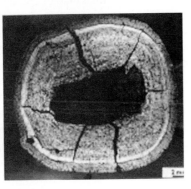

(b)

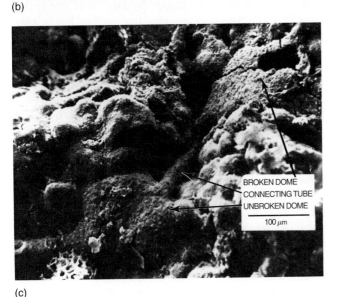

(c)

(a) Manganese nodules on the Pacific Ocean sea floor. (b) Cross-section of a manganese nodule showing the concentric layered structure and central nucleus. (c) SEM of the surface of a manganese nodule (×156). The small tube and dome structures were built by microscopic bacteria for shelter, thus nodules may owe their existence in part to biological processes. (*Courtesy of Scripps Institution of Oceanography, University of California, San Diego.*)

Phosphorites Phosphorus is a major element of apatite (calcium phosphate) minerals that precipitate as nodules and thin crusts at intermediate to shallow depths on continental shelves and slopes. Concentrations of phosphorus in such deposits commonly reach 30 percent by weight. The nodules have slow growth rates, ranging from 1 to 10 mm/1000 yr (0.04 to 0.4 in/1000 yr). Unlike manganese nodules, which grow outward equally in all directions, phosphorite nodules seem to grow downward into the sediment, so their tops are older than their bottoms. The source of phosphorus and the other elements in the nodules is water trapped in sediment pores on the ocean floor beneath areas of high surface productivity.

When plankton die, a lot of organic matter falls to the bottom along with the hard parts that we focused on above. In most places, bacteria in the water can consume all this organic detritus, converting it back into inorganic compounds, such as nitrate and phosphate ions, that re-dissolve in seawater. The photosynthetic plankton snap these ions right back up again—completing a fast, tight cycle that keeps phosphorus out of sediments. However, when productivity is high, the decomposer bacteria in the water column cannot keep pace; lots of organic matter gets past them and accumulates in the sediment. Here, bacteria living on sediment grain surfaces do the decomposing job—but the inorganic ions remain in the sediment pore waters instead of re-dissolving in the water where photosynthesizers use them. When phosphate ion levels become high enough, apatite precipitates. Over time, bottom currents may sweep away the finer grains that surround the nodules, increasing the fraction of nodules in the sediment and thereby enriching the total sediment phosphate content.

Glauconite *Glauconite* is a greenish, hydrous silicate mineral of variable composition. It apparently forms only in marine environments, perhaps by chemical alteration of biotite or clay on the sea floor. It is found as deep as 2500 m (8200 ft), but is most common on shelf areas near coastlines. Glauconite typically forms sand-sized pellets, coatings, or crusts in muddy or sandy sediments. When sediments contain enough glauconite to give them a greenish tint, they are called "green sands" or "green muds."

Evaporite minerals When seawater evaporates, *evaporite minerals* precipitate. As the volume of seawater decreases by evaporation, concentrations of dissolved ions increase in the remaining water to the point that minerals precipitate out. Evaporite minerals precipitate from seawater in the following order: calcite or aragonite (at 90 percent of the water remaining), gypsum (at 50 percent), halite (at 10 percent), and finally (at < 1 percent), minor amounts of assorted minerals collectively called *bittern salts*.

Evaporation of seawater to the extent necessary to form evaporites obviously does not happen in the open ocean. Areas near the edges of the oceans, where rainfall is low and topography is such that a drop in sea level could suddenly cut off seawater exchange, are good candidates. At the end of Chapter 7, you will find the story of how the Mediterranean was cut off from the Atlantic about 6 million years ago. At certain times in the geologic past, there have been many more marginal evaporative basins. The Permian Basin, Michigan Basin, and Alberta Basin are just a few of the places in North America that contain large masses of evaporite salts. Petroleum deposits also are found in these types of basins—another result of the chemical and physical conditions that evolve when circulation is restricted.

Note that it takes very little change in seawater volume to precipitate calcite or aragonite. Although most shallow marine calcium carbonate is considered to be produced by organisms, there are some enigmatic carbonate sediments in the geologic rock record and in modern environments. For example, the shallow offshore areas of the Bahama Islands, in addition to being home for many carbonate-precipitating organisms, contain muds composed of fine aragonite needles and sands composed of ooids. **Ooids** is not a misprint—these are small spheres of laminated calcium carbonate which look like onions when they are cut

open. The laminae form around sand grains, shell fragments, or quartz grains.

Are ooid laminae and aragonite needle muds biogenous or not? If you want to start a heated argument among scientists, bring up this topic at a scientific meeting of carbonate sediment specialists. Certain marine algae precipitate aragonite needles in their tissues, and when the organic matter decays after the algae die, these needles are contributed to sediments. However, there are phenomena known as *whitings* that occur frequently in the Bahama shallows. Seawater over the banks usually is crystal clear, but whitings are patches of cloudy seawater in which masses of aragonite needles are suspended. Scientists have subjected the seawater and suspended sediments in these whitings to microscopic, chemical, and isotopic analyses of every sort imaginable. The salinity and temperature of the seawater in these patches is slightly elevated, so evaporative precipitation of the aragonite needles certainly is not out of the question. The aragonite needles are chemically and morphologically identical to the needle muds on the banks. These facts seem to make a good argument for abiogenic precipitation—but, what if the needles went up instead of down? What if the needles in the water were eroded and re-suspended from mud that was already on the bottom? Bahamians call whitings *fish muds*, claiming that schools of fish stir up bottom muds while feeding. To test this idea, one scientist even set off some dynamite in a whiting to find out how many fish might be there—he didn't find any in that particular whiting. Other scientists have sighted fish in whitings; one of us once saw a small fish with a barracuda in hot pursuit, jump out of a whiting. Scientists advocating abiogenic whitings believe that fish simply take advantage of the whitings to hide from predators, if this is true, it obviously was not such a good idea for at least one fish.

Ooids form vast sand shoals in parts of the Bahamas and other places in the world. These sands are biological deserts because nothing can get a foothold in the shifting sand. Surely, this is an environment where biogenous carbonate precipitation is unlikely. However, high-powered microscopic examination of ooid surfaces shows that they are colonized by endolithic (endo = in, lithos = rock) algae. As we noted above, the photosynthetic and metabolic processes of algae create conditions ideal for calcium carbonate precipitation. So, it seems that we are still unsure whether ooids are biogenous or not.

In the geologic past, particularly in Precambrian sediments laid down before the evolution of organisms with hard parts, we find extensive deposits of carbonate mud and some ooid sands as well. Remember though, that photosynthetic bacteria and algae have been around for nearly 4 billion years and could well have played a role in creating these deposits.

Zeolites Some unusual minerals called zeolites form on the sea floor, some probably through alteration of the ocean crust during hydrothermal circulation of seawater. **Zeolites** are a group of hydrated framework silicate minerals that contain aluminum, calcium, sodium, and potassium cations.

They have the peculiar property of gradually losing their hydration water; they also possess the ability to substitute various metals for the alkaline metal cations originally present in their structures. *Phillipsite* is a white to reddish zeolite mineral that helps give abyssal red clay deposits their color. The total zeolite concentration in some deep-ocean sediments may exceed 50 percent.

Cosmogenous Particles

Cosmogenous particles, which originate from space, also occur in marine sediments. Common cosmogenous particles include **nickel-iron spherules** and **silicate chondrules** (Figure 5-7). Such particles were thought to be fragments of meteorites that broke up in the atmosphere or on impact. However, recent studies indicate that they may originate from collisions between bodies in our Solar System's asteroid belt, out beyond Mars. Iron spherules are generally about 30 μm (0.01 in) in diameter, and silicate chondrules occur in two size groups having average diameters of 30 μm and 125 μm (0.01 and 0.05 in). The overall size ranges of these smaller cosmic particles fall between mud and sand—10 to 640 μm (0.004 to 0.25 in).

Larger cosmic objects also fall into the ocean. We know of many meteorite impact craters on land, and given the larger surface area of the oceans, it is probably safe to assume that there have been many more impacts in the oceans. However, finding evidence of ocean impacts is quite difficult. Sediment cores taken in the Antarctic Basin 1400 km (870 mi) west of Cape Horn contain grains that indicate an asteroid meteorite struck the area 2.3 million years ago, near the start of the Pleistocene ice age. This is the only known impact of a meteor into a deep-ocean basin. The metallic element iridium (Ir) is found in greater concentrations in meteoric and mantle material than it is in Earth's crust. Therefore, layers of sediment that contain unusually high concentrations of iridium are usually accepted as evidence of meteoric impact unless there is evidence that volcanic events produced them. The concentration and spatial distribution of sediment cores with elevated levels of iridium suggest that the meteorite was between 0.5 and 1 km (0.31 and 0.62 mi) in diameter. It hit the ocean without producing a crater in the ocean floor. It is unlikely that this event caused the climatic cooling preceding the ice age, but it could have contributed to a rapid decrease in global temperatures because of the meteoric dust and water it sent into the stratosphere.

Another meteorite produced a crater 200 km (124 mi) southeast of Nova Scotia, Canada, near the edge of the shelf beneath 113 m (371 ft) of water. The crater is about 45 km (28 mi) in diameter and is believed to have been created by a stony meteorite or cometary nucleus 2 to 3 km (1.2 to 1.9 mi) in diameter. The impact is dated between 50 and 55 Ma in the early Eocene. Seismic and gravity data indicate the existence of an impact crater 90 km (56 mi) in diameter at the mouth of Chesapeake Bay. Buried by more recent sediment, this crater is about 35 million years old. Perhaps the most famous impact, which occurred on a shallow marine shelf off of the Yucatan Peninsula, occurred about 65 million years ago. The crater, called Chicxulub, is 180 km (110 mi) in diameter. Its impact created a tsunami that decimated shorelines and deposited iridium-enriched ejecta throughout the Caribbean. We will look at this event in more detail later in this chapter.

Classification of Marine Sediments

The classification of marine sediments is based on either the relative proportions of particles from the sources discussed above, or whether the sediments were deposited in the deep sea or on the continental margins. Neritic deposits are those found on continental margins; oceanic deposits are those found in the deep-ocean basins. Different processes control sediment accumulation in these environments.

Neritic Sediments

Neritic sediment deposits display a wide range of particle sizes and are composed primarily of lithogenous particles derived from the continents and deposited at rates in excess of 10 cm/1000 yr. Neritic sediments also contain biogenous, hydrogenous, and cosmogenous particles, but these constitute only minor percentages of the total sediment mass because they are deposited much more slowly than lithogenous particles. During the millions of years that sediments accumulated at the margins of continents, the continental shelf, slope, and rise developed into sedimentary wedges, some more than 10 km (6.2 mi) thick; together, they contain more than 75 percent of the sediment in the ocean basins. Less than 25 percent of marine sediment is found in the deep ocean even though that province makes up more than 80 percent of the ocean floor.

About 20 billion metric tons of sediment are carried to Earth's continental margins by rivers each year. Almost 80 percent is provided by runoff from Asia. As is shown in Figure 5-1, seven of the ten most important sediment-contributing rivers drain the Asian continent. The Hwang Ho, Yangtze, and Mekong empty into the marginal seas of the western Pacific Ocean; the Ganges, Brahmaputra, and Irrawaddy flow into the Bay of Bengal between India and Burma at the north end of the Indian Ocean. Also flowing into the Indian Ocean via the Arabian Sea is the Indus River. The three remaining large rivers drain the Americas. The sediment of the Colorado River remains near the northern end of the Gulf of California. The Mississippi and Amazon rivers are building major deltas of sediment in the Gulf of Mexico and on the South American shelf in the equatorial western Atlantic, respectively. The Mississippi and Amazon are typical of the long, low-gradient rivers that feed sediment onto broad continental shelves of passive continental margins.

Sediments, carried to the margins of the ocean by rivers and glaciers, settle out and fill estuaries, form deltas, or are spread across continental shelves by storms, tides, and

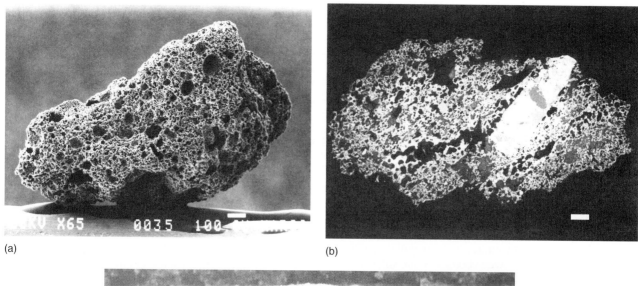

(a)

(b)

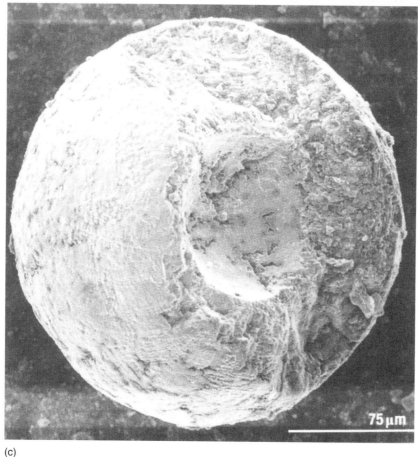

(c)

Figure 5-7 Cosmogenous Particles

(a) SEM of a vesicular silicate glass particle. Up to 50 percent of the volume of these particles is pore space. Bar length is 100 μm. (b) SEM of a vesicular particle with a basaltic inclusion. Bar length is 100 μm. (c) Iron spherule (Bar length is 75 μm)   *(Parts (a) and (b) courtesy of Frank T. Kyte, University of California, Los Angeles, (c) Reprinted by permission of the Open University Course Team,* Ocean Chemistry and Deep-Sea Sediments, *Butterworth-Heinemann, 1989.)*

turbidity currents. The rise in sea level caused by melting glaciers at the end of the last ice advance created estuaries from many flooded river valleys in coastal areas throughout the world. Until these estuaries are filled by sediments, they act as traps for river sediment that otherwise would reach the continental shelf. The sediments that cover about 70 percent of the continental shelves today are **relict sediments**, 3000 to 7000 years old, that have not been reworked by more recent sedimentary processes.

Sediments carried beyond estuaries are sorted by current and wave action, leaving coarser sand along the shore and finer silt and clay-sized particles farther out on the shelf. The greatest volume of sediment transport within the ocean is by longshore currents that move sand, brought into the oceans by rivers, along the shorelines (see Chapter 12). Some continentally derived sediment is carried beyond the shelf to the deep-ocean basin, but the particles do not settle out until waters are deep enough for the bottom to be below wave action.

The composition of neritic sediments is greatly influenced by latitude. Figure 5-8 was constructed by summing up the amounts of various sediment particle types occurring within latitudinal belts, but it does not give us any information on the compositions of individual sedimentary deposits. For example, from the graph you cannot tell if all the sediments at the equator are the same or of quite different compositions; they could be identical mixtures of particles in the percentages shown (3 percent rock and gravel, 20 percent sand, 50 percent silt and clay, 20 percent coral debris, and 7 percent shell fragments) or they could be deposits of a single particle type (3 percent of the total amount of deposits are pure rock and gravel, 20 percent are sand, etc.). The truth is that some deposits are purely

one grain type and others are mixtures of many grain types. The purpose of the graph is only to show the change in abundance of carbonate sediment particles (coral and shell debris) with latitude, and in grain size dominance for lithogenous particles with latitude.

Silt and clay are most abundant in equatorial and tropical latitudes, accounting for about 50 percent of total sediment at the equator, about 20 percent in the tropics, and about 15 percent at higher latitudes. Sand particles are most abundant at mid-latitudes, accounting for about 60 percent of sediment, decreasing to about 30 percent at the equator and to about 35 percent at high latitudes. Coarse, poorly sorted deposits (rock and gravel) are found primarily at high latitudes where they are deposited by glaciers and icebergs. At all latitudes, 6–7 percent of the sediment is composed of shell fragments. Coral reef debris is significant only at low latitudes because coral cannot survive in the cooler waters of latitudes higher than about 30°. At some locations near reefs, sediment may be composed entirely of reef debris; however, when summed with the rest of sediments found at these latitudes, coral debris accounts for only about 20 percent of the total.

Turbidites   Although it seems unlikely that storm waves, tides and ocean current systems could carry coarse material beyond the continental shelf into the deep-ocean basins, much neritic sediment is found at the base of the continental slope, forming the continental rise, and extending in gradually thinning layers toward the abyssal plain. These deposits, called **turbidites**, are deposited by turbidity currents that episodically move down the

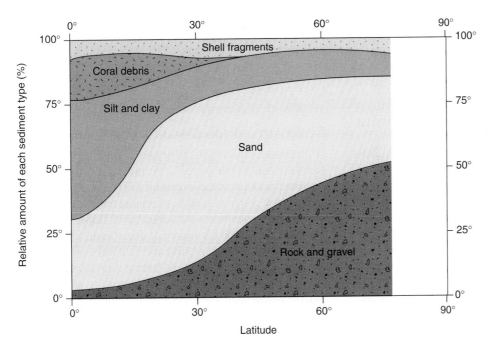

Figure 5-8   Sediment Distribution by Latitude

The percentage of various types of particles in neritic sediment varies by latitude. *(After M.O. Hayes, Marine Geology 5(2), 1967.)*

Figure 5-9 Turbidites and Graced Bedding

(a) Turbidity currents flow down submarine canyons and deposit sediment as deep-sea fans.
(b) Examination of the deep-sea fans shows that they are composed of a series of deposits called turbidites. Each turbidite represents the sediment deposited by one turbidity current event. Each turbidite displays graded bedding in which the sediment particles grade from coarse at the bottom to fine at the top. *(Reprinted with the permission of Macmillan Publishing Company from The Earth, 4th Edition, Figures 19.6 and 19.7, by Edward J. Tarbuck and Frederick K. Lutgens. Copyright © 1993 by Macmillan Publishing Company.)*

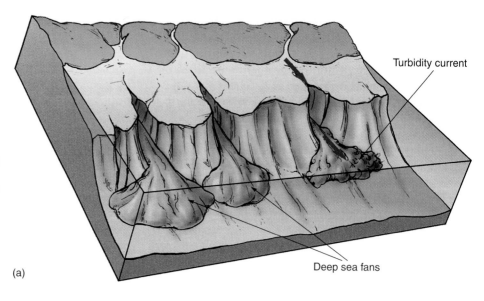

(a)

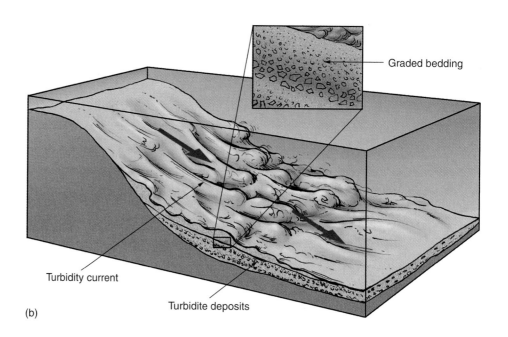

(b)

continental slopes through submarine canyons, carrying loads of neritic material (Figure 5-9).

Turbidites are characterized by graded bedding. This means that each deposit has coarser material at its base and progressively smaller particle sizes toward the top. This pattern results from the declining current velocity with time and within the current from the head to the tail. As the current moves down slope the head erodes bottom sediments and lower current velocities immediately behind the head allow the coarsest material to settle out. Smaller and smaller particles settle out as the waning current passes. Turbidite deposits are usually capped by the normal sediments that accumulate between currents, typically biogenous sediment.

Sediments of the Deep-Ocean Basin

As previously discussed, most neritic sediment is deposited on the continental margin forming the continental shelf, continental slope, and continental rise. Given that the continental rise is transitional between the continental margin and the even deeper open-ocean-basin floors, we might think that the rise is the source of deep ocean sediments in the same way that the continents are sources for the margin. After all, sediment should keep going downhill! Continental material does make it, via turbidity currents, out into the ocean basins near continental margins, but most of the ocean floor is not affected. Instead, wind is the major force bringing continental sediments to the open oceans. In light of all the high-energy processes required just to get sediment out to the continental margin, that statement may sound ridiculous. In truth, the wind deposits fine-grained particles almost everywhere; but everywhere except the deep-ocean basins, there is too much sediment from other sources for the wind-blown grains to amount to more than a very small fraction of the total. Even within the ocean basins, the wind has competition. Biogenous and hydrogenous sediments accumulate at faster or comparable rates, and if present, may swamp the wind-derived sediment component. For biogenous sediments to accumulate however, surface waters must be productive; hydrogenous sediments only form in special chemically modified settings and so much of the deep-ocean floor is dominated by wind-blown sediments.

To describe the relative percentages of biogenous and lithogenous material in deep sea sediments, scientists have developed a classification scheme that includes some rather odd names that you might not want to mention in polite dinner conversation. Sediments in the deep sea are divided into these two groups: (1) *pelagic deposits*, consisting mostly of material that has settled from the sea surface to the ocean floor and includes **oozes** and **abyssal clays**; and (2) *hemipelagic deposits*, which contain greater percentages of lithogenous material carried to the ocean floor by turbidity currents, or of sediment derived from volcanic activity. Table 5-2 shows the percentage by weight cutoffs for these various terms. The classification system first separates the sediment into particles more than 5 μm or less than 5 μm in diameter. For each of these size fractions, the relative percentages of sediment from various sources defines the name. For example, for a sediment to be pelagic, over 50 percent of the sediment mass must be composed of particles less than 5 μm in diameter, and less than 25 percent of the sediment fraction larger than 5 μm in diameter can be lithogenous. We will refer to these pelagic sediments as **oceanic sediment**.

Pelagic or Abyssal Clays   Both *pelagic* and *abyssal* are terms used to describe the clay particles that cover most of the ocean floor below about 4000 meters. Accumulating at rates of approximately 1 mm/1000 yr (0.04 in/1000 yr), these particles commonly are red-brown or buff in color from the oxidation of iron in clay minerals. The clay-sized

particles in these deposits are derived from the continents and carried to the oceans by winds. For a sediment to be called a *pelagic clay*, it must first fulfill the definition of pelagic, and in addition, contain less than 30 percent biogenous material by weight.

Oozes   At somewhat shallower depths, biogenous material makes up a significant portion of the sediment. A sediment is classified as a pelagic ooze if it meets the criteria for pelagic classification, and in addition, has greater than 30 percent biogenous material by weight. Oozes can form only where the deposition of lithogenous material occurs very slowly. The rates of deposition for biogenous sediment range from 1 to 15 mm/1000 yr (0.04–0.6 in/1000 yr).

Oozes are classified as calcareous or siliceous, depending on the mineralogy of the dominant skeletal material in the deposit. More specific names may be given to oozes to indicate the types of organic remains most abundant in the deposits. A diatom ooze is a siliceous ooze that contains mostly the remains of diatoms. Other descriptive names commonly used on this basis are radiolarian ooze (siliceous), foraminiferan ooze (calcareous), and pteropod ooze (calcareous).

Hemipelagic muds   More than 50 percent by weight of the particles in hemipelagic sediments are greater than 5 μm in size. The fraction of hemipelagic deposits greater than 5 μm in size, by definition, must contain more than 25 percent by weight of lithogenous material derived from current transport. Consequently, these deposits are restricted to regions close to the continental rise. Prefixes are applied depending on mineralogy and/or source. For example, calcareous muds contain more than 30 percent calcium carbonate material. Volcanogenic muds contain mostly ash and other volcanic debris.

Controls on Oceanic Sediment Accumulation

The distribution of sediments depends not only on their deposition rates, but also on the degree of preservation after deposition and the amount of input of other sediment types to the same piece of ocean floor. For example,

TABLE 5-2   Classification of deep-sea sediments as used by the Ocean Drilling Program

Name			% Particle origins
Pelagic deposits	Grains < 5 μm =	> 50% by weight	< 25% of > 5 μm is lithogeneous
Abyssal clays			< 30% biogenous
Oozes			> 30% biogenous
Hemipelagic deposits (muds)	Grains > 5 μm =	> 50% by weight	> 25% of > 5 μm is lithogenous
Calcareous			< 30% CaCO$_3$, < 50% lithogenous
Terrigenous			< 30% CaCO$_3$, 50% volcanogenic
Volcanogenic			

abyssal clays are deposited everywhere, but only become the dominant sediment type when the accumulation of other sediment types is negligible. Similarly, calcium carbonate accumulation must be negligible for siliceous oozes to dominate because calcareous oozes usually are deposited about 10 times more rapidly than siliceous oozes. In some areas, sediment accumulations include a variety of grain types and the percentage of biogenous grains is less than 30 percent. Thus, the sediment is not an "ooze" even though deposition rates of biogenous material may be equivalent to rates in areas where oozes are accumulating. This case is an example of the *dilution* effect. You should consider these processes when you look at the maps in this section.

Deposition: Getting to the Bottom

A problem that once puzzled marine geologists studying oceanic sediment distributions is that sediments on the deep-ocean floor very closely reflect the particle composition of the surface water directly above. This was difficult to understand because the calculated settling velocities of these small particles would result in them reaching the bottom only after 10 to 50 years. During this interval, a horizontal ocean current of only 1 cm/s (0.03 ft/s) (about 0.035 km/h, or 0.02 mi/h) could carry them 3000 to 15,000 km (1860 to 9300 mi) away.

These findings led scientists to look for a mechanism by which the small surface particles could be aggregated into larger particles that would settle faster. The mechanism, of course, becomes obvious if you consider the ocean food chain. Most particles in the surface water, living and non-living, organic and inorganic, are eaten at least once. Samples from sediment traps (Figure 5-10) deployed during the GEOSECS program contained numerous aggregated packages of deep-sea sediment particles. These packages are routinely formed in the intestinal tracks of marine animals and are called **fecal pellets** (Figure 5-11). Estimates made by those counting the pellets in the traps indicate that 99 percent of deep-sea particles have fallen to the ocean floor as parts of fecal pellets. Although they are only 50–100 μm (0.02–0.04 in.), the pellets are large enough to settle to the abyssal ocean floor in 10 to 15 days.

As it turns out, fecal pellets are also very important constituents of marine sediments at shallower depths. Extensive areas of mud deposited on the leeward sides of shallow carbonate banks have been repackaged as sand-sized fecal pellets by marine organisms. In these environments, the animals responsible are *deposit feeders*, animals that gain nourishment by eating mud to extract the

Figure 5-10   Sediment Trap

These funnel-shaped sediment traps are lowered to a specific depth in the ocean, allowed to remain for a specific time interval, and then retrieved. The amount and nature of the particles which have settled through the water column into the traps is analyzed.   (*Photo courtesy of Keith Bradley, Woods Hole Oceanographic Institution.*)

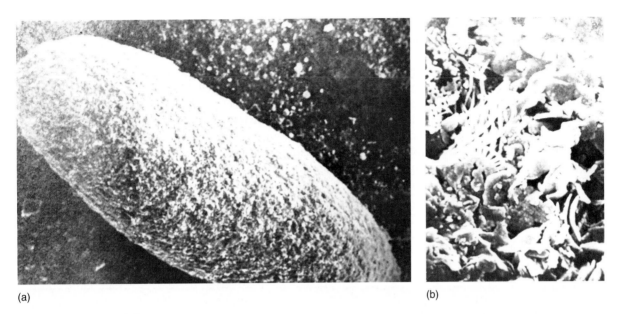

Figure 5-11 Fecal Pellets

(a) SEM of a fecal pellet produced by zooplankton. Length is about 200 μm. (b) SEM of the surface of the pellet showing the remains of phytoplankton and detritus. (*Photos courtesy of Susumu Honjo, Woods Hole Oceanographic Institution.*)

organic matter. Needless to say, this lifestyle leads to a lot of waste production. Pelleted lithogenous sediments are also found on muddy continental shelves.

Pelleting changes the hydrodynamic behavior of fine-grained sediment particles by allowing small particles to be transported as larger ones—requiring higher energy currents for erosion and shorter times for settling than they would individually. However, once deposited on the sea floor, the organic material that binds the pellet together disintegrates, and no record remains of the true hydrodynamic nature of the particles when transported.

Production

We will examine the controls on biological productivity in the oceans in much greater detail in Chapter 15, so our coverage here is only a quick summary. The bottom of any food chain is always the autotrophic organisms (algae and plants) that convert inorganic substances into organic matter. Thus, the population sizes of all heterotrophic organisms (consumers) depend on the success of the autotrophs. In the surface oceans, photosynthetic planktonic algae, diatoms, and coccolithophorids occupy the autotrophic niche. As we discussed in Chapter 2, photosynthesizers need only raw materials (the elements in organic matter: CO_2, water, nitrogen, phosphorus, etc.) and light.

The productivity of plankton that contribute most of the biogenous material to oceanic sediments is greatest in areas of upwelling along the margins of continents and in equatorial areas of the open ocean where surface currents diverge. In these settings, water from the deep oceans containing high levels of necessary raw materials (nutrients)

produced by the bacterial decay of organic matter comes to the surface. The general circulation patterns of deep-ocean water also influence the nutrient levels in the surface waters of each ocean, resulting in the Atlantic having generally higher productivity than the Pacific. We will explain why this occurs in Chapter 9.

Because photosynthetic organisms also rely on light, their productivity depends on how much sunlight reaches the ocean surface. This requirement creates latitudinal differences in productivity. At high latitudes, productivity depends mainly on the season; near the equator, productivity is nearly constant throughout the year.

Although sampling has been limited, there is evidence of a significant belt of springtime diatom productivity in the ocean surrounding Antarctica and at high latitudes of the North Pacific. Calcium carbonate-secreting organisms do not occur in these areas of nutrient-rich waters because water is too cold for them to survive. Populations of foraminifera decrease greatly north and south of about 50° latitude. The divergence of currents in the equatorial Pacific produces large populations of both siliceous and calcareous organisms. Coastal margins in middle and high latitudes are often limited to siliceous organisms because of the seasonal excursions of calcareous organisms to cold temperatures that occur in shallower waters.

Preservation or Destruction

Destruction of skeletal material occurs primarily through dissolution in seawater as the particles fall through the water column. Aggregation in fecal material, in addition to getting particles to the bottom faster, also gives biogenous particles some protection from dissolution during the trip. Once on the bottom, biogenous material continues to dissolve as long as it is in contact with seawater. Preservation of these sediment particles is only assured after they have been buried under another layer of sediment and the seawater trapped in sediment pores has lost its dissolving power.

At all depths, seawater will dissolve biogenous silica. Bacterial decomposition of the protective protein coating on diatom shells also aids the dissolution process (Figure 5-12). Siliceous sediments are only found where biological productivity is extremely high and planktonic populations are dominated by diatoms and/or radiolaria. In these places, the supply of siliceous material is great enough that some of it reaches the sea floor before dissolving. About 80 percent of the silica reaching the ocean floor dissolves back into the water before it can be buried under a sediment cover.

Calcium carbonate dissolves more readily with increasing CO_2 content in water, decreasing temperature, and increasing pressure. Photosynthesizers at the surface are continuously using up CO_2 from the air and water to

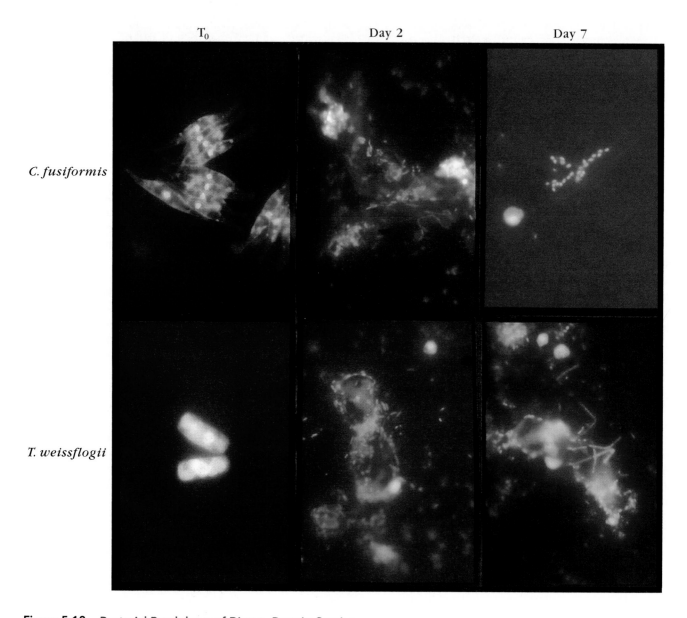

Figure 5-12 **Bacterial Breakdown of Diatom Protein Coatings**

From left to right, bacteria (stained bright blue) attack the protective protein coating on two species of diatoms (red), *Cylindrotheca fusiformis*, and *Thallassiosira weissflogii*, in lab cultures examined by epifluorescence microscopy. *(Permission and courtesy of K.D. Bidle, SIO. Reprinted by permission from* Nature *(www.nature.com) 397:6719, 475 (c)1999 Macmillan Magazines, Ltd.).*

make organic matter, but bacteria at depth are usually just as busy breaking down that organic matter and releasing the CO_2 into the water. In addition, pressure increases and temperature decreases with depth. At some point, the changes in these three factors result in transforming seawater from a solution that precipitates calcium carbonate at the surface into a solution that aggressively dissolves calcium carbonate at depth. Differences in temperature and CO_2 concentrations with depth result in the onset of carbonate dissolution at slightly different depths within different parts of the oceans. The depth at which dissolution begins is defined as the top of the *lysocline*. Throughout the depth interval of the lysocline, calcium carbonate continues to dissolve until a depth is reached where the entire supply settling down from the surface is dissolved, i.e., supply rate = dissolution rate. This depth is called the **carbonate compensation depth**, or **CCD**. Average CCDs occur at 4500–5000 m (15,000–16,000 ft), but may be as deep as 6000 m (20,000 ft) in parts of the Atlantic Ocean, or as shallow as 3500 m (11,000 ft) in regions of low biological productivity in the Pacific Ocean.

Do not forget that sea-floor spreading acts as a conveyor belt moving the ocean crust, with its carbonate sediment blanket, to greater depths over time. Even where carbonates reach the ocean floor above the CCD, they may dissolve after the ocean floor descends below the CCD. As long as they sit in contact with ocean bottom waters, they will continue to dissolve; they must be covered by other sediments in order to be preserved. About 50 percent of the total calcium carbonate supply that reaches the deep-ocean floor dissolves before it is protected by sediment. But the carbonate sediments that have been preserved, coupled with rate curves for sea-floor subsidence, give us a history of the position of the CCD through time, and of the seawater chemistry that controls the level of the CCD.

Distribution of Oceanic Sediments

Oceanic sediments cover about 75 percent of the ocean bottom. Figure 5-13 shows that calcareous oozes cover almost 48 percent of this area, abyssal clay accounts for 38 percent, and siliceous oozes account for 14 percent. If you look carefully at Figure 5-13, you will see that the percentage of calcareous ooze in the three ocean basins decreases as the average depth increases. The shallowest ocean, the Atlantic Ocean, has 68 percent of its floor covered with calcareous ooze; in the slightly deeper Indian Ocean, the coverage is 54.3 percent, and the deepest ocean, the Pacific, has a 36.2 percent coverage of calcareous ooze. Which of the controls previously mentioned is responsible?

A map of all deep-sea sediments is shown in Figure 5-14. Calcareous oozes are deposited primarily near

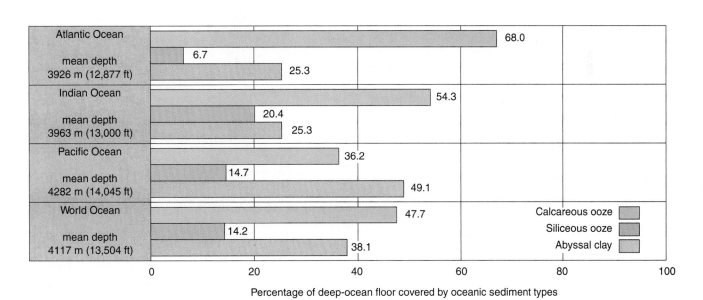

Figure 5-13   Distribution of Oceanic Sediment

The percentage of an ocean basin floor covered by calcareous ooze decreases with increasing mean depth of the basin probably because the deeper an ocean basin is, the greater is the percentage of its floor that lies beneath the carbonate compensation depth. The mean depths shown exclude the shallow adjacent seas where little oceanic sediment accumulates. Notice that the dominant oceanic sediment found in the deepest ocean basin, the Pacific Ocean, is abyssal clay, while calcareous ooze is the most widely deposited abyssal sediment in the shallower Atlantic and Indian Oceans.   *(After Sverdrup, Johnson, and Fleming, 1942.)*

Meteorite Impact and the Cretaceous–Tertiary Boundary Extinction Event

The 1980 report of the discovery of iridium-enriched sediments at the Cretaceous–Tertiary (K–T) boundary (65 Ma) in northern Italy initiated the search for the impact site of a meteor 10 km (6 mi) in diameter. Because iridium is found in greater concentrations in meteorites than in Earth's crust, deposits enriched in iridium are usually considered to be evidence of a meteorite impact. Scientists were intrigued by two pieces of evidence. First, iridium enrichments had been found in widely scattered deposits of the same age. Second, the age of the deposits corresponded to a major extinction event that had eliminated many groups of organisms from Earth at about the same time. This event led to the demise of the dinosaurs, making it possible for mammals of that day to eventually rise to the position of dominance they hold on Earth today.

Despite considerable opposition to the idea, the search for the site of the impact picked up steam. The only known coarse K–T boundary deposits were found in Deep Sea Drilling Project holes in the Caribbean (Figure 5A). Researchers believed the deposits were produced by tsunamis (giant waves) that were generated by a meteorite impact. This would have been possible only if the impact had occurred within a distance of about 1000 km (620 mi). Examination of a surface exposure of the K–T boundary on the southern peninsula of Haiti produced further support for a nearby impact. A 46-cm-thick layer of clay there contains an iridium anomaly, shocked quartz (known to be formed only by meteorite impact), and small glass spherules. Glass spherules are particles of rock melted upon impact and ejected through the atmosphere where they cool, solidify, and fall back to Earth. The composition of the spherules indicates they could have formed from a mixture of carbonate and silicate rocks. The Haitian deposit is 25 times thicker than any other K–T ejecta layer observed.

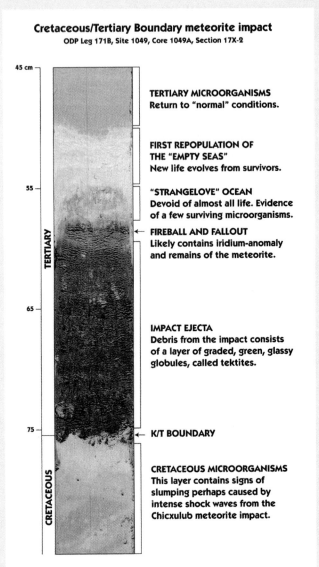

Cretaceous/Tertiary Boundary meteorite impact
ODP Leg 171B, Site 1049, Core 1049A, Section 17X-2

TERTIARY MICROORGANISMS
Return to "normal" conditions.

FIRST REPOPULATION OF THE "EMPTY SEAS"
New life evolves from survivors.

"STRANGELOVE" OCEAN
Devoid of almost all life. Evidence of a few surviving microorganisms.

FIREBALL AND FALLOUT
Likely contains iridium-anomaly and remains of the meteorite.

IMPACT EJECTA
Debris from the impact consists of a layer of graded, green, glassy globules, called tektites.

K/T BOUNDARY

CRETACEOUS MICROORGANISMS
This layer contains signs of slumping perhaps caused by intense shock waves from the Chicxulub meteorite impact.

Figure 5A   Cretaceous–Tertiary Sediment Core from the Caribbean Area

(Courtesy of the Ocean Drilling Program, Texas A & M University)

Figure 5B   Cretaceous–Tertiary Meteorite Impact Site

Oil company seismic surveying in the Gulf of Mexico during 1980 identified a crater 180 km (112 mi) in diameter along the north coast of the Yucatan peninsula near Chicxulub (Figure 5B). When this information finally became public knowledge in 1990, the site was immediately seized upon as a likely candidate for an impact. Rocks exposed at Chicxulub showed the characteristic carbonate and silicate composition. Once the Chicxulub crater and Haitian deposits were dated and appeared to be of the same age within the error of the dating method, many skeptics began to rethink the role of a meteorite in the extinction event at the K–T boundary. Now, some scientists are finding evidence that multiple impacts occurred over a period of time as a result of the breakup of a comet within the solar system at that time.

But, how would meteorite impacts lead to mass extinctions? Did darkening of the atmosphere by ejecta limit photosynthesis, or were wildfires responsible? Did the creation of nitrous oxides cause intense acid rain, or was there an intense greenhouse effect resulting from vaporization of carbonate rock and the release of huge amounts of carbon dioxide into the atmosphere? Why did some groups of organisms survive this catastrophe while nearly all species of other types of organisms were wiped out? The mystery of what type of event caused the K–T extinction appears to have been solved but the devil is in the details.

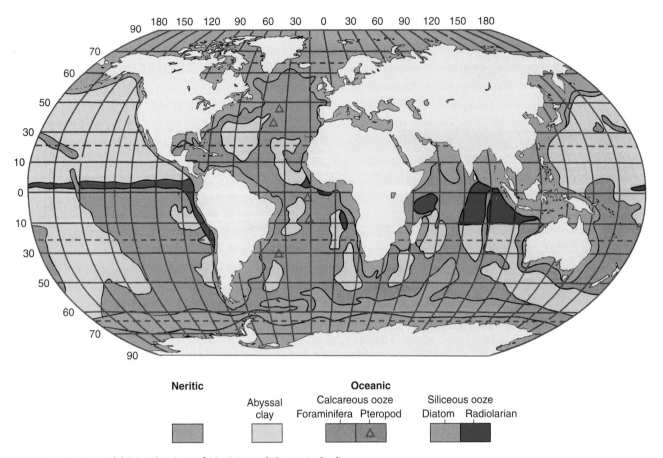

Figure 5-14 World Distribution of Neritic and Oceanic Sediments

Beyond the neritic deposits of the continental margins lie the oceanic marine deposits. Abyssal clays are found in the deeper ocean basins, where the bottom lies beneath the carbonate compensation depth. The best-developed calcareous oozes are found on the relatively shallow deep-ocean environments of the oceanic ridges and rises. Siliceous oozes are found beneath areas of unusually high biological productivity such as the Antarctic and equatorial Pacific Ocean. The red line in the Pacific Ocean shows the position of the cross section in Figure 5-17.

the oceanic ridges and rises that are generally above the carbonate compensation depth. The crests of these subsea mountain ranges usually lie at depths between 2500 and 3000 m (8200 and 9840 ft). The deep North Pacific basin is free of calcareous ooze except for the shallow Hawaiian Island Ridge.

Figure 5-15 shows the changes in concentration of biogenous calcareous particles in ocean floor sediments. From this map, it is easier to see that there are gradational changes in concentration moving away from the ridges. From the boundaries on this map, you should be able to approximate the CCD in each ocean. Figure 5-16 shows the distribution of biogenous silica (opal) in the surface sediment of the oceans; generally, significant accumulations are found only in areas of high productivity in equatorial and high-latitude regions. Because of the higher rates of calcium carbonate accumulation, siliceous oozes

are found only in areas below the carbonate compensation depth or at high latitudes.

The diatom ooze of the North Pacific does not have the high opal concentration of the Antarctic ooze because of dilution by lithogenous material. Radiolarian oozes develop in equatorial waters of the Atlantic, Indian, and Pacific Oceans. Equatorial surface waters are especially productive in the eastern Pacific Ocean, and the production rates of both calcium carbonate and opal particles are high (Figure 5-17). The ocean floor to the north deepens to below the CCD, preventing the siliceous ooze from being diluted by carbonate production. The southern limit of the band of radiolarian ooze is fixed by the intersection of the CCD with the ocean floor. To the north of this line, calcium carbonate particles do not reach the bottom, so radiolarian particles account for more than 30 percent of the sediment mass, the criteria for calling

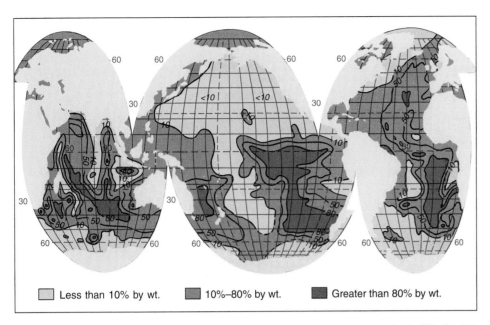

Figure 5-15 Calcium Carbonate in Deep-sea Sediments (on a per weight CaCO$_3$ basis)

Concentrations of calcium carbonate in ocean sediments are negatively correlated with ocean depth. The deeper the ocean floor, the less calcium carbonate found in the sediment. Calcium carbonate is also low in sediments accumulating beneath cold, high-latitude waters. Extensive areas of concentrations of calcium carbonate in the sediment in excess of 80 percent are associated with the relatively shallow ocean floor on the Carlsberg Ridge in the Indian Ocean, the East Pacific Rise, and the Mid-Atlantic Ridge. Note that the deepest ocean basin, the North Pacific, lies for the most part beneath the carbonate compensation depth and has very little calcium carbonate in its accumulating sediment.   *(After Biscaye et al., 1976; Berger et al., 1976; and Kolla and Biscaye, 1976.)*

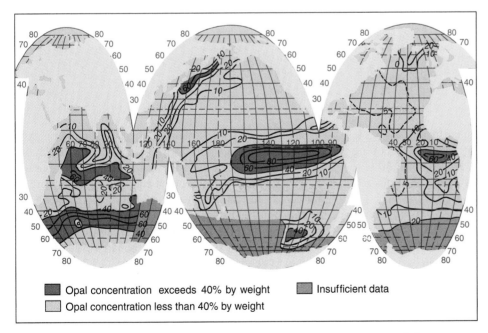

Figure 5-16 Biogenous Silica (SiO$_2$·nH$_2$O–Opal) in Surface Sediments of the World's Oceans (on a weight % total SiO$_2$ basis)

The distribution pattern of opal in surface sediments of the ocean shows maximum concentrations in areas of highest biological productivity. In the equatorial and northwest Pacific, opal is produced predominantly by radiolarians; elsewhere, diatoms are more abundant. The highly productive waters south of the Antarctic Convergence show up well in the south Indian Ocean and southeast Pacific Ocean. Data for the southern Atlantic Ocean and southwest Pacific Ocean are insufficient to determine the relationship between highly productive surface waters and opal concentrations in the sediment.   *(After Leinen et al., 1986.)*

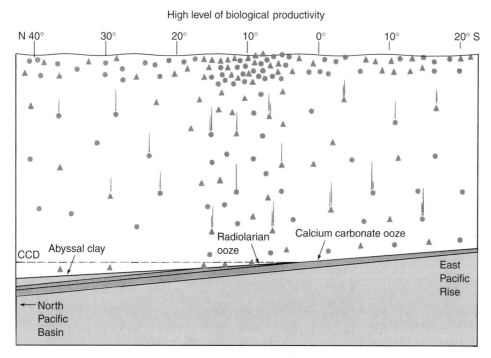

Figure 5-17   North-South Cross Section through the East Pacific Equatorial Region

High levels of biological productivity in the equatorial East Pacific region create an unusually large supply of biogenous particles. South of the equator, the ocean floor is above the carbonate compensation depth (CCD) and enough $CaCO_3$ particles reach the bottom to produce a $CaCO_3$ ooze. The ocean floor deepens to the north and descends beneath the CCD under the area of high biological productivity. No $CaCO_3$ particles reach the ocean floor here, but enough SiO_2 particles are deposited to produce a siliceous ooze. Further north, the productivity is lower and the ocean floor deeper. Not enough biogenous particles reach the ocean floor to produce an ooze, and the sediment is abyssal clay.

the sediment a radiolarian ooze. The northern limit of the radiolarian ooze is set by surface productivity declining to the point that radiolaria account for less than 30 percent of the sediment. Abyssal clay then becomes the dominant sediment type.

Abyssal Clay Distributions

A far greater percentage of the Pacific Ocean basin is covered by abyssal clay (49.1 percent) than are the Atlantic and Indian oceans (25.3 percent). Abyssal clay deposition becomes dominant when biological productivity is low and the ocean floor extends below the CCD. For example, the dominance of clay in the North Pacific results from low nutrient concentrations, a relatively shallow CCD that is often as shallow as 3000 m (9800 ft), and a large area of deep, old ocean floor (see Figure 5-14).

The abyssal clay distribution in the Atlantic Ocean is influenced by bottom topography and the CCD in a somewhat different way. The distribution of carbonate sediments across the mid-Atlantic Ridge is asymmetric even though the ocean crust has the same age-depth relationship on both sides of the ridge axis and the ocean floor has descended to the same depth at equivalent distances from the ridge axis. The greatest portion of the Atlantic Ocean bottom covered by abyssal clays is on the west side of the mid-Atlantic Ridge because the CCD is shallower on the west than on the east. Calcium carbonate is found in sediments to depths of more than 6000 m (19,700 ft) on the eastern side of the ridge, but deposition below 5000 m (16,400 ft) is rare in the basins west of the ridge. This difference results from restrictions in deep-water circulation patterns created by the ridge. Antarctic Bottom Water, originating in the Weddell Sea to the south, and North Atlantic Deep Water, originating to the north, flow into the area on the western side of the ridge. Both of these deep waters are very cold and have high carbon dioxide contents, a recipe for high dissolving capacity for calcium carbonate. The ridge protects the eastern side from the invasion of this CCD-raising water.

Summary

Sediment particles that accumulate on the ocean floor are classified by source as lithogenous (derived from rock), biogenous (derived from organisms), hydrogenous (derived from water), and cosmogenous (derived from outer space, beyond Earth's atmosphere). Sediment texture, determined by mineralogy, size, and sorting of sediment particles, is affected greatly by the type of transportation (water, wind, or ice), the duration of transport, and the amount of energy in the depositional environment.

Lithogenous sediment is composed of fragments of rocks released by weathering and transported by wind, water, or glaciers to the oceans. Volcanogenic, or pyroclastic, particles, deposited in the ocean via volcanic eruption, are a special class of lithogenous particles. Biogenous particles include particles of calcium carbonate ($CaCO_3$) from the remains of foraminifera, pteropods, and coccolithophorids, and of hydrous silica ($SiO_2 \cdot nH_2O$) from the remains of diatoms and radiolarians. Hydrogenous deposits include a wide variety of materials that precipitate from the water or form from interaction of substances dissolved in the water with materials on the ocean floor. Evaporites, manganese nodules, phosphorites, glauconites, carbonates (including ooids), and zeolites are examples. Cosmogenous particles include tiny nickel-iron spherules and silicate chondrules that originated in space, as well as debris from meteorite collisions.

Neritic sediment accumulates rapidly along the margins of continents. It is dominated by particles of lithogenous origin. More than 75 percent of all marine sediments are deposited within the thick sediment wedges that create the continental shelves, slopes, and rises. In high latitudes, these accumulations include poorly sorted glacial deposits. The percentages of neritic sediment composed of shell fragments, coral debris, silt and clay, sand, and gravel varies with latitude. The recent rise in sea level from melting glaciers has created estuaries from flooded river valleys of coastal areas throughout the world. River sediments are trapped in these estuaries rather than reaching the continental shelves; about 70 percent of shelf sediments are relict, 3000 to 7000 years old. Turbidity currents transport shelf sediment down submarine canyons and deposit turbidites on the continental rise.

Oceanic sediment accumulates at low rates on the floors of the deep-ocean basins far from the continents. Although it would take from 10 to 50 years for individual sediment particles in the ocean surface waters to settle to the bottom, most appear to be combined into larger aggregates as fecal pellets of small marine animals and reach the ocean floor in 10 to 15 days. This process of particle transport explains how the particle composition of the surface waters is similar to that of the sediment on the ocean floor immediately below.

In the deepest basins, where most biogenous sediment dissolves before reaching the bottom, abyssal clay deposits predominate. Oozes, composed of over 30 percent by weight of biogenous sediment, are found in shallower basins, on ridges and rises, and below surface waters of particularly high productivity. Rates of biological productivity, dissolution, and dilution determine whether biogenous sediments or abyssal clays predominate on the sea floor. Much of the dissolution of $CaCO_3$ occurs because the CO_2 content of the ocean water increases with depth. The carbonate compensation depth (CCD) is defined as the depth at which ocean water dissolves $CaCO_3$ at the same rate at which it settles from above, thus preventing accumulation on the ocean floor below that depth. Because the bottom water in the western Atlantic Ocean is colder, the carbonate compensation depth is shallower on the west side of the mid-Atlantic Ridge and abyssal clay in the western basins has a broader distribution than in the eastern Atlantic. In the Pacific Ocean, abyssal clay covers more of the ocean floor than in other oceans because the ocean floor is deeper (especially in the North Pacific) and lies beneath the carbonate compensation depth. Siliceous oozes are well developed in the North Pacific Ocean and around the southern margins of all oceans because little lithogenous sediment reaches these areas and foraminifera that produce $CaCO_3$ do not thrive in these cold waters. Manganese nodules are found on the deep-ocean floor where other sediments are accumulating at rates of less than 7 mm (0.3 in)/1000 yr. There is a relationship between the biological productivity of the overlying waters and the metallic content of the nodules.

Key Terms

Abyssal clays
Biogenous particles
Calcium carbonate ($CaCO_3$)
Carbonate compensation depth (CCD)
Coccolithophorids
Cosmogenous particles
Diatoms

Fecal pellets
Foraminifera deposits
Hydrogenous deposits
Lithogenous particles
Manganese nodules
Neritic sediment
Nickel-Iron spherules
Oceanic sediment

Ooids
Ooze
Opal ($SiO_2 \cdot nH_2O$)
Radiolaria
Relict sediments
Sediment maturity
Silica (SiO_2)

Silicate chondrules (SiO_2)
Stromatolites
Turbidites
Volcanogenic particles
Wentworth scale
Zeolites

Questions and Exercises

1. List the four basic sources of marine sediment particles.

2. What characteristics of marine sediment indicate increased maturity?

3. Why is a higher-velocity current required to erode clay-sized particles than larger, sand-sized particles?

4. List the two major chemical compounds of which most biogenous sediment is composed, and the organisms that produce them.

5. What are the chemical compositions of hydrogenous deposits of manganese nodules, phosphorite, glauconite, and carbonate? In which environments do they form? What is the relationship between a deposit's formation and its environment?

6. Describe the most common types of cosmogenous sediment and give the probable source of these particles.

7. Describe the basic differences between neritic sediment and oceanic sediment.

8. Discuss the processes by which sediments are carried to and distributed across the continental margin.

9. Discuss how productivity, dissolution, and dilution combine to determine whether an ooze or abyssal clay will form on the deep-ocean floor.

10. Explain why abyssal clay deposits on the floor of deep-ocean basins are commonly underlain by calcareous ooze.

References

Archer, D., and E. Maier-Reimer, 1994. Effect of deep-sea sedimentary calcite preservation on atmospheric CO_2 concentration. *Science* 367:6460, 260–63.

Berger, W. H., C. G. Adelseck, Jr., and L. A. Mayer, 1976. Distribution of carbonate in surface sediments of the Pacific Ocean. *Journal of Geophysical Research* 81:15, 2617–29.

Bidle, K.D., and F. Azam, 1999. Accelerated dissolution of diatom silica by marine bacterial assemblages. *Nature* 397:6719, 508–12.

Billett, D. S., R.S. Lampitt, A. L. Rice, and R. F. C. Mantoura, 1983. Seasonal sedimentation of phytoplankton to the deep-sea benthos. *Nature* 302:5908, 520–22.

Biscaye, P. E., V. Kolla, and K. K. Turekian, 1976. Distribution of calcium carbonate in surface sediments of the Atlantic Ocean. *Journal of Geophysical Research* 81:15, 2595–602.

Blum, J. D., and C. P. Chamberlain, 1992. Oxygen isotope constraints on the origin of impact glasses from the cretaceous-tertiary boundary. *Science* 257:5073, 1104–07.

Dravis, J. J., 1983. Hardened subtidal stromatolites, Bahamas. *Science* 219:4583, 385–86.

Heath, G. R., 1982. Manganese nodules: Unanswered questions. *Oceanus* 25:3, 37–41.

Howell, D. G., and R. W. Murray, 1986. A budget for continental growth and denudation. *Science* 233:4762, 446–49.

Kolla, V., and P. E. Biscaye, 1976. Distribution of calcium carbonate in surface sediments of the Atlantic Ocean. *Journal of Geophysical Research* 81:15, 2602–16.

Kyte, T.K., and D. E. Brownlee, 1985. Unmelted meteoric debris in the late pliocene anomaly: Evidence of the ocean impact of a nonchondritic asteroid. *Geochimica et Cosmochimica Acta* 49, 1095–1108.

Leinen, M., Cwienk et al., 1986. Distribution of biogenic silica and quartz in recent deep-sea sediments. *Geology* 14:3, 199–203.

McLane, M. D., 1995. *Sedimentology*. New York: Oxford University Press.

Sharpton, V. L., et al., 1992. New links between the chicxulub impact structure and the cretaceous-tertiary boundary. *Nature* 359:6398, 819–21.

Spencer, D. W., S. Honjo, and P. G. Brewer, 1978. Particles and particle fluxes in the ocean. *Oceanus* 21:1, 20–26.

Sverdrup, H. U., M. W. Johnson, and R. H. Fleming, 1942. Renewal 1970. *The Oceans: Their Physics, Chemistry, and General Biology*. Englewood Cliffs, NJ: PrenticeHall.

Tucker, M.E., and V. P. Wright, 1990. *Carbonate Sedimentology*. Oxford: Blackwell Scientific.

Udden, J. A., 1898. *Mechanical Composition of Wind Deposits*. Augustana Library Pub. 1. Rock Island, IL.

Wentworth, C. K., 1922. A scale of grade and class terms for clastic sediments. *Journal of Geology* 30, 377–92.

Suggested Reading

Sea Frontiers

Cox, V., 1994. *It's No Snow Job*. 40:2, 42–49. An overview of the work done by Alice Alldredge observing, collecting, and attempting to understand the role of *marine snow* in the ecology of the oceans.

Dietz, R. S., 1978. *IFOs (Identified Flying Objects)*. 24:6, 341–46. Discusses the source of Australasian tektites and microtektites.

Dudley, W., 1982. *The Secret of the Chalk*. 28:6, 344–49. This informative article discusses the formation of marine chalk deposits.

Dugolinsky, B. K., 1979. Mystery of Manganese Nodules. 25:6, 364–69. The mystery of origin, growth, and the environmental implications of manganese nodule mining are presented.

Feazel, C. T., 1986. *Asteroid Impacts, Seafloor Sediments, and Extinction of the Dinosaurs*. 32:3, 169–78. High concentrations of iridium and osmium are found in a marine clay deposited at the time dinosaurs and many other species died out 65 million years ago. This event may have been the result of a collision of a meteor, 6 miles in diameter, with Earth.

Feazel, C. T., 1989. *Inner space: Porosity of Seafloor Sediments.* 35:1, 49–52. A petroleum geologist discusses how porosity forms in marine sediments and the significance of the pores.

La Que, F. L., 1979. *Nickel from Nodules?* 25:1, 15–21. The economic and political problems related to nodule mining are discussed.

Shinn, E. A., 1987. *Sand Castles from the Past: Bahamian Stromatolites Discovered.* 33:5, 334–43. Descibes the formation of stromatolites, sand domes trapped by algal growth, in the Bahamas.

Wiley, J. P., Jr., 1992. *The Search for Missing* CO_2. 38:5, 40–43. Not all of the anthropogenic carbon dioxide that enters the atmosphere stays there. Is it removed by the terrestrial biosphere or does the ocean absorb it and store it in sediments?

Wood, J., 1987. *Shark Bay.* 33:5, 324–33. Summarizes the history of Shark Bay, Australia, since its discovery by Dirk Hartog in 1616. The stromatolites and "tame" dolphins that attract tourists are also discussed.

Scientific American

Erwin, D.E., 1996. The mother of mass extinctions. 275:1, 72–79. Eighty percent of Earth's animal species became extinct 250 million years ago, possibly resulting from a drastic lowering of sea level and/or a massive volcanic event.

Gehrels, T. T., 1996. Collisions with comets and asteroids. 274:3, 54–61. Reviews the nature of asteroids and comets. The age and size of some objects involved in major Earth-jarring impacts by extraterrestrial bodies, and the odds of impacts by different size ranges of objects are discussed.

Hollister, C.D. and S. Nadis, 1998. Burial of radioactive waster under the seabed. 278:1, 60–65. Outlines the many reasons why burying high level nuclear waste in ocean sediments may be a good idea.

Nelson H., and K. R. Johnson, 1987. *Whales and Walruses as Tillers of the Sea Floor.* 256:2, 112–18. The 200,000 walruses and 16,000 gray whales that feed by scooping sediment from the Bering Sea continental shelf suspend large amounts of sediment that is transported by bottom currents.

Pinter, N., and M. T. Brandon, 1997. How erosion builds mountains. Along with volcanic and tectonic activity, erosion and deposition of sediment at the margins of mountain chains affects their growth and shape.

Suess, E., G. Bohrmann; J. Greinert; and E. Lausch, 1999. Flammable Ice. 281:5, 76–83. Ice in sediments at depths below meters (1605 ft.) contains enough methanse to exceed the energy reserves of all known fossil fuel reserves of petroleum and coal.

Oceanography on the Web

Visit the Introductory Oceanography home page for on-line resources for this chapter. There you will find an on-line study guide with review exercises and links to ocean-ography sites to further your exploration of the topics in this chapter. Introductory Oceanography is at http://prenhall.com/thurman (click on the Table of Contents menu and select this chapter).

Sharks

- Would it matter if sharks were to disappear from the ocean?
- Is this an event that could happen?
- How do scientists determine the status of shark populations?
- What roles do sharks play in their environment?
- What are policy makers doing to protect sharks?

Background

There is one feature of the biology of sharks that has become legendary – their dominance as oceanic predators. Four hundred million years of evolution have endowed sharks with a suite of adaptations ranging from an array of acute senses to a jawfull of teeth that are replaced before they can dull. Like the big cats of Africa and Asia, sharks are supreme hunters. But evolutionary adaptations do not guarantee success. Sharks, perceived by most swimmers as their nightmarish nemesis at the beach, ironically have themselves become threatened with extinction (Figure 1).

By 1998, the number of sharks along the East coast of the U.S. had declined so precipitously that, in April of that year, the National Marine Fisheries Service (NMFS), in accordance with the Magnuson-Stevens Fishery Conservation and Management Act, implemented a revised Fishery Management Plan for 39 species of sharks (and other highly migratory fishes like swordfish and tuna). One year later, the plan was amended and the number of shark species protected was increased. These plans were designed to replenish stocks of sharks that had become severely depleted.

Other countries and international organizations are also taking actions to protect sharks. The Department of Fisheries and Oceans of Canada issued an Atlantic Shark Management Plan in 1997. In 1999, member countries of the Food and Agricultural Organization of the United Nations endorsed an international shark management agreement.

Most recently (October 27, 1999), the U.S. House Committee on Resources unanimously approved a resolution opposing the wasteful practice of shark finning after it came to light that 55,000 blue sharks were killed in 1998 for their fins, which are the key ingredient in shark-fin soup, an Asian delicacy. Days later the entire U.S. House of Representatives also unanimously endorsed the ban.

Why has there been so much concern for sharks? What events are responsible for the plight of sharks? What does "severe depletion" mean and can it lead to extinction? Would the consequences of shark extinctions be entirely negative, especially since they consume economically important fish and they also periodically "terrorize" swimmers? What roles do sharks play in their environment? In this issue we will address these questions and discuss shark ecology and conservation.

Characteristics of Sharks

Sharks are fish. Thus, like tuna, mackerel, salmon, etc., they are members of the Phylum Chordata and Subphylum Vertebrata. However, sharks and their relatives (skates and rays) are collec-

Figure 1

Tiger shark on a cart in Bombay India. Now a protected species along the U.S. Atlantic and Gulf coasts, this species is still threatened due to overfishing. *(BarnabasBosshart/CORBIS)*

tively known as elasmobranchs. They differ from tuna, mackerel, and salmon, which are bony fishes while sharks, skates, and rays are cartilaginous fishes. These terms distinguish the principal material comprising the skeleton.

The following list compares some of the characteristics of sharks (Figure 2) with bony fish.

- Sharks have 5–7 external bilateral gill slits, while bony fish have a single bilateral opening covered by bony opercula.
- Sharks may have spiracles (small openings on the top of the head leading to the gill chamber).
- Sharks have ventral mouths, that is, underslung jaws, with some exceptions, most notably the recently discovered shark known as megamouth. The mouths of bony fish can be at the front (terminal), ventral (facing downward) or dorsal (facing upward).
- Sharks have a tail fin whose upper lobe extends farther up and out than the lower lobe. Bony fish have a variety of symmetrical and asymmetrical tail shapes.
- Sharks have sandpaper-like skin due to the presence of placoid scales, which are structures resembling small teeth. Scales on bony fish are small bony plates.

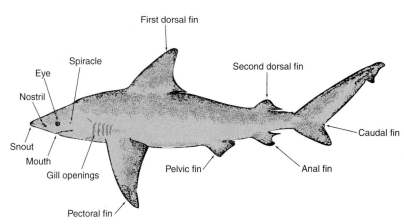

Figure 2

Shark external anatomy.

Labels: First dorsal fin, Spiracle, Eye, Nostril, Snout, Mouth, Gill openings, Pectoral fin, Pelvic fin, Second dorsal fin, Caudal fin, Anal fin

Figure 3

Oceanic white-tip shark caught in a Japanese driftnet in the Tasman Sea. *(Greenpeace/Grace)*

- Sharks employ internal fertilization via claspers, the primary male sexual accessory organ. Very few bony fish use internal fertilization. Instead they release sperm and eggs into the water, and fertilization is external.

- Sharks lack a swimbladder, an organ that regulates buoyancy.

- Sharks reduce the stress of living in a salty medium by conserving and storing urea, which is a waste product eliminated by other organisms.

Sharks and sharklike ancestors have been on the planet for nearly 400 million years, and there are over 400 extant species. Contrary to popular opinion, they are not "living fossils." While several features of sharks, especially the shark look, have been retained from their early evolutionary origin, many other characteristics, such as advanced jaw suspension, which enables them to feed more efficiently, and more flexible fin structure, which allows more maneuverability, have arisen more recently.

Ecology of Sharks

In coastal and oceanic ecosystems, larger sharks occupy a position at the top of the food web. Thus, they are apex, or top, predators. Within an ecosystem, apex predators exert a strong influence on the organisms below them in the food web. First, they play a major role in controlling both the diversity and abundance of other species within a community. Second, they influence the evolution of other species. In both cases they do this through predation on sick, injured, slow, weak, or otherwise less fit individuals.

Threats to the Survival of Sharks

For any species or population of organisms there exists a population size below which that group is believed to be doomed to extinction. This is known as a population's or species' **critical number.** When the number of adult individuals drops to this critical level, there is simply not a large enough reservoir of genetic

variability and potential for mating to allow the population or species to propagate and successfully face the rigors of its environment, including competition with other species. Genetic variation is central to the survival of a species because it is the raw material for natural selection. When a population's environment changes, genetic variation may produce some individuals which have the characteristics necessary for survival.

Critical numbers are difficult to establish and efforts to do so may not be undertaken until a species is near extinction. They also vary among species and are influenced by life history characteristics. The California condor, the big cats of Africa, India, and Asia, many whales, and the desert pupfish are examples of organisms that may be at or below their critical numbers.

Many species of sharks may be threatened with extinction because of overfishing and habitat alteration. Annual catches of sharks, skates and rays reached 800,000 metric tons (nearly 2 billion pounds) by 2000. Fishing pressure is likely to increase because most shark fisheries are still small-scale. As a transition to more modern, industrial-type fisheries is made, landings could increase substantially, at least for the short term.

Sharks are also frequently captured on longlines and in nets as bycatch (Figure 3), that is, as untargeted and hence unused catch. For example, in the 1991 Japanese squid driftnet fishery, which has subsequently ceased (although driftnets are still used in other fisheries), at least 11 kinds of sharks were taken as bycatch, including nearly 100,000 blue sharks.

Sharks are particularly vulnerable to pressures that can reduce their population because of their life history characteristics.

Go to our web site to continue the analysis.

See the complete issue at:
www.prenhall.com/thurman

HTTP://WWW.PRENHALL.COM/THURMAN •

6 The Physical Properties of Water and Seawater

Three delicate snowflakes. (Courtesy of Jim Zuckerman/Corbis)

The presence of liquid water on Earth has had a tremendous impact on the planet's past; the planet's future will depend critically on how humankind cares for this water, including the oceans, rivers, lakes, rain, and groundwater. To be good caretakers, we must not only study the water environments, but also water itself.

Water exists in these three phases on Earth: a solid, as ice or snow, a gas, as water vapor in the atmosphere, and most abundantly a liquid, as rivers, lakes, seas and underground reservoirs. The transformation of water from one phase to another and the movement of water from one place to another are major factors in the distribution of mass and energy on Earth. The relentless action of water has worn away many of Earth's great mountain ranges, reducing them to salt and dust, and carrying them to the sea. Yet water is also where life originated. The same chemical and physical properties that allow water to dissolve rock also have influenced the form and function of every organism on Earth since the first one evolved nearly 4 billion years ago.

The Water Molecule

Molecular Structure

Two hydrogen atoms and one oxygen atom combine to form one molecule of water. Oxygen has six electrons in its outermost shell, and hydrogen has one electron (Figure 6-1a). Oxygen shares two electrons, one with each hydrogen. Each hydrogen, in turn, shares its electron with oxygen. This electron sharing creates a covalent bond between hydrogen and oxygen. Oxygen also has two pairs of unbonded electrons in its outer orbital shell. Thus, there are four sets of electrons. Two sets are bonded (shared), and two unbonded sets are accommodated in the outer electron orbits for the molecule. Because of their negative charges, electrons repel each other, making the favored electron configuration one with the four electron orbits as far from each other as possible. This configuration results in a water molecule that resembles a tetrahedron, which is like a pyramid, except the base has only three sides (Figure 6-1b). Ideally, this configuration would result in the electron orbitals and

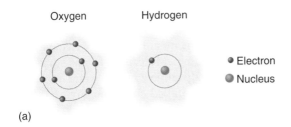

(a)

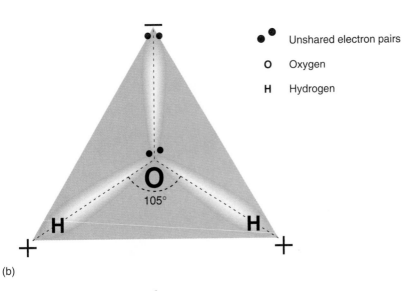

(b)

Figure 6-1   The Structure of Water

(a) The electron orbitals for oxygen and hydrogen. (b) The water molecule approximates a tetrahedral shape. The two hydrogens and the two unbonded electron pairs occupy the corners with the oxygen in the center. There is excess positive charge near the hydrogens and excess negative charge near the electron pairs (see Appendix V).

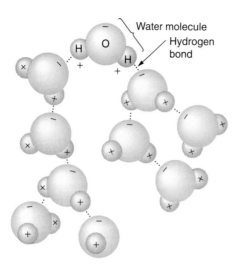

Figure 6-2 Hydrogen Bonding in Water

Attraction between the hydrogens of one water molecule and the unbonded electrons of another water molecule lead to hydrogen bonding between molecules, giving liquid water its structure.

hydrogen atoms all being 109.5 degrees apart; however, chemists have determined that the two covalent bonds with hydrogen are actually separated only by 104.5 degrees. The orbitals for the unbonded pairs are larger than the orbitals of the bonded pairs, pushing the bonded pairs closer together.

In a water molecule, the electrons tend to be found more often around the oxygen atom than around the hydrogen atoms, making the area around the oxygen negatively charged and the area around the hydrogens positively charged. This uneven distribution of charge gives the water molecule **polarity**. Polarity means that the water molecule will orient itself in an electric field, with its negative end attracted to the field's positive pole and the positive end directed toward the field's negative pole.

Hydrogen bonding

The polarity of the water molecules also means that the negatively charged end of one molecule will be attracted to the positively charged ends of other water molecules. This attraction results in the formation of **hydrogen bonds** between adjacent water molecules. These weak bonds form between the negatively charged end of one water molecule and the positively charged end of another water molecule (Figure 6-2). Hydrogen bonds are considerably weaker than covalent or ionic chemical bonds, but are much stronger than van der Waals forces. They give liquid water a structure that is responsible for a number of its unique and important properties.

Surface Tension and Viscosity

We can fill a container with water, not only to the brim, but above it (Figure 6-3a). Water can pile up a short distance above the container's rim, forming a convex (pushed-up) surface. This phenomenon is a result of water's high **surface tension,** a measure of the tendency of surface molecules of a liquid to cling together. With a low surface tension, water poured into a container would overflow as soon as the container's top was reached. Water droplets also demonstrate water's high surface tension; water beads up into drops that stand above a surface rather than spreading out as a thin surface film.

These phenomena are additional evidence of water's hydrogen bonding. Molecules at the surface of liquid water are strongly attracted to the water molecules around them, a property called cohesion (Figure 6-3b). The air above the water's surface has a very low density of molecules compared with that of the water, so the attraction between the surface water molecules and the air molecules is comparatively slight. Water molecules cluster together into drops, and pile up above a container's rim because the hydrogen bond attraction is just enough to offset gravity

Figure 6-3 Surface Tension and Capillarity

(a) The high surface tension of water causes it to pile up above the edge of a container. (b) Surface tension is created by hydrogen bonds holding water molecules together against the force of gravity. (c) Capillarity causes water to climb up the sides of containers. (d) Attraction between the positive areas of water molecules and the negatively charged surface of glass causes water molecules to cling to glass.

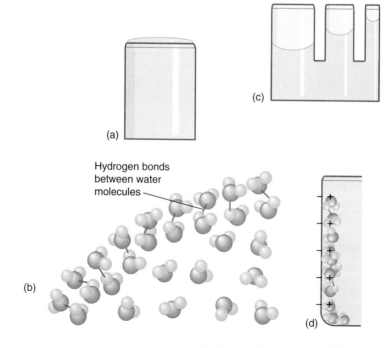

(a)

(c)

Hydrogen bonds between water molecules

(b)

(d)

pulling the droplets down flat to the surface or the water over the side of the container.

Many organisms take advantage of the high surface tension of water. Insects can often be seen walking across the surface of a quiet pool of water as though it were solid; the common water strider is one example. Other creatures are adapted to hang, bat-like, down into the water from the underside of the surface, taking advantage of surface tensional forces that counteract gravity pulling them downward. It is also possible to float small or thin objects that are much denser than water on water surfaces. A razor blade, as well as a metal paper clip or steel needle, although they are all about five times denser than water, will float if laid carefully on a quiet surface. Can you explain why, when water is in motion, the high surface tension is lost?

Water also clings to the surfaces of many substances. We refer to this characteristic as adhesion, or wetting. For example, when water is poured into a glass container, it strongly adheres to the sides, and the force of attraction between the water molecules and the glass molecules even cause the water to climb a little way up the sides (Figure 6-3c). The positively charged portions of the water molecules (the hydrogen ends) are attracted to negatively charged unbonded electrons in the oxygen atoms of the glass (Figure 6-3d). The hydrogen bond attraction between water molecules is too weak to counteract this attraction at the sides of the container. In very thin glass tubes, the area of contact between the water and the glass is maximized, and the water's free surface (where hydrogen-bonding forces are dominant) is minimized. Such tubes are called capillary tubes, and this property of liquids is called **capillarity**. Many organisms make use of the capillary properties of water; for example, plants use capillarity in transporting water upward in their stems.

Viscosity, a measure of the tendency of a substance to resist flow, is controlled by molecular structure, and thus, for water, is increased by hydrogen bonding. Although from our viewpoint, liquid water appears to flow quite readily (compared to substances such as honey or molasses with much higher viscosities), very small organisms living in the ocean find the viscosity of water to be a serious impediment to movement. For us to appreciate their problem, we would have to try swimming in a pool of honey! Organisms have evolved a number of anatomical features to cope with this, as we will see in Chapter 16.

The viscosity of water generally increases with decreasing temperature, increasing salinity and decreasing pressure. Higher viscosity results from increasing structure, that is, from increasing the number of hydrogen bonds or the frequency of hydrogen bond formation. In decreasing the temperature, the energy of each water molecule is lowered, which favors hydrogen bonding, that is, increased structure. The presence of dissolved ions in water (salinity) promotes formation of hydration spheres, which increases the degree of structure. Increasing pressure forces water molecules closer together, breaking apart hydrogen bonds and decreasing structure. Temperature has the most impact in changing the viscosity of seawater because salinity is nearly constant throughout the oceans and the pressure effect is relatively small. The effect of temperature on the viscosity of water is shown in Figure 6-4.

Light Transmission in Water

Light is electromagnetic energy with wavelengths from about 0.01 to 1000 micrometers (μm), ranging from short wavelength ultraviolet to long wavelength infrared. Light visible to the human eye covers only a very narrow band,

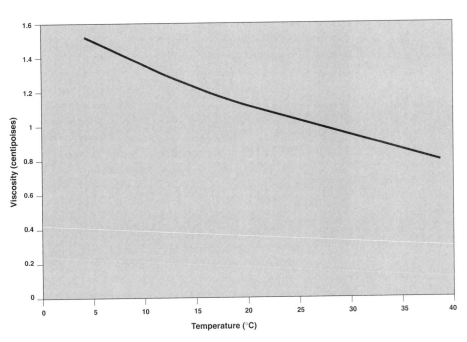

Figure 6-4   Water Viscosity vs. Temperature for Seawater

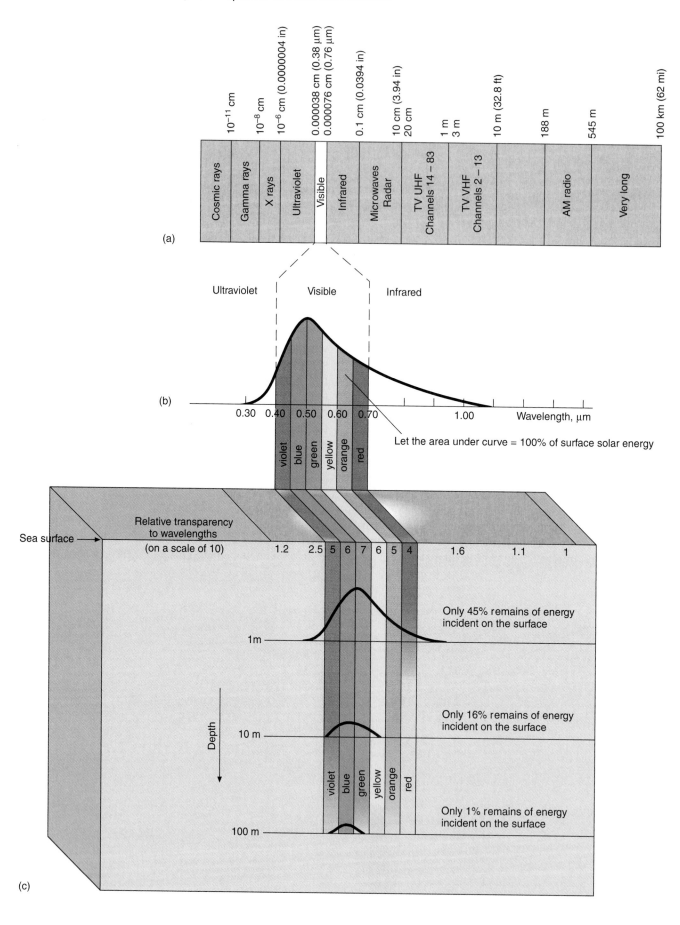

(a)

(b)

Let the area under curve = 100% of surface solar energy

(c)

from 0.38 to 0.76 μm in wavelength. Combined, these visible light waves produce white light, but if they are separated, as a prism does (or raindrops acting as prisms to make a rainbow), we see the separate wavelengths as colors. Violet corresponds to short wavelengths; blue, green, yellow, orange, and red correspond to progressively longer wavelengths.

Matter, depending upon its chemical composition, absorbs or reflects light energy to varying degrees. Objects appear to have color because they absorb or reflect different wavelengths of light to varying degrees. We see the object as having the color of the reflected wavelengths. For example, plants absorb most wavelengths of light except green and yellow, which are reflected, so leaves appear green. However, if certain wavelengths are missing from the light that falls on the object, those colors cannot be seen. If you are underwater, the true colors of objects can be observed only near the surface because the entire visible spectrum is transmitted there. However, with increasing depth, as the red end of the spectrum is absorbed by the water, objects reflect only the remaining blue and green light, thus appearing blue-green. You can even observe this effect to some degree in the deep end of a swimming pool. Underwater photographers carry lights and equip their cameras with special filters to compensate for this effect.

The gases in the atmosphere also absorb and reflect incoming solar energy. The portion of solar energy emission that reaches Earth's surface is shown in Figure 6-5b. The atmosphere is most transparent to the blue and green wavelengths of visible light. Absorption increases rapidly toward the ultraviolet such that only a small amount of ultraviolet is transmitted, but absorption increases somewhat more gradually toward the infrared, transmitting some infrared light up to about 1-μm wavelength.

Liquid water also absorbs light energy. In a column of water, over half the energy incident at the surface is absorbed within the first meter of depth. At 100 meters, only 1 percent of the surface energy remains (Figure 6-5c). Absorption is greatest at longer wavelengths, resulting in the selective loss of the red end of the spectrum with increasing depth. Within the top 10 meters, red wavelengths are absorbed; yellow is absorbed within 100 meters. The very low amount of light remaining below this depth consists of blue and green to about 250 meters; at greater depths, only

blue and a few green wavelengths are present. No light penetrates below 1000 meters.

The transmission of light also is affected by particles suspended in the water column. The depths given in Figure 6-5c for certain percentages of light transmission apply only to clear water. Suspended mud or plankton will scatter and absorb light, considerably reducing the amount transmitted. Thus, for coastal areas where suspended sediment is common, and for areas of high biological productivity, light transmission is considerably less. Much of the light is absorbed above 50 meters.

A Secchi disk is a simple device to measure visible light transmission in the ocean. The disk, which is 30 cm in diameter, is lowered into the water at the end of a cable and the depth at which it begins to disappear is used as a measure of the amount of suspended material in the water column. The shallower the depth, the greater the amount of suspended matter, measured as the turbidity of the water column (Figure 6-6). More quantitative measurements are made with spectrophotometric instruments that measure the amount of light absorption or scattering caused by the suspended particles.

The extent to which light penetrates water is a critical factor in photosynthesis. The algae that are the base of all ocean food chains can only photosynthesize when enough sunlight penetrates the water. Sunlight also is the heat source for the surface ocean and differential heating of the surface ocean layer is an important driving force for ocean circulation.

Sound Transmission in Seawater ....

Sound is transmitted much more efficiently through water than through air. Can you speculate why this is true? The efficiency of sound travel makes it possible to use sound technology to determine the position and distance of objects in the ocean. The basic technique is called **sonar** (Sound Navigation And Ranging). Sonar relies on echoes and measures the time it takes an echo to return to the sound source. Porpoises and whales use their own sonar in hunting and communication.

The average velocity of sound in the ocean is 1450 m/s (4750ft/s). This is more than four times faster than the velocity of sound in air, which is 334 m/s (1100ft/s) in a dry atmosphere at 20°C. Sound velocity in the ocean increases

Figure 6-5   The Electromagnetic Spectrum and Transmission of Visible Light in Water

(a) The electromagnetic spectrum runs from extremely short cosmic rays (left) through ultraviolet, visible light, infrared, microwaves, television and radio broadcasting frequencies, to extremely long radio waves (far right). (b) The narrow portion of the spectrum that is visible to humans as light. Violets and blues correspond to shorter wavelengths; oranges and reds correspond to longer wavelengths. (c) Light passing through water is selectively absorbed. At 1 m, only 45 percent of the original incident energy remains, and more of the red end of the spectrum has been absorbed. At 10 m, only 16 percent of the surface energy is left and the red end of the spectrum is gone. By 100 m, only 1 percent remains, limited to wavelengths in the blue region.

Figure 6-6   Secchi Disk Used to Measure Turbidity

(Photo courtesy of Patricia Deen.)

as temperature, salinity, and pressure increase (Figure 6-7a and b). Sound velocity can be measured directly by use of a velocimeter, or calculated from measurements of temperature, pressure, and salinity.

In many parts of the open ocean, between depths of about 100 and 1000 meters, there is a layer of water in which temperatures decrease rapidly with depth (Figure 6-7c). This is the thermocline layer (see Chapter 7). Because sound velocity decreases as temperature decreases, sound travels more slowly in this layer than in the warmer water above. Below the thermocline, water is of uniformly low temperature. As water pressure increases with depth, the effect of the low temperature in decreasing velocity is more than offset by the pressure effect, and the velocity increases again. The net result is a layer at a depth of about 1 km that has slower sound velocities than the water above and below. Sound waves bend or refract toward low-velocity layers; thus, sound becomes trapped in this layer and can be transmitted over long distances, in fact, across entire ocean basins. This layer is called the **SOFAR** (Sound Fixing And Ranging) **channel**. It is thought that marine mammals make use of this channel

for long-distance communication. It also is of special interest to those concerned with submarine surveillance and warfare.

Thermal Properties of Water

Heat and Changes of State

Water exists on this planet in three **states** or **phases**: solid, liquid, and gas. The temperature and pressure of the environment determine which phase water takes (Figure 6-8). Within each area on the figure, water is present in only one phase. At low temperatures and pressures to the left of lines B and C, water is solid; at high temperatures and low pressures to the right of lines A and C, water is a gas; in the middle region between lines A and B, water is liquid. On the lines forming the field boundaries, water is present in two phases—gas and liquid (line A), solid and liquid (line B), or gas and solid (line C). Water is present in all three phases only at the point (TP) where the three lines intersect. This is the **triple point** on a phase diagram. For water, this point is 0.0098°C and 0.006 atm (4.579 mm) pressure.

To change phase requires the input or removal of energy. If this energy is heat energy, we move left and right in the figure. Input of heat energy can raise temperature or break hydrogen bonds; removal of heat energy causes temperature decreases or hydrogen bond formation. Increasing pressure compresses the volume of space that the molecules occupy or causes hydrogen bond formation; decreasing pressure allows expansion of the volume occupied by the molecules or causes hydrogen bonds to break.

In oceanography, the major energy changes are heat inputs and outputs, therefore, the following sections will focus on the effects of temperature changes and phase changes related to heat energy. The sun is the original source of this energy and the effects of these energy and phase transformations on almost all processes at Earth's surface is truly amazing. Pressure effects become important in the deep ocean, and the organisms that live in or visit that environment have adaptations that allow them to deal with volume and phase changes (Chapter 16).

Heat Capacity

We measure the amount of heat energy added to or removed from substances in **calorie** units. By definition, 1 calorie is the amount of energy that must be added to 1 gram of liquid water to raise its temperature by 1 degree Celsius. The Calorie, the unit used to determine the amount of energy in food, is equal to 1000 calories. Adding energy to a substance either raises its temperature or does work, such as breaking molecular bonds.

The **heat capacity** of a substance is defined as the amount of energy required to raise the temperature of 1 gram of that substance by 1°C. The heat capacity of liquid water is 1 calorie/g/°C. The heat capacities of most other substances are much less than that of water, that is,

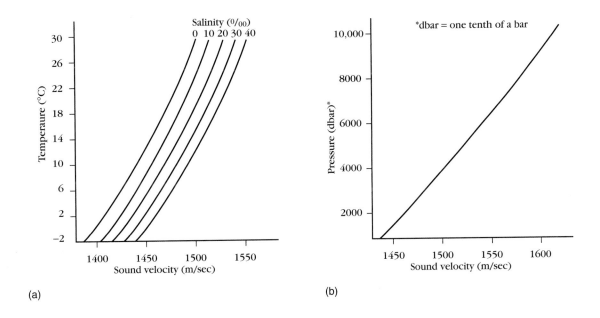

(a)

(b)

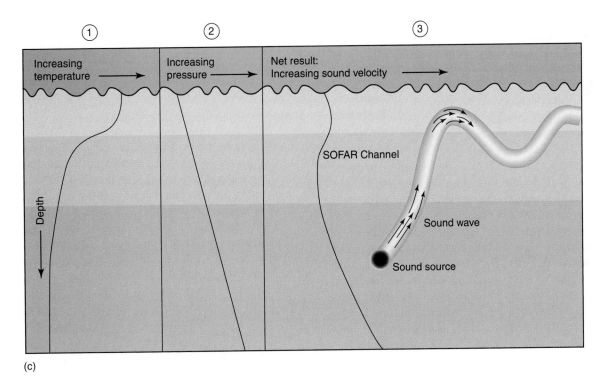

(c)

Figure 6-7   Sound Transmission in the Ocean

(a) Temperature effects on sound velocity at different salinities (b) Pressure effects on sound velocity. (c) Temperature and pressure changes with depth are shown for midlatitudes in the ocean in the first two panels. Resulting sound velocity and the path taken by a sound wave are shown in the third panel. Because water pressure increases steadily with depth, sound tends to travel faster; however, the rapid decline in temperature with depth offsets the effect of the increased pressure. The net result is the lower-velocity sound channel shown at the base of the thermocline. Sound waves bend (refract) into low-velocity areas. This produces shadow zones where velocity maxima exist. The SOFAR (Sound Fixing And Ranging) channel traps sound energy in the low-velocity zone. The sound source beneath the SOFAR channel illustrates how refraction bends sound. For the band of sound waves traveling to the right from the sound source in the bottom layer, the waves in the upper part of the band travel faster than those in the lower part due to increasing water pressure with depth (arrow lengths in diagram are proportional to velocity). The wave band bends upward into the SOFAR channel. As it passes through the SOFAR channel and into the upper layer, the top of the wave band moves faster than the bottom and will bend back down into the channel, trapping the sound in the SOFAR channel.

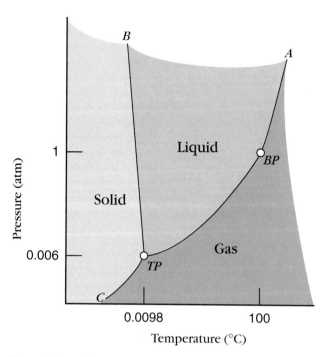

Figure 6-8 **Phase Diagram for Water**

Temperature and pressure cause changes in the phase or state of water. There is only one combination of temperature and pressure at which water can exist simultaneously as gas, liquid, and solid. This is the triple point (TP). The boiling point (BP) at 1 atm (760 mm) occurs at 100°C and is one point on a curve defining the boundary between liquid and gas (steam).

at the point where a phase change occurs, there is no temperature change even though energy is continuously added. The reason the temperature remains constant is that the heat energy is used entirely to break all the intermolecular bonds necessary to complete the phase change. Conversely, when a substance remains in one phase, the removal of energy results in a decrease in temperature except where a phase change occurs. At that point, the energy removed results in bond formation rather than temperature change. Only when all the bonds necessary to complete the phase change have formed does the temperature begin to decrease again.

The **latent heat of melting** is the amount of heat that must be added to 1 g of a substance, at its melting point, to change it completely from solid to liquid. Similarly, the **latent heat of vaporization** is the amount of heat that must be added to 1 g of a substance, at its boiling point, to change it completely from liquid to gas. When these processes are reversed, the terms are **latent heat of fusion** for a liquid becoming a solid, and **latent heat of condensation** for a gas becoming a liquid. Note, however, that the quantity of heat energy involved in these phase changes is the same regardless of the direction of the phase change.

latent heat of vaporization = latent heat of condensation

latent heat of melting = latent heat of fusion

The quantity of heat energy is the same because the same number of bonds must be formed or broken to complete the phase change regardless of the direction.

To help clarify the concept of latent heat, we will follow the phase change of water from a solid at −40°C, through liquid and vapor states, to a temperature above 100°C. In the lower left corner of Figure 6-10, we start with 1 g of ice at −40°C (this arbitrary starting point must be below the **freezing point**, 0°C). The addition of 1 calorie of heat raises the temperature of the ice by 2°C (the heat capacity of ice = 0.5 cal/g/°C); it takes 20 calories of heat to raise the temperature to 0°C. With the ice at 0°C, the addition of heat does not increase the temperature (the plateau in the graph) until 80 calories of energy have been added. This added energy goes into breaking intermolecular bonds in the ice. The heat required to convert 1 g of ice into 1 g of liquid water at 0°C is the latent heat of melting—80 calories.

After the change from ice to liquid water at 0°C, the addition of 1 calorie of heat to 1 g of water increases the temperature by 1°C (the heat capacity of water = 1 cal/g/°C). We must therefore add 100 calories before 1 g of water reaches 100°C, the **boiling point**. At this temperature, we reach another plateau where 540 calories must be added to convert liquid water to water vapor. Again, the added heat energy is used to break intermolecular bonds in the liquid water. The heat required to convert 1 g of liquid water to 1 g of water vapor at 100°C is the latent heat of vaporization —540 calories. After the change from liquid to vapor, any further addition of heat raises the temperature of the vapor

the same input of energy would increase their temperatures more. Put another way, for the same observed temperature change, water has the capacity to store much more energy than other substances. The difference in the heat capacities of ocean water and the rocks that make up the continents is strikingly illustrated by the difference in day and night temperatures for ocean and land areas (Figure 6-9). Throughout most of the ocean, day and night temperatures vary by less than 1°C because the high heat capacity of water easily absorbs the daily gains and minimizes the daily losses of heat energy. By contrast, the much lower heat capacity of continental rocks, soil, and vegetation can cause day and night temperature differences of 15 to 30°C. The high heat capacity of water moderates these temperature shifts in coastal regions.

Latent Heats of Melting and Vaporization

Closely related to water's unusually high heat capacity are its high latent heat of melting and its high latent heat of vaporization. *Latent* means hidden, and in this case, it refers to heat or energy stored in the molecules with no resulting temperature change. This energy is apparent only during a change of phase—when ice turns to liquid water or vice versa, and when liquid water changes to water vapor or vice versa—in other words, when bonds are being broken or formed. When a substance is in one state, any energy input results in a rise in temperature. However,

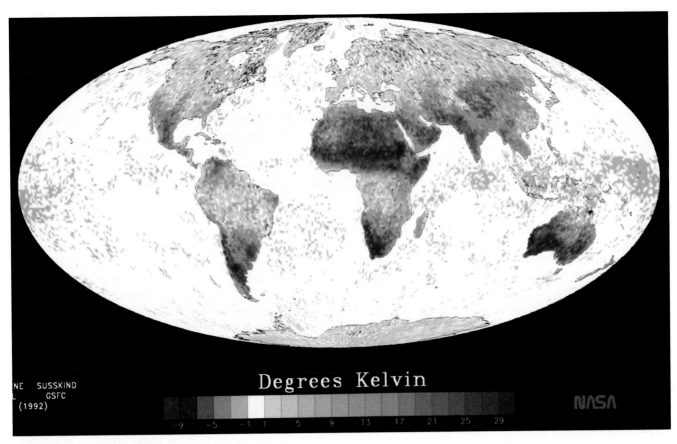

Figure 6-9   Day and Night Temperature Differences

This map shows differences between day and night temperatures, averaged for the month of January 1979. In white areas of the open ocean, there is only a 1°K difference in day and night temperatures (note Kelvin temperature scale). In land areas such as South America, Africa, and Australia, there is a difference of up to 30°K from day to night. This dramatically demonstrates the effects of the high heat capacity of water vs. the low heat capacity of land. The map was made by subtracting the temperatures at 2 a.m. from the temperatures at 2 p.m.   *(Courtesy of Moustafa T. Chahine, Jet Propulsion Laboratory, California Institute of Technology.)*

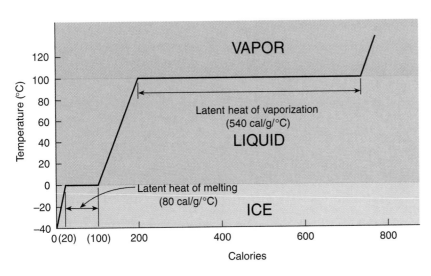

Figure 6-10   Latent Heats and Change of State of Water

Graph showing the change in temperature of water with energy input. During phase conversions, there is no temperature increase because the added energy is used to break intermolecular bonds. The latent heat of melting (80 cal/g/°C) is much less than the latent heat of vaporization (540 cal/g/°C).

at the rate of 2.3°C per calorie (the heat capacity of water vapor = 0.44 cal/g/°C).

We must look at bonding patterns among water molecules to understand why so much more heat is needed to convert 1 g of liquid water to water vapor (540 calories) than to convert 1 g of ice to liquid water (80 calories). In ice, each water molecule is bonded to four other molecules, with each oxygen attracting two hydrogens, and each hydrogen attracted to another oxygen (Figure 6-11a). This creates an open hexagonal lattice structure, manifested in the beautiful patterns of snowflakes (see Chapter 6 front photo). As ice melts, only a few hydrogen bonds have to be broken for the structure to collapse. Once the collapse has occurred, the phase change is complete even though many hydrogen bonds may be unbroken. In contrast to the rigidity of ice, the arrangement of these bonds in liquid water shifts continually such that clusters of water molecules form and disintegrate continuously. When water changes from the liquid to vapor state, much more energy is required because all of the hydrogen bonds in these clusters must be broken (Figure 6-11b and c).

What does all this have to do with the oceans? We know that the above descriptions of phase changes are not applicable to some of the most important ocean-atmosphere processes. For example, we know that water vapor in the atmosphere forms through evaporation of liquid water at Earth's surface. This phase change occurs at 1 atm pressure, but the temperature of the water is nowhere near 100°C! We also sometimes see ice changing directly to vapor, a process called *sublimation*. How do these phase changes occur?

The answer lies in the fact that all molecules contain **kinetic energy** and are in constant motion. On average, this motion acts to maximize the distance between neighboring molecules such that molecules in a liquid tend to escape to the vapor phase where the density of molecules is lower. The amount of kinetic energy each molecule contains increases with increasing temperature. At low temperatures, most molecules in the liquid do not have enough kinetic energy to escape, so they remain in the liquid phase. However, at 100°C and 1 atm. pressure, most all molecules in the liquid have enough kinetic energy to escape, or evaporate, from the surface of the liquid into the gas phase, and the input of 540 cal/g of water is required only to break the hydrogen bonds. At 20°C, 585 cal/g are required to convert liquid water to a vapor.

The Sun provides most of the energy input to the ocean surface, breaking bonds and increasing kinetic energy. The greater the kinetic energy, the more frequently neighboring molecules collide with each other, allowing transfer of energy between molecules. Water molecules that obtain this additional energy from their neighbors escape to the vapor phase. Important results from this process are: (1) evaporation leads to cooling of the residual molecules that have lost heat energy to their escaping neighbors; (2) the heat energy is transferred along with the escaped molecules from the liquid water to the gas phase (i.e., to the atmosphere). Sublimation occurs in the same way except that the required energy inputs are much higher.

Chemists measure the tendency of molecules in a liquid to escape into the gas phase as **vapor pressure**. Figure 6-12 shows how the vapor pressure of water depends on temperature. The vapor pressure describes the amount of water vapor (in pressure units) that air can hold at a given temperature. The concept of expressing amounts of gas as pressures is not so strange when we remember that, by definition, the atmosphere as a whole exerts a pressure of 1 atmosphere (atm) on Earth's surface. If a compound

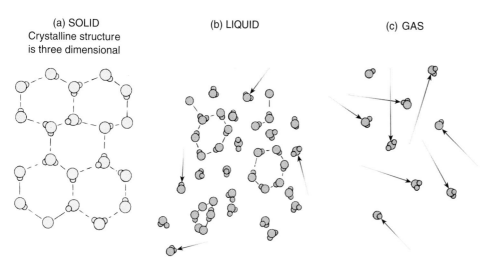

(a) SOLID
Crystalline structure
is three dimensional

(b) LIQUID

(c) GAS

Figure 6-11   Hydrogen Bonds in H₂O

(a) Below 0°C there is too little heat energy to permanently break the hydrogen bonds that bind water molecules to one another. (b) Between 0° and 100°C there is sufficient heat energy to make molecules active enough to break many of the hydrogen bonds. (c) Above 100°C heat energy makes molecules so active that hydrogen bonds cannot hold any molecules together.

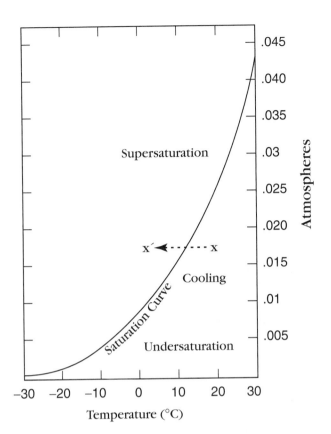

Figure 6-12   Vapor Pressure Dependence on Temperature

The dashed arrow (x to x') shows the path from undersaturation to supersaturation as air cools from 20 °C to 0 °C at a constant water vapor pressure (constant amount of water in air). Supersaturated air cannot hold the water, and it falls as rain or snow.

makes up 20 percent of the atmosphere, it is providing 20 percent of the pressure, or 0.2 atm. The curve in Figure 6-12 shows that, at 20°C the amount of water vapor in the atmosphere is about 0.017 atm or 1.7 percent. As the temperature of the air increases, (moving up the curve) the atmosphere can hold more water because the vapor pressure for water is higher. As temperature decreases (moving horizontally off the curve to the left), it rains because water must be removed from the atmosphere.

Evaporation from the oceans provides a critical moderating effect on the latitudinal heat gradient. Evaporation occurring at warm, low latitudes transfers heat to the atmosphere. Winds carry this moisture-laden air to higher latitudes where lower temperatures cause the water to condense and fall as rain or snow, thereby releasing the heat to the air.

The Water Cycle

Every year, the upper 1 meter of the ocean surface evaporates to the atmosphere. This water returns to Earth's surface as rain. Some rain lands in the oceans and some lands on the continents. The water that falls on land runs off into rivers

and returns to the ocean. This transfer of water from ocean to atmosphere to continents and back to the ocean is called the surface **water cycle**, or **hydrologic cycle** (Figure 6-13). The cycle consists of reservoirs and fluxes. A **reservoir** is a location where water is found; a **flux** is the reaction or pathway that takes water from one reservoir to another. Each reservoir has a size, given in mass units (e.g., grams); each flux is a rate given in mass change per unit time (e.g., grams added/year). When the total mass of the element does not change within the cycle, the cycle is said to be a **closed system**—that is, no mass is gained from or lost to the outside. Based on the assumption that no significant mass is gained from or lost to space or Earth's interior, scientists usually model the hydrologic cycle as closed.

Of the total yearly evaporation of water from Earth's surface, 86 percent evaporates from the oceans and 14 percent from the surface of the continents. When this water leaves the atmosphere as precipitation, 79 percent falls directly back into the oceans and 21 percent falls onto the continents. Except for the amount that returns directly to the atmosphere via evaporation, this continental water eventually returns to the oceans by river runoff or groundwater seepage.

The hydrologic cycle is obviously a dynamic process, but by taking a "freeze-frame," we can determine how much of the Earth's water is present in specific reservoirs of the hydrologic cycle. Today, we find that approximately 96 percent of the water on Earth is contained in the ocean basins, 3 percent is tied up in glaciers and polar ice, 1 percent is groundwater, about 0.01 percent is in rivers and lakes, and only 0.001 percent is in the atmosphere (Table 6-1). The rates at which water is transferred among these reservoirs are shown in Figure 6-13. These are the fluxes of the hydrologic cycle. Evaporation, precipitation, and river inflow to the oceans are examples.

Given that the oceans have remained about the same size, there have never been large differences in the hydrologic cycle through geologic time; however, small but

TABLE 6-1   Reservoirs of the hydrosphere

Reservoir	Volume (10^6 km³)	Percent of total
Oceans and sea ice	1400	95.96
Glaciers and land ice	43.4	2.97
Surface waters		
Lakes	0.125	0.009
Rivers	0.0017	0.0001
Subsurface waters		
Groundwater	15.3	1.05
Soil Moisture	0.065	0.0045
Atmosphere	0.0155	0.001
Biosphere	0.002	0.0001
Total	**1459**	**100**

(Berner and Berner, 1987, 1996; Drever, 1988)

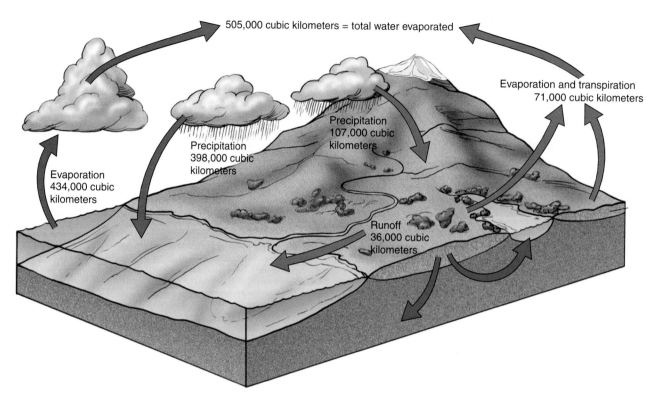

Figure 6-13 The Hydrologic Cycle

All of Earth's water is in motion. Water is removed from the liquid reservoirs (oceans, rivers, and lakes) by evaporation and it returns to them by precipitation. The atmosphere transports water vapor from the oceans to the continents, where it is deposited as precipitation and returns to the oceans by stream flow or groundwater runoff. Flux rates are /yr. *(Modified from The Earth, 4th ed. (Figure 10.2), by Edward J. Tarbuck and Frederick K. Lutgens. Copyright © 1993 by Macmillan Publishing Company).*

critical differences can be brought about by, or result in, large changes in Earth's climate. During glacial periods in Earth's history, sea level has decreased more than 120 meters, resulting from a transfer of 5 to 6 percent of the Earth's water from oceans to glaciers. If Earth's climate became warm enough to transfer all the present glacial reservoir to the sea, sea level would rise about 60 meters. Water vapor in the atmosphere has nearly double the effect of all the other greenhouse gases combined, therefore, changes in the size of the atmospheric reservoir may play a major role in driving climate change. We will explore these concepts further in Chapter 8.

Residence Time

Residence time is the average length of time that water, or another chemical substance, resides in a reservoir. It can be calculated from the formula:

$$\text{Residence time} = \frac{\text{Total amount of substance in reservoir}}{\text{Sum of the fluxes into or out of reservoir}} \quad (6\text{-}1)$$

The denominator assumes that the sum of all the influxes (processes that add water to the reservoir) are equal to the sum of all the outfluxes (processes that remove water from the reservoir). If additions were greater than removals, the reservoir would increase in size over time; similarly, if removals were greater than additions, the reservoir would shrink over time. Thus, another assumption in calculating residence time is that the size of the reservoir is constant.

From Table 6-2, the fluxes to and from the continental reservoir are:

TABLE 6-2 Annual fluxes between reservoirs	
Pathway	Volume (10^3 km^3 yr^{-1})
Atmosphere to ocean	398
Atmosphere to continent	107
Continent to atmosphere	71
Ocean to atmosphere	434
Continent to ocean	36

(Berner and Berner, 1987, 1996; Drever, 1988)

Flux to continents from atmosphere: 107×10^3 km³/yr

Flux from continents to atmosphere: 71×10^3 km³/yr

Flux from continents to ocean: 36×10^3 km/yr

The size of the continental water reservoir is constant because:

$$107 \times 10^3 \text{ input} = (71 + 36) \times 10^3 \text{ output} \qquad (6\text{-}2)$$
$$= 107 \times 10^3 \text{ output}$$

Using the data in Table 6-1 and 6-2, we can also calculate the residence time of water in the atmosphere. For the atmosphere, the sum of influxes and outfluxes are equal:

$$434 + 71 = 505 \text{ km}^3/\text{yr input} \qquad (6\text{-}3)$$

$$398 + 107 = 505 \text{ km}^3/\text{yr output} \qquad (6\text{-}4)$$

The residence time is:

$$\text{Residence time} = \frac{0.0155 \times 10^6 \text{ km}^3}{505 \times 10^3 \text{ km}^3/\text{yr}} \qquad (6\text{-}5)$$
$$= 0.0307 \text{ yr}$$
$$= 11.2 \text{ days}$$

Because influxes and outfluxes are not precisely known for every reservoir, some residence times can only be approximated. Residence times for water are shown in Table 6-3.

The reservoir size relative to the flux rates is the important control on residence times. Large reservoirs and small flux rates result in very long residence times, whereas small reservoirs and large flux rates result in very short residence times. In Table 6-3, we see that the deep ocean and deep groundwater have very long residence times (on the order of thousands of years), whereas water in the atmosphere has a very short residence time (about 11 days).

To illustrate this idea further, let's add something to the water—say a nonreactive pollutant. If the water containing the pollutant is in the atmosphere, it will only stay there an average of 11 days before moving to another reservoir. If it moves into a river, its average stay will also be about 11 days (0.03 years). But, if it moves into deep

TABLE 6-3	Residence times of water in reservoirs
Reservoir	**Residence time (years)**
Oceans and sea ice	
Shallow	100's
Deep	1000–2000
Glaciers and land ice	10–1000's
Groundwater	
Extractable	0.003–1000
Deep	10,000's
Lakes	1–3
Rivers	0.03
Soil moisture	0.04–1
Atmosphere	0.0307

(Berner and Berner, 1987, 1996; Drever, 1988)

groundwater or deep ocean water, the water and its dissolved pollutant will stay for thousands of years. This fact has motivated scientists to design ways of increasing the flux of CO_2 from the atmosphere to the deep ocean as a way of combating the increased atmospheric CO_2 from fossil fuel burning. Of course, the fact that CO_2 is very reactive seriously complicates the problem, unlike our simple example of a pollutant that is carried passively by water. More on this in Chapter 8!

Residence times are also useful for estimating the impact of pollution on isolated or semi-isolated bodies of water, such as lakes, estuaries, and coastal embayments. Residence times show roughly how long it takes for all the water to be replaced, or exchanged. If replacement happens rapidly, then the impact of a polluting event, such as an oil spill, will be less than if it happens slowly. For example, the impact of the oil spill in Prince William Sound in Alaska was less in those areas with short residence times (rapid exchange with the ocean) than in more protected areas. The oil spills in the Persian Gulf are particularly devastating because the gulf water has a very long residence time. We will explore these issues in greater detail in Chapter 18. See "Toxic Chemicals in Seawater" on page 186 for a discussion of the effects of manmade chemicals on marine organisms.

Summary

Water is very unusual because it exists on Earth in three states (solid, liquid, gas), and it has the capacity to store great amounts of heat energy. Hydrogen bonds (the bonds between water molecules resulting from their bipolar nature) play the major role in giving water its unusual properties. Hydrogen bonds account for the unusual thermal properties of water, such as its high freezing point (0°C) and boiling point (100°C), high heat capacity (1 cal/g/°C), high latent heat of melting (80 cal/g), and high latent heat of vaporization (540 cal/g). Hydrogen bonds are also a major factor in water's high surface tension and viscosity.

With increasing depth in the ocean, the colors of the visible spectrum are absorbed by ocean water in a way that removes red and yellow wavelengths at relatively shallow depths; blue and green wavelengths are removed at greater depths. Turbidity refers to the amount of suspended material in water; high turbidity greatly reduces the depth to which light can penetrate.

The velocity of sound transmission in the ocean increases with temperature, salinity, and pressure. A low-velocity SOFAR sound channel can conduct sound over great distances in the oceans.

The ocean contains about 96 percent of Earth's surface water. Water is recycled between the ocean, atmosphere, and continents by the hydrological cycle, which involves evaporation, precipitation, and surface and groundwater runoff from the continents.

Key Terms

Boiling point	Hydrogen bond	Light	States
Calorie	Hydrologic cycle	Phases	Surface tension
Capillarity	Kinetic energy	Polarity	Triple point
Closed system	Latent heat of condensation	Reservoir	Vapor pressure
Flux	Latent heat of fusion	Residence time	Viscosity
Freezing point	Latent heat of melting	SOFAR Channel	Water cycle
Heat capacity	Latent heat of vaporization	Sonar	

Questions and Exercises

1. Explain why water molecules are polar.

2. Explain how the hydrologic cycle transfers heat energy.

3. Explain how hydrogen bonding among water molecules affects the physical properties of water.

4. Why are the freezing and boiling points of water higher than would be expected for a compound of its molecular makeup?

5. How does the heat capacity of water compare with that of other substances? Describe the effect of this characteristic on climate.

6. Heat energy added as latent heat of melting and latent heat of vaporization does not increase water temperature. Why? Why is the latent heat of vaporization so much greater than the latent heat of melting?

7. How does hydrogen bonding produce the surface tension phenomenon of water?

8. How does the decrease in temperature at the base of the warm surface water produce a SOFAR, or sound channel, below the ocean's surface?

9. Explain the link between climate and the kinetic energy in water molecules.

10. Compare and contrast light and sound transmission in water.

References

Berner, E.K., and R. A. Berner, 1996. *Global Environment: Water, Air, and Geochemical Cycles*. Upper Saddle River, NJ: Prentice-Hall.

Davis, K. S., and J. S. Day, 1961. *Water: The Mirror of Science*. Garden City, NY: Doubleday.

Kuenen, P. H., 1963. *Realms of Water*. New York: Science Editions.

Kevelle, R., 1963. Water. *Scientific American* 209:93–108.

Suggested Reading

Baker, J. A., and D. Henderson, 1981. The fluid phase of matter. *Scientific American* 245:5, 130–39. The structure of gases and liquids is modeled using hard spheres.

Buckingham, M. J., J. R. Potter, and C. L. Epifanio, 1996. Seeing underwater with background noise. *Scientific American* 274:2, 86–91. A technique called acoustic-daylight imaging uses ambient sound in the ocean instead of light to form images. The resolution may be a bit less than for seeing in daylight, but the potential is very exciting.

Gabianelli, V. J., 1970. Water: The fluid of life. *Sea Frontiers* 16:5, 258–70. The unique properties of water are explained.

Covers hydrogen bonds, capillarity, heat capacity, ice, and solvent properties.

Gerstein, M. and M. Leavitt, 1998. Simulating water and the molecules of life. *Scientific American* 279:5, 100–105. Computer modeling of molecular interactions between water and biological substances such as proteins and DNA.

Wettlaufer, J. S., and J. G. Dasy, 2000. Melting Below Zero. 282:2, 50–53. Ice is coated with a layer of quasiliquid water as a result of a process of surface melting—even at temperatures well below freezing. The consequences of this condition are discussed.

Oceanography on the Web

Visit the Introductory Oceanography home page for on-line resources for this chapter. There you will find an on-line study guide with review exercises and links to ocean- ography sites to further your exploration of the topics in this chapter. Introductory Oceanography is at http://prenhall.com/thurman (click on the Table of Contents menu and select this chapter).

Global Warming and Sea Level Rise

- What factors cause changes in sea level?
- How can scientists estimate the magnitude of sea level rise given the thermal expansion of seawater and melting of ice associated with sustained global warming?
- How can we assess the severity of impact of sea level rise to shallow coastal communities and human populations given the best estimate of sea level rise over the next 100 years?

Introduction

2:06 am 02/22/2005; You wake up abruptly as police sirens begin blaring outside of your home. As you stumble to turn on the light, you hear something being said on a bullhorn about evacuation. A minute later, air raid sirens join in the frenzy. You turn on the television to catch the news, and there are images of Antarctica, and ice, and enhanced satellite images. What could possibly be the connection between the peaceful and serene images of the frozen continent and all of this noise? You listen carefully to the live news report. "At 11:53 PM yesterday evening—slightly more than two hours ago, a large portion of the West Antarctic Ice sheet slid violently into the adjoining polar ocean." You think, "Well that's great news, but what does it have to do with me, and why all of this fuss?" The news anchor continues, "All residents of coastal states living within 200 miles of the ocean or coastal bays are requested to evacuate immediately. The National Oceanographic and Atmospheric Administration has predicted that the water displacement generated by the slide of this ice sheet into the ocean will send a tsunami wave of 10 meters along the edge of the Atlantic Ocean basin. It is predicted that the South Atlantic regions of Florida, Georgia, and South Carolina will be impacted by this wave in roughly 16 to 17 hours. This evacuation will be permanent because the global sea level after the wave passes will be 6 meters above current sea level. We now transfer you to the press room of the White House for an emergency address by the President." You pinch yourself hard on the cheek thinking this must be a dream, this is impossible! But is it?

Background

It is thought that water began to accumulate on Earth's surface between 4.2 and 4.4 billion years ago, after the crust cooled below the boiling point of water. Water that now covers 71% of the earth's surface was derived from two main sources, outgassing of water-laden volcanic gases, and from fragments of comets impacting the upper atmosphere. Compared to the early Earth's history, the rates of input of water onto the earth's surface has slowed, and it is now thought that the volume of water (in all its states) on the surface has been relatively constant for several billion years. This implies that there are also mechanisms causing a slow loss of water away from Earth's surface. These losses could be due to recycling of Earth's crust and

water-laden ocean sediments, as well as the slow escape of water vapor from the upper atmosphere into space.

This relative constancy of water mass on the planet over the last several billion years does not, however, mean that sea level has remained constant. Still, the 220-meter range of sea level variation over this period is small compared to the mean depth of the ocean–3,800 meters.

Sea Level Changes

Changes in sea level, which have occurred over most of the earth's geologic history, are due to two processes. Eustatic processes change the absolute amount of liquid water within the ocean basins. The main mechanism driving eustatic changes in sea level is the re-proportioning of water between liquid and solid phases (ice) due to changes in global climate. Isostatic processes change the underlying topography of the sea floor. These changes can occur on either regional scales, as in the rebound of crust after deglaciation, or in the slow subsidence of deltas at passive continental margins, or on global scales, as in periods of marked increases in sea floor spreading rates. This increases the height of ridge and rise features throughout the global ocean basin, in turn causing displacement of water upward onto the coastal continental landscape.

Changes in sea level have been implicated directly and indirectly as contributing to mass extinction events that have occurred within the earth's geologic past. Rapid sea level decline has even been hypothesized to cause changes in atmospheric oxygen levels. In this case, rapid decay (oxidation) of shallow, newly exposed organic-rich marine deposits would remove oxygen from the atmosphere.

Sea level has been rising since the end of the last glaciation about 15,000 years ago. The rate of global sea level rise for the last 100 years has been 2 mm/year and 15–20 cm total (0.08 in/year, 6–8 in total). Estimates for global sea level rise for the next century suggest that this rate will double to 4 mm/year and 40–45 cm total (0.16 in/year, 16–18 in total). The observed rate of coastal sea level rise varies from region to region because of the variability in isostatic crustal movement (Figure 1). If future increases in sea level become more rapid, shallow water intertidal or subtidal communities may not be able to keep pace with sea level change and will die off as environmental conditions change beyond their limits of tolerance. Such a change is now occurring in intertidal salt marsh environments around the Mississippi River

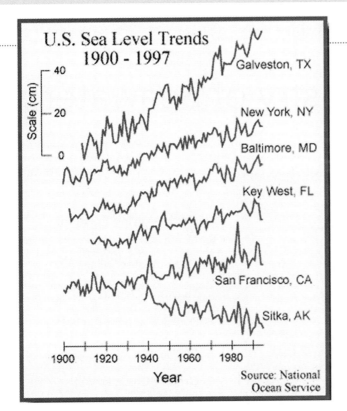

Figure 1

Trends of sea level rise within coastal city regions of the United States. Note the higher rate of sea level rise for Galveston, Texas which is experiencing coastal subsidence, whereas in Sitka, Alaska sea level appears to be falling because of coastal uplift.   *(NOAA National Ocean Service)*

delta in Louisiana, as the coastal landscape subsidence combined with global sea level rise exceeds the rate of vertical growth of the marsh community.

Impacts of Global Warming

The newfound concern about global warming impact stems from the virtual consensus among climatologists that the planet is experiencing a period of global warming.

Global warming may result in an increase in the rate of sea level rise due to thermal expansion of seawater and the melting of ice in glaciers and polar regions (Figure 2). The coefficient of thermal expansion of seawater is 0.00019 per degree Celsius, meaning that if a volume of seawater occupied 1 cubic meter of water (1000 L, 264.20 gal), after warming by 1°C, it would expand to 1.00019 m^3 (1000.19 L, 264.25 gal). Translated over the mean depth of the ocean (3.8 km, 2.4 mi), an increase in temperature of 1°C will cause a sea level rise of about 70 cm (28 in). Whereas thermal expansion acts upon water already in the basin, contributions from melting ice represent new, added water to the present ocean volume. The melting of ice that is currently perched upon land, as in the ice sheets of Greenland, Iceland, and the Antarctic, has the potential to raise sea level considerably, by about 80 meters (262 ft). Ice that is already floating in the ocean water, as in

Figure 2

Landsat 1 MSS digitally enhanced image of the Byrd Glacier in Antarctica, where it joins the Ross Ice Shelf. Melting of glaciated areas will directly increase sea level, whereas the melting of shelf ice will not cause a rise in sea level but may be an important indicator of regional climate change.   *(NOAA Central Library Photo Collection)*

the Arctic ice mass, Antarctic ice shelves, and much smaller icebergs, may melt but will not contribute to sea level rise since the mass of water contained in these features already displaces its equivalent water volume.

There is ample evidence that the increase in melting has begun. Monitoring of mountain glaciers throughout the world over the last two decades indicates that 75% of them are losing mass and that the rate of this loss has almost doubled, from 0.25 m to 0.50 m (0.82 to 1.64 ft) of water equivalent, in the last 20 years. In 1991, NASA reported that the extent of sea ice in the Arctic Ocean declined by 2% between 1978 and 1987. More recently, Norwegian scientist Ola Johannessen, of the Nansen Environmental and Remote Sensing Center, has presented satellite measurements of microwave emissions that show declines in permanent Arctic ice of 7% over each of the past two decades.

In the Antarctic, a series of ice shelves (Wordie, Larsen a, and Larsen b) have collapsed over the last decade spurred by the 2.5°C increase in the average temperature on the peninsula since the mid-1940s. There is some concern that loss of these ice shelves may destabilize continental ice sheets, with the Western Antarctic Ice Sheet being considered as potentially capable of slippage into the surrounding ocean.

To see which areas will be impacted and to continue the analysis, go to our web site.

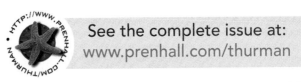

See the complete issue at:
www.prenhall.com/thurman

7 The Chemistry of Seawater

- The Dissolving Power of Water
- Ocean Water Salinity
- Seawater Density
- Dissolved Gasses in Water
- Acidity and Alkalinity
- Organic Chemistry of Seawater
- Variations in Seawater Composition with Depth and Latitude
- Source of Salts
- Biogeochemical Cycles
- Summary

One of the two rosettes carried aboard Scripps Institution of Oceanography's research vessel Melville *during GEOSECS Pacific operations shows nonmetallic bottles used in collecting seawater samples. The ends of each bottle can be triggered to close by the scientist operating the shipboard console. (Photo by Peter Wiebe courtesy Woods Hole Oceanographic Institution.)*

ater is often called the universal solvent because it dissolves so many substances in significant quantities. Ocean water contains substances dissolved out of the atmosphere, from rocks on the continents, and from rocks and sediments below the sea floor. Remarkably, the concentrations of the major chemical components of seawater are invariant throughout the world's oceans, but minor components can vary significantly. This constancy of composition indicates that the mixing of ocean water occurs more rapidly than the geologic processes that regulate the major components of ocean salinity. The patterns of variation of minor elements give clues to the more rapid processes that control their concentrations.

The Dissolving Power of Water

We are constantly reminded of how well water conducts electricity in the warnings to keep household electrical appliances away from the bath water. However, pure water is actually a very poor conductor of electricity because water molecules have no net charge. In the presence of an electrical field, water molecules simply rotate in place until they become oriented with their positively charged hydrogen nuclei toward the negative pole and their negatively charged oxygen nuclei toward the positive pole of the field. This orientation of the water molecules tends to neutralize the electrical field to 1/80th of its strength out of water. Only when other substances are dissolved in it does water conduct electricity because the dissolved salt ions move through the water and disrupt the ability of the water molecules to orient themselves.

Recall that chemical bonds involve electrostatic attraction between the component ions. When a salt is added to water, the electrostatic attraction between the ions that make up the salt is greatly reduced, and the salt dissolves. The dissolved salt ions are surrounded by water molecules oriented according to the charge of the ions in this way: negative oxygen ends of the water molecules point toward a positive ion, and positive hydrogen ends of the water molecules point toward a negative ion (Figure 7-1). This cover of water molecules, termed a **hydration sphere**, effectively shields the dissolved ions from other ions still in the solid salt, and thus, more salt ions travel into solution.

The high dissolving power of water is therefore a consequence of its ability to reduce the electrostatic attraction of ionic solids and to form hydration spheres around dissolved ions. Both of these abilities are the result of hydrogen bonding and the polarity of the water molecule.

Ocean Water Salinity

Ocean water tastes salty, but figuring out where the salt comes from and why the salt composition of the oceans is remarkably constant worldwide are very complex problems. The total amount of salt dissolved in ocean water averages about 35 parts per thousand (35‰). This means that there are 35 grams of salt in every 1000 grams of sea-

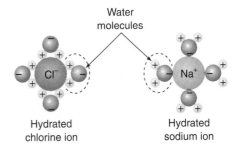

Figure 7-1   Structure of Water as a Solvent
Water molecules arrange themselves around ions in solution, forming hydration spheres. Negative ions, such as Cl^-, are surrounded by water molecules with hydrogens pointed inward; positive ions, such as Na^+, are surrounded by water molecules with hydrogens pointed outward.

water. (If you weighed out 1000 g of seawater and set it aside to evaporate, you would find that the salts left would weigh 35 g) The **salinity** of seawater is defined as the amount of dissolved solids in seawater. In our example, seawater would have a salinity of 35‰.

The major salt components of seawater are present in about the same amounts throughout the world's oceans. This **rule of constancy of composition** was established by English chemist George Forchammer in 1865, confirmed by William Dittmar in his analysis of the Challenger expedition samples, and has been verified over and over again in the countless seawater analyses done since. Six ions account for over 99 percent of the dissolved components in seawater: chloride, sodium, sulfate, magnesium, calcium, and potassium (Table 7-1 and Figure 7-2). Because of

TABLE 7-1 Composition of seawater.

Ion	% (by weight)	g/1000 g seawater (%)
Na^+	30.66	10.77
Mg^{2+}	3.65	1.29
Ca^{2+}	1.17	0.41
K^+	1.13	0.40
Sr^{2+}	0.023	0.008
Cl^-	55.02	19.35
SO_4^{2-}	7.71	2.71
HCO_3^-	0.30	0.12
Br^-	0.19	0.067
CO_3^{2-}	0.046	0.016
$B(OH)_4^-$	0.023	0.008
F^-	0.004	0.0014
OH^-	0.0004	0.0001
$B(OH)_3$	0.055	0.019
Trace components	0.02	0.007
Total	**100**	**35.0**

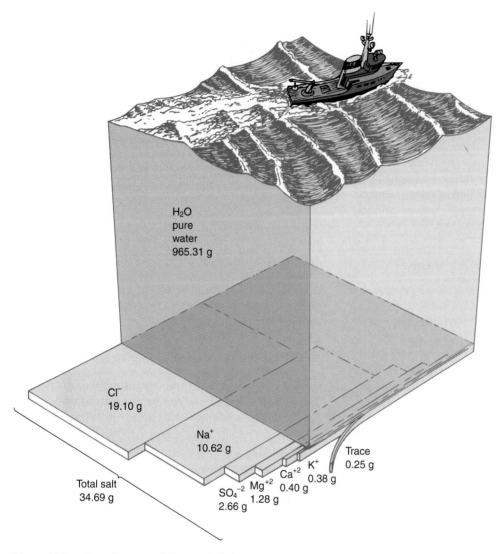

Figure 7-2　Constituents of Ocean Salinity

The components of 1 kg of seawater. Note that the 965.31g of pure water plus the 34.69 g of salts equal 1000.00 g. Rounding, average seawater contains about 965 parts per thousand (‰) water and 35 ‰ salt.

seawater's constancy of composition, we only need to measure the concentration of one major constituent to determine the salinity of a given water sample. The easiest ion to measure is chloride, and for many years, the salinity of seawater was calculated from the chloride ion concentration using the formula:

$$\text{Salinity (‰)} = 1.80655 \times \text{Cl}^- \text{ (‰)} \qquad (7\text{-}1)$$

Modern techniques utilize the fact that the electrical conductivity of water changes with the amount of dissolved salts. Thus, salinity can be easily and accurately determined by conductivity measurements. Modern salinometers can determine salinity by conductivity to within 0.003‰.

Seawater Density

Density refers to how tightly the molecules of a substance are packed together in a specified space, and it is usually expressed in units of mass per unit volume (for example, g/cm³). Density is important in considering the movement of water in the ocean because denser water sinks and less dense water rises. Furthermore, anything denser than water will sink, but a substance that is less dense than water will float.

Density is affected by state (gas, liquid, or solid), temperature, salinity, and pressure. The pressure effect for liquid water is very small, however, and for most purposes, ocean water is considered to be incompressible. Every 10 meters of water depth produces approximately

1 additional atm of pressure; thus, at a depth of 4000 meters, the pressure is about 401 atm (400 atm pressure from the water plus 1 atm from the overlying atmosphere). If 1 ml³ of seawater were lowered to this depth, it would only be compressed to about 98 percent of its volume at the surface.

The effects of temperature and salinity on seawater density are much greater than the effects of pressure. Most substances become denser as they become colder, so that their solid phases are denser than their liquid phases, which in turn are denser than their gas phases. This density increase occurs because the same number of molecules occupies less space as they cool down and lose energy. This condition, called *thermal contraction*, occurs in water. As pure water cools, it becomes denser, at least until it reaches a temperature of 4°C. Below 4°C, however, we observe that the density of liquid water decreases—in other words, the water stops contracting and actually ex-

pands. This is very unusual behavior for a chemical compound. The result is that the volume of a given quantity of water is greater in the solid phase (ice) than in the liquid phase, just the opposite of most other substances. Ice floats, whereas the solids of other compounds sink to the bottom of their liquids.

We can explain this curious behavior by considering water's molecular structure and the hydrogen bond. Figure 7-3a shows how the density of water changes with temperature. Starting at 20°C, at the left end of the curve, water is liquid and has a density of 0.9982 g/cm³. As the temperature of liquid water decreases, the amount of thermal agitation (kinetic energy) of the water molecules also decreases, and the unbonded water molecules therefore occupy less volume. The result is that liquid water contracts (becomes denser). At 15°C, the density has increased to 0.9991 g/cm³. As we approach 4°C (3.98°C to

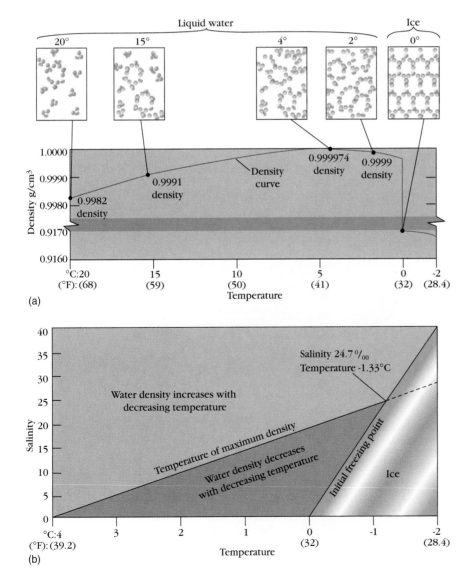

Figure 7-3 Temperature, Salinity, and Freezing of Water

Note that temperature decreases to the right in both diagrams. (a) Formation of ice clusters and density changes in freshwater as temperature falls from 20°C to −2°C. The density of freshwater starts at 0.9982 g/cm³ at 20°C, increases to a maximum of 0.999974 g/ml³ at 4°C, and then declines to less than 0.9170 g/ml³ as ice. (b) Diagram in (a) corresponds to 0 ‰ salinity in this diagram. With increasing salinity, both the temperature of maximum density and the freezing point decrease. These two trends define three fields. Ice occurs regardless of salinity below the freezing line. Between the freezing point and temperature of maximum density, the density is decreasing with decreasing temperature, equivalent to the declining part of the curve in (a) above. Above the maximum density line, the density is increasing with decreasing temperature, corresponding to the climbing part of the curve in (a). At 24.7 ‰ and −1.33°C, freezing occurs at the point of maximum density.

be more exact), we reach the maximum density of 0.999974 g/cm^3. Below this temperature, the density decreases as the temperature falls. At these low temperatures, some of the water molecules begin to organize into the six-sided lattice works that form the ice structure. These are bulky, open structures that take up more space than the freely moving unbonded water molecules. The density of liquid water continues to decrease. At 0°C, the temperature of the liquid-to-solid phase transition when liquid water actually freezes into solid ice, there is a dramatic density decrease to 0.9170 g/cm^3.

That water attains its maximum density at a temperature where it is still liquid is extremely important in the dy-

namics of circulation of freshwater bodies and to the survival of the aquatic organisms living in them (Figure 7-4). During the fall, as fresh surface waters are cooled, they become denser than the underlying water and sink, while the deep water rises to the surface. This new surface water then cools and sinks. The overturn continues until the entire freshwater column has reached 4°C. As cooling continues below this temperature, the water becomes less dense and therefore does not sink. Overturn ceases and the surface water cools until it freezes, forming a layer of ice over a still-liquid body of water. Organisms living within freshwater lakes live out the winter in relative hibernation in waters of 4°C. If ice were denser than cold water, the entire

Figure 7-4 Dynamics of Lake Stratification

(a) Cool air (4°C) in fall causes surface cooling of lake waters and the surface layer becomes more dense than the underlying warmer water, shown here as 15°C. This creates convective overturn and mixing of lake water. (b) With continued cooling and convection, the lake water reaches a uniform temperature of maximum density, about 4°C. (c) As winter approaches and air temperatures decrease further, additional cooling of the lake surface occurs, but this water is less dense than the underlying 4°C water. The surface freezes while the underlying water remains at 4°C.

(a)

4°C
Surface cooling

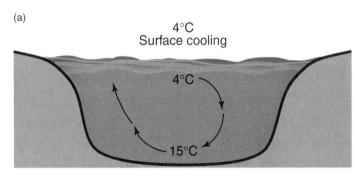

4°C

15°C

(b)

4°C

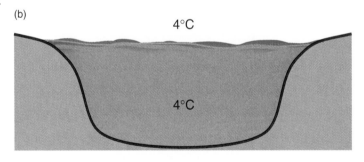

4°C

(c)

0°C

Ice

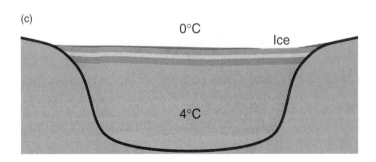

4°C

lake would freeze along with all its inhabitants. The presence of dissolved salts is one factor that prevents the same thing from happening in ocean basins.

The addition of dissolved salts lowers the temperature of maximum density. Increasing the pressure has the same effect. The reason is that both pressure and dissolved salts inhibit the formation of the open lattice works. Thus, to produce the same number of lattice works requires the removal of more energy, resulting in lower temperatures of maximum density. The point of the liquid-solid phase transition also shifts to a lower temperature. Figure 7-3b shows the effect of salinity on the temperature of maximum density and the freezing temperature of liquid water. Note that these two curves cross at $-1.33°C$ and $24.7‰$ S. To the left of this point, water behaves as previously described; it reaches its maximum density at a temperature higher than its freezing point. At the point of intersection, water is at its maximum density at the freezing point. To the right of this point, water freezes *before* it reaches maximum density.

The salinity effect in depressing the freezing point means that to freeze most seawater ($35‰$ average salinity), the water temperature would have to be lowered to less than about $-1.5°C$ for sufficiently long periods of time. Because of the high heat capacity of water and the transfer of warm, low latitude water to higher latitudes, the oceans do not freeze except very close to the poles.

At a constant temperature, the addition of salts increases the density of water. The added ions have mass, plus the electrostatic attraction of the dissolved ions draws the water molecules closer together. This effect can be observed by carefully marking the height of the water in a long-necked bottle, then noting the decrease in height when a large amount of salt is added.

Differences in the density of ocean water are the main driving forces in the vertical circulation of the world's oceans. Within estuaries and lagoons, wide seasonal temperature differences and the presence of water bodies with large differences in salinity can lead to density stratification and seasonal overturn similar to that previously described for lakes. The open oceans also are density-stratified, but the huge volumes of water and relatively small density contrasts lead to turnover times on the order of thousands of years.

Measuring Density

It is difficult to measure density directly, thus, temperature and salinity values are used to calculate the density. For very precise density measurements in the deep ocean, the effect of pressure must also be factored into the calculations. The force of pressure causes an increase in the energy content and temperature of the water. This pressure-induced rise in temperature must be factored out to get a correct picture of the component of temperature that affects density. These corrected terms are called *potential temperature* and *potential density*. Because the potential temperature is lower than *in situ* temperature, the potential density is greater than the *in situ* density.

The symbol used to express density is σ **(sigma)**. This symbol with a subscript T, σ_T or sigma-T, refers to *in situ* density, σ_θ or sigma-theta refers to potential density. This symbol denotes a dimensionless relative density that is calculated from the ratio of the density of the seawater sample to the maximum density of pure water. The equation is:

$$\sigma = 1000(ñ/ñ°(\text{max}) - 1) \qquad (7\text{-}2)$$

where ñ is the density of the seawater sample and ñ° (max) is the maximum density of pure water at 1 bar pressure and $3.98°C$ ($= 0.999974 \text{ g/cm}^3$).

Recall that seawater sample density must be calculated from temperature, salinity and pressure data; the equation used is the International Equation of State for Seawater:

$$ñ = ñ°[1/(1 - PK)] \qquad (7\text{-}2)$$

where ñ is the density at in situ pressure, ñ° is the density at in situ temperature and salinity, also a calculated value. K is a function of in situ temperature, salinity and pressure.

The effect of these complicated calculations is to convert density values to values without units (dimensionless numbers), and to eliminate the first two digits (1.0). For example, suppose we measure in situ temperature, pressure, and salinity for a water sample and, by plugging these numbers into the International Equation of State, we calculate a density value of 1.02567 g/cm^3. Using the equation for σ, we then calculate:

$$\sigma = 1000 \, (1.02567/0.999974 - 1) \qquad (7\text{-}3)$$
$$= 25.67$$

A far easier approach to observing how temperature and salinity affect density is to learn to interpret a graph constructed for a wide range of salinity and temperature combinations (Figure 7-5). With temperature on the vertical axis, and salinity on the horizontal axis, constant σ values plot as curves with values increasing to the right. We can use this graph to compare the densities of water samples simply by plotting them according to their salinities and temperatures. For example, a sample with a temperature of $20°C$ and a salinity of $35‰$ plots at point B on the figure, and we can estimate its σ value as 24.7 because it is about one-third the distance between the 24.0 and 26.0 σ curves. Another sample, at $0°C$ and $35‰$, plots at point D, which corresponds to a σ value of 28.1. Sample D is denser than sample B.

We can also use the graph to see how density changes with temperature. If sample B is warmed to $25°C$ (point A), the density decreases to 23.5 σ, a change of 1.2 σ units. If sample D is warmed to $5°C$ (point C), however, the density decreases by only 0.4 σ units. This difference shows that temperature change has a much greater effect on density in the high-temperature, low-latitude areas of the oceans than it does in polar regions, and is important in predicting seasonally induced mixing.

Figure 7-5 Seawater Density Variations with Temperature and Salinity

Density, plotted as σ contours, is shown as a function of temperature and salinity. A warm, low latitude water is shown at point A (25°C, 35 ‰), with a σ of 23.5. If this water were cooled to 20°C, point B, the density would increase, resulting in a σ of 24.7. If a cold, high latitude water, point C (5°C, 35 ‰) is cooled to 0°C, point D, σ changes from 27.7 to 28.1. Note that the 5 degree temperature change has a much greater effect on density for warmer waters than it does for colder waters.

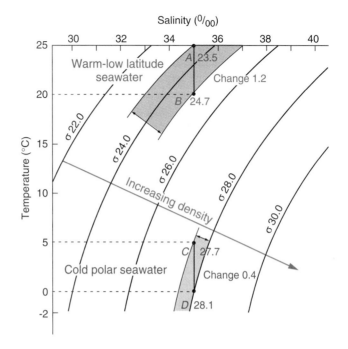

Density Structure of the Oceans

Figure 7-6 shows the average latitudinal variations in salinity, temperature, and density for the surface waters of the oceans. The graph illustrates that temperature has a much greater impact than salinity on the density distribution. Cold temperatures are the primary control on the density increases at high latitudes, and warm temperatures are the primary control on the density decreases at low latitude. From the salinity curve, you would expect low densities for surface waters at high latitudes and increasing density toward the equator, trends that are opposite to the observed densities.

Figure 7-7 shows the changes in density, temperature, and salinity for three vertical profiles at different latitudes. Again, these show the overriding control of temperature on density in the ocean. At high latitudes, density is close to constant with depth, as is temperature. At the equator, low surface water densities reflect high water temperatures. Again, the salinity curves suggest that trend in densities should be opposite to the observed values.

The equatorial depth profile shows a zone where density increases very rapidly. This region of rapid density change with depth is called the **pycnocline** (density slope). The pycnocline is a very stable barrier to mixing between the low-density water above and the high-density water below. The pycnocline has a high gravitational stability because it would require greater energy to move a mass of water up or down within the pycnocline, than it would to move an equal mass the same distance where density changes very slowly with increasing depth.

The pycnocline results from the combined effect of rapid vertical temperature change (the **thermocline** or temperature slope), and salinity change (the **halocline** or salinity slope). The relative magnitude of density difference dictated by the thermocline and halocline determines the degree of separation between the upper-water and deep-water masses.

The blue lines in Figure 7-7 show that the pycnocline is lacking in high latitudes. You can see that little difference exists between surface water density and bottom water

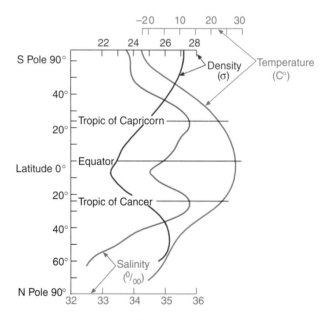

Figure 7-6 Average Temperature, Salinity, and Density of Surface Seawater by Latitude

Density (as σ) increases from about 22 near the equator to near 27 at high latitudes, while temperature is near 25°C near the equator and declines to −20°C at high latitudes. The mirror images of these curves indicate the overriding effects of temperature on density. Salinity maxima at tropical latitudes (23.5° N and S) have no apparent effect on the density curve.

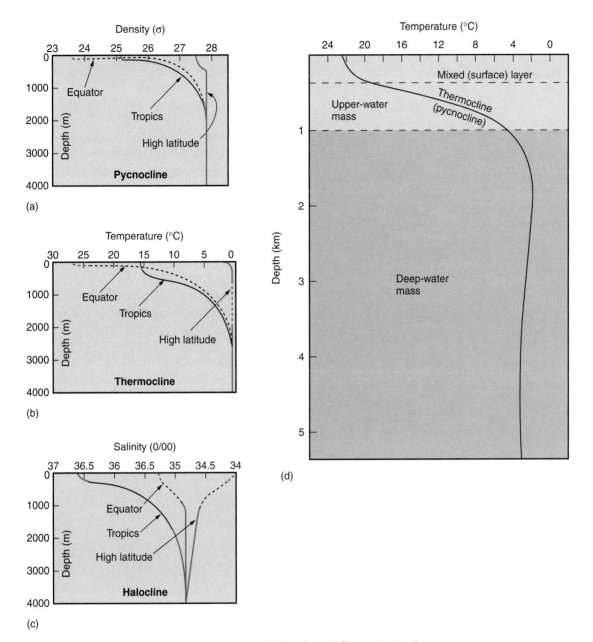

Figure 7-7 **Seawater Density Variations with Depth at Different Latitudes**

Note the uniformity in temperature, salinity, and density of deep ocean waters at all three latitudes. (a) Density changes with depth. Density changes rapidly from 0 to 1500 m at equatorial and tropical latitudes. At high latitudes, density is nearly constant. The depth interval over which density changes rapidly is termed the *pycnocline*. (b) Temperature changes with depth. Note that temperature decreases to the right. These curves mimic the density curves in (a). The depth interval with rapid temperature changes is termed the *thermocline*. (c) Salinity changes with depth. Note that salinity decreases to the right. High salinities occur in surface waters of the tropics and decrease rapidly. Equatorial and high latitude salinities show only small depth variations. The depth interval of rapid salinity changes is called the *halocline*. (d) The ocean temperature/depth profile. Mixing by waves creates the mixed surface layer of uniform temperature and density. The thermocline/pycnocline is the zone of rapid temperature and density change. It is well developed throughout equatorial and mid-latitude oceans. Under the upper water mass layer is deep water of uniformly cold temperatures and high density.

density in these polar areas. Thus, these water columns are more inclined to mix than are water columns possessing a pycnocline. In Chapter 9, we will consider the importance of low-stability water columns in polar latitudes to mixing ocean waters worldwide.

Dissolved Gases in Water

Take-a-Break, Grab a Soda

At this point, it would be instructive to find an unopened can or bottle of carbonated soda. This soda is an excellent example of water with dissolved gas in it (the sugar and flavoring don't noticeably affect the gas behavior for our purposes). Carbonated sodas are made by injecting CO_2 into liquid under excess pressure. If you have a clear glass bottle of soda, you should note how few bubbles are in the liquid prior to opening it.

By popping the top, you release the excess pressure, and the gas starts to bubble out of the liquid—you can feel the fizz escaping from the top if you stick your hand or nose over the opening. You can also fit a balloon over the opening and watch it inflate. After a short time, this mad rush of bubbles abates, but bubbles continue to escape more slowly. Now stir your soda or shake it a bit, and you'll produce a new rush of bubbles out the top. If you gently warm the soda, you will also increase the amount of gas that escapes. Eventually, most of the gas will be gone, and your soda will be flat.

From this little experiment you have learned that the gas content of water depends on pressure, temperature, and its ability to escape to the atmosphere. You decreased the gas content first by decreasing the pressure and then by increasing the temperature. By stirring, you increased the atmospheric escape. The final gas content of your soda is the equilibrium volume that can dissolve in water at room temperature and atmospheric pressure given the CO_2 content of the atmosphere.

Gas Exchange Between the Atmosphere and Natural Waters

At the surface of a water body, where the water is in contact with the atmosphere, a small amount of each atmospheric gas will dissolve into the water. Wave action enhances this air-water contact in the same way that you did when you stirred your soda. As wind velocities increase and waves grow in size, there is an increase in turbulence and the depth to which it extends. Thus, the amount of mixing of atmospheric gas into water is proportional to wind velocity.

The total amount of gas that can be dissolved eventually reaches an equilibrium concentration in the same way your soda did. The equilibrium concentration for any gas is proportional to the atmospheric concentration of that gas and depends on the temperature, pressure, and salinity of the water. The amount of gas that dissolves increases as pressure increases and as temperature and salinity decrease. Temperature is the most important controlling

factor in the ocean environment; the salinity effect is much smaller given the small range in salinity of the open ocean. Figure 7-8 shows how gas solubility in ocean water changes with temperature for the most abundant atmospheric gases and for the noble gases.

The most abundant atmospheric gases (by volume) are nitrogen (N_2, 78.1%), oxygen (O_2, 20.9%), argon (Ar, 0.9%), and carbon dioxide (CO_2, 0.03%). Argon and the other noble gases (helium, neon, krypton, and radon), although present in very small quantities in the atmosphere and oceans, are important tracers of ocean physical processes because they are chemically unreactive. Radon is radioactive (see front fly sheet) and thus also provides rate information on air-water gas exchange. Nitrogen (N_2) is converted to nitrate ions (NO_3^-) by bacteria, but because the amount converted is not large enough to appreciably alter the concentration, N_2 also can be used as a tracer. These tracer compounds, as well as other chemicals that are essentially unreactive in the ocean are termed **conservative tracers**.

In contrast to the previously mentioned gases, carbon dioxide and oxygen concentrations are altered appreciably by many biological and chemical processes in the sea. The amounts of dissolved oxygen and carbon dioxide contributed by the atmosphere at the sea surface are altered

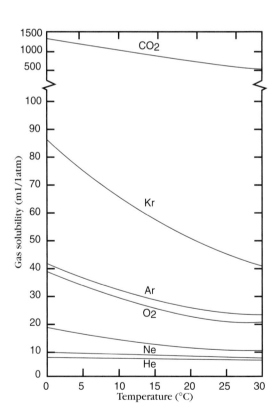

Figure 7-8 Relative Gas Solubility in Seawater with Temperature

Solubility of common atmospheric gases and noble gases, assuming all gases in seawater are present in equal amounts at one atm total pressure. Note that carbon dioxide is 10 times more soluble than the other gases.

by photosynthesis and respiration. Carbon dioxide levels are also affected by volcanic activity and sediment-water interactions on the sea floor. These compounds are **nonconservative tracers,** and they can be used to trace chemical and biological processes.

For all the gases except carbon dioxide, the amounts in the atmosphere constitute over 95 percent of the total global amount. Thus, the atmosphere dictates the ocean concentration for these gases. Because the amount of carbon dissolved in the ocean is much greater than the amount in the atmosphere, the ocean dictates the atmospheric carbon dioxide concentration. This is a critical point in understanding the ocean's role in global climate change, as we will see later.

Oxygen and carbon dioxide are produced and consumed by organisms. In plants and algae, **photosynthesis** uses energy from sunlight to convert carbon dioxide and water into organic matter and oxygen. In most living organisms, including bacteria, plants, and animals, **respiration** converts organic matter back into carbon dioxide using oxygen in the process. We can write simplified reactions to show these processes:

$$CO_2 + H_2O + \text{sunlight} \Rightarrow \text{organic matter} + O_2 \quad (7\text{-}4)$$
$$\text{photosynthesis}$$

$$\begin{array}{c}\text{organic} \\ \text{matter}\end{array} + O_2 \Rightarrow CO_2 + H_2O + \begin{array}{c}\text{metabolic} \\ \text{energy}\end{array} \quad (7\text{-}5)$$
$$\text{respiration}$$

These processes will be examined in more detail in Chapter 15.

Acidity and Alkalinity

An acid is a compound which, when dissolved in water, forms hydrogen ions. An alkaline compound, or base, takes up hydrogen ions. The **acidity** of a solution is indicated on a scale that measures the hydrogen ion concentration, the pH scale.

The Concept of pH

By definition, **pH** is equal to the negative logarithm of the hydrogen ion concentration. The equation is:

$$pH = -\log[H^+] \quad (7\text{-}6)$$

Technically, chemists want to measure the reactive concentration, called the *activity,* of hydrogen rather than the total concentration. You can understand the reasoning here more easily by considering the difference between counting all the students present at a dull lecture—the total concentration—and counting only those students that are awake—a better measure of the amount of student activity in the classroom. For our purposes, we will assume that all the ions are awake; so activity and concentration are the same.

By defining pH as a logarithm, we can refer to the exponent only (see Appendix I if you don't understand logarithms and exponents) and by putting the negative sign in the equation, we can express pH as a positive number. Thus, saying that something has a pH of 7 is just a much simpler way of saying that the hydrogen ion concentration is 10^{-7} mole/L. (A mole, like a dozen, is a unit that represents a fixed number of objects. Just as there are twelve objects in a dozen, there are 6.023×10^{23} molecules or ions in a mole.)

In liquid water, because of the attraction of the hydrogen bond to another molecule's oxygen, H_2O molecules can easily break apart, or dissociate, so that one hydrogen remains attached to another water molecule, forming H_3O^+ and leaving behind the OH^-, or hydroxide ion. We can write this reaction as:

$$H_2O + H_2O = H_3O^+ + OH^- \quad (7\text{-}7)$$

Or for the dissociated water molecule:

$$H_2O = H^+ + OH^- \quad (7\text{-}8)$$

In one liter (55.6 moles) of pure water, we find that about 10^{-7} moles have dissociated. Because for each H^+ produced, an OH^- is also produced, we have 10^{-7} moles/L of H^+ and 10^{-7} moles/L of OH^-. This is equivalent to a pH of 7, which is the neutral point for water on the pH scale at 25°C and 1 atm. The pH values for some common substances are shown in Figure 7-9.

If substances added to water produce more H^+, the water will have a lower pH (higher H^+ concentration) and be more acidic; if the added substances use up H^+, the water will have a higher pH (lower H^+ concentration) and be more basic, or alkaline. For example, if hydrochloric acid (HCl) is added to pure water, the hydrochloric acid molecules dissociate into H^+ and Cl^- ions, and the resulting solution has a lower pH. Conversely, a base such as sodium hydroxide (NaOH) dissociates into Na^+ and OH^- ions, and the added OH^- ions react with H^+ ions, producing a solution with a higher pH.

pH values of comon substances

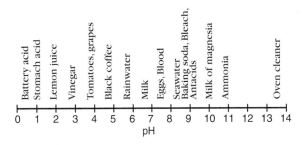

Figure 7-9   pH Values of Common Substances

Buffering and Alkalinity

The **alkalinity** of a solution is a measure of the amount of ions present that can react with, or neutralize, H^+. The higher the alkalinity of a solution, the more difficult it is to produce a pH change in the solution by adding acid. Therefore, the alkalinity of a solution is a measure of its acid **buffering** capacity.

In addition to OH^-, ions such as HCO_3^- (bicarbonate) and CO_3^{2-} (carbonate) are important components in the alkalinity of natural waters. HCO_3^- can take up one hydrogen ion and CO_3^{2-} can take up two hydrogen ions, in each case a molecule of H_2CO_3 (carbonic acid) is formed. We can write these chemical reactions as:

$$H^+ + HCO_3^- = H_2CO_3 \qquad (7\text{-}8)$$

$$2H^+ + CO_3^{2-} = H_2CO_3 \qquad (7\text{-}9)$$

The alkalinity of seawater comes predominantly from HCO_3^-, CO_3^{2-}, and OH^- ions. An equation for alkalinity is expressed by the sum of the ion concentrations multiplied by the number of hydrogen ions each can take up.

$$Alkalinity \ (A) = [HCO_3^-] + 2[CO_3^{2-}]$$
$$+ [OH^-] - [H^+] \quad (7\text{-}10)$$

The subtraction of the $[H^+]$ term at the end of the equation recognizes that addition of acid serves to lower the alkalinity, or acid-buffering capacity, of the solution.

The pH of surface seawater is about 8.2, meaning it has a $[H^+]$ concentration of $10^{-8.2}$ or about 7×10^{-9} moles/L (see Appendix I for information on non-integer exponents, if you need to). The alkalinity of surface seawater is typically 0.0023 moles/L (as HCO_3^-); thus, there are orders of magnitude more carbonate and bicarbonate ions than there are H^+ ions in seawater. This gives seawater a huge capacity to neutralize, or buffer, acids. If more H^+ ions are added to seawater, they combine with the CO_3^{2-} and HCO_3^- ions and there is no net change in pH.

The Influence of CO_2

As previously discussed, carbon dioxide is added to seawater through atmospheric exchange and as a product of respiration by ocean-dwelling organisms. The carbon dioxide molecules combine with water molecules to produce HCO_3^- and CO_3^{2-}. We can write these reactions as follows:

$$CO_2 + H_2O = HCO_3^- + H^+ \qquad (7\text{-}11)$$

$$CO_2 + H_2O = CO_3^{2-} + 2H^+ \qquad (7\text{-}12)$$

Note that although HCO_3^- and CO_3^{2-} ions are created, there is no net increase in alkalinity through these reactions, but there is a decrease in pH. The alkalinity does not change because for every molecule of HCO_3^- added in the alkalinity equation above, one molecule of H^+ is created that must be subtracted from the equation. Likewise, for

every molecule of CO_3^{2-} added, the two new molecules of H^+ must be subtracted. The new H^+ molecules, however, add to the concentration of acid, and the pH decreases. Remember that seawater has a lot of excess HCO_3^- and CO_3^{2-}. These ions quickly grab the H^+ molecules and neutralize them, by the reactions in Equations 7-8 and 7-9 above, and thus, raise the pH of seawater back to its previous value.

It is possible for carbon dioxide input to exceed the buffering capacity of seawater. The pH of seawater will then start to fall, but carbonate sediments on the ocean floor also may act as a buffer. If carbonate sediments are present, the decrease in pH will cause them to dissolve in the reaction as follows:

$$H^+ + CaCO_3 = Ca^{2+} + HCO_3^- \qquad (7\text{-}13)$$

This reaction introduces additional HCO_3^- into the seawater, increasing the alkalinity and the buffering capacity and bringing the pH back to its original value (Figure 7-10). The alkalinity and availability of carbonate sediments give the oceans a large buffering capacity for carbon dioxide. In fact, the oceans apparently have removed a large fraction of the CO_2 added to the atmosphere by fossil fuel burning without producing a significant change in seawater pH.

Organic Chemistry of Seawater

By weight, organic matter is made up mostly of carbon atoms. Each carbon has a 4^- charge and so must form four bonds with other atoms. The basic structure of organic material is carbon atoms bonded to each other in long chains,

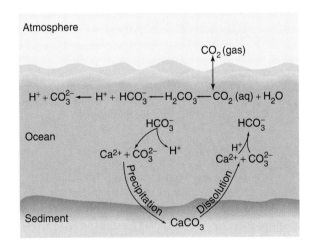

Figure 7-10   Carbonic Acid System

CO_2 exchanges between the atmosphere and surface ocean. The dissolved gas also dissociates into HCO_3^-, CO_3^{2-}, and H^+ ions, increasing the acidity of the ocean, but not the alkalinity. The dissolution of calcium carbonate from sea floor sediments can buffer the added H^+.

with hydrogen atoms bonded to the carbon atoms, completing the carbon valence. Oxygen, nitrogen, phosphorus, and sulfur may substitute for hydrogen or added to make a variety of organic compounds. Alcohols, for example, have $-OH$ groups; amines have $-NH_2$ groups. Metallic elements also are present in very small but critical amounts. Many of these elements are listed on your vitamin bottle: iron, zinc, magnesium, etc.

Organic matter in seawater is present as molecules dissolved in the water and as solid particulate matter. The solid particles are very tiny and remain suspended in the water for very long periods of time. Some of the largest dissolved molecules are not much smaller than the smallest solid particles and are very large in comparison to the simple inorganic ions we've been discussing. Consequently, there is a rather fuzzy boundary between what is actually dissolved (surrounded by a hydration sphere) and what is particulate. We follow the practical definition that uses the filter pore size most commonly used to filter seawater samples. Any organic matter retained by a 0.2 micrometer pore filter is particulate organic matter (POM); any organic matter that passes through the filter is dissolved organic matter (DOM). The terms **particulate organic carbon (POC)** and **dissolved organic carbon (DOC)**, are used frequently because with common analytical techniques, carbon quantities can be measured more easily than organic matter quantities.

One way or another, organic material in the oceans is derived from organisms. By far the most important source is organic matter synthesized by marine algae. The algae convert inorganic carbon (CO_2) into a wondrous array of organic molecules that together form living tissue. Every year about 2×10^{16} g of organic carbon is produced in this way. Land plants also contribute. Rivers carry about

4×10^{14} g of organic carbon from the continents to the oceans each year. There is some controversy about the amount of organic material that comes from the atmosphere to the oceans. Measurements of organic matter in rainwater suggest that about 2×10^{14} g of organic carbon are contributed annually. However, much of this may have originated in the oceans and is simply recycled through the atmosphere. Some organic matter may also be reintroduced to seawater from resuspended bottom sediments.

The organic matter that oceanographers measure in seawater includes living planktonic organisms, wastes produced by living organisms, and the remains of dead organisms. It exists as dissolved organic carbon (DOC) and particulate organic carbon (POC). Depth profiles for DOC and POC are shown in Figure 7-11. The relative amounts of the living and nonliving (detrital) organic matter change drastically with depth. As you might expect, both total organic matter concentrations and the amounts of living matter are highest in surface waters where active photosynthesis and feeding are taking place. Total values are lowest at the depth of the oxygen minimum and remain at these levels with increasing depth. Below the level of sunlight penetration, most organic matter is detrital, consisting of the waste products of living organisms. Bacteria living in the water column and on the surfaces of particles continue to break down this organic material, but some organic compounds resist the digestive abilities of the bacteria. Scientists call this indigestible part **refractory organic matter.** With depth, the total amount of organic matter decreases, but the fraction of refractory material increases as the bacteria continue breaking down the digestible fraction. Finally, a depth is reached, below which organic matter concentrations remain essentially constant.

Figure 7-11 Depth Profile of Dissolved Organic Carbon (DOC) and Particulate Organic Carbon (POC) in Seawater

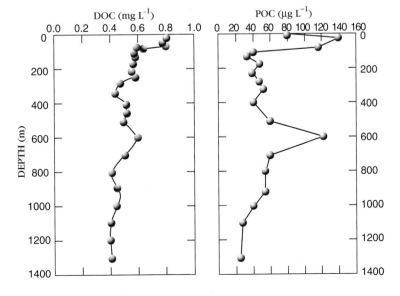

Variations in Seawater Composition with Depth and Latitude

Figure 7-12 shows the depth profiles of the North Atlantic and North Pacific for dissolved oxygen, dissolved carbon dioxide, pH, alkalinity, phosphate, nitrate, and silica.

Organisms that make siliceous shells (radiolaria and diatoms) deplete dissolved silica from the surface ocean. The increase in silica concentration with depth results from dissolution of the shells after the organisms die and sink toward the bottom. Because surface seawater is undersaturated in opaline silica, the siliceous skeletons dissolve immediately, and silica concentration increases continuously with depth. Values are much higher in the deep Pacific than in the deep Atlantic because of differences in water circulation patterns between the two oceans. In the Pacific, the deep waters have been isolated from the surface for much longer than the deep Atlantic waters and thus have had more time to accumulate silica.

Organisms that make carbonate shells (foraminifera and coccolithophorids) remove alkalinity from seawater. Alkalinity is returned to seawater by the dissolution of their shells after they die. Because the processes that remove and produce dissolved silica and alkalinity are the same, you would expect that the alkalinity depth profiles would be similar to the dissolved silica depth profiles. The profile for the North Pacific seems to fit this prediction, however, the alkalinity values in the North Atlantic are higher at the surface than in deep water. This makes no sense until you find out that the salinity of the North Atlantic is significantly higher than the North Pacific because of evaporation and that the curves need to be corrected to constant salinity. When this correction is done, the alkalinity profiles look very much like the silica profiles; alkalinity is low in the surface layers of both oceans and increases continuously with depth.

Oxygen values are highest at the surface because of photosynthetic production and exchange with the atmosphere. As you might expect, oxygen levels decrease rapidly with depth because oxygen is used up by respiration and there is no further production below the depth of sunlight penetration. Curiously, deep ocean oxygen contents are typically higher than values at intermediate depths. This makes sense once you find out that deep ocean waters come from sinking cold surface water at polar latitudes (see Chapter 9). Because it is so cold, this water contains more oxygen to start with than the intermediate waters that come from lower latitudes. Also, this cold, dense water sinks before respiration can deplete much of the oxygen.

Surface ocean carbon dioxide concentrations are in equilibrium with the atmosphere. As you would expect, the depth profiles of carbon dioxide look like mirror images of the oxygen profiles. Production of carbon dioxide by respiration causes the profiles to increase rapidly over the same depth interval at which oxygen is rapidly depleted. In the deep ocean, carbon dioxide values decrease again

for the same reason that the oxygen profiles increase—the concentrations are those the waters acquired at the surface at polar latitudes.

The depth profile of pH of seawater is the inverse of the carbon dioxide profile. The pH minimum coincides with the oxygen minimum and carbon dioxide maximum (see acid section). As carbon dioxide is added to seawater, the pH drops. The surface alkalinity values are low because some living organisms are using carbonate ions to synthesize their skeletal material. The amount of remaining alkalinity is insufficient to buffer the acidity caused by carbon dioxide from respiration. Only at a depth where biogenic carbonate starts dissolving to produce new alkalinity does the pH start to rise again.

The concentrations of phosphate and nitrate in seawater are very small. Phytoplankton need both of these compounds for growth and will take up all that is available. Consequently, the concentrations of these compounds in surface waters are very near zero. With conversion of organic matter back into CO_2, the phosphate and nitrate are released. Thus, the phosphate and nitrate depth profiles reach maxima at the same depth where oxygen is at a minimum and CO_2 is at a maximum. Values in deep water remain fairly constant at maximum values, again, a result of these waters being out of contact with the surface for long periods of time and the lack of further organic matter breakdown.

Source of the Salts

Recall from Chapter 2 that Earth has had oceans for about 4 billion years. Water vapor condensing in Earth's early atmosphere reacted with atmospheric dust and picked up volatile gases from volcanic eruptions. Among the most important gases that dissolved in the rainwater were chlorine, sulfur dioxide, and carbon dioxide. When these gases dissolve in water, they produce these acids: hydrochloric acid, sulfuric acid, and carbonic acid, respectively. When this acid rain fell to the solid surface of Earth, it reacted with the rocks, dissolving some of the minerals in the rock. Some rain fell into the ocean and some formed streams and rivers flowing into the sea. Through evaporation, water from the surface ocean returned to the atmosphere to begin the process again. Today, similar processes occur, although volcanic emissions are much lower than they were during Earth's early history and the compositions of rain and the atmosphere are significantly different (see Chapter 2). The average composition of rain today is shown in Table 7-2.

The reactions of water with rocks on Earth's surface are collectively termed **weathering**. Some common weathering reactions are shown in Table 7-3. Weathering results in addition of dissolved material to water, as well as loss of material precipitated out of water as new minerals. The major constituents of average river water are shown in Table 7-2.

Ocean water reacts with the rocks and sediments on the seafloor, gaining dissolved material and losing material that precipitates out as new minerals. These reactions include both high-temperature reactions of seawater with

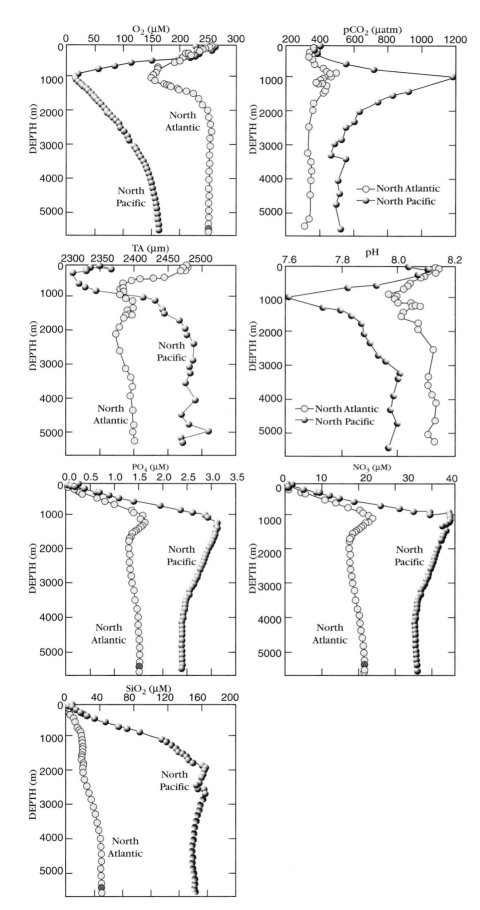

Figure 7-12   Depth Profiles of Dissolved Oxygen (O_2), Carbon Dioxide (pCO_2), pH, Alkalinity (TA), P (PO_4), N (NO_3), and Si (SiO_2)

TABLE 7-2 Composition of average rain and river water.

Ion	Rain (mg/L)	Rivers (mg/L)
Na^+	0.2 – 1	5.2
Mg^{2+}	0.05 – 0.5	3.4
Ca^{2+}	0.1 – 3.0	13.4
K^+	0.1 – 0.3	1.3
Cl^-	0.2 – 2	5.8
SO_4^{2-}	1 – 3	5.3
HCO_3^-	trace	52.0
$SiO_2(aq)$	trace	10.4
Total	**1.7–9.8**	**99.6**

(from Berner and Berner, 1996, *Global Environment: Water, Air and Geochemical Cycles*. Upper Saddle River: Prentice Hall, Inc.)

basalt at midocean ridges and low-temperature reactions with volcanic rocks and oceanic sediments (Table 7-3).

Note that there are large differences in the composition of rainwater, river water, and seawater (Tables 7-1 and 7-2). There is controversy as to the exact processes that control the composition of seawater. Obviously, it is much more complex than the simple accumulation and concentration of river water over time, as we shall see below.

You might think that over Earth's history, as the rivers dumped their loads into the sea and water evaporated from the ocean surface, that the oceans would have become increasingly salty. However, scientists have evidence that the salinity and composition of the oceans have remained relatively constant since very early in Earth's history. This suggests that the oceans do not simply accumulate salts brought to them by the rains and rivers. Instead, the salts are subject to cycles in which they are added to and removed from the oceans at equal rates such that the amount in the oceans stays the same. The residence times for some elements dissolved in ocean water are given in Table 7-4.

Biogeochemical Cycles

Biogeochemical cycles describe the movement of chemical elements or compounds on the Earth. As we discussed in Chapter 6 in the context of the hydrologic cycle, each geochemical cycle consists of reservoirs and fluxes. Figure 7-13 is a simple biogeochemical cycle with three reservoirs (shown as boxes) and six fluxes between the reservoirs (shown as arrows between the boxes). Each **reservoir** has

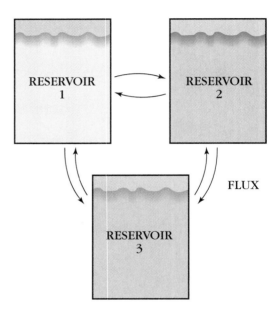

Figure 7-13   Simple Box Model of Biogeochemical Cycling

This schematic model consists of three reservoirs (boxes). Elements exchange in and out of the reservoirs at flux rates represented by the arrows.

TABLE 7-3 Common weathering reactions and hydrothermal reactions on the sea floor.

Weathering Reactions

$2H_2CO_3 + 9H_2O + NaAlSi_3O_8 \Rightarrow Al_2Si_2O_5(OH)_4 + 2Na^+ + 2HCO_3^- + 3H_4SiO_4$

carbonic acid + water + albite feldspar $\Rightarrow$ kaolinite clay + dissolved sodium + bicarbonate + silicic acid

$H_2CO_3 + CaCO_3 \Rightarrow Ca^{2+} + 2HCO_3^-$

Carbonic acid + calcite $\Rightarrow$ dissolved calcium + bicarbonate

Hydrothermal Reactions

$CaAl_2Si_2O_8 + 2Na^+ + 4SiO_2 \Rightarrow 2NaAlSi_3O_8 + Ca^{2+}$

Anorthite Feldspar + dissolved sodium + dissolved silica $\Rightarrow$ albite feldspar + dissolved calcium

$3Mg^{2+} + 2CaAl_2Si_2O_8 + 2H_2O \Rightarrow Mg_3Si_4O_{10}(OH)_2 + 2Al_2O_3 + 2Ca^{2+} + 2H^+$

dissolved magnesium + anorthite feldspar + water $\Rightarrow$ talc + aluminum oxide + dissolved calcium + acid

$2Mg^{2+} + 2SO_4^{2-} + 11Fe_2SiO_4 + 2H_2O \Rightarrow Mg_3Si_2O_5(OH)_4 + FeS_2 + 8SiO_2 + 7Fe_3O_4$

dissolved magnesium + dissolved sulfate + olivine + water $\Rightarrow$ serpentine + pyrite + dissolved silica + hematite

a size, given in mass units of the element (e.g., grams of carbon). Each **flux** is a rate given in mass change per unit time (e.g., grams added/year). The element cycles of the Earth usually are considered as closed cycles, based on the assumption that essentially no mass is gained or lost to space.

A summary diagram of the biogeochemical cycles of the elements carbon, nitrogen, phosphorus, sulfur and oxygen is shown in Figure 7-14. The cycling of these elements is key to the maintenance of life in the oceans and on land. It is also important to note that the cycles of these elements are strongly interdependent.

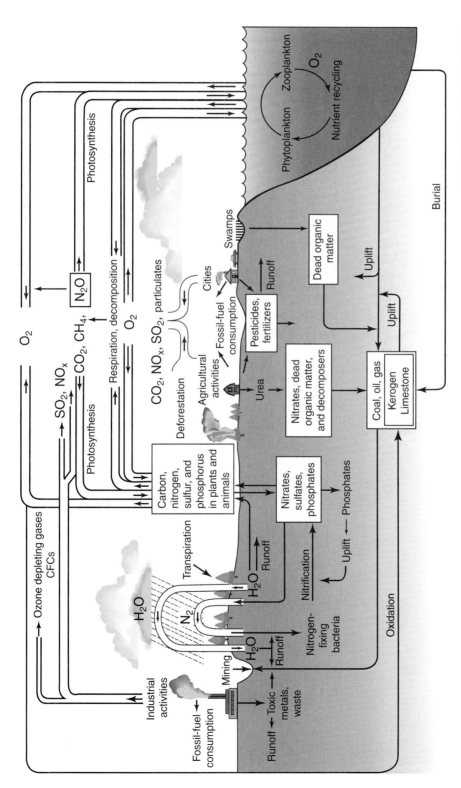

Figure 7-14

A summary diagram of the biogeochemical cycles of the key elements carbon, nitrogen, sulfur, and oxygen. The cycling of these elements is critical to the maintenance of life. (*From Mackenzie, F. T., 1998, Our Changing Planet: An Introduction to Earth System Science and Blobal Environmental Change, 2nd ed. Upper Saddle River: Prentice Hall.*)

When Circulation Stops:
Lessons From The Mediterranean

About 5.8 to 5.2 million years ago, during the Miocene Epoch, the eastern part of the Mediterranean Sea dried up because water inflow from the Atlantic Ocean stopped. The Mediterranean lies in an arid climatic belt and has few rivers flowing into it, making the Atlantic's inflow its only significant source of new water. At the beginning of the Pliocene (5 million years ago), seawater from the Atlantic reflooded the Mediterranean basin, bringing water levels up to their current position. There is some controversy as to how this happened. Some scientists think that tectonic forces uplifted the Mediterranean seafloor in the Miocene and then rapidly dropped it again in the Pliocene. Other scientists argue that only the Strait of Gibraltar, which connects the Mediterranean to the Atlantic, was uplifted, thereby blocking the exchange of water between the Mediterranean and Atlantic.

When the seawater in a basin evaporates, the salts in the water precipitate out, forming thick sequences of mineral salts in the bottom of the basin. The salt minerals present in significant amounts are gypsum ($CaSO_4$ $2H_2O$), anhydrite ($CaSO_4$), and halite ($NaCl$). Up to 4000 meters of salt are thought to have formed in many parts of the Mediterranean.

Since these salts were deposited, they have been buried by hemipelagic muds and terrigenous sediments that now form a cover about 100-200 meters thick over the salts. Because of the active compressional tectonic setting of the Mediterranean, however, the sediments and underlying salt minerals are being folded and fractured. Small sub-basins also are forming in areas of strike-slip motion. Seawater penetrating the fractures is causing the salt minerals to redissolve.

This salt dissolution creates very dense salty pools of water, in the bottoms of the basins. Because the brine is about 10 times saltier than normal seawater, it is much denser and consequently stays below the normal Mediterranean seawater. There are insufficient bottom currents to mix the brine with the seawater. Since this briny water doesn't circulate, once the oxygen it contains becomes depleted by bacterial processes, it cannot be replenished. Thus, these brines are very salty and anoxic. If something happened to cause this deep water to come to the surface, all of the surface dwelling creatures would die immediately in this anoxic water. These deep basins are lifeless, but do provide a perfect environment for preservation of the organic matter sinking from the surface. Over geologic time, if a large quantity of organic matter is buried in the sediments of an anoxic basin, it may undergo chemical changes that turn it into petroleum.

As you can see from the Mediterranean example, loss of contact with the open ocean can lead to drastic changes in the chemistry of marine waters. Salinity can increase drastically in arid climates, or decrease drastically in humid areas with high rainfall or high river inflow. In this extreme example, the sea dried up completely! The example also shows that the effects of this event are still felt in the modern Mediterranean through the redissolution of the Miocene salt. This process is now preventing recirculation of the deep Mediterranean waters, and these waters are acquiring chemistries that are very different from normal seawater and toxic to marine life. Other marginal seas, bays and estuaries where exchange with the ocean is restricted are also susceptible to large variations in seawater chemistry. We will look at these environments in greater detail in Chapter 13.

Element	Amount in Oceans (grams)	Residence Time (years)
Sodium, Na	147×10^{20}	2.6×10^{8}
Potassium, K	5.3×10^{20}	1.1×10^{7}
Calcium, Ca	5.6×10^{20}	8.0×10^{6}
Silicon, Si	5.2×10^{18}	1.0×10^{4}
Manganese, Mn	1.4×10^{15}	7.0×10^{3}
Iron, Fe	1.4×10^{16}	1.4×10^{2}
Aluminum, Al	1.4×10^{16}	1.0×10^{2}

TABLE 7-4 Residence times of some elements in the oceans.

(Source: Data from Broecker, 1974.)

The Carbon Cycle

Carbon is the building block for all life forms on Earth; it is also abundant in carbonate minerals. We categorize carbon as either organic (from living things) or inorganic (from minerals). Carbon in organic matter is in a reduced state, having an average oxidation state of 4^-, whereas the carbon in inorganic matter is oxidized, having an average oxidation state of 4^+. Carbon atoms change oxidation state through a number of chemical reactions and thereby can travel from the organic to the inorganic side or vice versa, thus, the inorganic and organic carbon cycles are linked.

Most of the carbon in the sea is inorganic, trapped in vast quantities of calcium carbonate sediments preserved on the seafloor. Inorganic carbon also includes the reservoir of dissolved HCO_3^- and CO_3^{2-} ions in seawater and CO_2 in the atmosphere. Most of the organic carbon on Earth is nonliving, tied up in rocks and sediments rather than in living organisms. The nonliving organic carbon occurs as organic matter in shales, petroleum, coal, and as dispersed organic particulates in ocean floor sediments and seawater. Note that the burning of fossil fuels and deforestation represent a large flux of carbon from organic to inorganic reservoirs which cannot be easily returned to organic carbon reservoirs.

Figure 7-15a shows the carbon cycle, excluding the small percentage that is lost to the ecosystem for millions of years by burial in marine sediments. This cycle involves the uptake of carbon dioxide by algae and plants for their use in the photosynthetic process. Carbon dioxide is returned to the ocean water primarily through respiration of algae, animals, and microbial decomposition.

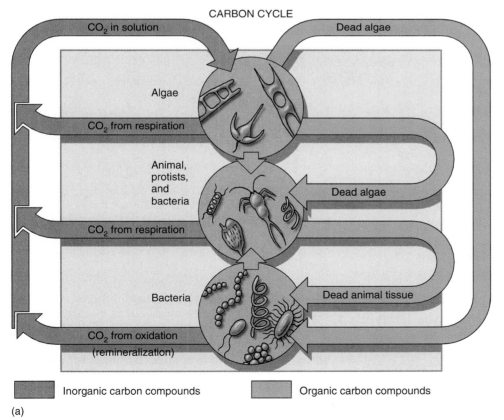

Figure 7-15   Carbon, Nitrogen and Phosphorus Cycles

NITROGEN CYCLE

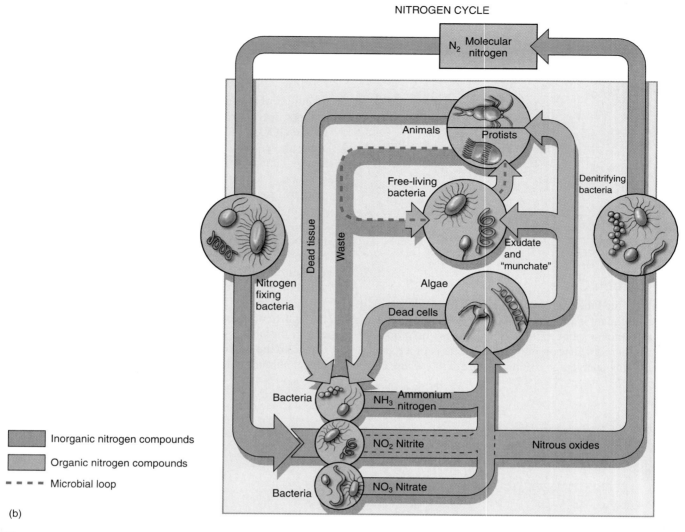

Inorganic nitrogen compounds

Organic nitrogen compounds

- - - - Microbial loop

(b)

Figure 7-15 (continued)

Nitrogen and Phosphorus Cycles

In addition to carbon, organisms need oxygen, nitrogen, and tiny amounts of sulfur and phosphorus in order to synthesize organic tissue. As we know from Section 7-4, oxygen dissolves in seawater, and provided exchange with the atmosphere occurs as rapidly as respiration, oxygen levels are sufficient to maintain life.

There is more sulfur than is needed by living things in the ocean. Phosphorus and nitrogen, however, are present in small quantities in the ocean and therefore are often the elements that limit phytoplankton growth. The ratio of carbon to nitrogen to phosphorus in the tissues of marine algae is on average 105/15/1, C/N/P. This relationship is named the Redfield ratio after Alfred C. Redfield who first discovered it. In surface seawater, the ratio is 1000/15/1, C/N/P. Thus, organisms are limited to using about one-tenth of the available carbon because of the availability of nitrogen and phosphorus. Because their concentrations are critical to ocean productivity, the phosphorus and nitrogen cycles are among the most studied in chemical oceanography.

Nitrogen compounds involved in photosynthetic productivity may be ten times the total nitrogen compound concentration that can be measured as a yearly average. This implies that the soluble nitrogen compounds must be cycled completely up to ten times per year. Available phosphates may be turned over up to four times per year.

The Nitrogen Cycle Figure 7-15b shows the nitrogen cycle. Nitrogen is essential in producing amino acids, the building blocks of proteins. The different bacteria involved in the nitrogen cycle make it somewhat complicated. Although some bacteria consume dissolved

PHOSPHORUS CYCLE

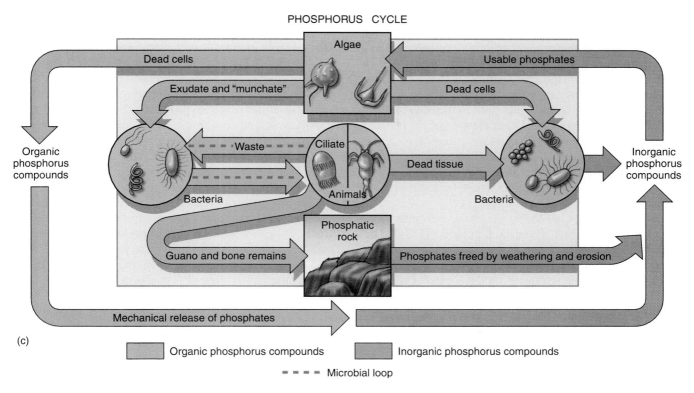

Figure 7-15   (continued)

organic matter or convert organic compounds to inorganic substances, others can bind molecular nitrogen (N_2) into useful nutrients such as nitrate (NO_3). These are **nitrogen-fixing bacteria**. Another special group is the **denitrifying bacteria**, whose metabolism depends upon the breakdown of nitrates and the liberation of molecular nitrogen.

Nitrogen availability is a limiting factor in marine productivity during summer in temperate oceans. Figure 7-15b shows that the conversion of organic nitrogen to the nutrient nitrate requires three stages of bacterial processing. These can take up to three months. By the time this conversion is completed much of the organic nitrogen that is being decomposed will have fallen below the euphotic zone.

Phosphorus Cycle   Figure 17-15c shows the phosphorus cycle. It is simpler than the nitrogen cycle. Enzymes within cells begin to break down organic molecules when some cells die. This autolytic breakdown along with the fact that only one stage of bacterial decomposition is required to convert organic phosphorus to inorganic nutrient form makes phosphorus more quickly available to phytoplankton.

Summary

The density of pure water increases with decreasing temperature, as do the densities of most substances. But water density increases only to 4°C (39.2°F), at which point it reaches its maximum density. Below 4°C, water density decreases with temperature, due to formation of bulky hydrogen-bonded lattices. At the point where water fully freezes, the density of the ice is only about 0.9 that of water at 4°C. Density increases with salinity and pressure. The unit, σ (sigma), is used to express seawater density.

The polarity of the water molecule, resulting from the uneven distribution of charge, gives water a very high dissolving power. Salinity is the amount of dissolved solids in ocean water, averaging about 35 g of dissolved solids per kilogram of ocean water. Salinity is usually expressed in parts per thousand (‰). Over 99 percent of the dissolved solids in ocean water come from six ions—chloride, sodium, sulfate, magnesium, calcium, and potassium. The rule of constancy of composition means that in any seawater sample, these ions are in the same proportion, so we can determine salinity by measuring the concentration of only one, usually the chloride ion.

Gases dissolve in seawater through atmospheric exchange processes and through the biological processes of photosynthesis and respiration. Salinity, temperature, and pressure affect the amount of gas that can be dissolved.

Depth profiles of many compounds in the ocean reflect atmospheric exchange, circulation patterns of the oceans, and reactions influencing ion concentrations. Two of the most important processes are photosynthesis and respiration. If atmospheric exchange processes and reactions are well known, then circulation patterns can be deduced from the depth profiles.

pH measures the concentration of hydrogen ions (protons) dissolved in water. The neutral point is pH = 7. The ocean has a great capacity to prevent changes in its pH because of the buffering capacity of its high alkalinity and the presence of carbonate sediments on the ocean floor. Much of the carbon dioxide added to the atmosphere by fossil fuel burning has apparently been taken up by the oceans.

The oceans also contain organic compounds in dissolved and particulate forms. Most of the organic matter derives from marine algae that convert oxidized carbon in CO_2 to reduced carbon in organic matter. Organisms that eat the algae and bacteria eventually recycle most of this organic matter back to CO_2.

Weathering reaction products contributed by river runoff, volcanic emissions, and basalt-seawater interactions are major processes contributing to the salt composition of the oceans. However, the ocean attained its present composition relatively early in its history and has maintained that composition through geologic time.

Biogeochemical cycling models Earth's environments and chemical reactions as reservoirs and fluxes, respectively. Biogeochemical cycles of particular importance in oceanography include the carbon, oxygen, nitrogen, sulfur and phosphorus cycles. Nitrogen and phosphorus availability often limit biological productivity.

Key Terms

Acidity	Dissolved organic carbon	Particulate organic carbon	Respiration
Alkalinity	(DOC)	(POC)	Salinity
Biogeochemical cycles	Flux	pH	σ (sigma)
Buffering	Halocline	Photosynthesis	Thermocline
Conservative tracers	Hydration sphere	Pycnocline	Weathering
Constancy of composition, rule of	Nitrogen fixing bacteria	Refractory organic matter	
Density	Nonconservative tracers	Reservoir	

Questions and Exercises

1. Explain how the polar nature of the water molecule makes it an effective solvent for ionic compounds.

2. As water becomes more saline, its temperature of maximum density and its freezing temperature decrease and converge. At what level of salinity does water cease to have a temperature of maximum density above its freezing temperature?

3. What condition of salinity makes it possible to determine chemically the total salinity of ocean water by measuring the concentration of only one constituent, the chloride ion?

4. If there is an estimated 18×10^{20} g of magnesium (Mg) in the ocean, and it is being added (or removed) at the rate of 56.6×10^{12} g/yr, what is the residence time of magnesium in the ocean?

5. What chemical compounds are the major focus of chemical oceanographers and why?

6. Why is the salinity of ocean water nearly constant?

7. How do biogeochemical cycles account for the concentrations of chemicals in the oceans?

8. Which processes seem to be most important in influencing the depth profiles of key chemical compounds?

9. Is understanding the structure of water and the nature of hydrogen bonding necessary to understand ocean chemistry? Why?

10. Describe the factors that control the amount of gas that dissolves in water.

References

Berner, E.K., and R.A. Berner, 1996. *Global Environment: Water, Air and Geochemical Cycles.* Upper Saddle River, NJ: Prentice-Hall.

Broecker, W.S., and T. Peng, 1982. *Tracers in the Sea.* New York: Eldigio Press-Columbia University.

Gregor, C.B., R.M. Garrels, F.T. Mackenzie, and J.B. Maynard, 1988. *Chemical Cycles in the Evolution of the Earth.* New York: J. Wiley and Sons.

Hammond, A. L., 1977. Oceanography: Geochemical tracers offer new insight. *Science,* 195:164–66.

Harvey, H. W., 1960. *The Chemistry and Fertility of Sea Waters.* New York: Cambridge University Press.

Mackenzie, F.T., 1998. *Our Changing Planet: An Introduction to Earth System Science and Global Environmental Change,* 2nd ed. Upper Saddle River, N.J.: Prentice-Hall, Inc.

Millero, F.J., 1997. *Chemical Oceanography.* Boca Raton, FL:CRC Press.

Packard, G. L., 1975. *Descriptive Physical Oceanography,* 2nd ed. New York: Pergamon Press.

Pilson, M.E.Q., 1998. *An Introduction to the Chemistry of the Sea.* Upper Saddle River, N.J.: Prentice-Hall, Inc.

Schlesinger, W.H., 1997, Biogeochemistry: An Analysis of Global Change. 2nd ed., San Diego, CA: Academic Press.

Suggested Reading

Behrenfeld, M.J., and Z.S. Kolber, 1999. Widespread iron limitation of phytoplankton in the south Pacific Ocean. *Science* 283: 840–843. Describes a study of iron limitation on phytoplankton populations and the use of fluorescence in measuring photosynthetic productivity.

Bidle, K.D., and F. Azam, 1999. Accelerated dissolution of diatom silica by marine bacterial assemblages. *Nature* 397: 508–512. Describes the role of bacteria in the silica cycle.

MacIntyre, F., 1970. Why the sea is salt. *Scientific American* 223:5, 104–15. A summary of processes that add and remove elements dissolved in the ocean.

Smith, F., and R. Charlier, 1981. Saltwater fuel. *Sea Frontiers* 27:6, 342–49. Considers potential for using the salinity difference between river water and coastal marine water to generate electricity.

Oceanography on the Web

Visit the Introductory Oceanography home page for on-line resources for this chapter. There you will find an on-line study guide with review exercises and links to ocean-ography sites to further your exploration of the topics in this chapter. Introductory Oceanography is at http://prenhall.com/thurman (click on the Table of Contents menu and select this chapter).

Toxic Chemicals in Seawater

- What are the effects of polychlorinated biphenyls and tributyl tin on marine organisms?
- What are their concentrations in seawater and in the tissues of marine organisms?

Introduction

Polychlorinated biphenyls (PCBs) are a group of over 200 synthetic chemicals of a type called **organochlorines**. Their formula is complex and their atomic weight depends on the number of relatively heavy chlorine atoms in the structure. PCBs do not exist naturally on earth; they were originally synthesized during the late nineteenth century. Because of their stability when heated, they were widely used in electrical capacitors and transformers. In the 1960s scientists began to report toxic effects on organisms exposed to PCBs, and by 1977, the manufacture of PCBs was banned in the U.S., the U.K., and elsewhere. By 1992, 1.2 million metric tons (2.6 billion pounds) of PCBs were believed to exist worldwide, while 370,000 metric tons (810 million pounds) were estimated to have been dispersed into the environment.

PCBs are relatively heavy molecules (average atomic weight of around 360 g/mole) and are relatively insoluble in water. Concentrations in seawater may reach 1 part per million (ppm), but PCBs typically concentrate in sediment. From there, they enter the food chain mainly through the activities of organisms called sediment- or deposit-feeders, which eat sediment, extract organic matter, and excrete the rest. PCBs tend to accumulate in the fatty tissues of animals and thus can become concentrated in animal tissue, a process called *bioaccumulation*. If animals eat the deposit-feeders, the PCBs move up the food chain and become concentrated, a process called *biomagnification*. While seawater concentration is usually below 1 ppm, concentrations exceeding 800 ppm have been measured in the tissues of marine mammals. According to the Environmental Research Foundation, this would qualify the creature for hazardous waste status!

PCBs have become widespread and serious pollutants, and have contaminated most terrestrial and marine food chains. They are extremely resistant to breakdown and are known to be carcinogenic. PCBs have been linked to mass mortalities of striped dolphins (Figure 1) in the Mediterranean, to declines in orca (killer whale; Figure 2) populations in Puget Sound, and to declines of seal populations in the Baltic.

PCBs have been shown to cause liver cancer and harmful genetic mutations in animals. PCBs may inhibit cell division, and they have been implicated in reduction of plant growth and even mortality of plants.

According to a report edited by Paul Johnston and Isabel Mc-Crea for Greenpeace UK,

"Since the rate at which organochlorines break down to harmless substances (has been) far outstripped by their rate of production, the load on the environment is growing each year.

Figure 1

Striped dolphins, whose mass mortalities in the Mediterranean Sea have been linked to PCBs. *(Photo by Erik Stoops, Corel Corporation)*

Organochlorines (including PCBs) are arguably the most damaging group of chemicals to which natural systems can be exposed."

PCBs and Orcas in Puget Sound

Even though soldiers during World War II used them for target practice, orcas have become a symbol of the Pacific Northwest.

In 1999, Dr. P.S. Ross, a research scientist with British Columbia's Institute of Ocean Sciences, took blubber samples from 47 live killer whales and found PCB concentrations from 46 ppm to over 250 ppm, up to 500 times greater than those found in humans. Ross concluded, "The levels are high enough to represent a tangible risk to these animals."

Ross compared the orca population he studied with the endangered beluga whale population of the St. Lawrence estuary of eastern North America, in which a high incidence of diseases have been linked to contaminants and which have shown evidence of reproductive impairment.

For the orcas, the PCBs are likely passed from generation to generation—an example of their persistence. PCBs are highly fat-soluble and are concentrated in mother's milk. Ross said, "...Calves are bathed in PCB-laden milk at a time when their organ systems are developing and they are at their most sensitive." While PCBs have been banned in the U.S. for 20 years, they are still being used in many developing countries. Moreover, PCB-laden waste has been transported from industrial countries to de-

Figure 2

The killer whale, an organism at the top of the food web in which high levels of PCBs have been found, an example of bioconcentration and biomagnification. *(Photo by Francois Gohier)*

veloping countries for "disposal." (PCBs can be removed from soil or sediment by incineration, but it is expensive—too expensive for most developing nations). Approximately 15% of PCBs reside in developing countries, mostly as a result of shipments from industrialized countries. Accordingly, Ross speculates that PCBs in the Pacific could be derived from East Asian sources and could end up concentrated in the tissues of migratory salmon, which are a prime food source for the orcas.

Ross' study, one of the most comprehensive on cetaceans (whales and their relatives) to date, was done in collaboration with the University of British Columbia, the Vancouver, B.C. Aquarium, and the Pacific Biological Station of British Columbia.

J. Cummins, in a 1988 paper in The Ecologist, stated that adding 15% of the remaining stock of PCBs to the ocean would result in the extinction of marine mammals.

Tributyl Tin

Tributyl tin (TBT) is the name given to a number of compounds with the general formula $(CH_3(CH_2)_3)Sn$. Typical variations are TBT chloride and TBT cyanide. A typical molecular weight is 320 g/mole. Added to marine paint, the compound inhibits the growth of barnacles and algae on the bottom of boats. However, tributyl tin kills not only organisms in direct contact with the boat (at sub-part per billion levels) but is also lethal to other marine organisms as it seeps out of the paint. TBT also bioaccumulates in shellfish. These discoveries led certain nations such as Australia, France, the United States, and others to ban TBT starting around 1980.

However, TBT is still widely used on large ocean-going vessels, so trace amounts of it are found in almost every harbor. For the impact of tributyl tin on shipyard workers, go to http://www.ban.org/ban_news/shipbreaking_is.html. For a review of the nature and impact of tributyl tin, try http://www.mbhs.edu/class/teched/matsci/stephen.html.

TBT in Japanese Tuna

A university research team has found that tuna and bonito in the waters around Japan have concentrations of TBT many times that found in fish in the South Pacific and Indian Oceans, but not believed high enough to harm humans. This TBT is presumably from anti-fouling paints used on ship hulls as well as from material used to protect fish nets. The researchers concluded that, in addition to contaminating local fish, the seas around Japan are now sources of TBT contamination of migratory fish like bonitos and tunas.

Shinsuke Tanabe, a professor at Ehime University's Agriculture School, and his colleagues collected 47 tuna and bonito from the central Sea of Japan, the waters off Kochi Prefecture, Papua New Guinea, the Indian Ocean and five other areas from 1983 through 1996. Tuna caught in the central Sea of Japan were found to have the highest concentration of TBT–320 nanograms (1 nanogram is equal to 1 billionth of a gram) as well as two other forms of tins. Tuna caught off Kochi Prefecture contained 310 nanograms, followed by 300 nanograms in bonito caught in the central Sea of Japan, according to the team. The researchers said those concentration levels are comparable to levels found in fish living in contaminated waters like Tokyo Bay or off the Italian coast. In the fish caught in the South Pacific or around the Philippines, the team found only 24 to 50 nanograms of tin. (Source: Environmental News Network, April 8, 1997.)

TBT in Dolphins

The February 8, 1997 issue of Science News contained a report of dolphins found dead along the coast of Japan which had accumulated a variety of butyl tin compounds, the breakdown products of tributyl tin.

Over the previous decade, dolphins along the Atlantic and Gulf Coasts as well as in the Mediterranean Sea experienced several mass die-offs. In the tissues of the dead dolphins scientists found TBT. While the finding doesn't prove that tributyl tin killed the coastal-dwelling bottle-nosed dolphins, there is other evidence that butyl tin compounds are potent immune system suppressors and they may have diminished the dolphins' ability to fight off the bacterial or viral infections thought to have caused the deaths. Exposure to TBT will continue because it persists in sediments and is still allowed on large vessels and aluminum hulls.

Now go to our web site to continue this analysis.

See the complete issue at:
www.prenhall.com/thurman

187

8 Air–Sea Interaction

- Solar Energy Received By Earth

- Heat Flow and the Coriolis Effect

- Heat Budget of the World Ocean and Climate Patterns

- Fog

- Ice Formation

- Energy From Ocean Thermal Gradients

- Summary

Satellite view of Hurricane Andrew which hit southern Florida in 1992 causing over $20 billion in damage (inset). (Courtesy of the National Hurricane Center, NOAA.)

Physical processes in the atmosphere create weather; similar "weather" occuring within the waters of the oceans as currents, turbulence, and mixing is driven by thermal gradients. The reason is simple: all fluids, which include gases and liquids, behave according to the same natural laws. For instance, the winds of the atmosphere correspond to currents in the oceans. The biggest difference is that seawater is about 1000 times denser than the atmosphere, so "weather" systems in the oceans develop more slowly and last longer than those in the atmosphere. Most of the currents and waves we observe in the surface ocean are created directly by atmospheric winds. Heating by solar energy creates the winds themselves. The ultimate source of energy for all of these forms of fluid motion, both in the atmosphere and in the ocean, is radiant energy from the Sun. The atmosphere and ocean constantly exchange this energy by reflection, absorption, evaporation, condensation, and precipitation.

Solar Energy Received By Earth ....

Most of the energy available at Earth's surface comes from the Sun. Solar energy strikes Earth at an average rate of 2 cal/cm^2/minute. Considering that the side of Earth facing the sun at any moment exposes about 255 trillion square centimeters to solar radiation, this is a tremendous amount of incoming energy. This energy drives the global ocean-atmosphere engine, creating pressure and density differences that create currents and waves in both air and sea. Solar energy is also the source of energy for almost all living things. However, if Earth's surface has

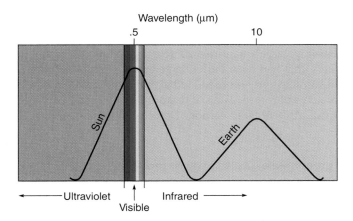

Figure 8-1   Comparison of Spectrums of Energy Radiated by the Sun and Earth

....
Most of the energy coming to Earth from the sun is within the visible spectrum and peaks at 0.48 μm (0.0002 in.). The atmosphere is transparent to most of this radiation, and it is absorbed at Earth's surface. When Earth reradiates this energy back toward space, it does so as infrared radiation with a peak at a wavelength of 10 μm (0.004 in.). The CO$_2$ and H$_2$O in the atmosphere absorb this infrared radiation to produce the greenhouse effect, which keeps the atmosphere's temperature higher than it otherwise would be.

maintained a constant average temperature over long periods, this energy must be re-radiating back to space at the same rate at which it is absorbed. This balance between energy input and output is called the **heat budget**.

The Heat Budget and the Atmospheric Greenhouse

The wavelength of energy radiated from a body is inversely proportional to the temperature of the body. Therefore, as is shown in Figure 8-1, the very hot Sun radiates energy most intensely at short wavelengths, averaging near 0.5 μm in the visible spectrum; Earth is a much cooler body and radiates at longer wavelengths, averaging around 10 μm, in the infrared range of the spectrum.

The average temperature of Earth's surface and the lower layer of the atmosphere is about 15°C (59°F). However, if the atmosphere contained no water vapor, carbon dioxide, methane, or other trace gases, Earth's average temperature would be far colder, about –18°C (–4°F). All of the water on Earth would be frozen. The more pleasant temperatures are made possible by the **greenhouse effect**, or insulating capacity of these gases. Water vapor is the most important greenhouse gas, accounting for about 24°C or 75 percent of the 33°C of total warming. Carbon dioxide accounts for 7°C, and methane contributes about 2°C.

It is the change of wavelength of energy radiating back from Earth's surface relative to the incoming solar radiation that is the key to the greenhouse effect. The greenhouse effect is somewhat analogous to the way a plant greenhouse operates. In a greenhouse, the short-wavelength solar energy passes through a transparent roof, striking the plants, floor, and other objects inside. It is absorbed by these objects and converted to heat. Heat raises the temperature in the greenhouse and also is re-radiated as heat energy at longer infrared wavelengths. It escapes the greenhouse at a magnitude equal to the incoming radiation at some elevated temperature.

Figure 8-2 shows the various components in Earth's heat budget and their estimated percentages of the total energy flux. In the upper atmosphere, much of the solar radiation within the visible spectrum (0.38–0.76 μm) penetrates the atmosphere to Earth's surface. About 47 percent of the solar radiation that is directed toward Earth is absorbed by oceans and land, about 23 percent is absorbed by the atmosphere and clouds, and about 30 percent is reflected into space by atmospheric backscatter, clouds, and Earth's surface.

Although Earth's atmosphere and clouds absorb only 23 percent of the shorter-wavelength incoming solar radiation, the outgoing longer-wavelength infrared radiation from Earth is absorbed in much greater quantities by water vapor, carbon dioxide, and other gases in the atmosphere. This occurs in the same way that air in a greenhouse is warmed by trapped heat energy. Some of the infrared energy absorbed in the atmosphere becomes reabsorbed by Earth and the process continues (the rest of the energy is

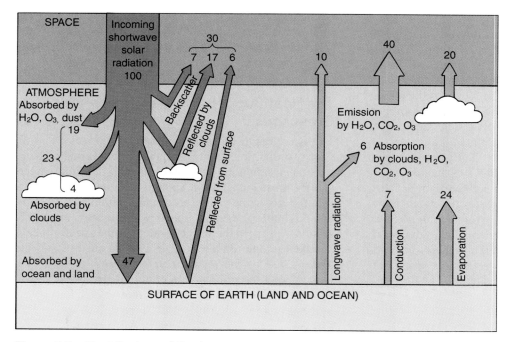

Figure 8-2   Heat Budget of Earth

Shortwave solar radiation from sun (mostly visible) is reflected, scattered, and absorbed by various components of the Earth-atmosphere system. All of the absorbed energy is radiated back into space by Earth as longwave infrared radiation. These values are estimates.

lost to space). Therefore, the solar radiation received is retained for a time within our atmosphere. It moderates temperature fluctuations between night and day and between seasons.

The Effect of Industrialization

Recent research has shown that the concentrations of some greenhouse gases in the atmosphere are increasing. These gases are listed in Table 8-1, in order of their potential contribution to increasing the greenhouse effect. Clearly, carbon dioxide dominates. The other trace gases—methane, nitrous oxide, tropospheric ozone, and chlorofluorocarbons—are present in far lower concentrations. However, they are im-

portant because, per molecule, they absorb far more infrared radiation than does carbon dioxide (Table 8-2). Thus, none of these gases should be ignored when considering the greenhouse effect.

In 1995, the U.N.-sponsored Intergovernmental Panel on Climate Change (IPCC) revealed unofficially that there was finally an international consensus that global warming "is unlikely to be entirely due to natural causes." Since the Industrial Revolution in the 1700s, humans have been burning large quantities of fossil fuels—coal, oil, and gas—that had been buried in sedimentary rocks. Mechanization in agriculture has led to the removal of thousands of acres of forests. Both of these are natural storage places for Earth's organic carbon. Deforestation and fossil fuel

TABLE 8-1	Estimated contributions of various greenhouse gases to increased greenhouse effect based on their present concentration and observed rate of increase in the atmosphere.		
Species	Concentration (ppbv)	Contribution to Greenhouse (%)	Increase (% per year)
CO_2	353000	60	0.5
CH_4	0.0017	15	1
N_2O	310	5	0.2
O_3*	10-50	8	0.5
CFC-11	0.28	4	4
CFC-12	0.48	8	4

*In the troposphere. ppbv = parts per billion by volume. Source: After H. Rodhe, 1990

TABLE 8-2 Comparison of increased greenhouse effect for known greenhouse gases.

The ability of one molecule of each of these gases to absorb infrared radiation is compared to one molecule of carbon dioxide. It is clear that all of these gases absorb infrared radiation much more efficiently than does carbon dioxide. Table 8-1 shows that their smaller overall contribution to increased greenhouse effect in the atmosphere is due to their low concentrations in the atmosphere compared to carbon dioxide.

Species	Mass basis (per kg)	Mole basis (per mole)
CO_2	1	1
CH_4	70	25
N_2O	200	200
O_3*	1800	2000
CFC-11	4000	12000
CFC-12	6000	15000

(Source: After H. Rodhe, 1990.)

burning releases previously stored carbon into the atmosphere. The basic reaction is chemically similar to respiration. For example, burning natural gas, or methane (CH_4), produces carbon dioxide and water:

$$CH_4 + 2O_2 = CO_2 + 2H_2O \qquad (8\text{-}1)$$

This process, of course, perturbs the natural carbon cycle. On a geologic time scale, the 300 years of human intervention amount to an extremely rapid change. Natural processes that buffer these perturbations act over much longer periods of time, on the order of thousands of years.

In 1958, Charles Keating began taking measurements of atmospheric gas compositions in remote locations. The data on carbon dioxide concentrations that he collected from a station on Mauna Loa in Hawaii were startling, showing nearly a 10-percent increase over a 34-year period (Figure 8-3a). To extend the record backward in time, scientists took ice cores from glaciers and analyzed the gases trapped in air pockets within the ice. Since the 1750s, atmospheric carbon dioxide had increased by 28 percent (Figure 8-3b). When trends for the other greenhouse gases—methane and nitrous oxide—were compiled by similar techniques, the picture was the same (Figures 8-3c and d). All of these gases are produced by industrial and agricultural processes.

Atmospheric scientists began to calculate the effects of these increases on the atmospheric greenhouse effect. A global increase in temperature was the obvious result—but when would it be noticeable, how rapidly would it rise, and how high would it go? To model climatic processes takes a lot of computing power and a lot of data. There are many interactive processes that must be accounted for to achieve reliable predictive models. Scientists test models by seeing how well they can predict historic data. There are two

atmospheric components that are particularly difficult to model: water vapor and small airborne particulates called aerosols. Water is a greenhouse gas; it absorbs infrared radiation quite effectively and, thus, can raise atmospheric temperatures. However, when water vapor condenses to form clouds, the white cloud surfaces reflect incoming radiation, causing cooling. Aerosols generally reflect energy, leading to cooling. Note, however, that aerosols may also form the nuclei on which water condenses, thus, the overall atmospheric effect depends on the interactions between these constituents. In addition, episodic events cause perturbations that must be factored out to test whether a longer-term model is a good match for historical trends. For example, the volcanic eruption of Mt. Pinatubo in 1991 emitted a large quantity of gas and aerosols into the atmosphere. This event may have cooled the climate by an average of one-half degree.

The Kyoto Protocol

The best climatic models presently available indicate that Earth's average surface temperature has risen by 0.5°C (0.9°F) in the last 100 years. However, there is no clear, simple proof that this has resulted from the increase in atmospheric carbon dioxide or other greenhouse gases. Many scientists are convinced that it has; others are not so certain. Also, some scientists who develop computer models of climate have evidence that greenhouse warming may evaporate more seawater, using up much of the excess heat.

The consequences of global warming are not just balmier weather. In many parts of the world, people would probably like to enjoy slightly warmer weather in the winter. However, computer models indicate that global warming will result in shifts in weather patterns, creating droughts in what are now prime agricultural regions, causing intensified rainfall and flooding in other areas, increasing the frequency and severity of hurricanes and typhoons, and melting glacial ice in polar regions. A global rise in sea-level is the consequence of glacial melting and the increased volume of the oceans resulting from warming. (Recall from Chapter 6 that the density of water above 4°C decreases with increasing temperature—in other words, the same mass of water will occupy a larger volume if you warm it up.) Given that about 75 percent of the world population live in low-lying coastal areas, even a 1-meter rise in sea level will have catastrophic consequences.

The mounting scientific evidence that global warming and some of its side-effects are occurring has led to international efforts to address the human contribution to the greenhouse effect. A number of international conferences have resulted in an agreement amongst 60 nations to voluntarily limit greenhouse gas emissions. This agreement, called the Kyoto Protocol because it was created at an international conference held in Kyoto, Japan, sets target reductions for each country. For example, the U.S. is committed to reducing emissions to 7 percent below 1992 emissions by 2007. The protocol also sets up processes by which technology

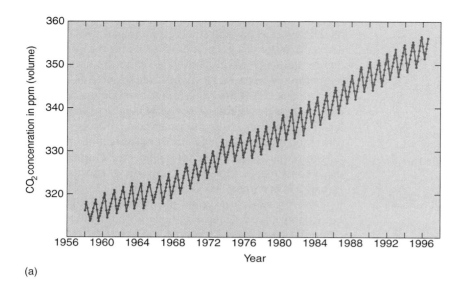

(a)

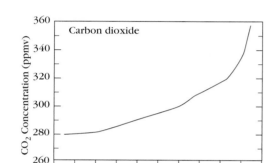

(b)

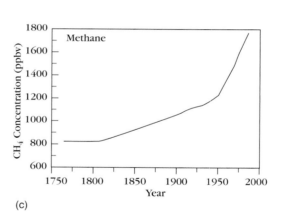

(c)

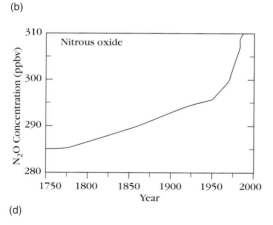

(d)

Figure 8-3   Atmospheric Gas Increases

(a) The concentration of atmospheric CO_2 at Mauna Loa Observatory in Hawaii, expressed in parts per million by volume of dry air (ppmv). The annual oscillation reflects the seasonal cycles of photosynthesis and respiration by biota in the Northern Hemisphere.   (Data courtesy of Charles Keeling and Timothy Whorf, Scripps Institution of Oceanograhy, University of California, San Diego.) (b, c and d) Historical trends in atmospheric CO_2, CH_4 and N_2O, respectively.   (*From Spiro, T.G. and Stigliani, W.M., 1996*, Chemistry of the Environment. *Saddle River: Prentice-Hall, Inc.*)

may be transferred to developing nations to enable them to industrialize without becoming producers of large quantities of greenhouse gases. See "Greenhouse Gases, Global CO_2 Emissions and Global Warming on pages 218–219.

Distribution of Solar Energy

The spherical shape of Earth results in an unequal distribution of solar energy at the surface. Consider a spherical surface; Figure 8-4 shows that a beam of light shining on a spherical surface strikes at an angle of 90° near the equator, forming a circle of light. If we keep the beam parallel to the equator, but move it up to shine on the surface above the equator, the angle of contact is less than 90° and the beam forms an ellipse that has a greater area than the circle of light we formed at the equator. At the pole, the light misses entirely (the light ray is tangent to Earth's surface). You can prove this to yourself with a flashlight, a round object, and some way to trace the beam outline on papers that you can overlay to compare areas.

In terms of the distribution of incident solar energy on Earth, if we consider a 1-km² (0.4-mi²) cross-sectional area of solar radiation that falls near the equator, this ray of light covers 1 km² of Earth's surface. By contrast, if we consider the same cross-sectional area falling on a high-latitude portion of Earth's surface, this 1-km² beam spreads out over considerably more than 1 km² of Earth's surface. The intensity of radiation available per unit of surface area is greatly decreased compared to that available in the lower latitudes.

The angle at which direct sunlight strikes the ocean surface is important in determining how much of the solar energy is absorbed and how much is reflected. If the sun shines down on a flat sea from directly overhead, only 2 percent of the radiation is reflected; however, 40 percent is reflected if the sun is only 5° above the horizon (Table 8-3).

Relative to its orbital plane around the sun, Earth's rotational axis is tilted. Thus, during different times of the year, the sun's rays are perpendicular to Earth's surface at latitudes above and below the equatorial plane (Figure 8-5). As Earth rotates on its axis, inclined at 23.5° from a perpendicular to the orbital plane, significant portions of its surface above 66.5°N latitude, the Arctic Circle, and 66.5°S

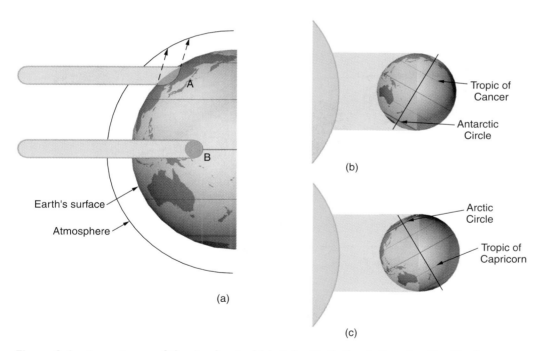

Figure 8-4   Importance of the Angle at which Solar Radiation Is Received by Earth

(a) The amount of solar energy received at higher latitudes is much less than that at lower latitudes for three reasons: (1) An area of solar radiation that strikes Earth at a high latitude, A, is spread over more surface than the same area of radiation that strikes Earth perpendicular to the surface at a lower latitude, B. (2) Radiation that strikes Earth at high polar latitudes also passes through a greater thickness of atmosphere. (3) More of the energy is reflected due to the low angle at which it strikes Earth's surface. (b) During the Northern Hemisphere winter, areas north of the Arctic Circle receive no direct solar radiation, and the Northern Hemisphere receives reduced radiation because of the low angle. However, areas south of the Antarctic Circle receive continuous radiation ("midnight sun") for months. (c) During the Northern Hemisphere summer, the situation is reversed. Throughout the year, the sun is directly overhead at noon at some latitudes between the tropics, affording intense, closer-to-vertical radiation to the hot equatorial area.

TABLE 8-3 Reflection and Absorption of Solar Energy Resulting from Different Angles of Incidence on a Flat Sea.					
Elevation of sun above horizon	90°	60°	30°	15°	5°
Percent of radiation reflected	2	3	6	20	40
Percent of radiation absorbed	98	97	94	80	60

latitude, the Antarctic Circle, spend up to 6 months in darkness. As the Earth progresses in its orbit, the direct rays of the sun migrate back and forth across the equator between the **Tropic of Cancer** (23.5°N) and the **Tropic of Capricorn** (23.5°S). At the beginning of Northern Hemisphere summer, the sun's rays are perpendicular to the Tropic of Cancer; at the beginning of Southern Hemisphere summer, the sun's rays are perpendicular to the Tropic of Capricorn.

The belt between the two tropics receives much more radiation per unit of surface area and undergoes less seasonal variation than the portions of Earth's surface to the north and south of this belt. At higher latitudes, there is much greater variation in the amount of solar energy, creating greater seasonal variation and generally cooler climates relative to tropical latitudes.

Heat Flow and the Coriolis Effect .....

Because of the low angle at which solar radiation strikes Earth's surface near the poles, and the very reflective ice cover in polar latitudes, more energy is reflected back into space than is absorbed. In contrast, because of the high angle at which sunlight strikes Earth at latitudes between 35°N and 40°S, more energy is absorbed than is radiated back into space. Figure 8-6 shows how this phenomenon is manifested in the average daily heat flow in the world ocean.

Figure 8-5   Attitude of Earth's Axis

As Earth orbits the sun during the year, its axis of rotation is constantly tilted at 23.5 degrees from perpendicular to the orbital plane (the plane of the ecliptic). This causes variation in the amount of solar energy that reaches each hemisphere through the year, resulting in the seasonal climatic variations we associate with spring, summer, autumn, and winter.

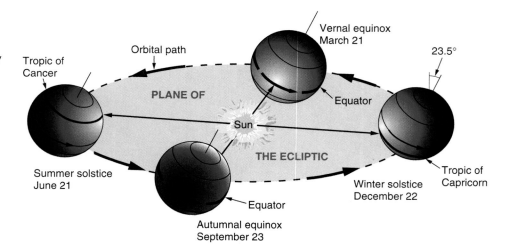

Figure 8-6   Heat Gain and Heat Loss from Oceans

Heat gained by the oceans at equatorial latitudes (red) equals heat lost at polar latitudes (blue). This creates a balanced heat budget for Earth's oceans.

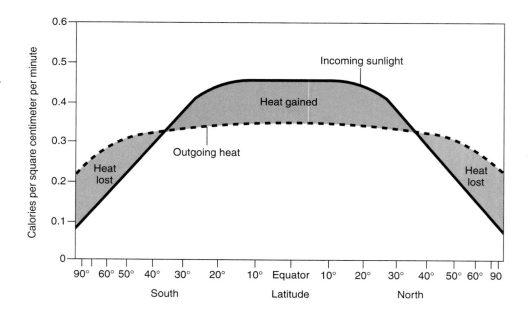

As a direct result of this condition, you might expect the equatorial zone to grow warmer as the years pass and the polar regions to become progressively cooler. Such is not the case. The polar regions are always considerably colder than the equatorial zone, but the temperature difference is not increasing. Clearly, excess heat at equatorial latitudes is being transferred toward the poles by some mechanism. This mechanism involves both the oceans and the atmosphere.

The Coriolis Effect

You might expect that patterns of atmospheric and oceanic circulation that transfer heat would be direct paths, forming straight, north-south arcs from areas of high heat to areas of low heat. However, the rotation of the Earth modifies these paths, diverting them to the east or west. Figure 8-7a shows that, as Earth rotates on its axis, points at different latitudes rotate at different velocities. The velocity depends on the distance from the point to Earth's axis of rotation. The surface velocity increases from 0 km/h at the poles to over 1600 km/h (nearly 1000 mi/h) at the equator. Any object that changes its latitude as it travels above Earth's surface is affected by this change in velocity. This is called the **Coriolis effect**. Calculating the Coriolis effect correctly is important to baloonists, airline pilots, anyone who launches rockets, meteorologists, and oceanographers.

To understand this concept better, let's launch some imaginary rockets. Imagine that you have three rockets, each of which will travel upward and then fall back to Earth in exactly one hour. You launch the first one at the North Pole (or South Pole), straight up. Where does it land an hour later? Earth is rotating beneath it, but you are on the axis of rotation, so it lands right where you launched it, on the pole. Now, again from the North Pole, let's launch the second rocket. This time, aim it toward a point along the 30°N latitude line, say the Canary Islands off the western coast of Africa. Where does the rocket land an hour later? Along the 30° latitude line, Earth is rotating eastward at about 1400 km/h (870 mi/h). Thus, the rocket will land in the mid-Atlantic, 1400 km west of its target. (Figure 8-7). Now let's fire the third rocket at the Canary Islands from a point on the equator directly south of the Canary Islands. The point on the equator from which the rocket is fired is moving east at 1600 km/h (1000 m), 200 km/h (124 mi) faster than the Canary Islands are moving. This means the rocket is also initially moving east 200 km/h faster than the Canary Islands. Thus, when the rocket returns to Earth one hour later at the latitude of the islands, it will land near the African Coast about 200 km east of the Canary Islands.

These examples demonstrate the apparent veering motion that the Coriolis effect causes. We can summarize the effect qualitatively. For the Northern Hemisphere, any object traveling horizontally above Earth's surface over a significant distance or time, will veer to the right of its direction of travel. You can easily remember this if you use your right hand, palm up with your fingers pointed in the direction of travel. Now, look at your thumb and you will see the direction of the Coriolis effect. In the Southern Hemisphere, an object will veer to the left of its direction of travel. You can easily remember this if you use your left hand in the same way we described above.

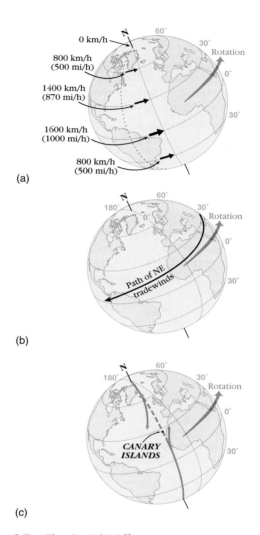

Figure 8-7 The Coriolis Effect

(a) Earth spins at a steady rate, but the actual velocity of any point varies with latitude. Because Earth measures about 40,000 km (about 25,000 mi) around at the equator, dividing that distance by 24 h in a day means that Earth's equator is traveling at more than 1600 km/h. Moving poleward, the velocity diminishes to 0 km/h at the poles. (b) To see the Coriolis effect, assume that an air mass is rotating with Earth at 30° latitude, which means it is traveling along with Earth's surface at 1400 km/h (870 mi/h). If it heads southward toward the equator at 32 km/h, it will reach the equator 100 h later. However, because the equator is rotating eastward faster than the air mass—some 200 km/h faster—the air will arrive at the equator at a point that is 20,000 km west of the point on the equator that was directly south of the air mass when it started moving (200 km/h × 100 h = 20,000 km). This air mass could represent the northeast trade winds, which follow a similar path. (c) Trajectories of two shot toward Tenenefe, Canary Islands; one from the North Pole and one from the equator. Dashed lines show the intended paths; solid lines show the trajectories would appear to follow as viewed from Earth's surface.

Atmospheric Circulation

Figure 8-8 shows how heat is transferred in the atmosphere. The greater heating of the atmosphere over the equator causes air to expand. This causes it to decrease in density and therefore to rise. As it rises, it cools by expansion, and the water vapor contained in the rising air mass condenses and falls as rain in the equatorial zone. After losing its moisture, this dry air mass travels some distance north and south of the equator. It then descends in the subtropical regions, around 30° latitude in both the Northern and Southern Hemispheres.

As the descending air approaches Earth's surface, it is warmed by compression. Upon reaching the surface it spreads away from the subtropics, moving both toward the equator and toward higher latitudes. This descending, denser air creates a high-pressure belt in the subtropics. Conversely, over the equator, the low density of the rising air creates a ring of low pressure that girdles the globe. The masses of air that move across Earth's surface from the subtropical high-pressure belts toward the equatorial low-pressure belt are the **trade winds**. Some of the air that descends in the subtropical regions moves along Earth's surface to higher latitudes as the wind belt called the **westerlies**. These masses rise over the dense, cold air moving away from the polar high-pressure caps at the subpolar

low-pressure belts located near 60°N and 60°S latitude. The air moving away from the poles produces the **polar easterlies**. The air that rises at the 60° latitudes cools, releases precipitation in these regions, and ultimately descends in the polar regions or in the subtropics.

These latitudinal patterns are a simplified picture of atmospheric circulation. In reality, the patterns are significantly altered by the uneven distribution of land and ocean over Earth's surface. The general processes of this idealized system are, however, clearly visible on a very broad scale.

Note in Figure 8-8 that Northern Hemisphere air masses move away from descending air in the subtropics both southward toward the equator and northward toward high latitudes. If Earth did not rotate, these would be simple north and south movements. However, the northeast trade winds do not blow directly from the north but from the northeast toward the southwest; likewise, the northward moving air, the westerlies, blow from the southwest to the northeast. These differences result from the Coriolis effect.

The trade winds blow from the northeast because, as the air mass moves southward from the subtropical region near 30°N latitude, it is also moving with the rotating Earth in an easterly direction at about 1400 km/h (870 mi/h). The air mass starts moving southward, toward a point on the equator at the same longitude. However, this point on the equator is moving eastward faster, at 1600 km/h

Figure 8-8 **Air Mass Circulation Around Earth**

Intense, year-round heating of the equatorial zone creates rising columns of warm air, developing a "permanent" low-pressure belt along the equator. In the 60°-latitude regions, winds riding up over cold polar air rise, create two more zones of low-pressure air. Descending cool, dry air produces high-pressure belts in the 30°-latitude regions (subtropics). Air movements are primarily vertical in these belts. Between the belts, however, strong lateral air movements occur, producing westerlies and trade winds. Due to cold, dense air masses overlying the polar regions, high-pressure conditions also exist there. The pattern shown here is general, but is extensively modified by the seasons and distribution of the continents.

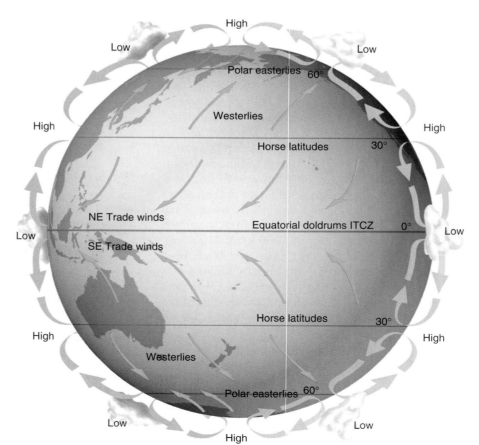

(about 1000 mi/h). The distance that the air mass must cover to reach the equator is about 3200 km (2000 mi). If it moves to the south at 32 km/h (20 mi/h), it will require 100 hours to arrive at the equator. The air mass is moving south at 32 km/h and east at 1400 km/h. For every hour that the air mass moves in the southerly direction, the point on the equator toward which it started moves farther east by 200 km (124 mi) than does the air mass (1600 km/h − 1400 km/h = 200 km/h). So, during the 100 hours it takes for the air mass to make the trip, the point on the equator toward which it started will be 20,000 km east of the air mass when it reaches the equator (100 h × 200 km/h = 20,000 km).

While the air mass was traveling 3200 km in a southerly direction, it appears as a result of Earth's rotation to have also moved 20,000 km in a westerly direction. Stated another way, as the air mass covered 30° of latitude from north to south, it appeared to move across 180° of longitude in a westerly direction. Certainly, if we were to encounter this air mass aboard a ship at some point between the equator and 30°N latitude, it would be coming out of the east-northeast.

A similar situation prevails for the winds that move northward from 30° latitude toward 60° latitude. These masses also are moving, along with Earth's surface, eastward at 1400 km/h. However, the point due north at 60° latitude, toward which the air mass is traveling, is moving in an easterly direction at only 800 km/h (500 mi/h). Unlike the situation for the trade winds, however, this westerly air mass is moving eastward at a velocity greater than that of the point at 60°N latitude toward which it started. For every hour that this mass moves northward, it also moves 600 km (375 mi) farther east than the point at 60° latitude toward which it started. If we were to encounter this air mass at some point between 30° and 60°N latitude, it would appear to come from the west-southwest. If the air mass moves northward at the same velocity as a trade wind mass moves southward, the deviation to the right will be greater for the westerlies than for the trade winds. This happens because the degree of Coriolis effect depends on the difference between rates that points at different latitudes rotate about Earth's axis.

We noted before that the rotational velocity of these points increases from 0 km/h at the poles to more than 1600 km/h at the equator. However, the rate of change with each degree of latitude is not constant. As we approach the pole from the equator, the rate of change of rotational velocity per degree of latitude increases. As a comparison, we note that there is a difference of 200 km/h in rotational velocity from the equator to 30°N latitude, but there is a difference of 600 km/h in rotational velocity from 30°N to 60°N latitude. From 60°N latitude to the pole, where the rotational velocity is zero, the difference is over 800 km/h. This explains why the Coriolis effect increases with latitude toward the poles. However, the magnitude of the effect depends mainly on the length of time that an air mass is in motion. Thus, even at low latitudes, a large Coriolis deflection is possible if an air mass is in motion for a long time.

Heat Budget of the World Ocean and Climate Patterns

As explained above, variations in latitude and absorption of solar energy make ocean temperatures vary from place to place and from time to time. Also, the temperature difference between polar and equatorial regions remains constant due to the transfer of heat energy from the equator to the higher latitudes by moving air and ocean masses. When wind blows over the ocean surface, friction drags the water, creating waves and currents. Consequently, large surface current systems, driven by the moving air masses shown in Figure 8-8, exist in the world's oceans. They play an important role in the transfer of heat energy.

Temperatures at various places in the ocean depend upon the rate at which heat flows in and out of these areas. Each ocean location has a heat budget, which describes the heat flow into and out of the area. The heat budget is expressed as:

$$\frac{\text{Rate of}}{\text{heat gain}} - \frac{\text{Rate of}}{\text{heat loss}} = \frac{\text{Net rate of}}{\text{heat loss or gain}} \quad (8\text{-}2)$$

In symbols, where Q is heat contributed or lost by the subscripted process, this equation becomes:

$$(Q_{\text{sun}} + Q_{\text{current}}) - (Q_{\text{radiation loss}} + Q_{\text{evaporation loss}} + Q_{\text{conduction loss}}) = Q_{\text{net}} \quad (8\text{-}3)$$

If Q_{net} is positive, it means the temperature of the ocean in that locality is rising. If it is negative, local temperature is falling. If it is zero, the temperature of the water is not changing. Q_{current} and Q_{net} should be zero if we consider the world ocean over a long period of time. Seasonal increases and decreases in temperature should average so that Q_{net} becomes zero. Therefore, the rate at which the world ocean gains heat through solar radiation should equal the rate at which it loses heat through radiation, evaporation, and conduction into the atmosphere (Figure 8-9).

The heat budget for the entire world ocean is expressed with these average values:

$$Q_{\text{sun}} = Q_{\text{radiation}} + Q_{\text{evaporation}} + Q_{\text{conduction}} \quad (8\text{-}4)$$
$$100\% = 41\% + 53\% + 6\%$$

Heat lost by radiation, primarily infrared, is greatest near the equator where the most solar energy is absorbed. Heat loss by evaporation is greatest in the dry subtropics. Heat lost through conduction reaches a maximum along the western margins of ocean basins where warm water currents carry water into regions where the atmosphere is much colder than the water.

Oceans, Weather, and Climate

At high and low latitudes, day-to-day weather may change little. Polar regions are usually cold and dry, regardless of the season. Near the equator also, day-to-day weather also is the same year-round; the air is warm,

Figure 8-9 Avenues of Heat Flow between the Ocean and the Atmosphere

Solar radiation (Q_{solar}) provides heat to the oceans. It is circulated within the ocean by currents ($Q_{current}$). Heat is lost from the ocean through evaporation ($Q_{evaporation}$), radiation ($Q_{radiation}$), and conduction ($Q_{conduction}$). Note that the heat loss totals 100 percent, equaling the heat gain from the sun.

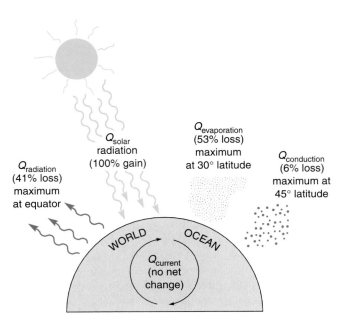

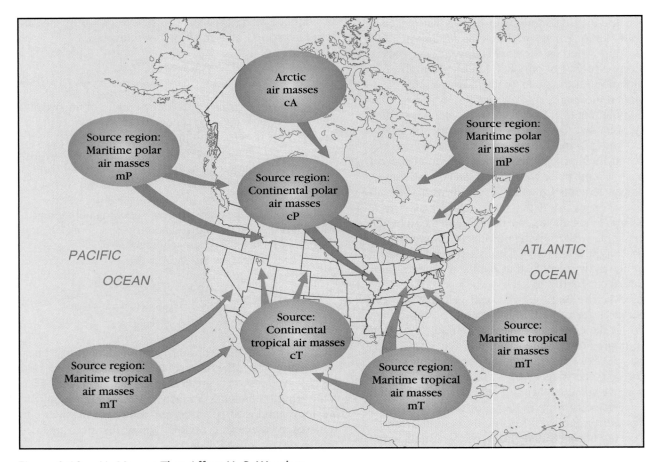

Figure 8-10 Air Masses That Affect U. S. Weather

Polar air masses (blue) are more likely to invade the United States during winter, and the tropical air masses (red) tend to move in from the south during summer. Air masses are classified on the basis of their source region. The designation continental (c) or maritime (m) indicates moisture content, whereas polar (P) and tropical (T) indicate temperature conditions. (*Reprinted by permission from Lutgens, F.K., and Tarbuck, E.J., The Atmosphere, 6th ed., Englewood Cliffs, N.J.: PrenticeHall, 1995, Figure 8-2.*)

damp, and still because air movement in the *doldrums* (the belt of maximum water vapor content) is primarily upward. Midday rains are common.

At midlatitudes weather gets interesting. Due to the seasonal change of pressure systems over continents, air masses from the high and low latitudes may move into the midlatitudes, where they meet and produce severe storms. In the United States, we are likely to be invaded by major polar air masses and tropical air masses with definite origins and distinctive characteristics (Figure 8-10). Note in the figure that some air masses originate over land (c = continental) and therefore are dryer; but most originate over the sea (m = maritime) and are moist. Some are cold (P = polar) and some are warm (T = tropical).

As polar and tropical air masses move into the midlatitudes, they also move gradually in an easterly direction. A *warm front* is a warm air mass moving eastward into an area occupied by cold air. A *cold front* is a cold air mass moving eastward into an area occupied by warm air (Figure 8-11). These fronts result from the movement of the **jet stream**, a fast-moving, easterly flowing air mass. It exists well above ground level, at about 10 km (6 mi) elevation above the midlatitudes. It usually follows a wavy path and may cause unusual weather by steering polar air masses far to the south or tropical air masses far to the north.

Regardless of whether the direction of colliding air masses makes the boundary between them a cold front or a warm front, the less-dense, warm air always rises above the denser cold air, where it cools, and the water vapor in it condenses as precipitation. A cold front is usually steeper, and the temperature differences across it are greater. Therefore, rainfall along a cold front is usually heavier and briefer than rainfall along a warm front.

Climate Patterns in the Oceans

We think of climate as applying to land areas where people live, but the term applies to regions of the ocean as well. We divide the open ocean into climatic regions that run generally east-west, and have relatively stable boundaries (Figure 8-12). In this section, we will look briefly at each region.

Figure 8-13 is a satellite view of water vapor over land and sea. Its patterns reveal that areas with high water vapor content indicate higher temperatures. In the equatorial region, the major air movement is vertical as the heated air rises, so surface winds are weak. Surface waters are warm, and the air is saturated with water vapor; this is quite noticeable in the figure. Heavy precipitation keeps salinity relatively low. Sailors once referred to this region as the *doldrums* because their sailing ships were becalmed by the lack of winds. Meteorologists refer to it as the **intertropical convergence zone** (ITCZ) because it is the region between the tropics where the trade winds converge (see Figure 8-8).

Figure 8-14 is a satellite view of wind speeds. Tropical regions, those between the Tropic of Cancer or Tropic of Capricorn, are characterized by strong northeasterly trade winds in the Northern Hemisphere and southeasterly trade winds in the Southern Hemisphere. These winds push the equatorial currents and create moderately rough seas. Relatively little precipitation falls at higher subtropical latitudes (about 30°), but precipitation increases toward the equator. Hurricanes and typhoons form in the doldrums and pass through the tropics as tropical storms. They carry large quantities of heat into higher latitudes.

Belts of high pressure occur over the subtropical regions. The dry air descending on the subtropics results in little precipitation and a high rate of evaporation, producing the highest surface salinities in the open ocean. Winds are weak in these parts of the open ocean, as are currents. However, strong boundary currents (along the east and west boundaries of the ocean basins) flow north and south, particularly along the western margins of the subtropical oceans.

The temperate regions are characterized by strong westerly winds blowing from the southwest in the Northern Hemisphere and from the northwest in the Southern Hemisphere. Severe storms are common, especially during

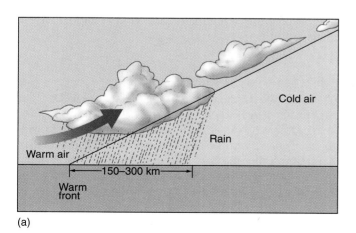

(a)

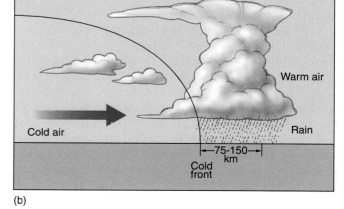

(b)

Figure 8-11   Warm and Cold Fronts

Cross sections through a warm front (a) and cold front (b).

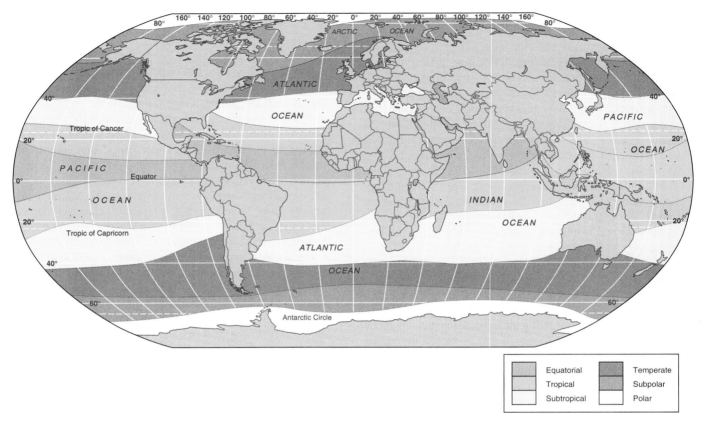

Figure 8-12 Climatic Patterns of the Open Ocean

The open ocean can be divided into climatic regions with relatively stable boundaries that generally run east–west.

winter, and precipitation is heavy. In fact, the North Atlantic is noted for fierce storms that have claimed many ships and many lives over the centuries.

The subpolar ocean is covered by sea ice in winter. It melts away, for the most part, in summer. Icebergs are common, and the surface temperature seldom exceeds 5°C (41°F) in the summer. Ocean surface temperatures remain at or near freezing in the polar areas and are covered with ice throughout most of the year. In these areas, which include the Arctic Ocean and the ocean adjacent to Antarctica, there is no sunlight during the winter, and no night during the summer.

The idealized pressure belts and resulting wind systems that we just discussed are significantly modified by two factors: (1) Earth's tilted axis of rotation and (2) the difference in heat capacities of water and land. As a result, during winter, continents usually develop atmospheric high-pressure cells due to the weight of cold air centered over them; during the summer, they develop low-pressure cells. As air moves away from high-pressure cells and toward low-pressure cells, the Coriolis effect comes into play. In the Northern Hemisphere, it produces a counterclockwise *cyclonic* flow of air around low-pressure cells and a clockwise *anticyclonic* flow of air around high-pressure cells (Figure 8-15). (The directions are reversed in the Southern Hemisphere.) Thus, wind patterns over continents may reverse themselves season-

ally as winter high-pressure cells are replaced by summer low-pressure cells.

Figure 8-16 shows average surface temperatures for the world ocean in August and February. These temperature belts migrate north and south with the seasons to some degree. Figure 8-17 shows how much greater the temperature range is for land than for coastal oceans and open oceans. Typical annual variations for open and coastal oceans at various latitudes are also shown. Figure 8-18 shows typical surface salinity patterns for the open ocean; they result from latitudinal variation in precipitation and evaporation.

Fog

There are two major sources of **fog** over ocean waters. The first forms over land on clear nights, which allows extensive heat loss by radiation. This fog usually forms in winter over marshy areas where the air has a high relative humidity during the day. The fog, which forms at night, may be carried by light winds out over coastal waters. The second variety forms over water when warm, moist air moves from a region with warmer surface-water temperatures to an area with colder water temperatures. Such fog usually occurs in spring or summer when air temperatures may be significantly higher than water temperatures. An example is the fog that forms when warm moist air moves

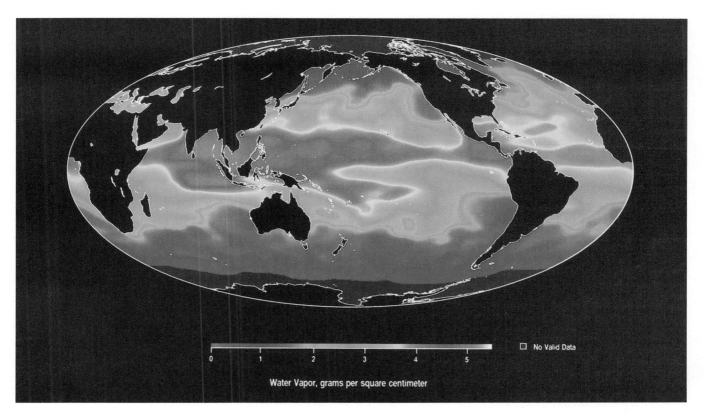

Figure 8-13   TOPEX/POSEIDON Water Vapor, October 3–12, 1992

A microwave radiometer aboard the TOPEX/POSEIDON satellites measures the water vapor in the atmosphere between the satellite and Earth's surface. It is expressed on the map in grams of water vapor per square centimeter. Most of the water vapor is in the lower atmosphere (troposphere). Because the ability of the atmosphere to hold water vapor increases with increased temperature, the greatest amount of atmospheric water vapor is found near the equator. This illustration reflects the great amount of heat being removed from the ocean by evaporation in the low latitudes (the doldrums). It is a region where warm air slowly rises and there are minimal horizontal air movements (wind) to propel sailing vessels or cool their crews. Because these data were taken in early fall, the great amount of heating in the Northern Hemisphere during the summer is still reflected in the shifting of the doldrums belt significantly north of the equator.   (*Courtesy of Jet Propulsion Laboratory, NASA.*)

from the Gulf Stream onto the Grand Banks off Newfoundland (Figure 8-19a).

Less common is **sea smoke**, or steam fog. It forms when cold air moves over ocean water that is about 10°C (18°F) warmer than the air. The lower layers of this cold air are warmed by water evaporated from the ocean surface and rise into the overlying cold air. Condensation occurs in a pattern that looks like smoke rising from the ocean surface (Figure 8-19b).

Ice Formation

Polar cold causes a permanent or nearly permanent ice cover on the sea surface. The term **sea ice** is used to distinguish such masses of frozen seawater ice from icebergs. **Icebergs** also are found at sea, but they are ice masses that originate by breaking off from glaciers, which form on land.

The freezing point of pure water is 0°C (32°F); however, the freezing point of seawater with a salinity of 35 parts per thousand (‰) is nearly 2 degrees lower, at 1.91°C (28.6°F). As water freezes at the ocean surface, the dissolved solids cannot fit into the crystalline structure of the ice. Thus, the salts are largely left dissolved in the surrounding seawater, and the salinity and density of the surrounding water increases. This greater salinity tends to lower the freezing point of the remaining water. However, the low-temperature, high-density water that is excluded from the ice tends to sink and be replaced by warmer, less dense water at the surface. This establishes a circulation that enhances the formation of sea ice, because freezing is aided by low salinity and calm water conditions. Sea ice is found throughout the year around the margin of Antarctica, within the Arctic Sea, and in the extreme high-latitude region of the North Atlantic Ocean.

Sea ice begins as small, needlelike, hexagonal (six-sided) crystals. These crystals eventually become so numerous that a slush develops. As the slush begins to form into a thin sheet, it is broken by wind stress and

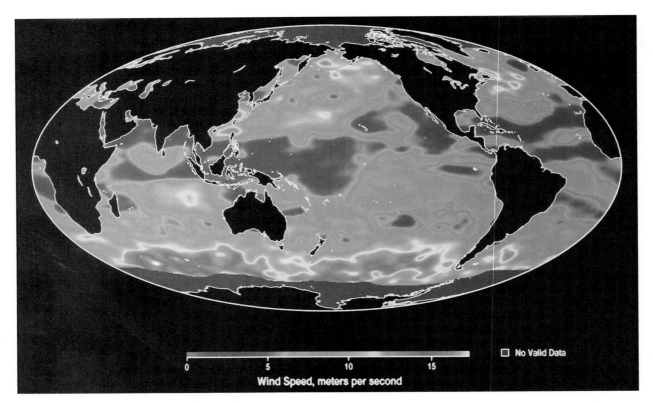

Figure 8-14 TOPEX/POSEIDON Wind Speed, October 3-12, 1992

This figure shows wind speed as measured by a radar altimeter aboard the TOPEX/POSEIDON satellites. A calm sea (low wind speed) returns a strong radar signal and a rough sea (high wind speed) scatters the radar signal sent down from the satellite and the return signal is weak. The lowest wind speeds occur in the doldrums, where the trade winds converge near the equator. The higher wind speeds occur in the westerly belts. Because there is less continental interference with the prevailing wind belts in the Southern Hemisphere, the strongest winds are in the westerly wind belt of that hemisphere. The highest wind speeds there are over 15 m/s (54 km/h or 34 mi/h). You can see why this region is often referred to as the "roaring forties." (*Courtesy of Jet Propulsion Laboratory, NASA.*)

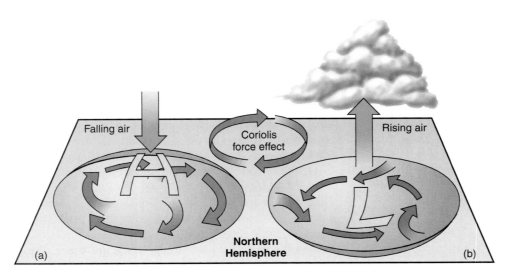

Figure 8-15 High-Pressure and Low-Pressure Cells and Air Flow

As air moves away from high-pressure cells (a) and toward low-pressure cells (b), the Coriolis effect causes the air masses (winds) to veer to the right in the Northern Hemisphere. This results in clockwise (anticyclonic) winds around high-pressure cells and counterclockwise (cyclonic) winds around low-pressure cells.

**Figure 8-16 (right)
Surface Temperatures
of the World Ocean
(°C)**

Average surface temperature distribution for August (a) and February (b). Note the warmest area, in the Pacific between Asia and Australia, and how temperatures migrate north–south with the seasons. (*After Sverdrup et al.*, 1942.)

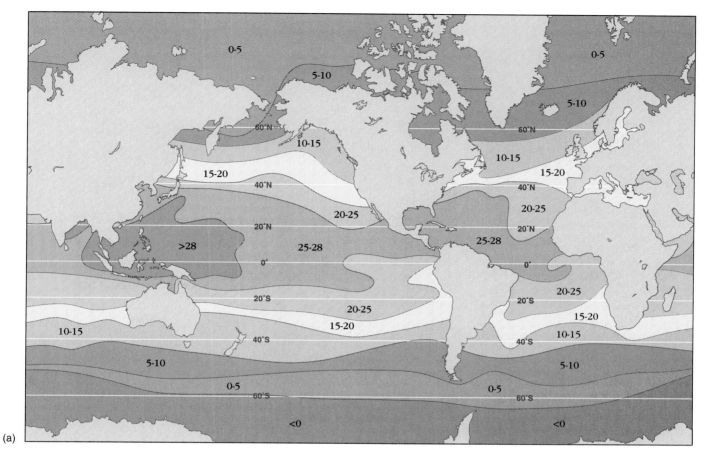

(a)

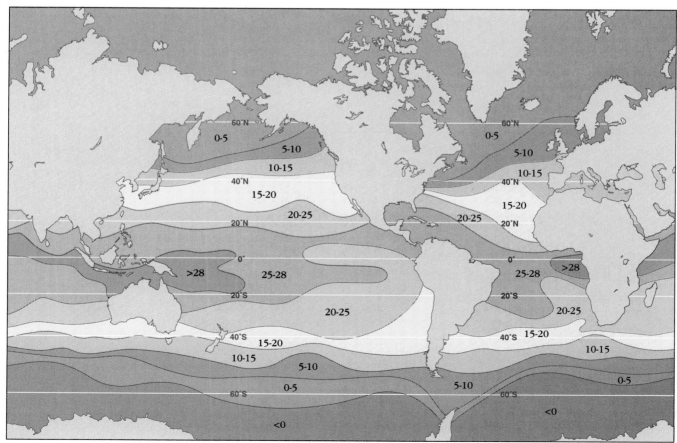

(b)

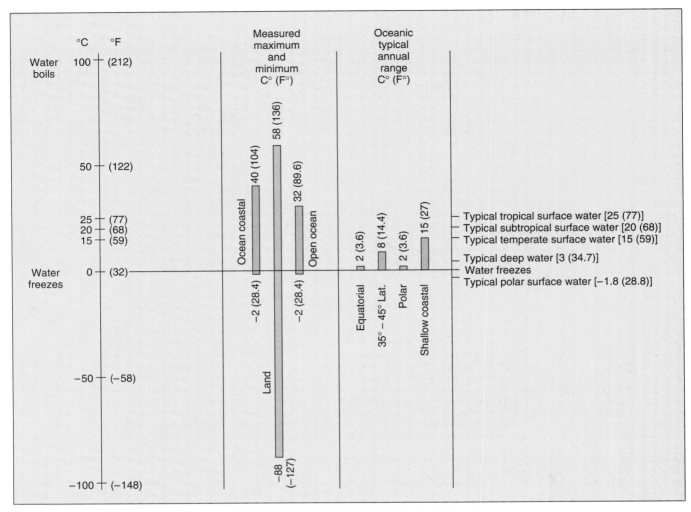

Figure 8-17   Comparison of Land and Ocean Temperature Ranges

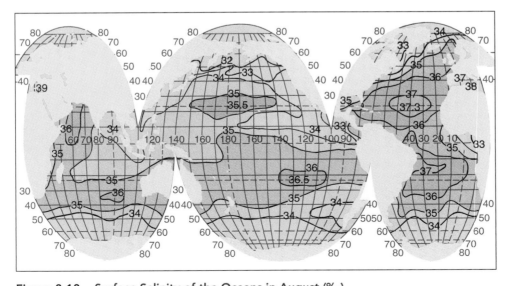

Figure 8-18   Surface Salinity of the Oceans in August (‰)

The dry air descending on the subtropics results in little precipitation and a high rate of evaporation, producing the highest surface salinities in the open ocean. (*After Sverdrup et al., 1942.*)

(a)

Figure 8-19 *Sea Fog and Sea Smoke*

(a) Water vapor and air are cooled by cold surface water and condense to produce sea fog. (b) Arctic sea smoke, formed when water evaporated from the warm ocean surface rises into overlying cold air. (*Part (a) Martin Bond, Science Photo Library; (b) Howard Bluestein, Science Source.*)

(b)

wave action into disk-shaped pieces called **pancake ice** (Figure 8-20a). As further freezing occurs, the pancakes coalesce to form **ice floes**.

The rate at which sea ice forms is closely tied to temperature conditions. Large quantities of ice form in relatively short periods when the temperature is very low, for example, $-30°C$ ($-22°F$). Even at low temperatures, the rate of ice formation slows as ice thickness increases due to the poor heat conduction of ice. Although newly formed sea ice contains little salt, it does trap a significant quantity of brine (salty water) during the freezing process.

Depending on the rate of freezing, newly formed ice may have a total salinity from 4 to 15 ‰. The more rapidly ice forms, the more brine it will capture and the higher the salinity will be. Over time, the brine will trickle down through the coarse structure of the sea ice, and the salinity of the ice will decrease. By the time it is a year old, sea ice has usually become relatively pure.

Pack Ice, Polar Ice, Fast Ice

In the Arctic Sea, ice that forms at the sea surface can be classified into one of three categories—pack ice, polar ice, or fast ice. **Pack ice** forms around the margin of the Arctic Sea (Arctic Ocean in some atlases). The ice extends through the Bering Strait into the Bering Sea on the Pacific side, and as far south as Newfoundland and Nova Scotia in the North Atlantic. Pack ice reaches its maximum extent during May, and its minimum extent when it breaks up in September (Figure 8-21). This ice, which can be penetrated by ships with reinforced hulls called *icebreakers*, reaches a maximum thickness of about 2 m (6.5 ft) in winter.

Pack ice is driven primarily by wind, although it also responds to surface currents which produce stresses that continually break and reform its structure. Pack ice formation is achieved by the expansion of floes that begin to raft onto one another as they expand (Figure 8-20c).

(a)   (b)

(c)   Rafted ice   Ridged ice   Weathered ridge ice   Hummocked ice

Figure 8-20   Sea Ice

(a) Pancake ice, freezing slush that is broken by wind stress and wave action into disk-shaped pieces. (b) Strong offshore winds blow across the Antarctic ice shelf near the Antarctic Peninsula. Ribbons of sea ice remain seaward of the shelf ice as the spring season begins (September in Antarctica). (c) Rafted ice. As ice floes expand, they slide over and beneath one another to become "rafted."   *(Part (a) Stephen J. Krasemann; (b) Courtesy of NASA.)*

The **polar ice** that covers the greatest portion of the Arctic Sea, including the polar region, attains a maximum thickness in excess of 50 m (over 160 ft). During summer, melting may produce enclosed water bodies called *polynyas*, although the polar ice never totally disappears. Its average thickness during summer is over 2 m (6.5 ft). Polar ice is constantly being exchanged, because floes from the adjoining pack ice are carried into the polar region during winter and floes break out of the polar ice and reenter the pack ice during summer. Circling in a clock-

wise direction around the Arctic Ocean, about one-third of the pack and polar ice is carried into the North Atlantic by the East Greenland Current each year.

Fast ice develops in the winter from the shore out to the pack ice and it completely melts during the summer. The fast ice, so-called because it is "held fast" or firmly attached to the shore, attains a winter thickness exceeding 2 m (6.5 ft).

In the Southern Hemisphere, where the polar region is not an ocean but a continental mass, we might characterize all the sea ice that forms around the Antarctic continental

Figure 8-21 (right)   Extent of Ice in Arctic Sea

These 1974 images show the percent of sea ice concentration in March (top) and September (bottom). They are accurate to within 15 percent for first-year ice (pack ice) and 25 percent for multiyear ice (polar ice cap). The scales of ice concentration along the right margins are for first-year ice. The other scale can be used for first-year ice (left side of scale) and multiyear-ice (right side of scale). The images were developed from the ESMR (Electrically Scanning Microwave Radiometer) aboard the *Nimbus 5* satellite.   *(Courtesy of NOAA.)*

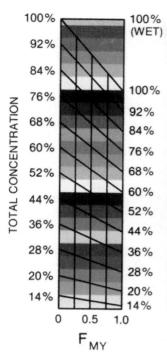

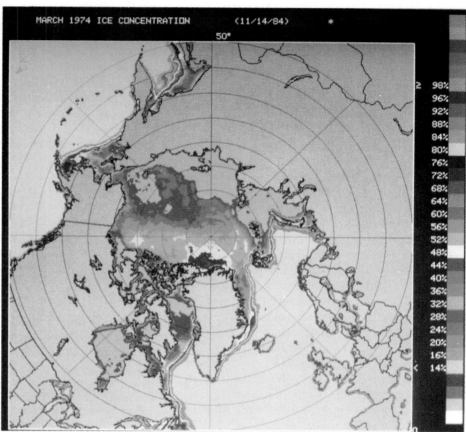

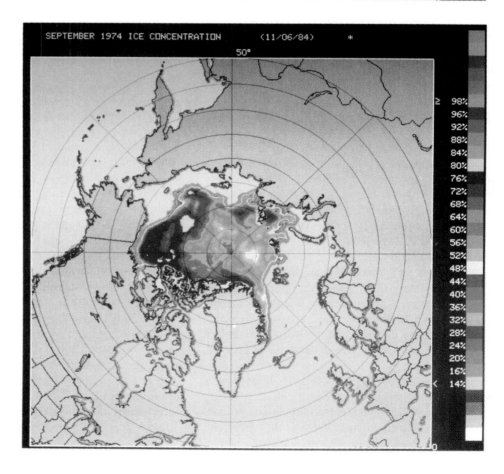

margin as pack ice and fast ice that have a rather temporary existence. The pack ice rarely extends north of 55°S latitude and breaks up rather completely from October to January except in the very quiet bays. Winds that are frequently very strong help prevent the formation of a greater pack ice accumulation around the Antarctic continent.

Icebergs

As previously noted, icebergs are quite distinct from sea ice because they derive from glaciers. In the Antarctic, source glaciers cover the continent. In the Arctic, glaciers cover most of the island of Greenland and part of Ellesmere Island (Figure 8-22). These vast ice sheets grow from snow accumulation on the landmass and spread outward until their edges push into the sea. There the ice sheets, which are less dense than seawater, are buoyed by the water and break up under the stress of current, wind, and wave action. This breakup, called *calving*, produces icebergs.

In the Arctic, icebergs originate primarily from glaciers that follows narrow valleys into the sea along the western coast of Greenland. Icebergs are also produced along the eastern coasts of Greenland and Ellesmere Island. The East Greenland Current and the West Greenland Current (arrows on map, Figure 8-22d) carry the icebergs up to 20 km (12.4 m) per day into the North Atlantic, where the Labrador Current may move them into North Atlantic shipping channels. They seldom are carried south of 45°N latitude, but during some seasons icebergs will move as far south as 40°N latitude and become shipping hazards.

Such an accumulation of icebergs had developed when *Titanic* sank in April 1912. After receiving repeated warnings of the iceberg hazard, the supposedly unsinkable 46,000-ton luxury ship proceeded with 2224 passengers at an excessive speed of 41 km/h (25.5 m/h) until it came to a grinding halt after hitting an iceberg with its starboard bow. Within hours, the Titanic sank at 41°46'N latitude, 50°14'W longitude near the Grand Banks, claiming 1517 lives.

The tragedy brought about the formation of the ice patrol that has since prevented further loss of life from such accidents. The U.S. Navy began this patrol immediately following the *Titanic* disaster; it became international in 1914. The patrol, maintained today by the U.S. Coast Guard, is concentrated between 40° and 48°N latitude and 43° and 54°W longitude.

A more recent major iceberg incident occurred in 1989. The Soviet cruise liner *Maxim Gorky* rammed an iceberg well north of the Arctic Circle between Greenland and one of the Spitzbergen islands (north of the area shown in the map). Quick action by the crew and the Norwegian Coast Guard prevented any loss of life among the more than 1300 people on board.

In Antarctica, the edges of glaciers form **shelf ice**, which pushes into the marginal seas and breaks off to produce icebergs. These icebergs have received less attention than the icebergs of the North Atlantic because they interfere less with shipping. Vast tabular icebergs with lengths exceeding 100 km (60 mi) break from the edges of the shelf ice (Figure 8-22c). They may stand up to 200 m (650 ft) above sea level, although most rise less than 100 m (330 ft) above sea level. In August, 1991, an iceberg the size of Connecticut (13,000 km² or 5000 mi²) broke loose from the ice shelf of the Weddell Sea. Two similar sized bergs formed in 1998 and 1999. Most calving occurs, as in the Arctic, during the summer months. When the sea ice breaks up, ocean swells driven by strong winds reach the edge of the shelf ice. Large icebergs are calved and move north into the warmer, rougher water, where they disintegrate. Carried by the strong West Wind Drift around Antarctica, these icebergs move east around the continent, but rarely migrate farther north than 40°S latitude. One of the effects of global warming and sea level rise may be an increase in calving rates of shelf ice.

Energy from Ocean Thermal Gradients

As we have discussed above, the sun imparts a great deal of energy to the sea and the atmosphere. Can this energy be harnessed to perform work for human purposes? The potential for extracting energy from the movement and heat distribution patterns of the atmosphere and ocean is attractive for several reasons:

1. Work can be achieved without significant pollution of air or water.
2. The amount of energy available at any time is far greater than that in fossil fuels (coal, oil, natural gas) and nuclear fuel (uranium).
3. The energy is renewable as long as the sun continues to radiate energy to Earth, and will not be depleted.

In order of decreasing energy potential, the sources of renewable energy are (1) heat stored in the oceans, (2) kinetic energy of the winds, (3) potential and kinetic energy of waves, and (4) potential and kinetic energy of tides and currents. In later chapters we will consider the use and potential of winds, waves, tides, and currents. Here we will consider only the renewable source with the greatest store of potential energy—the warm surface layers of the tropical oceans.

Ninety percent of Earth's surface between the Tropic of Cancer and Tropic of Capricorn is ocean. What makes this warm tropical surface water such an important source of energy is the presence of much colder water beneath the thermocline. With a temperature difference as small as 17°C (30.6°F), useful work can be done by **Ocean Thermal Energy Conversion (OTEC)** systems (Figure 8-23a). In an OTEC unit, warm surface water heats a fluid, such as liquefied propane gas or ammonia, that is under pressure in evaporating tubes. Heating vaporizes the fluid, and the vapor pushes against the blades of a turbine, which turns an electrical generator. After passing through the turbine, the fluid is condensed by cold water that has been pumped up from the deep ocean. It is again ready for heating by warm surface water that will cause it to vaporize and pass through

(a)

(b)

(c)

(d)

Figure 8-22   Icebergs

(a) Small North Atlantic iceberg. (b) Part of a large tabular Antarctic iceberg with sea ice in the foreground. (c) B-1 iceberg. This infrared image shows the huge iceberg that broke from the Ross Ice Shelf in Antarctica, October 1987. This iceberg was 136 km (84 mi) long, but two larger icebergs formed the year before in the Weddell Sea. (d) Map showing North Atlantic current and typical iceberg distribution (#).   *(Part (a) William W. Bacon, Rapho; (b) Joyce Photographics; (c) Infrared image courtesy of Robert Whritner, Manager, Antarctic Research Center, Scripps Institution of Oceanography, University of California, San Diego.)*

the turbine. Thus, heat energy stored in the ocean is converted to useful electrical energy. The system works in the reverse order of a typical refrigeration system and on a much larger scale.

Along the continental coast of the United States, the only region with good potential for Ocean Thermal Energy Conversion is a strip about 30 km (19 mi) wide and 1000 km (600 mi) long. It extends northward from southern Florida to St. Augustine, along the western margin of

the Gulf Stream. OTEC is also an attractive option for Hawaii, particularly because of that state's high expense of transporting fossil fuel to the islands. Testing at the Seacoast Test Facility, a shore-based OTEC installation on the island of Hawaii, demonstrated that OTEC power generation can be commercially successful.

Another proposed use of OTEC is for producing ammonia. A floating factory that could produce 586,000 tons of ammonia per year has been described in a feasibility report

Hurricanes and Typhoons— Nature's Safety Valves

The most spectacular examples of the significance of water's latent heat of evaporation to Earth's climate are those observed in the phenomena of hurricanes and typhoons. The following discussion of these storms focuses on some specific examples that have been remarkable for their size or for the destructive effects of their landfalls.

In both hemispheres, tropical storms usually develop near the end of the summer season (Figure 8A). They begin as low-pressure cells that break away from the equatorial low-pressure belt and grow as they pick up heat energy from the warm ocean. By means of the large heat capacity and latent heat of evaporation of the ocean's water, they transport great quantities of heat energy into heat-deficient higher-latitude regions by moving in curved paths around the western sides of the oceanic subtropical high-pressure cells (Figure 8B). When wind velocity within the storms reaches 120 km/h (74 mi/h), they are classified as hurricanes in the Western Hemisphere and typhoons in the Eastern Hemisphere. They are classified as to strength by the Saffir-Simpson Scale (Table 8A).

Hurricanes and typhoons can have diameters exceeding 800 km (497 mi), but more typically, they have a diameter measuring less than 200 km (124 mi). As air moves across the ocean surface toward the low-pressure eye of the hurricane and is drawn up to the outer reaches of the atmosphere, its sustained velocity increases to as much as

TABLE 8A	The Saffir-Simpson Scale			
Category	Pressure (hPa)	Wind Speed (m s^{-1})	(mi h^{-1})	Damage
1	Above 980	33–43	74–96	Minimal
2	979–965	43–50	96–112	Moderate
3	964–945	50–59	112–132	Extensive
4	944–920	59–69	132–155	Extreme
5	Below 920	70 or more	157 or more	Catastrophic

300 km/h (186 mi/h). Once near the eye, the air begins to spiral upward, and horizontal wind velocities may be negligible. Because of the Coriolis effect, the air flowing toward the eye of a storm veers to the right in the Northern Hemisphere and to the left in the Southern Hemisphere, producing counterclockwise and clockwise rotation, respectively (Figure 8B).

The energy contained in one hurricane is greater than that generated by all energy sources over the last 20 years in the United States. The energy released by an atomic bomb detonation in the center of a hurricane would have

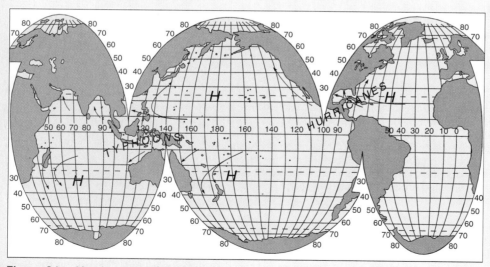

Figure 8A   Hurricanes and Typhoons

Hurricanes begin as low-pressure cells moving out of the equatorial low-pressure belt toward the higher latitudes in curved paths around the oceanic subtropical highs. They are called typhoons in the western Pacific and Indian Oceans and hurricanes in the western Atlantic and eastern Pacific Oceans.

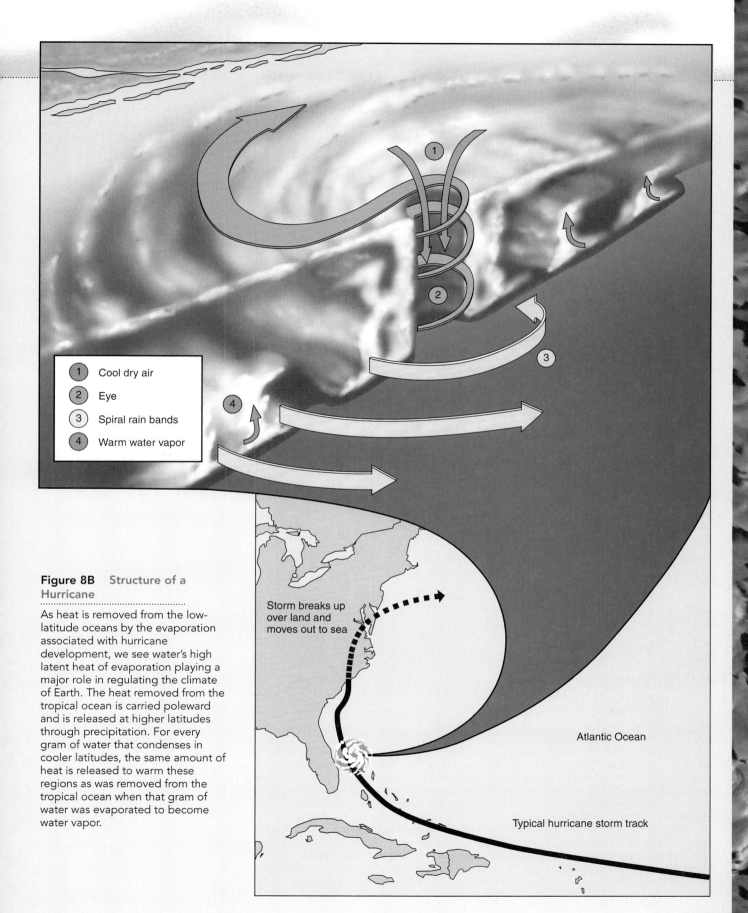

1. Cool dry air
2. Eye
3. Spiral rain bands
4. Warm water vapor

Figure 8B Structure of a Hurricane

As heat is removed from the low-latitude oceans by the evaporation associated with hurricane development, we see water's high latent heat of evaporation playing a major role in regulating the climate of Earth. The heat removed from the tropical ocean is carried poleward and is released at higher latitudes through precipitation. For every gram of water that condenses in cooler latitudes, the same amount of heat is released to warm these regions as was removed from the tropical ocean when that gram of water was evaporated to become water vapor.

Storm breaks up over land and moves out to sea

Atlantic Ocean

Typical hurricane storm track

no effect on the storm. It would quickly be sucked up through the eye and distributed throughout the system.

More than 30 percent of hurricanes are formed in the waters north of the equator in the western Pacific and are called typhoons. Another 18 percent form in the waters of the eastern Pacific, and 12 percent form in the Atlantic Ocean—mostly off the coast of Africa. The typhoons do much damage to islands and Asian coastal areas. The eastern Pacific storms usually die out in the vast expanse of cooler water between North America and the Hawaiian Islands. About once every 10 years, an eastern Pacific storm will make it to Hawaii. The greatest threat to life and prop-

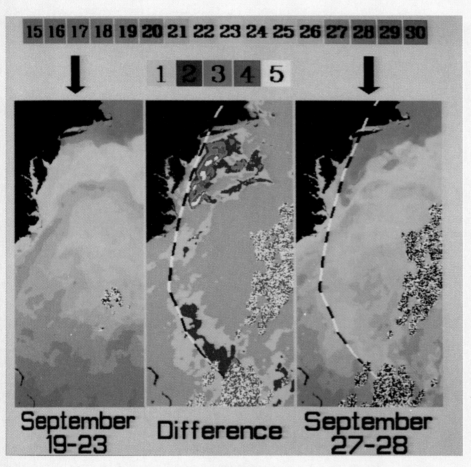

Figure 8C Sea-Surface Temperature (SST) Cooling—Hurricane Gloria, 1985

The path of hurricane Gloria is indicated by the black-and-white lines. The images to the left and right show the SST measured by the Advanced Very High Resolution Radiometer (AVHRR) aboard the *NOAA 9* satellite. The September 19–23 image reflects conditions just before the hurricane entered the area, and the September 27–28 image represents conditions just after the storm made landfall at Long Island. Note that the Gulf Stream, represented by the dark orange band just north of the center of the left-hand image, is broken up in the right-hand image. The temperatures represented by the colors on the end images are shown in the color band at the top. The center image shows the amount of cooling caused by Gloria, and the values in °C are shown on the short color band immediately above the center image. There was greater cooling along the north end of the track because cold subsurface water was closer to the surface there than to the south. The stippled areas represent cloud cover.   *(Courtesy of NOAA.)*

erty in the United States comes from the hurricanes of the Atlantic Ocean, which frequently find landfall along the North American coastline.

The worst natural disaster in U.S. history occurred on September 8, 1900, when a storm surge associated with a 135-km/h (84-mi/h) hurricane inundated the barrier island city of Galveston, Texas, killing 6000 people.

An illustration of the cooling effect of hurricanes on sea-surface temperature (SST) is the change that resulted from the passage of hurricane Gloria in September 1985. Its path out of the warm Sargasso Sea east of the Bahamas, along the East Coast, and across Long Island onto the mainland had the cooling effect shown in Figure 8C. The cooling of the ocean surface by as much as 5°C (9°F) resulted in part from evaporation and conduction, but much of the cooling resulted from cold water from beneath the surface mixing with the warmer surface water.

Hurricane Gloria was a strong hurricane, with a minimum central pressure of 919 millibars (mbar) compared to standard sea-level pressure of 1013 mbar. The lowest at-mospheric pressure ever recorded was 865 mbar in a typhoon off Okinawa in September 1945. The lowest pressure recorded in the Western Hemisphere was measured September 12, 1988, in hurricane Gilbert between Jamaica and the Cayman Islands. Gilbert at that point had sustained winds of 296 km/h (184 mi/h) and gusts of up to 320 km/h (199 mi/h). Figure 8D shows Gilbert two days later moving onto the Yucatán peninsula of Mexico, where it produced a 4-m (13-ft) storm surge. In all, 316 deaths were attributed to hurricane Gilbert. The island nations of Haiti, the Dominican Republic, and Jamaica suffered losses, as did Central American nations from Costa Rica to Guatemala. Mexico was hardest hit, with 202 deaths, most of which resulted from flooding caused by torrential rains as Gilbert came ashore near Monterrey. Property damage from Gilbert exceeded $10 billion, with Jamaica sustaining $6.5 billion of that total. Only $33 million in damage occurred in the United States. Most of the U.S. damage and the only U.S. death were from 29 confirmed tornadoes spawned by Gilbert in south Texas on September 16 and 17.

Figure 8D   Hurricane Gilbert

The strongest hurricane ever recorded in the Western Hemisphere, Hurricane Gilbert, moves onto the Yucatán peninsula of Mexico after having reached its fullest development.   (*Photo courtesy of Dave Clark, NOAA.*)

On August 24, 1992, Hurricane Andrew came ashore in southern Florida, doing more than $20 billion in damage before crossing the Gulf of Mexico and inflicting another $2 billion in damage. Andrew was by far the most destructive hurricane in history with regard to property damage. Winds reached 258 km/h (160 mi/h) as the hurricane crossed the Everglades ripping down every tree in its path. Fortunately, many people heeded the warnings to evacuate; 64 deaths resulted from that storm, which otherwise might have killed many more.

Even islands near the centers of ocean basins may be struck by hurricanes. The Hawaiian Islands were hit hard by Hurricane Dot in August 1959, and by Hurricane Iwa on November 23, 1982. Hurricane Iwa hit very late in the season with winds of up to 130 km/h (81 mi/h). Well in excess of $100 million in damage was done to the islands of Kauai and Oahu. Niihau, a small island occupied by 230 native Hawaiians, was directly in the path of the storm and suffered severe damage, but none of the population received serious injuries. Hurricane Iniki roared across the islands of Kauai and Niihau on September 11, 1992; with 210-km/h (130-mi/h) winds, it was the most powerful hurricane to hit Hawaii in the twentieth century. Property damage approached $1 billion.

Hurricanes will always be a threat to life and property. There is little hope that they will ever be controlled. Inhabitants of areas subject to a hurricane's destructive force need to be cautious of this important natural phenomenon. With present-day warning capabilities, prudent responses to hurricane warnings should help prevent much property damage and eliminate any great loss of life. By ignoring safety precautions, however, many lives will needlessly be lost in future hurricanes.

Reference: Thieler, E. R., and D. M. Bush. 1991. Hurricanes Gilbert and Hugo send powerful messages for coastal development. *Journal of Geological Education* 39:4, 291–98.

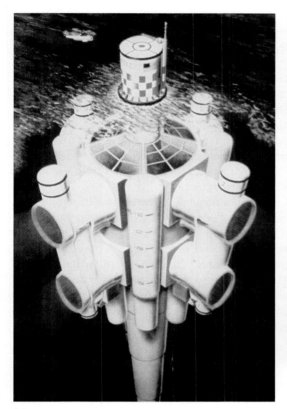

(b)

(a)

Figure 8-23   Proposed Uses of Ocean Thermal Power

(a) Ocean Thermal Energy Conversion (OTEC) system with crew quarters and maintenance facilities. Attached around the outside are turbine-generators and pumps. It is over 75 m (about 250 ft) in diameter, 485 m (about 1600 ft) long, and weighs about 300,000 tn. This unit is designed to generate 160 million watts of electrical power, enough to support a city of 100,000. (b) OTEC plant for ammonia production. Moving slowly through tropical waters, this plant could produce 1.4 percent of the U.S. annual ammonia requirement and save 22.6 billion ft^3 of natural gas.   *(Part (a) Courtesy of Lockheed Missiles and Space Co., Inc.; (b) Courtesy of U.S. Department of Energy.)*

from the Johns Hopkins Applied Physics Laboratory to the U.S. Maritime Administration. The proposed demonstration ship would weigh 68,000 tons, have a width of almost 60 m (200 ft), and a length of over 144 m (470 ft). It would move at less than 2 km/h (1.24 mi/h) through tropical waters and use energy derived from the temperature difference between warm surface waters and cold deep waters to produce ammonia. Each such plant could produce about 1.4 percent of our nation's ammonia requirements—75 percent of which is used for fertilizer—and save 22.6 billion ft^3 of natural gas each year (Figure 8-23b).

Summary

Radiant energy reaching Earth from the Sun is mostly in the ultraviolet and visible light range, whereas that radiated back to space from Earth is primarily in the infrared part of the spectrum. Water vapor, carbon dioxide, and other trace gases absorb the infrared radiation and heat the atmosphere. This phenomenon is called the greenhouse effect. Because human activities are increasing the concentrations of trace gases that enhance the greenhouse effect, there is concern that Earth's life zone may be warming.

The atmosphere is unevenly heated and set in motion because more energy is received than is radiated back into space at low latitudes, and because water vapor, which absorbs infrared radiation well, is unevenly distributed in the atmosphere. Belts of low pressure, where air rises, are generally found at the equator and at about 60° latitude. High-pressure regions, where dense air descends, are located at the poles and at about 30° latitude. The air at Earth's surface that is moving away from the subtropical highs produces trade winds moving toward the equator, and westerlies moving toward higher latitudes.

Because Earth's surface rotates at different velocities at different latitudes, increasing from 0 km/h at the poles to over 1600 km/h (1000 mi/h) at the equator, objects moving through the atmosphere tend to veer to the right in the Northern Hemisphere and to the left in the Southern Hemisphere. This is called the Coriolis effect.

In the heat budget of the world ocean, the only heat gain is from the Sun, whereas heat loss occurs through radiation, evaporation, and conduction into the atmosphere. The amount of heat lost by radiation, primarily of long infrared wavelengths, is greatest near the equator, where the most solar energy is absorbed. The greatest heat loss by evaporation is in the dry subtropics. Heat lost through conduction peaks along the western margins of oceans in temperate latitudes where warm water currents carry water into regions where the atmosphere is much colder than the water.

Earth's tilted axis of rotation and the distribution of continents both modify the idealized pressure belts. High-pressure cells form over continents in winter and are replaced by low-pressure cells in summer. There is a cyclonic (counterclockwise) movement of air around the low-pressure cells and a anticyclonic (clockwise) movement around high-pressure cells. At midlatitudes, cold air masses from higher latitudes meet warm air masses from lower latitudes and create cold and warm fronts that move from west to east across Earth's surface. Despite the modification of the idealized pressure belts, ocean climate patterns are closely related to them.

Sea ice forms when seawater is frozen in high latitudes. The process usually forms a slush, which breaks into pancakes that ultimately grow into floes. Sea ice develops as pack ice that forms each winter and melts almost entirely each summer, polar ice that is permanent in polar regions of the Arctic Sea, and fast ice that forms frozen to the shore during the winter. Icebergs form when large chunks of ice break off the large glaciers that form on Ellesmere Island, Greenland, and Antarctica.

From the motions and patterns of heat distribution in the atmosphere and oceans, renewable, nonpolluting energy can be exploited. These include heat stored in the ocean and potential and kinetic energy stored in winds, waves, tides, and currents. Ocean Thermal Energy Conversion (OTEC) is a process developed to use the difference in temperature between warm tropical surface water and the cold water below the thermocline to produce electricity and ammonia.

Key Terms

Coriolis effect	Iceberg	Pancake ice	Shelf ice
Fast ice	Intertropical Convergence Zone	Polar easterlies	Trade winds
Fog	Jet stream	Polar ice	Tropic of Cancer
Greenhouse effect	Ocean thermal energy	Sea ice	Tropic of Capricorn
Heat budget	conversion (OTEC)	Sea smoke	Westerlies
Ice floe	Pack ice		

Questions and Exercises

1. Explain why the atmosphere is heated primarily by re-radiation from Earth rather than by direct radiation from the Sun.

2. Discuss the greenhouse gases in terms of their relative concentrations and relative contributions to any increased greenhouse effect.

3. Describe the effect of the 23.5° tilt of Earth's axis of rotation.

4. As there is a net annual heat loss at high latitudes and a net annual heat gain at low latitudes, why does the temperature difference between these regions not increase?

5. Why are there high-pressure caps at each pole and high-pressure belts at 30° latitudes, and low-pressure belts in the equatorial region and at 60° latitudes?

6. Describe the Coriolis effect in the Northern and Southern hemispheres; include a discussion of why the effect increases with increased latitude.

7. Considering the heat budget of the world ocean over a long period, why should $Q_{current}$ (rate of heat gain or loss through ocean currents) and Q_{net} (net rate of heat gain or loss) be zero?

8. Describe the formation of sea ice from the initial freezing of the ocean surface water through the development of polar ice in the Arctic Ocean.

9. What is the difference in the average size and shape of icebergs in the Arctic and Antarctic? Why do these differences exist?

10. Construct your own diagram of how an Ocean Thermal Energy Conversion unit might generate electricity, or make a flow diagram presenting the steps of the process.

References

Changing climate and the oceans. 1987. *Oceanus*, 29:4, 1–93.

Charlson, R.J., et al., 1992. Climate forcing by anthropogenic aerosols. *Science* 255:5043, 423–30.

Deming, D., 1995. Climatic warming in North America: Analysis of borehole temperatures. *Science* 268:5217, 1576–77.

Ocean Energy. 1979. Oceanus 22:4, 1–68.

Oceans and Climate. 1978. Oceanus, 21:4, 1–70.

The Oceans and Global Warming. 1989. Oceanus 32:2, 1–75.

Pickard, G.L., 1975. *Descriptive Physical Oceanography*, 2nd ed. New York: Pergamon Press.

Rodhe, H., 1990. A comparison of the contribution of various gases to the greenhouse effect. *Science* 248:4960, 1217–19.

Spiro, T.G., and W.M. Stigliani, 1996. *Chemistry of the Environment*. Upper Saddle River: Prentice-Hall.

The Enigma of Weather—A Collection of Works Exploring the Dynamics of Meteorological Phenomena. 1994. *Scientific American*.

Suggested Reading

Sea Frontiers

Boling, G.R., 1971. *Ice and the Breakers*. 17:6, 363–71. An interesting history of people in icy waters and the development and improvement of icebreakers.

Charlier, R., 1981. *Ocean-fired Power Plants*. 27:1, 36–43. The potential of ocean thermal-energy conversion is discussed.

Houghton, R.A., and G.M. Woodwell, 1989. *Global Climate Change*. 260:1, 36–47. The history of global climate change.

Mayor, A., 1988. *Marine Mirages*. 34:1, 8–15. Discusses the nature of marine mirages and the research efforts that led to our understanding of them.

Scheina, R.L., 1987. *The Titanic's Legacy to Safety*. 33:3, 200–209. A brief summary of the Titanic sinking and an overview of the ice patrol and iceberg collision history subsequent to the sinking of that "unsinkable" luxury liner.

Smith, F.G.W., 1992. *Hurricane Special: An Inside Look at the Planet's Powerhouse*. 38:6, 28–31. Describes hurricanes, as they are called in the Atlantic Ocean, cyclones in the Indian Ocean, and typhoons in the Pacific Ocean—all huge cyclonic oceanic storms.

——. 1974. *Power from the Oceans*. 20:2, 87–99. A survey of the many tried and untried proposals for extracting energy from the oceans.

Sobey, E., 1979. *Ocean Ice*. 25:2, 66–73. The formation of sea ice and icebergs as well as their climatic and economic effects are discussed.

——. 1980. *The Ocean-Climate Connection*. 26:1, 25–30. The increasing knowledge of the effect of the oceans on Earth's climate may be used to predict climatic trends of the future.

Stuller, J., 1993. *Stormy Weather*. 39:4, 25–29. Describes the progress that has been made in predicting long-range changes in Earth's climates.

Scientific American

Herzog, H., Eliasson, B. and O. Kaarstad. 2000. Capturing Greenhouse Gases. 282:2, 72–79. We could reduce the global-warming effects of burning fossil fuels in carbon dioxide waste if we bury it deep underground or pump it into the ocean beneath the thermocline. Do we want to?

Jones, P.D., and T.M. Wigley, 1990. Global warming trends. 263:2, 84–91. Presents data relating to evidence of global warming over the past 100 years.

Karl, T.R., N. Nicholls, and J. Gregory, 1997. The coming climate. 276:5, 78–83. With Earth's temperature expected to rise 1–3.5°C by 2100, there may be more killer heat waves, tropical storms, and high-latitude precipitation.

Karl, T.R., and K.E. Trentberth, 1999. The Human Impact on Climate. 281:6, 100–05. We cannot quantify the effect humans are having on our future climate. We may be able to do so by 2050 if we commit to long-term climate monitoring now.

MacIntyre, F. 1974. The top millimeter of the ocean. 230:5, 62–77. Discusses processes that are confined to a thin film at the surface of ocean water and their role in the overall nature of the oceans.

Penney, T.R., and D. Bharathan, 1987. Power from the Sea. 256:1, 86–93. A prediction that generating electricity by ocean thermal energy conversion will become competitive with fossil fuel plants as the price of oil rises.

Pollack, H.N., and D.S. Chapman, 1993. Underground records of changing climate. 268:6, 44–53. Direct measurement of temperature shows Earth has warmed over the past 150 years. Data from boreholes may give us a more complete picture of the history of global climate.

Revelle, R., 1982. Carbon dioxide and world climate. 247:2, 35–43. Considers some of the possible effects of increasing atmospheric temperature due to carbon dioxide accumulation.

Stanley, S.M., 1984. Mass extinctions in the oceans. 250:6, 64–83. Geological evidence suggests most major periods of species extinction over the last 700 million years occurred during brief intervals of ocean cooling.

Stolarski, R.S., 1988. The Antarctic ozone holes. 258:1, 30–37. The discovery of the Antarctic ozone hole and its possible significance are discussed.

White, R.M., 1990. The great climate debate. 263:1, 36–45. Discusses the controversy over the degree of global warming we can expect in our future.

Oceanography on the Web

Visit the Introductory Oceanography home page for on-line resources for this chapter. There you will find an on-line study guide with review exercises and links to ocean-ography sites to further your exploration of the topics in this chapter. Introductory Oceanography is at http://prenhall.com/thurman (click on the Table of Contents menu and select this chapter).

Greenhouse Gases, Global CO_2 Emissions, and Global Warming

- What is the composition of the earth's atmosphere?
- What processes have influenced the atmosphere's composition?
- How does the earth's atmosphere interact with the ocean?
- What are greenhouse gases?
- What will be the impacts of global climate change associated with greenhouse gas increases?

Introduction

Earth's atmosphere is a relatively thin shell. About 95% of it is contained within 14 km (8.6 mi) of the earth's surface. All life on earth depends on the atmosphere, but despite our knowledge, humans are altering the atmosphere's composition. Scientific evidence suggests that emissions from the burning of fossil fuels, from industrial sources such as cement manufacture, and from deforestation are changing the make-up of our atmosphere. In addition, trace gases such as methane and chlorofluorocarbons (CFCs and HCFCs) are having an impact on the atmosphere wholly out of proportion to their concentration.

The Development of the Current Atmosphere

The earth's third and current atmosphere was produced partly by the metabolism of living organisms. Photosynthetic organisms, the cyanobacteria, appeared about 3.8 billion years ago and began to produce oxygen. Over the next 2 billion years, atmospheric O_2 concentrations rose and CO_2 concentrations fell as a direct result of photosynthesis. This increase in atmospheric oxygen then set the stage for two major evolutionary events on the planet: the evolution of aerobic (oxygen-using) life forms and the establishment of the ozone layer.

Increases in atmospheric and seawater levels of oxygen poisoned sensitive microorganisms and led to the evolution of other microorganisms capable of using oxygen to liberate energy from organic compounds. This new aerobic metabolism was much more efficient than the anaerobic metabolism it replaced. Today, natural anaerobic microbial communities are restricted in their habitats to deep marginal or isolated seas, organic-rich sediments, and regions beneath the earth's surface. The development of aerobic metabolism also permitted the evolution of multicellular organisms, which required more energy to support their increased biomass. All existing multicellular life forms employ aerobic metabolism.

Increases in atmospheric oxygen concentration eventually triggered the formation of the layer of ozone (O_3) that now exists in the upper atmosphere. Although considered a pollutant at ground-level, the ozone layer serves as an important filter of harmful ultraviolet radiation, which is responsible for sunburn and can cause skin cancer in humans.

Prior to the existence of the ozone layer, the earth's terrestrial and ocean surface were exposed to extremely high intensities of ultraviolet radiation. The surface ocean waters filtered out some of this radiation, and thus provided some protection to organisms, but it is likely that ocean primary production (that is, production of high-energy compounds from photosynthesis) was still limited by the high ultraviolet light intensities. The terrestrial surface fared worse and was effectively sterilized by this radiation.

The Atmosphere's Current Composition

The present-day atmosphere is composed primarily of N_2 gas (78.08% by volume), and oxygen (20.94% by volume). Also present, in quantities less than 1% by volume, are, in order: Ar, H_2O, CO_2, Ne, He, CH_4, NO_2, CO, NH_3, and O_3. The major controls upon the composition of the atmosphere and the cycling of these compounds are interactions with the Earth's biosphere (living matter) and lithosphere (rock and geological processes such as volcanism). Presently, O_2 levels are stable, but CO_2 levels are not and display seasonal and longer-term trends. Seasonal changes in CO_2 concentration are related to primary production changes due to changing light durations. Longer-term (decade–century) increases in CO_2 are due to a variety of anthropogenic (human-caused) inputs as well as changes in land use that reduce the ability of terrestrial biota to absorb CO_2.

Because the amount of CO_2 in the atmosphere is very small, the concentration is easily changed by the addition of CO_2 from various sources.

Atmospheric Functions

In addition to providing the oxygen needed by most of earth's life forms, the atmosphere provides a significant thermal insulation, preventing extreme changes in temperature over the daily light dark cycle. Unequal heating of the earth's atmosphere and terrestrial surface create long-term climate and short-term weather patterns. The winds that result from these heating differences and resultant pressure differences also drive ocean currents. The atmosphere also transfers heat.

	CO_2	CH_4	N_2O
Pre-industrial concentration	280 ppmv	700 ppbv	275 ppbv
Concentration in 1994	358 ppmv	1720 ppbv	312 ppbv[2]
Rate of concentration change[1]	1.5 ppmv/yr	10 ppbv/yr	0.8 ppbv/yr
Atmospheric lifetime (years)	50-200[a]	12[b]	120

ppmv = part per million by volume; ppbv = part per billion by volume

[1] Concentration increases in CO_2, CH_4, and N_2O are averaged over the decade beginning in 1984.

[2] Estimated from 1992-1993 data.

[a] No single lifetime for CO_2 can be defined because of the different rates of uptake by different processes.

[b] Defined as an adjustment time which takes into account the indirect effects of methane on its own lifetime.

Figure 1

Changes in the global concentration of greenhouse gases since the pre-industrial period.   *(IPCC, 1995)*

Changes Caused by Humans

After decades of research, scientists have finally concluded that humans have changed the composition of the earth's atmosphere. Since the beginning of the industrial revolution, atmospheric concentrations of the following greenhouse gases have changed: carbon dioxide has increased 30%; methane concentrations have more than doubled; and nitrous oxide concentrations have risen by about 15% (Figure 1). Greenhouse gases allow short wavelength radiation from the sun to pass through the atmosphere, but they absorb the longer wavelength radiation (i.e., heat) that is emitted by the earth. The need for energy to support industrial development, heat homes, cook food, watch television, and surf the internet, as well as the increased use of automobiles, has resulted in the burning of great stores of fossil fuel.

Fossil fuels, including coal, oil, and natural gas, were formed by the preservation and slow anaerobic decomposition of ancient plant and phytoplankton deposits. These deposits took tens of millions of years to form but we are now utilizing them at a rapid rate. A key by-product of fossil fuel consumption is CO_2. Since the industrial revolution, we have added CO_2 to the atmosphere more rapidly than it can be absorbed by its variety of sinks. This has led to a slow but steady increase in CO_2 concentration that will result in at least a doubling of pre-1860 atmospheric CO_2 content by the year 2150, if present trends continue. The increase in atmospheric CO_2 has occurred not only because of increased inputs into the atmosphere, but also due to changes in the landscape, such as deforestation, that result in less removal of CO_2 from the atmosphere.

Methane is another greenhouse gas whose concentration has also become elevated because of human activities. It is emitted by cows, and flooded farmlands (i.e., rice paddies), and both of these have increased dramatically within the last 2 centuries.

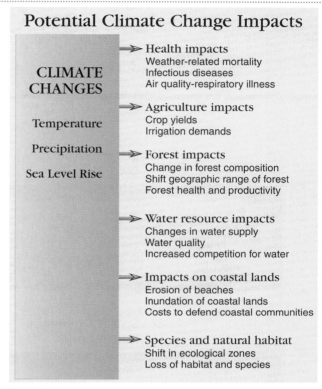

Figure 2

Impacts of climate change.   *(U.S. EPA)*

Global Warming Impacts

There is little controversy that human activities have caused these changes in the atmosphere, but there is intense debate regarding the magnitude of the problem that may result from these changes. However, knowledge that the 10 warmest years in this century all have occurred in the last 15 years has convinced most scientists that we are now seeing the direct response of the planet to the increase in greenhouse gases.

What will the impacts be of global climate change associated with greenhouse gas increases? Examine Figure 2 for an overview.

Now, go to our web site to continue this analysis.

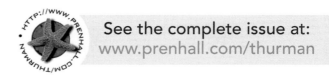

See the complete issue at:
www.prenhall.com/thurman

9 Ocean Circulation

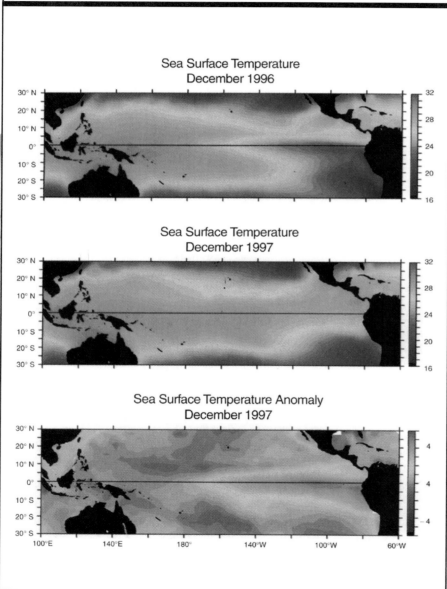

Computer images showing the development of the elevated surface temperatures in the eastern equatorial Pacific and the sea surface temperature anomalies associated with the 1997-1998 El Niño event. The 1997–1998 event was the most severe El Niño ever recorded.

Even though we think of the ocean as having great depth, ocean basins are very shallow compared to their widths. If the ocean basins were scaled down to be the width of this page, they would be no deeper than this page is thick! Yet within this thin shell of water, there is a rich, dynamic structure. Ocean currents can be categorized as being either wind-driven or density-driven. Wind-driven currents, set in motion by moving air, are horizontal and primarily affect the upper waters of the world ocean. Density-driven currents, created by temperature-salinity (*thermohaline*) differences between water masses, have significant vertical components and account for the formation and mixing of deep-ocean water masses. The effect of ocean currents on life is profound. Life in the deep sea is made possible only by a continuing supply of oxygen, carried there by cold, dense water currents that sink in subpolar regions and spread across the deep-sea floor. Ocean currents greatly influence the distribution of life throughout the oceans and, through their influence on climate, also powerfully affect the distribution of humans on land.

Horizontal Circulation

Horizontal circulation in surface waters develops from friction between the ocean and the wind. On a tiny scale, you can simulate this simply by blowing gently and steadily across a vessel of water, not hard enough to cause waves, but enough to start a water current flowing.

Boundary Currents

Trade winds blowing from the southeast in the Southern Hemisphere and from the northeast in the Northern Hemisphere are the principal drivers of the system of ocean surface currents. The trade winds thereby set in motion the water masses between the tropics. They develop the **equatorial currents** that travel along the equator worldwide. They are called *north* and *south* equatorial currents, depending on their position north or south of the equator. These currents move westward parallel to the equator. Deflection by the Coriolis effect and deflection along continental margins directs these currents away from the equator, as warm **western boundary currents,** that is, currents that travel along the western boundaries of their ocean basins. Figure 9-1 shows these warm currents as red arrows; the Gulf Stream is an example.

Between 30° and 60° latitude in both hemispheres, there are westerly winds that blow surface water in an easterly direction. These winds blow from the northwest in the Southern Hemisphere and from the southwest in the Northern Hemisphere. The Coriolis effect and continental barriers turn these waters toward the equator as cold **eastern boundary currents** that travel along the eastern boundary of their ocean basins. Eastern boundary currents are shown in blue in Figure 9-1; the Peru Current is an example.

The combination of western and eastern boundary currents produces the dominant feature of ocean basin surface circulation, the **subtropical gyre**, which rotates in a clockwise direction in the Northern Hemisphere (see North Atlantic Ocean in Figure 9-1) and in a counterclockwise direction in the Southern Hemisphere (see South Atlantic Ocean in Figure 9-1).

The polar easterlies are winds at subpolar latitudes that drive the surface currents in a westerly direction and, in combination with the Coriolis effect, produce **subpolar gyres** that rotate in a pattern opposite that of the adjacent subtropical gyres. Subpolar gyres are best developed in the Atlantic waters between Greenland and Europe and in the Weddell Sea off Antarctica (Figure 9-1).

Ekman Spiral and Transport

Why do water masses move as they do relative to wind direction? Fridtjof Nansen, during the voyage of the Fram (see Chapter 1), determined that the Arctic sea ice moved 20° to 40° to the right of the wind direction. He passed this information on to V. Walfrid Ekman, a physicist who developed the mathematics explaining this relationship. The circulation model that Ekman developed, the Wind Drift Model, is represented by the **Ekman spiral** (Figure 9-2). This model assumes that a uniform water column is being set in motion by wind blowing across its surface. Because of the Coriolis effect, the surface current moves in a direction 45° to the right of the wind in the Northern Hemisphere and to the left of the wind in the Southern Hemisphere. This surface mass of water moves with a velocity no greater than 3 percent of the wind speed. The water moves as a thin layer, and as it moves, it sets in motion another layer beneath it. As the wind energy put into the ocean surface is transmitted downward through successive layers, it is dissipated by water motion until none is left. Thus, with increasing depth, current speed decreases. The Coriolis effect causes a gradual change in current direction to the right.

Because each successive layer of water is set in motion at a slower velocity and in a direction to the right of the one above, a spiral pattern is created (Figure 9-2). At some depth, the momentum imparted by the wind to the moving water layers is so slight that no motion occurs as a result of wind at the surface. Although it depends on wind speed and latitude, motion normally ceases at a depth of about 100 m (330 ft), defining the base of the **Ekman layer**. The length of each arrow in Figure 9-2 is proportional to the velocity of the individual layer, and the direction of each arrow indicates its direction of movement. Under these idealized conditions, the surface layer should flow at an angle of 45° to the direction of the wind, as shown. However, the overall average transport of all the layers is a net water movement at 90° to the right of the wind direction. This is called **Ekman transport**.

Of course, the real ocean departs from idealized conditions. Frictional effects cause surface currents to move at an angle of less than 45° to the direction of the wind and, Ekman, transport to be at an angle of less than 90°

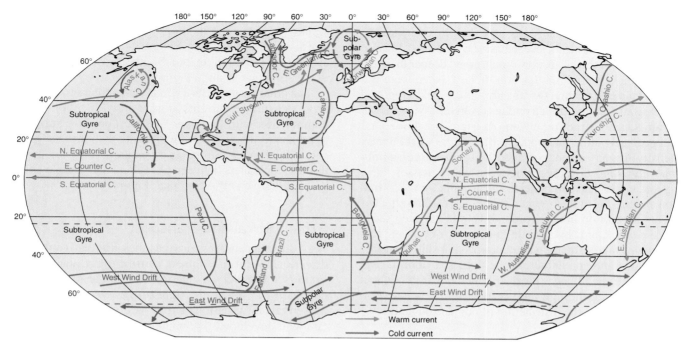

Figure 9-1   Wind-Driven Surface Currents in February and March

Major wind-driven surface currents of the world's oceans. Note the subtropical gyres that dominate the five major ocean basins: the North and South Pacific oceans, the North and South Atlantic oceans, and the South Indian Ocean. The gyres rotate clockwise in the Northern Hemisphere and counterclockwise in the Southern Hemisphere. The smaller subpolar gyres rotate in the reverse direction of the adjacent subtropical gyres.   *(After Sverdrup et al., 1942.)*

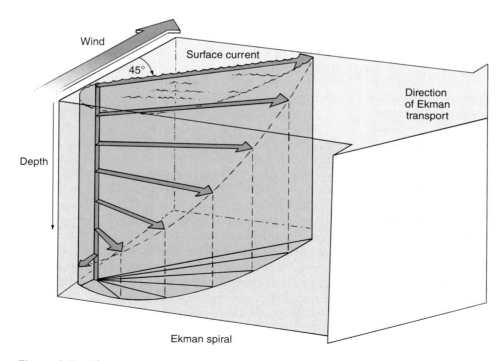

Figure 9-2   Ekman Spiral

Wind drives surface water in a direction 45° to the right of the wind in the Northern Hemisphere. Deeper water continues to deflect to the right and moves at a slower speed with increased depth. Ekman transport, which is the net water movement, thus is at right angles to the wind direction. This illustrates the principle, but in reality, the angles usually are somewhat less. The surface layer of water in which this spiraling occurs is called the Ekman layer.

to the direction of the wind. This is particularly true in shallow coastal waters, where all of the movement may be in a direction very nearly the same as that of the wind.

Geostrophic Currents

Let's consider Ekman transport in connection with the gyre in the North Atlantic Ocean. Ekman transport always turns water currents to the right in the Northern Hemisphere, and this clockwise rotation tends to produce a convergence of water in the middle of the gyre. Water literally piles up in the center. We find within all such ocean gyres a hill of water that rises as much as 2 m (about 6.6 ft) above the water level at the margins of the gyres.

As Ekman transport continually pushes water into this "hill" structure, gravity also acts to counter this effect, moving water down the surface slope. The Coriolis effect deflects the water flowing down the slope to the right in a curved path (Figure 9-3). The water piles up on these hills until the downslope component of gravity acting on individual molecules of water balances the Coriolis force. When these two forces balance, the net effect is a **geostrophic current** moving around the hill. The term *geostrophic* means Earth turning—the currents behave as they do because of the Coriolis force created by Earth's rotation.

What was just described is accurate, but idealized. Due to friction between water molecules, the water does converge and build up, but it gradually moves down the slope of the hill as it flows around it.

Westward Intensification   The apex of the hill formed within a rotating gyre is not in the gyre's center. The highest point of each hill is closer to the western boundary of the gyre. This is called **westward intensification**. Its causes are complex, but most of westward intensification can be explained by the Coriolis effect. Remember that the Coriolis effect increases toward higher latitudes (poleward). Thus, eastward-flowing, high-latitude water turns toward the equator more strongly than westward-flowing equatorial water turns toward higher latitudes. This causes a broad flow of water toward the equator across most subtropical gyres. In Figure 9-3b, we assume a steady state, with a constant volume of water rotating around the apex of the hill. The velocity of the water along the western margin is much faster than the velocity around the eastern side. This can be seen clearly in the North Atlantic and North Pacific gyres, where western boundary currents move northward in excess of 5 km/h (3.1 mi/h), but a flow along the eastern margin is more of a drift, moving well below 0.9 km/h (0.6 mi/h).

Western boundary currents commonly flow 10 times faster and to greater depths than eastern boundary currents. The warm western boundary currents are usually less than one-twentieth the width of the broad, cool drifts that flow toward the equator on the eastern side of subtropical gyres. Directly related to this difference in speed is the steepness of the hill's slope. The slope of the hill on the side of the slow-moving eastern boundary current is quite gentle; the western margin, on the other hand, has a comparatively steep slope corresponding to the high velocity of the western boundary current.

Dynamic Topography

Surface current velocity can be determined indirectly by measuring parameters controlling water density. The density distribution for an area is determined by collecting water samples at numerous locations over a depth interval. At each location, temperature and salinity are measured, and the average density of the water column is computed from the temperature and salinity characteristics measured above an arbitrary depth where the surface current is assumed to have died out (usually about 100 m, or 330 ft). No current is considered to exist at this depth and the pressures exerted downward by the weight of the overlying water columns are assumed to be equal at all locations. This assumption means that there is no horizontal gradient in pressure at our reference level, and that all measured water columns above this reference level must contain equal mass. A water column with a lower average density must have a greater volume, and therefore must stand higher above the datum than a water column of higher density. By computing for all locations the heights of water columns necessary to produce zero horizontal pressure gradients at the reference level, a topography of the ocean surface can be determined (Figure 9-4a).

Maps showing this topography can be prepared for any depth between the reference level and the ocean surface; they represent the **dynamic topography** that is used to compute current direction and velocity of the geostrophic current. The steeper the slope, the higher the velocity in a direction generally parallel to the topography contours (lines of equal height above the equal-pressure reference level). Rapid progress is being made in the use of radar altimeters mounted on satellites to map the dynamic topography of the oceans directly (Figure 9-4b).

Figure 9-4b is an image of average sea–surface elevation. The hills of water within the subtropical gyres of the Atlantic Ocean are clearly visible. In the Pacific Ocean, the hill in the North Pacific is easily seen, but the low equatorial elevations one might expect to see between the northern and southern subtropical gyres are missing. These data were recorded during a moderate El Niño event. Thus, the high stand of equatorial water is likely due to a well-developed equatorial countercurrent. The lower intensity of the South Pacific subtropical gyre also contributes to the lack of definition. Its lower intensity results from the great area it covers, plus its lack of confinement by continental barriers along its western margin. The South Indian Ocean hill is rather well developed, although its northeastern boundary stands high because of the influx of warm Pacific Ocean water through the islands of the East Indies.

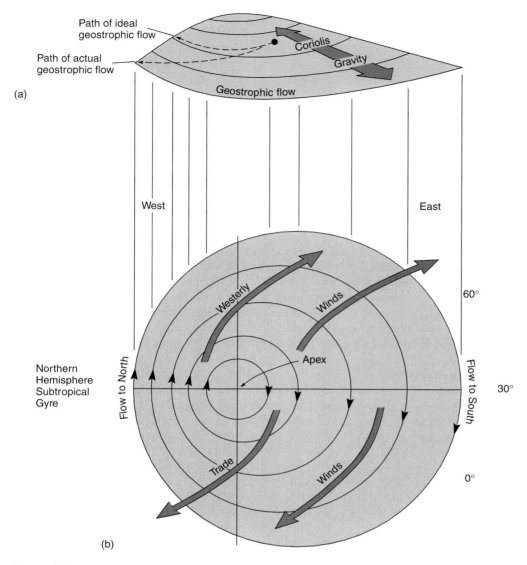

Figure 9-3   Geostrophic Current

(a) As Earth's wind systems set ocean water in motion, circular gyres are produced. Water literally piles up inside the subtropical gyre, forming a "hill" up to 2 m (6.6 ft) high. The apex of this hill is closer to the gyre's west margin because of Earth's rotation eastward. The theoretical geostrophic current would flow around the hill, in equilibrium between (1) the Coriolis force, which pushes water toward the apex through Ekman transport, and (2) gravity, which pulls the water downslope. However, friction between water molecules makes the current gradually run downslope. (b) The Coriolis effect is stronger on water farther from the equator. Thus, eastward-flowing, high-latitude water turns equatorward more strongly than westward-flowing equatorial water turns toward higher latitudes. This causes a broad, slow, equatorward flow of water across most of the subtropical gyre and forces the apex of the geostrophic hill toward the west. This phenomenon is referred to as westward intensification. Its main manifestation is a high-speed, warm, western boundary current that flows poleward along the hill's steeper western slope. The eastern slope has a slow drift of cold water toward the equator.

Equatorial Countercurrents

The large volume of water driven westward by the north and south equatorial currents piles up against continents because the Coriolis effect is minimal near the equator and does not turn the currents toward higher latitudes. Instead, the water piles up at the western margins of the ocean basins and very slowly flows back, creating **equatorial countercurrents**. This is particularly true in the western Pacific Ocean (see Figure 9-1) where a dome of equatorial water is trapped in the island-filled embayment between Australia and Asia. This dome of water has very weak current flow and displays the highest year-round sur-

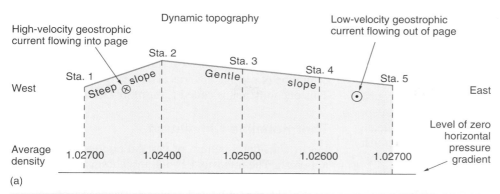

(a)

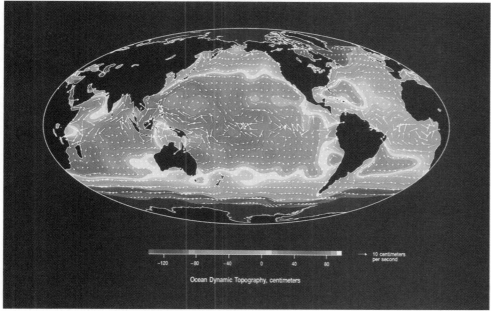

(b)

Figure 9-4   Dynamic Topography–Satellite View

(a) Comparison of average densities for the water columns above the arbitrary no-current datum tells us the apex of the hill is near station 2 because the lower the density of the water, the greater the volume a given mass will occupy. Northern Hemisphere geostrophic flow is indicated by circles beneath slopes. (b) TOPEX/POSEIDON Dynamic Topography, September 1992–September 1993. This radar altimeter determination of the dynamic topography of the world's oceans has a maximum relief of about 2 m (6.6 ft). The maximum elevation is in the western Pacific just seaward of the intense flow of the Kurishio Current. The westward intensification of the hills shows up very well in both the North Atlantic and South Atlantic gyres. The length of the arrows is proportional to current speed and helps demonstrate the relationship of ocean currents to the dynamic topography of the oceans.   *(Courtesy of Jet Propulsion Laboratory, NASA.)*

face temperature found anywhere in the world ocean (see Figure 8-16). The continued influx of water carried by the equatorial currents forces an eastward countercurrent of water that continues to South America.

Vertical Circulation

So far, we have discussed only horizontal currents in the ocean. However, vertical movements are essential in stirring the ocean, delivering oxygen to its depths, bringing nutrients to the surface, and distributing heat energy. Lat-

eral movements of water masses may bring about some shallow vertical circulation within the upper-water mass. We refer to this shallow vertical circulation system as *wind-induced circulation*. However, vertical circulation due to density changes at the ocean surface is of greater oceanwide importance in producing thorough mixing within the ocean. Density increases cause water masses to sink. Changes that increase water density are usually changes in temperature and salinity; this circulation is referred to as **thermohaline** (temperature–salinity) **circulation**.

Wind-Induced Circulation

There are regions throughout the world ocean where water masses rise or sink due to wind-driven surface currents that carry water away from or toward these regions. Wherever surface flow is away from an area, upwelling occurs. If horizontal surface flow brings insufficient replacement water into the area, then water must come from beneath the surface—there must be **upwelling** to replace water that was removed. The upwelled waters cause lower surface temperatures and higher concentrations of nutrients, making these areas higher in biological productivity.

This condition can occur in the open ocean or along the margins of continents. Winds from the east drive three westerly currents on either side of the equator: the North Equatorial Current of the Indian Ocean and the North and South Equatorial currents in the Atlantic and Pacific Oceans. The Ekman transport of water on the north side of the equator moves water to the right, toward a higher northern latitude; in the Southern Hemisphere it moves water to the left toward a higher southern latitude (see Figure 9-1). The net effect is a water deficiency at the surface between the two currents. Water from deeper within the upper-water mass comes to the surface to fill the void (Figure 9-5a). This phenomenon is called **equatorial upwelling**.

Coastal upwelling is common along the margins of continents. It occurs where surface waters adjacent to the continents are carried out to the open ocean via wind-driven Ekman transport (Figure 9-5b). The replacement of this water comes from depths between 100 and 200 m (330 and 660 ft). Coastal downwelling is the sinking of surface water caused by reversing the direction of the coastal winds that cause upwelling. Wind blowing in the opposite direction pushes water toward the shore, where it piles up and causes downwelling (Figure 9-5c).

Figure 9-6 shows the annual mean upwelling throughout the world's oceans. Calculated from wind stress data, vertical water motion at the base of the surface Ekman layer is depicted. Strong upwelling, in excess of 15 cm (6 in) per day, occurs in the equatorial regions of the Atlantic and Pacific oceans. Strong downwelling of similar magnitude occurs to the north and south of these regions of equatorial upwelling. Because of monsoonal (seasonal) changes in wind direction, the situation changes significantly on a seasonal basis in the Indian Ocean. There is a low rate of upwelling in the subpolar regions and a low rate of downwelling in the subtropical regions. Upwelling along the western margins of continents also is clearly shown.

Near the centers of the rotating gyres of the major ocean current systems, the winds are relatively weak and the water rotates at a very low rate. The weak winds that blow in these regions may be steady in direction and are known to set up convection cells in the upper-water mass. This phenomenon was first described scientifically by Irving Langmuir in 1938, based on his observations of the Sargasso Sea. He observed streaks, or straight rows, of seaweed parallel to the direction of the wind and concluded that the plants were trapped in zones of convergence be-

tween cells. Macroscopic algal material, microscopic organisms, and dissolved organic material are all concentrated in these regions. By contrast, in regions of divergence, where water surfaces as a result of convection, the concentration of organic material is relatively low (Figure 9-7). This phenomenon is called **Langmuir circulation**.

Thermohaline Circulation

Let us consider a pattern of vertical circulation that does not occur. The intensity of solar radiation is much greater in the equatorial region than at higher latitudes. Therefore, you might expect heated equatorial surface water to expand and move away from the equator toward the poles, spreading over the colder, denser, high-latitude waters. You might further expect some return subsurface flow from the high latitudes toward the equator. However, this exchange does not occur because the energy imparted to surface waters by winds greatly exceeds the effect of density changes due to heating. Thus, wind overcomes any tendency for such a pattern of circulation to develop. Instead, large-scale vertical circulation in the ocean results primarily from density differences in surface ocean waters. These density differences result from temperature and salinity differences.

Vertical mixing of ocean water is driven primarily by sinking cold water masses at subpolar latitudes. Surface masses sink to the ocean bottom and deep-water masses rise to the surface. The other variable that affects surface water density is salinity. Salinity appears to have a minimal effect on the movement of water masses in equatorial latitudes. Density changes due to salinity are important only at very high latitudes, where water temperatures remain relatively constant. For example, the highest-salinity water in the open ocean is found in the subtropical regions, but there is little or no density-driven sinking of water in these areas because water temperatures are high enough to maintain a low density surface-water mass. In such areas, a strong *halocline*, or salinity gradient, may develop. It features a relatively thin surface water layer where salinity may exceed 37‰. Salinity decreases rapidly with increasing depth to typical levels below 35‰.

Deep Water

Sinking cold surface waters become important deep-water masses in the subpolar regions of the Atlantic Ocean. In the North Atlantic, the major sinking of surface water is thought to occur in the Norwegian Sea. From there, it flows as a subsurface current into the North Atlantic. This flow becomes part of what is called **North Atlantic Deep Water**. Additional surface water may sink at the margins of the Irminger Sea off southeastern Greenland, and at the margins of the Labrador Sea. In the southern subpolar latitudes, the most significant area of deep-water mass formation is the Weddell Sea, where rapid winter freezing produces salt-laden, high-density water that sinks down the continental slope of Antarctica and becomes **Antarctic Bottom Water**, the densest water in the open ocean (Figure 9-8).

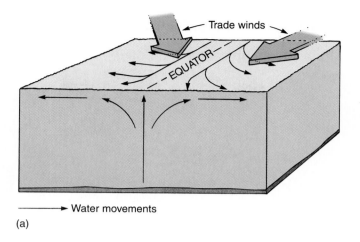

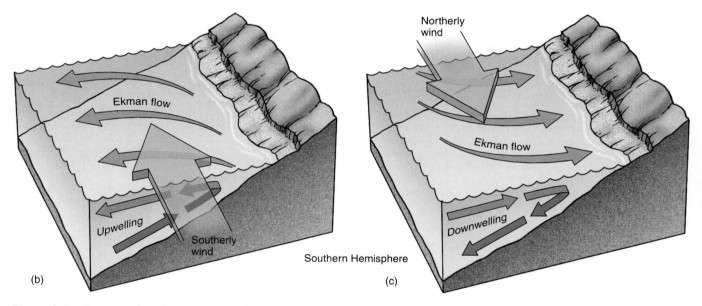

Figure 9-5   Equatorial and Coastal Upwelling and Downwelling

(a) Westward-flowing equatorial currents are driven by trade winds and partly steered by the Coriolis effect, which pulls surface water away from the equatorial region. This water is replaced by upwelling subsurface water. (b) Where wind-driven coastal currents flow along the western margins of continents and toward the equator, Ekman transport carries surface water away from the continent. An upwelling of deeper water replaces the surface water that has moved away from the coast. (c) A reversal of the direction of the winds that cause upwelling causes water to pile up against the shore, and forces downwelling.

On a broad scale, some surface-water masses converge, which may cause sinking. This convergence occurs within the subtropical gyres and in the Arctic and Antarctic. **Subtropical convergences** produce a low rate of downwelling. Major sinking does occur along the **Arctic Convergence (AC)** and **Antarctic Convergence (ANC)**, which is now called the **Antarctic Polar Front (APF)**. The major water mass formed from sinking at the APF is called the **Antarctic Intermediate Water** mass.

For every liter of water that sinks from the surface into the deep ocean, a liter of deep water is displaced and must therefore return to the surface. It is difficult to identify specifically where this vertical flow to the surface is oc-

curring. It may occur as a gradual, rather uniform return throughout the ocean basins. Although you might expect this return to the surface to be somewhat greater in low-latitude regions, where surface temperatures are higher, the data used to construct Figure 9-6 show that some major regions of upwelling are subpolar.

Figure 9-9 shows what oceanographers believe to be the general deep-water circulation of the world ocean. The most intense deep-water flows are the western boundary currents. Figure 9-10 integrates deep thermohaline circulation and surface currents to show the probable overall pattern of global circulation. This circulation provides the deep ocean with oxygen-rich water from sinking dense

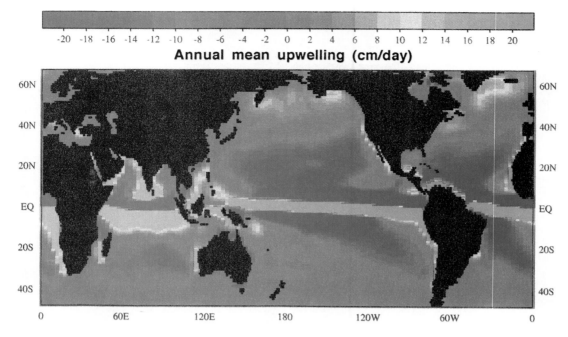

Figure 9-6 Upwelling and Downwelling in the World Ocean

This map depicts average upwelling and downwelling (negative on scale) at the base of the Ekman layer over the period from 1950 through 1988. Maximum rates of upwelling are shown in red and maximum rates of downwelling are shown in violet. In terms of yearly variation, upwelling in the north Pacific and Atlantic oceans intensifies in January and is greatly reduced in July. The reverse is true for the southern polar oceans. Due to the monsoonal wind changes, the belts of upwelling and downwelling seen in the equatorial Indian Ocean switch places during the northeast monsoon (winter). *(Image courtesy of William Hsieh, University of British Columbia.)*

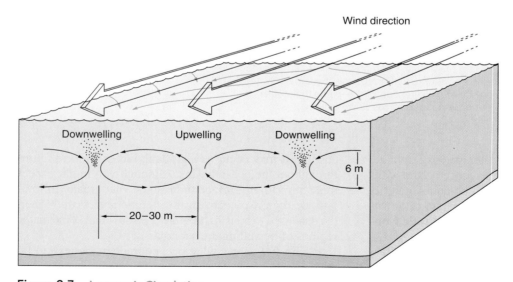

Figure 9-7 Langmuir Circulation

Steady winds blowing across the ocean surface create convection cells with alternate right- and left-hand circulation. The axes of these cells run parallel to wind direction. Organic debris accumulates in downwelling zones of convergence and produces streaks that run parallel to the wind direction for great distances.

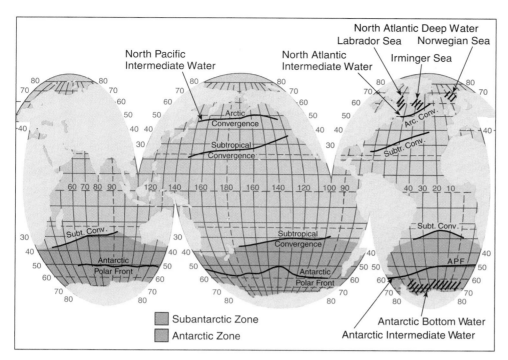

Figure 9-8   Regions of Sinking Intermediate and Deep-Water Masses

The only regions where deep and bottom water form are in the subpolar Atlantic Ocean. North Atlantic Deep Water forms north of the Arctic Convergence, and Antarctic Bottom Water forms near Antarctica. Best developed of all intermediate water masses, Antarctic Intermediate water forms at the Antarctic Polar Front. The subtropical convergences all are centered within subtropical gyres.

surface water, primarily in the subpolar North Atlantic Ocean. Here, a volume of water equal to 100 Amazon rivers begins its long journey into the deep basins of all of the world's oceans (gray arrows in Figure 9-10).

Deep-water masses move southward through the Atlantic, across the South Indian Ocean, and into the North Pacific Ocean. As they do so, dissolved oxygen concentrations decrease because respiration by living organisms and decomposition of organic matter both consume oxygen. Decomposition also increases the concentration of nutrients in the water. Dissolved oxygen decreases from 4.5 to 6.5 ml/l of water in the North Atlantic, where the water recently descended, down to 3.5 to 4.5 ml/l in the North Pacific, near the end of the water's subsurface journey. Conversely, nutrient levels increase in deep-sea water from the North Atlantic Ocean to the North Pacific Ocean. The return limb of this conveyor belt-like system begins with surface water flowing out of the northeastern Pacific Ocean (red arrow off the west coast of the United States in Figure 9-10). Upwelling water is added as the surface flow crosses the Indian Ocean and Atlantic Ocean on its return to the subpolar North Atlantic Ocean.

This conveyor belt-like pattern of return within the surface water may also help to explain the salinity pattern observed in the world ocean; the surface salinity of the North Pacific Ocean is lowest and that of the North Atlantic is highest. The return flow begins in the North Pacific as

low-salinity water, and its salinity increases with time by evaporation. Cooling of this high-salinity water produces a density great enough to initiate sinking and formation of the North Atlantic Deep Water by the time it returns to the subpolar North Atlantic Ocean.

The importance of sinking surface waters in the subpolar Atlantic Ocean to life on Earth cannot be overstated. Without this sinking and the return flow from the deep sea to the surface, distribution of life in the sea would be considerably different, and life would be absent in the deep ocean because of the lack of oxygen.

Though it seems unlikely that these conveyor belt-like circulation patterns have ever completely stopped, there is evidence of variations in sinking rates of the surface waters. Studies of Atlantic Ocean sediments and computer simulations appear to agree that, during the last glacial maximum, North Atlantic Deep Water flow was reduced to about 60 percent of its present volume and flowed south at a shallower depth. The deep North Atlantic basin appears to have been filled with Antarctic Bottom Water, which increased in flow and pushed farther north than at present.

Antarctic Circulation

The **Antarctic Circumpolar Current** dominates the movement of water masses in the southern Atlantic, Indian, and Pacific oceans south of 50°S latitude. This latitude

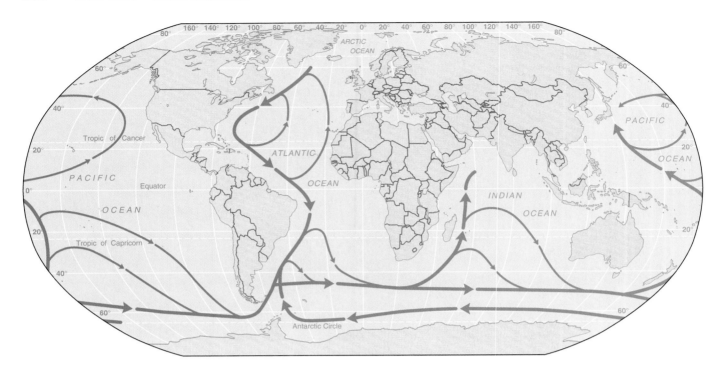

Figure 9-9 Stommel's Model of the Deep-Water Circulation of the World Ocean

(a) Based on information obtained to date, this highly schematic model of deep-water circulation developed by Henry Stommel in 1958 appears to be reasonably correct. Heavy lines mark the major western boundary currents. They result from the same forces that produce the more intense western boundary currents in the surface circulation. (After H. Stommel, 1958.) (b) Dr. Henry M. Stommel (1920–1992) entered the field of physical oceanography as a research assistant at Woods Hole Oceanographic Institution in 1944. Although his formal education consisted of no more than a B.S. in physics, he soon became prominent in the field. Many of his colleagues consider him to have been the world's preeminent physical oceanographer. *(Photo courtesy of Woods Hole Oceanographic Institution.)*

may be considered the northern boundary of what is unofficially referred to as the Southern Ocean. Although the current is not speedy (its maximum surface velocity is about 2.75 k/h or 1.65 mi/h), it does transport more water than any other ocean current, an average of about 130 sv. (The unit of ocean current volume, the **sverdrup (sv)**, used in Figure 9-11 and throughout the rest of this chapter, is equal to 1 million m³/s.) The Antarctic Circumpolar Current is driven by the westerly winds, which are very strong throughout much of the year. The surface portion of this flow is named the **West Wind Drift**.

The Antarctic Circumpolar Current meets its greatest restriction as it passes through the Drake Passage (named for Sir Francis Drake) between the Antarctic Peninsula and the southern islands of South America. The passage is

about 1000 k (600 mi) wide. A narrow surface current called the **East Wind Drift**, so-called because it is propelled by the polar easterlies, moves in a westerly direction around the margin of the Antarctic continent. The East Wind Drift is most extensively developed east of the Antarctic Peninsula in the Weddell Sea region, and in the Ross Sea (Figure 9-11).

Thus, we have two currents flowing around Antarctica in opposite directions. Recall that the Coriolis effect deflects moving masses to the left in the Southern Hemisphere; therefore the East Wind Drift is deflected toward the continent, and the West Wind Drift is deflected away from it. At about 60°S, this creates a zone of divergence, the **Antarctic Divergence**, or the 60° Divergence, between the two currents.

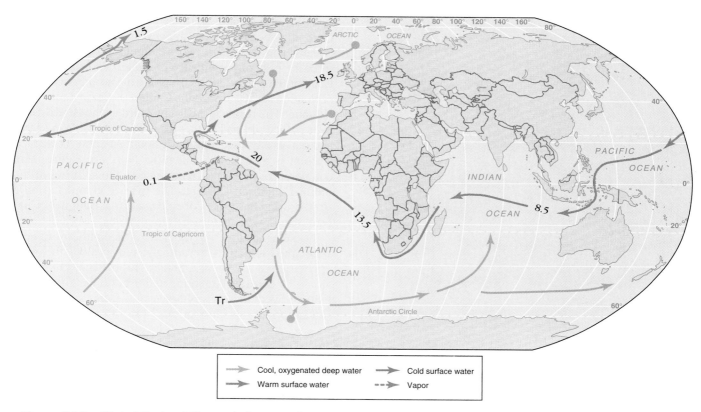

Figure 9-10   Global Cycle of Thermohaline Circulation

The deep ocean is replenished with oxygen by the sinking of the North Atlantic Deep Water (NADW). NADW is composed of water that sinks in the Norwegian Sea and is joined by water from Baffin Bay and the Mediterranean Sea. The total volume of this flow may be about 20 sv as it enters the South Atlantic. This deep flow is recooled in the Antarctic and flows on into the deep Indian Ocean and Pacific Ocean. A broad pattern of upwelling returns the bottom water to the surface in all oceans. A surface flow of warmer, low-salinity water of the North Pacific starts the return journey to the North Atlantic. A flow of 8.5 sv from the North Pacific increases to 13.5 sv after crossing the Indian Ocean and reaches a total of 18.5 sv by the time it is carried into the North Atlantic. A small amount of cold surface water enters the Atlantic from the Pacific through the Drake Passage, and 1.5 sv flows into the Atlantic through the Bering Strait. At least 0.1 sv of water is transferred from the Atlantic Ocean to the Pacific Ocean as water vapor across the Isthmus of Panama.   *(After Gordon, 1986.)*

At about 50°S latitude, the surface water in the circumpolar current begins to sink beneath warmer water from the north, creating a convergence, known as the Antarctic Polar Front (APF). Continuing north to about 40°S, we encounter the Subtropical Convergence, which forms the northernmost boundary of the Antarctic Circumpolar Current. The two convergences divide the Southern Ocean into two zones, the Antarctic Zone from the continent to the APF, and the Subantarctic Zone from the APF to the Subtropical Convergence.

Especially during winter, a mixture of Antarctic Circumpolar Water and shelf water from the Weddell Sea produces the densest water in the world ocean, the Antarctic Bottom Water. This mixture of shelf water with circumpolar water produces a water mass with a temperature of nearly –2°C and salinity of 34.7 resulting in a σ value of 27.9 (density of 1.0279 g/cm^3). Because of its high density, Antarctic Bottom Water flows down the Antarctic continental shelf northward toward the Atlantic Ocean. As it moves along the bottom, it is carried eastward by the Antarctic Circumpolar Current into the Indian and Pacific portions of the Southern Ocean.

At the APF convergence, Antarctic surface water sinks, mixes with North Atlantic Deep Water, and becomes the subsurface mass called Antarctic Intermediate Water, which has a temperature of about 2°C (35.6°F) and salinity of 34.3‰. This combination results in a σ value of 27.4. The water mass can be identified by its relatively low salinity and relatively high dissolved oxygen content as a tongue about 500 meters thick, extending northward with its core at about 900 m (2950 ft) depth at 40°S latitude (Figure 9-12).

Sandwiched between the Antarctic Intermediate and the Antarctic Bottom water masses at 40°S latitude is North Atlantic Deep Water, which moves southward with its core at a depth of about 2200 m (7216 ft). It is characterized by a temperature between 2° and 3°C (36° and 37°F) and average

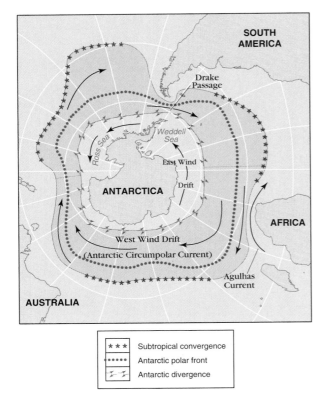

★ ★ ★	Subtropical convergence
• • • •	Antarctic polar front
⸗ ⸗	Antarctic divergence

Figure 9-11 Antarctic Surface Circulation

salinity of 34.7‰. North Atlantic Deep Water can be identified by its temperature, which is slightly above that of the Intermediate Water and the Antarctic Bottom Water. At 40°S, this deep water has been away from the surface for at least 300 years and is sandwiched between two water masses that have more recently left the surface; thus, it is also possible to identify it on the basis of dissolved oxygen minimum. It has a very positive effect on the biological productivity of the Antarctic Zone because of its high concentration of nutrients, accumulated from organic material drifting down during the hundreds of years it has been beneath the photosynthetic zone.

Atlantic Ocean Circulation

Atlantic Ocean surface currents are shown in Figure 9-13. It is helpful to study the currents on this map as they are described. In the Atlantic, the basic surface circulation pattern is that of two large gyres.

The North and South Atlantic Gyres

In response to the Coriolis effect, the North Atlantic gyre rotates clockwise and the South Atlantic gyre rotates counterclockwise. Both rotations are driven by the trade and westerly winds. Each gyre consists of a poleward-moving warm current (red) and an equatorward-moving cold return current (blue). The two gyres are partially separated by the Atlantic Equatorial Countercurrent.

The South Atlantic gyre includes the South Equatorial Current, which reaches its greatest strength just below the equator and is split in two by topographic interference from the eastern prominence of Brazil. Part of the South Equatorial Current moves off along the northeastern coast of South America toward the Caribbean Sea and the North Atlantic. The rest is turned southward as the **Brazil Current**, which ultimately merges with the West Wind Drift and moves eastward across the South Atlantic. The Brazil Current is much smaller than its Northern Hemisphere counterpart, the Gulf Stream, due to the splitting of the South Equatorial Current. The gyre is completed by a slow-drifting movement of cold water, the **Benguela Current**, that flows equatorward along Africa's western coast.

Outside the gyre, a significant northbound flow of cold water, the **Falkland Current**, also moves along the western margin of the South Atlantic. This cold current moves along the coast of Argentina as far north as 25° to 30°S latitude, wedging its way between the continent and the southbound Brazil Current.

The Gulf Stream The Gulf Stream moves northward along the East Coast of the United States, warming coastal states and moderating winters there (Figure 9-13). It is interesting to trace the origins and behavior of this important current.

The North Equatorial Current moves parallel to the equator in the Northern Hemisphere, where it is joined by that portion of the South Equatorial Current that is shunted northward along the South American coast. This flow then splits into two masses, the Antilles Current that flows along the Atlantic side of the West Indies and the Caribbean Current that passes through the Yucatán Channel into the Gulf of Mexico. The current then proceeds out of the Gulf through the Florida Straits where it reconverges with the Antilles Current and becomes the **Florida Current**.

The Florida Current flows close to shore over the continental shelf, carrying a volume that at times exceeds 35 sv. As the Florida Current moves off North Carolina's Cape Hatteras and flows across the open ocean in a northeasterly direction, it is called the **Gulf Stream**. It flows up to 9 km/h (5.6 mi/h), the fastest current in the world ocean.

The western margin of the Gulf Stream can frequently be defined as a rather abrupt boundary that periodically migrates closer to and farther away from the shore. Its eastern boundary becomes very difficult to identify, as it is usually masked by filamentous meandering water masses that change their position continuously.

The Gulf Stream water mass gradually merges with the Sargasso Sea. The **Sargasso Sea** is the water that circulates around the rotation center for the North Atlantic gyre. Its name derives from the seaweed *Sargassum* that floats on its surface. A volume transport of over 90 sv off Chesapeake Bay indicates that a large volume of Sargasso Sea water has been added to the flow provided by the Florida Current. However, by the time the Gulf Stream nears Newfoundland, its volume has been reduced to 40 sv. This indicates that most of the Sargasso Sea water that joined the

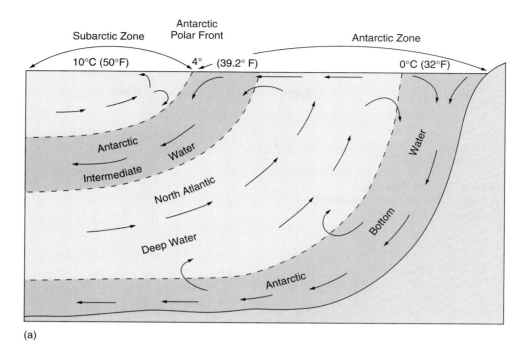

(a)

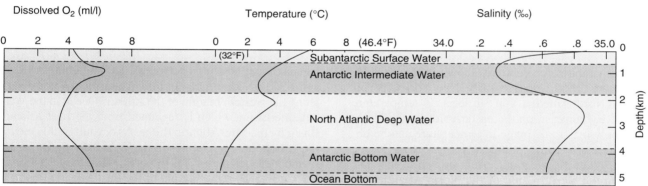

(b) Major subsurface water masses may be identified at 40° S. latitude by observing vertical temperature and salinity profiles as follows:

1. Antarctic Intermediate Water—Salinity minimum and oxygen maximum identifies core of mass at about 900 m (2952 ft).

2. North Atlantic Deep Water—Temperature and salinity maxima and dissolved oxygen minimum identify NADW. Core at 2200 m (7216 ft) approaches 34.9‰.

Figure 9-12 Antarctic Subsurface Water Masses

Florida Current to make up the Gulf Stream has returned to the diffuse flow of the Sargasso Sea. Just how this loss of water occurs has yet to be determined. However, much of it may be caused by meanders (snakelike bends in the current) that pinch off to form large eddies (whirlpools) (see the Special Feature at the end of this chapter).

Southeast of Newfoundland, the Gulf Stream continues in an easterly direction across the North Atlantic. Here the Gulf Stream breaks into numerous branches. Two branches combine water through the mixing of the cold **Labrador**

Current and the warm Gulf Stream. These are the **Irminger Current**, which flows along Iceland's west coast, and the **Norwegian Current**, which moves northward along Norway's coast. The other major branch crosses the North Atlantic and turns southward as the cool **Canary Current** passing between the Azores Islands and Spain. This broad, diffuse, southward flow eventually joins the North Equatorial Current, completing the gyre.

The effects of the westward intensification of current flow in the North Atlantic Ocean can be seen from the

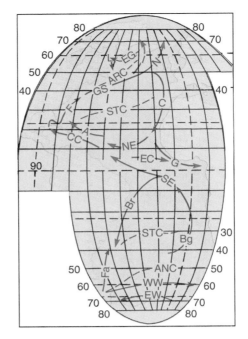

CONVERGENCES
ARC–Arctic
STC–Subtropical
ANC–Antartic

Cold →

Warm →

CURRENTS

A–Antilles
Bg–Benguela
Br–Brazil
C–Canary
CC–Caribbean
EG–East Greenland
EW–East Wind Drift
EC–Equatorial Counter
Fa–Falkland
F–Florida

G–Guinea
GS–Gulf Stream
I–Irminger
L–Labrador
NE–North Equatorial
N–Norwegian
SE–South Equatorial
WW–West Wind Drift

Figure 9-13   Atlantic Ocean Surface Currents

surface temperatures for February shown in Figure 8-16. From latitudes 20°N (Cuba) to 40°N (New York City) off the coast of North America, you can see a 20° temperature difference. By contrast, on the eastern side of the North Atlantic, only a 5° to 6° range in temperature can be observed between the same latitudes. Northwestern Europe is warmed by the Norwegian Current branch of the Gulf Stream. On the western side of the North Atlantic, the southward-flowing cold Labrador Current keeps Canadian coastal waters much colder. Also, during the Northern Hemisphere winter, North Africa's coastal waters, cooled by the southward-flowing cold Canary Current, are cooler than waters off Florida and the Gulf of Mexico.

Atlantic Deep-Water Masses

Throughout the South Atlantic and to some extent in the North Atlantic, surface and near-surface water masses are underlain by the Antarctic Intermediate Water. This intermediate water mass begins to lose its characteristics in the North Atlantic, and we find that **Mediterranean Water** that has passed into the Atlantic Ocean over the Gibralter Sill underlies much of the Central Water Mass in the North Atlantic. Figure 9-14 shows the temperature, salinity, and dissolved oxygen characteristics of the Atlantic water masses.

The North Atlantic Deep Water, transporting 14 sv, can be identified as far south as 40°S latitude on the basis of temperature and salinity maxima observed at depths around 3000 m (9840 ft). Antarctic Bottom Water, transporting 7 sv, can be identified as far north as 40°N latitude by its low temperature and salinity and high oxygen content. Mediterranean Water is most easily identified by its salin-

ity maximum at depths between 1000 and 2000 m (3280 and 6560 ft). Antarctic Bottom Water moves northward along the western margin of the Atlantic Ocean in the West Atlantic Basin, which is separated from the East Atlantic Basin by the Mid-Atlantic Ridge. As a result of this movement, bottom temperatures in the West Atlantic Basin are maintained well below 2°C (35.6°F), while they are around 2.5°C (36.5°F) at the bottom of the East Atlantic Basin.

Although the exact volumes involved are difficult to determine, it is evident that there is a large transfer of deep water across the equator between the North and South Atlantic. The residence time of deep water in the Atlantic Ocean is approximately 275 years. It appears that all of the deep-water and bottom-water masses throughout the world ocean originate in the Atlantic Ocean.

Changes in the Deep Ocean   Using historical oceanic observations for the 50-year period from 1948 to 1998, an analysis of the data from the surface to a depth 3000 m (9843 ft) has revealed an average warming of the top 3000 m of the world ocean by an average of 0.06°C (0.11°F). Compared to the average temperature and heat content for the years 1970-74, values reached a negative minimum in the early 50s, peaked in the late 70s, dropped to just above the comparison average in the late 80s and reached their highest levels in the 90s in all ocean basins. Since 1955, the oceans have gained about 20×10^{22} joules of heat—an amount equal to the average change in heat content of Southern Hemisphere oceans each year with the change of seasons. The distribution of this heat is about equally distributed between the top 300 m (984 ft) where the average temperature increase is 0.31°C (0.56°F) depths

(a)

Vertical Temperature Distribution (°C)

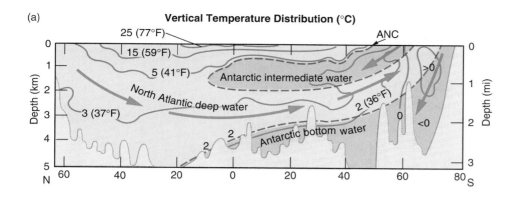

(b)

Vertical Salinity Distribution (⁰/₀₀)

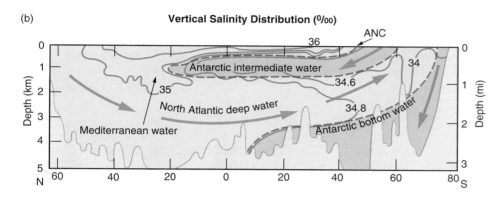

(c)

Vertical Dissolved Oxygen Distribution (ml/l)

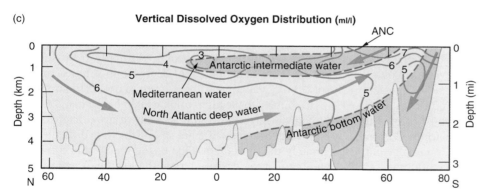

Figure 9-14   Atlantic Ocean Subsurface Circulation as Revealed by Temperature, Salinity, and Dissolved Oxygen.

(Adapted from Sverdrup, Johnson, and Fleming, The Oceans: Their Physics, Chemistry, and General Biology, *1970. Reprinted by permission of Prentice Hall, Inc., Englewood Cliffs, NJ)*

between 300 m and 3000 m. The warming could be due to natural variations, anthropogenic activities, or a more likely combination of both.

Pacific Ocean Circulation

Surface Circulation

The surface circulation in the Pacific Ocean is generally similar to that of the Atlantic Ocean, except that the Equatorial Countercurrent is much better developed in the Pacific Ocean (Figure 9-15). Southeast trade winds drive the

South Equatorial Current, which extends from about 10°S to 3°N latitude. Embedded within the South Equatorial Current and flowing in an easterly direction along the equator is a thin, ribbonlike **Equatorial Undercurrent**. This undercurrent extends for more than 6000 km (3726 mi) from the western Pacific to the Galapagos Islands. Only 0.2 km (0.12 mi) thick and approximately 300 km (186 mi) wide, the Equatorial Undercurrent at depths of 70 to 200 m (230 to 656 ft) has a volume transport of approximately 40 sv.

The South Equatorial Current becomes the western boundary current of the South Pacific gyre—the **East Australian Current**. This current, which is relatively

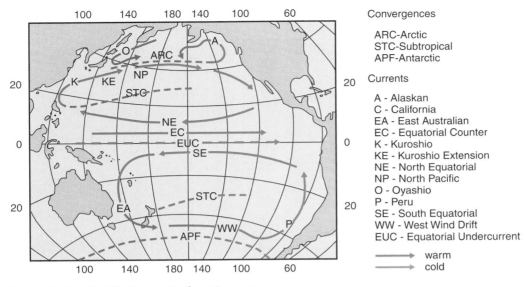

Figure 9-15   Pacific Ocean Surface Currents

weak for a western boundary current, joins the West Wind Drift that carries water across the southern Pacific. This water is pushed up along the western coast of South America as the **Peru Current**, which completes the gyre by turning west near the equator to join the South Equatorial Current.

In the North Pacific, the North Equatorial Current flows from east to west between latitudes of 8° and 20°N. As the North Equatorial Current approaches the western boundary of the North Pacific, most of the water moves north to form the **Kuroshio Current**, the North Pacific counterpart of the Florida Current and southern Gulf Stream in the North Atlantic. Part of the North Equatorial Current water turns south and joins the Equatorial Countercurrent. After the Kuroshio Current, flowing at a maximum velocity of 9 km/h (5.6 mi/h), leaves the coast in an easterly direction, it is called the **Kuroshio Extension** to about 170°E. This segment of the North Pacific gyre corresponds to the Gulf Stream of the North Atlantic and has a volume transport of about 65 sv.

The Kuroshio Extension is met by the cold **Oyashio Current**, which carries water south from the Bering Sea and the Sea of Okhotsk. Beyond this juncture the flow becomes the **North Pacific Current** that moves toward the North American continent and divides; part turns southward as the **California Current**, which completes the gyre by joining the North Equatorial Current, and the remainder flows north to become the Alaskan gyre.

Pacific Deep-Water Masses

The surface-temperature distribution in the Pacific Ocean is similar to that observed in the Atlantic Ocean, but surface salinity in the Pacific is lower than in the Atlantic, the South Pacific water being slightly more saline than that in the North Pacific (see Figure 8-18).

The bottom water of the Pacific Ocean is a mixture of North Atlantic Deep Water and Antarctic Bottom Water. It is called the **Oceanic Common Water** and is introduced into the Pacific Ocean by the Antarctic Circumpolar Current. It moves very slowly northward with a volume transport of about 25 sv. The northward movement of this water mass can be identified by observing the temperature of the bottom water, which increases gradually from south to north. There is also a slight decrease in salinity and dissolved oxygen content from south to north (Figure 9-16).

In contrast to the high-volume interchange in the Atlantic Ocean, the rates and volume of water transport between the North and South Pacific are so low that it is difficult to accurately measure them. One of the main criteria for concluding that the deep waters in the Pacific Ocean move at very low rates and are old is the relatively low dissolved oxygen content of these waters. Compared to the Atlantic deep-water masses, where dissolved oxygen content is from 4.5 to 6.5 ml/l, the Pacific deep-water dissolved oxygen content is between 3.5 and 4.5 ml/l (Figure 9-16). The residence time of deep water in the Pacific Ocean is 510 years.

As indicated in Figure 9-9, the general deep-water circulation in the Pacific Ocean is expected to be from west to east. However, in 1981 it was discovered that hydrothermal vent water from the East Pacific Rise at about 15°S latitude was emitting excess helium that could be traced 2000 km (1242 mi) west. Henry Stommel, who had developed the general deep-water circulation pattern, predicted that the East Pacific Rise might emit enough heat to overcome the general circulation and reverse the direction of flow. In 1987, floats released to track the flow confirmed that the current flows west at about 0.5 cm/s (0.2 in/s) and is probably driven by heat energy from the East Pacific Rise. The current extends west as far as the Society Islands. Although the existence of this vent-driven flow is still controversial, it is intriguing to consider that

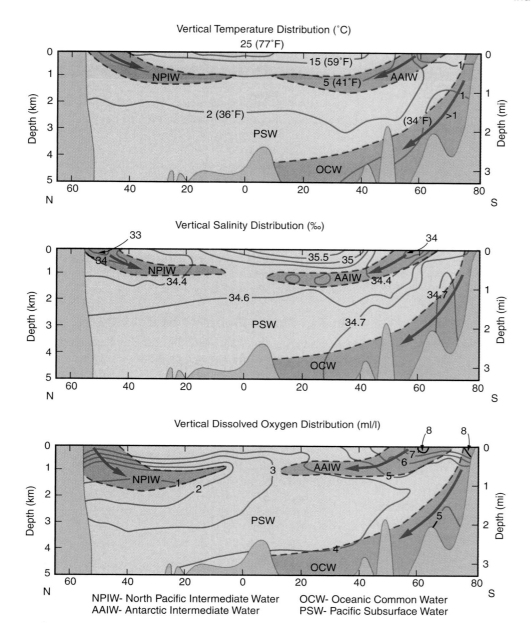

Figure 9-16   Pacific Ocean Subsurface Circulation as Revealed by Temperature, Salinity, and Dissolved Oxygen

(From Sverdrup, Johnson, and Fleming, The Oceans: Their Physics, Chemistry, and General Biology, *1970. Reprinted by permission of PrenticeHall, Englewood Cliffs, NJ.)*

energy from Earth's interior may play a small role in perturbing solar-driven ocean circulation.

Indian Ocean Circulation

Surface Circulation

Surface circulation in the Indian Ocean varies considerably from that in the Atlantic and Pacific. From November to March, the equatorial circulation is similar to that in the other oceans, with two westward-flowing equatorial currents (North and South Equatorial Currents) separated by an eastward-flowing Equatorial Countercurrent. However,

in contrast to the Atlantic and Pacific wind systems, which shift northward of the geographical equator, in the Indian Ocean the meteorological equator is shifted southward. The Equatorial Countercurrent flows between 2° and 8°S latitude, bounded on the north by the North Equatorial Current (which extends as far as 10°N latitude), and on the south by the South Equatorial Current (which extends to 20°S latitude).

The winds of the northern Indian Ocean have a seasonal pattern and are called **monsoon** winds (from the Arabic word meaning "season"). During winter, the typical northeast trade winds are called the northeast monsoon. They are

(a)
WINTER: November–March,
Northeast monsoon
wind season

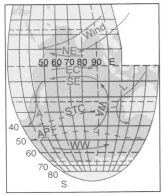

(b)
SUMMER: May–September,
Southwest monsoon
wind season

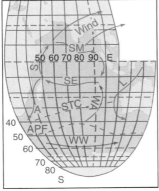

From November to March,
Indian Ocean equatorial
circulation is similar to other
oceans, but Equatorial
Countercurrent shifts
southward.

Low pressure over mainland
during summer draws Southwest
Monsoon winds over the North
Indian Ocean. This wind
produces the Southwest
Monsoon Current which flows
easterly and replaces the North
Equatorial Current.

Warm ——→   Cold ——→

CURRENTS **A**—Agulhas **EC** —Equatorial Countercurrent
L—Leeuwin **NE**—North Equatorial **S**—Somali
SE—South Equatorial **SM**—Southwest Monsoon
WA—West Australian **WW**—West Wind Drift

CONVERGENCES **STC**—Subtropical **ANC**—Antarctic

Figure 9-17   Indian Ocean Surface Currents

reinforced because rapid cooling of air over the Asian main-land during winter creates a high-pressure cell, which forces atmospheric masses off the continent and out over the ocean, where air pressure is less (green arrows in Figure 9-17a). During summer, the Asian mainland warms faster than the oceanic water. (Recall that this is due to the lower heat capacity of continental rocks and soil compared with water.) As a result, a summer low-pressure cell develops over the continent, allowing higher-pressure over the ocean to move air from the Indian Ocean onto the Asian landmass. This gives rise to the southwest monsoon (green arrows in Figure 9-17b). This may be thought of as a continuation of the southeast trade winds across the equator. During this season, the North Equatorial Current disappears and is re-placed by the Southwest Monsoon Current. It flows from west to east across the North Indian Ocean. In September or October, the northeast trade winds are re-established, and the North Equatorial Current reappears (Figure 9-17).

Surface circulation in the southern Indian Ocean is sim-ilar to the counterclockwise circulation observed in other southern oceans. When the northeast trade winds blow, the South Equatorial Current provides water for the Equator-ial Countercurrent and the **Agulhas Current**, which flows southward along Africa's eastern coast and joins the West Wind Drift. Turning northward out of the West Wind Drift is the **West Australian Current**, which completes the gyre by merging with the South Equatorial Current. Dur-

ing the southwest monsoon, a northward flow from the equator along the coast of Africa, the **Somali Current**, de-velops with velocities approaching 4 km/h (2.5 mi/h).

The eastern boundary current in the southern Indian Ocean is unique. Other eastern boundary currents of sub-tropical gyres are cold drifts flowing toward the equator that produce arid coastal climates, typically having less than 25 cm (10 in) of rain per year. However, in the southern In-dian Ocean, the West Australian Current is displaced off-shore by a southward-flowing current called the **Leeuwin Current**. This current is driven southward along the Aus-tralian coast from the warm-water dome piled up in the East Indies by the Pacific equatorial currents. The Leeuwin Cur-rent produces a mild climate in southwestern Australia, which receives about 125 cm (50 in) of rain per year. This current is weakened during ENSO events, discussed in the next section, and contributes to the Australian drought con-ditions associated with these events.

Indian Ocean Deep-Water Masses

Antarctic Intermediate Water lies directly beneath the sur-face and near-surface water of the Indian Ocean, and is identifiable by its low temperature and high dissolved oxy-gen content (Figure 9-18). **Red Sea Water**, a high-salinity mass with low oxygen content, is fed by outflow of the Red Sea and descends to a depth between 1000 and 1500 m (3280 and 4920 ft) in the northern Indian Ocean.

The main deep water mass in the Indian Ocean is Oceanic Common Water, a mixture of North Atlantic Deep Water and Antarctic Bottom Water. It has an average tem-perature of 1.5°C (34.7°F), an average salinity of 34.7‰, and enters the Indian Ocean at a rate of 20 sv, and has a residence time of 250 years.

El Niño–Southern Oscillation Events

For years, fishermen of the coastal waters of Peru knew that an influx of warm water every few years would result in a reduced population of anchovy, their main fishery and the source of food for coastal birds. This warm-water phenom-enon usually occurred around Christmas and was given the name **El Niño**, meaning "the child" in reference to the Christ child. In addition to its effects on the Peruvian an-chovy fishery, El Niño events are associated with extreme anomalies in weather patterns throughout the world. For ex-ample, hurricane frequency increases in the Pacific but de-creases in the Atlantic, catastrophic flooding occurs along the Pacific coast of South America, while severe drought afflicts Australia and Indonesia, and winter storms increase in frequency and severity in the southwestern United States. The 1997–1998 El Niño event resulted in 23,000 deaths and $33 billion in damages around the world.

What causes El Niño events? During the 1920s, G. T. Walker identified what he called the **Southern Oscilla-tion (SO)**, a condition in which the summer high-pressure system in the southeastern Pacific occurs in conjunction

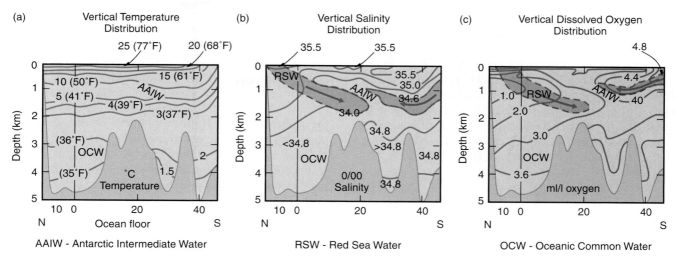

(a) Vertical Temperature Distribution

(b) Vertical Salinity Distribution

(c) Vertical Dissolved Oxygen Distribution

AAIW - Antarctic Intermediate Water

RSW - Red Sea Water

OCW - Oceanic Common Water

Figure 9-18   Indian Ocean Subsurface Circulation as Revealed by Temperature, Salinity, and Dissolved Oxygen

(From A. J. Clowes and G. E. R. Deacon, The Deep-Water Circulation of the Indian Ocean, 1935. Adapted by permission of Nature, Macmillan Journals, and G. E. R. Deacon.)

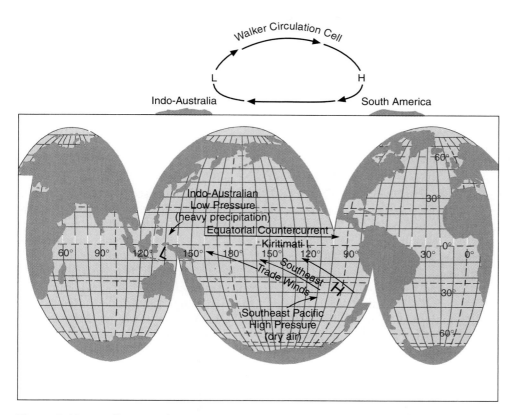

Figure 9-19   Walker Circulation

Normal oceanic and atmospheric conditions that become altered by El Niño–Southern Oscillation events.

with a low-pressure system over the Indo-Australian region (Figure 9-19). This pressure difference lessens when trade winds diminish, the temperature of surface waters in the eastern Pacific increases, and the Equatorial Countercurrent flow increases. When it was discovered that the El Niño seemed to be related to the Southern Oscillation, the occurrences began to be referred to as *El Niño–Southern Oscillation* (ENSO) events. We now realize that **ENSO events** are cycles of roughly 3 to 4 years in length that consist of warm El Niño events and cold La Niña events.

These cycles are superimposed on longer-scale variations in the complex ocean-atmosphere climate patterns.

The intensity of ENSO cycles apparently depends upon when they occur within the longer term climate patterns. There have been about a dozen ENSO events over the last 50 years; of these, two of the most intense occurred recently, in 1982–1983 and 1997–1998. There is some indication that these events coincided with a period when tropical Pacific and Indian Ocean waters were warmer. Global warming from anthropogenic greenhouse gas buildup may play a role.

ENSO events and Ocean-Atmosphere Circulation

Normally, in years without ENSO events, the Southern Oscillation circulation cell, called *Walker Circulation*, affects the southeast trade winds. These winds converge on the Indo-Australian low-pressure cell, rise, and produce high rates of precipitation in the low-pressure area. A large mass of the warmest water in the world ocean (29–30°C) lies beneath this low-pressure cell, having been pushed there by the trade winds. The thermocline in this warm-water area begins to develop only below a depth of 100 m (328 ft); in contrast, the thermocline develops as shallow as 30 m (98 ft) in the eastern equatorial Pacific where the waters are relatively cooler (22–24°C). These cooler temperatures are the result of upwelling of cold deep waters caused by the trade winds displacing surface waters toward the west. Dry air descends within the southeastern Pacific high-pressure cell off the west coast of South America, creating high rates of evaporation and the normal arid climate of the region.

However, these characteristics change during an ENSO event. One precursor of an ENSO event is the movement of the Indo-Australian low-pressure cell to the east beginning in October or November of the year previous to an El Niño. In extreme cases, severe droughts can occur in Australia because the low-pressure cell moves so far east. Concurrent with the eastward shift of the Indo-Australian low-pressure cell, the meteorological equator, or *intertropical convergence zone (ITCZ)*, where the northeast trade winds and southeast trade winds meet and rise, moves south. Its normal seasonal migration is from 10°N latitude in August to 3°N in February, but during ENSO

events, it may move south of the equator in the eastern Pacific. Associated with this shift are weak trade winds, a decreased upwelling in the eastern equatorial Pacific, and a deepening of the eastern Pacific thermocline, which results in a thickening layer of warm water.

Prolonged westerly wind bursts, persisting for one to three weeks, replace the normally weak easterly winds in the western Pacific. These wind bursts perturb the upper ocean and promote the eastward development of large-scale Kelvin waves. These are internal waves (see Chapter 10) with wavelengths of thousands of kilometers which form at the thermocline. Kelvin waves suppress the normal upwelling of water in the eastern Pacific.

As the ENSO develops, the weakened trade winds and anomalous warmth of surface waters observed in the western Pacific spread toward the east. The arrival of unusually warm surface waters at Kiritimati (Christmas Island; 2°N, 157°W) can be predicted by increased surface temperatures off the coast of Peru. The event is fully developed by January (Figure 9-20). Heavy rainfall from the southward shift of the ITCZ strikes the coasts of Ecuador and Peru, which are usually arid, and spreads west across the tropical Pacific. The intense eastward flow of the Equatorial Countercurrent causes a rise in sea level along the western coast of the Americas that progresses poleward in both hemispheres. The event ends 12 to 18 months after it starts, with a gradual return to normal conditions that begins in the southeastern tropical Pacific and spreads west. The below-average temperatures in the eastern Pacific at the end of an El Niño, as shown in Figure 9-20, mark the beginning of the La Niña part of the cycle.

Recent ENSO Events

The 1982–1983 ENSO event caused a severe drought in Australia and Indonesia. This event was anomalous in that it was initially confined to the central and western tropical Pacific and spread to the east late in its development. In November 1982, virtually all of the 17 million adult birds that normally inhabited Kiritimati had abandoned their nestlings, probably because the lack of upwelling nutrients prevented the normal algal bloom and caused the fish the birds normally feed on to move to other feeding grounds. Figure 9-21 shows the warming that occurred off southern California from January 1982 to January 1983 as

Figure 9-20 (right)   Sea Temperature Anomaly Resulting From Averaging of ENSO Event Temperature Anomalies from 1950–1973

(a) After onset: average of March, April, and May temperature anomalies. Condition starts in previous December–January and is well developed by March–May period shown. Note the abnormally high surface water temperatures extending into the eastern Pacific off the coasts of Ecuador and Peru. (b) Maximum development: average of December, January, and February anomalies. The condition expands westward out from along the coast of South America and also extends northward. (c) Ending of event: average of May, June, and July anomalies. Conditions return to near normal, with the exception of a significant negative temperature anomaly in the eastern Pacific. *(Data courtesy of NOAA.)*

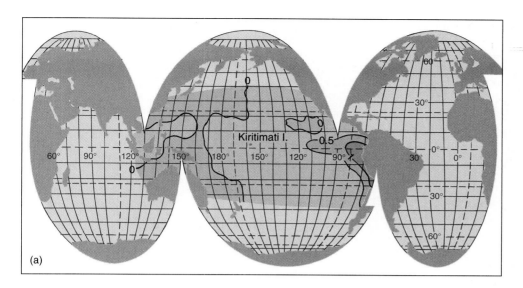

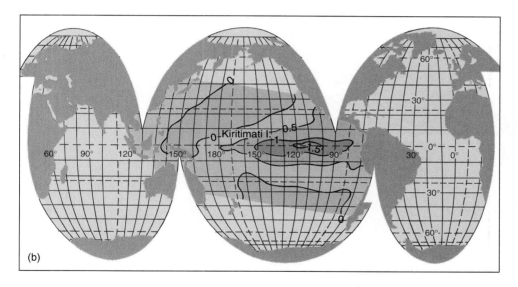

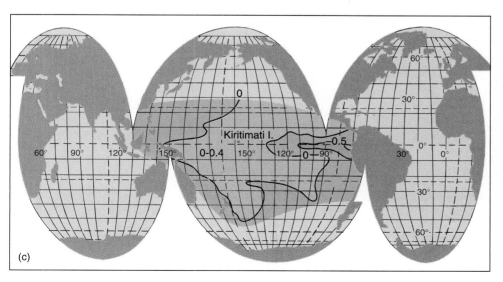

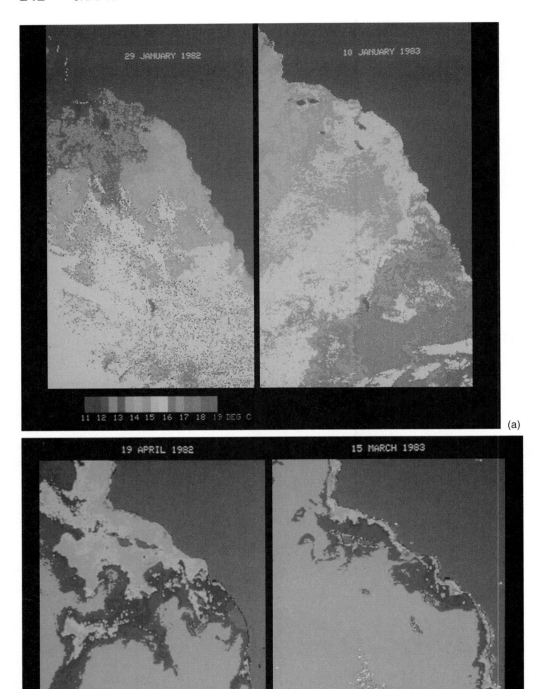

Figure 9-21   Water Temperatures and Phytoplankton Pigment Concentrations Off Southern California

(a) The advanced, very-high-resolution radiometer sea-surface temperature image shows that waters off southern California in January 1983 are an average of 1.72°C (3.1°F) warmer than a year before. The coastal-zone color-scanner phytoplankton pigment images show a reduction in plant productivity over the same time span. One of the strongest El Niño events ever recorded, the 1982–1983 event, pushed a layer of warm water north from the equator. Coinciding with this event was a reduction of upwelling of nutrient-rich, cold water along the coast. (b) This reduction in the nutrient level of the surface water reduced the level of biological productivity. Spectacular physical phenomena such as this and their biological consequences help highlight the close relationship between the physical and biological phenomena of the oceans.   *(Courtesy of Paul C. Fiedler/National Marine Fisheries Service.)*

a result of the 1982–1983 ENSO. This warming was accompanied by a major reduction in biological productivity, as is shown by the Coastal Zone Color Scanner phytoplankton pigment images of April 1982 compared with those of March 1983.

The 1997–1998 ENSO event was the most severe on record by some measures. Temperatures in the eastern equatorial Pacific exceeded normal ranges by record amounts (Figure 9-22), averaging more than 4°C above normal, and reaching more than 9°C above normal in some locations. The event also developed more rapidly than any previous event on record. Monthly changes in sea surface temperatures in the eastern equatorial Pacific were more rapid than for any previously recorded El Niño (see chapter opening photo).

Predicting ENSO Events

One of the major accomplishments of the Tropical Ocean Global Atmosphere (TOGA) program from 1985 to 1994 was the deployment of the ENSO observing system, consisting of a systematic array of satellites and moored buoys

designed to make coordinated measurements of atmospheric and oceanic conditions. This system allowed collection of the most comprehensive data set to date for the 1997–1998 El Niño event. The occurrence and severity of the 1997–1998 event was not forecast correctly by any of the existing ENSO models, but as the event developed, the data being collected was used to adapt these models so that they successfully predicted how the event subsequently developed. It is increasingly apparent that the success of predictive modeling depends on understanding atmospheric and oceanic phenomena that occur on time scales of tens or hundreds of years and their relationships to ENSO events.

Power From Winds and Currents ......

Solar energy drives the winds. The winds, in turn, have driven ocean currents for billions of years and have powered society's machines for centuries. Westerly winds off the coast of New England are a nearshore wind resource that is reasonably sustained and contains a large amount of energy. There is some optimism that Offshore Windpower Systems (OWPS) could generate electricity to meet

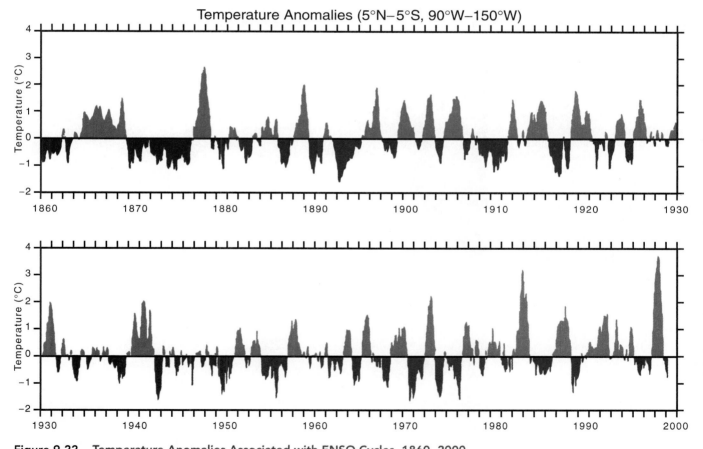

Figure 9-22 Temperature Anomalies Associated with ENSO Cycles, 1860–2000

The variation in sea surface temperatures from normal is expressed as an anomaly above or below the norm which is equivalent to 0°C variation. El Niño events are characterized by periods when temperatures are above the norm (red areas); La Niña events are characterized by periods when temperatures are below the norm (blue areas).

Warm Core and Cold Core Rings of the Gulf Stream

Superimposed on the large scale circulation patterns we have covered in this chapter, oceanographers have been discovering smaller scale circulation phenomena in the oceans. Of these, the most interesting are akin to the eddies and meanders of rivers. These eddies form within major surface currents. Once formed, the eddies or meanders often break off from the main current and form rings of current water surrounding a center of water having different properties. We will look at the rings forming off the Gulf Stream as examples of this process.

The Gulf Stream travels up the eastern coast of North America, separating the cooler continental shelf and slope waters on its north side from the warmer waters of the Sargasso Sea on its south side (Figure 9B). The Gulf Stream itself is warmer than the surrounding waters because it transfers water from the equatorial Atlantic and Gulf of Mexico northward. When the current flow develops meanders, some of the meanders close off completely, capturing the water type on the inside of the meander. Thus, meanders to the north enclose Sargasso Sea water, and meanders to the south enclose cold shelf and slope water. The resulting rings are called warm-core rings and cold-core rings, respective-

ly. Rings vary in diameter from 100 to 500 kilometers and have lasted for up to 2 years. Warm-core rings may be up to 1 km in depth; cold-core rings may extend to the ocean floor, influencing seafloor sediment features. Eventually, the rings are consumed by the Gulf Stream or dissipate within the waters beside it.

While these rings exist, they maintain not only unique temperature characteristics, but also unique chemical characteristics and biological populations. Thus, the rings may contain isolated colonies of warm-water organisms in a cold ocean, or conversely, colonies of cold-water organisms in a warmer ocean. Phytoplankton pigment concentrations recorded by the Coastal Zone Color Scanner (CZCS) on the *NOAA 7* satellite in April of 1982 clearly show a warm-core ring. Both the Gulf Stream and the Sargasso Sea waters to the south are characterized by low biological productivity (blue in Figure 9A) with pigment concentrations of less than 1 mg/m^3. Shelf and slope waters are much more productive, having concentrations up to 3-7 mg/m^3 (orange and red in Figure 9B). The warm-core ring show up as a blue, low-pigment area within the orange area of productive waters.

Figure 9A   CZCS image of the Gulf Stream

Image taken on April 24, 1982 of the Gulf Stream. Phytoplankton pigment concentrations, in mg/m^3, range from 0.05 (darkest blue) to more than 3.0 (darkest brown). The cooler waters of the shelf (b) and slope (c) show the highest pigment concentrations, 2.8 to 3.0. The warmer waters of the warm-core ring (d) and Gulf Stream (a) have lower pigment concentrations of 0.3 to 0.5. *(Courtesy of Otis Brown and Robert Evans, University of Miami/RSMAS.)*

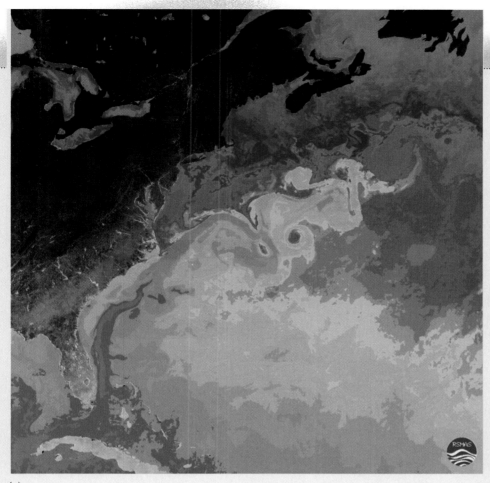

(a)

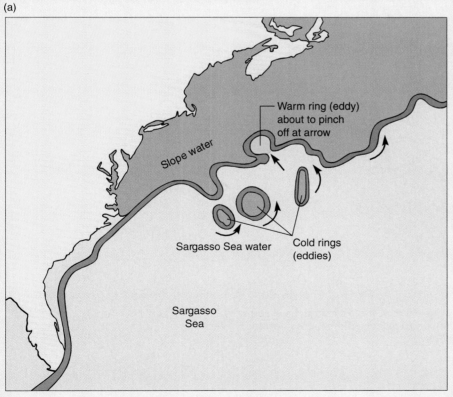

(b)

Figure 9B   The Gulf Stream and Sea-Surface Temperatures

(a) The northwestern Atlantic Ocean off the U.S.-Canadian East Coast, featuring the warm Gulf Stream. Sea–surface temperature data gathered by a NOAA satellite were processed to produce this false-color image. The warm waters of the Gulf Stream are shown in orange and red. Colder nearshore waters are shown in green, blue, and purple. Warm water from south of the Gulf Stream is transferred northward as warm core rings (yellow) surrounded by cooler (blue and green) water. Cold nearshore water spins off to the south of the Gulf Stream as cold core rings (green) surrounded by warmer (yellow and red-orange) water. (b) This overlay of the satellite image in (a) clarifies the process of ring formation. The red band is the core of the Gulf Stream flow. Note the three cold rings that have formed by the pinching off (closure) of meanders. Also note the warm ring that is being created by pinching off the current. The cold rings have separated from the main Gulf Stream flow. The forming warm ring will also separate from the Gulf Stream.   *(Part (a) courtesy of Dr. Charles McLain/Rosenstiel School of Marine and Atmospheric Science, University of Miami.)*

the needs of large areas of the U.S. North Atlantic coast (Figure 9-23).

Many have considered the great amount of energy in the Florida-Gulf Stream Current System and dreamed of harnessing it to power society's machines. Scientists and engineers met in 1974 to consider this possibility and concluded that some 2000 MW of electricity could be recovered along the east coast of southern Florida. Devices proposed for extraction range from underwater windmills to a Water Low Velocity Energy (WLVEC), which is operated by parachutes attached to a continuous belt. Again, calculations indicate such systems can be economically competitive.

The Coriolis Program was proposed in 1973 by William J. Moulton of Tulane University. Large hydroturbines are envisioned that could generate 43 MW of electricity from the movement of ocean currents (Figure 9-24). An array of 242 units covering an area 30 km wide (18.6 mi) and 60 km long (37 mi) could generate 10,000 MW annually—the equivalent of burning 130 million barrels of oil or about one-third of the present domestic inventory of crude oil in the United States.

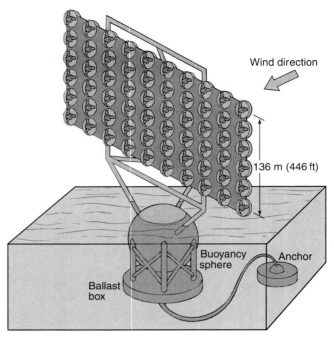

Figure 9-23   Offshore Windpower System (OWPS)

There is some optimism that an offshore windpower system could generate electricity to meet the needs of large areas of the U.S. North Atlantic coast.   *(Courtesy of Woods Hole Oceanographic Institution.)*

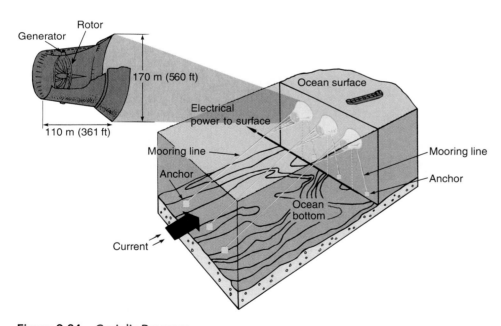

Figure 9-24   Coriolis Program

Buoyant Coriolis hydroturbines anchored in ocean current. It is estimated that an array of 242 such units placed in the Gulf Stream could provide 10 percent of the present electricity needs of Florida.   *(Courtesy of U.S. Department of Energy.)*

Summary

Horizontal currents set in motion by wind systems are characteristic of the surface waters of the world ocean. According to the Ekman spiral concept, winds set the surface waters in motion in a direction 45° to the right of the wind direction in the Northern Hemisphere, and the net water movement is at right angles to the wind direction. As a result of the Ekman spiral phenomenon, water is pushed in toward the center of clockwise gyres in the Northern Hemisphere and counterclockwise gyres in the Southern Hemisphere, forming hills of water. As water in the Northern Hemisphere runs downslope on the hills of water that result, the Coriolis effect causes it to turn right and into the clockwise flow pattern. As a result, a geostrophic current flowing parallel to the contours of the hill is maintained. The apex of the hill is always located to the west of the geographical center of the gyre. Using dynamic topography, oceanographers can map the surface currents in ocean basins by determining the average density of water columns at locations throughout the basin and mapping this distribution. The speed of the current increases with the increased slope of the hills, which is normally steepest on the west side.

Vertical circulation produced by wind is confined to the surface-water mass. An example is equatorial or coastal upwelling, which is created by Ekman transport pulling wind-driven water masses away from the equator or coastal areas. Water surfaces from a depth of no more than a few hundred meters to replace the surface water. An even shallower vertical movement of water is represented by Langmuir circulation of long, shallow convection cells running parallel to the gentle, steady winds that create them. These cells extend no deeper than 6 m (19.7 ft). Mixing of water in the deepest ocean basins is caused by thermohaline circulation, initiated by the sinking of dense surface water in high latitudes. North Atlantic Deep Water and Antarctic Bottom Water carry most of the oxygen-bearing water to the deep ocean basins, thereby maintaining their life-sustaining capacity.

The Circumpolar Current flows in a clockwise direction around the continent of Antarctica. It is the largest current in the world ocean, transporting over 130 sv of water. The surface portion of this flow is often referred to as the West Wind Drift because it is driven by strong westerly winds. Two major deep-water masses form in Antarctic waters, Antarctic Bottom Water, the densest water mass in the oceans, forms primarily in the Weddell Sea and sinks along the continental shelf into the Atlantic Ocean. Farther north, at the Antarctic Polar Front, low-salinity Antarctic Intermediate Water sinks to a depth of about 900 m (2952 ft). Sandwiched between these two masses is North Atlantic Deep Water, rich in algal nutrients after hundreds of years in the deep ocean. The highest-velocity current in the oceans is the Gulf Stream as it flows along the southeast Atlantic coast. The best development of an equatorial countercurrent system is found in the Pacific, where a large eastern movement of water is achieved by the Pacific Countercurrent and the Equatorial Undercurrent, a ribbon-like subsurface flow. The Kuroshio Current in the North Pacific is the counterpart of the Gulf Stream. The formation of deep-water masses is not pronounced in the Pacific Ocean as there appears to be no deep sinking of surface water. The Indian Ocean circulation is dominated by the monsoon wind systems that blow out of the northeast in the winter and the southwest in the summer.

ENSO events are important anomalies in world climate caused by changes in ocean-atmosphere dynamics. ENSO cycles consist of warm El Niño and cold La Niña events initiated by changes in winds and ocean circulation patterns in the equatorial Pacific which cause the movement of warm surface waters normal for the western equatorial Pacific to shift eastward. The eastward shift of the warm surface water and development of internal Kelvin waves at the thermocline prevent upwelling of cold, nutrient-rich waters off the Pacific coast of South America. These events cause severe damage to biological ecosystems, to local economies that depend on fisheries, and severe drought or flood conditions in many parts of the world.

Key Terms

Agulhas Current
Antarctic Bottom Water
Antarctic Circumpolar Current
Antarctic Convergence
Antarctic Divergence
Antarctic Intermediate Water
Antarctic Polar Front
Arctic Convergence
Benguela Current
Brazil Current
California Current
Canary Current
Coastal upwelling
Dynamic topography
East Australian Current

East Wind Drift
Eastern boundary currents
Ekman layer
Ekman spiral
Ekman transport
El Niño
ENSO event
Equatorial Countercurrent
Equatorial Current
Equatorial Undercurrent
Equatorial Upwelling
Falkland Current
Florida Current
Geostrophic current
Gulf Stream

Irminger Current
Kuroshio Current
Kuroshio Extension
Labrador Current
Langmuir circulation
Leeuwin Current
Mediterranean Water
Monsoon
North Atlantic Deep Water
North Pacific Current
Norwegian Current
Oceanic Common Water
Oyashio Current
Peru Current
Red Sea Water

Sargasso Sea
Somali Current
Southern Oscillation
Subpolar gyre
Subtropical convergence
Subtropical gyres
Sverdrup (sv)
Thermohaline circulation
Upwelling
West Australian Current
West Wind Drift
Western boundary currents
Westward intensification

Questions and Exercises

1. Compare the forces directly responsible for causing horizontal and deep vertical circulation in the oceans. What is the ultimate source of energy driving both circulation systems?

2. Diagram and discuss how Ekman transport produces the hill of water within major ocean gyres that causes geostrophic current flow. As a starting place on the diagram, use the prevailing wind belts, the trade winds, and the westerlies.

3. What causes the apex of these geostrophic hills to be offset to the west of the center of the ocean gyre systems?

4. Explain how oceanographers determine the dynamic topography of the ocean waters.

5. Describe the relationships among the wind, the surface current it creates, and the development of equatorial and coastal upwellings.

6. Discuss why thermohaline vertical circulation is driven by sinking surface water that occurs only at high latitudes.

7. The largest current in the world ocean in terms of volume transport is the Antarctic Circumpolar Current. Explain why its surface portion is referred to as the West Wind Drift. What is its maximum volume transport compared to the maximum volume transport of the Gulf Stream?

8. Why does the North Atlantic Deep Water enhance biological productivity in the Antarctic Zone as it surfaces between the Antarctic Intermediate Water and the Antarctic Bottom Water?

9. What evidence is there that the bottom water in the Pacific Ocean has been away from the surface much longer than that of the Atlantic Ocean?

10. Describe a typical El Niño event and the effects they have on climate and ecosystems.

References

Alayne, F., A. F. Street-Perrott, and R. Perrott, 1990. Abrupt climate fluctuations in the tropics: The influence of atlantic ocean circulation. *Nature* 343:6259, 607–12.

Behringer, D., *et al.* 1982. A demonstration of ocean acoustic tomography. *Nature* 299, 121–25.

Boyle, E., and A. Weaver, 1994. Conveying past climates. *Nature* 299, 121–125.

Clowes, A. J., and G. E. R. Deacon, 1935. The deep-water circulation of the Indian Ocean. *Nature* 136:936–38.

Fichefet, T., S. Hovine, and J.C. Duplessy, 1994. A model study of atlantic thermohaline circulation during the last glacial maximum. *Nature* 372:6503, 252–55.

Goldstein, R., T. Barnett, and H. Zebker, 1989. Remote sensing of ocean currents. *Science* 246:4935, 1282–86.

Gordon, A. L., 1986. Interocean exchange of thermocline water. *Journal of Geophysical Research* 91:C4, 5037–46.

Kerr, R., 1999. Big El Niños ride the back of slower climate change. *Science* 283:1108–1109.

Kunzig, R., 1991. Can Earth's internal heat drive ocean circulation? *Science* 252, 1620–21.

Levitus, S., et al., 2000. Warming of the World Ocean. *Science* 287:5461, 2225–29.

Liusen, X., and W.W. Hsieh, 1995. The global distribution of wind-induced upwelling. *Fisheries Oceanography* 4:1, 52–67.

McPhaden, M.J., 1999. Genesis and evolution of the 1991–1998 El Niño. *Science* 283:950-953.

Montgomery, R., 1940. The present evidence of the importance of lateral mixing processes in the ocean. *American Meteorological Society Bulletin* 21:87–94.

Parrilla, G., et al., 1994. Rising temperatures in the subtropical north atlantic ocean over the past 35 years. *Nature* 369:6475, 48–51.

Pedlosky, J., 1990. The dynamics of the oceanic subtropical gyres. *Science* 248:4935, 316–22.

Philander, S.G., 1990. *El Niño, La Niña, and the Southern Oscillation.* San Diego: Academic Press.

Pickard, G. L. and W.J. Emery., 1990. *Descriptive Physical Oceanography: An Introduction,* 5th ed. New York: Pergamon Press.

Stommel, H., 1987. *A View of the Sea.* Princeton, NJ: Princeton University Press.

——. 1958. *The Abyssal Circulation. Letters to the Editor.* Deep Sea Research 5. New York: Pergamon Press.

Stuiver, M., P. D. Quay, and H. G. Ostlund., 1983. Abyssal water carbon-14 distribution and the age of the world oceans. *Science* 219:4586, 849–51.

Sverdrup, H. U., M. W. Johnson, and R. H. Fleming., 1942. *Renewal 1970. The oceans: Their physics, chemistry, and general biology.* Englewood Cliffs, NJ: Prentice Hall.

Travis, J., 1994. Taking a bottom-to-sky "Slice" of the Arctic Ocean. *Science* 266, 1947–48.

Webster, P.J. and T.N. Palmer, 1997. The past and the future of El Niño. *Nature* 390:562-564.

Suggested Reading

Sea Frontiers

Alper, J., 1991. *Munk's Hypothesis*. 37:3, 38–43. Discusses Walter Munk's proposal to monitor global warming by transmitting sound signals across vast ocean reaches in SOFAR channel.

Frye, J., 1982. *The Ring* Story. 28:5, 258–67. Discusses the Gulf Stream, its rings, and how the rings are studied.

Miller, J., 1975. Barbados and the Island-Mass Effect. 21:5, 268–72. Waters around tropical islands are much more productive than the surface waters of the open ocean.

Smith, F. G. W., 1972. *Measuring Ocean Movements*. 18:3, 166–74. Discusses some practical problems related to current flow and some methods used to determine current direction, speed, and volume.

Smith, F. G. W., and R. Charlier, 1981. *Turbines in the Ocean*. 27:5, 300–305. Discusses the potential of extracting energy from ocean currents.

Sobey, E., 1982. *What is Sea Level?* 28:3, 136–42. Discusses the factors that cause sea level to change.

Scientific American

Baker, D. J., Jr. 1970. Models of ocean circulation. 222:1, 114–21. Observations of a model depicting a segment of the surface of Earth over which fluids move, and an explanation of geostrophic flow within ocean gyres. Some knowledge of basic physics is necessary for full comprehension of this material.

Hollister, C. D., and A. Nowell., 1984. The dynamic abyss. 250:3, 42–53. Submarine storms are associated with deep, cold currents flowing away from polar regions toward the equator.

Munk, W., 1955. The circulation of the oceans. 191:96–108. A young physical oceanographer, who has gone on to be highly honored by the scientific community, presents his views of ocean circulation.

Stewart, R. W., 1969. The atmosphere and the ocean. 221:3, 76–105. The exchange of energy between the atmosphere and ocean and the resulting phenomena, currents, and waves are covered in this readable, comprehensive article.

Spindel, R. C., and P. F. Worcester, 1990. Ocean acoustic tomography. 263:4, 94–99. Theoretical basis and practical application of OAT are clearly presented.

Stommel, H., 1955. The anatomy of the atlantic. 190:30–35. This vintage article is by another young oceanographer who went on to great achievements.

Webster, P. J., 1981. Monsoons. 245:5, 108–19. The mechanism of the monsoons and their role in bringing water to half of Earth's population are discussed.

Weibe, P. H., 1982. Rings of the Gulf Stream. 246:3, 60–79. The biological implications of the large cold-water rings are considered.

Oceanography on the Web

Visit the Introductory Oceanography home page for on-line resources for this chapter. There you will find an on-line study guide with review exercises and links to ocean-ography sites to further your exploration of the topics in this chapter. Introductory Oceanography is at http://prenhall.com/thurman (click on the Table of Contents menu and select this chapter).

10 Waves

The Ocean Drilling Program's research vessel, Resolution, experiencing a North Atlantic storm in September, 1995. The ship barely survived the 185 k/h (115 mi/h) winds and 21 m (70 ft) waves. (Courtesy of the Ocean Drilling Program, Texas A&M University.)

Stirred by the wind into waves and pulled up and down by tidal forces, the ocean surface is always restless, always in motion. Sailors relish the speed their ships achieve when the wind comes up, but the wind also arouses the sea. Wind-produced waves have splintered ships in two, rolled them over, and ripped them apart. Some very famous ships have been wrecked by waves, including the *Edmund Fitzgerald*, which sank in Lake Michigan; many yachts have perished to waves in the cup races, and countless others have doubtless been wrecked, leaving no survivors to tell the tale.

Waves are among the most common and easily observed ocean phenomena, yet it was not until well into the nineteenth century that some scientific understanding of waves developed. Waves are periodic, predictable phenomena if the amount of energy applied to the ocean is known and predictable. Wind, earthquakes, turbidity currents, landslides, friction, and the pull of the Sun and Moon all generate waves. The Sun and Moon generate the largest waves of all: the tides.

Definition of a Wave

Waves are energy transmitted through matter. The medium (solid, liquid, or gas) does not actually travel as the energy passes through it, but its constituent particles vibrate or oscillate in place as the energy passes through them. Try thumping your fist on a table; the energy you generate travels through the table and is felt as a vibration by someone sitting at the other end, but the table does not otherwise move.

Waves move through matter in three dimensions. Simple progressive waves (in which the waveform can be observed progressing or traveling as in water) are shown in Figure 10-1a. Progressive waves may be longitudinal, transverse, or orbital.

In **longitudinal waves** (push-pull), such as sound waves, the particles that are in vibratory motion push and pull in the same direction the energy is traveling, like a spring whose coils are alternately compressed and expanded. The waveform travels through the medium by compressing and decompressing. Energy may be transmitted through all states of matter—gaseous, liquid, or solid—by this longitudinal movement of particles.

In **transverse waves** (side-to-side), energy travels at right angles to the direction of particle vibration. An example is the wave created when one end of a rope is fastened and the other end is moved up-and-down or side-to-side. A waveform is set up in the rope and progresses along it, as energy is transmitted from the motion of the hand to the fastened end. If you watch any segment of the rope, you can see it move up-and-down (or side-to-side) at right angles to a line drawn from the fastened end to the hand that is putting energy into the rope. Generally, this type of wave transmits energy only through solids, because only in a solid are particles strongly bound to one another. Longitudinal and transverse waves are called *body waves* because they transfer energy through a body of matter.

Waves that transmit energy along an interface between two fluids (including gasses) of different densities have particle movements that are neither longitudinal nor transverse. The best example of such an interface is that between air and water (atmosphere and ocean). We may say that the movement of particles along such an interface involves components of both longitudinal and transverse waves, because the particles move in circular orbits. Thus, waves at the ocean surface are progressive **orbital waves**, or *interface waves*.

Wave Characteristics

When we observe the ocean surface, we see waves of various sizes moving in various directions. This produces a complex wave pattern that constantly changes. To introduce the concepts needed to understand waves, we will look at a simple, idealized waveform (Figure 10-1b). This waveform represents the transmission of energy from a single source that travels along the ocean–atmosphere interface and is not influenced by friction with the ocean bottom. Such a wave has very uniform characteristics.

As our idealized progressive wave passes a permanent marker, such as a pier piling, you will notice a succession of high parts of the waves, or **crests**, separated by low parts, or **troughs**. If we mark the water level on the piling when the troughs pass and then do the same for the crests, the vertical distance between the marks is the **wave height**, H. The horizontal distance between any two corresponding points on successive waveforms, such as from crest to crest or from trough to trough, is the **wavelength**, L. The ratio of wave height to wavelength, H/L, is called **wave steepness**. The time (in seconds) that elapses during the passing of one full wave, or wavelength, is the **wave period**, denoted as T, for time. Figure 10-2 shows the relationships among wavelength, period, and speed. Because the period is the time required for the passing of one wavelength, if either the wavelength or period of a wave is known, the other can be calculated:

$$\text{Wavelength } (L) \text{ / Period } (T) = \text{Speed } (S) \quad \text{10-1}$$

For example, if we have a wave with a wavelength of 156 meters and a period of 10 seconds, we can calculate its speed as:

$$156 \text{ m}/10\text{s} = 15.6 \text{ m/s} \quad \text{10-2}$$

Another characteristic related to wavelength and speed is frequency. **Wave frequency** (f) is the number of wavelengths that pass a fixed point per unit of time and is equal to $1/T$. If six wavelengths pass a point in 1 min, and we have the same wave system as in our previous example, then:

$$\begin{aligned} \text{Speed } (S) = \text{L} \times f &= 156 \text{ m} \times 6/\text{min} \\ &= 936 \text{ m/min} \end{aligned} \quad \text{10-3}$$

To convert to meters per second, we have to divide by 60:

$$936 \text{ m/min} \times 1 \text{ min}/60 \text{ s} = 15.6 \text{ m/s} \quad \text{10-4}$$

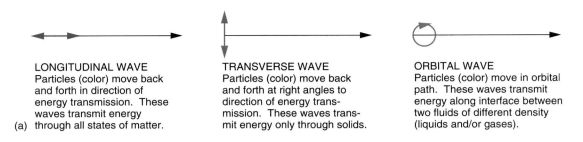

LONGITUDINAL WAVE
Particles (color) move back
and forth in direction of
energy transmission. These
waves transmit energy
(a) through all states of matter.

TRANSVERSE WAVE
Particles (color) move back
and forth at right angles to
direction of energy transmission. These waves transmit energy only through solids.

ORBITAL WAVE
Particles (color) move in orbital
path. These waves transmit
energy along interface between
two fluids of different density
(liquids and/or gases).

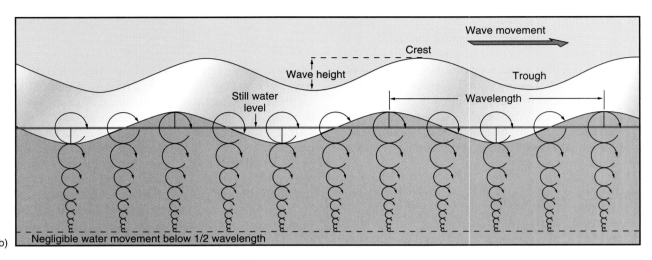

(b)

Negligible water movement below 1/2 wavelength

Figure 10-1   Types of Progressive Waves

(a) Types of progressive waves. (b) A simple sea wave and its components: height,
wavelength, and depth.   *(Part (b) from the Tasa Collection:* Shorelines. *Macmillan Publishing
Co., New York. Copyright © 1986 by Tasa Graphic Arts, Inc. All rights reserved.)*

Figure 10-2   Speed of Deep-Water Waves

Ideal relations among wave speed,
wavelength, and period for deep-
water waves. Speed (m/s) equals the
wavelength (m) divided by the
period (s), or $S = L/T$. For example
(shown with red lines), a wave with a
period of 8 s has a wavelength of
about 100 m and a speed of about
12.5 m/s ($L(m) = 1.56(m/s)T^2$).

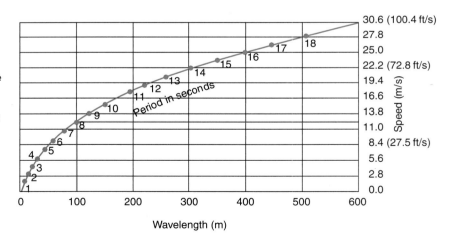

Because the speed and wavelength of ocean waves are
such that less than one wavelength passes a point per second, the preferred unit of time is period (rather than frequency), when calculating the speed of ocean waves.

Returning to the particle motion of ocean waves, the
circular orbits followed by water particles at the surface
have a diameter equal to the wave height. While a particle
is in the crest of a passing wave, it is moving in the direction of energy propagation. While it is in the trough, it is
moving in the opposite direction. The trough-half of the
orbit occurs at a lower speed than the crest-half of the
orbit. Therefore, there is a small net transport of water in
the direction the waveform is moving. This condition results from the fact that particle speed decreases with increasing depth below the still-water line. Also, the
diameters of particle orbits decrease with increased depth
until particle motion associated with our idealized wave
ceases at a depth of one-half wavelength, $L/2$.

Deep-Water Waves

Ocean waves with the ideal characteristics just described belong to a category called **deep-water waves**. Such waves travel across the ocean where the water depth (d) is greater than one-half the wavelength (Figure 10-3a). These waves are not affected by the ocean floor. Included in deep-water waves are all wind-generated waves as they move across the open ocean. As you can see in Figure 10-2, the speed of deep-water waves is equal to the wavelength (L) divided by the period (T):

$$\text{Speed (m/s)} = L/T \qquad \text{10-5}$$

The easier of these characteristics to measure is period (the time required for one wave to pass). Thus, the following equation is more commonly used for computing wave speed (g is acceleration due to gravity):

$$\text{Speed (m/s)} = gT/2\pi = 9.8 \text{ m/s}^2 \times T(s)/2\pi \\ = 1.56 \, T \qquad \text{10-6}$$

Shallow-Water Waves

Waves in which depth (d) is less than 1/20 of the wavelength ($L/20$) are classified as **shallow-water waves**, or *long waves* (Figure 10-3b). Included in this category are wind-generated waves that have moved into shallow, nearshore areas, tsunamis (seismic sea waves) generated by earthquakes in the ocean floor, and tide waves, generated by gravitational attraction of the sun and moon. In these waves, the wavelength is very long relative to water depth, and wave speed increases with increasing ocean depth (d):

$$\text{Speed (m/s)} = \sqrt{gd} = 3.1\sqrt{d(m)} \qquad \text{10-7}$$

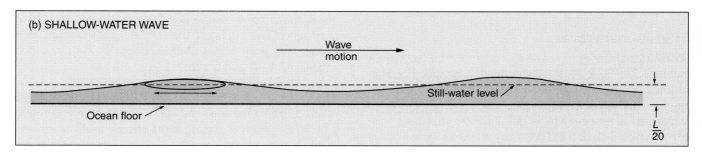

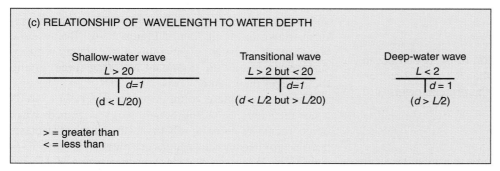

Figure 10-3   Deep-Water and Shallow-Water Waves

(a) Wave profile and water-particle motions of a deep-water wave. Note the diminishing size of the orbits with increasing depth below the surface. (b) Motions of water particles in shallow-water waves. Water motion extends to the ocean floor. (c) Relationship of wavelength to water depth.

Figure 10-4 Wave Speed

This graph shows the relation of water depth to wave speed. It is divided into three zones (red, purple, and green) because different factors affect wave speed at different water depths. The blue lines represent different wavelengths. Three examples of how to read the graph are shown by the red, green and purple circles and lines connecting to the axes. Red example is a shallow-water wave (red zone): To determine wave speed, find the water depth on the bottom scale, then move up to the "shallow-water waves" line, then over to the wave-speed scale. In the example, a wave in 50 m deep water travels about 21 m/s. Green example is a deep-water wave (green zone): Water depth is not a factor and so, to determine wave speed, simply extend the wavelength line (blue) to the left until it intersects the wave-speed scale; example is for a 50- m wavelength, showing a wave speed of about 9 m/s. Purple example is a transitional wave (purple zone): Combine the shallow-water and deep-water procedures for transitional waves; example is for a 1000- m wavelength and 200- m water depth, giving a speed of about 35 m/s. Note that the transitional waves have a slower wave speed than they would have had as deep-water waves.

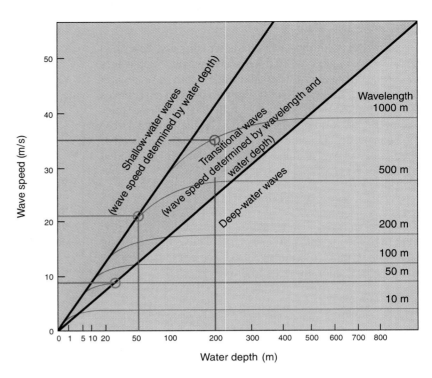

Particle motion in shallow-water waves occurs in a very flat elliptical orbit that approaches horizontal oscillation. The vertical component of particle motion decreases with increasing depth. The presence of shallow-water waves can be detected at the ocean bottom and therefore can affect the bottom.

Transitional Waves

Transitional waves have wavelengths greater than 2 times but less than 20 times the water depth. The speed of transitional waves is controlled partially by wavelength and partially by water depth (Figure 10-3c and 10-4). Deep-water waves generated by winds at the ocean surface usually have periods of 6 to 10 s. They maintain these periods after encountering shallow coastal waters.

Making Waves

Figure 10-5 shows the concentration of energy in ocean waves that have periods from less than 0.1 s to more than 1 day. The illustration also shows the principal causes of waves of different periods.

Wind-Generated Waves

Wind-generated waves have a history that includes their origin in windy regions of the ocean, their movement across great expanses of open water without subsequent aid of wind, and their termination when they break and release their energy, either in the open ocean or against the shore. As the wind blows over the ocean surface, it cre-

ates pressure and stress that deform the ocean surface into small, rounded waves with extremely short wavelengths—less than 1.74 cm (0.7 in.). Such waves are commonly called *ripples*, but oceanographers call them **capillary waves**. The name comes from the term capillarity, in reference to the surface tension of water. Surface tension is the dominant **restoring force**, that works to destroy these tiny waves and to restore the smooth ocean surface once again. Capillary waves characteristically have rounded crests and V-shaped troughs (Figure 10-6).

As capillary wave development increases, the sea surface takes on a rougher character. This catches more of the wind, allowing the wind and ocean surface to interact more efficiently and transferring more energy to the water. As more energy is transferred to the ocean, **gravity waves** develop. They have wavelengths exceeding 1.74 cm (0.7 in.) and are shaped more like a sine curve (the familiar waveform shown in the middle of Figure 10-6). Because the waves reach a greater height at this stage, gravity replaces surface tension as the dominant restoring force, giving these waves their name.

The length of these young waves is generally 15 to 35 times their height. As additional energy is gained, wave height increases more rapidly than wavelength. The crests become pointed and the troughs are rounded (Figure 10-6, right side). When the steepness reaches a ratio of 1/7, the speed with which the waves travel becomes 1.2 times that of typical deep-water waves.

Energy imparted by the wind increases wave height, length, and speed. When a wave encounters a wind with speed equal to that of the wave, neither wave height nor

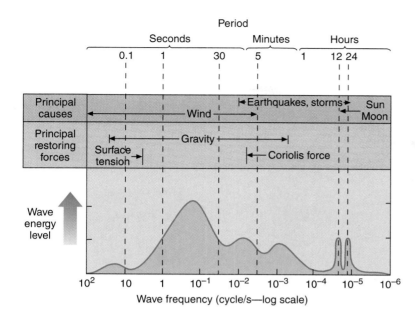

Figure 10-5   Distribution of Energy in Ocean Waves

This figure shows that most of the energy possessed by ocean waves is in wind-generated waves with a period of about 10 s. The long-period peaks on either side of the 5-minute period mark represent tsunami, while the two sharp peaks to the right represent ocean tides with their semidaily and daily periods.   *(After Blair Kinsman, Wind Waves: Their Generation and Propagation on the Ocean Surface. © 1965.)*

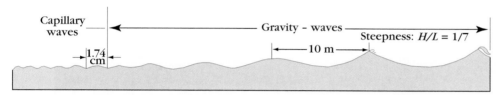

Figure 10-6   Capillary and Gravity Waves

As energy is put into the ocean surface by wind, small rounded waves with V-shaped troughs develop (capillary waves). As the water gains energy, the waves increase in height and length. When they exceed 1.74 cm (0.7 in) in length, they take on the shape of the sine curve and become gravity waves. Increased energy increases the steepness of the waves. The crests become pointed and the troughs rounded. As the steepness reaches 1/7, the waves become unstable, and whitecaps form as they break. Not drawn to scale.

length are changed because there is no net energy exchange; the wave is at its maximum size.

Figure 10-7 shows important relations between wind and the area where waves are initiated. The area where wind-driven waves are generated is simply called, sea. For clarity in this discussion, we will call it the sea area. It is characterized by choppiness and short wavelengths, with waves moving in many directions and having many different periods and lengths. The variety of wave periods and wavelengths is caused by frequent changes in wind speed and direction.

Factors important in increasing the amount of energy transferred to the waves are (1) *wind speed*, (2) the *duration* of time that the wind blows in one direction, and (3) the *fetch*, a term that refers to the distance over which the wind blows in one direction.

The energy in a wave is directly related wave height. Wave heights in a sea area are usually less than 2 m (6.6 ft), although it is not uncommon to observe wave heights of 10 m (33 ft) and periods of 12 s. As sea waves gain energy, their steepness increases. When steepness reaches a crit-

ical value of 1/7 ($H = 1/7 L$), open ocean breakers form; called *whitecaps,* they release the energy put into the waves by the wind. The appearance of a sea surface as it changes from calm to the condition that results from hurricane-force winds is described in Table 10–1. Figure 10-8 is a satellite image of average wave heights worldwide as they existed during the period, October 3–12, 1992.

The largest wind-generated wave authentically measured was 34 m high (112 ft), with a period of 14.8 s. It was seen in the North Pacific in 1935 by the crew of the U.S. Navy tanker *U.S.S. Ramapo.* For a given wind speed, there is a maximum fetch and duration of wind beyond which the waves cannot grow. When both maximum fetch and duration are attained for a given wind speed, a fully developed sea has been created. The reason it can grow no further is that waves are losing as much energy as breaking whitecaps as they are receiving from the wind.

Table 10–2 shows the fetch and duration of wind required to produce a fully developed sea for given wind speeds. Table 10–3 describes the fully developed sea at given wind speeds in terms of average height, length, and

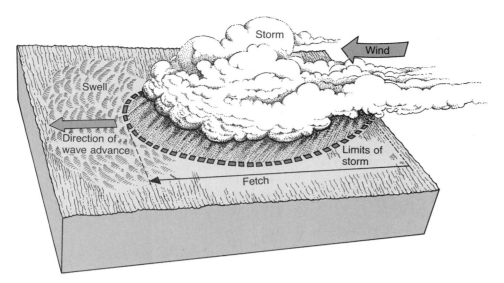

Figure 10-7   The Sea and Swell

As wind blows across the sea area, wave size increases with increased wind speed, fetch, and duration. As waves advance beyond the sea area, they continue to advance across the ocean surface as swell, free waves that are not driven by the wind but sustained by the energy they obtained in the sea.

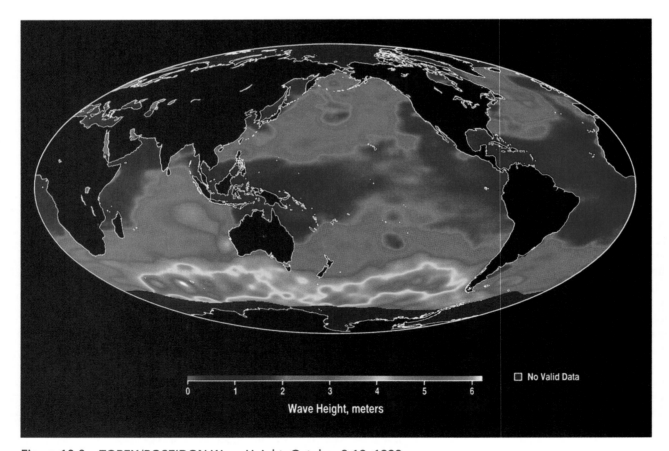

Figure 10-8   TOPEX/POSEIDON Wave Height, October 3-12, 1992

The TOPEX/POSEIDON satellite receives a return of a stronger signal from calm seas and a weaker signal from seas with large waves. Based on these data, this image shows there is a high correlation between wave height and wind speed (see Figure 8-14). The largest average wave heights are in the westerly wind belt of the Southern Hemisphere where the highest wind speeds were recorded. The highest average wave heights of over 6 m (20 ft) occur in the southern Indian Ocean.   *(Courtesy of Jet Propulsion Laboratory, NASA.)*

TABLE 10-1 Beaufort Wind Scale and the State of the Sea.

Number	Descriptive Term	(m/s)	(mi/h)	Appearance of the Sea
0	Calm	—	—	Like a mirror
1	Light air	0.3–1.5	1–3	Ripples with the appearance of scales, no foam crests
2	Light breeze	1.6–3.3	4–7	Small wavelets; crests of glassy appearance, no breaking
3	Gentle breeze	3.4–5.4	8–12	Large wavelets; crests begin to break, scattered whitecaps
4	Moderate breeze	5.5–7.9	13–18	Small waves, becoming longer; numerous whitecaps
5	Fresh breeze	8.0–10.7	19–24	Moderate waves, taking longer form; many whitecaps, some spray
6	Strong breeze	10.8–13.8	25–31	Large waves begin to form, whitecaps everywhere, more spray
7	Near gale	13.9–17.1	32–38	Sea heaps up and white foam from breaking waves begins to be blown in streaks
8	Gale	17.2–20.7	39–46	Moderately high waves of greater length, edges of crests begin to break into spindrift, foam is blown in well-marked streaks
9	Strong gale	20.8–24.4	47–54	High waves, dense streaks of foam and sea begins to roll, spray may affect visibility
10	Storm	24.5–28.4	55–63	Very high waves with overhanging crests; foam is blown in dense white streaks, causing the sea to appear white; the rolling of the sea becomes heavy; visibility reduced
11	Violent storm	28.5–32.6	64–72	Exceptionally high waves (small and medium-sized ships might for a time be lost from view behind the waves), the sea is covered with white patches of foam, everywhere the edges of the wave crests are blown into froth, visibility further reduced
12	Hurricane	>32.7	>73	The air is filled with foam and spray, sea completely white with driving spray, visibility greatly reduced

(Source: After Bowditch, 1958.)

TABLE 10-2 Fetch and Duration Required To Produce Fully Developed Sea for a Given Wind Speed.

Wind Speed km/h (mi/h)	Fetch km (mi)	Duration h
20 (12)	24 (15)	2.75
30 (19)	77 (48)	7
40 (25)	176 (109)	11.5
50 (31)	380 (236)	18.5
60 (37)	660 (409)	27.5
70 (43)	1093 (678)	37.5
80 (50)	1682 (1043)	50
90 (56)	2446 (1517)	65.25

period of waves. Also, the average height of the highest 10 percent of the waves is given.

Swell

Waves generated in a sea area do not all have the same characteristics because they are generated at different times in the storm from winds blowing at different speeds. As these waves move toward the margins of the sea area, where wind speeds diminish, they eventually are moving faster than the local wind. When this occurs, the wave steepness decreases, and the waves become long-crested waves called **swell**. Swell move with little loss of energy over large stretches of the ocean surface, transporting energy away from the sea area where wind energy is input. Most of the energy of the swell is eventually released along the continental margins where it is a major cause of shoreline erosion.

Waves with longer wavelengths travel faster and leave the sea area first. They are followed by the shorter, slower waves. This progression illustrates the principle of **wave dispersion**, a sorting of waves by wavelength. In the wave-generating area, waves of many wavelengths are present. In deep water, wave speed is a function of wavelength (Figure 10-2), so the longer waves outrun the shorter ones. Swell can travel huge distances without significant depletion of energy. Swell originating from Antarctic storms has been recorded breaking along the Alaskan coast after traveling more than 10,000 km (over 6000 mi).

As a group of waves leaves a sea area, it becomes a swell **wave train**. This group of waves moves across the ocean surface with a group speed (the speed at which the wave energy progresses) of only one-half the speed of an individual wave in the group. Because it takes time for the

TABLE 10-3	Description of Fully Developed Sea for a Given Wind Speed			
Wind Speed km/h (mi/h)	Average Height M (ft)	Average Length m (ft)	Average Period s	Highest 10% Waves m (ft)
20 (12)	0.33 (1)	10.6 (34.8)	3.2	0.75 (2.5)
30 (19)	0.88 (2.9)	22.2 (72.8)	4.6	2.1 (6.9)
40 (25)	1.8 (5.9)	39.7 (130.2)	6.2	3.9 (12.8)
50 (31)	3.2 (10.5)	61.8 (202.7)	7.7	6.8 (22.3)
60 (37)	5.1 (16.7)	89.2 (292.6)	9.1	10.5 (34.4)
70 (43)	7.4 (24.3)	121.4 (398.2)	10.8	15.3 (50.2)
80 (50)	10.3 (33.8)	158.6 (520.2)	12.4	21.4 (70.2)
90 (56)	13.9 (45.6)	201.6 (661.2)	13.9	28.4 (93.2)

wave energy to cause the water to oscillate (make a wave), the leading wave in the train keeps disappearing. However, the same number of waves always remains in the group because, as the leading wave disappears, a new wave replaces it at the back of the group. The cause of this phenomenon is the energy moving more slowly than the individual waves; its slower transmission generates the wave at the rear.

Interference Patterns Because swells may move away from a number of storm areas in any ocean, it is inevitable that swell from different storms will run together, and the waves will clash, or interfere, with one another. This gives rise to a special feature of wave motion: interference patterns, produced when two or more wave systems collide. The resulting interference pattern is the algebraic sum of the disturbances that all the colliding waves would have produced individually. The result may be larger or smaller troughs or crests, depending on conditions (Figure 10-9). When swells from two storm areas collide, the interference pattern may be constructive or destructive, but it is more likely to be mixed. Ideally, **constructive interference** results when wave trains having the same wavelength come together in phase, meaning crest-to-crest and trough-to-trough. If we sum the displacements that would result from each wave individually, we find that the resulting interference pattern is a wave having the same length as the two converging wave systems, but with a wave height equal to the sum of the individual wave heights (Figure 10-9, left). **Destructive interference** results when wave crests of one swell coincide with the troughs of a second swell. If the waves have identical characteristics, the algebraic sum of the crest plus the trough is zero, and the energy of these waves cancel each other.

However, it is more likely that the two swell systems possess waves of varied heights and lengths and thus come together with a mixture of destructive and constructive interference (Figure 10-10). This more complex **mixed interference pattern** explains the occurrence of the varied sequence of higher and lower waves and other irregular wave patterns we observe when swell approaches the seashore. Interference also occurs in the sea area where

waves of various lengths, directions, and heights combine to produce a confused or choppy surface.

You can experience this effect in another fluid, the air, by listening to a steady, high-pitched sound in a room. An example is a steadily blown note on a brass instrument. As you move your head, the sound will grow loud (constructive interference), or virtually disappear (destructive interference), but mostly it will vary (mixed interference).

Free and Forced Waves In the swell, we see what may be referred to as a **free wave**, which is moving with the momentum and energy imparted to it in the sea area but is not experiencing a maintaining force that keeps it in motion. In the wave-generating area, free and forced waves are present. A **forced wave** is one that is maintained by a force that has a periodicity coinciding with the period of the wave. This force is the wind. Because of the variability of the wind, many wave systems in the sea area alternate between forced waves and free waves. Tides are an example of a forced wave that is always maintained by the gravitational attraction of the Moon and Sun.

Rogue Waves

One of the ocean's mysteries is the cause of **rogue waves**, massive waves that can reach 10 stories in height. Such waves result from rare coincidence in ordinary wave behavior. On average, in the open ocean, 1 wave in 23 will be over twice the average wave height, 1 in 1175 will be three times the average wave height, and only 1 in 300,000 will be four times the average wave height. The chances of truly monstrous waves are 1 in billions, but they do happen. Of course, this statistical information is useless in predicting specifically when and where a rogue wave will arise. Lloyd's of London is the world's leading marine insurance underwriter. They report that during the 1980s, storms claimed an annual average of 46 ships weighing over 500 tons (for example, a 15 m-long, or 50- ft, fishing vessel). The total of vessels of all sizes lost may reach 1000/yr. Some are the victims of rogue waves (Figure 10-11).

Although it is impossible to predict their occurrence, a main cause of rogue waves is theorized to be extraordinary

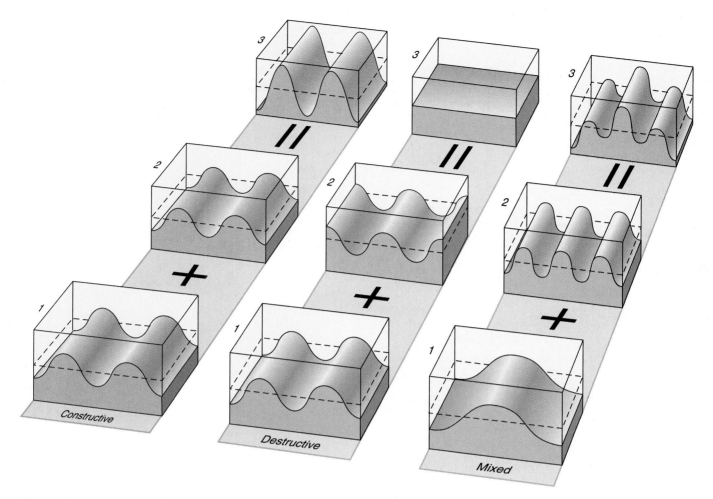

Figure 10-9   Wave Interference

As wave trains come together from different sea areas (1 and 2), three possible interference patterns may result (3). Very rarely, if the waves have the same length and come together in phase (crest to crest and trough to trough), totally constructive interference occurs. In this case, the amplitudes simply add to produce waves of the same length but of greater height. If two sets of waves have identical characteristics but come together exactly 180° out of phase, the result is complete cancellation, or destructive interference, with no remaining wave at all. More commonly, waves of different lengths and heights encounter one another and produce a complex, mixed interference pattern.

constructive interference. It has been determined that rogue waves tend to occur more frequently in locations that are downwind from islands or shoals, and where storm-driven waves move against strong ocean currents, such as the Agulhas Current off the southeastern coast of Africa. This stretch of water, where Antarctic storm waves drive northeast into the Agulhas Current, is probably responsible for sinking more ships than any other place on Earth.

Oceanographers learned more about ocean waves during the 3-month life of the *Seasat* satellite, than from centuries of gathering data by ship, buoy, and aircraft. Newer satellites, such as Poseidon and TOPEX, have provided much additional data (see Figure 10-8) but it is very unlikely that it will be possible to predict rogue waves any time soon, if ever.

Interaction with the Shore

As waves reach the shore, they transfer their contained energy to the shoreline. The loss of energy turns the waves to surf, and the energy released results in the varied erosive and depositional coastal processes that characterize that environment. We will learn more about coastal processes in Chapter 12; here, we will focus on what happens to the waves themselves.

Surf

Most waves generated in the sea area by storm winds move across the ocean as swell. They release their energy along the margins of continents in the **surf zone**, where the swell forms breakers. As deep-water waves of the swell toward the continental margin over gradually shallower

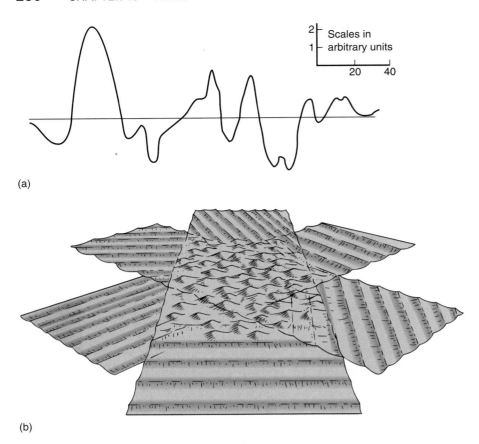

(a)

(b)

Figure 10-10   Mixed Interference Pattern

(a) Cross-section through the mixed interference pattern created when the three swell systems shown in (b) come together. Constructive interference creates the highest peaks and lowest troughs; where the trace is near the horizontal line (zero datum), destructive interference is dominant.

Figure 10-11   Rogue Waves

This illustration shows an oil tanker encountering a rogue wave. A deep trough, or "hole in the ocean" as it is called, drops 15 m (50 ft) below the still-water line. The bow of the ship drops into the trough, and the coming crest, which may be up to 15 m above the still-water line, crashes onto the bow. This huge mass of water overcomes the structural capacity of the ship, and so it sinks. Rogue waves have severely damaged a number of tankers and sunk others.

water, they eventually encounter water depths that are less than one-half of their wavelength (Figure 10-12). When the water depth becomes less than 1/20 the wavelength, waves in the surf zone begin to behave as shallow-water waves (see Figure 10-3). These shallow depths interfere with water particle movement at the base of the wave. The wave slows, and the wavelength decreases. Since the energy in

the wave remains the same, wave height increases. The crests become narrow and pointed and the troughs become wide curves, a waveform previously described for high-energy waves in the open sea. The increase in wave height accompanied by a decrease in wavelength increases the steepness (H/L) of the waves. As the wave steepness reaches 1/7, the waves break as surf.

If the surf is swell that has traveled from distant storms, breakers will develop relatively near shore in shallow water. By the point at which waves break, they have become shallow-water waves. The horizontal water motion associated with such waves moves water alternately toward and away from the shore, in a sort of oscillation. The surf is characterized by parallel lines of relatively uniform breakers. However, if the surf is composed of waves that have been generated by local wind, the waves may not have been sorted into swell. The surf may be more nearly characterized by unstable, deep-water, high-energy waves with steepness already near 1/7. They will break shortly after feeling bottom some distance from shore, and the surf will be rough, choppy, and irregular.

Ideally, when water depth is about 1.3 times breaker height, wave crests break, producing surf. Particle motion is greatly hampered by the bottom, and a significant

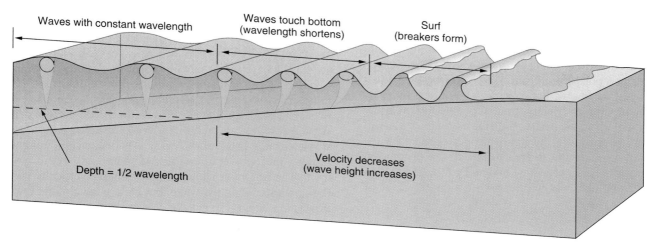

Waves with constant wavelength

Waves touch bottom
(wavelength shortens)

Surf
(breakers form)

Depth = 1/2 wavelength

Velocity decreases
(wave height increases)

Figure 10-12   Surf Zone

As waves approach the shore and encounter water depths of less than one-half wavelength, friction removes energy from the waves. The waves slow, stretching out so that wavelength decreases; wave height increases because the water must go somewhere, and the only way is up. When the water depth becomes 1.3 wave height, the wave reaches a steepness of 1/7 and breaks on the shore.   *(The Tasa Collection:* Shorelines. *Macmillan Publishing, New York. Copyright © 1986 by Tasa Graphic Arts, Inc. All rights reserved.)*

(a)

Figure 10-13   Breakers

(a) Plunging breaker. (Frances Bretague, Photo Researchers.) (b) Spilling breakers off Cape of Good Hope, South Africa.   *(Photo by Carl Purcell.)*

(b)

transport of water toward the shoreline occurs. Breaking of surf results from severe restriction of particle motion near the wave's bottom. This slows the waveform, but at the surface, individual water particles that are orbiting are not slowed because they have no contact with the bottom. Thus, the top of the waveform leans shoreward as the bottom of the waveform "drags bottom." The entire waveform slows and due to increasing wave height, water moves faster in the elliptical orbits near the surface. This causes water particles at the ocean/atmosphere interface to move toward shore faster than the waveform itself.

This motion can be seen particularly well in plunging breakers (Figure 10-13a). **Plunging breakers** have a curl-ing crest that moves over an air pocket. This happens when the curling particles literally outrun the wave, and there is nothing beneath them to support their weight. Plunging breakers form on moderately steep beach slopes.

The more commonly observed breakers are **spilling breakers** (Figure 10-13b) which result from a relatively gently sloping ocean bottom that extracts energy from the wave more gradually. This produces a turbulent mass of air and water that runs down the front slope of the wave instead of producing a spectacular cresting curl. Because energy is gradually extracted from spilling breakers, they have a longer life span and, thus, give surfers a longer but less exciting ride than plunging breakers.

The particle motion of ocean waves, shown in Figure 10-1, indicates that optimal surfing is found in front of the crest, where the water particles are moving up into the crest. This force, along with buoyancy, helps maintain a surfer's position in front of a cresting wave. When this upward motion of water particles is interrupted by the wave passing over water that is too shallow to allow this movement to continue, the ride is over. A skillful surfer, by positioning the board properly on the wave front, can regulate the degree to which the propelling gravitational forces exceed the buoyancy forces, and high speeds can be obtained while moving along the face of the breaking wave.

Wave Refraction

As we have seen, waves begin to bunch up and wavelengths become shorter as swell begins to "feel bottom" upon approaching the shore. However, it is seldom that swell approaches a straight shoreline at a perfect right angle (90°). Some segment of the wave can be expected to feel bottom first, and therefore it will slow before the rest of the wave. This bending of the wave front, called **refraction**, occurs as waves approach the shore. As shown in Figure 10-14a, waves approaching at an angle to a straight shoreline feel the bottom first close to the shoreline. In Figure 10-14b, an irregular shoreline slows portions of wavefronts at different distances from shore.

Wave refraction unevenly distributes wave energy along the shoreline. We can construct lines that are perpendicular to the wave fronts, and space them so the energy between lines is equal at all times. These are called orthogonal lines or *wave rays* (Figure 10-14). They help us see how energy is distributed along the shoreline by breaking waves. The orthogonal lines and wavefronts always bend toward shallow water.

Orthogonals indicate the direction that waves travel. You can see that waves converge on headlands that jut into the ocean, but the same waves diverge in bays. Thus, a concentration of energy is released against the headlands, while energy is more dispersed in bays. The result is heavy erosion of headlands, whereas deposition may occur in bays. The greater energy of waves breaking on headlands is reflected in an increased wave height.

Wave Diffraction

Wave **diffraction** results from wave energy being transferred around or away from barriers impeding its forward motion. Waves move past barriers into harbors because their energy moves laterally along the wave crest (Figure 10-15), and the wave behind the barrier goes out in all directions, forming an arc-shaped crest. This bending is on a much smaller scale and is less easily explained than the bending discussed in refraction, which is a simple response to changes in velocity. Diffraction occurs because any point on a wave front is a source from which energy can propagate in all directions.

Wave Reflection and Standing Waves

Not all of a wave's energy is expended as it rushes onto the shore. A vertical barrier, such as a seawall, can reflect swell back into the ocean with little loss of energy. For this ideal reflection to occur without energy loss, the wave must strike the barrier at a right angle. Such a condition is rare in nature. Nonetheless, less ideal reflections produce standing waves, a special interference pattern. **Standing waves** are the product of two waves of the same length moving in opposite directions, so there is no net movement. The water particles continue to move vertically and horizontally within the wave, but there is no circular motion as in a progressive wave. When standing waves travel within an enclosed basin, their motion is characterized by lines or points in space called *nodal lines* or *nodes*, respectively, at which there is no vertical movement. *Antinodes*, points at which wave crests alternately become troughs, represent the places where the greatest vertical movement occurs with the development of a standing wave (Figure 10-16). We will consider standing waves further when we discuss the tides in the next chapter. Under certain conditions, the development of standing waves significantly affects the tidal character of coastal regions.

Reflection of wind-generated waves from coastal barriers occurs at an angle equal to the angle at which the wave approached the barrier. This type of reflection may produce small-scale interference patterns similar to those previously discussed. An outstanding example of reflection phenomena is the Wedge, a wave feature that develops west of the jetty at Newport Harbor, California (Figure 10-17a). As incoming waves strike the jetty and are reflected, a constructive interference pattern develops. When the crests of incoming waves merge with crests of reflected waves, plunging wedge-shaped breakers may exceed 8 m (26 ft) in height (Figure 10-17b). These waves present a fierce challenge to the most experienced body surfers, attracting some who did not survive the challenge—the Wedge has killed or severely injured many who have come to try it.

Storm Surges

Large cyclonic storms create low atmospheric pressure at Earth's surface. When such a storm develops over the ocean, the low atmospheric pressure, relative to that over surrounding areas of the sea, produces a low hill of water. As the storm migrates across the open ocean, the hill moves with it. As the storm approaches shallow water near shore, the portion of the hill over which the wind is blowing shoreward will produce a **storm surge**, a mass of elevated, wind-driven water that produces an increase in sea level.

A storm surge can be extremely destructive to low-lying coastal areas, especially when it occurs at high tide. The coincidence of a storm surge with high tide in areas that normally have particularly high tides frequently produces major catastrophes, with great property damage and loss of life (Figure 10-18). One of the most remarkable storm surges occurred at a lighthouse at Dunnet Head, Scotland.

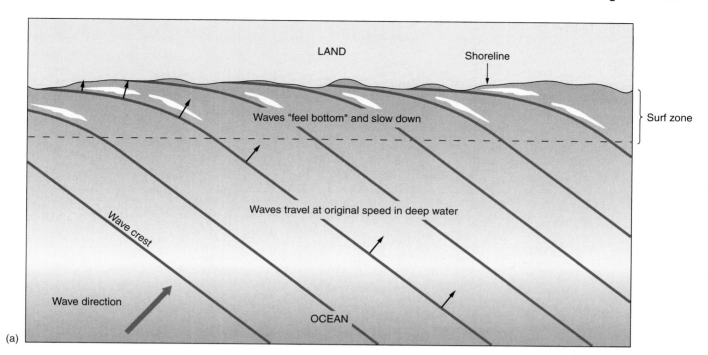

(a)

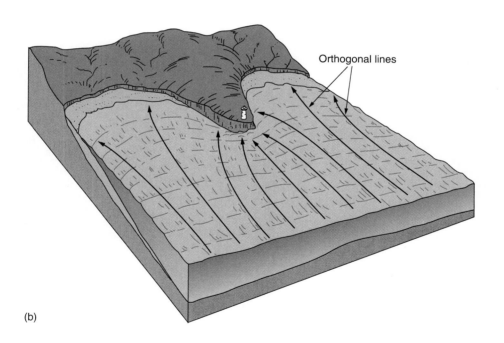

(b)

Figure 10-14   Refraction
....
(a) Waves approaching a straight shoreline at an angle "feel bottom" first close to shore. The part of the wave crest closer to shore slows more quickly than the part still in deeper water; the crest bends, or refracts. (b) As the waves "feel bottom" in the shallow areas off the headlands, they are slowed. The segments of the waves that move through the deeper water leading into the bay are not slowed until they are well into the bay. As a result, the wave crests are refracted (bent) and the release of wave energy is concentrated on the headlands. Erosion is active on the headlands, whereas deposition occurs in the bay, where the energy level is low. Orthogonal lines, spaced so that equal amounts of energy are between any two adjacent lines, help to show the distribution of energy along the shore.   *(Adapted with the permission of Macmillan Publishing Company from* Earth Science, *Sixth Edition, Figure 11-14, by Edward J. Tarbuck and Frederick K. Lutgens. Copyright © 1991 by Macmillan Publishing Company.)*

Figure 10-15   Diffraction

Diffraction is bending caused by waves passing an obstacle; it is not related to refraction. By diffraction, wave energy may spread to the most protected areas within a harbor.

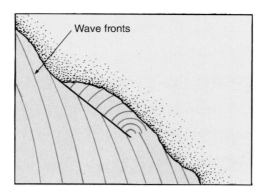

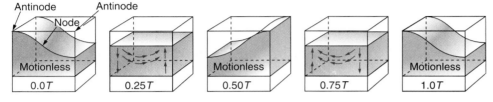

Figure 10-16   Reflection–Standing Waves

An example of water motion viewed at four points during a wave cycle. Water is motionless when antinodes reach maximum displacement. Water movement is maximum when the water is level. Movement is totally vertical beneath the antinodes, and maximum horizontal movement occurs beneath the node. The circular motion of particles that occurs in progressive waves does not exist in standing waves.

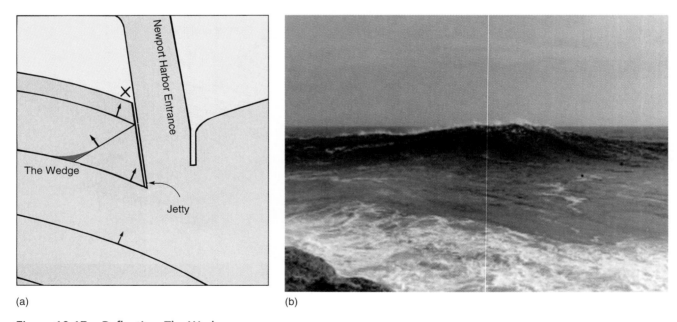

(a)   (b)

Figure 10-17   Reflection–The Wedge

(a) The Wedge, a wedge-shaped crest that may reach heights in excess of 8 m (26 ft), develops as the result of interference between incoming waves and reflected waves near the jetty protecting the entrance to Newport Harbor, California. (b) A view of a wedge crest taken from the landward end of the jetty. The three dots in the water in front of the wave are the heads of body surfers waiting to catch the wave.

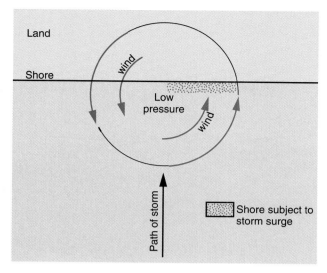

(a)

(b)

Figure 10-18   Storm Surge

(a) As a cyclonic storm in the Northern Hemisphere moves ashore, the low-pressure cell around which the storm winds blow and the onshore winds associated with the storm will produce a high-water phenomenon called a storm surge. (b) Storm waves generated by Hurricane Felix in August 1995 surge ashore in the Belmar/Avon, New Jersey, area. *(Photo by Bob Stovall.)*

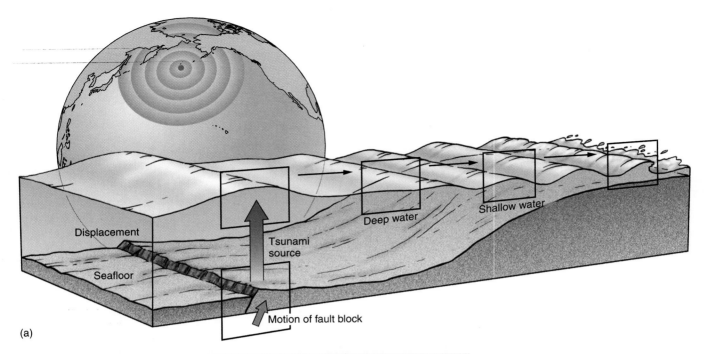

(a)

(b)

Figure 10-19   Tsunami

(a) Abrupt vertical movement along a fault in Earth's crust pushes up the ocean water column above the fault. The energy released is then distributed laterally along the atmosphere–ocean interface in the form of massive long, low waves called tsunami. The energy is transmitted across the open ocean by these waves, which exceed 200 km (125 mi) in length, but are only about 0.5 m high (1.6 ft), and thus are generally undetected. They release their energy upon reaching the shore, where they develop heights that can exceed 30 m (100 ft). Parts (b), (c), and (d) show the results of a 1983 tsunami in northern Japan, which killed 100 and injured 104.   *(Photos courtesy Kyodo News Agency.)*

(c)

(d)

Figure 10-19 (continued)

The lighthouse sits atop a cliff some 90 m high (about 300 ft). As storm surge waves broke against the cliff, they tossed stones that broke windows in the lighthouse.

Tsunami

The Japanese call the large, destructive waves that occasionally roll into their harbors **tsunami**, or harbor waves. Tsunami are unusual waves that originate from earthquakes. News media often mistakenly call them tidal waves, which implies they are related to the tides, but no such relationship exists.

Tsunami are usually caused by fault movement, or displacement, in Earth's crust along a fracture. This not only causes an earthquake, but also a sudden change in water level at the ocean surface above. Secondary events, such as underwater avalanches produced by the faulting, may also produce tsunami (Figure 10-19a). The ocean most plagued by tsunami is the Pacific. It is ringed by a series of trenches that are unstable margins of lithospheric plates, along which large-magnitude earthquakes occur.

Because the wavelength of a typical tsunami exceeds 200 km (125 mi), it is obviously a shallow-water wave, the speed of which is determined by water depth. Moving at great speeds in the open ocean, well over 700 km/h (435 mi/h), these waves only have heights of approximately 0.5 m (1.6 ft) in the open ocean. Thus, tsunami are not readily observable until they reach shore. In shallow water, they slow, and the water begins to pile up. Tsunami may have crests that exceed 30 m height (100 ft), and rush landward with destructive results (Figure 10-19b, c and d). Unlike a hurricane, whose high winds and waves threaten ships at sea and send them to the protection of a coastal harbor, a tsunami sends ships from their coastal moorings into the calm of the open ocean. A tsunami may consist of a single wave, but multiple waves are much more likely, depending on how the earthquake releases energy.

One of the most destructive tsunami ever generated came from the greatest release of energy from Earth's interior observed during historical times. On August 27, 1883, the volcanic island of Krakatau in Indonesia exploded and essentially disappeared. The sound of the explosion was heard an incredible 4800 km (2981 mi) away at Rodriguez Island in the western Indian Ocean. Dust from the explosion ascended into the atmosphere and circled Earth on high-altitude winds. This dust produced unusual and beautiful sunsets for nearly a year. Not many people were killed by the outright explosion of the volcano, but the displacement of water from the explosion was enormous. The resulting tsunami exceeded 30 m (100 ft). It devastated the coastal region of the Sunda Strait between Sumatra and Java, taking more than 36,000 lives. The energy carried by this wave was detected in the English Channel, after the wave crossed the Indian Ocean, passed the southern tip of Africa, and moved north in the Atlantic Ocean.

Before today's tsunami warning system existed, the first sign of a tsunami to most observers was the rapid recession of sea level. This recession, created by the trough preceding the first tsunami wave crest, is followed in minutes by one or more destructive waves. Such a recession was observed in the port of Hilo, Hawaii, on April 1, 1946. It was the beginning of a tsunami from an earthquake in the Aleutian Trench off the island of Unimak, Alaska, over 3000 km (1850 mi) away. The recession was followed by a wave that elevated the water nearly 8 m (26 ft) above normal high tide. The tsunami also struck Scotch Cap, Alaska, on Unimak Island. The island's lighthouse, the base of which stood 14 m (46 ft) above sea level, was destroyed by a wave estimated to have attained a height of 36 m (118 ft). This wave, produced by a disturbance in the Aleutian Trench, was recorded throughout the coastal Pacific at tide-recording stations.

This event prompted the creation of the International Tsunami Warning System (ITWS). Until 1948, it was impossible to warn of coming tsunami in time for people to avoid the destructive waves. When a seismic disturbance is recorded that has the potential to produce a tsunami, the closest tide-measuring stations are monitored for any indication of the rapid sea-level recession indicative of a tsunami. Because tsunami are shallow water waves that reach the bottom far offshore, pressure gauges installed on the sea floor are also used for early detection. If a tsunami is detected, warnings are sent to all the coastal regions that a destructive wave might reach. If there is sufficient time before the wave arrives, this warning allows for evacuation of people from low-lying areas and removal of ships from harbors.

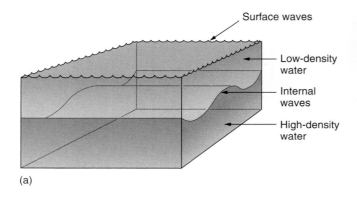

(a)

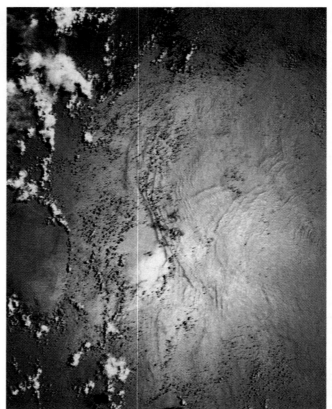

(b)

Figure 10-20   Internal Wave

(a) A simple internal wave moving along the density interface below the ocean surface. (b) This view of internal waves in the Indian Ocean was taken by the crew of the space shuttle Columbia in January 1990. The waves are diffracting around shoals south of the Seychelles and recombining to form interference patterns.   *(Photo courtesy of NASA.)*

Internal Waves

We have discussed waves that occur at the interface between the atmosphere and the ocean, where obvious density discontinuities exist; such density discontinuities also exist within the ocean water column. These are sites for generation of internal waves.

Just as energy can be transmitted along the interface between the atmosphere and the ocean, it also can be transmitted along density interfaces beneath the ocean surface in what are called **internal waves**. Internal waves may have heights exceeding 100 m (330 ft) (Figure 10-20). The greater the difference in density between the two fluids, the faster the waves will move.

Much remains to be learned about internal waves, but their existence is well documented and many of their causes have been identified. Some internal waves have periods that suggest that they are caused by tidal forces. Underwater avalanches in the form of turbidity currents, wind stress, and energy put into the water by moving vessels may also cause internal waves. Parallel slicks seen on surface waters may overlie troughs of internal waves. The slicks are caused by a film of surface debris that accumulates behind wave crests and dampens surface waves.

Internal waves reach greater heights from a smaller energy input than do the waves resulting from very large energy input observed at the ocean's surface. This is because they move along interfaces, such as the pycnocline, across which the density difference is considerably less than that between the ocean surface and the atmosphere. Internal waves are thought to move as shallow-water waves at speeds considerably less than those of surface waves, with periods of 5 to 8 minutes and wavelengths of 0.6 to 0.9 km (0.37 to 0.56 mi).

Harnessing Wave Energy

There is great energy in moving water, which is why we have hydroelectric power plants on rivers. Even greater energy exists in ocean waves; however, to harness it efficiently significant problems must be overcome. Where waves refract and converge, as they do around headlands, energy is focused, creating a potential setting for power generation. Such a system might extract up to 10 MW of power per kilometer of shoreline. This would be comparable to the electricity consumed by 20,000 households for 1 month. However, it could produce significant power only when large storm waves broke against it. Such a system would operate only as a power supplement, and a series of perhaps a hundred such structures along the shore would be required. Structures of this type could have a serious effect on natural processes, and might lead to serious coastal erosion in areas deprived of sediment. Along shores with suitable ocean-floor topography, internal waves could be effectively focused by refraction and induced to break against an energy-conversion device.

Lockheed Corporation has developed an interesting device called Dam-Atoll to extract energy from ocean waves (Figure 10-21). Waves enter the top of the unit at the ocean surface. Water spirals into a whirlpool within a central core. The swirling water turns a turbine, the unit's only moving part, which can provide a continuous electrical power output of 1 to 2 MW according to inventors Leslie S. Wirt and Duane L. Morrow. In particularly good wave areas, such as the Pacific Northwest, they think 500 to 1000 units could provide power in quantities comparable to those provided by Hoover Dam, a major hydroelectric power-generating facility on the Colorado River.

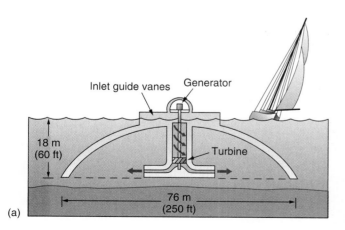

(a)

(b)

Figure 10-21 Dam-Atoll

Lockheed Corporation's Dam-Atoll designed to generate electricity from wave action. (a) Water entering at the surface spirals down an 18 m (600 ft) central cylinder to turn a turbine located at the bottom of the cylinder. Each unit is designed to produce from 1 to 2 MW. (b) View from above. *(Courtesy of Lockheed Corporation.)*

The Fastnet Disaster—
Wave Interference Creates A Monster Sea

On Tuesday, August 14, 1979, the worst disaster in the history of yachting struck 303 vessels entered in the Fastnet Race, the finale of the five-race Admiral Cup Series. The 1000-km (621 mi) race started on August 11 from the Isle of Wight off the south coast of England; the course required rounding Fastnet Rock off the southern tip of Ireland with a return to Plymouth, England (Figure 10A).

On the first day of the race, a low-pressure system that had spawned tornadoes and thunderstorms from Ohio to New England, weakened and moved into the Atlantic off the coast of Nova Scotia (Figure 10B). It was of no concern to the participants in Fastnet. By 1200 Greenwich Meridian Time (GMT) the following day, August 12, the system had accelerated into the mid-Atlantic to 40°E longitude and 48°N latitude. Having become no more than a minor low-pressure trough with a minimum pressure of 1007 millibars (as compared to standard sea level pressure of 1013 millibars), it was relatively unnoticed as it reached 19°E longitude and 48°N latitude at 1200 GMT August 13 (Figures 10B and 10C-1).

The storm immediately began to develop, and by 1800 GMT, August 13, the pressure had dropped to 996 millibars (Figure 10C-2). At 2100 GMT, with the storm center just off the southwest coast of Ireland, the pressure had dropped to 983 millibars (Figure 10C-3). Gusts up to 100 km/h (62 mi/h) were being felt south of Ireland. These west-southwest winds represented a force of 10 on

Figure 10A

The yachts sail from the Isle of Wight, around Fastnet Rock, then back to Plymouth.

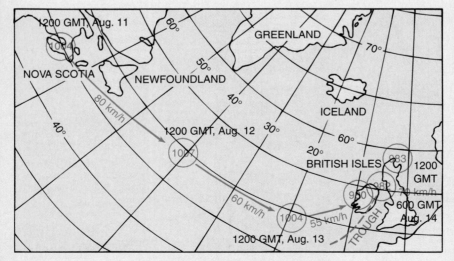

Figure 10B Course of the Storm

A weakening storm center moves off the coast of Nova Scotia at 1200 GMT (noon) Saturday, August 11. It races at 80 km/h (50 mi/h) into the Mid-Atlantic by 1200 GMT Sunday. Slowing slightly, the center moves to 19°E longitude and 48°N latitude by 1200 GMT Monday. Here the storm slows appreciably and deepens as it moves northeast toward Ireland. When the storm reaches the Irish Sea at 0600 GMT Tuesday, August 14, the trough crosses the race fleet and the west-southwest winds change to northwest winds. The storm has reached its peak. As the storm moves to the northnortheast, the seas in the race course recede. The circled numbers indicating the storm's location are atmospheric pressure in millibars. Velocity values along the course of the storm are the rate at which the storm moved across the Atlantic Ocean. A detailed view of the storm's development from 1200 GMT Monday, August 13, to 1200 GMT Tuesday, August 14, is presented in Figure 10C.

the Beaufort Wind Scale (Table 10). Although the larger, leading yachts had already rounded Fastnet Rock and had the wind on their stern-starboard beam, most of the fleet was struggling into a rising wind and sea.

By 0000 GMT on August 14, the storm center was in Galway Bay in central-western Ireland with a pressure of 980 millibars (Figure 10C-4). The trough of the storm moved into the fleet at 0600 GMT, and the winds shifted from west-southwest to northwest (Figure 10C-5). This right-angle change in the winds caused the tragedy. As the crests of incoming waves from the northwest merged with the crests of waves created by west-southwest winds, constructive interference produced very short, steep waves as high as 15 m (49.2 ft). Gusts approaching 145 km/h (90 mi/h) were reported. By 1200 GMT, the storm center

had moved to Moray Firth in northeastern Scotland, and the seas subsided dramatically along the course of the race (Figure 10C-6).

Considering the short duration of the winds—about 12 h for the west-southwest winds and less than 6 h for the northwest winds—the sea was abnormally developed. The compact, steep, rogue waves resulted from constructive interference between the two wave systems as they came together to produce a cross-sea condition (Figure 10D). When they took a yacht near their crests and on their leading edges, these short, elevated, massive waves could probably not have been managed by even the best designed yacht. As the yachts slid down the leading edge of the breaking wave, they rolled over (Figure 10E). The water rising into the wave crest pushed the keel up and rotated

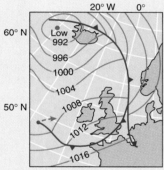

1. **1200 GMT, August 13.** From this location, the weak low-pressure cell (1004 millibars) begins to develop into the Fastnet storm, the path of which is marked by the red arrow.

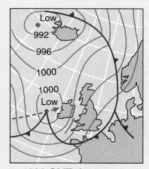

2. **1800 GMT, August 13.** Storm center pressure drops to 996 millibars, bringing gusts upward to 100 km/h (62 mi/h) in southwest winds south of Ireland. Wind direction is indicated by the orange arrow.

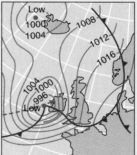

3. **2100 GMT, August 13.** Pressure falls to 983 millibars as west-southwest winds increase.

Figure 10C   Development of the Storm

(Courtesy of NOAA.)

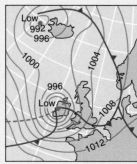

4. **0000 GMT, August 14.** Storm reaches maximum intensity as pressure drops to 980 millibars.

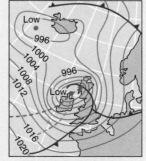

5. **0600 GMT, August 14.** Trough (orange dashed line) moves over the race course and winds shift from the west-southwest to northwest, producing a cross-sea.

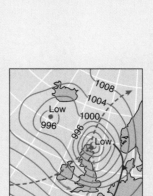

6. **1200 GMT, August 14.** Storm moves off to the northeast and winds diminish over the race course.

271

the mast into the sea. The following wave struck the keel, rotating it back into the water and completing the revolution. Many of the yachts reported being rolled a number of times in such a fashion. To survive, the crew would have to have been wearing safety harnesses. Some of the yachts may have been thrown onto others as the sea tossed them around.

Fifteen people died and 23 yachts were sunk or abandoned during the fierce encounter. Although weather forecasters were criticized for not giving sufficient warning to allow the yachts to seek harbor, they did not have adequate advance knowledge that the conditions would be so severe. At any rate, with the outcome of the race undetermined, it is unlikely that many yachts would have abandoned the race.

The Fastnet disaster is a recent example of the unpredictability of the oceans and the intense fury that they can develop. During a storm, one is likely to consider only the waves that may develop from a single wind direction. Although all mariners are aware of the potential effects of a cross-sea on wave conditions, it probably did not seem important at the time. Even as they came, most looked upon the rough seas as just another challenge to be met. Such is revealed in the words of Ted Turner, skipper of the race winner, *Tenacious*: "But we weren't really concerned with the conditions, we were concerned with winning."

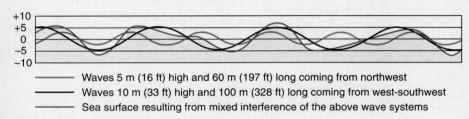

——— Waves 5 m (16 ft) high and 60 m (197 ft) long coming from northwest
——— Waves 10 m (33 ft) high and 100 m (328 ft) long coming from west-southwest
——— Sea surface resulting from mixed interference of the above wave systems

Figure 10D   Constructive Wave Interference

Although the exact size of the waves created by each of the wind regimes has not been determined, it seems unlikely that the 15-m (49.2 ft) waves reported could have developed without constructive wave interference. This condition would result from the combination of the west-southwest wind with the later northwest wind. A possible pattern of interference shows the production of crests of 7.5 m (24.6 ft) above the still-water line and troughs 7.5 m below it. This gives a maximum height of 15 m (49.2 ft) for the resulting waves.

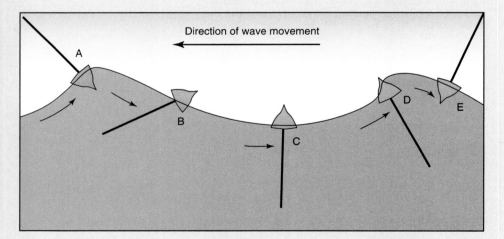

Figure 10E   Yachts Rolled Over

As the steep wave takes the yacht near the crest on its leading edge (A), the yacht begins to roll over, aided by the upward motion of water particles moving into the wave crest that push the keel up while the mast descends. As the wave passes, the downward movement of water particles into the trough helps rotate the mast down (B and C). The next wave strikes the upturned keel while water particles moving into the crest help push the mast upward (D). Finally, as the wave passes, the rotation is completed (E).

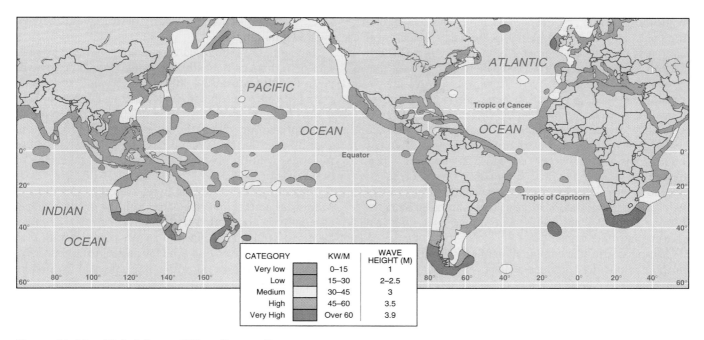

Figure 10-22   Global Coastal Wave Energy Resources

This map shows the effect of west-to-east movement of storm systems in the temperate latitudes. The western coasts of continents are struck by larger waves than eastern coasts, meaning that more wave energy is available along western shores. Furthermore, larger waves appear to be associated with the westerly wind belts between 30° and 60° latitudes in both hemispheres. kW/m is kilowatts per meter; every meter of "red" shoreline is a potential site for generating over 60 kW of electricity.   (Map constructed from data in U.S. Navy Summary of Synoptic Meteorological Observations, SSMO. Adapted from Sea Frontiers (July–August 1987) "Sea Secrets," International Oceanographic Foundation, Vol. 33, No. 4, 260–261; produced by The National Climatic Data Center with support from the U.S. Department of Energy.)

These units have many potential uses other than the generation of electricity. They could be used in cleaning up oil spills, protecting beaches from wave erosion, creating calm harbors in the open sea, and desalinating seawater through reverse osmosis. Economic conditions in the future may enhance the appeal of converting wave energy to electrical energy. The coastal regions with the greatest potential for exploitation are shown in Figure 10-22. A major barrier to the use of any device to harness wave energy is the monumental engineering problem of preventing the devices from being destroyed by the wave force they are built to exploit.

Summary

Wave phenomena transmit energy through various states of matter by setting up patterns of oscillatory motion in the particles that make up the matter. Progressive waves are longitudinal, transverse, or orbital, depending on the pattern of particle oscillation. Particles in ocean waves move primarily in orbital paths.

Characteristics used to describe waves are wavelength (*L*), wave height (*H*), wave period (*T*), and wave speed (*S*). If water depth is greater than 1/2 wavelength, a progressive wave travels as a deep-water wave with a speed that is directly proportional to wavelength. If water depth is less than 1/20 of the wavelength, the wave moves as a shallow-water wave, whose speed increases with increasing water depth.

As wind-generated waves form in a sea area, capillary waves with rounded crests and wavelengths less than 1.74 cm (0.7 in) form first. As the energy of the waves increases, gravity waves form, with increasing wave speed, wavelength, and wave height. Energy is transmitted from the sea area across the ocean by low, rounded waves called swell.

Swell releases its energy in the surf as waves that break in the shallow water near shore. If waves break on a relatively flat surface, the result is usually a spilling breaker, while those forming on steep slopes have spectacular curling crests and are called plunging breakers. When swell approaches the shore, segments of the waves that first encounter shallow water are slowed, whereas other parts that have not been affected by shallow water move ahead, causing the wave to refract, or bend. Refraction concentrates wave energy on headlands, whereas wave energy is generally dispersed in bays. Diffraction occurs when waves pass barriers or constrictions. Reflection of waves off seawalls or other barriers can cause an interference pattern called a standing wave. In standing waves, crests do not move laterally as in progressive waves but form alternately with troughs at locations called antinodes. Separating the antinodes are nodes, where there is no vertical movement of the water.

During storms, the combination of low air pressure and onshore winds may produce a storm surge that raises the water level at the shore many meters above normal sea level. Such surges are particularly destructive if they coincide with high tide.

Tsunami, or seismic sea waves, are generated by seismic disturbances beneath the ocean floor. Such waves have lengths exceeding 200 km (125 mi) and travel across the open ocean with undetectable heights of about 0.5 m (1.6 ft) at speeds in excess of 700 km/h (435 mi/h). On approaching shore, they may increase in height to over 30 m (100 ft). Tsunami have been known to cause millions of dollars worth of damage and take tens of thousands of lives.

Internal waves are not well understood but are thought to form at density interfaces beneath the ocean surface, especially in connection with the pycnocline. They may be up to 100 m (330 ft) high, with periods from 5 to 8 min. Ocean waves could be harnessed to produce hydroelectric power, but significant problems must first be solved.

Key Terms

Capillary wave	Internal wave	Spilling breaker	Tsunami
Constructive interference	Longitudinal wave	Standing wave	Wave dispersion
Crest	Mixed interference	Storm surge	Wave frequency (f)
Deep-water wave	Orbital wave	Surf zone	Wave height (H)
Destructive interference	Plunging breaker	Swell	Wavelength (L)
Diffraction	Refraction	Transitional wave	Wave period (T)
Forced wave	Restoring force	Transverse wave	Wave steepness
Free wave	Rogue wave	Trough	Wave train
Gravity wave	Shallow-water wave		

Questions and Exercises

1. Discuss longitudinal, transverse, and orbital wave phenomena, including the states of matter in which they transmit energy.

2. Calculate the speed (S) for deep-water waves with the following characteristics:
   A. $L = 351$ m, $T = 15$ s
   B. $L = 351$ m, $f = 4$ waves/min
      Express speed (S) in meters per second (m/s).
   C. $T = 11$ s

3. Calculate the speed with which a shallow-water wave will travel across an ocean basin 4 km (2.5 mi) deep.

4. Describe how the shape of waves change as they progress from capillary waves to increasingly large gravity waves until they reach a steepness ratio of 1/7. A change in which of these variables will make gravity the dominant restoring force, H, L, S, T, or f?

5. Waves from separate sea areas move away as swell and produce an interference pattern when they come together. If the waves from Sea A and Sea B have wave heights of 1.5 m (5 ft) and 3.5 m (11.5 ft), respectively, what would be the height of waves resulting from constructive interference and destructive interference? Illustrate your answer with a diagram (refer to Figure 10-9).

6. Describe changes, if any, in wave speed (S), length (L), height (H), and period (T) that occur as waves move across shoaling water to break on the shore.

7. Using orthogonal lines, illustrate how wave energy can be distributed along the shore. Identify areas of high and low energy release.

8. On the basis of the fundamental characteristics of standing waves shown in Figure 10-16, construct a similar diagram of a standing wave in which two nodes and three antinodes exist.

9. How long will it take for a tsunami to travel across 2000 km (1242 mi) of ocean if the average depth of ocean is 4500 m (14,760 ft)?

References

Bowditch, N., 1958. *American Practical Navigator*, rev. ed. H. O. Pub. 9. Washington, D.C.: U.S. Naval Oceanographic Office.

Pickard, G. L., 1975. *Descriptive Physical Oceanography: An Introduction.* 2nd ed. New York: Pergamon Press.

Changery, M. J., and R. G. Quayle., 1987. Coastal wave energy. *Sea Frontiers* 33:4, 260–61.

Gill, A. E., 1982. Atmosphere-ocean dynamics. *International Geophysics Series*, Vol. 30. Orlando, FL.: Academic Press.

Kinsman, B., 1965. *Wind Waves: Their Generation and Propagation on the Ocean Surface*. Englewood Cliffs, NJ: PrenticeHall.

Knamori, H., and M. Kikuchi., 1993. The 1992 Nicaragua earthquake: A slow tsunami earthquake associated with subducted sediments. *Nature* 361:6414, 714–16.

Melville, W., and R. Rapp., 1985. Momentum flux in breaking waves. *Nature* 317:6037, 514–16.

Stewart, R. H., 1985. *Methods of Satellite Oceanography*. Berkeley, CA .: University of California Press.

Sverdrup, H. U., M. W. Johnson, and R. H. Fleming., 1942. Renewal 1970. *The Oceans: Their Physics, Chemistry, and General Biology*. Englewood Cliffs, NJ: PrenticeHall.

Tarbuck, E. J., and F. K. Lutgens., 1991. *Earth Science*. New York: Macmillan.

Van Arx, W. S., 1962. *An Introduction to Physical Oceanography*. Reading, Mass.: Addison-Wesley.

Suggested Reading

Sea Frontiers

Barnes-Svarney, P., 1988. *Tsunami: Following the Deadly Wave*. 34:5, 256–63. Covers the origin of tsunamis, a history of major occurrences, and the methods used to detect them. .

Changery, M. J., 1987. *Coastal Wave Energy*. 33:4, 259–62. Considers the wave energy resources of the world's shores.

Ferrell, N., 1987. *The Tombstone Twins: Lights at the Top of the World*. 33:5, 344–51. Presents a short history of the two lighthouses on Unimak Island, Alaska.

Land, T., 1975. *Freak Killer Waves*. 21:3, 139–41. The British design a buoy that will gather data in areas where 30 m (98 ft) waves, which may be responsible for the loss of many ships, occur.

Mooney, M. J., 1975. *Tragedy at Scotch Cap*. 21:2, 84–90. The events resulting from an earthquake off the Aleutians on April 1, 1946, are recounted. The resulting tsunami destroyed the lighthouse at Scotch Cap, Alaska.

Pararas-Carayannis, G., 1977. *The International Tsunami Warning System*. 23:1, 20–27. The history and operations of the International Tsunami Warning System.

Robinson, J. P., Jr., 1976. *Newfoundland's Disaster of '29*. 22:1, 44–51. A tsunami that struck Newfoundland on November 18, 1929, caused massive destruction.

——. 1976. *Superwaves of Southeast Africa*. 22:2, 106–16. The formation and destruction caused by large waves that strike ships off the southeast coast of South Africa are discussed.

Smail, J., 1982. *Internal Waves: The Wake of Sea Monsters*. 28:1, 16–22. This informative discussion looks at the causes of internal waves and their effect on ships and submarines.

Smail, J. R., 1986. *The Topsy-Turvy World of Capillary Waves*. 32:5, 331–37. Capillary waves are clearly described, and their role in transmitting wind energy to the motion of waves and currents is discussed.

Smith, F.G.W., 1970. *The Simple Wave*. 16:4, 234–45. This is a very readable explanation of the nature of ocean waves. It deals primarily with the characteristics of deep-water waves.

——. 1971. *The Real Sea*. 17:5, 298–311. This discussion of wind-generated waves is comprehensive and easy to read.

——. 1985. *Bermuda Mystery Waves*. 31:3, 160–63. The author proposes a possible source of large waves that struck Bermuda on November 12, 1984.

Truby, J. D., 1971. Krakatoa—the killer wave. *Scientific American* 17:3, 130–39. Describes events leading up to the 1883 eruption of Krakatoa and the tsunami that followed.

Bascom, W., 1959. *Ocean Waves*. 201:2, 89–97. An informative discussion of the nature of wind-generated waves, tsunamis, and tides.

Koehl, M. A. R., 1982. *The Interaction of Moving Water and Sessile Organisms*. 274:6, 124–35. The author describes how benthic shore-dwelling animals have adapted to the stresses of strong currents and breaking waves.

Oceanography on the Web

 Visit the Introductory Oceanography home page for on-line resources for this chapter. There you will find an on-line study guide with review exercises and links to oceanography sites to further your exploration of the topics in this chapter. Introductory Oceanography is at http://prenhall.com/thurman (click on the Table of Contents menu and select this chapter).

Catch of the Day: ~~Slimehead~~, Orange Roughy, ~~Patagonian Toothfish~~, Chilean Sea Bass

- What is the state of global fisheries?
- What are the trends in "harvesting" of wild marine fish?
- What is bycatch?
- What is the environmental impact of commercial fishing?
- How much seafood do we eat?
- How important is seafood as a protein source?
- Is "sustainable fisheries" an oxymoron?

Introduction

Wild Oats Markets, a nationwide chain of more than 75 grocery stores, issued a press release on August 11, 1999 stating that it would no longer sell North Atlantic swordfish, marlin, orange roughy, or Chilean sea bass because these species are endangered due to overfishing.

Paul Gingerich, Meat and Seafood Purchasing Director, commented, "Floods, drought and overgrazing that affect food sources on land can be readily seen and measured. The effects of overfishing cannot be easily seen in the oceans. We need to be proactive to save these species for future generations."

Will the other 246,000 U.S. grocery stores follow suit? Are the species in question, and others, really endangered?

The Impact of Global Fisheries

Even though less than 1% of global caloric intake comes from fish, the importance of fisheries to the global and many national economies cannot be overstated. Consider the following information on fisheries worldwide:

- The value of the international fish trade for 1994 was $47 billion.
- The combined value of canned, fresh, and frozen fishery products in the U.S. in 1996 was over $2.9 billion.
- Nearly 85,000 people were employed in processing and wholesale jobs alone in the U.S. in 1995.
- The economies of many countries such as Iceland, Peru, and Norway depend heavily on fish product exports.
- In eastern Canada, the closure of the cod fishery cost at least 40,000 jobs, in a country with a population one-tenth that of the U.S.
- Of the $752 the average American spent on meat in 1995, $97 (13%) was for seafood (Figure 1).

Although the contribution of fish to the diet of humans may seem small if we simply count calories, it becomes critical if we consider protein: 16% of global animal protein is provided by fish, while in the far east, where most humans live, nearly 28% of animal protein comes from fish. In developing countries worldwide, where population growth rates are ominously high, 950 million people depend on fish as their primary source of protein.

Figure 1

An all-you-can-eat seafood bar, featuring marine organisms from around the world, is a mainstay at many oceanside resorts. *(Courtesy of Sunny Day Guide)*

In addition to serving as a basic food source, fish is increasingly considered by affluent westerners to be a "health food." Fatty fish like salmon and mackerel have relatively high levels of omega-3 fatty acids, which have been shown in clinical studies to reduce the risk of heart attack by 50–70%.

Finally, fish and fish by-products representing as much as one third of wild-caught fish, are a mainstay of the pet food industry and are used as a constituent of animal feed as well.

Environmental Costs of Fishing

Commercial fishing can be an environmentally costly activity. First, many fishing methods destroy habitat. Consider trawling, which is typically done for shrimp and other bottom-associated species like Atlantic cod and plaice. In this method, a 10 to 130 m (33–426 ft) long net is scraped across large areas of the bottom, collecting virtually everything in its path, including endangered sea turtles. Trawling, which became popular with the advent of the diesel engine in the 1920s, is practiced worldwide on virtually every different bottom type. Saturation trawling, in which the net is repeatedly fished in an area until virtually no fish or shrimp are left, has been compared to clearcutting a forest. The compar-

Figure 2

Bycatch includes sea turtles, mammals, birds, invertebrates, and fish, like this oceanic sunfish caught in a Japanese driftnet in the Tasman Sea. *(Courtesy of Greenpeace)*

ison is appropriate: trawling heavily damages sessile benthic organisms like sponges, hydroids, and tube-dwelling worms and displaces associated fauna like fish and crustaceans. Complete recovery in both cases may take decades or longer. The comparison between trawling and forest clearcutting breaks down when one considers the area involved annually. Approximately 100,000 km^2 (38,000 mi^2) of forest are lost annually, whereas an area 150 times as large is trawled!

Innovations like TEDs (Turtle Excluder Devices) shunt large objects like sea turtles out of the net. The use of TEDs, however, is not universal and is not entirely successful either.

Extremely destructive methods like dynamiting and poisoning still are used in some areas.

Commercial overfishing threatens fish stocks, which are already under stress from coastal environmental degradation due to overdevelopment and industrial, municipal, and agricultural pollution. A 1994 United Nations Food and Agricultural Organization (FAO) analysis of marine fish resources concluded that 35% of 200 top marine fisheries were overexploited (i.e., yielded declining landings); 25% were mature (and thus on the verge of endangerment if stressed); and 40% (largely in the Indian Ocean) were still developing.

Significantly, *no* major fisheries, according to the study, were undeveloped.

Additionally, at least 25% of the commercial catch is unused. This quantity is known as "bycatch" and refers to undersized, low-value, and non-target species (fish, crabs, etc.). Bycatch is often returned to the water dead, or dies soon after (Figure 2). The fishing activity that possibly generates the most bycatch is trawling for shrimp. In addition to damaging the ocean bottom, as much as 90% of the trawl contents may be non-target and hence unused species, sometimes called "trash fish" by fishers.

Finally, the environmental cost of commercial fishing extends to the pollution associated with the manufacture, transportation, and use of equipment (like fishing boats) and supplies; fuel spills; and transportation and refrigeration of fishery products.

Figure 3

The erstwhile slimehead (top), successfully marketed as the orange roughy. The fish formerly known as the Patagonian toothfish (bottom), now available in Western restaurants and seafood shops as the Chilean sea bass. It is not at all closely related to the group of fishes commonly known as sea basses. *(Top—Jean-Paul Ferrero/Jacana, Photo Researchers, Inc. Bottom—W. Savory, Center for Food Safety and Applied Nutrition)*

Case Study: The Slimehead and Patagonian Toothfish

Would you eat a fish called a slimehead? Or a Patagonian toothfish? Probably not, so clever marketing specialists transformed these into popular items by renaming them as "orange roughy" and "Chilean sea bass" (Figure 3). The former was popular in the mid to late 1980's, whereas the latter's popularity is just peaking.

Unfortunately, renaming the fish changed neither their biology nor their fate. The orange roughy is a classic example of misunderstanding the importance of gaining a complete understanding of a species' biology before exploiting it as a fishery. The Chilean sea bass is yet another reminder that we refuse to learn from our mistakes. Both cases demonstrate the power and environmental destructiveness of effective marketing.

To find out more about global fisheries and endangered fish stocks, go to our website.

See the complete issue at:
www.prenhall.com/thurman

11 Tides

View of Mont-Saint-Michel in the Gulf of Saint-Malo, France, where the tidal range exceeds 10 m (33 ft). At extreme low tide, the shoreline moves 15 km (9 mi) from the high-tide shoreline of the mainland. Mont-Saint-Michel is built on a granitic outcrop about 1.7 km (1 mi) from the high-tide shoreline of the mainland and is surrounded by ocean water during high tide.   (Porterfield-Chickering/Photo Researchers, Inc.)

Humans undoubtedly have observed the rise and fall of the tides since first settling by the sea. However, there are no surviving written accounts of tides prior to those of Herodotus (circa 450 B.C.), who concluded that the tides of the Mediterranean Sea were related to the motion of the Moon, for both followed a similar cyclic pattern. It was not until Sir Isaac Newton (1642–1727) developed his universal law of gravitation, however, that the relationship between the tides, Moon, and Sun could be explained adequately.

Tides are the ultimate manifestation of shallow-water wave phenomena. Tide waves possess lengths measured in thousands of kilometers and heights ranging to more than 15 m (50 ft). They are generated by the gravitational attraction of the Sun and Moon on the ocean masses, affecting every particle of water, from the surface to the deepest parts of the ocean basins.

Generating Tides

Tides are generated through a combination of gravity and the motions of the Earth, Moon, and Sun. We will now look at the interacting forces on these three celestial bodies to develop an understanding of the ocean's daily rhythms.

Newton's Law of Gravitation

Sir Isaac Newton published his *Philosophiae naturalis principia mathematica* (*Philosophy of Natural Mathematical Principles*) in 1686, stating in his preface:

"... I derive from the celestial phenomena the forces of gravity with which bodies tend to the Sun and several planets. Then from these forces, by other propositions which are also mathematical, I deduce the motions of the planets, the comets, the Moon, and the sea."

What followed from his thinking was our first understanding of why tides behave as they do.

Newton's law of gravitation states: every particle of mass in the universe attracts every other particle of mass. This occurs with a force that is directly proportional to the product of their masses and inversely proportional to the square of the distance between them. The greater the mass of the objects and the closer they are together, the greater the gravitational attraction. Mathematically **gravitational force** can be expressed as:

$$\text{Gravitational force} = Gm_1m_2/r^2$$

where G is the universal gravitational constant, m_1 and m_2 are the masses, and r is the distance between the masses. For a spherical body, all of the mass can be considered to exist at the center of the sphere, and thus, r will always be the distance between the centers of bodies being considered.

In addition to ocean tides, gravitational attraction also causes tides in Earth's atmosphere and lithosphere. However, the atmospheric and lithospheric tides are not "sensible"—we cannot see or feel them with our unaided senses.

Gravitational Attraction: Mass Versus Distance

It is gravity that tethers the Sun, its planets, and their Moons together. It is gravity that tugs every particle of water on Earth toward the Moon and the Sun, creating the tides. This part is not difficult to understand. What takes some effort to grasp is how tidal patterns are controlled by Earth's tilted axis, Earth's rotation, and the path of Earth and the Moon moving as partners through space.

The **tide-generating force** is the difference between the gravitational force of the tide-generating object acting on a mass at the Earth's surface and at the Earth's center. This force varies inversely as the *cube* of the distance from the center of Earth to the center of the tide-generating object instead of varying inversely to the *square* of the distance as does the gravitational attraction. Therefore, the tide-generating force, although derived from gravitational attraction, is not linearly proportional to it; distance is a more highly weighted variable:

$$\text{Tide-generating force} \propto m_1m_2/r^3$$

Although the gravitational attraction between Earth and the Sun is more than 177 times greater than between Earth and the Moon, the Moon dominates the tides. Figure 11-1 illustrates the massiveness of the Sun compared to the Moon. Because the Sun is 27 million times more massive than the Moon, it should, solely on the basis of mass, have a tide-generating force 27 million times greater than that of the Moon. However, we must consider the distance factor. The Sun is 390 times farther from Earth than the Moon (see the scale in Figure 11-1). So, the tide-generating force is reduced by 390^3, or about 59 million times compared to that of the Moon. This condition results in the Sun's tide-generating force being 27/59, or about 46 percent, that of the Moon.

The Earth-Moon System

You probably were taught that the Moon orbits Earth, but it is not that simple. The two bodies actually travel through space as a system. You can visualize this by imagining Earth and the Moon as ends of a sledge hammer, flung into space, tumbling slowly end-over-end. Thus, as Earth and the Moon orbit the Sun together, they rotate around the center of mass of the Earth–Moon system, which is called the *barycenter*. The barycenter is not located in the space between Earth and the Moon, as you might think. Instead, it is within Earth's mantle, at a point about 4700 km (2900 mi) from our planet's center. This is because Earth is a much larger mass than the Moon, and the two are close together in space. The barycenter follows a smooth orbit around the Sun, while Earth and the Moon themselves follow wavy paths (Figure 11-2).

The tidal pattern we see on Earth primarily results from this "sledge hammer" rotation of the Earth–Moon system around its center of mass. Figure 11-3a shows the path followed by the center of Earth as it rotates around the barycenter of the Earth–Moon system.

Figure 11-1 Why the Moon Generates Bigger Tides than the Sun

Earth has a diameter of 12,682 km (7876 mi). The diameter of the Moon is 3478 km (2160 mi), roughly 1/4 that of Earth. The diameter of the Sun is 1,392,000 km (864,432 mi), which is 109 times the diameter of Earth. Their relative sizes are shown to scale. The tabular data compares the masses of the Sun and Moon and their distances from Earth. These factors are important in determining their relative tide-generating effect.

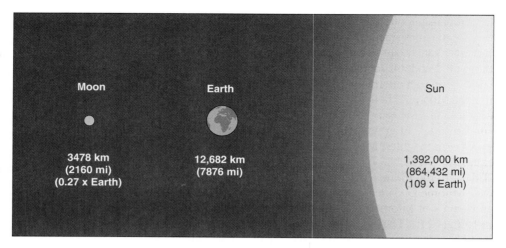

Tide-Generating Body	Distance from Earth (avg)	Mass (Metric tons)	Relative Tide-Generating Effects
Moon	384,835 km (234,483 mi)	7.3×10^{19}	Based on relative masses, the sun is 27 million times more massive than the moon and has 27 million times the tide-generating effect. However, since the sun is 390 times farther than the moon from Earth, its tide-generating effect is reduced by 390^3, or 59 million times.
Sun	149,758,000 km (93,016,845 mi)	2×10^{27}	

Moon

Earth ←— Distance of moon and sun from Earth shown approximately to scale —→ Sun

DETERMINATION OF TIDE-GENERATING FORCE OF SUN RELATIVE TO MOON

$$\text{Tide-generating force} \propto \frac{\text{Mass}}{(\text{Distance})^3} \propto \frac{\text{Sun–27 million times more mass}}{(\text{Sun–390 times farther away})^3}$$

$$(390)^3 = 59,000,000 \quad \text{Thus,} \quad \frac{27 \text{ million}}{59 \text{ million}} = 0.46 \text{ or } 46\%$$

The sun has 46% the tide-generating force of the moon.

Tide-Generating Force

To understand the tide-generating force, you need to understand centripetal force. **Centripetal force** pulls an orbiting body toward the center of its orbit. For example, if you tie a string to a rock and swing the tethered rock around your head (Figure 11-3b), the string pulls the rock toward your hand. The string provides a centripetal force to the rock, forcing the rock to seek the center of its orbit. If the string breaks, the force is gone, the rock no longer can maintain its circular orbit, and it will fly off in a straight line, tangent to the circle (Figure 11-3b). The same is true

of planets orbiting the Sun. They are pulled, not by string, but by gravity. Thus, gravity operates as a centripetal force. If the gravity of the Sun and its planets were abruptly shut off, centripetal force would vanish, and all the planets would fly off into space along straight-line paths.

Since the Moon is the dominant force producing tides on Earth, we will first consider the tide-generating forces resulting from the Earth–Moon system only. We will later discuss the modifications in this dominant pattern that result from the tide-generating force of the Sun.

Because the daily rotation of Earth on its polar axis has no tide-generating effect, we will ignore this motion in

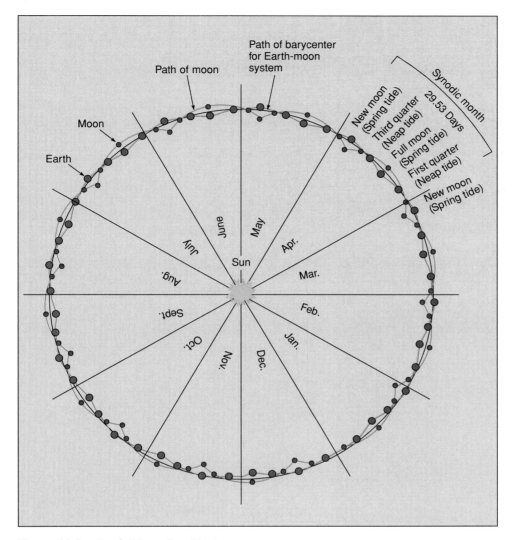

Figure 11-2   Earth-Moon-Sun Motion

As the Earth–Moon system orbits the Sun, the center of each body follows its own wavy path, because both also rotate about the barycenter of the Earth–Moon system (see Figure 11-3a). This center of mass is not in the middle of this system, because Earth is so much larger than the Moon; instead, it is shifted toward Earth's end, so much so that the center of mass actually lies within Earth's mantle. During a month, the Moon moves from a position between the Sun and Earth (new Moon) to a position that puts Earth between the Moon and the Sun (full Moon), and then back again.

our initial considerations. The tide generated on Earth is primarily the result of the rotation of Earth and the Moon about their common center of mass, the barycenter (Figure 11-4a). As the Earth and Moon rotate around the barycenter, all particles that make up Earth follow circles of equal radii. If we divide Earth into a great number of particles of equal mass, the centripetal force required to keep each particle of Earth following an identical orbit is the same (Figure 11-4b). The required centripetal force for each particle is supplied by its gravitational attraction to the Moon. Although the average gravitational attraction per unit of mass must equal the *average* centripetal force for different particles of Earth's mass to keep Earth

in its proper path, the two are not equal for all points on Earth. The centripetal force required for all particles is the same and is directed toward the center of each particle's orbit.

For all particles on Earth, the gravitational attraction of the Moon is directed toward the center of the Moon, however, the strength of this gravitational attraction is greater for particles closer to the Moon. For example, for particles on a line from the center of the Moon to the center of the earth, the attraction would be stronger for particles on Earth's surface, which would be closer to the Moon, than for particles near Earth's center. Figure 11-5a shows the required centripetal force and gravitational attraction of

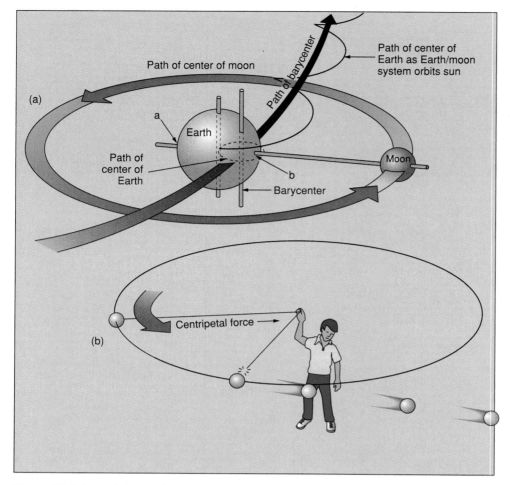

Figure 11-3   Earth-Moon System Rotation

Earth and its Moon move around a common center of gravity (barycenter) that is inside Earth. (a) The dashed circle is the path followed by the Earth's center as it moves around the barycenter— the common center of gravity—of the Earth–Moon system. The circular paths followed by points *a* and *b* have the same radii as that followed by Earth's center. (b) If you tie a string to a rock and swing it in a circle around your head, it stays in the circular orbit because the string exerts a centripetal (center-seeking) force on the rock. This force pulls the rock toward the center of the circle. If the string breaks with the rock at the position shown, the rock will fly off along a straight path, tangent to the circle.

the Moon acting on points along Earth's surface as vector lines. The arrow points in the direction of the force, and the length of the vector is proportional to the magnitude of the force. The difference in the magnitude and direction of the gravitational attraction and the required centripetal force may be determined by vector subtraction. The tides result from the horizontal component of this difference, the residual force.

The Moon is at zenith (directly overhead) at point *Z*, and at nadir (on the opposite side of Earth) at point *N* (Figure 11-5b). The residual forces at both of these points are vertical and away from Earth's surface (Figure 11-5a). This force is extremely small, however, and if you were to step on scales at the location of the Moon's zenith, your weight would be reduced by only about 0.11 mg

(0.000035 oz). By itself, this would have little effect in producing changes in the level of the ocean. However, the residual horizontal components of the forces present at other points on the surface, which direct water toward these points, results in the apexes of the two tidal bulges occurring at these two positions.

Figure 11-5b shows the distribution of horizontal tidal forces. A plane passed through the center of Earth, perpendicular to a line through the centers of Earth and the Moon, intersects Earth's surface along a circle where tide-generating forces are zero. The small residual forces are directed down toward the center of Earth along this circle. There is also no horizontal tide-generating force at the points on Earth where the Moon is at its nadir and zenith. At all other points on Earth's surface, there is a horizontal

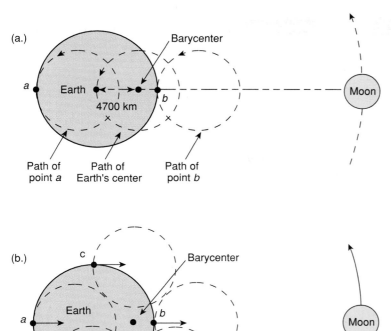

Figure 11-4   Earth-Moon Rotation
....
(a) The dashed line through the center of Earth is the path of Earth's center as it moves around the barycenter of the Earth–Moon system. The circular paths followed by points a and b have the same radius as that followed by Earth's center. (b) Arrows from points a, b, c, d, and e to the center of their circular orbits represent the magnitude and direction of centripetal force required to hold objects at these points in their orbital paths. Note that all of the arrows have the same length, indicating they represent the same amount of force. The direction of the required force is also the same for all the points— parallel to a line connecting the centers of Earth and the Moon.

component to the residual force. These tiny forces, the tide-generating forces, push the water into the bulges at the zenith and nadir. The magnitude of the tide-generating forces increases from zero along the circle to a maximum at an angle of 45° on either side of the circle and decreases again to zero at the zenith and nadir. The two bulges centered on the apex and nadir provide the framework within which the equilibrium tide operates (Figure 11-5b).

The more distant from Earth the tide-generating bodies, the smaller the residual forces (shorter arrows in Figure 11-5a) will be. Thus, the tide generating forces which are the horizontal component of the residual forces will also be smaller. This is why the Moon controls tides far more than the Sun.

Equilibrium Theory of Tides ....

In the above discussion, we have considered the forces that form the basis for the **equilibrium tide theory**. This theory ignores some of the complexities of real tides, but provides a good model of gross tidal phenomena. Some of the simplifying assumptions made by the theory are: (1) Earth has two equal tidal bulges, one toward the Moon and one away from the Moon; (2) the ocean covers the whole Earth at a uniform depth; (3) there is no friction between seawater and the seafloor; (4) the continents have no influence. Because these simplifications do not actually apply to real situations on Earth, we cannot use

equilibrium theory to predict the real tides at specific locations, but we can predict the general behavior of tides in the world's ocean basins. Newton made these same assumptions mathematically when he first developed the equilibrium theory of tides. Later in this chapter, we will consider the **dynamic tide theory** that accounts for variable ocean depth, the presence of continents that interrupt the ocean's continuity, and friction between the ocean water and ocean floor. However, we will first consider how a number of factors affect the equilibrium theory tidal bulges.

The Rotating Earth

Our ideal, uniformly deep ocean is modified only by the tide-generating forces that cause bulges on opposite sides of Earth, as shown in Figure 11-6. Let us assume that a stationary Moon is aligned with Earth's equator so that the maximum bulge occurs at the equator on opposite sides of Earth. Earth requires 24 hours for one complete rotation and so, at the equator, we experience two equal high tides and two equal low tides each day. The time that elapses during a complete tidal cycle (from one high tide to the next, or from one low tide to the next), is the tidal period, 12 hours in this case. If we move to any latitude north or south of the equator, we experience a similar tidal period except that the high tides are less high because we are not at the apex of the bulge. This pattern is called a **semidiurnal tide**; that is, there are two tidal cycles per 24-hour day.

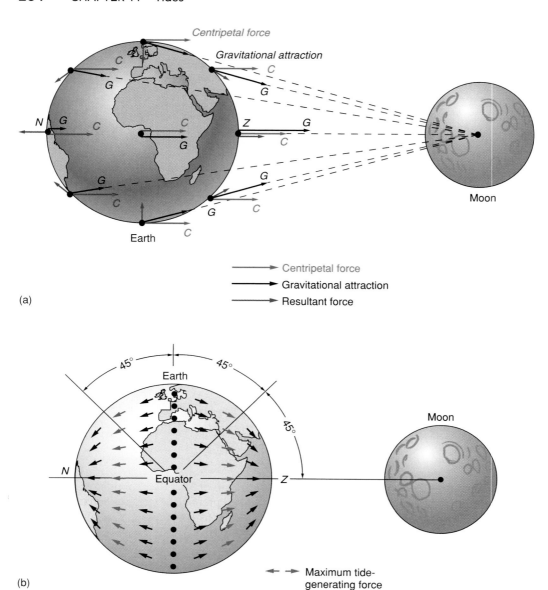

(a)

(b)

Figure 11-5 Tide-Generating Force

(a) The length and direction of the arrows represent the magnitude and direction of the forces related to each of nine identical particles located at each black point. The pink arrows labeled *C* show the relative strength and direction of the centripetal force required to keep each particle mass in the circular orbit it follows as a result of the Earth–Moon rotation system. They are all equal in magnitude are parallel to one another. The black arrows labeled *G* show the gravitational attraction between these particles and the Moon. This attractive force of gravity provides the required centripetal force to keep the particles in orbit, but the gravitational force is identical to the centripetal force only for the particle at the center of Earth. The gravitational force is in the required direction only for particles on a line connecting the centers of Earth and the Moon. For all particles on the half of Earth facing the Moon, the gravitational force is larger than required, and for the particles on the half of Earth facing away from the Moon, it is smaller. For all particles except at the center of Earth, there is a resulting residual force (blue arrows) because the gravitational force varies from the required centripetal force. This force is small —averaging about 10^{-7} of the magnitude of Earth's gravity. Therefore, where forces act perpendicular to Earth's surface, as does gravity, they do not have any tide-generating effect. However, where they have a significant horizontal component —tangent to Earth's surface —they aid in producing tidal bulges on Earth. Since there are no other large horizontal forces on Earth with which they must compete, these small, ever-present residual forces can push water across Earth's surface. (b) Along the intersection of Earth's surface and a plane running through Earth's center perpendicular to a line connecting the centers of Earth and the Moon, the residual forces are directed toward the center of Earth and have no horizontal component, and the tide-generating forces are zero (dots). There is also no horizontal component at the zenith (*Z*) and nadir (*N*) because the residual forces point away from the center of Earth. Elsewhere on Earth's surface, there are horizontal tide-generating forces. The magnitude of the tide-generating force varies. It reaches maximum values along two circles that lie 45° on either side of the previously described circle of zero tide-generating force.

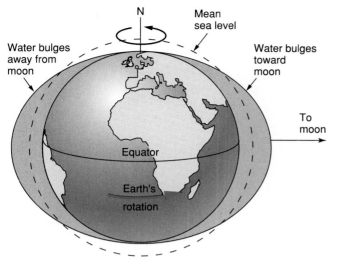

Figure 11-6   Idealized Equilibrium Tide

Assuming an ocean of uniform depth covering Earth, and the Moon aligned with the equator, tide-generating forces produce two bulges in the ocean surface. One extends toward the Moon and the other away from the Moon. As Earth rotates, all points on its surface (except the poles) experience two high tides daily because Earth rotates beneath the two bulges.

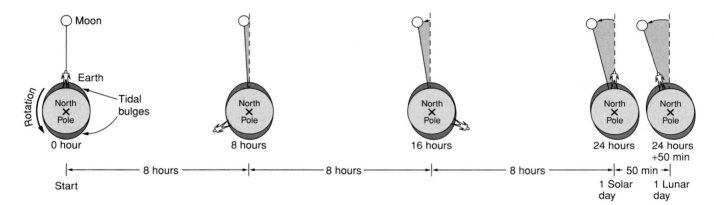

Figure 11-7   The Lunar Day

A lunar day is the time that elapses between successive appearances of the Moon on the meridian of a stationary observer. As Earth rotates, the Earth–Moon system rotation moves the Moon in the same direction (toward the east). During one complete rotation of Earth (the 24-hour solar day), the Moon moves eastward 12.2°, and Earth must rotate an additional 50 min to place the observer in line with the Moon again.

However, high tides do not occur every 12 hours on Earth's surface. Instead, they occur every 12 hours and 25 minutes. Twice this period is called the **lunar day**. The lunar day is the time that elapses between two successive passages of the Moon over any longitude line on Earth. If we observe the time at which the Moon rises on successive nights, we find that it rises 50 minutes later each night (Figure 11-7). Where does the additional 50 minutes come from? During the time that Earth is making a full rotation in 24 hours, the Moon has continued moving another 12.2° to the east. Thus, Earth must rotate an additional 50 minutes (24.83 h) to catch up and place the Moon directly over the observer's longitude. If we measured time using the lunar day, each **lunar hour** would be about 1 hour and 2 minutes of solar time (24.83 h/24 h = 1h 2min).

Combined Effects of Sun and Moon

So far, we have considered the effects of Earth's rotation and the revolution of the Earth–Moon system about its center of mass. We have ignored the effect of the Sun. Figure 11-2 shows how the Earth–Moon system revolves around the Sun. The oscillating paths of Earth and the Moon can be seen as they rotate together around their barycenter. The upper right portion of the figure shows that the Moon cycles through its phases every 29.53 days (the lunar, or synodic, month). When the Moon is between Earth and the Sun, it cannot be seen for a few days, and is called the new moon. When the Moon is on the opposite side of Earth from the Sun, its entire disk is brightly visible, and we call it the full Moon. A quarter Moon results when the Moon is at right angles to the Sun

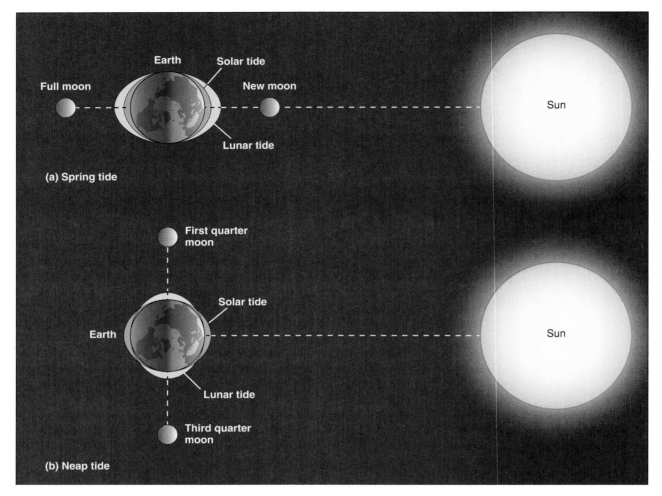

Figure 11-8   Earth-Moon-Sun Positions and the Tides

(a) When the Moon is in the new or full position, the tidal bulges created by the Sun and Moon are aligned, producing constructive interference and a therefore larger bulges, which "spring forth" as spring tides. (b) When the Moon is positioned halfway between the new and full phases (called the first and third quarters), the tidal bulge produced by the Moon is at right angles to the bulge created by the Sun. The bulges tend to cancel each other (destructive interference), and the resulting bulges are smaller, called neap tides. New Moon and full Moon phases produce spring tides with maximum tidal ranges, while the first and third quarter phases of the Moon produce neap tides with minimal tidal ranges.   *(From The Tasa Collection:* Shorelines. *Macmillan Publishing Co., New York. © 1986, by Tasa Graphic Arts, Inc. All rights reserved.)*

relative to Earth. (Viewed from Earth, it looks like half a Moon.)

Figure 11-8a shows that when the Sun and Moon are aligned, either with the Moon between Earth and the Sun (new Moon) or with the Moon on the side opposite the Sun (full Moon), the tide-generating forces of the Sun and Moon are added together. During these times of the month, we experience the highest and lowest tides, that is, the vertical difference between high and low tides, **tidal range**, is at a maximum. This maximum tidal range is called the **spring tide**, because the tide surges, or springs forth. (The name has no connection with the spring season.)

When the Moon is at **quadrature**, in either quarter phase (Figure 11-8b), the tide-generating force of the Sun is working at right angles to the tide-generating force of the Moon, and we experience minimum tidal range. This is called **neap tide**. (To help you remember this name, think of it as a nipped tide, that is, small.) The time that elapses between successive spring tides (full Moon and new Moon) or neap tides (first and third quarters) is a little more than 2 weeks (29.53/2 = 14.765 days).

In Figure 11-9, the tides generated by the Sun (yellow curve) and Moon (dashed curve) and their sum (blue curve) are shown over a 24-hour period for three different phases of the Moon. In Figure11-9a, we see the spring-tide condition

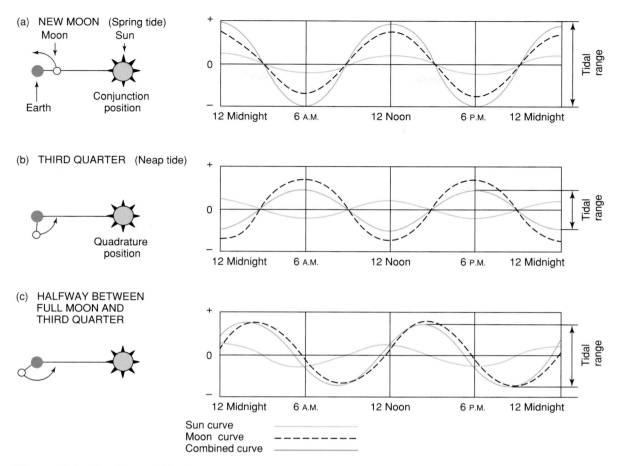

Figure 11-9   Sun-Moon Tide Curves

(a) Spring tide. During a new Moon, the tidal bulges of the Sun and Moon are centered at the same longitude, so their effects add together to produce maximum high tides and minimum low tides. This produces maximum tidal range. (b) Neap tide. During a third-quarter Moon, the Sun and Moon tidal bulges are at right angles to each other. Thus the Sun tide reduces the effect of the Moon tide. For example, at 6 A.M. the observer is passing through the Moon bulge and the Sun trough. This produces a high tide in which the effect of the Moon bulge is reduced by an amount equal to the Sun trough. At 12 noon, the Moon trough is reduced by the Sun bulge, so a reduced low tide occurs. This produces minimum tidal range. (c) Where the Sun and Moon bulges are separated by an angle of 45° the Sun bulge is alternately added to and subtracted from the Moon bulge. This produces a tidal range halfway between that of the spring and neap tides.

associated with the new Moon (the same conditions exist with a full Moon). Because the Moon and Sun are on the same side of Earth, maximum tide heights occur at midnight and noon. We have extremely high tides and extremely low tides with this arrangement of Earth, Moon, and Sun.

Figure 11-9b shows the tide that results from Earth-Moon-Sun positions during quadrature. During first-quarter and third-quarter Moons, the tide-generating effect of the Moon is greatest at 6:00 A.M. when it is at zenith, and 6:00 P.M. when it is at nadir in relation to the observer. The Sun is at nadir at 12:00 midnight and zenith at 12:00 noon. These are the times of maximum tide-generating effect of the Sun. In this instance, we can see that the maximum lunar tides correspond with the minimum solar tides, producing a tide pattern with relatively small tidal range—a neap tide.

Figure 11-9c shows the situation halfway between neap-tide and spring-tide alignments of Earth, Moon, and Sun, when the Moon is in a position halfway between full and the end of the third quarter. When the Moon is at zenith at 3:00 A.M., the observer still has to rotate 135° before the Sun will be at zenith at 12:00 noon. The Sun and Moon are at nadir at 12:00 midnight and 3:00 P.M., respectively. The resulting tidal pattern shows that the tidal range is less than that of spring tides, but greater than that of neap tides. "Grunion and the Tides," the special feature at the end of this chapter, discusses the close association between the spawning of these small fish and the spring–neap tide sequence.

Declination of the Moon and Sun

Up to this point, we have assumed that the Moon and Sun always remain aligned over the equator, but, of course, this is not the case. Most of the year, the Sun is north or south of the equator because of Earth's axial tilt relative to its orbital plane. Also, the orbital plane of the Moon does not correspond to Earth's equatorial plane. This angular distance of the orbital plane of the Sun or Moon above or below Earth's equatorial plane is called **declination**.

Earth revolves around the Sun along an invisible ellipse in space. Imagine a plane in space that includes this ellipse. This plane is called the **ecliptic**. Earth's axis of rotation is not perpendicular (upright) on the ecliptic, but is tilted 23.5° from perpendicular. The tilt of Earth's axis relative to its plane of revolution around the Sun (ecliptic) is shown in Figure 11-10. You can see that Earth's tilted axis always points the same direction throughout the yearly cycle. This tilt causes the seasons —spring, summer, fall, and winter.

For the Northern Hemisphere, spring begins at the **vernal equinox**, about March 21, when the Sun is directly above the equator at noon. On about June 21, summer begins with the **summer solstice** when the Sun reaches its most northerly point in the sky, directly above the Tropic of Cancer, which is at 23.5°N latitude. The Sun then moves southward in the sky each day, and on about September 23 it is directly above the equator again, producing the **autumnal equinox** and the beginning of fall in the Northern Hemisphere. During the next three months the Sun is more southerly in the sky until the **winter solstice**

on about December 22, when the Sun is above the Tropic of Capricorn, at 23.5°S latitude. For the Southern Hemisphere, the summer solstice is about December 22 and winter begins around June 21. Without Earth's 23.5° tilt, seasonal differences would disappear. Because of the tilt, the Sun's declination varies between 23.5° north and 23.5° south of the equator on a yearly cycle.

To further complicate matters, the plane of the Moon's orbit is at an angle of 5° relative to Earth's ecliptic. This results in the plane of the Moon's orbit intercepting the plane of the Earth's equator at an angle of 28.5° (5° + 23.5° = 28.5°) (Figure 11-11a). The plane of the Moon's orbit also precesses, or rotates, while maintaining this 5° angle. This **precession** completes a cycle every 18.6 years.

The Moon's declination changes from 28.5° south to 28.5° north and back to 28.5° south of the equator in one month (Figure 11-11a). Figure 11-11b shows the relationship of the ecliptic, the plane of the Moon's orbit, and the plane of Earth's equator after one-fourth of the precession, or 4.65 years later. At this time, the maximum declination of the Moon's orbit relative to Earth's equator still approaches 28.5°. However, in Figure 11-11c, one-half precession is completed; this is 9.3 years after the condition observed in Figure 11-11a, and the maximum declination of the Moon relative to Earth's equator is now 18.5°.

Taking declination into account, we must alter our previous concept of the predicted equilibrium tide. We must now expect that tidal bulges will rarely be aligned with the equator and will occur for the most part north and south of the equator. Since the Moon is the dominant

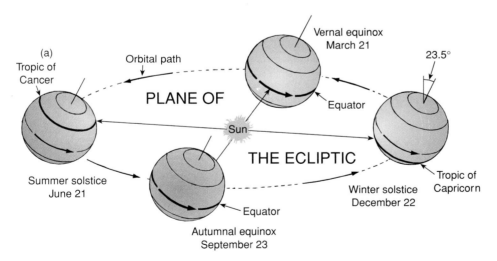

Figure 11-10   Declination of the Sun

As Earth orbits the Sun during one year, its axis of rotation constantly tilts 23.5° from perpendicular, relative to the plane of the ecliptic. The Sun shines directly overhead along the Tropic of Cancer (23.5°N) on the summer solstice, about June 21. Three months later, the Sun shines directly overhead along the equator (0°) during the autumnal equinox, about September 23. Three months later, the Sun shines directly overhead along the Tropic of Capricorn (23.5°S) on the winter solstice, about December 22. Three months later, the Sun shines directly overhead along the equator on the vernal equinox, about March 21. Three months later, the yearly orbit is completed, and the Sun again shines directly overhead along the Tropic of Cancer.

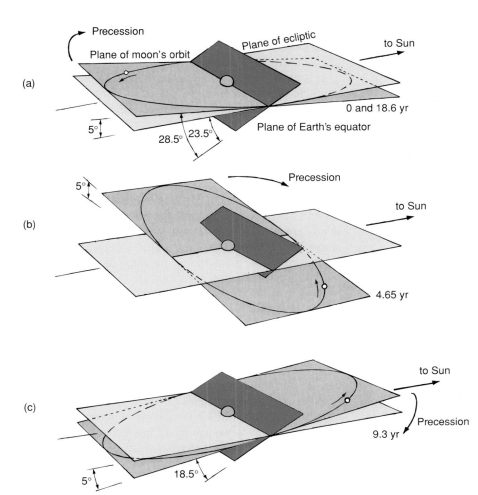

Figure 11-11   Declination of the Moon

The plane of the Moon's orbit (gray plane) is tilted at an angle of 5° relative to the plane of the ecliptic (yellow plane) and rotates with a clockwise precession that has a period of 18.6 years. (a) The declination of the Moon's orbit is the sum of the angle of intersection of the plane of Earth's equatorial plane (blue plane) with the plane of the ecliptic (23.5°) plus the angle of intersection between the plane of the Moon's orbit and the ecliptic (5°). This produces the maximum declination of the Moon relative to Earth's equator of 28.5°. (b) Positions of the planes 4.65 years later when the Moon has achieved one-fourth of its precessional rotation. (c) Relative positions of the planes after 9.3 years or one-half of the precession. The maximum declination of the Moon relative to Earth's equator is now 18.5°, or 23.5° less 5°.   *(From C. Hauge, Tides, Currents, and Waves, California Geology, July 1972. Reprinted by permission of the author.)*

force that creates tides in Earth's oceans, we would expect the bulges to follow the Moon as it moves on its monthly journey, with a maximum declination of 28.5° north and south of the equator (Figure 11-12).

Effects of Changing Distance

Additional considerations that affect the tide-generating forces of the Sun and Moon on Earth are their changing distances from Earth due to elliptical orbits. Earth ranges from **perihelion**, when the distance to the sun is 148.5 million km (92.2 million mi) during the Northern Hemisphere winter, to **aphelion**, when the distance to the Sun is 152.2 million km (94.5 million mi) during the Northern Hemisphere summer. The Moon moves from **perigee**, 375,200 km (233,000 mi) from Earth to **apogee**, of 405,800 km (252,000 mi) from Earth and back to perigee in 27.5 days (Figure 11-13).

Given that tide-generating forces vary inversely with the cube of the distance from the center of Earth to the center of the tide-generating body, it is not surprising that these changes in distance affect the tides. Because of the shorter distance to the Sun, tides worldwide have greater ranges during the Northern Hemisphere winter than in summer. Also, as the result of the Moon being closer to Earth, tidal ranges become greater at perigee.

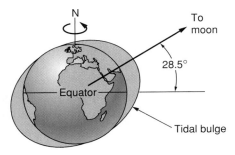

Figure 11-12   Maximum Declination of Tidal Bulges from Equator

The apexes of the tidal bulges may lie at any latitude from the equator to a maximum of 28.5° on either side of the equator, depending on the season of the year (solar angle) and the Moon's position at the moment.

Prediction of Equilibrium Tides

To predict tidal patterns for an idealized, water-covered Earth, let's return to the effect of declination. Let's assume that the declination of the Moon, which determines the alignment of the tidal bulges, is 28° north of the equator, then observe the sequence of tides that occur at a fixed

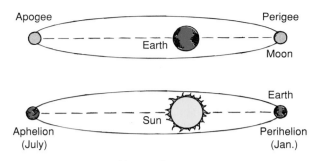

Figure 11-13 The Effects of Elliptical Orbits

The Moon moves from perigee of 375,200 km (233,000 mi) to apogee of 405,800 km (252,000 mi) (top). Greater tidal ranges are experienced during perigean tides. Perihelion brings Earth within 148.5 million km (92.2 million mi) of the Sun. Aphelion distance is 152.2 million km (94.5 million mi) (bottom). Greater tidal ranges are experienced during perihelion tides.

point at that latitude (Figure 11-14). When the Moon is directly over the observer's head, high tide occurs. Low tide is six lunar hours later, followed by another high tide in six hours, but this high tide is much lower than the first high tide. At the end of a 24-lunar-hour period, we have passed through a complete lunar-day cycle of two low tides and two high tides, a semidiurnal pattern of one tidal cycle in half of a lunar day.

A representative curve of the tide observed is shown in Figure 11-14e. The tidal curves for the same period recorded at the equator and at 28°S latitude are also provided. The difference in heights of successive high tides or successive low tides occurring at a given location as a result of the declination of the Moon and the Sun relative to Earth's equator is called the *diurnal inequality*. Diurnal inequality is greatest when the Moon is at maximum northern declination and again, about two weeks later, when it is at its maximum southern declination. Since the Moon is generally above the tropics at these times tides are called *tropical tides*. Occurring between the tropical tides are tides that have a minimum diurnal inequality when the Moon is over the equator—these are called *equatorial tides*.

The equilibrium tidal theory predicts that we should generally observe two high tides and two low tides per lunar day. Except for the rare occasions when the Sun and Moon are simultaneously above the equator, we would expect that: (1) Because of the changing declination of the Moon and the Sun neither the two high tides nor the two low tides would be of the same height; (2) there would be yearly and monthly cycles of tidal range related to the changing distances of Earth from the Sun and Moon; (3) every two weeks (half a lunar month), we would experience spring tides or neap tides (Figure 11-14).

Considering the great number of variables involved in the tides, it is interesting to consider when conditions might be right to produce the maximum possible tide-generating force. The maximum tidal range should be produced when the Sun is at perihelion, in conjunction or opposition with the Moon at perigee, and when both the Sun and Moon are at zero declination. This condition occurs only once each 1600 years. The last occurrence was about 1700 and the next occurrence is predicted for A.D. 3300.

It is clearly a real tidal phenomenon, and the following example illustrates how the effects of Earth's perihelion even nearly coinciding with the perigee of the Moon during a spring tide can be dramatic, particularly when combined with unusual weather events. During January 1983, storms in the North Pacific arose from slow-moving low-pressure cells that developed over the Aleutian Islands, Alaska. The resulting strong northwest winds blew across the ocean from Kamchatka to the continental U.S. coast (Figure 11-15a). Averaging about 50 km/h (30 mi/h), the winds produced a near fully developed 3-m (10-ft) swell along the coast from Oregon to Baja California. This storm surge would have been trouble enough under average conditions, but the situation was amplified by 2.25-m- (7.4- ft-) high spring tides and a slight increase in sea level along the coast from an influx of warm water caused by the 1982–1983 El Niño (see Chapter 9). The unusually high spring tides occurred because Earth was still near the perihelion of its orbit, which occurred on January 2, and the Moon was at its closest approach to Earth, reaching perigee on January 28.

Some of the largest waves came ashore January 26–28, causing over $100 million in damage (Figure 11-15b). At least a dozen lives were lost, 25 homes were destroyed, and over 3500 homes were seriously damaged. Many commercial and municipal piers collapsed and many boats were lost or damaged. With each such occurrence, we learn more of the cost of developing the shore. We will explore this topic further in Chapter 12.

Dynamic Theory of Tides

The dynamic theory of tides considers the effects of ocean basins of varying depth separated by continents, and friction. The tidal bulges of the equilibrium theory of tides are directed toward the Moon and away from the Moon on opposite sides of Earth. Thus, as Earth rotates, the bulges (or wave crests) are separated by a distance of one-half Earth's circumference (about 20,000 km or 12,420 mi). We might expect the bulges to move across Earth at about 1600 km/h (1000 mi/h) in order for the bulge to circle Earth in one lunar day.

The tides are an extreme example of shallow-water waves. As such, their speed is directly proportional to water depth. For a tide wave to travel at 1600 km/h (1000 mi/h), the ocean would have to be 22 km deep (13.7 mi)! However, the mean depth of the ocean is only 3.9 km (2.4 mi). Thus, tidal bulges move as forced waves, their velocity determined by ocean depth.

Based on mean ocean depth, the mean speed at which tide waves can travel across the open oceans is about 700 k/h (435 mi/h). With this limitation on the speed of tide waves, the idealized bulges that simply point toward and away from the tide-generating body cannot exist. Instead, they break up into a number of cells.

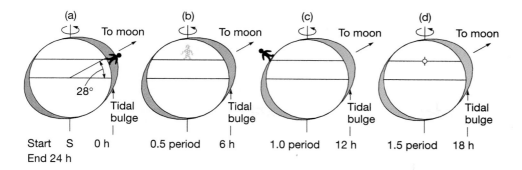

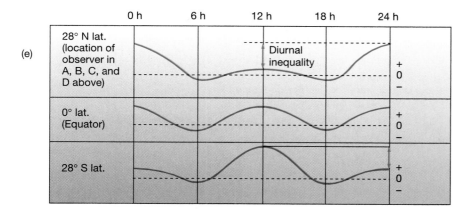

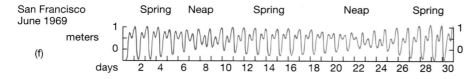

Figure 11-14   Predicted Equilibrium Tides

(a) Follow the "tidal experience" you would have if you stood at 28°N latitude, for example, along a Florida beach near Cape Canaveral. At the start of the lunar day, you are at the center of the tidal bulge, and experience a high tide. (b) Six lunar hours later (0.5 period), you experience low tide while the Moon is located on the back side of Earth. (c) After 12 lunar hours (1.0 period or 0.5 lunar day), you again experience high tide. It is much lower than the first high tide, however, because you now are passing through the edge of the tidal bulge. (d) You experience low tide again, 18 lunar hours (1.5 period) after start. At the end of one lunar day (24 h, 50 min), you return to (a) and experience a high tide. (e) Tide curves for 28°N, 0°, and 28°S latitudes when the declination of the Moon is 28°N. (All curves for the same longitude). Note that tide curves for 28°N and 28°S have identical highs and lows, but are out of phase by 12 h. This results from the fact that the bulges in the two hemispheres occur on opposite sides of Earth. (f) Along with the unequal heights for the two high and low tidal extremes occurring each lunar day, we expect to observe spring and neap tides, which are controlled by the Sun–Moon–Earth alignment. Thus, the tide curve for June in San Francisco demonstrates the general character of the predicted equilibrium tide.   *(From Anikouchine and Sternberg. The World Ocean: An Introduction to Oceanography, 1973. PrenticeHall, Englewood Cliffs, NJ.)*

In the open ocean, the crests and troughs of the tide wave actually rotate around a point near the center of each cell. It is called an **amphidromic point** (which means running around). There is essentially no tidal range at this point. Radiating from this point, we can draw on a map **cotidal lines** that connect points along which high tide will simultaneously occur. Figure 11-16 shows cotidal lines, labeled to indicate the time of high tide in hours after the

Moon crosses the Greenwich Meridian. The times indicate that the rotation of the tide wave is counterclockwise in the Northern Hemisphere and clockwise in the Southern Hemisphere. The wave makes one complete rotation during the tidal period. The size of the cells is limited by the fact that the tide wave must make one complete rotation during the period of the tide (usually 12 lunar hours). Within an amphidromic cell, low tide is 6 hours behind

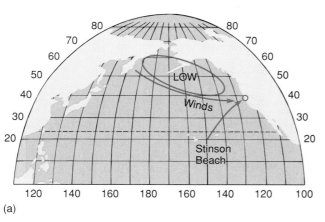

(a)

(b)

Figure 11-15   High Tides of January 1983

(a) January 1983, storm winds blow uninterrupted across the North Pacific Ocean from the Kamchatka Peninsula of Russia to the west coast of the United States. (b) Homes threatened by storm waves and unusually high tides on January 27, 1983, at Stinson Beach north of San Francisco.   *(Wide World Photos.)*

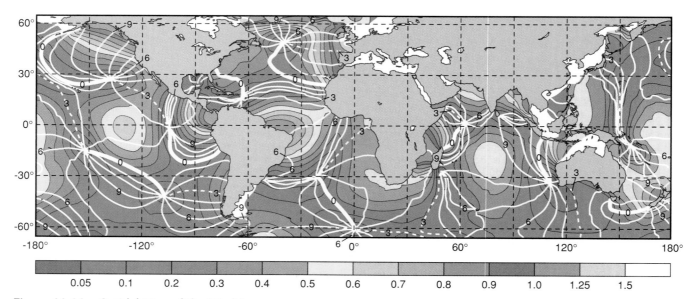

Figure 11-16   Cotidal Map of the World

Because tides must move as forced shallow water waves through the world's oceans, the two bulges we expected from the equilibrium tide theory cannot exist. Instead, the tide waves rotate around fifteen amphidromic points —seven in the Pacific Ocean and four each in the Atlantic and Indian oceans. This map depicts the amphidromic rotation of the M2 (main lunar) partial tide (see Figure 11-17). It represents 56 percent of the semidiurnal tide amplitude. Each amphidromic point has twelve lines radiating from it. The zero (0) line shows the location of the M2 tide crest as the Moon is over the Greenwich meridian. The other lines show the position of the tide crest at successive lunar hours during the 12 lunar hour semidiurnal period. The amplitude (one-half the range) of the M2 tide is shown by the color pattern on the map and scale beneath the map in centimeters. The tidal range increases with increasing distance from the amphidromic point. In the Gulf of Mexico, Caribbean Sea, and southeast Asian seas, there is a diurnal period to the tides and the amplitude is minimal. This map was constructed with data obtained from the TOPEX/POSEIDON satellite altimeter data from August 10, 1992, through August 10, 1993.   *(After Le Provost et al., 1995.)*

high tide. For example, if high tide is occurring along the cotidal line labeled 10, low tide is simultaneously occurring along the cotidal line labeled 4.

We must also consider the effect of the continents, which interrupt the free movement of the tidal bulges across the ocean surface. The ocean basins between continents have free standing waves set up within them. The character of these waves modifies the forced astronomical tide waves that develop within the basin.

It is impossible for us to explain here all the tidal phenomena that occur throughout the world. For instance, high tide rarely occurs at the time the Moon is at its highest point in the sky. The elapsed time between the passing of the Moon and the occurrence of high tides varies from place to place, a result of the many factors that determine the characteristics of the tide at any given location. To develop an optimal understanding of tides, a combination of mathematical analysis and observation is required.

Just as the sea is composed of multiple wave systems, the tides are also composed of multiple tide waves called **partial tides**. Harmonic analysis, a mathematical approach to the study of tides that embraces all the tide-generating variables that possess a periodicity (cyclic pattern), identifies almost 400 of them. The actual tide observed at any given location is the combined effect of all the partial tides at that point.

To simplify things somewhat, a relatively accurate model of the actual tide can be computed considering only the seven major partial tides. They are listed with their characteristics in Figure 11-17. Combining the periods of each of the partial tides with the amplitudes and phases that can be obtained from observation, relatively accurate predictions of the tide at any location can be made. To make the predictions as accurate as possible, the observations must be made throughout a period of at least 18.6 yr, the period of the precession of the plane of the Moon's orbit through the ecliptic.

Types of Tides

Ideally we expect two high tides and two low tides of unequal heights during a lunar day. However, due to modification from varying depths, sizes, and shapes of ocean basins, tides in many parts of the world exhibit different patterns: diurnal tide (daily), semidiurnal tides (twice daily), or mixed tides. These are shown in Figure 11-18.

Diurnal tides have a single high and low water each lunar day. These tides are common in the Gulf of Mexico and along the coast of Southeast Asia. Such tides have a tidal period of 24 h, 50 min. **Semidiurnal tides** have two high and two low waters each lunar day. The levels of successive high waters and successive low waters are approximately the same. Since tides are always growing higher or lower at any location due to the spring–neap tide sequence, successive high tides and successive low tides are never exactly the same at that location. Semidiurnal tides are common along the Atlantic Coast of the United States. The tidal period is 12 h 25 min.

Symbol		Period in solar hours	Amplitude $M_2 = 100$	Description
Semidiurnal tides	M_2	12.42	100.00	Main lunar (semidiurnal) constituent
	S_2	12.00	46.6	Main solar (semidiurnal) constituent
	N	12.66	19.1	Lunar constituent due to monthly variation in moon's distance
	K_2	11.97	12.7	Soli-lunar constituent due to changes in declination of sun and moon throughout their orbital cycle
Diurnal tides	K_1	23.93	58.4	Soli-lunar constituent
	O	25.82	41.5	Main lunar (diurnal) constituent
	P	24.07	19.3	Main solar (diurnal) constituent

(a)   The seven most important partial tides

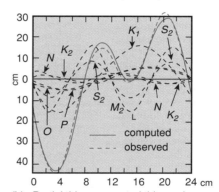

(b)   Partial tides, computed tide, and observed tide at Pula, Yugoslavia (January 6, 1909). Note close fit of computed and observed tides.

Figure 11-17   Partial Tides

The semidiurnal (twice daily) and diurnal (daily) partial tides described in (a) are shown in tide curve form in (b), along with the computed and observed tides that resulted from their combined effects at Pula, Yugoslavia.   *(From A. Defant,* Ebb and Flow, *1958. The University of Michigan Press, Ann Arbor.)*

Mixed tides have characteristics of both diurnal and semidiurnal tides. Successive high tides and/or low tides will have significantly different heights. Mixed tides commonly have a tidal period of 12 h 25 min, which is a semidiurnal characteristic, but also possess diurnal periods for a few days per month. This is the tide that is most common throughout the world and is the type found along the Pacific coast of the United States. Note that the tide curve for Los Angeles, shown in Figure 11-19, is mixed because

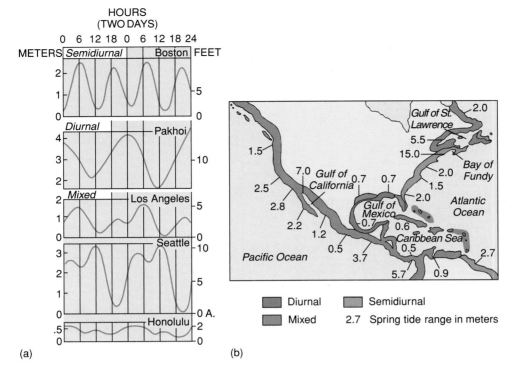

Figure 11-18   Types of Tides

(a) In a semidiurnal (twice daily) tide, there are two highs and lows during each lunar day, and the heights of each successive high and low are about the same. (Note that the diagrams show two days —not one.) In the diurnal (daily) type of tide, there is only one high and one low each lunar day. In the mixed type of tide, both diurnal and semidiurnal effects are detectable, and the tide is characterized by a large difference in high water heights, low water heights, or both, during one lunar day. Even though a tide at a place can be identified as one type, it still may pass through stages of one of both of the other types. (b) Types of tides observed along North and South American coasts. The numbers give the spring tide range in meters and are therefore near the maximum tidal range that can be expected. Storm waves, lower barometric pressure, ocean currents, and coincidence with extremes of distance from the Sun could increase the range.   *(After C. Hauge, Tides, Currents, and Waves. California Geology, July 1972.*

it is semidiurnal in period during all days except September 15, 16 and 17 when the period is diurnal. If you spend some time studying Figure 11-19, you will see the effects on the tides of the phases and declination of the Moon, distance to the Moon, as well as the position of the Sun at autumnal equinox.

Tides in Lakes

In general, lake basins are too small to have any appreciable tide set up by the tide-producing forces, but such forces may be significant where the long axis of the basin runs parallel to lines of latitude. In any such basin, very small standing waves may be generated with a period equal to that of the tide-generating force. This wave type is referred to as a *forced standing wave* (Figure 11-20).

Of much greater importance is the free standing wave initiated by strong winds at the surface or by some seismic disturbance that may be set up in the basin. The period of a free standing wave is determined by the length and depth of the basin, and this period is termed the *characteristic*

period for the basin. If the characteristic period of the free wave is very near that (or a multiple) of the period of a forced wave resulting from the tide-generating forces, the oscillations may reinforce each other, and produce a **resonance tide**. Lake Ontario displays such tides. Free oscillations of the type just described were given the name **seiche** by the inhabitants of the region near Lake Geneva, Switzerland. For seiches that have a single nodal line, the formula for the period in Figure 11-20 gives a close approximation of the period that actually develops. For rectangular basins with two and three nodal lines, the periods would be approximately one-half and one-third those of a mononodal seiche.

Tides in Narrow Basins Connected to the Ocean

As previously discussed, even under conditions of resonance between free and forced oscillations in small closed bodies of water, the maximum tides are not great. Seldom do they exceed a few centimeters. By contrast, in similarly

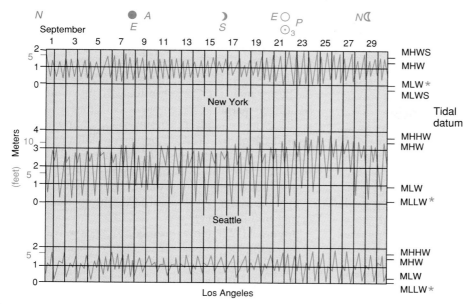

Figure 11-19   Tidal Curves for Three Locations

The declinations of the Moon and the Sun, phase of the Moon, and position of the Moon in its orbit as noted across the top of the chart all contribute to the tidal variations depicted. The position of Earth in its orbit around the Sun and the configuration of the sea bottom and basin boundaries also affect the tides. Some names of water levels are listed along the right margin. Those that are used as the chart datum for a place are marked with an asterisk. MHWS (mean high water springs) is the average height of the high water of the spring tides; MHW (mean high water) is the average height of all the high tides at a place; MLW (mean low water) is the average height of all the low tides at a place; MLWS (mean low water springs) is the average height of all low waters of the spring tides; MHHW (mean higher high water) is the average height of the higher high tides at a place where the tide is the mixed type and displays an inequality during a tidal day; and MLLW (mean lower low water) is the average height of the lower low tides at a place where the tide is of the mixed type.   *(From C. Hauge. Tides, Currents, and Waves, California Geology, July 1972.)*

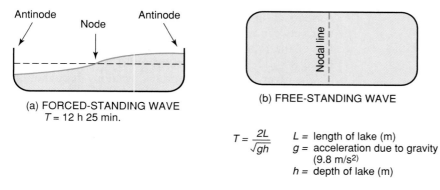

Figure 11-20   Tides in Lakes

(a) A forced standing wave generated by tide-generating forces. It will have a period of 12 h 25 min. (b) A mononodal free standing wave (seiche) generated by an atmospheric disturbance. Its period is determined by the size and shape of the basin of the lake. The period (T) is related to length and depth of the lake as shown by the equation in the figure. If the lake has dimensions that produce a period for the seiche that is approximately equal to or a multiple of the forced tide-generated wave, they will be resonant and produce greater displacements at the antinodes.

sized basins that are open at one end to the ocean, the tides may become relatively large. Why the difference?

To help answer this question, let us consider a rectangular bay aligned east-west with one end open to the ocean (Figure 11-21). Even though one end of the basin is open, we still have to consider the effect of free standing waves developing and being reflected from the open end.

At the open end of the bay, the water must always be at the same level as the ocean at that location. Therefore, the tidal range is of greater magnitude than it is in closed basins. If we think of the free standing wave, caused by wave energy reflected from the closed end of the basin, as being in resonance with the forced wave produced by tidal forces, we can see that combining the energy of these waves produces a standing wave with increased amplitude within the basin.

Tides in Broad Open Basins

In large, broad basins, Earth's rotation causes a circular movement of the wave crest around the margins of the embayment. Since there is a significant horizontal transport of water associated with standing waves, the Coriolis effect becomes important in embayments that are wide enough to allow lateral movement of water.

Let us assume an embayment in the Northern Hemisphere has developed a mononodal standing wave and the crest is developing at its closed end (Figure 11-22a). As the water begins to move toward the open end of the embayment (Figure 11-22b), the Coriolis effect diverts the

water to the right, forcing it to pile up along the left-hand side of the embayment as we look at it from the open end. One-half period after the crest at the closed end of the basin begins to fall, a crest will form at the open end (Figure 11-22c). As the water moves toward the closed end of the basin, again the Coriolis effect will pile it up along the right-hand margin of the embayment (Figure 11-22d). This produces an amphidromic point, a point of no tidal range, around which the crest of the wave rotates in a counterclockwise direction, instead of a mononodal line as might have developed in a narrow basin (Figure 11-22e). Of course, this amphidromic system is but a miniaturized version of the systems we described for the open ocean basins earlier in this chapter.

Figure 11-22f shows that as progressive waves move through broad basins, there is a much greater tidal range on the right-hand side of the basin, looking into the basin, than on the left. Assuming a theoretical high and low-tide level for the system, and a tilt in the surface of the wave caused by the Coriolis effect as it moves into and out of the basin, it can be seen that along the right-hand side of the basin the tidal range is equal to the theoretical tidal range plus $2x$. The x value is the result of the Coriolis force tilting the wave as it moves through the basin. On the left-hand side of the basin, the tidal range, as a result of this tilting, is equal to a theoretical difference between high and low tide of minus $2x$.

Coastal Tide Currents

The current accompanying the rotating tide crest in a Northern Hemisphere basin turns in a counterclockwise direction, producing a rotary current in the open portion of the basin. Near shore, except where the shore is straight and steeply sloping, the rotary current is changed to an alternating current, or **reversing current**, that moves in and out rather than along the coast as would a rotary current. These reversing currents are of the greatest concern to navigators, since they can reach high velocities. For example, currents of 44 km/h (27.3 mi/h) occur in restricted channels between islands of coastal British Columbia; velocities of the rotating tidal currents in the open ocean are usually well below 1 km/h (0.62 mi/h).

The reversing tidal current that develops throughout a lunar day in a mixed tide is depicted in Figure 11-23. Beginning at high tide, the current velocity is zero because the water has just reached its highest stage and is about to begin its outward flow. Following this high slack water, the lowering of the tide begins, and the **ebb current** (seaward flow of water) velocity increases and reaches a maximum about 3 lunar hours after high slack water. The velocity decreases and eventually reaches zero again at the first low slack water. Following the change in current velocity associated with the tidal phase, it can be seen that maximum current velocity is reached midway through the ebb current that occurs between the higher high water and lower low water. Figure 11-23 shows that the lower low water (LLW) is below the tide datum of the chart, the zero

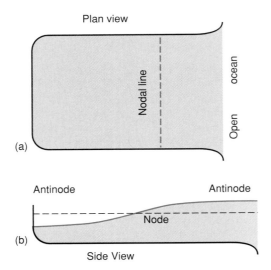

Figure 11-21   Tides in Seas, Bays, and Gulfs

In seas, bays, and gulfs, the forced standing waves have a greater height than those created in lakes. This is because the height of the tide at the open end of the basin must be the same as the open ocean. Therefore, the development of a resonant condition between the free and forced standing waves in such basins produce much greater displacements at the antinodes.

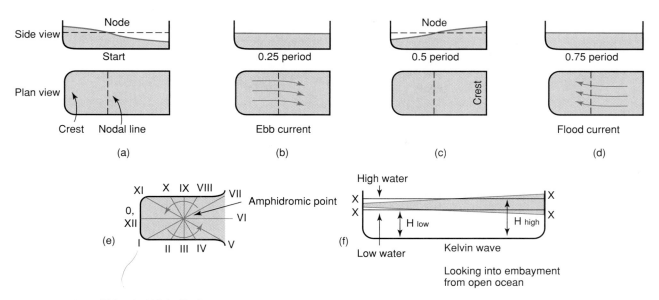

Figure 11-22   Tides in Wide Embayments

(a-d) As the crest forms in the head of the bay, there is no horizontal water movement. In (b), the surface is in an equilibrium (flat) condition, and water moves horizontally toward the open ocean. The Coriolis force causes the water to veer to the right in the Northern Hemisphere. After the crest forms at the mouth of the bay (c), the same effect causes water to veer toward the opposite side of the bay when water moves toward the head of the bay (d). In (e), for bays with sufficient width, the nodal line of the standing wave disappears and is replaced by an amphidromic point, a point of no tide. The standing wave has taken on a progressive character, and the crest moves in a counterclockwise path around the bay. The lines labeled with Roman numerals are cotidal lines that connect all points experiencing high tide at the indicated lunar hour. The tide crest rotates around the amphidromic point, and tidal range increases with increasing distance along the cotidal lines from the amphidromic point. In (f), a Kelvin wave —along with the effects shown in (a-e), the Coriolis force acting on a progressive wave moving along any channel or embayment produces a tilted water surface instead of a flat high and low water at tidal extremes. This results is a greater tidal range on one side of a basin than the other. On the right side, the tidal range is $H_{high} - H_{low} + 2x$. On the left side, the range is $H_{high} - H_{low} - 2x$.

mark. How can the tide be less than zero? The datum that is commonly used for mixed tides is the mean lower low water (MLLW), the average height of the lower low tides at the locality. Since the lower low tide that we are recording is below the average low tide for this locality, this tide has a negative value. In areas where mixed tides are not observed, the tide datum is very commonly the mean low tide recorded at that place—the average low tide. Thus most tide extremes recorded, even low tides, have positive values (see Figure 11-19). Only during spring tides are negative low tides observed.

Tides Observed Throughout The World

Bay of Fundy

The Bay of Fundy has an extremely wide opening into the Atlantic Ocean—258 km (160 mi). The bay splits into two narrow basins at its northern end, Chignecto Bay and Minas Basin (Figure 11-24). The period of free oscillation in the Bay of Fundy is very nearly that of the tidal period. This condition brings about resonance, which along with the narrowing and shoaling toward the north end of the bay, produces maximum tidal ranges in the extreme northern end of Minas Basin. The maximum perigean tidal range of about 17 m (56 ft) occurs at the north end of Minas Basin, and the minimum tidal range of about 2 m (6.6 ft) is found at the opening into the bay. There are no nodes within the bay, and the tidal range increases from the mouth of the bay northward. Along with the standing wave phenomenon, there is a rotational component that causes a greater tidal range on the south shore than on the north shore.

North Sea

The North Sea is a very broad basin that is relatively shallow and open to the Atlantic Ocean over a great distance along its northern margin. Bounded on the west by the British coast and to the south and east by the coast of

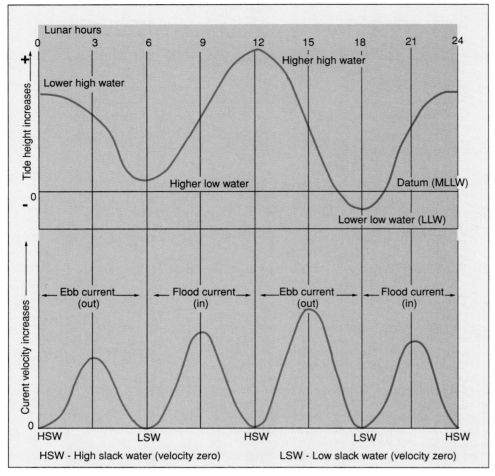

Figure 11-23   **Reversing Current**
Note that tidal current velocity is zero at high and low tidal extremes. Maximum tidal current velocity occurs midway between tidal extremes.

northern Europe, the North Sea experiences maximum tidal ranges approaching 5 m (16 ft). As the semidiurnal tide in the form of a progressive wave enters the North Sea from the Atlantic Ocean, the maximum tidal range within the sea is produced along the Scottish and English coasts. In contrast to the 3 to 4 m (9.8 to 13.1 ft) tidal range along the western coast of the North Sea, the tides recorded at the southern tip of Norway typically range less than 25 cm (10 in).

Because of the Earth's rotation, the mass of water that moves with the progressive wave is deflected to the right as it enters the North Sea. The wave is reflected off the coast of Belgium and the Netherlands and travels north pushing the mass of moving water against the coasts of Denmark and Norway as it moves toward the Atlantic Ocean. The reason tidal ranges are so much greater along the coast of Scotland and England is twofold. First, the wave possesses more energy as it moves in from the Atlantic Ocean. As the wave moves southward toward the European coast, it gradually loses energy to the gently shoaling bottom, and the reflected wave contains considerably less energy than that possessed by the southbound wave. Second, the Kelvin wave that develops in any situation where a progressive wave travels through a broad channel contributes to the effect. The effect of the loss of energy as the wave moves into and out of the North Sea, and the tilting of the Kelvin wave can be seen in the **corange lines** that indicate the value of tidal ranges throughout the basin (Figure 11-25a).

Theoretically, in a basin with the dimensions of the North Sea, the tidal pattern should be approximately that shown in Figure 11-25b. Two rotary tides should develop around amphidromic points, one in the north end and one in the south end of the basin. The tidal wave should travel from the north along the western margin, around the southern end of the basin, which is closed, then back along the eastern margin to the north. The actual pattern that develops in the North Sea, shown in

(a)

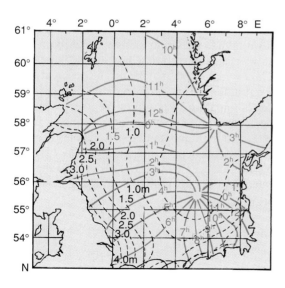

(b)

Figure 11-24 Bay of Fundy

The largest tidal range in the world occurs at the north end of the Minas Basin in the Bay of Fundy. Because of its dimensions, this bay has a natural free-standing-wave period about equal to that of the forced-tide wave. This combined with the fact that the bay narrows and becomes shallower toward its head causes a maximum tidal range at the northern end of 17 m (56 ft). High tide (a) and low tide (b) in Minas Basin. (c) map. *(Nova Scotia Department of Tourism.)*

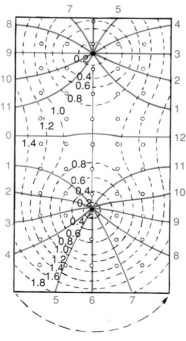

(c)

(a) Cotidal and corange lines of the North Sea *M₂* tide, —— : time of high water after moon's transit through Greenwich meridian; - - -: mean corange lines.

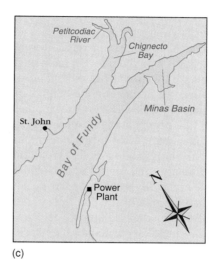

(b) Theoretical cotidal lines and corange lines of the semi-diurnal tide in a bay whose length is twice its breadth. The arrow shows the direction of the earth's rotation.

Figure 11-25 North Sea Tides

(a) Cotidal lines show a much higher tidal range along the coast of the British Isles than along the coasts of Denmark and Norway. Also, the amphidromic points are offset to the east, because the tide wave enters the North Sea from the North Atlantic, and the incoming Kelvin wave is much more energetic than the outgoing Kelvin wave that moves north along the coasts of Denmark and Norway. Much of the incoming wave energy is lost as it moves into the shallow water at the south end of the North Sea. (b) Theoretical pattern that would develop without the energy loss described in a bay the size and shape of the North Sea. *(From A. Defant, Ebb and Flow, © 1958. The University of Michigan Press, Ann Arbor.)*

Grunion and the Tides

Along the beaches of southern California and Baja California from March through September, a very unusual fish-spawning behavior occurs. Shortly after the maximum spring tide, small silvery fish come ashore to bury their fertilized eggs in the sand. They are the grunion (*Leuresthes tenuis*), slender little fish 12 to 15 cm (4.7 to 6 in.) in length. The name grunion comes from the Spanish gruñon, which means grunter. The early Spanish settlers gave the fish this name because of the faint noise they make during spawning.

The type of tide that occurs along southern California and Baja California beaches is a mixed tide. On most tidal days (24 h 50 min), there are two high and two low tides. There is usually a significant difference in the heights of the two high tides that occur each day. During the summer months, the higher high tide occurs at night. As the higher high tides become higher each night as the maximum spring-tide range is approached, sand is eroded from the beach (Figure 11A). After the maximum height of the spring tide is reached, the higher high tide that occurs each night is a little lower than the one of the previous night. During this sequence of decreasing tide heights as neap-tide conditions approach, sand is deposited on the beach. The grunion depend greatly on this pattern of beach sand deposition and erosion in their spawning process.

Grunion spawn only after each night's higher high tide has peaked on the three or four nights following the night of the highest spring high tide. This behavior assures that the eggs will be covered deeply by sand deposited by the receding high tides. The fertilized eggs buried in the sand are ready to hatch 9 days after spawning. By this time, another spring tide is approaching, and the higher high tide that occurs each night will be higher than that of the previous night. This condition causes the beach sand to erode, exposing the eggs to agitation by waves that break ever higher on the beach. The eggs hatch about 3 minutes after being freed in the water. Tests done in laboratories have shown that the eggs will not hatch until agitated in a manner that simulates the agitation of the eroding waves.

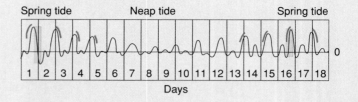

 Grunion deposit eggs in beach sand during early stages of the ebb of higher high tides on the three or four days following maximum spring tidal range.

Flood tides erode sand and free grunion eggs during higher high tide as maximum spring tidal range is approached.

Maximum spring tidal range.

Figure 11A   Grunion and the Tidal Cycle

Eggs deposited in the beach during ebb tides on days 3, 4, and 5 will be ready to hatch when eroded by the flood tides of days 13, 14, 15, and 16. The lines separating days on the tidal chart represent midnight.

The spawning begins as the grunion come ashore immediately following an appropriate high tide, and it may last from 1 to 3 hours. Spawning activity usually peaks about an hour after its start and may last from 30 minutes to an hour. Thousands of fish may be on the beach at this time. During a run, females move high on the beach. If no males are near, a female may return to the water without depositing her eggs. In the presence of males, she drills her tail into the sand until only her head is visible. The female continues to twist, depositing her eggs 5 to 7 cm (2 to 3 in) below the surface. The male curls around the female's body and deposits his milt against it (Figure 11B).

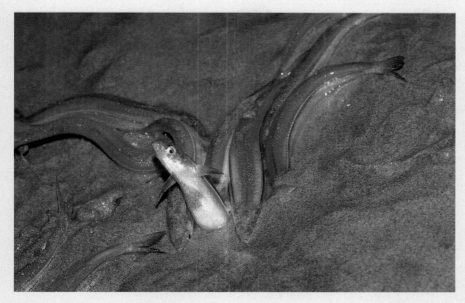

Figure 11B   Grunion Spawning

Male grunion swarm around a female that has burrowed into the sand to deposit her eggs below the surface. The males release sperm-laden milt that runs down the female's body to fertilize the eggs.   *(Photo by Eda Rogers.)*

The milt runs down the body of the female to fertilize the eggs. The spawning completed, both fish return to the water with the next wave.

As soon as the eggs are deposited, another group of eggs begins to form within the female. They will be deposited during the next spring-tide run. Large females are capable of producing up to 3000 eggs for each series of spawning runs, which are separated by the 2-week period between spring-tide occurrences. Early in the spawning season, only older fish spawn, but by May, even the 1-year-old females are in spawning condition.

Young grunion grow rapidly and are about 12 cm (5 in) long at one year of age and ready for their first spawning. They usually live 2 or 3 years, but 4-year-olds have been recovered. The age of grunion can be determined by their scales. After growing rapidly during the first year, they grow very slowly. There is no growth at all during the 6-month spawning season, which causes marks to form on each scale.

How the grunion are able to time their spawning behavior so precisely is not known. Some investigators believe the grunion are able to sense very small changes in hydrostatic pressure caused by the changing water levels associated with the tides. Certainly some very dependable detection mechanism keeps the grunion accurately informed of the tidal conditions, because their survival depends on a spawning behavior precisely tuned to tidal motions.

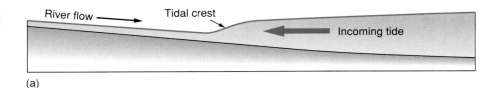

(a)

(b)

Figure 11-26   **River Bores**

(a) As the tidal crest moves upriver, it develops a steep forward slope through resistance to its advance by the river flowing to the ocean. Such crests (bores) may reach heights of 5 m (16.4 ft) and move at speeds up to 22 km/h (13.7 mi/h). (b) A tidal bore moves up a river flowing into Chignecto Bay on the coast of New Brunswick, Canada.   *(New Brunswick Department of Tourism.)*

Figure 11-25a, is similar to the predicted tidal pattern except that the amphidromic points are offset considerably to the east.

Tides in Rivers

The Amazon River probably possesses the longest estuary (river mouth) affected by oceanic tides. Tides can be measured as far as 800 k (500 mi) from the river's mouth, although the effects are quite small at that distance. Tide waves that move up river mouths lose their energy due to the decreasing depth of water and the flow of the river water against the tide. As a tide-crest wave moves up the river, it becomes more and more asymmetrical, developing a steep front (Figure 11-26). This front produces a rapidly rising tide that falls slowly.

An extreme development of this type produces a **tidal bore** in which a very steep wave front surges up the river.

In the Amazon, it is called "pororoca" and appears as a waterfall up to 5 m (16.4 ft) in height moving upstream at speeds up to 22 k/h (13.7 mi/h)! Other rivers that experience tidal bores are the Chientang in China, where bores may reach 8 m (26 ft); the Petitcodiac in New Brunswick, Canada; the Seine in France; and the Trent in England.

Tides as a Source of Power

The history of human efforts to harness tidal energy dates back to the Middle Ages at least. Pursuit of this renewable energy source waned with the availability of cheap fossil fuels, but today, as we realize that cheap fossil fuels eventually will run out, there is increased interest in generating electricity with tidal energy.

The most obvious benefit of generating electrical power with tidal energy is reduced operating costs, compared to

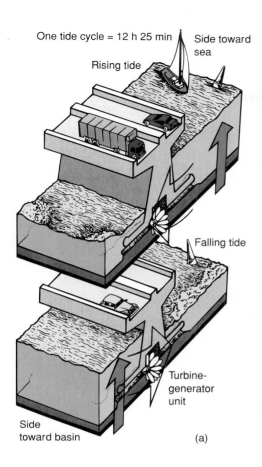

One tide cycle = 12 h 25 min

Rising tide

Side toward sea

Falling tide

Turbine-generator unit

Side toward basin

(a)

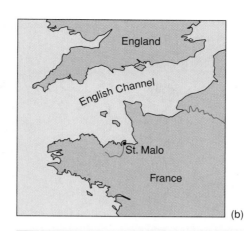

(b)

(c)

Figure 11-27   La Rance Tidal Power Plant at St. Malo, France

Diagram shows the barrier between the open ocean to the right and the La Rance estuary to the left. The relative water levels during rising and falling oceanic tides also are shown.   *(Photo courtesy of Phototeque/Electricite de France.)*

conventional thermal power plants that require fossil fuels or radioactive isotopes. Even though the initial cost of building a tidal power-generating plant may be higher, there would be no ongoing fuel bill. A negative consideration involves the periodicity of the tides. Power could be generated only during a portion of a 24-hour day. Because people operate on a solar period and tides operate on a lunar period, the energy generated from the tides would coincide with need only part of the time. To get around this, power would have to be distributed to a point of need at the moment—an expensive transmission problem. Power could be stored, but this presents a large and expensive technical problem. Furthermore, electrical turbines (generators) must run at constant speed, yet use the changing velocity and flow direction of the tidal current. This requires a special design that would allow both advancing and receding water to spin the turbine blades.

La Rance Tidal Power Plant

A successful tidal power plant is operating in the estuary of the La Rance River off the English Channel in France (Figure 11-27). This estuary has a surface area of approximately 23 km² (9 mi²), and a maximum tidal range of 13.4 m (44 ft). The deepest water ranges from just over 12 m (39 ft) at low tide to more than 25 m (82 ft) at high tide. Usable tidal energy is proportional to the area of the basin and to the amplitude of the tide. A barrier was built across the estuary about 3 km (2 mi) upstream, where the estuary is 760 m (2500 ft) wide. To allow water to flow through the barrier when the generating units are shut down, sluices (artificial channels) were built into the barrier. Twenty-four electrical generating units operate beneath the power plant. Each unit can generate 10 million watts (10 MW) of electricity, enough to serve over 1500 homes for a year.

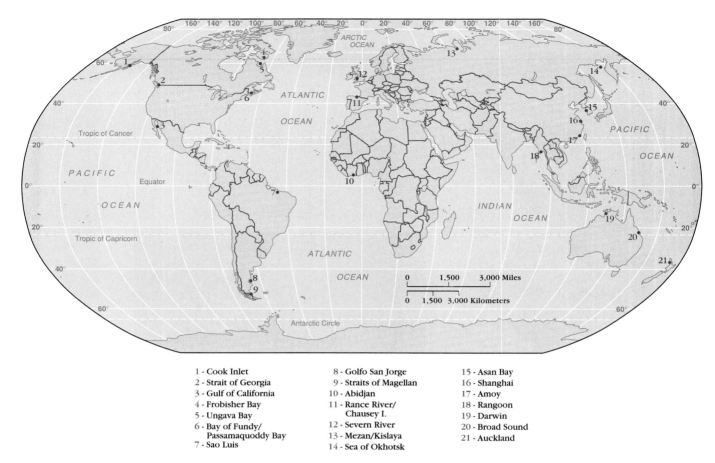

1 - Cook Inlet
2 - Strait of Georgia
3 - Gulf of California
4 - Frobisher Bay
5 - Ungava Bay
6 - Bay of Fundy/
   Passamaquoddy Bay
7 - Sao Luis

8 - Golfo San Jorge
9 - Straits of Magellan
10 - Abidjan
11 - Rance River/
   Chausey I.
12 - Severn River
13 - Mezan/Kislaya
14 - Sea of Okhotsk

15 - Asan Bay
16 - Shanghai
17 - Amoy
18 - Rangoon
19 - Darwin
20 - Broad Sound
21 - Auckland

Figure 11-28 Sites with Major Potential for Tidal Power Generation

Twenty-one locations worldwide where tidal ranges are great enough to have potential for generating electricity. Where a large area exists for storing water behind a dam, a tidal range of 3 m (10 ft) would suffice. Where smaller storage areas exist, greater tidal ranges would be required.

The plant generates electricity only when there is sufficient difference in water height between the basin side and the ocean side; this occurs only during about one-half of the tidal period. Annual power production of about 540 million kWh without pumping can be increased to 670 million kWh by using the turbine generators as pumps at the proper times.

Other Tidal Power Plants

Within the Bay of Fundy, the province of Nova Scotia has constructed a tidal power plant that has generated up to 40 million kWh per year since completion in 1984. It is built on the Annapolis River, where the maximum tidal range is 8.7 m (26 ft). Some engineers think that a tidal power plant could generate electricity constantly if located on the Pas-

samaquoddy Bay near the U.S.–Canadian border at the south end of the Bay of Fundy. Others are less optimistic. Potentially, the usable tidal energy seems great compared with La Rance, because the flow volume is about 117 times greater.

Whether or not tidal power stations ever are constructed on a large scale, this potential source of energy will receive increased attention as the cost of generating electricity by conventional means increases. Figure 11-28 shows locations of some sites that have potential for tidal power generation. However, any use of coastal waters for the tidal generation of electricity will have environmental costs resulting from the modification of current flow. It will interfere with many traditional uses of coastal waters, including boating and fishing.

Summary

The tides of Earth are caused by the gravitational attraction of the Sun and Moon. The Moon has about twice the tide-generating effect of the Sun because the Moon is so much closer. Small horizontal forces tend to push water into two bulges on opposite sides of Earth, one directly facing the tide-generating body, and the other directly opposite. Since the tidal bulges due to the Moon's gravity are dominant, the tides we observe on Earth have periods dominated by lunar motions. They are modified by the changing position of the solar bulges.

If Earth were a uniform sphere covered with an ocean of uniform depth, tides would be easily predicted. Such tides would have a period of 12 h 25 min, or half a lunar day. Tides with maximum tidal range (spring tides) would occur each new Moon and full Moon, and tides with minimum range (neap tides) would occur with the first and third quarter phases of the Moon.

Since the Moon has a declination up to 28.5° north or south of the equator, and the Sun is directly over the equator only two times per year, tidal bulges usually are located to create two high tides of unequal height per lunar day. The same can be said for the low tides. Tidal ranges are greatest when Earth is nearest the Sun and Moon.

Since Earth has an irregular surface, with continents dividing the world ocean into irregularly shaped basins, the tides we observe are explained by the dynamic theory of tides. Open-ocean tides rotate counterclockwise around an amphidromic point, a point of zero tidal range, in the Northern Hemisphere.

The basic types of tides observed on Earth are a diurnal tide (period of 1 lunar day); a semidiurnal tide (period of half a lunar day, like that predicted for the equilibrium tide); and a mixed tide with characteristics of both. Mixed tides are usually dominated by semidiurnal periods and display significant diurnal inequalities which are greatest when the Moon is over the tropics, and least when it is over the equator.

All basins have a characteristic free-standing wave, or seiche. It usually is not very large for small basins, but if it is in phase with the forced-standing wave created by the tide-generating bodies, its height can be significant. If the basin is wide enough, the standing wave may be converted to a progressive wave rotating in a counterclockwise direction (Northern Hemisphere) around an amphidromic point, a point of zero tidal range. Tidal currents follow this rotary pattern in open-ocean basins but are converted to reversing currents at the margins of continents. The maximum velocity of reversing currents occurs during ebb and flood currents when the water is halfway between high and low-standing waters.

The effects of shoaling and narrowing on the tide can be observed in the Bay of Fundy. The development of multiple amphidromic points can be observed in the North Sea. Tidal bores, tide waves that force their way up rivers, are common in such rivers as the Amazon, Chientang, Seine, and Trent.

Because tides can be used to generate power without the use of fossil or nuclear fuel, the possibility of constructing such generating plants has always been attractive to engineers. One such plant is located in the estuary of La Rance River in France.

Key Terms

Amphidromic point	Dynamic tide theory	Partial tides	Spring tide
Aphelion	Ebb current	Perigee	Summer solstice
Apogee	Ecliptic	Perihelion	Tidal bore
Autumnal equinox	Equilibrium tide theory	Precession	Tidal period
Centripetal force	Gravitational force	Quadrature	Tidal range
Cotidal lines	Lunar day	Resonance tide	Tide-generating force
Corange lines	Lunar hour	Reversing current	Vernal equinox
Declination	Mixed tide	Seiche	Winter solstice
Diurnal tide	Neap tide	Semidiurnal tide	

Questions and Exercises

1. Explain why the Sun's influence on Earth's tides is only 46 percent that of the Moon, even though the Sun exerts a gravitational force on Earth that is 177 times greater than that of the Moon.

2. Describe how the centripetal force required to keep particles of Earth rotating in their identical orbits within the Earth–Moon system varies from the centripetal force provided by the gravitational attraction between the particles and the Moon. Note: Consider strength and direction of the force.

3. Construct a diagram of vector arrows to show how the horizontal tide-generating force results from the difference in the required centripetal force, C, for a particle on Earth's surface at 45° from the zenith, and the centripetal force provided by the gravitational attraction between the Moon and that particle, G.

4. Explain why the maximum tidal range (spring tide) occurs during new and full-Moon phases and the minimum tidal range (neap tide) with quadratures.

5. Discuss the length of cycle and degree of declination of the Moon and Sun relative to Earth's equator. Include a discussion of the effects of precession of the plane of the Moon's orbit through the ecliptic.

6. Describe the effects of the declination of the Moon and Sun on the world ocean tides.

7. Diagram the Moon's orbit around Earth and Earth's orbit around the Sun. Label the positions on the orbits where the Moon and Sun are closest to and farthest from Earth, stating the terms used to identify them. Discuss the effects that the Moon and Earth being in these positions have on Earth's tides.

8. Describe the period and diurnal inequality of the following: diurnal tide, semidiurnal tide, and mixed tide.

9. What forces produce forced- and free-standing waves in lakes and narrow ocean embayments?

10. In narrow embayments, standing waves may develop as a result of tides entering from the ocean. In wider embayments, a rotating progressive wave may develop instead. Discuss how the increased width of the embayment allows this rotary wave to develop.

References

Anikouchine, W. A., and R. W. Sternberg. 1973. *The World Ocean: An Introduction to Oceanography.* Englewood Cliffs, N.J.: Prentice-Hall.

Clancy, E. P. 1969. *The Tides: Pulse of the Earth.* Garden City, N.Y.: Doubleday.

Defant, A. 1958. *Ebb and Flow:* The Tides of Earth, Air and Water. Ann Arbor: University of Michigan Press.

Gill, A. E. 1982. Atmosphere and ocean dynamics. *International Geophysics Series,* Vol. 30. Orlando, Fla.: Academic Press.

Hauge, C. 1972. Tides, currents, and waves. *California Geology,* July.

Le Provost, C., A. F. Bennett, and D. E. Cartwright. 1995. Ocean tides for and from TOPEXP/POSEIDON. *Science* 267:5198, 639–42.

Pond, S., and G. L. Pickard, 1978. *Introductory Dynamic Oceanography.* Oxford: Pergamon Press.

Sverdrup, H. U., M. W. Johnson, and R. H. Fleming. 1942. Renewal 1970. *The Oceans: Their Physics, Chemistry, and Biology.* Englewood Cliffs, N.J.: Prentice-Hall.

von Arx, W. S. 1962. *An Introduction to Physical Oceanography.* Reading, Mass.: Addison-Wesley.

Suggested Reading

Sea Frontiers

Sobey, J. C. 1982. *What is Sea Level?* 28:3, 136–42. The role of tides and other factors in changing the level of the ocean surface is discussed.

Zerbe, W. B. 1973. *Alexander and the Bore.* 19:4, 203–8. Alexander the Great encounters a tidal bore on the Indus River.

Scientific American

Goldreich, P. 1972. *Tides and the Earth-Moon system.* 226:4, 42–57. The tide-generating force of the Sun and Moon on Earth, and the effect of transfer of angular momentum from Earth to the Moon as a result of tidal friction are discussed. Also considered are theories of lunar origin.

Greenberg, D. A. 1987. *Modeling Tidal Power.* 257:5, 128–31. The effects of using the large tidal ranges experienced in the Bay of Fundy to generate electricity are modeled on a computer.

Lynch, D. K. 1982. Tidal bores. 246:4, 146–57. The tidal bore phenomenon is explained.

Oceanography on the Web

Visit the Introductory Oceanography home page for on-line resources for this chapter. There you will find an on-line study guide with review exercises and links to ocean-ography sites to further your exploration of the topics in this chapter. Introductory Oceanography is at http://prenhall.com/thurman (click on the Table of Contents menu and select this chapter).

Beaches or Bedrooms? The Dynamic Coastal Environment

- What factors affect beach location and shape?
- How has coastal development influenced the beach environment?
- What are the pros and cons of beachfront stabilization?

Background

Beaches are dynamic zones that change in shape over both space and time. Beaches may stretch for thousands of km along passive margin shorelines (i.e., those without seismic activity such as the U.S. Atlantic coast), but extensive beaches are rare in tectonically active uplift areas along active margin shorelines, for example, the Pacific coast of North and South America.

The shape of a beach may vary over distances of a kilometer to hundreds of kilometers and is based on the balance between processes that promote erosion and processes that favor deposition of sediment. These in turn are influenced by local physical and geologic characteristics.

Over time, the shape of beaches is influenced by short-term storm events (days); normal seasonal changes in wave intensity (month to year); longer-term (years to decades) climatic phenomena like El Niño and La Niña and periods of high hurricane activity; and very long-term (century to millennium) changes in global sea level.

In the last 18,000 years, sea level has risen about 120 meters (400 ft) and many coastal shorelines have moved dramatically inland. Off the East Coast of the United States geologists estimate that the shoreline has migrated inland more than 80 km (50 mi) during this period, alternating between active migration and static periods. Based upon a rough average of typical beach slopes, for each unit rise in sea level, the beach will migrate 1000 to 2,000 units inland.

All 30 states bordering an ocean or the Great Lakes have erosion problems, and 26 are presently experiencing net loss of their shores. As an example of this erosion and shoreline migration process, consider the Cape Hatteras lighthouse in North Carolina and its location relative to the shoreline. In 1870, the lighthouse was situated about 450 m (1475 ft) feet from the ocean. By 1919, the ocean had advanced to within 100 m (325 feet) of the tower, and by 1935, it was within 30 m (100 ft). After a series of strategies designed to hold back the coastal erosion failed (not a surprise to coastal geologists), the lighthouse was finally moved nearly 900 m (2950 ft) inshore in 1999. Attempts to save the neighboring Morris Island lighthouse in South Carolina are currently in progress.

Erosion is a natural and sometimes predictable process. Still, despite this knowledge, the movement of shorelines inland has been a serious problem to coastal residents and businesses, especially in the middle and southern Atlantic and Gulf regions of the United States. This issue is further examined on our website.

Composition and Structure of Beaches

Many factors affect the shape, composition and structure of beaches. The shape varies with sand supply, sea level change, and wave size. When any of these factors change, the other factors respond accordingly. For example, if the supply of sand is reduced and sea level and wave size remain constant, the beach profile will change because the diminished sand supply creates an erosional environment, leading to the landward movement of the beach's intertidal region. Decreasing wave energy, holding the other physical factors constant, will cause either reduced erosion or natural accretion (building) of the beach.

The slow but unrelenting assault of sea level rise upon coastlines is difficult to recognize on a daily basis, but storm or hurricane-induced erosion, property damage, and loss of life capture our immediate attention. Fortunately, increased capabilities in weather forecasting have resulted in a decrease in the loss of life from storms over the last century in the United States. However, increasing population growth and development in coastal regions has led to an increase in property damage from extreme weather events over this same time period (Figure 1; http://hurricanes.noaa.gov/prepare/).

It has been predicted that global sea level will rise by approximately 1 meter in the next century, meaning that along the U.S. Atlantic and Gulf coasts, the shoreline will migrate approximately 2000 meters (6560 ft) inland. Compound this phenomenon with the anticipated increased hurricane activity in the Atlantic, and the long-term prognosis for coastal development begins to look precarious.

To further complicate matters, human attempts to slow beach erosion often exacerbate the problem. Construction of seawalls, jetties, revetments, and groins may stabilize local portions of the beach but can cause problems in adjoining areas. Seawalls, which parallel the shoreline, may temporarily protect property but eventually lead to the loss of the beach, a process called "Newjerseyization," named for the state where seawall construction has been a common practice in postponing property loss (Figure 2).

Beach Renourishment

Although many states no longer permit construction of hardened structures such as sea walls and groins, the alternative practice of beach "renourishment" has become increasingly common in

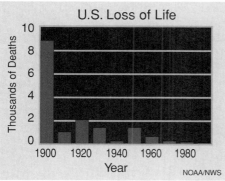

Figure 1

The graphs above from the NOAA National Weather Service illustrate the great increase in property damage but reduction in the loss of life associated with U.S. hurricanes since 1900. (NOAA)

Figure 2

Newjerseyization is evident at high tide in North Myrtle Beach South Carolina as erosion has removed the beach up to the seawall. Inset, erosion has excavated the foundation to a modern beachfront high rise in North Myrtle Beach South Carolina. *(Photo by Daniel Abel)*

the last several decades. Beach renourishment involves bringing in sand, by dredge, barge, or truck, to build up the height and breadth of disappearing beaches. The sand may come from inland sources, or from subtidal shelf regions. This practice is expensive and temporary, but it provides a key benefit compared to constructing hard structures in that it allows the beach to continue to exist. This benefit of beach renourishment, combined with a significant federal subsidy, is an especially appealing alternative in areas which depend on tourism.

The U.S. Army Corps of Engineers is responsible for designing and implementing renourishment operations. Much controversy exists regarding the efficacy and accuracy of the Corps' renourishment practices. One of the most common problems is that the Corps' estimated renourishment interval tends to be considerably longer than the actual interval before the next required renourishment; that is, the Corps overestimates how long a beach will last after renourishment. As of 1987, 400 million cubic yards (365×10^6 m^3) of sand have been used to renourish beaches in the U.S. By 1995, the total cumulative national cost of continued renourishment approached $5 billion.

Renourishment practices have led to a false sense of security for many beachfront residents, and the continued subsidy of renourishment by the federal government has had the effect of stimulating additional beachfront development. Technically, we

may be able to continue in the near future to repair the localized storm impact erosion to prime tourism locales (win the battle), but the reality of global sea level rise means that in the future we will eventually have to abandon current beachfront property (lose the war), or install an extremely expensive series of dikes, as they have done in Holland, to protect our shoreline. The Dutch have spent around $15 billion during the 90's, which is around $1,000 for each citizen, suggesting that we might have to spend $270 billion to "protect" our coasts!

Beach renourishment may have both direct and indirect effects upon flora and fauna of the sandy shore. The direct effects of beach renourishment include impacts that occur during dredging of subtidal materials, as well as those that occur during emplacement. Many coastal dredging operations which obtain sand material for beach renourishment impact species in the process. Certain types of dredges can cause high mortalities of female sea turtles during the nesting season. Many of the subtidal invertebrate organisms collected in the dredge material die after emplacement into the intertidal environment.

Go to our website to continue the analysis of this issue.

See the complete issue at:
www.prenhall.com/thurman

12 The Shore

The beach at Malibu, California. (Photo by Harold V. Thurman.)

Throughout our history, we have been attracted to seacoasts for their moderate climate, seafood, recreational opportunities, and commercial benefits. Most of the world's population lives near coastal areas. In the United States over the last 50 years, populations have shifted toward the coasts so that over 80 percent of residents now live within easy access. About 50 percent of the world's population lives within 200 km (125 mi) of the shoreline.

Because development has been carried out with little understanding of the dynamics and ecology of coasts, there is constant tension between nature and human structures. In a few years, development can destroy or irreparably damage fragile ecosystems; likewise, natural processes that shape the coastlines, can, in a few short hours, cause billions of dollars of damage and the loss of many lives.

Smarter land use policies, in harmony with coastal processes, are key to ensuring that our coastlines are preserved for the enjoyment of future generations. Unfortunately, current policies encourage land misuse. In the United States, the National Flood Insurance Program (NFIP) was meant to encourage people to build safely inland of flood-threatened areas in return for federally subsidized flood insurance. Instead, the program has resulted in subsidized overdevelopment of coastal areas that is risky, but lucrative. If global climate change results in accelerated sea level rise, costs and losses will escalate.

Plate Tectonics and the Coasts

As the waves beat against the rocks at the shore, they cause erosion. This erosion produces sediment that is transported along the shore and deposited in areas where wave energy is low. All shores undergo some degree of erosion and deposition. Shores that are often described as primarily erosional have well-developed cliffs and are generally in areas where tectonic uplift, or emergence, of the coast is occurring. Much of the U.S. Pacific Coast is classified as primarily erosional. In contrast, along the U.S. southeastern Atlantic Coast and Gulf Coast, sand deposits and offshore barrier islands are common. These coasts are submerging because of a gradual tectonic subsidence of the shore. Although we classify such coasts as primarily depositional, erosion can be a major problem when human developments interfere with natural coastal processes. The most dramatic examples of sea level changes during the past 3000 years have been due to tectonic processes. To understand the underlying tectonic cause of emergence or submergence, we need to return briefly to the concept of global plate tectonics.

Atlantic-Type (Passive) Coasts

We can readily understand this emergence–submergence pattern of tectonic plates if we consider the breakup of Pangaea that initiated the formation of the Atlantic Ocean. Recall that the lithosphere thickens, the ocean deepens, and heat flow decreases as lithospheric plates cool with age or with increasing distance from the spreading centers (see

Chapter 3). As a result of this process, the ocean floor of this passive margin, the Atlantic-type margin, is thought to have subsided about 3 km (2 mi) over the past 150 million years. Furthermore, erosion of the continent has produced a tremendous volume of sediment that has accumulated to a maximum thickness of 15 km (9.3 mi) along the Atlantic Coast. Such a thick deposit has been made possible by the plastic asthenosphere, which adjusts isostatically to allow the lithosphere to flex downward to accommodate its sediment burden. This thick sedimentary wedge underlying the continental shelf, slope, and rise has been exploited for large petroleum reserves in the Gulf of Mexico.

Sea-floor spreading and isostatic sinking of the crust account for the subsidence of the Atlantic and Gulf coasts. Although the rate of subsidence due to cooling decreases with time and the rate of subsidence due to sediment loading depends on sediment supply, subsidence continues until tectonic or isostatic conditions change (Figure 12-1a).

Pacific-Type (Active) Coasts

In contrast to passive, subsiding, Atlantic-type coasts, the Pacific-type coast, or active coast, is characterized by intense tectonic activity (Figure 12-1b). Along such margins, volcanoes erupt and earthquakes rumble. The thick, broad sediment wedge characteristic of the Atlantic Coast is poorly developed along a Pacific-type coast because ocean crust is being destroyed, the lithospheric plate is subducting, and compressive forces push crust upward, causing an emergent coast.

Along active margins, mountain ranges are characteristically aligned parallel to the continental margin. In all cases where continental margins are associated with plate convergence, the compressive forces produce uplift. Along the California coast, the process has been further complicated by the development of the San Andreas Fault. This fault has made possible the local subsidence of the depositional basins in the Los Angeles area.

Emergent and Submergent Features

Along some coasts are flat platforms, called **marine terraces**, backed by cliffs. Stranded beach deposits and other evidence of marine processes may exist many meters above the present shoreline. These features characterize **emerging shorelines**. They have reached their present positions relative to the existing shoreline by an uplift of the continent, a lowering of sea level, or a combination of the two.

In other areas, we find underwater **drowned beaches** and submerged dune topography. These features, along with the drowned river mouths along the present shoreline, indicate **submerging shorelines**, those that are sinking below sea level. This submergence must be due to subsidence of the continent, a rise in sea level, or a combination of the two.

Attempts to determine the causes of change in relative levels of the ocean and the continent have not met with great success. Whether the shoreline has emerged or submerged because of a falling or rising sea level or an

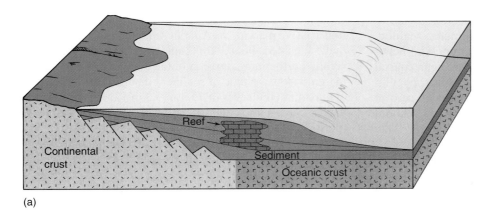

(a)

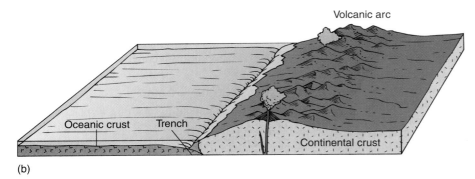

(b)

Figure 12-1   Atlantic- and Pacific-type Coasts

(a) Atlantic-type margin. This schematic shows why the U.S. Atlantic coast is slowly subsiding. When a continent is moving away from a spreading center like the Mid-Atlantic Ridge, its trailing edge subsides due to cooling and the weight of accumulating sediment. A low level of tectonic deformation, earthquakes, and volcanism accompanies such conditions, making the coast far more quiet and stable than a Pacific-type margin. (b) Pacific-type margin. This schematic shows why the U.S. Pacific coast is slowly rising, or emerging. Uplift is typical of continental margins where plate collisions occur. Earthquakes, volcanic action, and mountain chains paralleling the coast are common characteristics of such margins.

emerging on subsiding continent in a particular region cannot be determined by examining the coastal features in that area, since both processes produce the same end results (Figure 12-2).

Coastal Terminology

We usually speak of the shore, generally or perhaps even poetically, to describe the boundary between land and sea. Technically, the **shore** is the zone that lies between the lowest tide level (low tide) and the **coastline**, the highest elevation on land that is affected by storm waves. From this point landward is the **coast**, which extends inland as far as ocean-related features can be found (Figure 12-3). The width of the shore varies between a few meters and hundreds of meters. The width of the coast may vary from less than 1 km (0.6 mi) to many tens of kilometers.

The shore is divided into the **foreshore** (also called the intertidal or littoral zone), that portion exposed at low tide

and submerged at high tide, and the **backshore**, which extends from the normal high tide to the coastline. The **shoreline** is the water's edge and migrates back and forth with the tide. The **nearshore** zone is that region between the low-tide shoreline and breakers. Beyond the low-tide breakers is the **offshore** zone.

A **beach** is a sediment deposit on the shore area. It consists of wave-worked particles that are transported along the coast. If a coast was previously erosional, a **wave-cut bench,** a flat, wave-eroded surface, would underlie the beach. A beach may continue from the coastline across the nearshore region to the line of breakers.

All coastal regions follow the same general developmental path. As long as there is no change in the elevation of the landmass relative to the ocean surface, cliffs continue to erode and retreat until the beaches widen sufficiently to prevent waves from reaching them. The eroded material is carried from high-energy areas and deposited in low-energy areas.

(a)

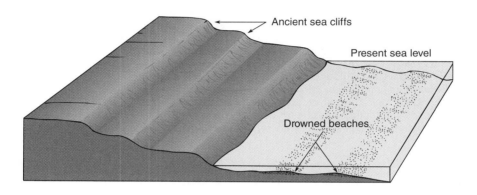

(b)

Figure 12-2   Evidence of Changing Shoreline Levels
....
(a) Marine (wave-cut) terraces on San Clemente Island south of Los Angeles, California. Once at sea level, the highest terraces in the background are now at an elevation of about 400 m (1320 ft). (b) Marine terraces bounded by ancient seacliffs mark ancient shorelines as do drowned beaches that lie below sea level.   *(Part (a) by John S. Shelton)*

Wave Erosion

Due to wave refraction (see Chapter 10), wave energy is concentrated on **headlands,** while energy reaching the shore in bays is reduced. As wave energy is concentrated on the headlands, erosion occurs and the shoreline retreats; the reduced wave energy in bays leads to deposition. On a day-to-day basis, the greatest concentration of wave energy is in the foreshore region. However, during rare periods when storm waves batter the shore, more erosion may occur across the entire shore in one day than occurs from average wave conditions over a period of years.

The cliffs shown in Figures 12-2 and 12-3 are called **wave-cut cliffs** because they are formed by wave action undercutting the base of the cliff. The cliff develops as the upper portions collapse after being undermined. The undermining may be evident in the form of a notch at the cliff base that also may be characterized by **sea caves.** Such caves are most commonly cut into hard rocks whereas softer rocks collapse. Further wave action may develop the caves into openings running through the headlands,

called **sea arches,** and these eventually crumble to leave **sea stacks** (Figure 12-4). Such remnants rise from the relatively smooth wave-cut bench that is formed into the bedrock by wave erosion (Figure 12-3).

The rate of wave erosion is determined by the following variables:

1. Degree of exposure of the coastal region to the open ocean; fully exposed coasts receive higher-energy wave action.

2. Tidal range—given the same amount of wave energy, a small tidal range will erode much more rapidly than one with a large tidal range. The smaller tidal range intensifies wave energy over a much narrower shore. (Although high-velocity tidal currents may develop in areas where a large tidal range exists, such currents are of limited importance in eroding the coastline.)

3. Composition of coastal bedrock—crystalline igneous rocks such as granite, metamorphic rocks, and hard

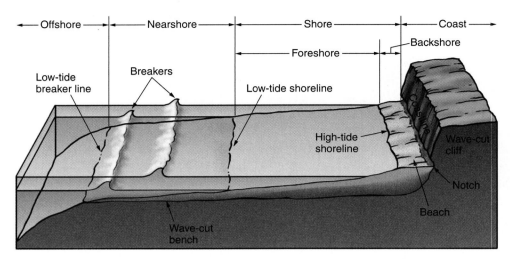

Figure 12-3 Landforms and Terminology of Coastal Regions
The coastline marks the most landward evidence of direct erosion by ocean waves. It separates the shore from the coast. The shore extends from the coastline to the low-tide shoreline (water's edge). It is divided into the backshore, above the high-tide shoreline, which is covered with water only during storms, and the foreshore (intertidal or littoral zone). Never exposed but affected by waves that touch bottom is the nearshore, which extends seaward to the low-tide breaker line. The greatest amount of sediment transport as beach deposit occurs within the shore and nearshore zones. Beyond the nearshore lies the offshore region, where depths are such that waves rarely affect the bottom.

sedimentary rocks are relatively resistant to erosion and produce a rugged shoreline topography. Weak sedimentary rocks such as sandstone and shale are more easily eroded; they produce a gentler topography and are associated with more extensive beach deposits.

Longshore Drift

Along most coastlines, waves strike the shore at an angle. This sets up a longshore movement of water, called the **longshore current**, which moves parallel to the shore between the shoreline and the breaker line. Coastal erosion produces large amounts of sediment. Sediment delivered to the ocean by rivers adds to the supply. All of this sediment is distributed along the continental margin by the longshore current.

At the landward margin of the surf zone, the **swash**, a thin sheet of water, moves sediment up onto the exposed beach at an angle. However, gravity pulls the backwash with its sediment load straight down the beach face. As a result, swash-transported pebbles and sand grains move in a zigzag pattern along the shore in the same direction as the longshore current within the surf zone (see the zigzag longshore drift line in Figure 12-5). The net direction of annual longshore currents is southward along both the Atlantic and Pacific shores of North America. The velocity of the longshore current increases with increased beach

slope, angle of breakers to the beach, and wave height. The current's velocity decreases with increasing wave period.

Longshore drift refers to the movement of sediment by the process just described. The amount of longshore drift in any coastal region is determined by an equilibrium between erosional and depositional forces. Any interference with the movement of sediment along the shore destroys this equilibrium, causing a new erosional and depositional pattern to form.

Rip Currents

As noted, longshore current water flows up onto the shore and then runs back into the ocean. This backwash of water generally finds its way into the open ocean as a thin "sheet flow" across the ocean bottom. However, some of this water flows back in local **rip currents**, strong narrow currents occurring either perpendicular to, or at an angle to the coast where topographic lows or other conditions allow their formation. Rip currents may be less than 25 m (80 ft) wide and can attain velocities of 7 to 8 km/h (4–5 mi/h). They do not travel far from shore before they break up. If a light-to-moderate swell is breaking, numerous rip currents of moderate size and velocity may develop. A heavy swell usually produces fewer, more concentrated rip currents (Figure 12-5). Rip currents that occur during heavy swell are a significant hazard to coastal swimmers. However, swimmers who realize they are caught in a rip current

Figure 12-4   Sea Arch and Sea Stack along the Coast of Iceland

If one ever doubted the erosive power of ocean waves, here is ample evidence.   *(Bruce F. Molnia, Terra-photographics/BPS.)*

and do not panic can escape by swimming parallel to the shore for a short distance (simply swimming out of the current). Another option is to ride out the current until it becomes less intense, because rip currents extend only a short distance from shore.

Beach Composition and Slope

The sediment deposited on a beach may be derived from a local source or it may be transported by longshore drift. Where the sediment is provided by coastal mountains, beaches are composed of mineral particles from the rocks of those mountains. This sediment may be relatively coarse in texture. Sediment in coastal regions provided primarily by rivers that drain lowland areas is normally finer in texture. Often, mud flats develop along the shore because only tiny clay-sized and silt-sized particles are emptied into the ocean. In low-relief, low-latitude areas such as southern Florida, where there are no mountains or other sources of rock-forming minerals nearby, most beach material is derived from the remains of the organisms that live in the coastal waters. Beaches in these areas are composed predominantly of shell fragments and the remains of microscopic animals. Many beaches on volcanic islands in the open ocean are composed of dark fragments of the basaltic lava that makes up the islands and of coarse debris from coral reefs that develop around islands in low latitudes.

The slopes of beaches are closely related to the size of the particles of which they are composed (Table 12-1). Waves washing onto the beach carry sediment, thereby increasing the slope of the beach. If the backwash returns as much sediment as the waves carried in, the beach has reached equilibrium and will not steepen. A beach composed of fine-grained sand that is relatively angular will have a gently sloping, firm surface. Because these small grains interlock closely, little of the swash sinks down among the grains. Most of it runs back down the slope to the ocean, possessing enough energy to maintain equilibrium on a gentle slope. The backshores of such beaches are usually nearly horizontal. Beaches composed of coarse sands or pebbles usually contain more rounded particles that are more loosely packed. The swash quickly percolates into such deposits as it moves up the beach slope. Deposition of particles by the swash continues until the

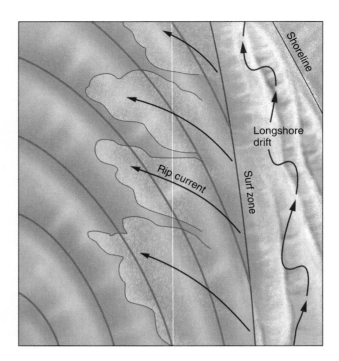

Figure 12-5   Longshore Currents and Rip Currents

In this example, as waves (curving blue lines) approach shore from a southerly direction, they produce a longshore current and longshore drift that flow northward. Water in the current follows a zigzag path; waves push it up the beach slope from the direction of approach, but it runs straight back down the face of the slope under the influence of gravity. The longshore drift of sediment particles carried by the longshore current follows a similar zigzag path. Water of the longshore current finds its way offshore by passing through topographic lows as strong seaward flows called rip currents. These currents are visible in the photograph as jets of water that appear light in color because of the turbidity from sediment they have resuspended from the ocean floor. The arrows in the diagram show the path of the four rip currents.   *(Courtesy of Scripps Institution of Oceanography, University of California, San Diego)*

beach slope becomes steep enough that the speed of the gravity-driven backwash running to the ocean is equal to the speed of the wave-driven swash up the beach. Such

beaches are usually much less firm than beaches composed of finer material, and the backshores slope significantly toward the coastline.

Special Geomorphic Features

Numerous depositional features are partially or wholly separated from the shore. These are deposited by the longshore drift and through other processes that are not well understood. Some of these are shown in Figures 12-6 and 12-7.

Spits and Tombolos   A **spit** is a linear ridge of sediment attached to land at one end. The other end points in the direction of longshore drift and ends in the open water (Figure 12-6). Spits are simply extensions of beaches into the deeper water near the mouth of a bay. Spits grow because particles are deposited when longshore currents slow as they turn corners at the end of the existing beach or point of land. The open-water end of the spit normally curves into the bay as a result of current action. If tidal currents or currents from river runoff are too weak to keep the mouth of the bay open, the spit may eventually extend across the bay and tie to the mainland, thus completely cutting off the bay from the open ocean. The spit then has become a **bay barrier**.

TABLE 12-1	The Relationship of Particle Size to Beach Slope.	
Wentworth Particle Size	(mm)	Mean Slope of Beach
	256	
Cobble		24°
	64	
Pebble		17°
	4	
Granule		11°
	2	
Very coarse sand		9°
	1	
Coarse sand		7°
	0.5	
Medium sand		5°
	0.25	
Fine sand		3°
	0.125	
Very fine sand		1°
	0.063	

Source: After Table 9, "Average beach face slopes compared to sediment diameters" from *Submarine geology,* 3rd ed. By Francis P. Shepherd, p. 127. Copyright © 1948, 1863, 1973 by Francis P. Shepherd. Reprinted by permission of Harper & Row, Publishers, Inc.

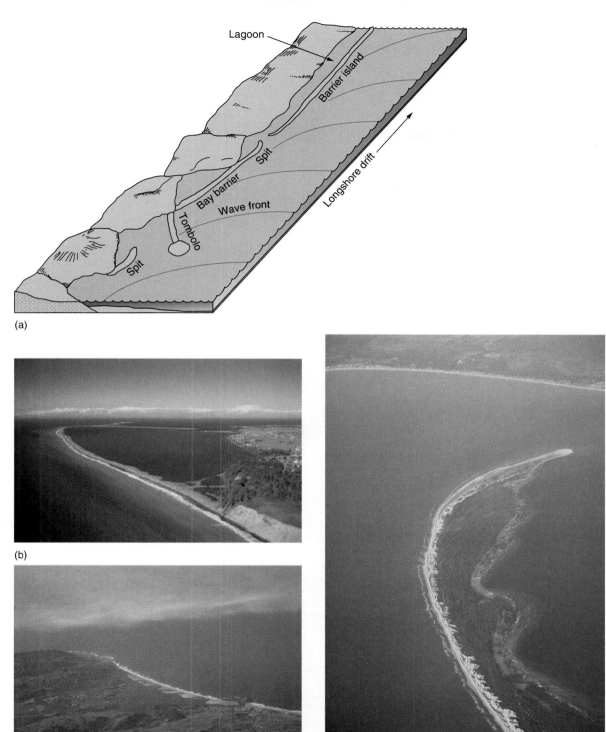

(a)

(b)

(c)   (d)

Figure 12-6   Spits, Tombolos and Barrier Islands

(a) Diagram showing basic structures of spits, tomobolos, bay barriers, and barrier islands. (b) Dungeness Spit, Dungeness National Wildlife Refuge, WA.   (*F. Stuart Westmoreland, courtesy Photo Researchers, Inc.*) (c) Bay barrier, Morro Bay, CA.   (*Eda Rogers*) (d) Barrier Island off Gulf Coast of Florida.   (*Eda Rogers*)

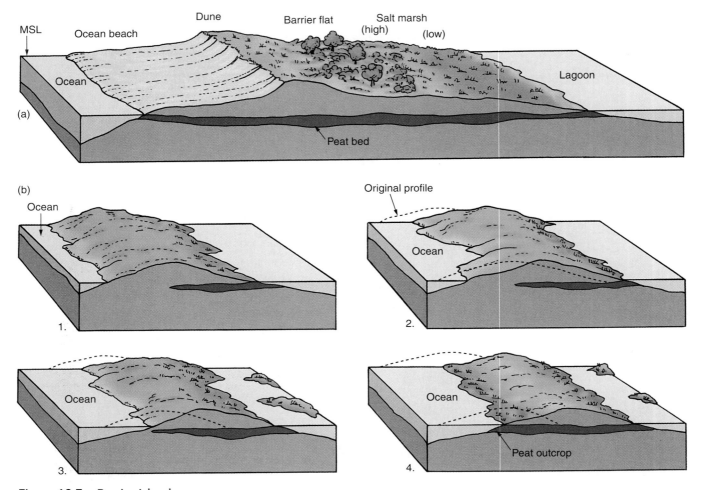

Figure 12-7   Barrier Islands

(a) Cross section through a barrier island. The major physiographic zones of a barrier island are the ocean beach, dunes, barrier flat, high salt marsh, and low salt marsh. MSL is mean sea level. The peat bed represents ancient marsh environments that have been covered by the island as it migrates toward the mainland with rising sea level. (b) 1. Barrier island before overwash, with salt marsh development on mainland side toward the lagoon to the right. 2. Overwash erodes seaward margin of barrier island, cuts inlets, and carries sand into the lagoon, covering much of the existing marsh. 3. New marsh forms in the lagoon that has retreated toward the mainland 4. With continued rise in sea level and migration of the barrier island toward the mainland, peat formed from previous marsh deposits may be exposed by erosion of the foreshore of the ocean beach.

A **tombolo** is a sand ridge that connects an island either to another island or to the mainland (Figure 12-6). Tombolos are usually aligned at a large angle to the main shore because they are usually sand deposits laid down in the wave-energy shadow of the island. Thus, their alignment is indicative of the average direction of wave approach.

Barrier Islands   Long offshore deposits of sand lying parallel to the coast are called **barrier islands** (Figure 12-6d). They are well named for they do constitute a barrier to storm waves that otherwise would severely assault the mainland shore. The origin of barrier islands is complex, and several explanations have been proposed for their existence. However, it appears that many barrier islands developed during the worldwide rise in sea level that began with the melting of the most recent major glaciers, some 18,000 years ago.

Barrier islands are nearly continuous along the Atlantic Coast of the United States. They extend around Florida and along the Gulf of Mexico coast, extending well south of the Mexican border. Barrier islands may exceed 100 km length (60 mi) and have widths of several kilometers. Examples of such features are Fire Island off the New York coast, North Carolina's Outer Banks, and Padre Island off the coast of Texas.

A typical barrier island has the following physiographic features, shown in Figure 12-7a. From the ocean landward, they are: (1) ocean beach, (2) dunes, (3) barrier flat,

(4) high salt marsh, (5) low salt marsh, and (6) lagoon between the barrier island and the mainland. We will look briefly at each portion.

The ocean beach is typical of the beach environment discussed earlier. During the summer, as gentle long-wavelength waves carry sand to the beach, it widens. Higher-energy, steeper winter waves carry sand offshore and produce a narrower beach.

Winds blow sand inland during dry periods to produce dunes, which are stabilized by dune grasses. These plants can withstand salt spray and burial by sand. Dunes are the lagoon's primary protection against excessive flooding during storm-driven high tides. Numerous passes exist through the dunes, particularly along the southeastern Atlantic Coast, where dunes are less well developed than to the north.

Behind the dunes, the barrier flat forms as the result of deposition of sand blown from the dunes by onshore winds or washed through inlets and over dune tops during storms. These flats are quickly colonized by grasses. During storms, these low barriers are washed over by seawater. If the frequency of storm overwash decreases, the plants undergo natural biological succession, the grasses being successively replaced by thickets, woodlands, and even forests.

Salt marshes are intertidal grasslands that typically lie inland of the barrier flat. They are divided into the low marsh, extending from about mean sea level to the high neap-tide line, and the high marsh, extending to the highest spring-tide line. The low marsh is by far the most biologically productive part of the salt marsh. New marshland is formed as overwash carries sediment into the lagoon, filling portions that become intermittently exposed by the tides. Marshes may be poorly developed on parts of the island that are far from floodtide inlets. Their development is greatly restricted on barrier islands where people perform artificial dune enhancement and fill inlets, activities that prevent overwashing and flooding.

Because of subsidence, the shoreline and barrier islands along the southeastern North American coast are migrating landward. This is clearly visible to those who build structures on these islands. Other evidence for such migration can be seen in peat deposits, remnants of old marshes that lie beneath the barrier islands. As salt marsh is the only barrier island environment in which peat forms, the island must have moved inland over previous marsh development. A possible cycle of salt marsh formation and destruction by landward migration of a barrier island is presented in Figure 12-7b.

Deltas Some rivers carry more sediment than can be distributed by the longshore current. Such rivers dump sediment at their mouths, forming a **delta** deposit (Figure 12-8). One of the largest such features on Earth is that produced by the Mississippi River where it empties into the Gulf of Mexico. Deltas are fertile, flat areas that are subject to periodic flooding. Delta formation begins when a river has filled its mouth with sediment. Once the delta forms, it grows through the distribution of sediment by forming **distributaries**, branching channels that radiate out over the delta (Figure 12-8a, b). Distributaries lengthen as they deposit sediment and produce fingerlike extensions to the delta. When the fingers get too long, they become choked with sediment. At this point, a flood may easily shift the distributary course and provide sediment to low-lying areas between the fingers. Where depositional processes dominate over coastal erosion and transportation processes, a bird's foot, Mississippi-type delta results.

Where erosion and transportation processes exert a significant influence, a delta shoreline will be smoothed to a gentle curve, like that of the Nile River Delta in Egypt (Figure 12-8c). The Nile Delta is presently eroding due to human intervention—sediment is being entrapped behind the Aswan High Dam, completed in 1964. Prior to building of the dam, generous volumes of sediment were carried by the Nile into the Mediterranean Sea. Erosion has had both negative and positive effects. On one hand, the delta is being destroyed. Reduced sediment availability may create significant problems along the eastern end of the delta, where subsidence has occurred at a rate of 0.5 cm (0.2 in.)/yr for 7500 yr. At the present rate of inundation, the sea may move inland 30 m (100 ft) during the next century. On the other hand, the erosion may have made possible the discovery of the ruins of the ancient Greek city of Alexandria beneath the Mediterranean waters at the edge of the delta.

Classification of Coasts

According to the classification of coasts developed by Francis P. Shepard, all coasts can be divided into two basic categories, primary and secondary. **Primary coasts** are young features that have been formed by nonmarine processes. **Secondary coasts** have aged enough that physical or biological marine processes have obliterated the nonmarine character of the coast.

Table 12-2 shows the nonmarine processes that produce primary coasts. Land erosion coasts such as drowned rivers (Chesapeake Bay) or drowned glacial erosion coasts (Puget Sound) were formed by an ocean that began rising 18,000 years ago as the last Pleistocene ice advance began to wane. Rising water invading these nonmarine landforms reached its present level about 3000 years ago, and sea level has been relatively stable since then. Subaerial deposition coasts, glacial deposition coasts, volcanic coasts, coasts shaped by earth movements, and ice coasts all reflect coastlines that have been formed recently by nonmarine geological processes (Figure 12-9).

With time, exposure of primary coasts to the action of ocean waves destroys all evidence of the nonmarine process that produced them, converting them to secondary coasts. For instance, an irregular primary coast resulting from earth movements may be eroded sufficiently by wave action to produce a relatively straight cliffed coast if the

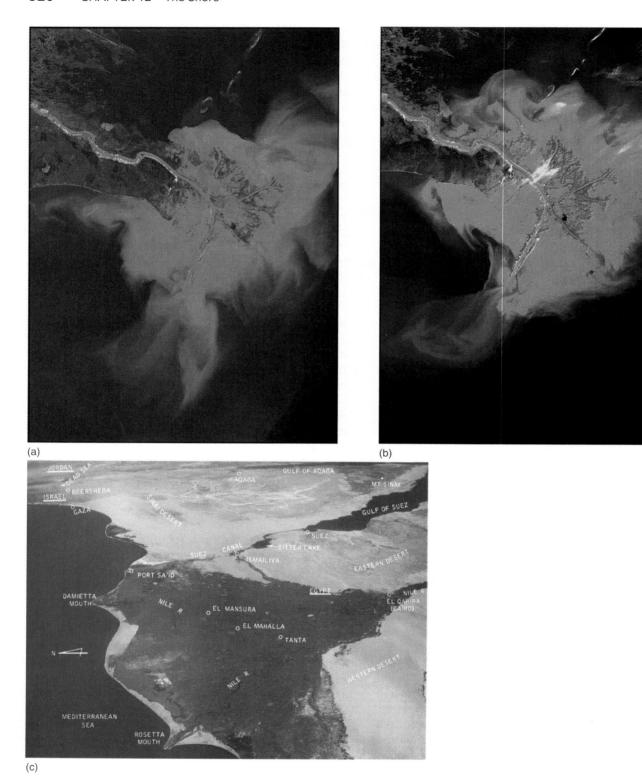

(a)

(b)

(c)

Figure 12-8 Deltas

(a, b) Digitate (fingerlike) structure of the Mississippi River Delta results from a low-energy environment. These images show the growth of the delta at the mouth of the Mississippi River's main channel during the 16 years between January 16, 1973 (a), and March 3, 1989 (b). (c) The relatively smooth, curved shoreline of the Nile River Delta dominates this view. The Mediterranean Sea is to the left, and the Gulf of Suez is in the upper right.   (*Part (a) GEOPIC, Earth Satellite Corporation; (b) Courtesy of NASA.*)

TABLE 12-2 Classification of Coasts.

I. Coasts Shaped by Nonmarine Processes (Primary Coasts)

    A. Land erosion coasts

        1. Drowned rivers

        2. Drowned glacial-erosion coasts

            a. Fjord (narrow)

            b. Trough (wide)

    B. Subaerial deposition coasts

        1. River deposition coasts

            a. Deltas

            b. Alluvial plains

        2. Glacial-deposition coasts

            a. Moraines

            b. Drumlins

        3. Wind deposition coasts

            a. Dunes

            b. Sand flats

        4. Landslide coasts

    C. Volcanic coasts

        1. Lava flow coasts

        2. Tephra coasts

        3. Coasts formed by volcanic collapse or explosion

    D. Coasts shaped by earth movements

        1. Faults

        2. Folds

        3. Sedimentary extrusions

            a. Mud lumps

            b. Salt domes

    E. Ice coasts

II. Coasts Shaped by Marine Processes or Marine Organisms (Secondary Coasts)

    A. Wave erosion coasts

        1. Straightened coasts

        2. Irregular coasts

    B. Marine deposition coasts (prograded by waves, currents)

        1. Barrier coasts

            a. Sand beaches (single ridge)

            b. Sand islands (multiple ridges, dunes)

            c. Sand spits (connected to mainland)

            d. Bay barriers

        2. Cuspate forelands (large projecting points)

        3. Beach plains

        4. Mud flats, salt marshes (no breaking waves)

    C. Coasts formed by biological activity

        1. Coral reef, algae (in the tropics)

        2. Oyster reefs

        3. Mangrove coasts

        4. Marsh grass

        5. Serpulid reefs

(a)

Figure 12-9   Examples of Primary and Secondary Coasts

Primary coasts: (a) Land erosion coast —drowned river. Chesapeake Bay and Delaware Bay formed when the rising ocean flooded the valley of the Susquehanna River and Delaware River   (*Courtesy Earth Satellite Corporation*)

underlying bedrock is of uniform strength. If the resistance of the bedrock varies along the coast, it will be irregular. Much of the Pacific Coast of the United States is classified as a wave erosion coast. From Massachusetts south, most of the Atlantic Coast is classified as a marine deposition coast. Coasts formed by marine biological activity are quite restricted along the U.S. shoreline; some of them are discussed elsewhere in the text.

Tectonic and Isostatic Movements of Earth's Crust

Changes in sea level relative to a continent may occur because of movement of the land, which is tectonic movement. Such movement could mean large-scale uplift or large-scale subsidence of major portions of continents or ocean basins, or it could mean a more localized deformation

(b)

(c)

(d)

(e)

Figure 12-9   (continued)

Primary coasts: (b) Subaerial deposition coast —delta. Satellite view of the Mississippi River Delta clearly shows the characteristic "bird's foot" shape.   (*NASA*) (c) Volcanic coast —lava flow. Lava from the Kilauea east rift flowing into the sea changes the contour of the island of Hawaii.   (*U.S. Geological Survey*) (d) Coast shaped by earth movements —faults. Fault movements associated with the Alaskan earthquake of 1964 uplifted the wave-cut bench to become a wave-cut terrace.   (*U.S. Geological Survey*) (e) Coast shaped by earth movements —faults. Satellite view of San Francisco Bay shows a coastline controlled by vertical movements of fault blocks within the San Andreas Fault system.   (*NASA*)

(f)

(g)

(h)

Figure 12-9 (continued)

Secondary coasts: (f) Wave erosion coast —irregular. View of coast along the island of Santa Cruz, California. (*James R. McCullagh*) (g) Marine deposition coast—barrier coast. Spit and bay barrier along the coast of Martha's Vineyard, Massachusetts. (*USD-ASCS*) (h) Coast formed by biological activity—mangrove coast. Mangrove trees on Isabela Island in the Galopagos extend the coast seaward. (*James R. McCullagh*)

of the continental crust involving folding, faulting, and tilting. Earth's crust also responds isostatically, by sinking under the accumulation of heavy loads of ice, sediment, or lava, or rising when such a load is removed.

There is evidence that during the last 2.5 to 3 million years, at least four major accumulations of glacial ice developed at high latitudes. Although Antarctica is still covered by a very large, thick glacial accumulation, much of the ice cover that once existed over northern Asia, Europe, and North America has disappeared. The most recent pe-

riod of melting began about 18,000 years ago. Accumulations of ice that were up to 3 km (2 mi) thick have disappeared from northern Canada and Scandinavia.

While these areas were beneath the thick ice sheet, the crust beneath them sank. Today, they are still recovering, slowly rising after the melting of the ice. Some geologists believe that by the time this isostatic rebound is finished, the floor of Hudson Bay, which is now about 150 m (500 ft) deep, will be above sea level. There is evidence in the Gulf of Bothnia between Sweden and Finland of 275 m

Figure 12-10   Extent of Ice Coverage During Most Recent Glacial Age

(a) Arrows show the general direction of ice flow. Coastlines are shown as they may have been at that time, when sea level may have been 120 m (390 ft) lower than at present. The present Hudson Bay and the Gulf of Bothnia are areas that were depressed beneath the weight of the glacial ice. With the ice now melted, these areas still are rebounding slowly, and eventually may rise above sea level—if another ice advance does not occur before rebound is complete. (b) Sea Level Change During the Most Recent Advance and Retreat of Pleistocene Glaciers. Sea level dropped worldwide by about 120 m (400 ft) as the last glacial advance removed water from the oceans and transferred it to continental ice sheets. About 18,000 years ago, sea level began to rise as the glaciers melted and water was returned to the oceans.

(a)

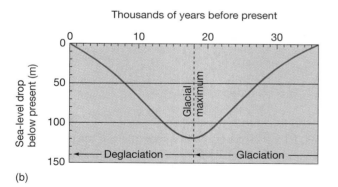

(b)

(900 ft) of isostatic rebound during the last 18,000 years (Figure 12-10a). Generally, tectonic and isostatic changes in the level of the shoreline are local or regional, confined to a segment of the shoreline of a given continent.

Eustatic Changes in Sea Level

A worldwide change in sea level can be caused by an increase or decrease of seawater volume, or by an increase or decrease in ocean basin capacity. Such a change in sea level is called **eustatic**. This term refers to a highly idealized phenomenon in which all of the continents remain static while the sea rises or falls. For example, small eustatic changes in sea level could be created by formation or destruction of large inland lakes. More important is the effect of glaciers. As they freeze, large glaciers tie up vast volumes of water, causing a eustatic lowering of sea level. During interglacial stages, like the present, glaciers melt, releasing great volumes of water that drain to the sea, causing a eustatic rise of sea level.

Changes in sea-floor spreading rates also can change sea level. Fast spreading produces larger rises, like the East Pacific Rise, that displace more water than slow-spreading ridges like the Mid-Atlantic Ridge. Thus, fast spreading produces a rise in sea level. Significant changes of sea level resulting from changes in spreading rate typically occur over hundreds of thousands or millions of years.

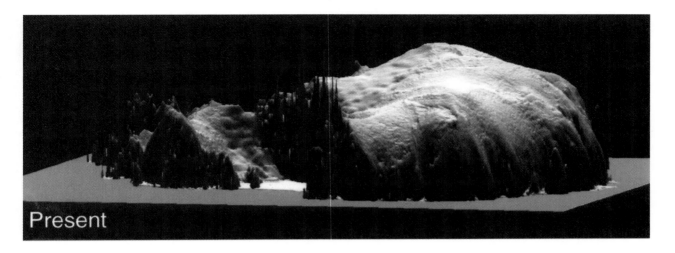

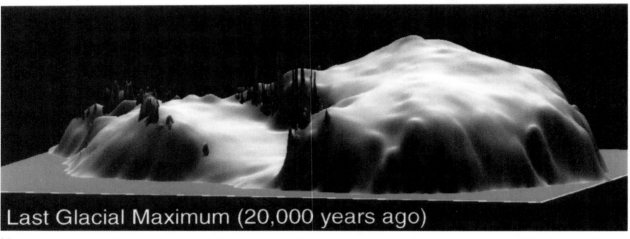

(c)

Figure 12-10   **(continued)**

(c) The Antarctic Ice Sheet at present and 20,000 years ago. The West Antarctic ice sheet (left side) has lost two-thirds of its mass.   *(Robert Bindschadler)*

During the Pleistocene Epoch, when glacial growth occurred, the amount of water in the ocean basin fluctuated considerably. We know this from the present elevations of marine features. Because the climate was colder during glacial times, we might account for some of the lowering of the shoreline by the contraction of the ocean volume as its temperature decreased. It has been calculated that for every 1°C (1.8°F) decrease in the mean temperature of the ocean water, sea level would drop 2 m (6.6 ft). Temperature indications derived from study of fossils in Pleistocene ocean sediments suggest that ocean surface temperature may have been as much as 5°C (9°F) lower than at present. Therefore, contraction of the ocean water may have lowered sea level by about 10 m (33 ft).

Although it is difficult to definitely state the range of shoreline fluctuation during the Pleistocene, there is cause to believe that it was at least 120 m (400 ft) below the present shoreline (Figure 12-10b). It is also estimated that if all the remaining glacial ice on Earth were to melt, sea level would rise another 60 m (200 ft). This would give a possible range of sea level during the Pleistocene of 180 m

(600 ft), most of which must be explained through the capture and release of Earth's water by glaciers. Figure 12-10c shows the fluctuations in the extent of the Antarctic Ice Sheet over the last 20,000 years. The West Antarctic ice sheet has lost about two-thirds of its mass, which is sufficient to account for 11 m of sea level rise.

During the last 18,000 years, ocean volume has increased due to the expansion of seawater resulting from warmer global temperatures and the melting of glacial ice; however, during the last 3000 years, sea level has probably not risen enough to impact coastal character. Local and regional tectonic changes may mask the effect of eustatic changes in sea level. Coastal regions rarely can be classified as purely emergent or submergent because many areas show evidence of both emergence and submergence in the recent past.

At present, there is much discussion about the results of historical research and ongoing monitoring programs. For example, there has been an apparent global warming of about 0.5°C (0.9°F) and a eustatic rise in sea level of 10 cm (4 in) since 1880 (Figure 12-11). In addition, atmospheric carbon dioxide levels have increased by 28 percent since

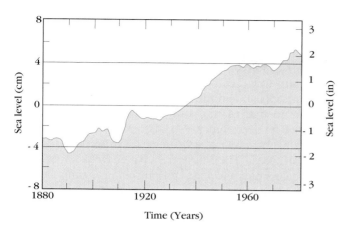

Figure 12-11   Sea Level Change from 1880 to 1980

Tide-gauge data, averaged over 5-yr periods, show that global mean sea level has risen about 10 cm (4 in) over the past 100 years. However, more recent data that remove the effect of isostatic recovery indicate that global sea level has risen 1.75 mm/yr during the last 50 yr. This rate is 1.5 times the rate shown in this figure.   *(After V. Gornits, S. Lebedeff, and J. Hausen, 1982. Science 215:1611–14.)*

1958, when monitoring began. The obvious question is: Are these observations related, and are they the result of an increased greenhouse effect? The answer is unknown, but the great lesson for humankind is that we cannot dominate nature; we must live with it. To do otherwise is futile. Some of the consequences of a rising sea level for coastal communities in the United States are discussed in the following sections and in "Global Warming and Sea Level Rise" on page 162.

U.S. Coastal Conditions

Whether the dominant process along a coast is erosion or deposition depends on the combined effect of the variables we have been discussing—degree of exposure to ocean waves, tidal range, composition of coastal bedrock, tectonic subsidence or emergence, isostatic subsidence or emergence, and eustatic sea level change.

In 1971, the U.S. Army Corps of Engineers reported on erosion of the U.S. coastline, which is 135,870 km long (84,240 mi, including Alaska but not Hawaii). They found over 24 percent of it to be "seriously eroding." Subsequent

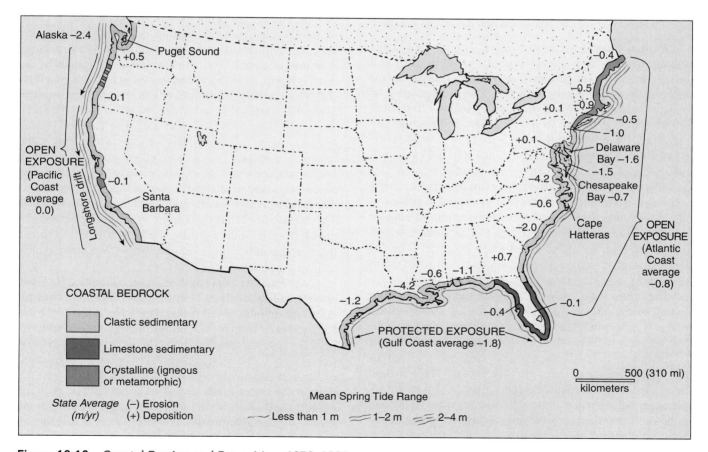

Figure 12-12   Coastal Erosion and Deposition, 1979–1983

For the Pacific, Atlantic, and Gulf Coasts, this map shows coastal bedrock type (red, yellow, blue), the mean spring-tide range (light blue lines), degree of exposure, and rates of erosion/deposition (m/yr). Net erosion (−) or net deposition (+) is shown for specific locations, along with average condition for each coast.

studies supported by the U.S. Geological Survey along the shores of the contiguous coastal states and Alaska determined the rates of shoreline change (Figure 12-12).

The Atlantic Coast

The U.S. Atlantic coast is quite a study of coastal processes in action (refer to Figure 12-12):

- Most of the Atlantic coastline is exposed to storm waves from the open ocean. However, barrier islands from Massachusetts southward protect the mainland from large storm waves.
- Tidal ranges generally increase from less than 1 m (3.3 ft) along the Florida coast to more than 2 m (6.6 ft) in Maine.
- The bedrock of most of Florida is a resistant limestone. However, from Florida northward through New Jersey, most bedrock is poorly consolidated sedimentary rocks formed in the recent geologic past. These rocks have poor resistance to erosion. As they erode, they become sand sources for barrier islands and other depositional features common along the coast.
- From New York northward, continental glaciers affect the coastal region directly. Many coastal features, including Long Island and Cape Cod, are glacial deposits, moraines left behind when the glaciers melted.

North of Cape Hatteras in North Carolina, the coast is subject to very high energy conditions during fall and winter when storms called *nor'easters* (northeasters) blow in from the North Atlantic. The great energy of these storms is manifested in waves up to 6 m (20 ft) high with a 1 m (3.3 ft) rise in sea level that follows the low pressure as it moves northward. Such high energy conditions significantly change coastlines that are predominantly depositional and cause considerable erosion.

Most of the Atlantic coast appears to be experiencing a sea level rise of about 0.3 m (1 ft) per century, primarily as a result of coastal subsidence. This condition may be reversed in northern Maine because of isostatic rebound due to the melting of the Pleistocene ice sheet. The Atlantic Coast has an average annual rate of erosion of 0.8 m (2.6 ft). In everyday terms, this means that the sea is migrating landward each year by about the length of your legs. The majority of East Coast states are slowly losing real estate. Virginia leads the way with a 4.2- m (13.7- ft) loss per year, which is largely confined to barrier islands.

Erosion rates for Chesapeake Bay are about average for the Atlantic coast, but Delaware Bay shows average erosion rates about twice the Atlantic coast average (1.6 m/yr). Of the observations made along the Atlantic coast, 79 percent indicated some degree of erosion. Delaware, Georgia, and New York are net depositional coasts, despite serious erosion problems in these states as well.

The Gulf Coast

The erosion/deposition story of the Gulf Coast is simpler. The Louisiana–Texas coastline is dominated by the Mississippi River Delta, which is being deposited in a microtidal and generally low energy environment. The tidal range is normally less than 1 m (3.3 ft). Except for the hurricane season (June–November), wave energy is generally low. Tectonic subsidence occurs throughout the Gulf Coast, and the average rate of sea level rise is similar to that of the southeast Atlantic coast, about 0.3 m (1 ft) per century.

The average rate of erosion is 1.8 m (6 ft) per year in the Gulf Coast. The Mississippi River Delta experiences the greatest rate; the state of Louisiana is losing an average of 4.2 m (13.7 ft) per year. With erosion made worse by the dredging of barge channels through marshlands, Louisiana has lost more than one million acres of delta since 1900. The Pelican State is now losing marshland at a rate exceeding 130 km^2 (50 mi^2) per year.

Although all Gulf states show a net loss of land and the Gulf Coast has a greater erosion rate than the Atlantic Coast, only 63 percent of the shore is receding because of erosion. The high average rate of erosion reflects the heavy erosion loss in the Mississippi River Delta area.

The Pacific Coast

In general, the Pacific Coast is experiencing less erosion than the Atlantic and Gulf Coasts. Along the Pacific coast, relatively young, weak marine deposits dominate the bedrock, with local outcrops of more resistant granite, metamorphic rock, and volcanic rock. Tectonically, the coast is rising, as evidenced by wave-cut terraces. Sea level still shows at least small rates of rise, except for a segment along the coast of Oregon and the Alaskan coast. The tidal range is mostly between 1 and 2 m. The Pacific Coast is open to large storm waves. High-energy waves may strike the coast in winter, with 1 m (3.3 ft) waves being normal. Frequently, the wave height will increase to 2 m, and a few times per year, 6 m (20 ft) waves hammer the shore. These high energy waves erode sand from many beaches. The exposed beaches, which are composed primarily of pebbles and boulders during the winter months, regain their sand as low energy summer waves beat more gently against the shore.

Many Pacific Coast rivers have been dammed for flood control and hydroelectric power generation. This has reduced the amount of sediment supplied by rivers to the sea for longshore transport. As a result, some areas now experience a severe erosion threat. With an average erosion rate of only 0.005 m/yr (0.016 ft/yr), and only 30 percent of the coast showing erosion loss, the Pacific Coast appears to be under much less of an erosion threat than the Atlantic and Gulf coasts. Yet, all along the coast, some areas experience high rates of erosion. Alaska is losing an average of 2.4 m/yr (7.9 ft/yr). Only Washington shows a net deposition. The long, protected Washington shoreline within Puget Sound helps skew the Pacific Coast figures. Although the average erosion rate for California is only 0.1 m (0.33 ft)/yr, over 80 percent of the coast is experiencing erosion, with local rates of up to 0.6 m/yr (2 ft/yr).

TABLE 12-3	Rates of Accretion + and Erosion − for Coastal Landform Types.

Region	Change (m/yr)
Mud Flats	−2.0
Florida	−0.1
Louisiana–Texas	−2.1
Gulf of Mexico	−1.9
Sand Beaches	−0.8
Maine–Massachusetts	−0.7
Massachusetts–New Jersey	−1.3
Atlantic Coast	−1.0
Gulf Coast	−0.4
Pacific Coast	−0.3
Barrier Islands	−0.8
Louisiana–Texas	−0.8
Florida–Louisiana	−0.5
Gulf of Mexico	−0.6
Maine–New York	+0.3
New York–North Carolina	−1.5
North Carolina–Florida	−0.4
Atlantic Coast	−0.8
Rock Shorelines	+0.8
Atlantic Coast	+1.0
Pacific Coast	−0.5

Erosion Rates and Landform Types

The variability we see in erosion rates along coasts is closely correlated with coastal landform types (Table 12-3).

- Deltas and mud flats composed of fine-grained sediment erode most rapidly, with mean erosion rates of 2 m (6.6 ft)/yr. This is not surprising, and it accounts for the fact the Gulf Coast is eroding at a greater rate than other coasts.
- Coarser deposits of sandy beaches and barrier islands erode more slowly (0.8 m/year or 2.6 ft).
- The rock shore of Maine, which is experiencing isostatic rebound, shows great resistance to erosion, with a deposition rate of 1 m/yr (3.3 ft/yr).
- On the Pacific Coast, the rocks are not very durable and erode at an average rate of 0.5 m/yr (1.6 ft/yr).

Coastal Development

The National Flood Insurance Program, established in 1968, has clearly accelerated coastal development. During the decade prior to passage of the plan, hurricanes killed 186 people and did $2.2 billion in damage. However, in the decade following its passage, after coastal de-velopment boomed, there were 411 deaths and $4.7 billion in property damage. This is a 121 percent increase in deaths and a 114 percent increase in damage. The reason, of course, was that the NFIP had encouraged overdevelopment, placing more people and buildings in harm's way.

The problem continues to worsen. Studies conducted by the General Accounting Office in 1983 into the soundness of the program showed that many insurance premiums should be increased by 800 percent. However, only new properties were subject to these higher rates. Rates have increased further in recent years, but the government still pays out billions with each flooding disaster in the form of grants and loan subsidies —to cover damage that would not have happened if our government had a sensible policy on seacoast development. The insurance remains a real bargain. For about $1000, the owner of a $200,000 coastal home can purchase insurance that would cost $18,000 from a private insurance company.

In addition to these benefits, which help owners replace structures built in locations that are clearly in danger of storm damage, the federal government subsidizes development of these areas by paying most of the cost of infrastructure development. (Infrastructure refers to water systems, sewage treatment facilities, highways, and bridges.) Further, the federal government assumes from 50 to 70 percent of the cost of erosion-control programs—and erosion is what coastal geology is all about.

In 1982, Congress passed the Coastal Barrier Rezoning Act. It excludes construction in undeveloped coastal areas from receiving federal assistance. The act also encourages states to develop setback regulations that require construction to be set back from the shore, realistically incorporating the threat of storm damage into all coastal building programs. See "Beaches or Bedrooms? The Dynamic Coastal Environment" on page 308–309.

Artificial Barriers

People continually modify coastal sediment erosion and deposition in attempts to improve or preserve their property (Figure 12-13). One device for doing so is a jetty, a structure built at an angle to the shore. Jetties constructed to protect harbor entrances from wave action are the most common barriers to the longshore transport of sediment. Many examples can be cited of beach destruction resulting from jetties. Along the southern California coast, excessive beach erosion related to the development of local harbor facilities is especially common. California's longshore drift is predominantly southward, so construction of a jetty traps sediment on its northern side, causing increased erosion on its southern side.

The harbor at Santa Barbara, California, provides an interesting example of damage caused by jetties. In Figure 12-13a, you can see the jetty constructed as a **breakwater**, to protect the harbor. The jetty greatly disturbed the equilibrium of this coastal region. The breakwater on the western side of the harbor caused an accumulation of the sand that otherwise migrated eastward along the coast. The beach to the

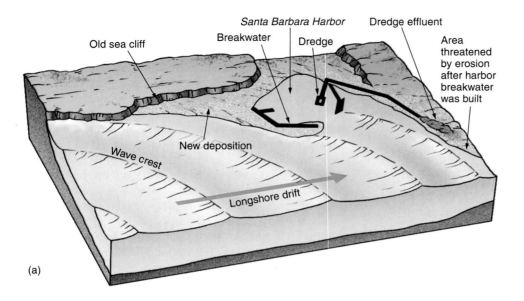

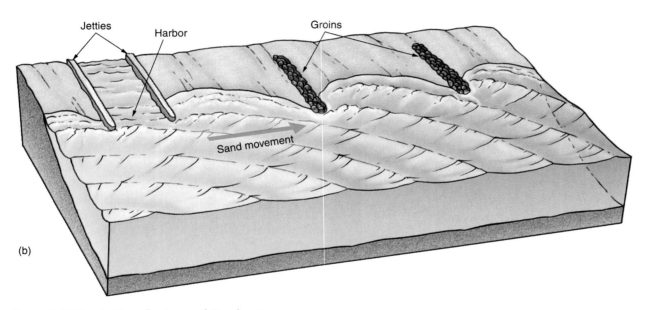

Figure 12-13 Jetties, Groins, and Breakwaters

(a) Santa Barbara Harbor. Construction of a breakwater to the west of Santa Barbara Harbor interfered with the eastward-moving longshore drift, creating a broad beach. Sand being deposited against the breakwater was no longer available to replace sand being removed by wave erosion to the east. As the beach extended around the breakwater into the harbor, dredging operations were initiated to keep the harbor open and to put the sand back into the longshore drift. This helped reduce coastal erosion east of the harbor. *(U.S. Coast and Geodetic Survey Chart 5161)* (b) Jetties and groins trap sand that would otherwise be moved along the shore by wave action. *(From Tarbuck, E. J., and Lutgens, F. K., Earth Science, 5th ed. Columbus: Merrill, 1988. Reprinted by permission.)*

west of the harbor continued to grow until finally the sand moved around the breakwater and began to fill in the harbor. While abnormal deposition occurred on the west, erosion proceeded at an alarming rate east of the harbor. The wave energy east of the harbor was no greater than before, but the sand that had formerly moved down the coast was no longer available because it was entrapped behind the break-

water. Dredging keeps the harbor from filling in, and to compensate for the deficiency of sand downcurrent from the harbor, dredged sand is pumped down the coast so it can re-enter the longshore drift and replenish the eroded beach. The dredging operation has stabilized the situation, but at a considerable expense. It seems obvious that any time people interfere with natural processes in a coastal region, they must

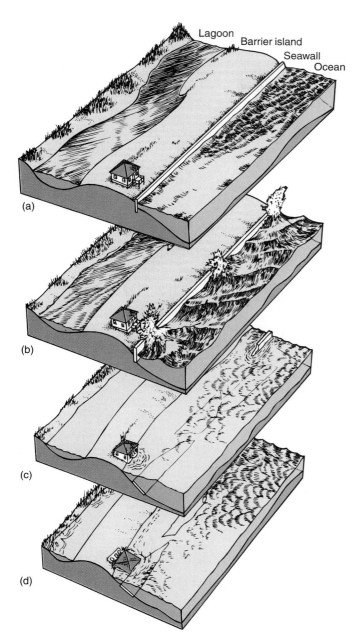

Figure 12-14 Seawalls

(a) A seawall is built along a beach to protect beachfront property. (b) The first large storm removes beach from the seaward side of the wall and steepen its seaward slope. (c) Eventually, the wall is undermined and falls into the sea. (d) The property is lost as the oversteepened beach slope advances landward in its effort to reestablish a natural slope angle.

provide the energy needed to replace that which they have misdirected through modification of the shore environment.

Along many coasts, small jettylike structures called *groins* have been constructed at intervals along the beach (Figure 12-13b). They help beach development by causing sand to accumulate on their upcurrent sides. Being shorter than jetties built to protect harbors, groins eventually allow sand to migrate over their tops and around their ends. An equilibrium may be reached that allows sufficient sand transport along the coast before excessive erosion occurs downcurrent from the last groin. Although some serious erosional problems have developed, groins usually cause much less beach erosion than a single large jetty.

One of the most destructive structures built to "stabilize" eroding shores is the **seawall** (Figure 12-14). Instead of extending out into the longshore current, as do groins and jet-

ties, seawalls are built parallel to the shore. The idea is to protect developments from the action of ocean waves. However, once waves begin breaking against a seawall, turbulence generated by the abrupt release of wave energy quickly erodes the sediment on its seaward side, causing the seawall to collapse into the surf. Where seawalls have been used to protect property on barrier islands, the seaward slope of the island beach is steepened, and the rate of erosion increased.

Although most local laws state that beaches belong to the public, government action allows owners of shore property to destroy beaches in an effort to provide short-term protection to their property. To add to the public insult, taxpayers subsidize property owners in many shore developments by helping to rebuild their storm-damaged properties at great public expense through the programs discussed earlier in this chapter.

(a)

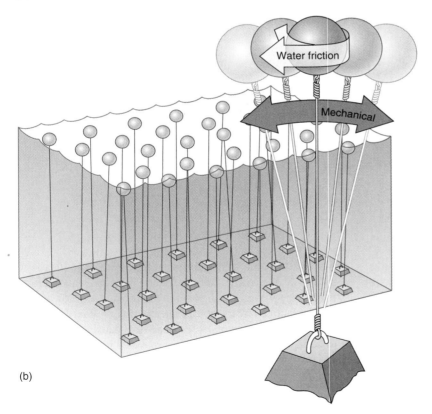

(b)

Figure 12-15 Tethered-float Breakwaters

(a) Deployment of a tethered-float breakwater in San Diego Harbor. (b) Artist's sketch of a tethered breakwater system incorporating 1.5 m (4.8 foot) spheres of steel placed 1.5 m apart. The spheres are anchored to the ocean floor and held just beneath the ocean surface. This system has the advantage over traditional breakwater structures of removing energy from waves without completely stopping the transport of sediment along the shore.
(Part (a) Scripps Institution of Oceanography, University of California, San Diego.)

There is an interesting alternative to the solid structures used to protect local shore areas from ocean waves. It is the tethered-float breakwater developed at Scripps Institution of Oceanography (Figure 12-15). Tested successfully in San Diego Harbor, these inverted pendulums protect against coastal wave erosion by absorbing energy from passing waves, while allowing the sediment to pass through unhindered. If they prove durable, they will be a welcome substitute for traditional breakwaters.

Summary

The region of contact between the oceans and the continents is marked by the shore, lying between the lowest low tides and the highest continental elevations affected by storm waves. The coast extends inland from the shore as far as marine-related features can be found. The shore is divided into the foreshore, extending from low to high tide; and the backshore, extending beyond the high-tide line to the coastline. Seaward of the low tide are the nearshore zone, which extends to the breaker line, and the offshore zone beyond.

Wave erosion of the shore produces a wave-cut cliff that constantly retreats, leaving behind such features as wave-cut benches, sea caves, sea arches, and sea stacks. Wave erosion increases with greater exposure of the shore to the open ocean, decreasing tidal range, and decreasing strength of bedrock.

As waves break at an angle to the shore, a longshore current is set up that produces a longshore drift of sediment along the shore. Deposition of sediment thus transported may produce beaches, spits, tombolos, and barrier islands. Viewed from ocean side to lagoon side, barrier islands commonly have an ocean beach, dunes, barrier flat, and salt marsh. Deltas form at the mouths of rivers that carry more sediment to the ocean than can be distributed by the longshore current.

Coasts can be classified as primary coasts or secondary coasts. Primary coasts include those developed by nonmarine processes; they include land erosion coasts, subaerial deposition coasts, glacial deposition coasts, volcanic coasts, coasts shaped by earth movements, and ice coasts. Secondary coasts include those where the nonmarine character has been destroyed by marine processes; they include wave erosion coasts, marine deposition coasts, and coasts formed by marine biological activity.

A sea level drop may be indicated by ancient wave-cut cliffs and stranded beaches well above the present shoreline. A sea level rise may be indicated by old submerged beaches, wave-cut cliffs, or drowned river valleys. Such sea level changes may result from tectonic processes causing local movement of the land mass, or from eustatic processes changing the amount of water in the oceans or the capacity of ocean basins. Melting of continental ice caps during the past 18,000 years has caused a eustatic rise in sea level of about 120 m (400 ft).

Along the Atlantic Coast, sea level is rising about 0.3 m (1 ft) every century along most of the coast; the average annual erosion rate is 0.8 m (2.6 ft). Along the Gulf Coast, sea level is rising at 0.3 m (1 ft) per century, and the average rate of erosion is 1.8 m (6 ft) per year. The Mississippi River Delta is eroding at a rate of 4.2 m (13.7 ft) per year, resulting in a large annual loss of wetlands. Along the Pacific Coast, the average erosion rate is only 0.005 m (0.016 ft) per year. Based on coastal landform types, mud flats erode more rapidly than sand beaches, barrier islands, or rocky shores.

Structures built along the shore, such as jetties to protect harbors and groins used to widen beaches, trap sediment on their upcurrent side, and erosion may then become a problem downcurrent. A less-damaging replacement for fixed breakwater structures might be the tethered-float breakwater.

Since passing the National Flood Insurance Program (NFIP) in 1968, the federal government has encouraged development of shores that are subject to storm damage. The result has been increased death, destruction, and cost. The direction of this and complementary programs must be reversed to improve the situation. One helpful piece of legislation was the Coastal Barrier Rezoning Act, which excludes construction in undeveloped coastal areas from receiving federal assistance.

Key Terms

Backshore	Drowned beach	Offshore	Shore
Barrier island	Emerging shorelines	Primary coasts	Shoreline
Bay barrier	Eustatic	Rip current	Spit
Beach	Foreshore	Salt marsh	Swash
Breakwater	Headland	Sea arch	Submerging shorelines
Coast	Longshore current	Sea cave	Tombolo
Coastline	Longshore drift	Sea stack	Wave-cut bench
Delta	Marine terrace	Seawall	Wave-cut cliff
Distributary	Nearshore	Secondary coasts	

Questions and Exercises

1. Why has the National Flood Insurance Program had a negative impact on the nation's shores?

2. Define the characteristics of the two major categories of coast classification, and list the subcategories of each.

3. Discuss the formation of such erosional features as sea cliffs, sea caves, sea arches, and stacks.

4. List and discuss three factors that affect the rate of wave erosion.

5. What is longshore drift, and how is it related to the longshore current?

6. List and define the depositional features spit, tombolo, bay barrier, and barrier island. Include discussion of how some barrier islands develop peat deposits running through them from ocean shore to marsh.

7. Discuss why some rivers have deltas and others do not. Also, include the factors that determine whether a "bird's foot" or a smoothly curved Nile-type delta will form.

8. Compare the causes and effects of tectonic vs. eustatic changes in sea level.

9. How does a Pacific-type margin differ from an Atlantic-type margin?

10. How does placement of breakwater or jetty, across the longshore current and drift, affect erosion and deposition?

References

Bindschadler, R. 1998. Future of the west Antarctic ice sheet. *Science* 282:5388, 428–29.

Bird, E. C. F., 1985. *Coastline Changes: A Global Review.* Chichester, U.K.: John Wiley & Sons.

Burk, K., 1979. The edges of the ocean: An introduction. *Oceanus* 22:3, 2–9.

Kuhn, G. G., and F. P. Shepard. 1984. *Sea Cliffs, Beaches, and Coastal Valleys of San Diego County: Some amazing histories and some horrifying implications.* Berkeley: University of California Press.

Leatherman, S. P. 1983. Barrier dynamics and landward migration with holocene sea-level rise. *Nature* 301:5899, 415–17.

May, S. K., et al. 1982. The coastal erosion information system. *Shore Beach* 50, 19–26.

Peltier, W. R., and A. M. Tushingham. 1989. Global sea level rise and the greenhouse effect: might they be connected: *Science* 244:4906, 806–10.

Roemmich, D., 1992. Ocean warming and sea level rise along the southwest U.S. Coast. *Science* 257:5068, 373–75.

Sahagian, D. L., F. W. Schwartz, and D. K. Jacobs. 1994. Direct anthropogenic contributions to sea level rise in the twentieth century. *Nature* 367:6458, 54–56.

Shepard, F. P. 1973. *Submarine Geology*, 3rd ed. NY: Harper & Row.

———. 1977. *Geological Oceanography.* N.Y.: Crane, Russak & Company.

Stanley, D. J., and A. G. Warne. 1993. Nile Delta: Recent geological evolution and human impact. *Science* 260:5108, 628–34.

Suggested Reading

Sea Frontiers

Carr, A. P., 1974. *The Ever-Changing Sea Level.* 20:2, 77–83. A well-presented discussion of the causes of sea-level change.

Emiliani, C., 1976. *The Great Flood.* 22:5, 256–70. An interesting discussion of the possible relationship of the rise in sea level resulting from the melting of glaciers 11,000–8,000 years ago and biblical and other ancient accounts of a great flood.

Feazel, C., 1987. *The Rise and Fall of Neptune's Kingdom.* 33:2, 4–11. A discussion of factors that change the level of the sea, including atmospheric conditions, currents, climate, ocean topography, and sea-floor spreading.

Fulton, K., 1981. *Coastal Retreat.* 27:2, 82–88. The problems of coastal erosion along the southern California coast are considered.

Goldsmith, V., 1992. *New York City is Flooding ... NOT.* 38:4, 24–31, 62. Focusing on New York, the author provides an interesting insight into the problem of sea level rise.

Grasso, A., 1974. *Capitola Beach.* 20:3, 146–51. The destruction of the beach at Capitola, California, shortly after the construction of a harbor by the Army Corps of Engineers at Santa Cruz to the north.

Mahoney, H. R., 1979. *Imperiled Sea Frontier: Barrier beaches of the east coast.* 25:6, 329–37. The natural alteration of barrier beaches is considered.

Pilkey, O. H., 1990. *Barrier Islands.* 36:6, 30–39. The origin, evolution, and types of barrier islands are discussed.

Schumberth, C. J., 1971. *Long Island's Ocean Beaches.* 17:6, 350–62. A very informative article on the nature of barrier islands. The specific problems observed on the Long Island barriers serve as examples.

Tucker, C., 1993. *Beach Sweeping and Ocean Keeping—A Practical Guide to Fighting Marine Pollution.* 39:4, 30–36. The author suggests ways for people to become involved in the fight against beach and ocean pollution.

Wanless, H. R., 1989. *The Inundation of our Coastlines: Past, present and future with a focus on South Florida.* 35:5, 264–7. An overview of evidence useful in evaluating the past and future movements of the shorelines.

Wanless, H. R., and L. P. Tedesco, 1988. *Sand Biographies.* 34:4, 224–32. Discusses how we can determine where beach sand came from and how it traveled to its present location.

Westgate, J. W., 1983. *Beachfront Roulette.* 29:2, 104–9. The problems related to the development of barrier islands are discussed.

Scientific American

Bascom, W., 1960. Beaches. 203:2, 80–97. A comprehensive consideration of the relationship of beach processes, both large- and small-scale, to release of energy by waves.

Broecker, W. S., and G. H. Denton, 1990. What Drives Glacial Cycles? 262:1, 48–107. Includes the most up-to-date information on the causes of glacial cycles.

Dolan, R., and H. Lins, 1987. Beaches and Barrier Islands. 257:1, 68–77. A discussion of the futility of trying to develop and protect structures on beaches and barrier islands.

Fairbridge, R. W., 1960. The Changing Level of the Sea. 202:5, 70–79. A discussion of the causes of changing sea level, which seems to be related mostly to the formation and melting of glaciers and changes in the ocean floor.

Oceanography on the Web

 Visit the Introductory Oceanography home page for on-line resources for this chapter. There you will find an on-line study guide with review exercises and links to ocean-ography sites to further your exploration of the topics in this chapter. Introductory Oceanography is at http://prenhall.com/thurman (click on the Table of Contents menu and select this chapter).

Human Impacts on Estuaries

- How are estuaries formed?
- What contributes to the high biological productivity in estuaries?
- Why are estuaries important to human society?
- Why are estuaries exceptionally prone to human impacts?

Introduction

Today's estuaries are all less than 12,000 years old, reflecting the geologically recent rise of sea level since the last glacial maximum. Estuaries are typically ephemeral coastal features with youthful ecosystems.

Ecological Function of Estuaries

The shallow, nutrient-rich waters of estuaries and associated wetlands create a highly productive environment for plants and animals. In fact, estuarine environments are among the most productive on earth. The high concentration of nutrients and shallow depth support phytoplankton, seagrasses, macroalgae, emergent grasses and, in tropical environments, mangroves.

Animal communities in estuarine sediments are of low diversity, since many species cannot tolerate the extreme fluctuations in temperature and salinity. Animals that feed upon benthic (bottom-associated) organisms are in turn food for larger consumers, which include larger fishes as well as birds and mammals.

Many subhabitats within estuaries, such as salt marshes and seagrass beds, serve as nursery grounds for the juvenile stages of commercially valuable fish and shellfish. The production of the higher plants (seagrasses, marsh grasses, and mangroves) is in great surplus of what is directly consumed, and the decomposition of this material leads to a detrital food web where a large portion of the nutrition to the consumers derives from the microbes that grow on the dead plant material.

The wetlands that fringe many estuaries also perform valuable services for human society. Estuaries provide habitat for more than 75% of America's commercial fish catch and for 80–90% of the recreational fish catch. Water draining from the uplands carries sediments, nutrients, and pollutants. As the water flows through fresh and salt marshes, much of the sediments and pollutants is filtered out. Wetland plants and soils also act as a natural buffer between the land and ocean, absorbing flood waters and dissipating storm surge. This protects upland organisms as well as valuable real estate from storm and flood damage.

Population Growth in Coastal Regions

Features of estuaries which favor their high productivity, namely their semi-enclosed nature, riverine inflow of nutrients, and extensive intertidal plant communities, also attract humans. Many major cities are located along estuaries (Boston, Baltimore, Charleston, Norfolk, San Francisco, Seattle, etc.). Historically, these cities were, or are still major ports for commercial shipping.

River connection to estuaries expanded the ability of cities far from the estuary mouth to develop port operations and dependent industry. The floodplain region of estuaries and rivers possesses rich soils that enhance agricultural production, and having nearby ports for marketing these products encouraged agriculture throughout the watershed region of the estuary.

The aesthetics of rivers, estuaries, marshes, and beaches have also encouraged a dramatic growth both of residents and tourists. It is therefore not surprising that population growth has been much greater in coastal and estuarine regions of the U.S. compared to the inland regions (Figure 1). More than 139 million people—about 53% of the national total—reside along the narrow coastal fringes. This coastal population is increasing by 3,600 people per day. The increase in human population and their associated activities within estuarine watersheds have caused significant declines in water quality and changes in the structure and function of estuarine ecosystems.

Human Effects Upon Estuarine Water Quality

Human activities have altered estuarine watersheds throughout the world. Increases in nutrients (eutrophication) occur naturally in aquatic systems as they age, but human population growth has accelerated the accumulation of these nutrients many orders of magnitude faster than what would occur naturally. High rates of nutrient input into estuaries can contribute to fish disease, toxic and nontoxic algae blooms, low dissolved oxygen, and marked change in plankton and benthic community structure.

Toxicants and pathogens also enter estuaries. Toxicants, including organic substances and metals, may exert direct and acute effects upon the aquatic community, or they may be bioconcentrated as these materials are passed up the food web, which selectively impacts the higher trophic level species. Exposure to toxicants or pathogens in coastal waters, or the consumption of undercooked or raw seafood harvested from those waters, can cause severe illness or death. Coastal population growth has led to an increasing flux of pathogens (viruses, bacteria, and parasites) to coastal waters, primarily from sewage.

Illness and an occasional death still result from human pathogens in coastal waters. In the case of toxic phytoplankton, coastal monitoring occurs on a local basis, often by local health departments, as well as by the Food and Drug Administration.

Seafood safety is a significant concern to the nation. Paralytic, diarrhetic, neurotoxic, and amnesic shellfish poisonings are all

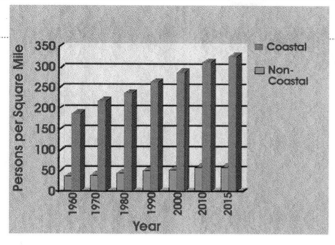

Figure 1

Population growth of coastal and non-coastal constituents of the U.S. from 1960 to estimated values in 2015. Data from the NOAA State of the Coast Report, 1998. *(NOAA)*

Figure 2

Sores on menhadden associated with an outbreak of *Pfiesteria piscicida*. The incidence of harmful algae blooms in coastal and estuarine environments have been increasing world wide in the last several decades. *(J. Purkholder)*

Figure 3

Hundreds of pigs fight for their lives as flooding engulfs a large hog farm in North Carolina. Upwards of 100,000 hogs drowned as a result of Hurricane Floyd. *(The News and Observer/Chris Seward)*

caused by biotoxins accumulated from algae. Outbreaks of poisoning due to various toxins accumulated by shellfish and fish such as the recent *Pfiesteria* outbreaks in Chesapeake Bay have occurred several times in the past few years (Figure 2).

Some human activities directly modify the habitat of resident estuarine species. These include: conversion of open land and forest for commercial development, agriculture, forestry, highway construction, marinas, dredging and filling, and damming. The timber harvesting operations in the U.S. Pacific Northwest include expansive road construction and forest clearcutting. These activities generate silt that can smother the eggs of salmon. The precipitous decline of a variety of the West Coast salmon is probably due, in part, to this habitat loss.

Changes in the hydrology of watersheds draining to the coast have also occurred as a result of channelization and damming, and diversion to other drainage basins.

Dredging channels to promote port operations enhances the transport of salty oceanic water into bays and estuaries, changing the salinity structure, circulation, flushing, and water residence times of these semi-enclosed coastal systems. Such changes can have dramatic effects on biological productivity and ecosystem structure and function.

The exploitation of living and nonliving resources can affect coastal ecosystem health. Fishing provides many benefits to society, including food, employment, business opportunities, and recreation. However, like many human activities, fishing also can have deleterious ecological effects. In Chesapeake Bay, the Atlantic oyster (*Crassostrea virginica*) used to cover extensive regions of the subtidal bottom, being capable of filtering the entire bay water volume over several days. The overharvesting of oysters during the last century has now eliminated commercial oyster reefs. The organic matter that used to be filtered by oysters now falls to the bottom and is consumed primarily by benthic microbes, leading to increasing frequency, extent, and duration of low oxygen (anoxia and hypoxia) conditions.

Living near the coast, although highly desirable to most of our population, does carry additional threats of natural disasters like hurricanes (Figure 3), storm erosion, and tsunamis.

Now, go to our web site to continue the analysis.

See the complete issue at:
www.prenhall.com/thurman

13 Coastal Waters and Marginal Seas

A fjord, an estuary formed by sea level flooding of a glacially cut valley. (Photo by H. P. Merton)

Estuary and wetland environments are among the most biologically productive environments on Earth. They are the nursery grounds for many species of marine organisms that inhabit the open ocean. They also are the conduits through which land-derived compounds must pass to reach the open ocean. There are many chemical, physical, and biological processes, occurring in the sediments and waters of these environments, that protect the quality of the water in the open ocean. Human activities, however, are increasingly altering conditions beyond the capacity of these processes to adapt. As these ecosystems decline and collapse, the balance of the whole ocean comes into question.

Coastal Water Circulation

The primary physical difference between coastal waters and the open ocean is depth. The shallowness of coastal waters and coastal geography that limits rapid exchange with the open ocean allow river runoff and tidal currents to create significant changes in seawater chemistry in time and space.

Salinity

In general, salinity is lower in coastal regions than in the open ocean due to the runoff of fresh water from the continents (Figure 13-1a, c). Along the coast where tidal and wave action is low, river runoff reduces the salinity of the surface layer. Large volumes of runoff from major river systems may reduce salinity throughout the water column even where extensive mixing occurs. Where precipitation on land is mostly rain, river runoff peaks in the same season that precipitation does. However, if runoff is largely fed by melting snow and ice, runoff always peaks in summer.

Prevailing offshore winds counteract the salinity-reducing effect of runoff in some coastal regions. These winds lose most of their moisture over the continent and evaporate considerable water as they move across the surface offshore. Increased in evaporation in these areas tends to increase surface salinity (Figure 13-1b).

Temperature

In coastal regions where the water is relatively shallow, temperatures may vary greatly in the course of the year. Seasonal changes in temperature can be most easily observed in coastal regions of the midlatitudes, where surface temperatures are coolest in winter and warmest in late summer. Sea ice forms in many high-latitude coastal areas where temperatures are determined by the water's freezing point, generally warmer than −2°C (28.4°F) (Figure 13-1d). Maximum surface temperatures may approach 45°C (113°F) in low-latitude coastal waters where circulation with the open ocean and strong mixing are restricted (Figure 13-1e). Figures 13-1f and g shows how strong thermoclines may develop where mixing does not occur. Very high- and low-temperature surface water may form a relatively thin layer. In contrast, mixing reduces or increases the surface temperature by distributing the heat through a greater vertical column of water, thus pushing the thermocline deeper and making it less pronounced.

Tidal currents can significantly influence vertical mixing in shallow coastal water, and thus temperature. Prevailing offshore winds can significantly affect surface temperatures. These air masses usually have relatively high temperatures during the summer, increasing ocean surface temperatures and seawater evaporation rates. During winter, they are much cooler than the ocean surface and absorb heat, cooling surface water near shore.

Coastal Geostrophic Currents

Geostrophic currents, which we discussed in Chapter 9 with respect to ocean basins, also develop in coastal waters. Wind blowing parallel to the coast piles up water along the shore. Gravity then pulls this piled-up water back downslope toward the open ocean. As it runs downslope away from the shore, the Coriolis effect causes it to turn. Along continents of the Northern Hemisphere, currents veer northward along western coasts and southward along eastern coasts. In the Southern Hemisphere, the directions are reversed. Also producing geostrophic flow along continental margins is high-volume fresh water runoff gradually mixing with ocean water. This produces a slope of surface water away from the shore. The seaward slope is enhanced by salinity and temperature gradients that cause the density of seawater to increase seaward. These currents often vary in intensity with the seasons because their strength depends on the wind and amount of runoff. Their seaward limits are the boundary currents resulting from the open-ocean gyres. Coastal geostrophic currents frequently flow in a direction opposite the boundary current. Such is the case with the Davidson Current that develops along the northwestern coast of North America in winter. Heavy precipitation occurs in the Pacific Northwest during winter, and southwesterly winds are strongest then. Their combined effect produces a relatively strong northward-flowing geostrophic current. It flows between the shore and the southward-flowing California Current, which is the eastern boundary current in the North Pacific (Figure 13-2).

Estuaries

Origin of Estuaries

The most common estuary is found where a river empties into the sea. Many bays, inlets, gulfs, and sounds are also considered to be estuaries. Thus, we define an **estuary** as a partially enclosed coastal body of water in which salty ocean water is significantly diluted by fresh water from land runoff. The mouths of large rivers form the most economically significant estuaries. Many have been seaports and centers of ocean commerce for hundreds or even thousands of years—Cairo in Egypt, Baltimore, New York City, and San Francisco in the United States, Buenos Aires in Argentina, and London in England are but a few examples. Many estuaries support important

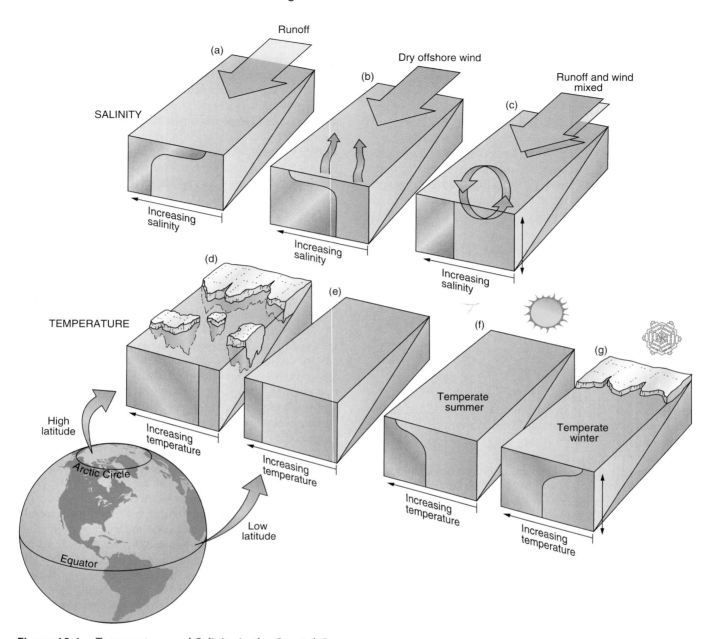

Figure 13-1 **Temperature and Salinity in the Coastal Ocean**

(a) Freshwater runoff from streams and rivers is not mixed into the water column, but instead forms a low-salinity surface layer and well-developed halocline. (b) Warm, dry offshore winds cause a high rate of evaporation that may offset the effects of runoff. This situation produces a halocline with a gradient that is the reverse of that in (a). (c) Runoff is mixed with deeper water, producing an isohaline ("equally salty") water column of generally lower salinity than the open ocean. (d) In high latitudes where sea ice is forming or thawing throughout the year, the temperature of coastal water remains uniformly near freezing. (e) Water in shallow, low-latitude coastal regions, protected from free circulation with the open ocean, may become uniformly warm. (f) In midlatitudes, coastal water is significant warmed during summer, and a strong seasonal thermocline may develop. (g) Winter may produce a low-temperature surface layer. The thermocline shown may develop. However, cooling may cause surface water to sink by increasing its density, causing a well-mixed isothermal water column. Mixing from strong winds may drive the thermoclines in (f) and (g) deeper and even mix the entire water column, producing an isothermal condition.

commercial fisheries as well. Environmental changes caused by heavy commercial use of estuaries have only recently become a major concern.

Essentially all estuaries in existence today owe their origin to sea level rise over the past 18,000 years. Sea level has risen approximately 120 m (400 ft) due to extensive melting of major continental glaciers. These glaciers covered portions of North America, Europe, and Asia during the Pleistocene Epoch, often referred to as the Ice Age. Four major classes of estuaries can be identified on the basis of their origins (Figure 13-3):

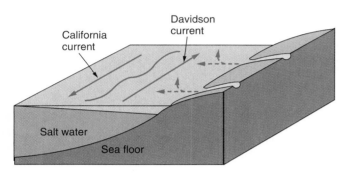

Figure 13-2   Davidson Current

During the winter rainy season, fresh runoff water produces a seaward slope away from low-salinity surface water near the shore. There is a surface flow of low-salinity water from the shore toward the open ocean. As the flow of surface water is acted on by the Coriolis effect, it veers to the right. This creates a north-flowing current (Davidson Current) along the coast of Washington and Oregon between the shore and the southbound California Current.

1. **Coastal plain estuaries**: formed as rising sea level causes the oceans to invade existing river valleys. These estuaries are appropriately called *drowned river valleys*. **Chesapeake Bay** off Maryland and Virginia is an example.

2. **Fjords**: (pronounced fee-yord) formed from flooded glacially carved valleys. Unlike water-carved valleys which are V-shaped, glacially carved fjords are U-shaped valleys with steep walls. They usually have glacial sediment deposits that form mounds called **sills** near the ocean entrance. Fjords are common along the coasts of Norway (the word "fjord" is Norwegian), Canada, Alaska, and New Zealand.

3. **Bar-built estuaries**: formed by sandbars that separate shallow lagoons from the open ocean. The sandbars are deposited parallel to the coast by wave action. Examples abound in the United States, including Laguna Madre along the Texas coast, Pamlico Sound in North Carolina and Chincoteague Bay in Maryland.

4. **Tectonic estuaries**: produced by faulting or folding of rocks that creates a restricted down-dropped area into which rivers flow. San Francisco Bay is, in part, a tectonic estuary.

Water Mixing in Estuaries

Generally, freshwater runoff flows into the *head* of an estuary and moves as an upper layer of low-density water toward the *mouth* of the estuary near the open ocean. Beneath this upper layer, an opposing inflow of denser, salty seawater occurs. Mixing takes place at the contact between these water masses. This is the classical pattern of **estuarine circulation**. Often, no single mixing pattern applies

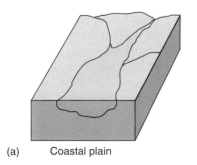

(a)   Coastal plain

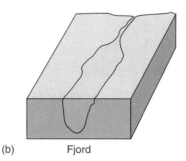

(b)   Fjord

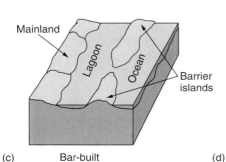

(c)   Bar-built

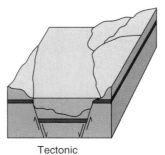

(d)   Tectonic

Figure 13-3   Classification of Estuaries on the Basis of Origin

Four classes of estuaries are (a) coastal plain (for example, Chesapeake Bay, Maryland and Virginia), (b) fjord (Strait of Juan de Fuca, Washington), (c) bar-built (Laguna Madre, Texas), and (d) tectonic (San Francisco Bay, California).

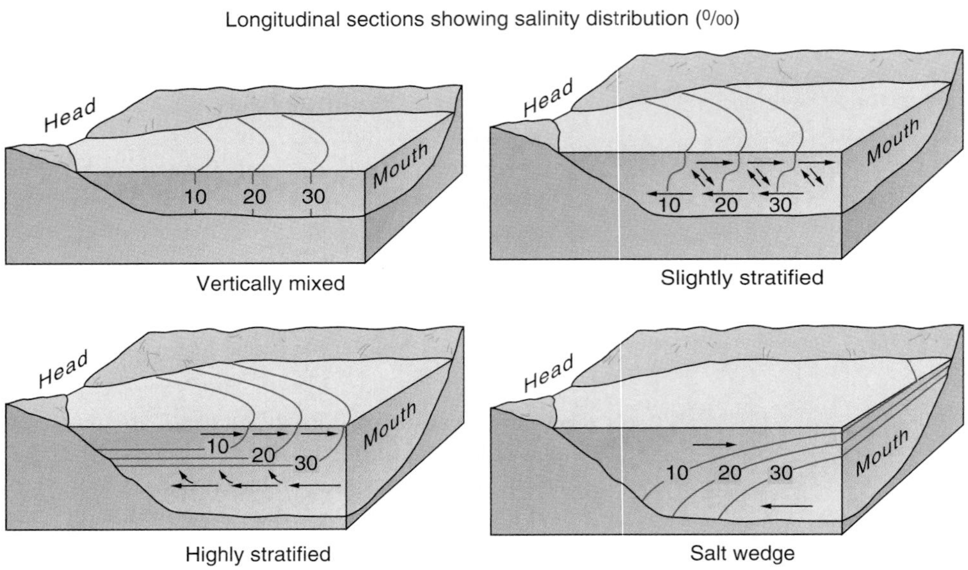

Longitudinal sections showing salinity distribution (⁰/oo)

Vertically mixed

Slightly stratified

Highly stratified

Salt wedge

Figure 13-4   Classifying Estuaries by Mixing

The basic flow pattern in an estuary is a surface flow of less dense freshwater toward the ocean, and an opposite subsurface flow of salty seawater into the estuary. The dimensions of each flow, and the degree of mixing between the two, depends on specific conditions in each estuary. Note that surface freshwater flow toward the mouth of the estuary extends seaward much farther along the right-hand shore as one faces seaward. This is due to the Coriolis effect. The opposite side of the estuary experiences a greater marine (saltwater) influence. In most estuaries, the marine water inflow occurs in the subsurface.

to an estuary as a whole, and mixing may also change with distance from the mouth of the estuary, season, or tidal conditions. Estuarine mixing is classified into four general patterns (Figure 13-4).

1. **Vertically mixed estuary**: This is a shallow, low-volume estuary where the net flow always proceeds from the river head of the estuary toward the estuary's mouth. Salinity at any point in the estuary is uniform from the surface to the bottom because the river water mixes evenly with the ocean water at all depths. Salinity simply increases from the river head to the mouth of the estuary, as shown in the figure. Note that the Coriolis effect skews the salinity contours.

2. **Slightly stratified estuary**: This is a somewhat deeper estuary in which salinity increases from the head to the mouth at any depth, as in vertical mixing. However, two basic water layers can be identified, less saline, less dense upper water provided by the river, and denser, deeper marine water. These two layers are separated by a zone of mixing. In this type of estuary, we begin to see classical estuarine circulation patterns develop because there is a net surface flow of low-salinity water toward the ocean and an opposite net subsurface flow of marine water toward the head of the estuary.

3. **Highly stratified estuary**: This is a deep estuary in which upper-layer salinity increases from the head to the mouth, reaching a value close to that of open-

ocean water. The deep-water layer has a rather uniform marine salinity at any depth throughout the length of the estuary. Net flow of the two layers is similar to that in a slightly stratified estuary, but mixing at the interface of the upper water and the lower water is such that net movement is from the deep-water mass into the upper water. Less saline surface water simply moves from the head toward the mouth of the estuary, growing more saline as water from the deep mass mixes into it. Relatively strong haloclines develop in such estuaries at the contact between the upper and lower water masses. It is not unusual for these haloclines to extend over 20‰ near the head of the estuary at maximum river flow.

4. **Salt wedge estuary**: This is an estuary in which a saline wedge of water intrudes from the ocean beneath the river water, typical of the mouths of deep, high-volume rivers. No horizontal salinity gradient exists at the surface in these deep estuaries. Water is essentially fresh throughout the length of, and even beyond, the estuary. There is, however, a horizontal salinity gradient at depth and a very pronounced vertical salinity gradient manifested as a strong halocline at any station throughout the length of the estuary. This halocline is shallower and more highly developed near the mouth of the estuary.

Chesapeake Bay exemplifies the complexity of a real estuary in comparison with these simplified models.

Chesapeake Bay is one of the most highly researched bodies of water on the planet, but the data often seem to highlight how poor our understanding is rather than to verify it. Some examples of observations based on circulation data collected over an extended time period are:

1. Researchers have recorded intervals of upstream surface flow accompanied by downstream deep flow, just the opposite of the expected pattern.

2. Periods of total downstream flow have been observed.

3. Even more complex patterns involve a surface and bottom flow in one direction separated by a mid-depth flow in the opposite direction, or landward flows along the shores and seaward flows in the central portions of the estuary.

Estuaries and Human Activities

Estuaries are vital natural ecosystems that have evolved over millennia. Their health is essential to world fisheries and to coastal environments worldwide. However, they are exploited heavily by economic activities that do not depend on the health of the estuary. Estuaries are sites of shipping, logging, manufacturing, waste disposal, and other activities that often damage the estuarine environment. Estuaries are most threatened where the human population is high and growing. However, even where population pressure is modest, estuaries can be severely damaged. See Human Impacts on Estuaries on page 000.

A good example is the **Columbia River estuary**, which forms most of the border between Washington and Oregon. The Columbia River estuary is flushed by tides that drive a salt wedge 42 km (26 mi) upstream and raise its water level by over 3.5 m (12 ft) (Figure 13-5). When the tide falls, fresh water flowing at up to 28,000 m³/s (1,000,000 ft³/s) pushes out hundreds of kilometers into the Pacific Ocean. Salmon returning to spawn need to find this plume of water, swim upstream and lay their eggs in the gravel of their home streams (see Chapter 16).

In the late nineteenth century, farmers and dairymen moved onto the rich floodplains of the Columbia River and eventually erected dikes to prevent inevitable natural flooding. Now these lands are removed from the natural ecosystem and, ironically, bad farming practices have made them unsuitable for agriculture. As the principal conduit for the logging industry, which has dominated the region's economy through most of its modern history, the river has also been subjected to extensive pollution by soil erosion and chemicals. But hydroelectric dams have dealt the river system its most serious and permanent blow. Many of these dams did not include salmon ladders, which help fish "climb" in short vertical steps around the dams to reach the headwaters of their tributary home streams. This closed off many spawning grounds. In addition, the presence of dams and removal of water from the system for irrigation reduces the freshwater volume reaching the ocean. As is occurring in many cosatal areas, exotic species are reducing the species diversity within the indigenous estuarine community. See Illegal Immigration: Ballast Water and Exotic Species on page 390–391.

As is the case with most interactions between nature and humans, the dams were almost essential to the development of the region, but a high price was paid. At present, in an effort to diversify the local economy, shipping facilities are being developed, with the accompanying dredges and increased pollution potential. If problems have developed in such a sparsely populated areas as the Columbia River estuary, what must be the conditions at more highly stressed estuaries?

Chesapeake Bay is a classic example of a coastal plain estuary. It was produced by the drowning of the Susquehanna River valley and valleys cut by its tributaries

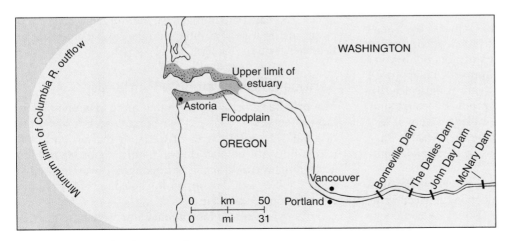

Figure 13-5   Columbia River Estuary

This high-volume salt wedge estuary has suffered more from interference with its flow by hydroelectric dams and diked floodplains than from chemical pollution. Like the Mississippi River, the estuary's large flow volume aids it in reducing the effects of chemical pollution.

(Figure 13-6). Most of the fresh water entering the bay does so along the western margin via rivers that drain the slopes of the Appalachian Mountains. Figure 13-6 shows the salinity of the Bay. Note that salinity increases oceanward and also reflects the Coriolis effect, acting to deflect eastward the seawater entering the bay from the south and to deflect westward the southward flow of river water.

With maximum river flow in the spring, a strong halocline (pycnocline) develops, preventing mixing of the fresh surface water and saltier deep water. Beneath the pycnocline, which can be as shallow as 5 m (16 ft), waters may become anoxic (without oxygen) from May through August. Blooms of microscopic algae, and the eventual decay of this organic matter as it sinks into deep water outstrips the supply of oxygen in the bottom layer. Major kills of commercially important blue crab, oysters, and other bottom-dwelling organisms occur during these events. For unknown reasons, the degree of stratification and kills of bottom-dwelling animals have increased over the last 40 years. The Bay has experienced intense pressure from human activities, and increased nutrients from sewage and fertilizers have been added to the bay as a result. These extra nutrients probably promote larger algal blooms and also may contribute to blooms of toxic dinoflagellates in coastal waters up and down the eastern coast of North America (see Chapter 18) as is discussed in *Pfiesteria,* Morphing Peril to Fish on page 000.

Estuary Circulation and Plankton

Plankton are microscopic algae, protists, and animals that abound in nutrient-rich waters, where they drift with the current. Phytoplankton are microscopic algae which make their own food through photosynthesis and are the foundation of the food chain in an estuary. These tiny floaters are carried about by currents, and their distributions correlate with estuarine circulation patterns.

Plankton communities were studied in four estuarine environments and the adjacent continental shelf areas on the east coast of the United States (Figure 13-7). These studies showed that, as plankton were carried seaward by estuarine flow, their rates of photosynthetic production increased sixfold because of improved water clarity, but out beyond 5 km (3 mi), production gradually dropped with approach to the nutrient-poor Gulf Stream waters (Figure 13-7b). Seasonal changes in photosynthetic production were similar in estuaries, in nearshore shallow shelf waters of less than 20 m depth, and in deeper offshore shelf waters. Photosynthetic productivity in all regions peaked in late summer (green curve in Figure 13-7c). .

Unexpectedly, in shelf waters, similar seasonal changes were observed in phytoplankton production (green curve in Figure 13-7c) and the abundance of fish larvae and copepods (purple and red curves in Figure 13-7c). Zooplankton populations should lag behind phytoplankton populations, yet the data showed simultaneous peaks. Apparently, both populations developed in the estuaries and were subse-

quently flushed into the shelf waters by estuarine outflow. If this is true, seasonal circulation patterns in estuaries also have a powerful influence on the ecosystems of the continental shelf. The populations of many of the fish and mammals living on the shelf are linked to these plankton through complex food webs (see Chapter 15).

Wetlands ....

Wetlands, which border estuaries and other shore areas protected from the open ocean, are strips of land delicately in tune with natural shore processes. They are very biologically productive. Two types of wetlands exist, salt marshes and mangrove swamps (Figure 13-8). Both are within the tidal zone, intermittently submerged by ocean water, and rich in life forms adapted to large changes in salinity and moisture as seawater and fresh waters compete. Bottom sediments are organic-rich, oxygen-poor mud and peat deposits. Marshes, characteristically inhabited by various grasses, occur from the equator to latitudes as high as 65°, but **mangrove swamps** are restricted to latitudes below 30°, both north and south of the equator (Figure 13-9). Once mangroves colonize an area, they normally outgrow and replace marsh grasses.

Undisturbed wetlands have a very high economic value. **Salt marshes** are believed to serve as nurseries for over half the species of commercially important fish in the southeastern United States. Other fish, such as flounder and bluefish, use marshes for feeding and over-wintering. Fisheries for oysters, scallops, clams, and fishes, such as eels and smelt, are located directly in the marshes.

A very important characteristic of wetlands is their ability to filter or purify polluted water. Wetlands remove nutrient compounds and metals from waters polluted by industrial, agricultural, and residential sources. Bacterial processes remove nutrients; most metal removal is probably achieved through attachment to clay-sized particles in the wetlands. Nutrient compounds containing nitrogen and phosphorus are decomposed by bacteria or utilized by plants. As plants die in marshes, their organic material is either incorporated into the sediment to become peat or becomes food for bacteria, fungi, or fish.

Serious Loss of Valuable Wetlands

You may be surprised to learn that more than one-half of the wetlands in the United States have been lost. Of the original 215 million acres of wetlands that once existed (excluding Alaska and Hawaii), only about 90 million acres remain. Both marsh and mangrove wetlands have been filled in and developed for housing, industry, and agriculture in coastal areas.

To help prevent the loss of remaining wetlands, the U.S. Environmental Protection Agency established an Office of Wetlands Protection (OWP) in 1986. At that time, wetlands were being lost to development at a rate of 300,000 acres per year. The OWP is actively enforcing regulations against

Figure 13-6 Chesapeake Bay

(a) The map shows average surface salinity. The dark blue area is a region of anoxic (oxygenless) waters, from Baltimore to the mouth of the Potomac River. (b) Comparison of dissolved oxygen levels for the summers of 1950 and 1980. Deeper midbay waters are essentially without oxygen in 1980 (*After Officer et al., 1984*).

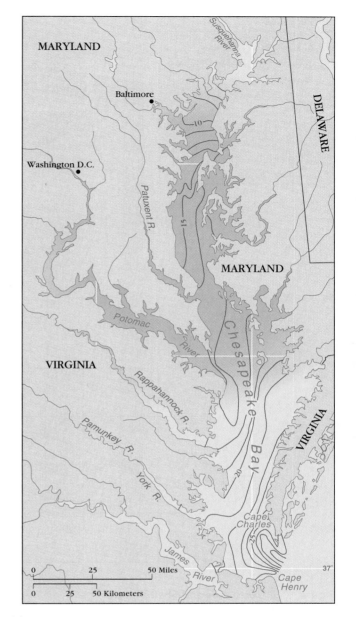

(a)

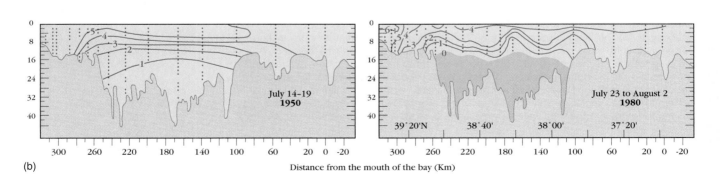

(b)

Distance from the mouth of the bay (Km)

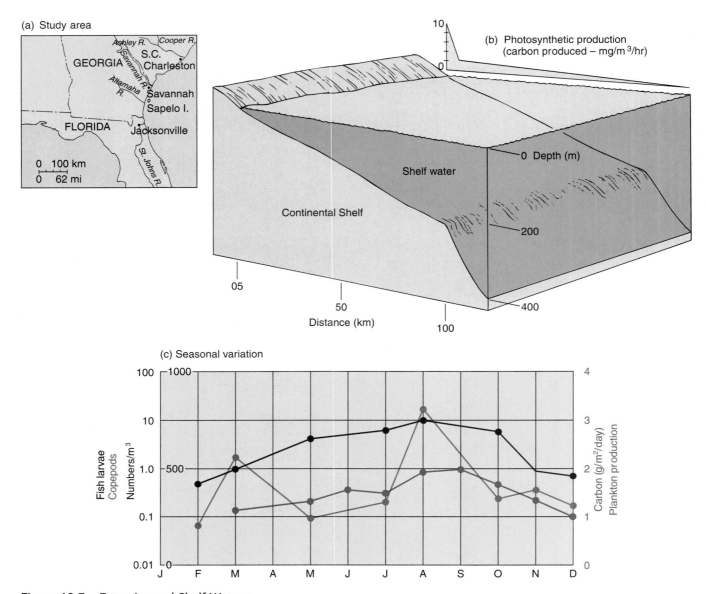

Figure 13-7 Estuaries and Shelf Waters

Along the southeastern coast of the United States, a study of the relationship between estuaries and the coastal ocean had the following results: (a) Along this coastal area, relatively small-volume rivers flow onto a gently sloping continental shelf at least 100 k (60 mi) wide. At the four locations shown, transects from estuaries across the continental shelf revealed photosynthetic production (measured as the amount of carbon in phytoplankton) and plankton abundance. (b) The graph (upper right) simulates the average photosynthetic production pattern along the transects. The graph aligns with the continental shelf view below it. Zero on the distance scale is the nearshore shelf waters adjacent to the estuary mouth. (c) Seasonal changes in three biological indicators: photosynthetic production, copepods (an important population of small zooplankton that eat photosynthetic cells), and fish larvae. The similar pattern for these quite varied life forms indicates they may be produced in the estuary and flushed together into the shelf waters by estuarine outflow. *(After Turner, R. E., et al., 1979).*

(a)

(b)

Figure 13-8   Salt Marsh and Mangrove Trees

(a) Salt marsh on San Francisco Bay at Shoreline Park, California.   *(Courtesy of Eda Rogers)*
(b) Mangrove trees on Lizard Island, Great Barrier Reef, Australia.   *(Courtesy of R. N. Mariscal/*
Bruce Coleman, Inc.)

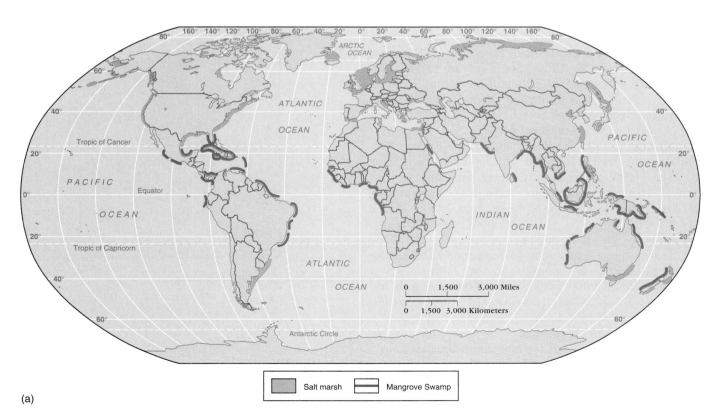

(a)

Figure 13-9   World Distribution of Salt Marshes and Mangrove Swamps

Note the dominance of mangrove swamps in low latitudes and salt marshes in high latitudes.

wetlands pollution and is identifying the most valuable wetlands that can be protected or restored.

Lagoons

Lagoons are protected, shallow bodies of water landward of barrier island salt marshes. Recall that a lagoon is a bar-built estuary. Because of very restricted circulation between lagoons and the ocean, three distinct zones usually can be identified within a lagoon. Typically, we find a freshwater zone near the mouths of rivers that flow into the lagoon, a transitional middle zone of brackish water formed by mixing fresh water and ocean water, and a saltwater zone close to the entrance, where the maximum tidal effects within the lagoon can be observed. Away from the salt-water zone near the mouth of the lagoon, tidal effects diminish and usually are undetectable in the freshwater region. Figure 13-10 shows the characteristics of a typical lagoon.

Ocean water flows through the lagoon entrance during a warm, dry summer to compensate for the volume of water lost through evaporation. In latitudes that have seasonal variations in temperature and precipitation, lagoon salinity is determined by a factor other than nearness to the sea. This flow increases salinity in the lagoon. Lagoons may become hypersaline (excessively salty) in arid regions where seawater inflow is too small to keep pace with surface evaporation. During the rainy season, the lagoon becomes much less saline as freshwater runoff increases.

Laguna Madre

Laguna Madre, on the Texas coast between Corpus Christi and the mouth of the Rio Grande River, is one of the best known lagoons in the United States (Figure 13-11). This long, narrow body of water is protected from the open ocean by Padre Island, an offshore island 160 km (100 mi) long. The lagoon probably formed about 6000 years ago as sea level rose to its present height.

Much of the lagoon is less than 1 m (3.3 ft) deep. The tidal range of the Gulf of Mexico in this area is about 0.5 m (1.6 ft). The inlets at each end of the barrier island are quite small. There is, therefore, very little tidal interchange of water between the lagoon and the open sea. Laguna Madre is a hypersaline lagoon. The shallowness of its waters creates a very great seasonal range of temperature and salinity. In this semiarid region, water temperatures are above 27°C (80.6°F) in the summer and may fall below 5°C (41°F) in winter. Salinities range from 100‰ during dry periods, to less than 2‰ when infrequent local storms provide large volumes of fresh water. High evaporation generally keeps salinity well above 50‰.

Because even the salt-tolerant marsh grasses cannot withstand such high salinities, the marsh is replaced by an open sand beach on Padre Island. At the inlets, ocean inflow occurs as a surface wedge over the denser water of the lagoon. In turn, the outflow from the lagoon occurs as a subsurface flow. This is just the opposite of the typical circulation of estuaries.

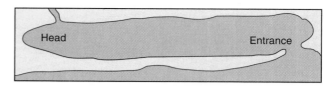
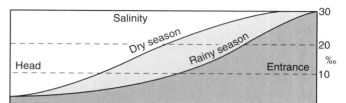
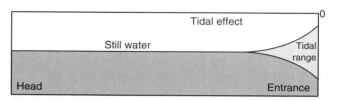

Figure 13-10   Lagoons

Typical configuration, salinity conditions, and tide conditions of a lagoon.

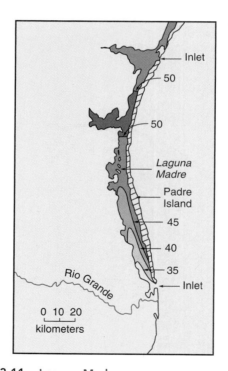

Figure 13-11   Laguna Madre

Typical summer surface salinity distribution (‰).

Marginal Seas

At the margins of the ocean are relatively large bodies of water called **marginal seas**. Most of these seas result from tectonic events that have isolated pieces of ocean crust between continents, such as the Mediterranean, or behind volcanic island arcs, such as the Caribbean. These waters have varying degrees of exchange with the open oceans, depending on climate and geography; thus, salinities and temperatures depart from those of typical seawater. We will look at some of the more unusual and important marginal seas that border the Atlantic, Pacific, and Indian oceans.

Marginal Seas of the Atlantic Ocean

Mediterranean Sea The **Mediterranean Sea** (Figure 13-12a) is a most unusual body of water. It is actually a number of small seas connected by narrow necks of water into one large sea. It is the remains of an ancient sea (the Tethys Sea) that separated the former continents of Laurasia and Gondwanaland about 200 millions years ago.

The Mediterranean, bounded by Europe and Asia Minor on the north and east, and by Africa to the south, is essentially surrounded by land. Only three narrow straits connect it to other water bodies—to the Atlantic Ocean through the Strait of Gibraltar (about 14 km or 9 mi wide, 244 m or 800 ft deep); to the Black Sea through the Bosphorus Strait (roughly 1.6 km or 1 mi wide); and to the Red Sea via the Suez Canal, an artificial waterway 160 km (100 mi) long.

The Mediterranean has two major basins separated by an underwater sill (ridge) extending from Sicily to the coast of Tunisia at a depth of 400 m (1300 ft). Strong currents run between Sicily and the Italian mainland through the Strait of Messina. The Mediterranean also has a very irregular shape, which divides it into subseas like the Aegean Sea and Adriatic Sea. Each of these seas has a separate circulation pattern.

Atlantic Ocean water enters the Mediterranean through the Strait of Gibraltar as a surface flow. It enters to replace water that rapidly evaporates in the very arid eastern end of the sea. The water level in the eastern Mediterranean is generally 15 cm (6 in) lower than at the Strait of Gibraltar. The surface flow follows the African coast throughout the length of the Mediterranean, spreads northward across the sea and eastward to Cyprus. Here, during winter, it sinks to form a water mass called Mediterranean Intermediate Water, with a temperature of 15°C (59°F) and a salinity of 39.1‰. This water flows westward, back along the African coast, at a depth of 200 to 600 m (650 to 2000 ft), then passes into the North Atlantic as a subsurface flow through the Strait of Gibraltar.

By the time Mediterranean Intermediate Water passes through the Strait of Gibraltar, its temperature has dropped to 13°C (55°F) and its salinity to 37.3‰. It is still much denser than the surface water in the Atlantic Ocean, and so it moves downward along the continental slope until it reaches a depth of approximately 1000 m (3300 ft), where it encounters Atlantic Ocean water of the same density. At this level, it spreads in all directions into the Atlantic water and becomes a distinct water mass characterized by its high salinity over a broad area.

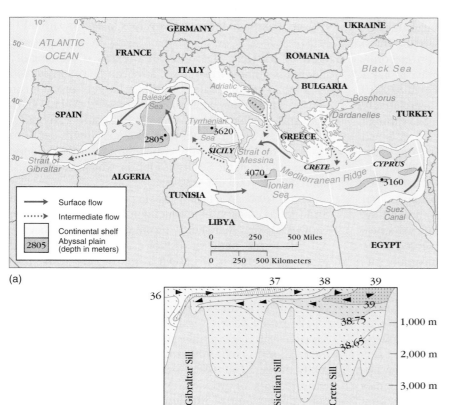

(a)

(b)

Figure 13-12 Mediterranean Sea–Subseas, Depths, Circulation, and Salinity

(a) The map shows the subseas, depths, sills (underwater ridges), surface flow, and intermediate flow. (b) Vertical distribution of salinity (‰). Most of the mass of Mediterranean water has salinities between 38 and 39 ‰. Maximum salinities (exceeding 39‰) occur in surface waters at the eastern end. After sinking and moving as Intermediate Water toward the Atlantic, salinity is reduced to about 37.3‰ as it flows over the sill at Gibraltar. *(Part (a) Adapted from Encyclopedia of Oceanography, Ed. Rhodes Fairbridge, © 1966. Reprinted by permission of Dowden, Hutchinson, & Ross, Inc., Stroudsburg, PA.)*

Circulation in the Mediterranean Sea is typical of closed, restricted basins where evaporation exceeds precipitation. Such restricted basins always lose water rapidly to surface evaporation, and this water must be replaced by surface inflow from the open ocean. Evaporation of seawater greatly increases salinity. This denser, saltier water eventually sinks and returns to the open ocean as a subsurface flow (Figure 13-12b).

Note that this circulation pattern is opposite to the pattern characteristic of estuaries, where surface freshwater flow goes into the open ocean and saline subsurface flow enters the estuary. The Mediterranean Sea is the *type area* for such circulation, which is called **Mediterranean circulation.** Its opposite, estuarine circulation, develops between a marginal water body and the ocean when fresh water input exceeds the water loss to evaporation.

Caribbean Sea The **Caribbean Sea** is separated from the Atlantic Ocean by an island arc called the Antillean Chain. Composed of the islands of Cuba, Hispaniola, Puerto Rico, and Jamaica, the **Greater Antilles** form the northern boundary of the Caribbean Sea. The **Lesser Antilles** extend in an arc from the Virgin Islands to the continental shelf of South America. The deepest connection between the Caribbean and the Atlantic Ocean is the Anegada Passage east of the Virgin Islands, with a maximum depth near 2300 m (7455 ft). Other channels through which circulation occurs have depths ranging from 1500 to 2000 m (4920 to 6560 ft). The Caribbean Sea is divided into four major basins from east to west —Venezuela, Colombia, Cayman, and Yucatán —all of which reach depths in excess of 4000 m (13,120 ft) (Figure 13-13a).

The Guiana Current, entering the Caribbean through channels between the Lesser Antilles islands, represents a portion of the South Equatorial Current that moves northwest along the Guiana Coast of South America. It has a temperature between 26°C and 28°C (78.8°F and 82.4°F) and a salinity between 35.0‰ and 36.5‰. This relatively thin mass of water passes into the Caribbean Sea through the shallow channels north and south of St. Lucia Island and mixes in a 1:3 ratio with North Atlantic Water. Becoming the Caribbean Current, which travels 200-300 km (124 and 186 mi) north of the Venezuelan coast, the current continues generally west over the deepest portion of the Caribbean Sea. It turns north and passes through the Yucatán Strait into the Gulf of Mexico. Surface velocities as high as 4.5 km/h (2.8 mi/h) have been observed in the main axis of the Caribbean Current, but the maximum surface velocities usually average less than 2 km/h (1.2 mi/h). The current can be detected to depths of 1500 m (4920 ft) in the eastern basins. Volume transport into the Caribbean Sea by the Caribbean Current is about 30 sv.

The easterly component of the trade winds blowing along the coast of Venezuela and Colombia sets up a surface flow away from the coast that produces a shallow upwelling. Most of the water rising to the surface comes from depths of less than 250 m (820 ft). The upwelling water contains high concentrations of nutrients and low temperatures that result in relatively high biological productivity in the coastal region.

There are four readily identifiable water masses in the Caribbean Sea. Two are relatively warm surface masses found above 200 m (656 ft) depth, and two are deeper masses characterized by lower temperature. The salinity of the Caribbean Surface Water mass is determined by the rate of evaporation versus the amount of precipitation and runoff; it is generally above 36‰ during the winter, partially because of the upwelling of high-salinity water. Moving northward, the surface salinity decreases to values generally less than 35.5‰.

Extending from the southeast to the northwest of the Caribbean Sea near the Yucatán Strait is a thin, sheetlike, high-salinity layer —the Subtropical Underwater. Located at depths as shallow as 50 m (164 ft) in the southeast, it dips to a depth of 200 m near the Yucatán Strait. The maximum salinity within this sheet seems to follow the axis of flow for the Caribbean Current and exceeds 37‰ in the Yucatán Strait. Salinity decreases away from the flow axis. Directly beneath the Subtropical Underwater, following the main flow axis, is a low-salinity water mass, the Subtropical Intermediate Water. A salinity minimum that falls below 34.7‰ in the southeast and becomes less detectable at the Yucatán Strait identifies this mass (Figure 13-13b).

North Atlantic Deep Water enters the Caribbean primarily through the Anegada Passage between the Virgin Islands and the Leeward Islands of the Lesser Antilles, and through the Windward Passage between Cuba and Hispaniola. Characterized by a salinity slightly less than 35‰ and temperature just above 2°C (35.6°F), this water spreads as Caribbean Bottom Water and can be identified by an oxygen maximum that reaches values in excess of 5 ml/l.

Gulf of Mexico The Gulf of Mexico is tectonically much less complex than the Caribbean Sea. Surrounded by a broad continental shelf, this relatively broad basin reaches a maximum depth in excess of 3600 m (11,800 ft). The **Gulf of Mexico** is connected to the Caribbean Sea by the Yucatán Strait, which reaches a maximum depth of 1900 m (6232 ft); its only connection with the Atlantic Ocean is through the Straits of Florida, which reach depths approaching 1000 m (3280 ft).

The Caribbean Current passing through the Yucatán Strait loops clockwise, producing a dome of water in the Gulf of Mexico that stands 10 cm (4 in) higher than the Atlantic water southeast of Florida. This hydraulic head forces an intense flow known as the *Florida Current* to pass through the Straits of Florida. It joins with water carried north by the Antilles Current and flows north along the coast of Florida (see Figure 13-13a).

The *Loop Current* is a significant feature of the Gulf of Mexico's surface circulation. Figure 13-14a and b show the relationship of the current flow direction to the topography of the 20°C (68°F) isotherm surface in the southeastern Gulf of Mexico. Much of the surface water entering the Gulf of Mexico through the Yucatán Strait loops around the temperature contour in a clockwise flow

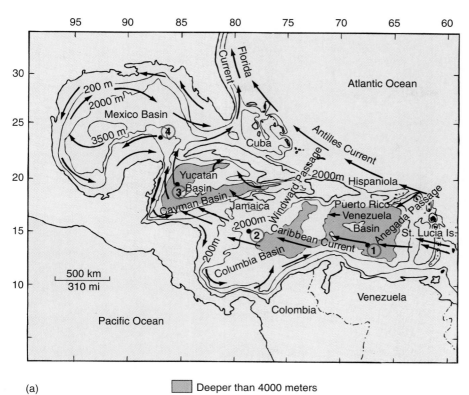

Figure 13-13   The Caribbean and Gulf of Mexico

(a) Surface currents and bathymetry. (b) The cores of the Subtropical Underwater and the Subtropical Intermediate Water can be identified on vertical salinity profiles as maxima and minima, respectively, at the stations 1, 2, 3, and 4 in (a). The Caribbean Surface Water salinity is generally above 36‰, whereas the Caribbean Bottom Water salinity is uniformly between 34.9 and 35.0‰. The saltier surface water does not sink because of its high temperature.

(a)   ▢ Deeper than 4000 meters

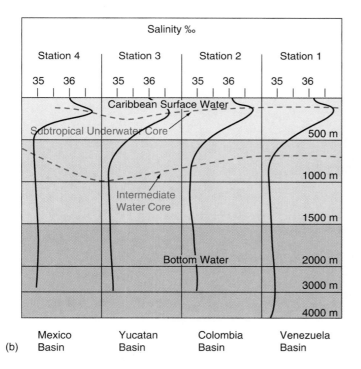

(b)

and heads toward the Straits of Florida. After passing through the Yucatán Strait, surface-water characteristics can be identified to a depth of 90 m (295 ft) during the winter, and to 125 m (410 ft) during the summer, this being the depth through which seasonal temperature changes extend. Generally, the surface temperature just off the Yu-

catán coast ranges from 24° to 27°C (75° to 81°F), whereas temperatures along the northern Gulf Coast range between 18° and 21°C (64° and 70°F) (Figure 13-14c).

The salinity of the surface water generally ranges from 36.0‰ to 36.3‰. Runoff from the Mississippi River significantly decreases salinity to depths of 50 m (164 ft) and

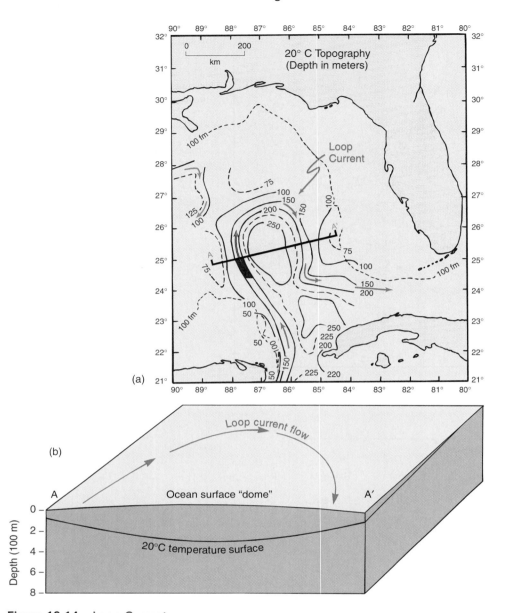

Figure 13-14   Loop Current

(a) The position and strength of the Loop Current can be deduced from the depth (m) to the 20°C (68°F) temperature surface. Flowing clockwise, its velocity increases with increased slope of the temperature surface. What may be a clockwise-rotating eddy is indicated northwest of the main loop current.   *(Courtesy of NOAA.)* (b) Block diagram showing flow of Loop Current in a clockwise pattern around the 10-cm-high surface "dome" of water that directly overlies the "bowl" representing the depth of the 20°C temperature surface. The vertical scale is greatly exaggerated. (c) NOAA 9 satellite infrared image of sea-surface temperature in the Gulf of Mexico. The 20°C temperature surface that forms the "bowl" in (b) intersects the ocean surface where the two lightest shades of blue meet. The Loop Current loops clockwise around the warm water that "fills" the "bowl," then moves off through the Straits of Florida. Note that the warm water that forms the "dome" reaches temperatures in excess of 26°C (79°F) (red).   *(Courtesy of NOAA.)*

as far as 150 km (93 mi) from the coast. The influence of the runoff can be identified by salinity values as low as 25.0‰ near the surface.

The Subtropical Underwater is still identifiable as a salinity maximum north of the Yucatán Strait at a depth of 100 to 200 m (328 to 656 ft). The boundary between the upper water (containing the surface water and Subtropical Underwater) and the deep water is marked by the 16°C (61°F) isotherm at a depth of about 200 m. Intermediate Water entering through the Yucatán Channel below 850 m (2788 ft) can be identified by a salinity minimum throughout the Gulf of Mexico. The core of the Intermediate Water

Figure 13-14   (continued)

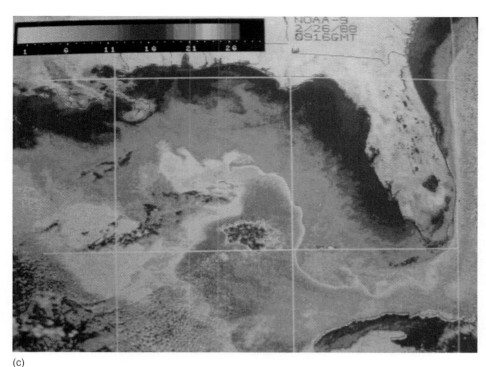

(c)

can be identified throughout the Gulf at depths as shallow as 550 m (1804 ft), but its identity is lost passing through the Straits of Florida. Salinity and temperature increase very slightly toward the bottom of the basin.

Marginal Seas of the Pacific Ocean

Gulf of California   This narrow, northwest-southeast trending sea extends from the Tropic of Cancer at its open south end to the mouth of the Colorado River, which is the main river system draining into it (Figure 13-15a). Some half-dozen smaller rivers also empty into the Gulf from the east, carrying water from the Sierra Madre Occidental across a broad coastal plain that forms the western margin. From this coastal plain, which is on the North American Plate, a continental shelf extends out from 5 to 50 km (3 to 30 mi). The shelf terminates at an average depth in excess of 100 m (328 ft). This depositional feature does not exist on the western side of the gulf, which is characterized by steep rocky slopes and is on the Pacific Plate. The Colorado River has developed a significant delta at the northern end, with depths rarely exceeding 200 m (656 ft). Two exceptions are the 1500-m (4920-ft) and 550-m (1804-ft) basins on the west and east side, respectively, of Angel de la Guarda Island. Depths gradually increase to the south through a series of basins, the deepest of which is 3700 m (12,136 ft). These basins are separated by sills that generally extend from 100 to 400 m (328 to 1312 ft) above the basin floor. The sills represent east–west–trending offsets of segments of spreading centers that are responsible for separating Baja California from the Mexican mainland.

The character of the Gulf of California has changed somewhat since the completion of Hoover Dam in 1935. Until that construction was completed, the Colorado River provided an annual average flow of almost 18 billion m³ of water, which carried 161 million tons of suspended sediment to the Gulf. Since 1935, the average annual flow has been reduced to less than 8 billion m³ of water and about 15 million tons of sediment. The drainage systems on the eastern side of the basin have relatively small flow volumes, and many are intermittent owing to the arid nature of their drainage basins.

Seasonal winds control the surface circulation. A low-pressure atmospheric system in summer located over the northern end of the peninsula develops winds that drive the surface water from the Pacific into the Gulf. This flow produces upwelling along the steep rocky coast of the peninsula. During the winter months, the low-pressure system is found on the mainland east of the Gulf. These winter winds produce upwelling along the mainland side. Upwelling produces a rich plankton bloom of diatoms and dinoflagellates which is the basis of a thriving biological community throughout the year. The Gulf supports a large fish population and is the nursery for whales that migrate there from the North Pacific.

The tidal range increases from about 1 m (3.3 ft) in the south to more than 10 m (33 ft) during spring tides at the mouth of the Colorado River. The tidal currents that develop in the north, along with convective mixing, produce an isothermal water column during the winter. Temperatures may drop as low as 16°C (61°F), as opposed to summer surface temperatures that may reach 30°C (86°F). A high rate of evaporation produces a marked stratification, with surface

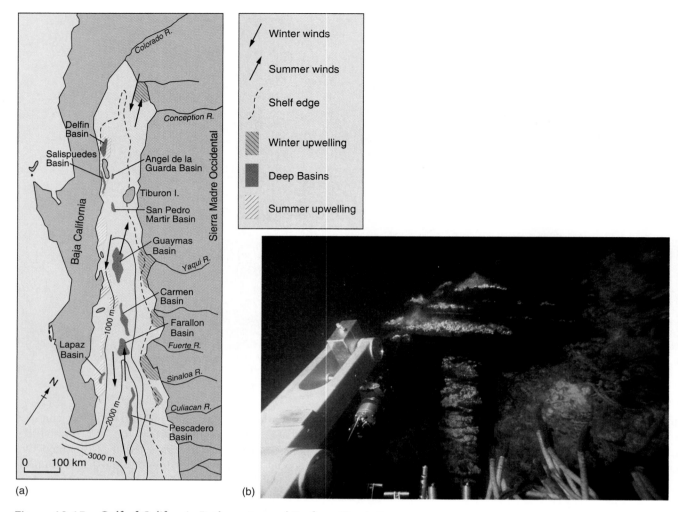

Figure 13-15 **Gulf of California Bathymetry and Surface Circulation**

(a) Basins increase in depth from the 980 m (3214 ft) depth of the Delfin Basin in the north to more than 3700 m (12,136 ft) in the Pescadero Basin. Winds that reverse on a seasonal basis produce upwelling on the east side of the gulf in the winter and on the west side during the summer. *(From Encyclopedia of Oceanography, Ed. Rhodes Fairbridge, © 1966. Reprinted by permission of Dowden, Hutchinson, & Ross, Inc., Stroudsburg, PA.)* (b) Pagodalike structure that vents 315°C (599°F) water atop sediment mound that covers a spreading center of the Guaymas Basin. Alvin's "arm" and vestimentiferan tube worms can be seen in the foreground. *(Robert Brown, Scripps Institution of Oceanography, University of California, San Diego.)*

water being warmer and more saline. The water below the thermocline in the central southern portion of the Gulf of California possesses an oxygen minimum as low as 0.01 ml/l between the depths of 400 m (1312 ft) and 800 m (2624 ft).

A joint Mexican–American expedition identified a hydrothermal vent biological community in the Guaymas Basin during the summer of 1980. In 1982, the biological community was observed and sampled during a dive of the submersible Alvin. Although they are covered by sediment, the vents are on a segment of a spreading center (Figure 13-15b).

Bering Sea The **Bering Sea**, on the northern margin of the Pacific Ocean, extends to 66°N latitude and has the shape of a triangle with a curved base formed by the

volcanic island arc of the Aleutians (Figure 13-16). A broad continental shelf with depths less than 200 m (656 ft) extends off the Siberian and Alaskan coasts, but most of the rest of the basin is deeper than 1000 m (3280 ft). The deepest part of the basin is located in the western half of the sea, and fully 90 percent of the sea is either less than 200 m in depth or more than 1000 m in depth. Except where it is cut by the Bering Canyon at the end of the Alaska Peninsula, the continental shelf slopes off very abruptly at 4° to 5° into the deep basin.

The major surface flow of water into the Bering Sea occurs between the Komandorski Islands of Russia and Attu Island, the most westerly of the Aleutian Islands of Alaska. Here the Alaskan Stream, flowing in a westerly direction south of the Aleutian chain, converges with

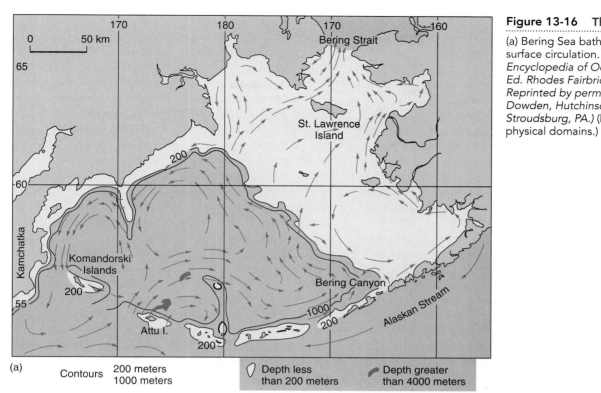

(a)

Contours   200 meters
           1000 meters

Depth less than 200 meters

Depth greater than 4000 meters

Figure 13-16   The Bering Sea

(a) Bering Sea bathymetry and surface circulation.   *(From Encyclopedia of Oceanography, Ed. Rhodes Fairbridge, © 1966. Reprinted by permission of Dowden, Hutchinson, & Ross, Inc., Stroudsburg, PA.)* (b) Bering Sea physical domains.)

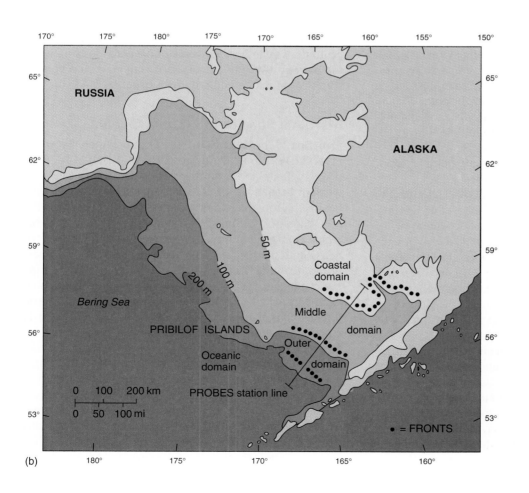

(b)

northward-moving water of the western Pacific and flows through the passages into the Bering Sea. A counterclockwise gyre is set up north of the Komandorski Islands, while a clockwise rotation develops to the north of the eastern end of the Aleutians. The main flow passes on to the east until it reaches the broad Alaskan shelf, then follows this shelf edge to the north. A portion of this northward flow passes between St. Lawrence Island and the east Siberian coast before crossing the Bering Strait into the Arctic Sea. Tidal currents dominate the shallow Alaskan shelf region, but a persistent northward-flowing current resulting from the runoff of fresh water from the Alaskan coast flows at speeds up to 10 km/h (6 mi/h) along the coast and through the Bering Strait (Figure 13-16a).

The continental shelf of the Bering Sea is divided into three domains by fronts where water masses moving in onshore and offshore directions converge (Figure 13-16b). The coastal domain extends seaward to a depth of about 50 m (164 ft) and ranges in width from 80 to 150 km (50 to 93 mi). The middle domain extends seaward to near the 100 m (328 ft) depth, which occurs at distances of 350 to 400 km (217 to 248 mi) from shore. The outer domain extends to the shelf break, at a depth of about 170 m (558 ft), about 500 km (311 mi) from shore.

The coastal domain is vertically mixed by tidal currents and wind waves (Figure 13-17a). Freshwater runoff and salt water are well mixed, although the surface water is slightly less salty than deeper water. The middle domain has enough water depth to separate the effects of tidal mixing (most effective on bottom) from wind and wave mixing (most effective on top). This produces two distinct layers of water. A high-salinity, low-temperature (predominantly marine) layer underlies a warmer, less-saline surface layer. The greater water depth in the outer domain produces a three-layered water structure. The top and bottom layers are similar to those of the middle domain although they are slightly more saline. Separating these layers is a zone of mixing where alternating thin layers of offshore and shelf water can be identified.

The circulation patterns described above are critical to the biological productivity of this sea. The Bering Sea possesses a large fisheries resource. About 5 percent of the world's catch is taken from the waters overlying the Bering Sea continental shelf, which represents 45 percent of the Bering Sea area. Most of the shelf lies in U.S. water. While U.S. fishermen have concentrated on the king and tanner crabs and salmon, the large stocks of Alaska pollock and Yellow-fin sole have been exploited primarily by Japan, Korea, and Russia.

Upwelling and enhanced biological productivity are associated with all of the fronts that separate shelf domains. This enhancement is most pronounced at the middle front, between the middle and outer domains (Figure 13-17b). The patterns of phytoplankton and zooplankton distribution are organized around the inner, middle, and outer fronts. Primary productivity within the middle domain is about 400 gC/m²/yr, or about twice that of the outer domain (200 gC/m²/yr). Most of the outer-domain phytoplankton is grazed by relatively large copepods and krill. These large grazers are prevented from entering the middle domain by their sensitivity to changes in the physical properties of water that occur at the middle front (Figure 13-17c). In deep water off the shelf, where food is scarce in winter, large outer-shelf copepods reproduce only once a year. To take advantage of the available food, juveniles migrate into the surface waters before the spring bloom begins (usually in May). They graze heavily on the phytoplankton and keep the bloom from attaining a large population.

The large crustaceans of the outer shelf are accompanied by smaller species of copepods that inflict much smaller grazing pressure on the phytoplankton than do the larger forms. These smaller forms are found across the entire shelf and are the dominant grazers of the middle domain. The smaller copepods produce several broods per year, but reproduction does not occur until after the spring bloom has developed when food becomes available. Thus a large spring bloom of phytoplankton is able to develop in the middle domain, and the small copepods do not harvest a large amount of it. Because of the continual heavy grazing by the larger outer-domain herbivores, the phytoplankton population is kept at a low level. Little of this productivity is made available to bottom dwellers. However, the large number of phytoplankton unharvested by smaller herbivores in the middle domain is available to bottom dwellers. Three times as much phytoplankton remains ungrazed in the middle domain as in the outer domain. The Alaskan pollock, sablefish, and other fish that live in the upper water column are found in greater concentrations near the shelf break of the outer domain, while the bottom-dwelling tanner crab, king crab, and yellow-fin sole dominate middle-domain fisheries.

Marginal Seas of the Indian Ocean

Red Sea The **Red Sea** extends more than 1900 km (1180 mi) north of the narrow Strait of Bab-el-Mandeb (Gate of Tears) to the northern tip of the Gulf of Suez, the western branch of the Red Sea that separates the Sinai Peninsula from the African mainland. Forming the eastern boundary of the Sinai Peninsula is the Gulf of Aqaba, which is an eastern branch at the northern end of the Red Sea (Figure 13-18). The Red Sea is the beginning of a new ocean forming as the plates containing the Arabian Peninsula and the African mainland separate. Extending from 12° to 30°N latitude, the Red Sea lies in a highly arid region and is characterized by surface waters of unusually high temperature and salinity. Broad reef-covered shelves no more than 50 m (164 ft) deep drop off sharply to a gently sloping surface at about 500 m (1640 ft), which eventually leads into a central trough of 1500 m (4920 ft) to more than 2300 m (7545 ft) depth in the central region.

Geological evidence indicates the Red Sea formed primarily during the past 20 million years by intracontinental rifting. The rifting may have begun as early as 180 million years ago, about the time Pangaea separated to form the Atlantic. Oceanic crust underlies the Red Sea and the Arabian

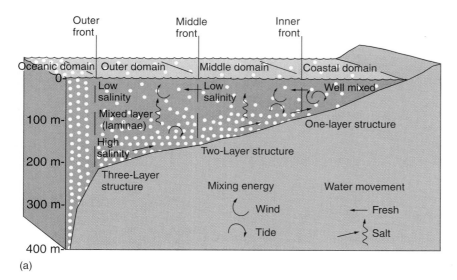

(a)

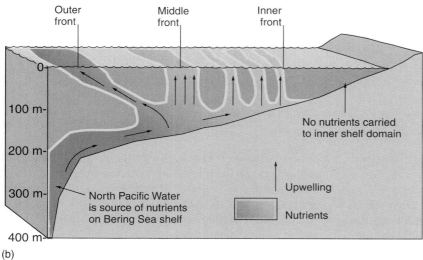

(b)

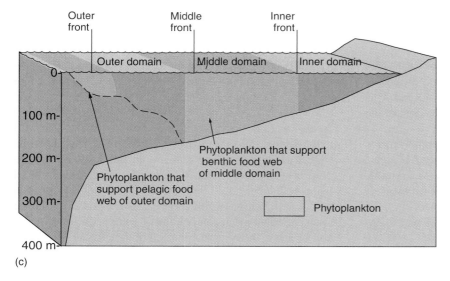

(c)

Figure 13-17   Biological Productivity and Circulation in the Bering Sea

(a) Water structure in the southeast Bering Sea. (b) Nutrient supply from upwelling
(c) Areas of large phytoplankton biomass.

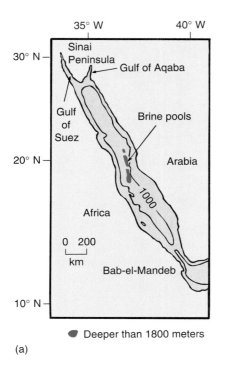

(a)

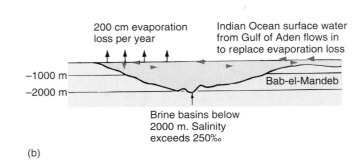

(b)

Figure 13-18   The Red Sea

(a) Bathymetry of the Red Sea. (b) Circulation between the Red Sea and the Indian Ocean.

Peninsula is slowly separating from the African mainland through the process of sea-floor spreading.

Sill depth at the Strait of Bab-el-Mandeb is only 125 m (410 ft) compared to a maximum depth of more than 1000 m (3280 ft) over parts of the Red Sea. Across this shallow sill, the basic circulation seems to be dominated by a high rate of evaporation, which exceeds 200 cm/yr (79 in/yr). The surface water loss in the Red Sea must be replaced by Indian Ocean water from the Gulf of Aden, which enters as a surface flow. As this surface flow moves north, its density increases as evaporation increases its salinity. The dense water sinks and returns as a subsurface flow to the sill and out into the Gulf of Aden. This outflowing, warm saline water sinks rapidly until it finds its equilibrium depth then spreads out into the Indian Ocean. This pattern of circulation (Figure 13-18b) is similar to that of the Mediterranean Sea.

Because of the arid conditions in the region, the surface water in the Red Sea reaches a salinity of 42.5‰ and a temperature of 30°C (86°F) during the summer months. Below a depth of 200 m (656 ft), a very uniform mass of water extends to the bottom throughout most of the Red Sea. This deep-water mass has a temperature of 21.7°C (71°F) and a salinity of 40.6‰.

In 1966, investigators aboard the Woods Hole Oceanographic Institution vessel *Chain* studied a series of deep basins in the central Red Sea region that had been previously noted to contain extremely high-salinity and high-temperature water masses. Between 21°15' and 21°30'N latitude, two major basins were found—the Discovery Deep to the south and the Atlantis II Deep to the north. These and other similar basins contain brines with temperatures in excess of 36°C (96.8°F) and salinities as high as 257‰. These

brines, because of their high salinity, have densities great enough to keep them in their basins and prevent them from mixing with the overlying Red Sea water. The top of the brine accumulations is slightly below 2000 m (6650 ft).

The brines may be formed from hydrothermal circulation of seawater in the hot, porous oceanic crust. The sediments associated with the brines have concentrations of salts and metals that give them a great potential for future mining. Quite probably the enrichment in the crust beneath the sediment is also great and will enhance the economic potential of this area. The Discovery Deep contains the first hydrothermal springs to be discovered. This discovery provided the impetus to initiate the search that led to the discovery of the hydrothermal vents in the eastern Pacific.

Arabian Sea   The **Arabian Sea** is the northward extension of the Indian Ocean between Africa and India (Figure 13-19). Surface currents in the Arabian Sea are dominated by monsoon winds that blow from the northeast from November until March, when the southwest monsoon begins to develop. The air carried onto the continent during this southwest monsoon contains large quantities of water and produces heavy precipitation in the coastal regions.

During the northeast monsoon, the surface current moves south along the west coast of India and turns west at about 10°N latitude. Here, some of the surface water flows into the Gulf of Aden and the rest turns south along the Somali coast to converge with the North Equatorial Current. When the southwest monsoon begins, the North Equatorial Current disappears, and a portion of the South Equatorial Current flows north along the Somali coast as the Somali Current. This strong seasonal current flows with velocities in excess of 11 km/h (7 mi/h) and continues along the coast of Arabia and India in a clockwise pattern until it reaches 10°N latitude. Here, it becomes the Southwest Monsoon Current, and replaces the North Equatorial Current. Because of the alignment of winds relative to the African and Arabian coasts, significant upwelling occurs during the southwest monsoon.

Surface salinities north of 5°N latitude in the Arabian Sea are generally above 36‰ during the northeast monsoon. The Somali coastal region surface salinity may fall below 35.5‰ because of dilution by the South Equatorial Current and upwelling during the southwest monsoon. During the rainy season, surface salinities of less than 35‰ can be found as a result of dilution due to precipitation and runoff. Surface temperatures in the central region reach a maximum of 28°C (82°F) in June and a minimum temperature of 24°C (75°F) in February.

Below 200 m (656 ft), the salinity decreases to values near 35‰ until an abrupt salinity maximum of 35.4 to 35.5‰ develops at depths above 1000 m (3280 ft). This salinity maximum represents the flow of water into the Arabian Sea from the Persian Gulf and the Red Sea. The temperature of the Red Sea water ranges from 9°C to 10°C (48°F to 50°F). Red Sea water can also be identified on the basis of 0.45 ml/l concentration of dissolved oxygen at about 790 m (2591 ft) off the Somali coast. Hydrogen sulfide has been observed on the continental slope at depths where this oxygen minimum occurs in the northern Arabian Sea.

Bay of Bengal The **Bay of Bengal** is bounded by India and the island of Sri Lanka on the west, and by the Malay Peninsula and the Andaman-Nicobar Island Ridge to the east (Figure 13-19). A significant continental influence is exerted on this body of water by the runoff from the Ganges and Brahmaputra rivers at the extreme north end of the bay.

The currents in the Bay of Bengal are also dominated by the monsoon wind system. During the southwest monsoon, a clockwise rotation is established within the Bay; this circulation is accompanied by upwelling along the east Indian coast. With the development of the northeast monsoon in November, the circulation reverses and forms a counterclockwise gyre. At Chittagong in southeast Bangladesh, a seasonal change in the sea level of 1.2 m (4 ft) occurs as a result of monsoon wind changes.

The surface salinity in the Bay of Bengal seldom exceeds 34‰. During the southwest monsoon, particularly during late summer when the rainfall is the greatest, the runoff from the Ganges, Brahmaputra, and other rivers along the coast of Burma and India dilute the water and reduce surface salinity. Values as low as 18‰ may be observed at the extreme north end of the Bay. The major influence of the dilution is observed along the Indian coast; it is less significant away from shore.

Of all the nations on Earth, Bangladesh is under the greatest threat if a significant sea level rise occurs. More than 80 percent of this nation is built on delta, and it is regularly devastated by storm surges that occur during the southwest monsoon. Any rise in sea level will only make things worse. See "Coastal Population Growth: A Global Ecosystem at Risk" on page xviii–1 for a more detailed discussion of the peril Bangladesh may face.

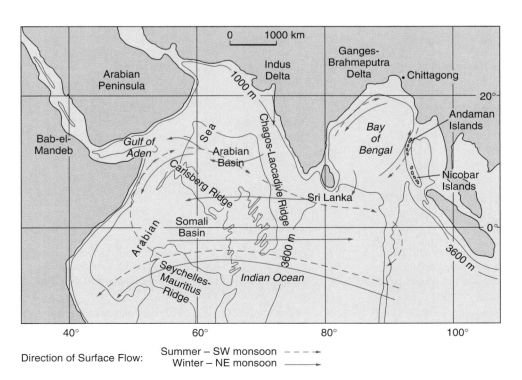

Figure 13-19 Bathymetry and Surface Circulation of the Arabian Sea and the Bay of Bengal

Summary

Characteristics of the coastal ocean, such as temperature and salinity, vary over a greater range than the open ocean. This is because the coastal ocean is shallow and experiences river runoff, tidal currents, and seasonal changes in solar radiation. Coastal geostrophic currents are produced as a result of fresh water runoff and coastal winds.

Estuaries are semi-enclosed bodies of water where fresh water runoff from the land mixes with ocean water. Estuaries are classified by their origin as coastal plain, fjord, bar-built, or tectonic. Mixing patterns of fresh and saline water in estuaries include vertically mixed shallow estuaries; slightly stratified; highly stratified; and salt wedge, which is characteristic of large-volume river mouths. Typical estuarine circulation consists of a surface flow of low-salinity water toward the mouth, and a subsurface flow of marine water toward the head. Chesapeake Bay is a classic example of a coastal plain estuary with most of its stream inflow entering along its western margin. This fact, along with the Coriolis effect acting on the estuarine circulation, makes the eastern side of the bay saltier.

Wetlands are of two types—salt marsh and mangrove swamp. These biologically productive regions are alternately covered and exposed by the tide. Marshes scrub the water of land-derived pollutants before they reach the ocean. Despite their ecological importance, wetlands are fast disappearing from our coasts as they are destroyed by human activities.

Long offshore deposits called barrier islands protect marshes and lagoons. Some lagoons have very restricted circulation with the ocean and may exhibit a great range of salinity and temperature conditions as a result of seasonal change.

Marginal seas are relatively large semi-isolated bodies of marine water often formed by tectonic processes that isolate patches of ocean crust or that are forming new ocean basins. For example, the Mediterranean is a remnant of the ancient Tethys Sea, which has been closing up as the plates containing Africa and Europe move closer together. The Red Sea is the beginning of a new ocean forming as the plates containing the Arabian Peninsula and the African mainland separate.

Marginal seas typically are shallower than the open ocean and have varying degrees of restricted circulation. The Mediterranean Sea has a classic circulation characteristic of restricted bodies of water in areas where evaporation greatly exceeds precipitation. It is the reverse of estuarine circulation. In other marginal seas, salinities may fall below those of normal seawater because of large influxes of river runoff.

Seasonal wind patterns like those found in the Gulf of California, cause coastal upwelling and high biological productivity. An extensive shelf on the northeast side and a deep basin to the southwest are characteristic of the Bering Sea north of the Aleutian Islands. Many important commercial fisheries exist in this sea. The Arabian Sea to the west of India, and the Bay of Bengal to the east have circulation patterns determined by the monsoon winds.

Key Terms

Arabian Sea	Estuarine circulation	Lagoon	Salt marshes
Bar-built estuaries	Estuary	Lesser Antilles	Salt wedge estuary
Bay of Bengal	Fjords	Mangrove swamps	Sill
Bering Sea	Greater Antilles	Marginal seas	Slightly stratified estuary
Caribbean Sea	Gulf of California	Mediterranean circulation	Tectonic estuaries
Chesapeake Bay	Gulf of Mexico	Mediterranean Sea	Vertically mixed estuary
Coastal plain estuaries	Highly stratified estuary	Red Sea	Wetlands
Columbia River estuary			

Questions and Exercises

1. Discuss the effects of offshore winds and freshwater runoff on salinity distribution in coastal oceans where deep mixing does not occur. How do the winter and summer seasons affect the temperature distribution in the water column?

2. How does coastal runoff of low-salinity water produce a longshore geostrophic current?

3. Describe the difference between vertically mixed and salt wedge estuaries in terms of salinity distribution, depth, and volume of river flow. Which displays the more classic estuarine circulation pattern?

4. Name the two types of wetland environments and the latitude ranges where each most commonly develops. How do wetlands contribute to the biology of the oceans and the cleansing of polluted river water?

5. Describe the circulation between the Atlantic Ocean and the Mediterranean Sea, and explain how and why it differs from estuarine circulation.

6. Describe how coastal upwelling in the Gulf of California is related to seasonal winds.

7. Discuss the unusual depth distribution of the Bering Sea and its relationship to biological productivity.

8. Compare and contrast the circulation between the Red Sea and the Indian Ocean to that between the Mediterranean Sea and the Atlantic Ocean.

9. Explain why the Red Sea water that flows into the Indian Ocean at a depth of 125 m (410 ft) sinks to 1000 m (3280 ft) before spreading throughout the Arabian Sea.

10. Describe the relationship between the surface circulation and monsoon winds in the Bay of Bengal.

References

Cherfas, C., 1990. The fringe of the ocean—under siege from land. *Science* 248:4952, 163–65.

Cooper, S. R., and G. S. Brush, 1991. Long-term history of Chesapeake Bay anoxia. *Science* 254, 992–96.

Fairbridge, R. W., ed. 1966. *Encyclopedia of Oceanography*. New York: Van Nostrand Reinhold.

Molinari, R. L. et al., 1977. Winter intrusions of the loop current. *Science* 198:505–6.

Officer, C. B., 1976. Physical oceanography of estuaries. *Oceanus* 19:5, 3–9.

Pickard, G. L., 1964. *Descriptive Physical Oceanography: An Introduction.* New York: Macmillan.

Turner, R. E., S. W. Woo, and H. R. Jitts, 1979. Estuarine influences on a continental shelf plankton community. *Science* 206:218–220.

Valiela, I. and S. Vince, 1976. Green borders of the sea. *Oceanus* 19:5, 10–17.

Suggested Reading

Sea Frontiers

Baird, T. M., 1983. *Life in the High Marsh.* 29:6, 335–41. The ecology of the salt marsh is discussed.

Baker, R. D., 1972. *Dangerous Shore Currents.* 18:3, 138–43. Various types of currents found near the shore can be hazardous to swimmers.

de Castro, G., 1974. The Baltic—to Be or Not to Be? 20:5, 269–73. Reviews the pollution threat to life in the Baltic Sea and what has been done to solve the problem.

Edwards, L., 1982. *Oyster Reefs: Valuable to More than Oysters.* 28:1, 23–25. The value of oyster reefs to various estuarine lifeforms is detailed.

Heidorn, K. C., 1975. *Land and Sea Breezes.* 21:6, 340–43. The causes of land and sea breezes are discussed.

Iverson, E. S., and D. E. Jory, 1990. *Arms Race on the Grass Flats.* 35:5, 304–11. The arsenal of weapons used by crustaceans, turtles, and octopuses to subdue their molluscan prey on the marsh grass flats is revealed.

Olson, D. B., 1990. *Monsoons and the Arabian Sea.* 36:1, 34–41. The monsoons of the Arabian Sea are discussed in terms of the physical conditions that cause them and their effects on coastal inhabitants.

Osing, O., 1974. The Meeting of the Seas. 20:1, 21–24. Conditions at the northern tip of Denmark, where the North Sea and the Baltic Sea meet, are described.

Schaefer, F. S., 1993. *To Harvest the Chesapeake.* 39: 1,36–49. A history of Chesapeake Bay's fisheries provides the basis for explaining why the oysters, blue crabs, and softshell clams are becoming scarce in the bay.

Sefton, N., 1981. *Middle World of the Mangrove.* 27:5, 267–73. The life-forms associated with Cayman Island mangrove swamps are described.

Smith, F., 1982. *When the Mediterranean Went Dry.* 28:2, 66–73. The author details how global plate tectonics caused the Mediterranean Sea to dry up some 5 million years ago.

Scientific American

Degens, E. T., and D. A. Ross, 1970. The Red Sea hot brines. 222:4, 32–53. The source of the hot brine pools in the Red Sea and the economic potential resulting from metal enrichment associated with the brine deposits are considered.

Holloway, M., 1994. Nurturing nature. 270: 4, 98–108. The Florida Everglades are drying up because they receive only 20 percent of the water they received before humans began to modify them. They are the focus of a mammoth restoration effort.

Hsu, K. H., 1972. *When the Mediterranean Dried Up.* 227:6, 26–45. Evidence is presented that shows the Mediterranean Sea was a dry basin 6 million years ago.

Kusler, J. A., W. J. Mitsch, and J. S. Larson, 1994. Wetlands. 270: 1. The nature of seven types of wetlands, their ecological significance, and the efforts to save or restore them are discussed.

Oceanography on the Web

Visit the Introductory Oceanography home page for on-line resources for this chapter. There you will find an on-line study guide with review exercises and links to ocean-ography sites to further your exploration of the topics in this chapter. Introductory Oceanography is at http://prenhall.com/thurman (click on the Table of Contents menu and select this chapter).

Bycatch: Dolphin-Safe Tuna[1] and Turtle-Safe Shrimp

- What is bycatch?
- What do consumers think "dolphin-safe" means?
- What is the exact meaning of "dolphin-safe?"
- What are the social, economic, and environmental costs and benefits of dolphin-safe fishing methods?
- What is an ecologically acceptable impact for commercial fishing?

Introduction

One of the main problems with large scale "harvesting" of wild marine organisms for human consumption is that most commercial fishing techniques are indiscriminate, that is, they cannot selectively capture only the target species. As a result, as much as 25% of the total global commercial catch is wasted or unused.

This quantity is known as "bycatch" and refers to undersized fish, low-value, and non-target species It may include benthic organisms like sponges, worms, sea stars, crabs, etc, and also sharks, dolphins, whales, sea turtles, and even birds. Bycatch may die in nets or on longlines, or is often returned to the water dead or dying.

Shrimp Trawling

Among the most harmful fishing activities is trawling for shrimp. In addition to damaging the ocean bottom (trawling has been compared to clearcutting a forest!), as much as 90% of the trawl contents may be non-target and hence unused species, sometimes called "trash fish" by fishers, and at times including endangered sea turtles. (Stop and consider how the phrase "trash fish" strike you. Does it show respect for life?)

Shrimp trawling is widespread and may cause extreme environmental damage. While as many as 25,000 boats ply U.S. waters and the U.S. imports wild-caught shrimp from nearly 40 countries, consumers are virtually unaware of the dimensions of its destructiveness. We'll discuss aspects of this issue on our web site.

In contrast to this ignorance of the environmental impact of shrimp trawling is a bycatch issue that most canned tuna consumers and others are well aware of—the capture of dolphins by tuna fishers.

In this issue we will focus on the multifaceted issue of bycatch in the tuna fishing industry and evaluate the costs and benefits of bycatch-reduction techniques.

Dolphins and Tuna in the ETP

The eastern tropical Pacific Ocean (ETP), an area of approximately 8 million square miles (21,000,000 km^2), is one of the world's richest sources of commercially important tunas. The ETP fishery for yellowfin tuna (*Thunnus albacares*), in fact, has been called one of the most important fisheries in the world. Yel-

Figure 1

School fishing for tuna.   (© Hulton-Deutsch Collection/CORBIS)

lowfin and skipjack tunas (*Katsuwonus pelamis*) are mainstays of the canned light meat tuna industry. The ETP fishery for albacore (*Thunnus alalunga*), whose flesh is the basis of the white-meat tuna industry, is small by comparison.

Two methods have been widely used to catch yellowfin and skipjack tunas in large-scale fisheries in the ETP. In *school fishing* (Figure 1), a technique no longer practiced in the ETP, rugged commercial fishers used stout rods to catch tunas, which frequently bit unbaited hooks during their feeding frenzy. Worldwide, according to Bumblebee Seafoods, 40% of the world's commercial tuna are caught on pole and line. A more productive method of catching yellowfin and skipjack tunas is *purse seining*. Globally, 30% of the world's commercial tuna are caught in purse seines. (Long-lining, in which hooks are set at intervals along a horizontal line stretching for miles, accounts for 30% of world commercial tuna catch, essentially albacore, which are also caught commercially by trolling).

In purse seining (Figure 2), a school of fish is encircled by speed boats with a net that may be 2 km (1.2 mi.) long and 200 meters (660 ft) deep. A purse line attached to the bottom of the net is then pulled in, trapping the tunas and other organisms unfortunate enough to be in the same location. Vessels from 12 nations, including the U.S., purse seine in the ETP for tuna.

Figure 2

A tuna purse seiner with spread net. When the circle is complete, the bottom of the net is closed, trapping all sea creatures within. *(Photos by Glenn Oliver/Visuals Unlimited)*

Figure 3

These photos of dead dolphins being hauled on board the Panamanian tuna boat "Maria Luisa" were captured from a video taken by a marine biologist who went undercover for five months on the boat to document the dolphin killings. The video from which these images were taken was broadcast on U.S. television and resulted in a huge public outcry. *(AP LaserPhoto)*

In the ETP, tunas frequently congregate around floating objects, such as tree trunks[2], and also along with two kinds of dolphins, northern offshore spotted (*Stenella attenuata*) and eastern spinner (*Stenella longirostris*), a fact discovered by the U.S. tuna fleet nearly 3 decades ago. This relationship is thought to benefit the tunas, which can easily follow dolphins and take advantage of the latter's superior prey-finding abilities. Setting nets around dolphins typically catches the largest tunas and is thus the more desirable method.

When modern tuna seiners enter an area, they can spot aggregations of tunas and dolphins fairly easily, especially by helicopter, because dolphins are noisy and disturb the sea surface. The netting process, which can take 2–3 hours, does not discriminate between the tunas and the dolphins, which stay together throughout the process. A number of dolphins can die due directly to entanglement and drowning (Figure 3), and more may die later due to the delayed effects of severe trauma. It is estimated that the purse-seine fishery for tuna killed more then 1.3 million eastern spotted dolphins in the ETP between 1959 and 1990. As many as 5 million dolphins were killed during the first 14 years of purse seining in the ETP.

Policies to curb dolphin mortality

There have been several legislative and international attempts to curb the killing of dolphins during tuna seining. The first of these was an agreement reached with the Inter-American Tropical Tuna Commission (IATTC) in 1976 (but not funded until 1979). This program sought (1) to determine dolphin mortality, (2) to reduce it such that dolphin populations were not threatened and accidental killing was avoided, and (3) to maintain a high level of tuna production. The chief result of this effort was the placement of observers on one-third of all vessels fishing in the ETP. As a result, the first reliable estimates of dolphin mortality were made.

This was followed by a series of policies designed to lower dolphin mortality, including new provisions of the Marine Mammal Protection Act (MMPA) in 1988 and 1990; the Dolphin Protection Consumer Information Act of 1990; the Agreement for the Conservation of Dolphins of 1992 (more commonly called the La Jolla Agreement); the Panama Declaration of 1995; and the International Dolphin Conservation Program Act (IDCPA) of 1999.

These resulted in 100% observer coverage of ETP tuna seiners and established international limits of fewer than 5000 dolphins killed by 1999. Moreover, criteria were instituted for labeling canned tuna as "dolphin-safe". As we will see, the success of the "dolphin-safe" labeling program as a deterrent to killing dolphins is unsettled. However, as a marketing tool it is unequivocal: people buy the product. For 1996, domestic canned tuna sales approached $1 billion (Statistical Abstract of the U.S. 1998, p. 697, Table 1164)

There is no question that dolphin mortality has decreased in the ETP as a result of conservation measures. But the issue remains controversial and repercussions have been felt ecologically, economically, socially, and politically, as you will see below and on our web site, where you should now proceed.

See the complete issue at:
www.prenhall.com/thurman

---

[1] The "dolphin-safe" portion of this issue was written with Robert Young of Coastal Carolina University.

[2] Surprisingly, enough such objects enter the ocean to be worthwhile to commercial fishers.

14 The Marine Environment

A comb jelly (Mertensia ovum) trails red tentacles to attract and capture prey in Arctic waters. (Photo by Norbert Wu, courtesy Peter Arnold, Inc.)

The distribution of life on our planet is far from uniform. On land, life is most abundant and diverse in areas where water is plentiful—consider the lush vibrant environment of the tropical rainforest compared to the sparse, barren desert. In the oceans, water supply is obviously not a problem, and the physical and chemical conditions are generally far more constant than in the terrestrial environment. However, the distribution of life in the oceans is extremely uneven; there are drastic variations in the diversity and abundance of marine life both with depth and area. Ocean water temperatures, pressures, chemical compositions, and circulation patterns are all important influences on the distribution of living things.

Most organisms found in the open ocean do not have highly specialized regulatory systems to combat sudden changes that might occur in their environment. They are, therefore, affected to various degrees by quite small changes in salinity, temperature, turbidity, and other environmental variables. On the other hand, organisms that dwell in coastal waters have had to adopt various strategies to cope with salinity and temperature changes, with turbidity, and, in intertidal zones, with exposure to the air at low tide. All marine organisms face a number of challenges to their success. Locomotion through water requires far more energy than moving through air, maintaining bodily fluid composition against osmotic effects is critical, and competition is fierce.

Marine Organisms and the Chemistry of Their Environment

Water constitutes over 80 percent of the mass of protoplasm, the substance of living matter. Over 65 percent of your weight, and 95 percent of a jellyfish's weight, is water (Table 14-1). Water is the transport medium for gases, nutrients, and wastes in living organisms. Water itself is a raw material for photosynthesis. Land plants and animals have developed very complex plumbing systems to distribute water and the vital substances it carries throughout their bodies and to prevent its loss.

Salinity and Osmosis

When water-permeable membranes, such as those that surround living cells, separate water solutions of unequal salinity, water molecules diffuse through the membrane from the lower-salinity solution into the higher-salinity solution. This process is called **osmosis**, from a Greek word meaning "to push" (Figure 14-1). *Osmotic pressure* is the pressure that must be applied to the more saline solution to prevent passage of water molecules into it from a supply of pure water.

Osmosis is important to organisms that live in water, either marine or fresh. These organisms have body parts that are semipermeable membranes separating internal body fluids from seawater. If the salinity of an organism's body fluid equals that of the ocean, the organism is said to be **isotonic**, in reference to the fact that the osmotic pressure inside and outside the cell is equal. Thus, no net transfer of water occurs through the membrane in either direction.

The higher the salinity difference between the internal and external solutions, the higher the osmotic pressure is for the system. We can compare the osmotic pressure of solutions by comparing their concentrations of dissolved solids—their salinities (Figure 14-2). This comparison is readily made by determining their freezing points since the addition of dissolved solids to an aqueous solution lowers its freezing point (see Chapter 6).

If water outside an organism's cells has a lower salinity than the fluid within the cell, there will be a net movement of water through the cell membranes into the cells (always moving into the more concentrated solution). This organism is **hypertonic** (saltier) relative to its environment. If the salinity in an organism's cells is less than that of the surrounding water, water passes out of the cells through the cell membranes, into the more concentrated solution. This organism is **hypotonic** (less salty) relative to the external medium.

It may be easier to view osmosis from the perspective of the concentration differences of water rather than of the salts dissolved in it. If you consider the relative concentration of water molecules on either side of the semipermeable membrane, then osmosis is simply the net transfer of water molecules from the side with the greater concentration of water molecules, through the membrane, to the side with the lesser concentration of water molecules.

OSMOSIS

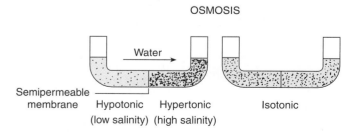

Semipermeable membrane Hypotonic (low salinity) Hypertonic (high salinity) Isotonic

Figure 14-1   Osmosis

Osmosis occurs when two water solutions of different salinities are separated by a semipermeable membrane. The membrane allows water molecules to pass through it, but not the ions of dissolved substances. Water molecules diffuse through the membrane from the less concentrated (hypotonic) solution (left) into the more concentrated (hypertonic) solution (right). If the salinities of the two solutions are the same (isotonic), there is no net movement of water through the membrane.

TABLE 14-1 Water Content of Organisms	
Organism	Water content (%)
Human	65
Herring	67
Lobster	79
Jellyfish	95

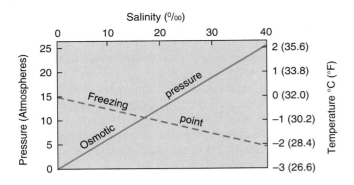

Figure 14-2 Osmotic Pressure and Freezing vs. Salinity
Osmotic pressure increases and the freezing point decreases with increased salinity.

In the case of marine invertebrates, the body fluids and the seawater in which they are immersed are in a nearly isotonic state. Since no significant difference exists, these creatures do not have special mechanisms to maintain their body fluids at a specific salinity. Thus, they have an advantage over their freshwater relatives, whose body fluids are hypertonic relative to the low-salinity freshwater (Table 14-2).

In contrast to marine invertebrates, marine fish have body fluids that are only about one-third as saline as ocean water. They are, therefore, hypotonic (less salty) with respect to the surrounding seawater. This low level may be the result of their having evolved in low-salinity coastal waters. This difference creates a problem in that ocean-dwelling fish, without some means of regulation, would dehydrate by osmosis, with loss of water from their body fluids into the surrounding ocean. However, they counter this loss by drinking seawater and excreting the salts through *chloride cells* in their gills. They also maintain their body water by discharging only very small amounts of highly concentrated urine (Figure 14-3). Osmosis also explains why we humans shouldn't drink seawater. The salinity of our own body fluids is about one-fourth that of seawater. To drink seawater would cause a drastic loss of water by osmosis, through the membranes surrounding our digestive tract, causing eventual dehydration.

Freshwater fish are hypertonic (more salty) relative to the very dilute water in which they live. The osmotic pressure of their internal fluids may be 20 to 30 times greater than that of the fresh water that surrounds them. Freshwater fish avoid taking on water by osmosis, which would eventually rupture cell membranes, by not drinking water and by having cells that can absorb salt. They dispose of excess water as large volumes of very dilute urine (Figure 14-3).

Sensitivity to salinity varies among species. For example, some oysters live in estuarine environments at the mouths of rivers and are capable of withstanding a considerable range in salinity. Organisms that inhabit coastal regions have evolved a tolerance for a wide range of salinity conditions. They are called **euryhaline,** from the Greek root word *eury-,* meaning wide or broad; they are "widely salty." By contrast, other marine organisms, particularly those that

TABLE 14-2 Osmotic Pressure of Body Fluids in Various Organisms.

A comparison of the freezing points of the body fluids of some marine and freshwater organisms and the waters in which they live. The higher the salinity and osmotic pressure, the lower the freezing point. Marine invertebrates and sharks are essentially isotonic, whereas most marine vertebrates are hypotonic. Essentially all freshwater forms are hypertonic. The freezing points of ocean water corresponding to invertebrate specimens vary because the specimens were taken from coastal waters, where water salinity varied.

Freezing Points (°C)			Freezing Points (°C)		
Animals	Body Fluid	Water	Animals	Body Fluid	Water
Marine Invertebrates			Freshwater Invertebrates		
Annelid worm	−1.70	−1.70	Mussel	−0.15	−0.02
Mussel	−2.75	−2.12	Water flea	−0.20	−0.02
Octopus	−2.15	−2.13	Crayfish	−0.80	−0.02
Lobster	−1.80	−1.80			
Crab	−1.87	−1.87			
Marine Vertebrates			Freshwater Vertebrates		
Cod	−0.74	−1.90	Eel	−0.62	−0.02
Shark	−1.89	−1.90	Carp	−0.50	−0.02
Turtle	−0.50	−1.90	Water snake	−0.50	−0.02
Seal	−0.50	−1.90	Manatee	−0.50	−0.02

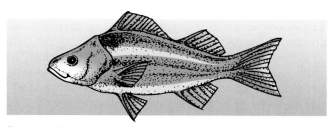

FRESH WATER FISH
(Hypertonic)

Do not drink
Cells absorb salt
Large volume of dilute urine

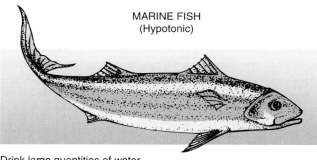

MARINE FISH
(Hypotonic)

Drink large quantities of water
Secrete salt through special cells

Figure 14-3   Freshwater (Hypertonic) and Marine
(Hypotonic) Fish

inhabit the open ocean, are seldom exposed to much salinity variation. Consequently, they are adapted to steady salinity and can withstand only very small changes. These organisms are called **stenohaline**, from the Greek root word *steno-*, meaning narrow; they are "narrowly salty."

Dissolved Silica and Carbonate

Some phytoplankton and many animals extract dissolved compounds from seawater to construct the hard parts of their bodies that serve as supportive skeletons or protective coverings. These hard materials are most commonly made of silica ($SiO_2 \cdot nH_2O$) or calcium carbonate ($CaCO_3$). Silica is used by a very important phytoplankton population in the ocean—diatoms—and by microscopic protozoa called radiolaria. Calcium carbonate is used by the coccolithophorids, foraminifera, most mollusks, corals, and many benthic (bottom-dwelling) algae.

Nutrients and Sunlight

The distribution of life throughout the oceans depends on the availability of nutrients, the raw materials needed to make organic matter, and sunlight needed by the phytoplankton that form the first link in ocean food chains. Where and when sunlight is available and the conditions are right for the supply of large quantities of nutrients, marine populations reach their greatest sizes.

How do nutrients get into, and waste products get out of cells? In addition to being water-permeable, the membrane surrounding a living cell allows diffusion of many small molecules dissolved in the water. **Diffusion**, like osmosis, is driven by concentration differences (Figure 14-4). Both processes tend to eliminate differences in the concentrations of dissolved compounds in fluids. Diffusion refers to the movement of dissolved compounds in water, with or with out a membrane; osmosis refers to the movement of water molecules through a membrane separating waters of different salt concentrations.

Nutrient compounds are relatively plentiful in seawater. They diffuse through the cell membrane into the cytoplasm, where nutrients tend to be less concentrated because the cell rapidly converts them into cell material, energy, or waste products. Waste products diffuse out of the cell through the cell membrane, driven by the higher concentrations within the cell than in the surrounding fluid medium.

Note that these diffusional processes reach an equilibrium, with no net movement of molecules across the cell membrane, if the water outside the cell, which is waste rich and nutrient poor, is not replaced by fresh nutrient-rich water. Large complex animals have developed intricate internal circulatory systems, but small single-celled organisms and those without control of their own mobility must rely on the physical motion of the water they live in to perform this task. For this reason, there are very few aquatic organisms that can live in stagnant pools. Waves, tides, and currents perform a critical role in providing nutrients and removing wastes for the organisms in many marine ecosystems.

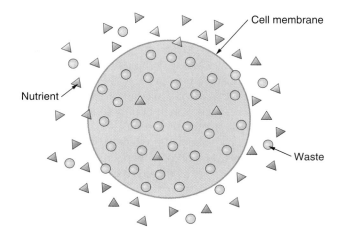

Figure 14-4   Diffusion

If a water-soluble substance is piled on the bottom of a container of water, it eventually becomes evenly distributed throughout the water. This is due to random molecular motion. Diffusion means that molecules of a substance move from an area of higher concentration (of the substance) to an area of lower concentration until the distribution of molecules is uniform. Nutrients (triangles) are in higher concentration outside the cell in this example. They diffuse into the cell through the cell membrane. Wastes (circles) are in higher concentration inside the cell and diffuse out of the cell through the membrane.

To summarize, these three things happen simultaneously across the cell membrane:

1. Water molecules move from the side of greater water concentration toward the side of lesser water concentration (higher salinity) by osmosis;
2. Nutrient molecules or ions move by diffusion into the cell, where they are rapidly used up in metabolic processes;
3. Waste molecules move by diffusion from the cell into the surrounding water.

Availability of Nutrients In Chapters 2 and 5, we discussed the role of water in eroding the continents, carrying the eroded material to the oceans, and depositing it as sediment at the margins of the continents. During this process, water accumulates many dissolved compounds. Among these are some of the nutrients that organisms need to make their cell material and their hard parts, as well as compounds to be broken down for energy. Because the continents are the major sources of these nutrients, you might expect the greatest concentrations of marine life to be at the continental margins—and this is the case. As we head for the open sea from the continental margins, we travel through waters containing progressively less marine life.

Deep-ocean water is rich in nutrients produced by the bacterial decomposition of organic matter in the water column and in bottom sediments. Wherever this deep water returns to the surface by upwelling, we find large populations of organisms. As we discovered in Chapter 9, upwelling can be highly seasonal in many locations and also is affected by El Niño events.

Availability of Solar Radiation Phytoplankton carry on photosynthesis, but photosynthesis cannot proceed without light energy provided by the Sun. Sunlight penetrates the atmosphere quite readily, so land plants would seem to have an advantage over those in the ocean. There is almost never a shortage of solar radiation to drive the photosynthetic reactions for terrestrial plants. In contrast, ocean water is a significant barrier to the penetration of solar radiation. In the clearest water, a minute amount of solar energy is detectable to depths of about 1 km (0.6 mi), but the amount is too small to power photosynthetic reactions. Photosynthesis in the ocean is restricted to a very thin surface layer, approximately the upper 100 m (330 ft), the depth to which sufficient light energy for photosynthesis penetrates. Near the coast, the photosynthetic zone is much thinner because the water contains more suspended material, which restricts light penetration.

Marine Organisms and the Physical Properties of Their Environment

Importance of Organism Size

Marine phytoplankton are quite simple compared to the specialized plant forms that exist on land. Why have phytoplankton remained as single-celled organisms instead of becoming multicellular? The major requirements of phytoplankton are that they stay in upper ocean water where solar radiation and nutrients are available. These requirements lead us to consider phytoplankton design—specifically their size and shape.

Phytoplankton in the upper layers of the ocean have no roots and no means of swimming, yet they maintain their general position in the water. How they do so is closely related to the ratio between their surface area and body mass. Greater surface area per body mass means a greater resistance to sinking because more surface is in frictional contact with the surrounding water. An examination of Figure 14-5 will help you see how this ratio of surface area to mass increases with decreasing size. You can see that it benefits single-celled organisms, which make up the bulk of photosynthetic marine life, to be as small as possible.

Also related to the ratio of surface area to mass, is the efficiency with which photosynthetic cells take in nutrients from the surrounding water and expel waste through their cell membranes. Both the intake of nutrients and waste disposal depend partly on diffusion, so eventually, increased size would reduce the ratio of surface area to body mass to the point where the cell could not function properly and would die. This is why we find cells in all plants and animals to be microscopic, regardless of the overall size of the organism, from bacteria to whales. Each cell must take in nutrients and dispose of wastes rapidly enough to maintain life processes.

Diatoms are an important group of phytoplankton. You can see representative diatoms in Figures 14–14c-l. Examining diatoms carefully, we see that they commonly have long, needlelike extensions. These increase a diatom's ratio of surface area to mass. Other members of the microscopic marine community display a similar strategy.

Small organisms use another contrivance to stay in the upper layers of the ocean—some produce a drop of oil, which lowers their overall density and increases buoyancy. Despite these adaptations to improve floating ability, the organisms are still denser than water and tend to sink, if ever so slowly. However, this is not a serious handicap, for the tiny cells are carried readily with water movements. Near the surface, wind causes considerable mixing and turbulence. Turbulence dominates the movement of these small organisms, keeping them positioned to bask in the solar radiation needed to photosynthesize, producing the organic compounds and oxygen needed by other members of the marine community.

Physical Support

Land plants have vast root systems that securely anchor the plant to the earth. In the Animal Kingdom, a number of support systems are used. Each requires some combination of appendages—legs, arms, fingers, toes—that must support the entire weight of the land animal. In the ocean, the water that so lavishly bathes the organisms and supplies their needed gases and nutrients serves as their physical support as well. Organisms that live in the open

CUBE *A*

Linear dimension – 1 cm
on each side

|←→|1 cm|←→|

Area – 6 cm² (6 sides)
Volume – 1 cm³

Ration of surface area to volume:

$$\frac{6 \text{ cm}^2}{1 \text{ cm}^3} = 6:1$$

Cube *A* has 6 units of surface area
per unit of volume.

CUBE *B*

Linear dimension – 10 cm
on each side

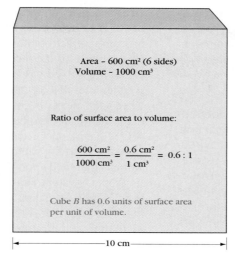

Area – 600 cm² (6 sides)
Volume – 1000 cm³

Ratio of surface area to volume:

$$\frac{600 \text{ cm}^2}{1000 \text{ cm}^3} = \frac{0.6 \text{ cm}^2}{1 \text{ cm}^3} = 0.6:1$$

Cube *B* has 0.6 units of surface area
per unit of volume.

|←————— 10 cm —————→|

Figure 14-5   The Importance of Size

Obviously, cube A is far smaller than cube B. However, compared with cube B, cube A has 10 times the surface area per unit of interior volume. In practical terms, if cubes A and B were plankton, cube A would have 10 times greater resistance to sinking per unit of mass than cube B. Therefore, it could stay afloat by exerting far less energy than cube B. If cube A were a planktonic alga, it could take in nutrients and dispose of waste through its cell membrane 10 times more efficiently than an alga with the dimensions of cube B.

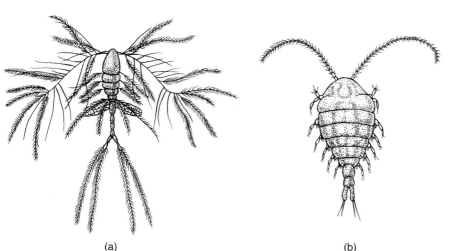

(a)   (b)

Figure 14-6   Water Temperature and Appendages

(a) The copepod *Oithona* displays the ornate plumage characteristic of warm-water varieties. (b) The copepod *Calanus* displays the less ornate appendages found on temperate- and cold-water forms. *(After Sverdrup, Johnson, and Fleming, 1942.)*

ocean depend primarily upon buoyancy and frictional resistance to sinking to maintain their desired positions. This is not to say that they maintain their positions without expending energy. Some have special adaptations that increase their efficiency in positioning themselves. We will discuss these adaptations in this and succeeding chapters.

You know that fluids flow with varying degrees of ease. For example, syrup flows less readily than water. This property is viscosity, which is defined as internal resistance to flow. It is a characteristic of all fluids—water, syrup, motor oil, air, and molten lava. Loosely speaking, viscosity is the thickness of a fluid; the higher the viscosity, the thicker the fluid. Viscosity is particularly affected by temperature (compare the viscosity of refrigerated syrup to heated syrup).

The viscosity of ocean water is affected by two variables—temperature and salinity. Greater salinity increases viscosity. However, temperature has an even greater effect, with lower temperatures thickening seawater. As a result of this temperature relationship, phytoplankton and animals that float in colder waters have less

need for extensions to aid them in floating because the water is more viscous. In fact, members of the same species of floating crustaceans are very ornate with featherlike appendages where they occupy warmer waters, but lack such appendages where they occupy colder, more viscous waters (Figure 14-6). Thus, cold, high-viscosity water benefits floating organisms, helping them maintain their position near the surface.

Not all aquatic organisms are one-celled floaters. With increasing organism size and swimming ability, viscosity ceases to enhance survival and becomes an obstacle instead. This is particularly true for larger animals that swim freely in the open ocean. They must pursue prey or flee a predator, which means they must displace water to move forward. The faster they swim, the greater is the stress on them. Not only must water be displaced ahead of the swimmer, but water must also move in behind it to occupy the space that the animal has vacated. These are important considerations in streamlining.

The familiar shape of free-swimming fish and marine mammals like whales and dolphins exemplifies the

streamlining adaptations that enable organisms to move with minimum effort through water. A common shape that serves this purpose is a flattened body, which presents a small cross section at the front end, with a gradually tapering back end. This form is characteristic of fish. It reduces stress from movement through water and allows minimal energy to be expended in overcoming viscosity (Figure 14-7).

Temperature

Of all the conditions we can measure, none is more important to life in the ocean than temperature. The marine environment is much more stable than the land environment, and a comparison of temperature ranges illustrates this. Ocean temperatures remain within a far narrower range, and change far more slowly, than land temperatures (see Figure 8-17). The minimum surface temperature observed in the open sea is seldom below −2°C (28.4°F), and the maximum surface temperature seldom exceeds 32°C (89.6°F). In contrast, record continental temperatures are -88°C (-127°F) and 58°C (136°F).

Daily ocean surface variations rarely exceed 0.3°C (0.5°F), although in shallower coastal waters they may vary daily by 2–3°C (3.6–5.4°F). In contrast, continental temperatures vary widely from day to day and season to season. Annual ocean temperature variations are also small. They range from 2°C (3.6°F) at the equator to 8°C (14.4°F) at 35° and 45° latitude and decrease again in the higher latitudes. Annual temperature variations in shallower coastal areas may be as high as 15°C (27°F). In contrast, continental temperatures can vary dramatically over a year. Temperature variations in surface waters are reduced with depth. In the deep ocean, daily or seasonal temperature variation becomes insignificant. Throughout the deeper parts of the ocean, the temperature remains uniformly low. At the bottom of the ocean basin, where depth

exceeds 1.5 km (about 1 mi), temperatures hover around 3°C (37°F), regardless of latitude.

Recall from Chapter 6 that reducing temperature increases density and viscosity, and it increases the capacity of water to hold gases in solution. All these changes significantly affect organisms inhabiting the ocean. A direct consequence of the increased capacity of colder water to contain dissolved gases is seen in the vast phytoplankton communities that develop in high latitudes during summer, when solar energy becomes available for photosynthesis. A major factor in this phenomenon is the abundance of dissolved gases, specifically, carbon dioxide, which phytoplankton need for photosynthesis, and oxygen, needed by animals that feed upon the phytoplankton.

Tropical waters support more species, but in colder waters, floating organisms are physically larger. The total biomass of floating organisms in colder, high-latitude planktonic environments greatly exceeds that of the warmer tropics. The fact that organisms in the tropics are smaller than those observed in the higher latitudes may be related to the lower viscosity and density of tropical seawater. Being smaller, tropical species expose more surface area per unit of body mass.

Warmer temperatures increase the rate of biological activity, which more than doubles with an increase of 10°C (18°F). Tropical organisms apparently grow faster, have a shorter life expectancy, and reproduce earlier and more frequently than their counterparts in the colder waters.

In spite of the general belief that species diversity is greatest in higher temperature waters, recent studies of the diversity of planktonic foraminifera in the Atlantic Ocean have shown that diversity peaks in subtropical waters (25°-30° latitude) and begins to decrease when sea surface temperatures (SST) exceed 27°C (80.6°F) (Figure 14-8). It is speculated that this decrease results from temperatures decreasing more gradually and terminating at greater depth in the subtropical thermocline than in the equatorial thermocline, thereby providing more niches for speciation.

Some animal species can live only in cooler waters, whereas others can live only in warmer waters. Many of these can withstand only very small temperature change, and are called **stenothermal** (narrowly thermal). Other varieties are apparently little affected by temperature and can withstand changes over a large range. These are classified as **eurythermal** (widely thermal). Stenothermal organisms are found predominantly in the open ocean at depths where large ranges of temperature do not occur. Eurythermal organisms are found in both shallow coastal waters, where the largest ranges of temperature are found, and in surface waters of the open ocean.

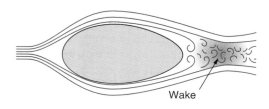

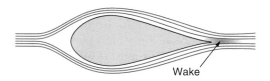

Figure 14-7   Streamlining
Due to the viscosity (thickness, or resistance to flow) of water, any body moving rapidly through it must produce as little stress as possible as it displaces the water through which it moves. After the water has moved past the body, it must flow in behind the body with as little eddy action as possible.

The Distribution of Life ....

Because of its immense volume and the paucity of our knowledge, it is difficult to describe the degree to which the sea is inhabited. We know that some populations fluctuate greatly each season increasing the difficulty of describing the extent to which the marine environment is populated.

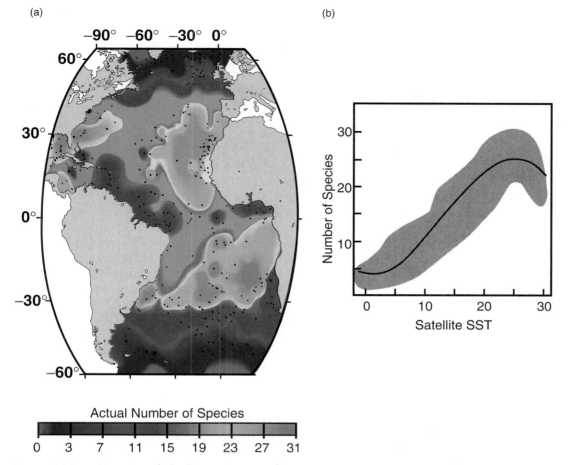

(a)   (b)

Figure 14-8   Diversity of Planktonic Foraminifera vs. Temperature and Latitude

(a) Distribution of species of planktonic foraminifera as measured from 351 core-top sediment samples from the floor of the Atlantic Ocean. Note that diversity increases from the subpolar latitudes to the subtropics and decreases again toward the equator. (b) This curve shows that there is a positive relationship between sea surface temperature (SST) and diversity from -2°C to 27°C and a negative relationship above 27°C.

However, we may compare the marine and terrestrial environments by comparing the number of marine species to land species. Figure 14-9 shows the distribution of Earth's known species. Well over 1,410,000 species have been described, and less than 17 percent of these live in the ocean. Many biologists believe there may be from 3 to 10 million additional unnamed species living on Earth. A large number of these probably inhabit the oceans, yet we may expect fewer species of marine life than of terrestrial life.

If the ocean is a prime habitat for life and life originated in such an environment, why do we now see such a small percentage of the world's species living there? This lesser number may well exist because the marine environment is more stable than the terrestrial environment. The relatively uniform conditions of the open ocean do not produce pressures for organisms to adapt. Also, once we get below the surface layers of the ocean, temperatures not only are stable, but also relatively low. Chemical reactions are retarded by this lower temperature, and this, in turn, may reduce the rate of reproduction and tendency for variation to occur.

Considering the great variety of organisms on the continents, we can assume that this development was the product of a less stable environment, one presenting many opportunities for natural selection to produce new species to inhabit varied new niches. At least 75 percent of all land animals are insect species that have evolved to inhabit very restricted environmental niches. If we ignore the insects, the oceans contain over 45 percent of the remaining animal species.

Water Color and Life in the Oceans

Usually you can determine areas of high and low organic production in the ocean from the color of the water. Neritic, or coastal, waters are almost always greenish in color because they contain more large particulate matter. This material disperses solar radiation in such a way that the most scattered wavelengths are those of greenish or yellowish light. This condition is also partly the result of yellow-green microscopic marine algae in these coastal

Figure 14-9   Distribution of Species on Earth

The entire large cube represents the 1,410,000 species known to be living on Earth. The blue-shaded portion indicates the proportion (16.7 percent) that lives in the oceans. Of the marine species, 98 percent live on or in the ocean floor, and 2 percent are plankton or nekton.

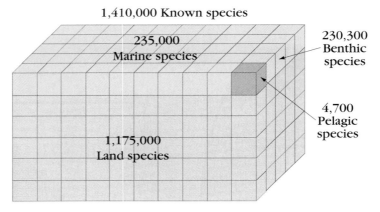

waters. In the open ocean, where particulate matter is relatively scarce and marine life exists in low concentration, the water appears blue due to the size of the water molecules and the scattering they effect on the solar radiation.

Green color in water usually indicates the presence of a lush biological population, such as might correspond to the tropical forests on the continents. The deep indigo blue of the open oceans, particularly within the tropics, usually indicates an area that lacks abundant life and could be considered a biological desert.

Divisions of the Marine Environment

We can readily divide the marine environment into two basic units. The ocean water itself is the **pelagic environment** (pelagic means *of the sea*). Here, floaters and swimmers play out their lives in a complex food web. The ocean bottom constitutes the **benthic environment** (benthic means *bottom*). Plants, algae, and animals that neither float nor swim (or swim poorly) spend their lives there.

Based on our very incomplete knowledge of marine species diversity, the distribution of the 235,000 known marine species shows that only about 4700 (about 2 percent) live above the ocean floor in the pelagic environment. The other 98 percent are benthic, inhabiting the ocean floor. These proportions will certainly change as recent discoveries indicate the existence of many more benthic species in the deep sea than previously thought.

There are many terms for the different divisions of the pelagic environment. You may not remember all of them, but it is important to understand that each labels a zone in the ocean that has distinctive physical and biological characteristics. All of these divisions are shown in Figure 14-10a. The pelagic environment is divided into two provinces. The **neritic province** extends from the shore seaward, including all water overlying an ocean bottom that is less than 200 m (660 ft) in depth. The word neritic means shallow. Seaward, where depth increases beyond 200 m (660 ft), is the **oceanic province**, which includes water that ranges in depth from the surface to the bottom of the deepest ocean trenches.

Neritic Environments

The combined availability of solar radiation and nutrients results in maximum biomass concentrations in shallow coastal waters. Biomass decreases with increasing distance from shore and with greater depth. The coastal region has many physical conditions that appear, at first, to be deterrents to the establishment of life:

1. Waters are shallow, allowing seasonal variations in temperature and salinity that are much greater than in the open ocean.
2. Water depth varies in the nearshore region as a result of tide movements that periodically cover and uncover a thin strip along the margins of the continents.
3. Breaking waves at the shoreline release large amounts of energy.

Such a great concentration of biomass in an environment that contains so many stress factors highlights the importance of the evolutionary process in developing new species by natural selection. Over the eras of geologic time (billions of years), new life forms have developed to fit every imaginable biological niche. Many of these life

Figure 14-10 (right)   Oceanic Biozones

(a) Biozones of the pelagic and benthic environments. (b) Distribution of oxygen (O_2) and the nutrient phosphate (PO_4) in the water column at mid-to-low latitudes. In surface water, oxygen is abundant due to continual mixing with the atmosphere and continual algal photosynthesis; nutrient content is low due to continual uptake by algae. At the very top of the chart, below the euphotic zone, oxygen content abruptly decreases and nutrient content abruptly increases. At or near the base of the mesopelagic zone, an oxygen minimum and nutrient maximum are recorded. Nutrient levels remain high to the bottom. Oxygen increases with depth as deep-water and bottom-water masses carry it into the deep ocean.

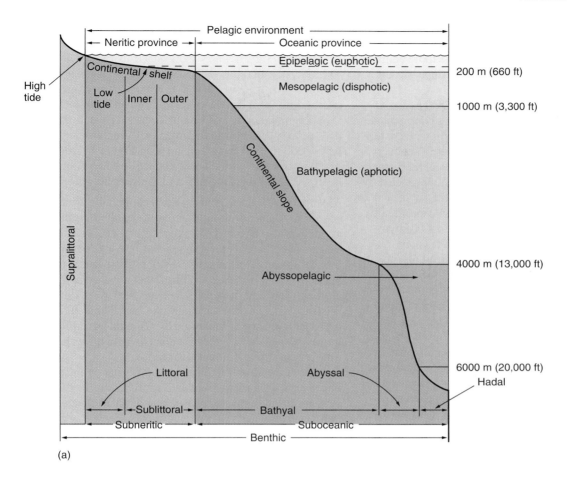

(a)

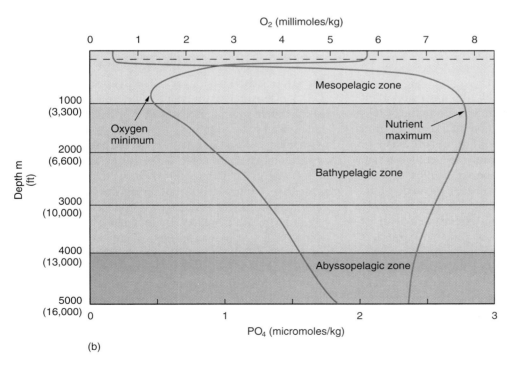

(b)

forms are adapted to adverse conditions under which life can exist as long as nutrients are available.

Along continental margins, some areas have more abundant life than others. What characteristics produce this uneven distribution? Again, we need consider only those basic requirements for the production of food for the answer. If you measure the properties of the water along continental coasts and compare these measurements from place to place, you will find that the areas with the greatest biomass are those with lower water temperatures. The colder water keeps more oxygen and carbon dioxide in solution than does warmer water. Of particular importance is the increased availability of carbon dioxide. This again is an example of the greater availability of the basic requirements of phytoplankton affecting the distribution of life in the oceans.

In certain areas of the coastal margins we find an additional factor that enhances conditions for life—upwelling. Upwelling, as the name implies, is a flow of subsurface water toward the surface. This phenomenon occurs along the western margins of continents in areas where surface currents are moving toward the equator (Figure 14-11a). Along the western margins of continents in both Northern and Southern Hemispheres, the Ekman transport moves surface water away from the coast. As this happens, water from depths of 200 to 1000 m (660 to 3300 ft) rises to replace it (Figure 14-11b). This water comes up from depths where no photosynthesis occurs. It is rich in nutrients because there are no phytoplankton in deeper waters to use them. This constant replenishing of nutrients at the surface enhances the conditions for life in areas of upwelling. Upwelling water is usually of low temperature, which produces the additional benefit of having a high capacity for dissolved gases (Figure 14-11c).

Oceanic Environments

The availability of light is probably the most important factor in determining the distribution of life in the oceanic province. The **euphotic zone** (good light) extends from the surface to a depth where enough light still exists to support photosynthesis, which is rarely deeper than 100 m (330 ft). The euphotic zone is, of course, also found in the neritic province. Below the euphotic zone is the **disphotic zone** (apart from light), where there is a small, but measurable, quantity of light. This zone extends to a depth of about 1000 m (3280 ft). Below about 1000 m (3300 ft), is the **aphotic zone** (without light) where no light reaches.

The oceanic province is further subdivided into four zones. The **epipelagic zone** (top of ocean) extends from the surface to a depth of 200 m (660 ft); the **mesopelagic zone** (middle depth of ocean), from 200 to 1000 m (660 to 3280 ft); the **bathypelagic zone** (deep ocean), from 1000 to 4000 m (3300 to 13,000 ft); and the **abyssopelagic zone** (without bottom) includes all the deepest parts of the ocean below 4000 m (13,000 ft) depth.

Epipelagic Zone

The upper epipelagic zone is the only oceanic biozone where there is sufficient light to support photosynthesis. The boundary between it and the meso-

pelagic zone, at 200 m depth (660 ft), is also the approximate depth at which the level of dissolved oxygen begins to decrease significantly (red curve in Figure 14-10b). This decrease in oxygen occurs because no algae are found below about 150 m (500 ft), and dead organic tissue descending from the biologically productive upper waters is undergoing decomposition by bacterial oxidation, which uses up oxygen. Nutrient content also increases abruptly below 200 m (600 ft) (green curve in Figure 14-10b). This depth also serves as the approximate bottom of the mixed layer, seasonal thermocline, and the surface water mass.

Mesopelagic Zone Within the mesopelagic zone, a dissolved oxygen minimum occurs at depths ranging from about 700 to 1000 m (2300 to 3280 ft) (Figure 14-10b). The intermediate-water masses that move horizontally in this depth range often possess the highest levels of algal nutrients in the ocean. Within the mesopelagic zone, although sunlight from the surface is very, very dim, there is still evidence that animals can sense this light. The mesopelagic zone is inhabited by fish that have unusually large and sensitive eyes, capable of detecting light 100 times weaker than humans can sense.

Other important inhabitants of this zone are the bioluminescent organisms, which literally glow in the dark. This group includes many species of shrimp, squid, and fish. Approximately 80 percent of mesopelagic organisms carry light-producing photophores, glandular cells containing luminescent bacteria surrounded by dark pigments. Some organisms also possess lenses to amplify the light radiation. This cold light, called **bioluminescence**, is produced by a chemical process involving the compound luciferin. Molecules of luciferin are excited and emit photons of light in the presence of oxygen. This system is similar to that used by fireflies and glowworms.

The Deep Scattering Layer The **deep scattering layer** (DSL), discovered inadvertently when the U.S. Navy was testing sonar equipment early in World War II, indicates that marine life migrates between the epipelagic and mesopelagic layers. The Navy observed a sound-reflecting surface that changed depth daily; it was found at a depth of 100 to 200 m (330 to 660 ft) during the night, but was as deep as 900 m (3000 ft) during the day. The Navy determined that this surface was produced by masses of migrating marine plankton. These organisms moved closer to the surface at night and then to greater depths during the day. The DSL appeared to respond to the changing intensity of light (Figure 14-12).

Research with plankton nets and submersibles has found the DSL to contain layers of various organisms, including small fish. A very small concentration of fish is sufficient to reflect sonar signals. Fish are predators so they are likely following the organisms on which they prey; it is these organisms that are actually responding to the light changes.

Bathypelagic and Abyssopelagic Zones The aphotic (lightless) bathypelagic and abyssopelagic zones represent over 75 percent of the living space in the oceanic province.

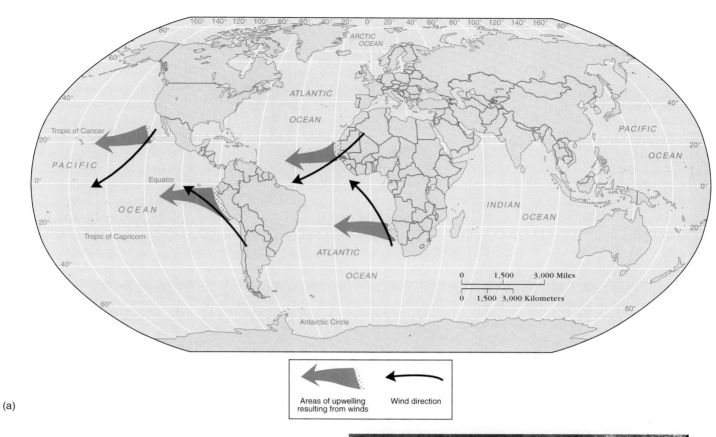

(a)

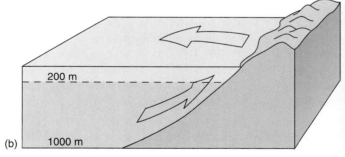

(b)

Figure 14-11   Upwelling

(a) Coastal winds (black arrows) usually drive currents southward along the western margins of continents in the Northern Hemisphere. In the Southern Hemisphere, such currents usually are driven northward. The Ekman transport moves water away from the western margins of the continents. The net current directions are shown with blue arrows. (b) As the surface water moves away from the continent, water rises to replace it from depths of 200 to 1000 m (660 to 3300 ft). Because this water comes from below the zone of photosynthesis, it is cold (unheated by sunlight) and rich in nutrients. (c) In this false-color Coastal Zone Color Scanner (CZCS) image from the Nimbus-7 satellite, wind-driven upwelling along the coast of northwestern Africa results in high phytoplankton biomass and productivity. Highest productivity = red and orange, decreasing through yellow, green, and blue.   *(Image courtesy of NASA/Goddard Space Flight Center.)*

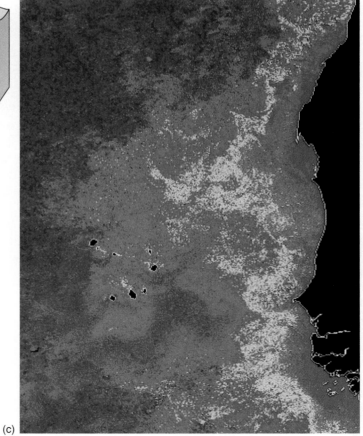

(c)

Figure 14-12 Deep Scattering Layer

The deep scattering layer (DSL) scatters and reflects sonar signals well above the bottom. It varies in depth from 100 to 200 m in daylight to as deep as 900 m at night. It may be caused by large numbers of euphausids and lanternfish (myctophids). They are predators that feed on smaller planktonic organisms, like the copepod shown, which daily migrate vertically in the water column.

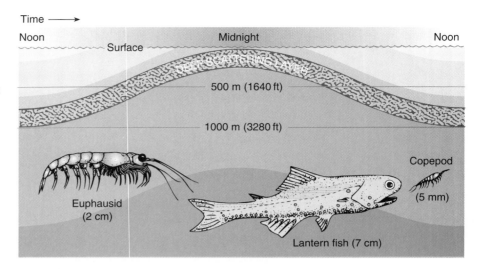

Figure 14-13 Deep-Sea Anglerfish

The anglerfish, *Lasiognathus saccostoma*, attracts two deep-water shrimp to its bioluminescent "lure." The fish is about 10 cm (4 in) long. *(Photo by Peter Arnold, Inc.)*

In this region of total darkness, there are many totally blind fish. Bizarre-looking small predaceous species also are present. Many species of shrimp that normally feed on detritus become predators at these depths, where the food supply is greatly reduced. Animals that live in these deep zones feed mostly on one another. They have evolved impressive warning devices and unusual apparatus that make them more efficient predators. They are characterized by small, expandable bodies, extremely large mouths relative to body size, and very efficient sets of teeth (Figure 14-13).

Oxygen content increases with depth, as deep- and bottom-water masses carry oxygen from cold surface waters where they formed to the deep ocean. The abyssopelagic zone is the realm of bottom-water masses that commonly move in the opposite direction of the overlying deep-water masses of the bathypelagic zone.

Benthic (Sea Bottom) Environment

The benthic, or sea-floor, environment is subdivided into two units that correspond to the neritic/oceanic provinces of the pelagic environment above them. These units are the **subneritic** (underneritic) **province** and the **suboceanic province** (see Figure 14-10a).

Subneritic Province   The subneritic province extends from the spring high-tide shoreline to a depth of 200 m (660 ft), approximately encompassing the continental shelf. The word *littoral* comes from a Latin word meaning shore, and there are several areas or zones that incorporate the term—supralittoral, sublittoral, and others. A transitional region from land to sea floor is above the high-tide line and is called **supralittoral** (higher than the shore) **zone**. Commonly called the *spray zone*, it is covered with water only during periods of extremely high tides and when tsunamis or large storm waves break on the shore. The *intertidal zone* (zone between high and low tides) is the **littoral zone** (shore zone). From low tide shoreline out to 200 m (660 ft) water depth, the subneritic province is called the **sublittoral zone** (lower than the shore), or *shallow subtidal zone*, and is essentially the continental shelf. The **inner sublittoral zone** includes the sublittoral to a depth of approximately 50 m (160 ft). This seaward limit varies considerably because it is set at the depth below which we find no growing plants or algae attached to the ocean bottom. This depth is controlled primarily by the amount of solar radiation that reaches the bottom. This means that beyond the inner sublittoral region, all photosynthesis is carried on by microscopic floating planktonic algae. The **outer sublittoral zone** includes that portion of the sublittoral zone from the inner sublittoral zone out to a depth of 200 m (660 ft) or the shelf break, which is the seaward edge of the continental shelf. (See Chapter 17 for the effects of these zonations on benthic life forms).

Suboceanic Province   The suboceanic province includes all the benthic environment deeper than 200 m (660 ft). The **bathyal zone** extends from a depth of 200 to 4000 m (660 to 13,000 ft), generally corresponding to the continental slope. The **abyssal zone**, covers over 60 percent of the surface area of the benthic environment or almost 43 percent of Earth's surface, stretching from 4000 to 6000 m (13,000 to 20,000 ft) in depth. Here, the ocean floor is covered by soft oceanic sediment, primarily abyssal clay. The tracks and burrows of animals that live in this sediment are frequently recorded in bottom photographs (Figure 14-14). The **hadal zone** is a restricted environment, including all depths below 6000 m (20,000 ft); this zone is found only in deep trenches along the margins of continents. The isolation of these deep, linear depressions allows the development of animal communities unique to each trench.

Classifying Organisms

Marine organisms commonly are categorized by their means of locomotion: plankton (floaters), nekton (swimmers), and benthos (bottom dwellers). Chances are that you probably have never seen living plankton, and you may not even have heard of them before you took this course. Most of Earth's biomass is plankton. Furthermore, the volume of Earth's space inhabited by plankton and nekton greatly exceeds that occupied by all animals that live on land or on the ocean's bottom.

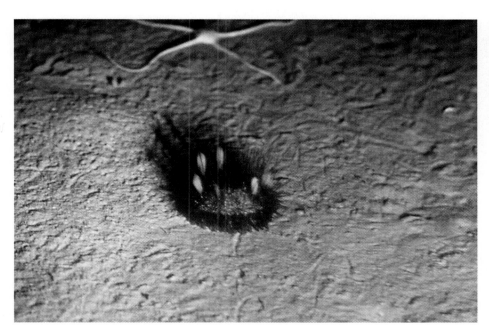

Figure 14-14   Tracks of Benthic Organisms Cover the Deep Ocean Floor

Sea urchin (center) and brittle star (top) are two examples of benthic organisms that leave tracks or impressions as they move or rest on the sediment surface.

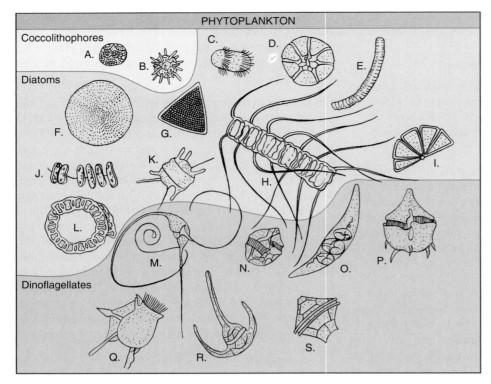

PHYTOPLANKTON

Coccolithophores

Diatoms

Dinoflagellates

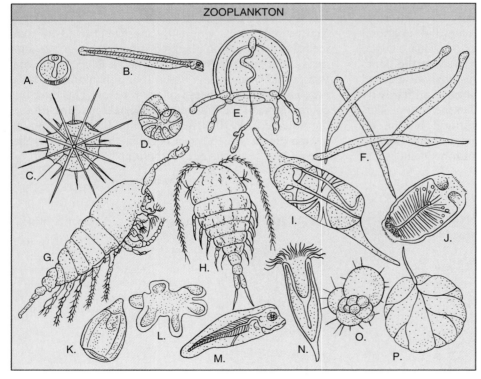

ZOOPLANKTON

Figure 14-15 Phytoplankton and Zooplankton

Plankton comes from a Greek word meaning to drift, roam, or wander. (Maximum dimensions in parentheses.) Phytoplankton: (A) and (B), coccolithophoridae (15 μm, or 0.0006 in). (C)–(L) Diatoms (80 μm, or 0.0032 in): (C) *Corethron*, (D), (E) *Rhizosolenia*,(F) *Coscinodiscus*, (G) *Biddulphia favus*, (H) *Chaetoceras*, (I) *Licmophora*, (J) *Thalassiorsira*, (K) *Biddulphia mobiliensis*, (L) *Eucampia*. (M)–(S) Dinoflagellates (100 μm, or 0.004 in): (M) *Ceratium recticulatum*, (N) *Goniaulax scrippsae*, (O) *Gymnodinium*, (P) *Goniaulax triacantha*, (Q) *Dynophysis*, (R) *Ceratium bucephalum*, (S) *Peridinium*. Zooplankton: (A) fish egg (1 mm, or 0.04 in), (B) fish larva (5 cm, or 2 in), (C) radiolaria (0.5 mm, or 0.02 in), (D) foraminifera (1 mm, or 0.04 in), (E) jellyfish (30 m, or 98 ft), (F) arrow worms (3 cm, or 1.2 in), (G), (H) copepods (5 mm, or 0.2 in), (I) salp (10 cm, or 4 in), (J) doliolum, (K) siphonophore (30 m, or 98 ft), (L) worm larva (1 mm, or 0.04 in), (M) fish larva (5 cm, or 2 in), (N) tintinnid (1 mm, or 0.04 in), (O) foraminifera (1 mm, or 0.04 in), (P) dinoflagellate (*Noctiluca*) (1 mm, or 0.04 in).

Plankton (Floaters)

Plankton includes many types of organisms—algae, protozoa , animals, and bacteria—that drift with ocean currents. An individual planktonic organism is called a plankter. Plankton drifting does not mean they are all unable to swim. Many plankton can swim, but either move only weakly or are restricted to vertical movement and

cannot determine their horizontal position within the ocean.

We typically classify plankton as phytoplankton, zooplankton, or bacterioplankton. Algae are called **phytoplankton**, protozoa and animals (including larvae and other immature forms) are called **zooplankton**. Representative members of each group are shown in Figure 14-15. Plankton also include bacteria. It has recently

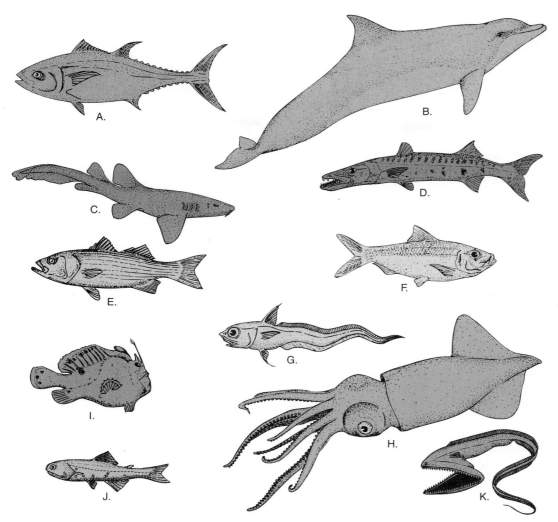

**Figure 14-16
Nekton (Swimmers)**

Not drawn to scale; typical maximum dimension is in parentheses. (A) bluefin tuna (2 m, or 6.6 ft); (B) bottle-nosed dolphin (4 m, or 13 ft); (C) nurse shark (3 m, or 10 feet); (D) barracuda (1 m, or 3.3 feet); (E) striped bass (0.5 m, or 1.6 feet); (F) sardine (15 cm, or 6 in); (G) deep-ocean fish (8 cm, or 3 in); (H) squid (1 m. or 3.3 ft); (I) anglerfish (5 cm, or 2 in); (J) lanternfish (8 cm, or 3 in); (K) gulper (15 cm, or 6 in).

been discovered that free-living **bacterioplankton** are much more abundant in the plankton community than previously thought. Having an average dimension of only 0.5 μm (0.00002 in), they were overlooked in earlier studies because of their small size.

Plankton range greatly in size. They include large floating animals such as jellyfish and large algae, such as the seaweed Sargassum. These large floating organisms are called **macroplankton**, and they measure from 2 to 20 cm (0.8 to 8 in). At the other extreme are bacterioplankton that are too small to be filtered from the water with a fine silk net. They must be removed by other types of microfilters. These very tiny floaters are called **picoplankton**, measuring 0.2 to 2 μm (0.000008 to 0.00008 in.).

An additional scheme for classifying plankton is based on the portion of their life cycle spent within the plankton community. Organisms that spend their entire life as plankton are *holoplankton*. Many organisms we normally consider to be nekton or benthos, because they spend their adult life in one of these living modes, actually spend a portion of their life cycle as plankton. Many nekton and very nearly all the benthos are plankton during their larval stages. These organisms, which as adults either sink to the bottom to become benthos or begin to swim freely as nekton, are called *meroplankton*.

Nekton (Swimmers)

Nekton include all those animals capable of moving independently of the ocean currents, by swimming or other means of propulsion. They are not only capable of determining their own positions, but are also capable, in many cases, of long migrations. Included in the nekton are most adult fish and squid, marine mammals, and marine reptiles (Figure 14-16). When you go swimming in the ocean, you become nekton, too.

Most freely moving nekton are somewhat limited in their horizontal movement across the ocean. Certain nonvisible barriers effectively limit their lateral range. These barriers are gradual changes in temperature, salinity, viscosity, and availability of nutrients. For example, many stenothermal fish are killed by temporary horizontal shifts of water masses which cause shifts in the vertical temperature distribution. Vertical ranges of nekton are normally determined by water pressure.

Fish appear to be everywhere, but normally are considered to be more abundant near continents and islands and in colder waters. Some fish, such as salmon, swim into freshwater rivers to spawn. Many eels do just the reverse, growing to maturity in fresh water, then descending the streams to breed in the great depths of the ocean.

Pfiesteria, the Morphing Peril to Fish

In 1995, a fish kill of mostly Atlantic menhaden occurred in North Carolina's Neuse Estuary. Fifteen million fish were killed in what appeared to be an outbreak of *Pfiesteria piscicida,* a toxic dinoflagellate known to have caused a similar kill in Pamlico Sound in 1991. This and related dinoflagellate species kill millions of fish yearly in North Carolina's coastal waters (Figure 14A). This complex of toxic microorganisms is referred to as *Pfiesteria.* They have been found in coastal waters from Delaware to Alabama, where they undermine the ability of fish to reproduce and resist disease; however, they have been implicated in fish kills only in North Carolina and Maryland.

P. piscicida is known to change into as many as 24 distinct forms within a life cycle that can be initiated by the presence of fish (Figure 14B). In water that is at least 26°C, cysts can lie dormant in sediments then produce nontoxic zoospores that feed on other microorganisms present in the water. When fish are present, the nontoxic zoospores become toxic and head for the fish where the toxins drug the fish and destroy their skin. Disease-causing bacteria and fungi can then more easily attack the fish. The spores then feed on the material oozing from sores that develop on the skin. The presence of fish cause the cysts to produce more nontoxic zoospores which become toxic. Toxic zoospores reproduce asexually and also produce gametes that merge

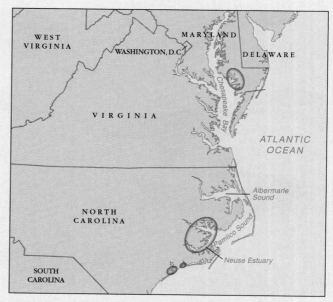

Figure 14A   Fish Kills Related *Pfiesteria* Outbreaks

One billion fish died in the first outbreak of *Pfiesteria* that occurred in 1991 in Pamlico Estuary. Since then smaller fish kills related to the *Pfiesteria* complex have occurred in Chesapeake Bay and along the North Carolina coast. Fish kills have occurred within the circled areas.

Figure 14B (right)   Scanning electron micrographs of stages of *Pfiesteria piscicida*

(1). The most lethal stage, a *toxic zoospore* that releases the toxin that kills fish (10 μm). Peduncle retracted (arrow). (2). While fish is dying, peduncle (arrow) is extended to feed on fish tissue (10 μ). (3). Toxic zoospore transforming to a *filipodial amoeba* that feeds on dead tissue. Pseudopods of the forming amoeba can be seen developing as knobs on the spherical surface in the right half of the image (diameter of main sphere is 5.6 μm). (4). A fully developed filipodial amoeba (diameter of central sphere is 4 μm). (5). When the dead fish sinks to the bottom, filipodial amoeba may transform into *lobos amoeba* that consume the dead animal (maximum left-right thickness of central lobe is 20 μm). (6). During the process represented by the above transformations, others occur. Within hours after the last fish dies without being replaced by live fish, some toxic zoospores transform into *cysts* covered by long-bristled scales and settle into the sediment (diameter of sphere without extensions 10 μm). (7). By one month after fish death, the bristles have been lost from the cyst surface. Upon the presence of live fish, they will transform into *nontoxic zoospores* that will become toxic zoospores that begin the cycle over again.   (Credit: (1) JoAnn Burkholder. (2) Howard Glasgow and JoAnn Burkholder. (3)–(5) JoAnn Burkholder and Howard Glasgow. (6) Joann Burkholder, et al. With permission from Nature. (7) Karen A. Steidinger.)

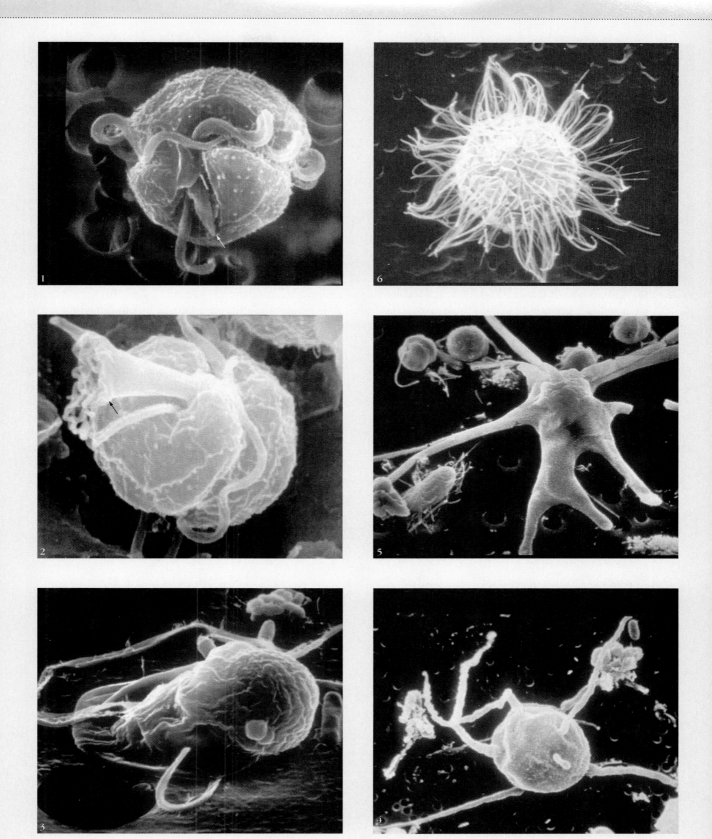

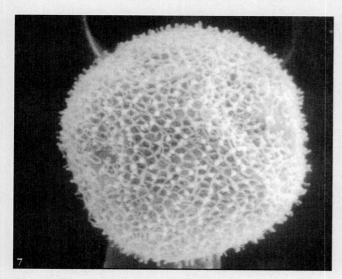

Figure 14B (Continued)

to produce planozygotes that resemble the zoospores. When the fish dies, the zoospores and planozygotes change into amoenae that gorge on the dead animal. In colder water, amoebae rise from the sediment when fish are sensed, attack the fish with toxins, and return to the bottom to feed on the fish when it dies. Unable to perform photosyntheses on their own, zoospores can save chloroplasts from algae they ingest in their nontoxic form and use them for weeks to generate food.

Pfiesteria outbreaks are initiated by an overabundance of nutrients such as nitrogen that allow a proliferation of algae that serve as food source for zoospores that reproduce rapidly and are ready to attack fish when they arrive on the scene.

Humans can suffer from *Pfiesteria* also. However, the disease is not contracted by eating fish from *Pfiesteria*-contaminated water. Instead, harm comes to humans from getting toxin-laden water on their skin or from breathing air over the toxin-laden waters. Symptoms of the human illness are shortness of breath, itchy or burning eyes, headaches, and forgetfulness.

The *Pfiesteria* outbreaks are part of a generally worsening pattern of increased rate of harmful algal blooms and ciguatera. Ciguatera is caused by dinoflagellate toxins accumulating in tropical fish, including barracuda, red snapper and grouper, and causes more human illness than any other form of seafood poisoning. Water quality in our streams is poor due to runoff of nutrients and other chemicals, wetlands that serve to clean the water from these streams are being destroyed at a high rate, and the density of human populations in coastal areas is increasing. Sickness in our coastal waters must be taken seriously, as it is certain that the quality of life for fish in our coastal waters is linked to the quality of life for humans.

Figure 14-17   Benthos (Bottom-Dwellers)–Some Intertidal and Shallow Subtidal Forms

Not drawn to scale. (Typical maximum dimension shown in parentheses.) (A) sand dollar (8 cm, or 3 in); (B) clam (30 cm, or 12 in); (C) crab (30 cm, or 12 in); (D) abalone (30 cm, or 12 in); (E) sea urchins (15 cm, or 6 in); (F) sea anemones (30 cm, or 12 in); (G) brittle star (20 cm, or 8 in); (H) sponge (30 cm, or 12 in); (I) acorn barnacles (2.5 cm, or 1 in); (J) snail (2 cm, or 0.8 in); (K) mussels (25 cm, or 10 in); (L) gooseneck barnacles (8 cm, or 3 in); (M) sea star (30 cm, or 12 in); (N) brain coral (50 cm, or 20 in); (O) sea cucumber (30 cm, or 12 in); (P) lamp shell (10 cm, or 4 in); (Q) sea lily (10 cm, or 4 in); (R) sea squirt (10 cm, or 4 in).

Benthos (Bottom Dwellers)

Benthos are organisms that live on or within the ocean bottom. **Epifauna** live on the surface of the sea floor, either attached to rocks or moving over the bottom. **Infauna** live buried in the sand, shells, or mud. Some benthos not only live on the bottom, but also move with relative ease through the water above the ocean floor. They are called *nektobenthos*. Examples of benthos are shown in Figure 14-17.

The littoral (shore) zone and inner sublittoral zone have a great diversity of physical and nutritive conditions.

Animal species have developed in great numbers within this nearshore benthic community as a result of the variations existing within the habitat. As you move across the bottom from the shore into the deeper benthic environments, the number of benthic species per square meter may remain relatively constant. However, you can observe an inverse relation between the distance from shore and the number of benthic individuals and biomass.

The littoral and inner sublittoral zones are the only areas where we find large algae (seaweeds) attached to the bottom because these are the only benthic zones where sufficient light penetrates. Throughout most of the benthic environment, animals live in perpetual darkness where no photosynthesis can occur. These animals must feed on each other or on whatever nutrients fall from the productive zone near the surface. The deep-sea bottom is an environment of coldness, stillness, and darkness. Under these conditions, life moves slowly. For animals that move around on the deep bottom, the streamlining that is so important to nekton is of little concern.

Organisms that live in or on the deep-ocean floor are normally widely distributed because physical conditions vary little, even over great distances. A few species appear to be extremely tolerant of pressure changes in that members of the same species may be found in the littoral province and at depths of several kilometers.

Hydrothermal Vent Biocommunities

In 1977, the first **biocommunity** at a hydrothermal vent was discovered in the Galápagos Rift off South America. This discovery showed that high concentrations of relatively large, deep-ocean benthos are possible. It appears that the primary limiting factor for life on the deep-ocean floor is the availability of food.

At these hydrothermal vents, food is abundant. It is produced not by photosynthesis, for no sunlight is available, but by chemosynthesis in bacteria. The size of individuals and the total biomass in the hydrothermal communities far exceeds that previously known for the deep-ocean benthos. These biocommunities will be discussed in detail in Chapter 17.

Seamounts and the Distribution of Benthos

A seamount is an interesting place to examine the effects of changing physical conditions on the benthos because significant depth changes occur over small horizontal distances. Jasper Seamount and Volcano 7 in the eastern Pacific have been investigated to determine the effects of current conditions and the oxygen minimum zone on benthic communities (Figure 14-18a). On Jasper Seamount, there are eight peaks ranging in depth from less than 600 m (2000 ft) to 1200 m (4000 ft); the seamount rises over 3600 m (11,800 ft) from the 4200 m (13,800 ft) deep abyssal ocean floor. A Deep Tow photographic survey showed the benthos to include sponges, black corals, gorgonian corals, anemones, and tunicates. The survey indicated that black corals, which grew at depths above 1150 m (3770 ft), were up to three times more abundant near peaks than at the same depth on midslope (Figure 14-18b). Gorgonian corals, which grew between the depths of 1170 and 1250 m (3838 and 4100 ft), were five times more abundant near peaks than at the same depth on midslope.

Current-meter readings indicated that current speed was about twice as fast near peaks than at midslope (4.9 cm/s vs. 2.7 cm/s, or 1.9 in/s vs. 1.1 in/s). This acceleration of currents over the peaks is analogous to the acceleration of airflow over the curved top of an airplane wing. The accelerated currents may increase the density of benthos by carrying in more larvae ready to settle and more particulate food per unit of time to areas near the peaks.

Volcano 7 rises from a depth of 3400 m (11,150 ft) to 730 m (2390 ft). In the overlying water, the base of the euphotic zone is at 85 m (280 ft). The oxygen minimum zone is well developed, and there is less than 0.2 ml of oxygen per liter of seawater from 288 to 1077 m (945 and 3533 ft) depth. There are few bacterioplankton throughout this range of depth, so dead organic matter from the euphotic zone falls undegraded to the seamount surface where it produces a thick, green **flocculent** rich in organic matter (Figure 14-18c). No large benthos is found near the peak above 750 m. The flocculent concentration decreases with depth, and the concentration of

Figure 14-18 (right)   Seamounts and Benthos Zonation

(a) Locations of Jasper Seamount and Volcano 7 in the eastern Pacific Ocean. (b) A high-density bed of spiral black corals, *Stichopathes* sp., near a peak on Jasper Seamount. A deep-sea sponge is visible on the right side of the photo.   *(Photo courtesy of Paul K. Dayton, Scripps Institution of Oceanography, University of California, San Diego.)*
(c) On the upper summit of Volcano 7 (746 m), rock and foraminiferal sand are covered by protozoan mat and flocculent detritus. White dot is a small coelenterate. (d) On the lower summit of Volcano 7 (780 m), rock and foraminiferal sand support two large and a number of smaller white sponges, brachyuran crabs (average length is 5 cm), shrimp, sea stars, and white serpulid worm tubes. The green, fluffy material is flocculent detritus. Rattail fish swim near the right margin of image. (e) On the lower summit of Volcano 7 (819 m), rocky area supports crabs, shrimp, serpulid worm tubes, sea stars, two stalked barnacles in lower center, and many sea anemones.   *(Photos (c)–(e) courtesy of Karen Wishner, University of Rhode Island.)*

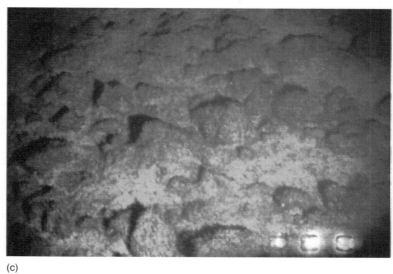

(c)

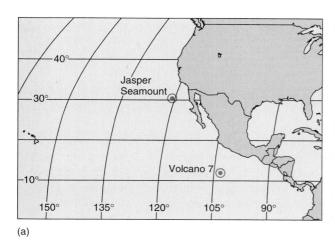

(a)

(d)

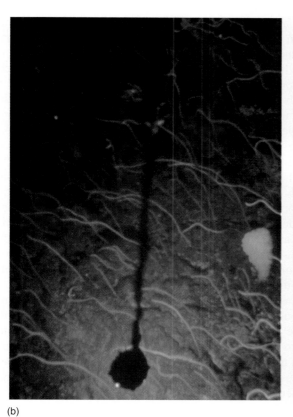

(b)

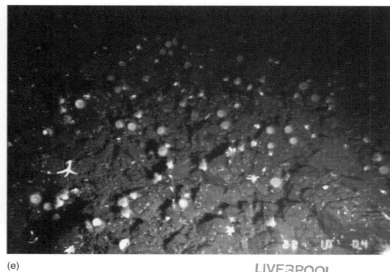

(e)

large benthos increases to 19 individuals/m^2 at 810 m (Figure 14-18d and e). Below that depth the concentration of large benthos decreases and is less than 0.5 individual/m^2 below 1000 m.

Considering the observations at Jasper Seamount, we would have expected the large filter-feeding benthos to have been more abundant near the peak of Volcano 7. It appears that the peak's penetration of the very well-developed oxygen minimum allowed too much organic flocculent to settle onto the peak, and the large filter-feeding benthos could not cope with the high organic sedimentation rate. At greater depths, the concentration of flocculent decreased, and the sponges, tubeworms, anemones, and barnacles were able to establish themselves while crabs, seastars, and shrimps fed on the rocky surface and in the near-bottom water.

Summary

Organisms living in the oceans have developed special adaptations to the chemical and physical characteristics of their environment. These adaptations are quite different than those required for life on land. If the body fluids of an organism are separated from ocean water by a membrane that allows water molecules to pass through, problems related to osmosis may result. Osmosis is the passage of water molecules through a semipermeable membrane from a region in which they are in higher concentration into a region where they are in lower concentration. Marine invertebrates and sharks are essentially isotonic, having body fluids with a salinity similar to that of ocean water. Most marine vertebrates are hypotonic, having body fluids with a salinity lower than that of ocean water, and they tend to lose water through osmosis. Freshwater organisms are essentially all hypertonic, having body fluids much more saline than the water in which they live, so they must compensate for a tendency to take water into their cells through osmosis. Euryhaline organisms can live in waters of variable salinity; stenohaline organisms require constant salinity.

The availability of nutrients and sunlight are important controlling factors in the distribution of all organisms because the basic producers of food are algae. The requirements of algae must be met first if food is to be plentiful for all.

Pelagic marine organisms depend primarily on buoyancy and friction to avoid sinking. The viscosity of water decreases with increasing temperature and decreasing salinity. Planktonic organisms in warmer waters have adapted to lower-viscosity by evolving feathery appendages. Stenothermal organisms require environments with stable temperatures, whereas eurythermal organisms can live in waters with a range of temperatures. Algae must stay in surface water to receive sunlight. The small animals that feed on them do not have effective means of locomotion so they depend on their small size and other adaptations to give them a high ratio of surface area per unit of body mass, which results in a greater frictional resistance to sinking. Large animals that swim freely face an altogether different problem and generally have streamlined bodies to reduce frictional resistance to motion.

Since solar radiation is available only in the surface water of the ocean, algal life is restricted to a thin layer of surface water, usually no more than 100 m (330 ft) deep. Nutrients derived ultimately from continental erosion are much more abundant near continental features. Biomass concentration decreases away from the continents and with increased depth. The color of the oceans ranges from green in highly productive regions to blue in areas of low productivity. Compared to life in colder regions, organisms living in warm water tend to be individually smaller, comprise a greater number of species, and constitute a much smaller total biomass. Warm-water organisms also tend to live shorter lives and reproduce earlier and more frequently than their cold water counterparts. For planktonic foraminifera, there is evidence that diversity of species peaks at 27°C.

The marine environment is divided into two basic units—the pelagic (ocean water) and the benthic (ocean bottom) environments. These regions, which are further divided on the basis of depth, are inhabited by organisms we can classify into three categories on the basis of life style—plankton (free-floating forms with little power of locomotion), nekton (free swimmers), and benthos (bottom-dwellers).

Research on seamounts shows that filter-feeding benthos are more abundant on peaks where currents are up to twice as swift as on slopes. However, if peaks extend into the oxygen minimum zone where oxygen concentrations drop to below 0.2 ml/l, filter-feeding benthos are rare because they cannot cope with the large amount of organic matter that reaches the bottom.

Key Terms

Abyssal zone	Disphotic zone	Isotonic	Plankton
Abyssopelagic zone	Epifauna	Littoral zone	Stenohaline
Aphotic zone	Epipelagic zone	Macroplankton	Stenothermal
Bacterioplankton	Euphotic zone	Mesopelagic zone	Streamlining
Bathyal zone	Euryhaline	Nekton	Sublittoral zone
Bathypelagic zone	Eurythermal	Neritic province	Subneritic province
Benthic environment	Flocculent	Oceanic province	Suboceanic province
Benthos	Hadal zone	Osmosis	Supralittoral zone
Biocommunity	Hypertonic	Outer sublittoral zone	Zooplankton
Bioluminescence	Hypotonic	Pelagic environment	
Deep scattering layer	Inner sublittoral zone	Phytoplankton	
Diffusion		Picoplankton	

Questions and Exercises

1. Define the terms euryhaline, stenohaline, eurythermal, and stenothermal. Where in the marine environment will organisms displaying a well-developed degree of each characteristic be found?

2. Describe the relationships between osmotic pressure, salinity, and freezing point of a solution.

3. What problem requiring osmotic regulation is faced by hypotonic fish in the ocean? How have these animals adapted to meet this problem?

4. An important variable in determining the distribution of life in the oceans is the availability of nutrients. Describe the relationship between the continents, nutrients, and the concentration of life in the oceans.

5. A major determinant of plant productivity is the availability of solar radiation. Why is biological productivity relatively low in the tropical open ocean where the penetration of sunlight is greatest?

6. Discuss the characteristics of the coastal waters where unusually high concentrations of marine life are found.

7. Compare the ability to resist sinking of an organism with an average linear dimension of 1 cm (0.4 in) to that of an organism of the same density with an average linear dimension of 5 cm (2 in). Discuss some adaptations other than size used by organisms to increase their resistance to sinking.

8. Changes in water temperature significantly affect the density, viscosity, and ability of water to hold gases in solution. Discuss how decreasing water temperature changes these variables and may affect marine life.

9. Construct a table listing the subdivisions of the benthic and pelagic environments and the physical factors used in assigning their boundaries.

10. Describe the vertical distribution of oxygen and nutrients in the oceanic province, and discuss the factors affecting this distribution. (Refer back to Chapter 7 also.)

References

Burkholder, J. M., et al. 1992. New "phantom" dinoflagellate is the causitive agent of major esturaine fish kills. *Nature* 358:3685, 407–10.

Burkholder, J. M., and H. B. Glasgow, Jr., 1997. *Pfiesteria piscicida* and other *Pfiesteria*-like dinoflagellates: Behavior, impacts, and environmental controls. *Limnology and Oceanography* 42(5, part 2), 1052–75.

Borgese, E. M., N. Ginsburg, and J. R. Morgan. 1994. *Ocean Yearbook 11*. Chicago: The University of Chicago Press.

Genin, A., et al. 1986. Corals on seamount peaks provide evidence of current acceleration over deep-sea topography. *Nature* 322:6074, 59–61.

Grassle, J. F., and N. J. Maciolek. 1992. Deep-sea species richness: Regional and local diversity estimates from quantitative bottom samples. *American Naturalist* 139:2, 313–41.

Hedpeth, J., and S. Hinton. 1961. *Common Seashore Life of Southern California*. Healdsburg, Calif.: Naturegraph.

Isaacs, J. D. 1969. The nature of oceanic life. *Scientific American* 221:65–79.

Mann, K. H., and J. R. N. Lazier. 1991. *Dynamics of Marine Ecosystems. Biological-Physical Interactions in the Oceans.* Boston: Blackwell Scientific Publications.

May, R. M. 1988. How many species are there on Earth? *Science* 241:4872, 1441–48.

Pimm, S. L., et al. 1995. The future of biodiversity. *Science* 269:5222, 347–50.

Rutherford, S., S. D'Hondt, and W. Prell, 1999. Environmental Controls on the Geographic Distribution of Zooplankton Diversity. Nature 400: 6746, 749–53.

Sieburth, J. M. N. 1979. *Sea Microbes.* New York: Oxford University Press.

Sumich, J. L. 1976. *An Introduction to the Biology of Marine Life.* Dubuque, Iowa: Wm. C. Brown.

Sverdrup, H., M. Johnson, and R. Fleming. 1942. *Renewal 1970. The Oceans.* Englewood Cliffs, NJ : Prentice-Hall.

Thorson, G. 1971. *Life in the Sea.* New York: McGraw-Hill.

Wilson, E. O. 1992. *The Diversity of Life.* Belknap Press of Harvard University Press, Cambridge, Mass.

Wishner, K., et al. 1990. Involvement of the oxygen minimum in benthic zonation on a deep seamount. *Nature* 346:6279, 57–59.

Suggested Reading

Sea Frontiers

Burton, R. 1977. Antarctica: Rich around the edges. 23:5, 287–95. Discusses the high level of biological productivity around the continent of Antarctica.

Gruber, M. 1970. Patterns of Marine Life. 16:4, 194–205. Describes how the form and size of many varieties of life in the ocean fit them for life in a particular environmental niche.

Hammer, R. M. 1974. Pelagic adaptations. 16:1, 2–12. This comprehensive discussion explains how pelagic organisms have adapted to reduce the energy required to maintain their positions in the open ocean.

Patterson, S. 1975. To be seen or not to be seen. 21:1, 14–20. Discusses the possible role of color in the protection and behavior of tropical fishes.

Perrine, D. 1987. The strange case of the freshwater marine fishes. 33:2, 114–19. The author explains how marine crevalle jacks are able to inhabit the fresh waters of Crystal River, Florida.

Schellenger, K. 1974. Marine life of the Galápagos. 20:6, 322–32. Examines the unique life forms of the Galápagos Islands, 950 km (589 mi) from South America.

Thresher, R. 1975. A place to live. 21:5, 258–67. How bottom-dwelling animals compete for space on the ocean floor is the subject of this interesting discussion.

Williams, L. B., and E. H. Williams, Jr. 1988. Coral reef bleaching: Current crisis, future warning. 34:2, 80–87. Corals and related reef animals underwent bleaching along the Pacific Central American coast in 1983 and in the Caribbean Sea in 1987. Possible causes are discussed.

Wu, N. 1990. Fangtooth, Viperfish, and Black Swallower. 36:5, 32–39. Strange adaptations help fish survive in the food-scarce and dark waters below 1000 meters' depth.

Scientific American

Denton, E. 1960. The buoyancy of marine animals. 203:1, 118–29. Discusses means by which some marine animals reduce the energy expenditure required to live in the ocean water far above the ocean floor.

Eastman, J. T., and A. L. DeVries. 1986. Antarctic fishes. 255:5, 106–14. Explains how one group of fish survived when the Antarctic turned cold.

Horn, M. H., and R. N. Gibson. 1988. Intertidal fishes. 258:1, 64–71. Intertidal fishes have undergone remarkable adaptation to survive this physically harsh environment.

Isaacs, J. D. 1969. The nature of oceanic life. 221:3, 146–65.

The conditions for life in the ocean as they relate to the variety and distribution of marine life-forms are presented in this well developed survey.

Isaacs, J. D., and R. A. Schwartzlose. 1975. Active animals of the deep-sea floor. 233:4, 84–91. Automatic cameras dropped to the ocean bottom show a surprisingly large population of large fishes on the deep-sea floor.

Palmer, J. D. 1975. Biological clock and the tidal zone. 232:2, 70–79. Investigation of the mechanism of biological clocks found in organisms from diatoms to crabs set to the rhythm of the tides.

Partridge, B. L. 1982. The Structure and function of fish schools. 246:6, 114–23. Considers the benefits of schooling and the means by which fish maintain contact with the school.

Robinson, B. H. 1995. Light in the ocean's midwaters. 273:1, 60–65. A scientist from the Monterey Bay Aquarium Research Institute describes animals observed from submersibles in the twilight of the mesopelagic waters.

Vogel, S. 1978. Organisms that Capture Currents. 239:2, 128–39. The manner in which sponges use ocean currents is an important part of this discussion

Oceanography on the Web

Visit the Introductory Oceanography home page for on-line resources for this chapter. There you will find an on-line study guide with review exercises and links to ocean-ography sites to further your exploration of the topics in this chapter. Introductory Oceanography is at http://prenhall.com/thurman (click on the Table of Contents menu and select this chapter).

Illegal Immigration: Ballast Water and Exotic Species

- What are exotic species?
- How are they able to cross oceans?
- What enables them to colonize foreign environments?
- Do they represent a threat to coastal ecosystems?
- What can policy makers do to protect estuaries from non-native species?

NEWS FLASH!

In September 1999, U.S. President Bill Clinton issued an executive order that directed the Departments of Agriculture, Interior, and Commerce, the Environmental Protection Agency, and the U.S. Coast Guard to develop an alien species management plan within 18 months to blunt the economic, ecological, and health impacts of invasive species.

Agriculture Secretary Dan Glickman promised "a unified, all-out battle against unwanted plants and animal pests." But senior administration officials acknowledged that the task poses difficulties and may succeed only through greater international cooperation.

Officials acknowledge that the United States also has species that cause adverse impacts when they are carried to other countries. Interior Secretary Bruce Babbitt said the long-term answer is to resolve these issues through international agreements that would benefit all countries.

Environmentalists, meanwhile, complain that the Clinton Administration has been slow in regulating ballast discharges from freighters—one of the major pathways for exotic aquatic organisms such as the Chinese mitten crab (annual economic cost unknown); green crab (annual economic cost $44 million); and Asian clam (annual economic cost, $1 billion!), which are threatening native marine life in San Francisco Bay and as far north as Washington state. Another such animal is the veined rapa whelk (Figure 1), which has been recently discovered in Chesapeake Bay.

Since the issue involves interstate and international commerce, individual states and counties cannot, under the U.S. Constitution, regulate ballast water. The West Coast invaders, says Linda Sheehan of the Center for Marine Conservation in San Francisco, are driving out native crabs and clams and threatening local oysters—even burrowing into and weakening flood control levees, which could potentially result in huge losses from property damage during floods.

A coalition of environmental groups and the Association of California Water Agencies asked the EPA to regulate freighter ballast water discharges under the Clean Water Act.

Introduction

Silently, almost imperceptibly, the planet's oceans, seas, estuaries, and lakes are being invaded by plants, animals, bacteria, and even viruses from distant climes. These organisms are called alien, exotic, or invasive species. Sometimes their impact is negligible, rarely is it beneficial, and often it borders on the disastrous. At a January 1999 meeting of the American Association for the Advancement of Science, Cornell University ecologist David Pimentel estimated the total cost of invasive species at $123 billion a year.

Instead of remaining in an ecosystem in which all members have evolved and interacted over time, in the relative blink of an eye invasive species may be transported beyond their natural range into the presence of other organisms with which they will immediately begin to interact, and perhaps compete.

Once thrust into a new environment, an organism faces a whole new set of conditions. To survive, all living organisms must live long enough to bear offspring and thus ensure the future of their gene pool. The 'aim' of exotic species is not to take over an estuary or clog a factory's water pipes, but rather to simply survive and reproduce.

Scientists believe that most non-native organisms fail to survive in their new environment long enough to become established. And that's a very good thing. But occasionally the introduced organism finds its new home completely livable, sometimes even ideal. Successful invasive species usually share a similar set of characteristics, according to the U.S. Coast Guard:

- They are *hardy,* indicated by their surviving a trip inside a ship for perhaps thousands of miles.
- They are *aggressive,* with the capacity to outcompete native species.
- They are *prolific breeders,* and can take quick advantage of any new opportunity, and
- They *disperse rapidly.*

Rapid dispersal is facilitated by having a planktonic larval stage, which allows the juveniles to be carried far and wide by currents. Such an introduced species often spreads rapidly, especially when predators and pathogens normally encountered in its home range are absent from the new environment, or when they are better able to feed than their new neighbors. (Or if they find their new neighbors especially tasty!)

In the preceding scenario, alien species flourish and potentially can reach astonishingly high population levels. Often, native species are displaced, or 'outcompeted' by the invaders. Then the situation is often called an *invasion.*

Have You Seen This Animal?

The Veined Rapa Whelk
(Rapana venosa)

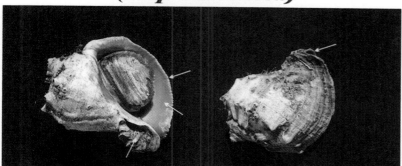

The Virginia Institute of Marine Science (VIMS) is interested in any sightings of this large snail in Virginia waters. The veined rapa whelk is native to the Sea of Japan, reaching sizes of 5 to 7 inches in length. There are several distinguishing characteristics that are highlighted by arrows in the above pictures. Note the small teeth along the edge of the shell and the orange coloration along the inner edge of the shell. Other characteristic features are a pronounced channel (columella) and the ribbing at the lower end of the shell.

Figure 1

The veined rapa whelk. Scientists at the Virginia Institute of Marine Science are tracking this invader from the Sea of Japan. As of February 15, 2000, 650 confirmed observations of this species have been recorded. U.S. east coast estuaries have favorable temperatures and ample prey (bivalves) for the rapa whelk, but they lack predators. One novel but ironic way of managing the invasion is to develop a fishery for the whelks as a means of controlling their population. *(Juliana Harding, Molluscan Ecology Program, Virginia Institute of Marine Science)*

Invasive species can inflict damage on ecosystems by:

- outcompeting native species,
- introducing parasites and/or diseases
- preying on native species, and/or
- dramatically altering habitat, e.g., rearranging the spatial structure of an ecosystem

Most invasive species are brought to new shores in the ballast water of ships, but animals dumped into an estuary from aquariums or accidental releases from aquaculture facilities may also contribute.

What is ballast water? Ballast water is carried by ships in special tanks to provide stability, optimal steering, and efficient propulsion. According to the U. S. Coast Guard, the use of ballast water varies among vessel types, among port systems, and with cargo and sea conditions.

How much ballast water is involved? The National Oceanographic and Atmospheric Administration (NOAA) has calculated that 40,000 gallons (150,000 liters) of foreign ballast water are dumped into U.S. harbors each *minute*.

The problem with ballast water is very simply stated: ballast water is taken up by a ship in ports and other coastal regions, in which the waters may be usually rich in planktonic (small, floating, or weakly swimming) organisms. It may be released at sea, in a lake or a river, or in the open ocean along coastlines—wherever the ship reaches a new port. As a result, a myriad of organisms is transported and released around the world within the ballast water of ships. Here are two examples.

Scientists studying an Oregon bay counted 367 types of organisms released from ballast water of ships arriving from Japan over a four-hour period! Another study documented a total of 103 aquatic species introduced to or within the United States by ballast water and/or other mechanisms, including 74 foreign species.

Now, go to our web site to continue the analysis.

See the complete issue at:
www.prenhall.com/thurman

15 Biological Productivity and Energy Transfer

Killer whale hunting sea lions near Valdes, Argentina. (Francois Gohier, courtesy Photo Researchers, Inc.)

Plants, algae, and other organisms that produce their own food by photosynthesis from carbon dioxide, water, and sunlight are called producers. This process, which uses solar energy and dissolved nutrient compounds from seawater to create organic matter, is the basis for most nutrition in the ocean. The ocean's producers are the foundation of feeding relationships in the ocean. The primary producers of the oceans are marine algae. They synthesize most of the organic matter used to support the marine biological community.

When you think of marine algae, you probably picture the large macroscopic seaweeds that grow near shore in many coastal areas. However, these large algae play only a minor part in the production of food for the ocean population as a whole. Instead, marine organisms depend primarily on the small, planktonic algae that inhabit the near-surface sunlit water of the world's oceans. These algae are microscopic, but they are the fundamental biological community in the marine environment—the phytoplankton.

In addition to the photosynthesizing algae living near the ocean surface, another food-production method operates in the deep ocean. In the total darkness of the deep sea, where no measurable sunlight penetrates, certain bacteria synthesize organic matter using energy released from the oxidation of hydrogen sulfide or methane. This process, called chemosynthesis, supports a vast array of unusually large deep-sea benthos. The chemical energy stored by both of these producers of organic matter—surface phytoplankton and deep-sea bacteria—is passed to the various populations of animals that inhabit the oceans through a series of feeding relationships called food chains and food webs.

Classification of Organisms

All living things belong to one of the three domains of life, **Archaea**, **Bacteria**, and **Eukarya** (Figure 15-1). The Archaea were first discovered by Carl Woese in 1976 when he examined some microbes thought to be bacteria that produced methane. They did not have some of the genetic characteristics he had found in all other bacteria, and he designated them *archaebacteria*—and later archaea. The Eukarya include the four kingdoms shown in Figure 15-2—Protists, Fungi, Plantae, and Animalia.

Archaea and **bacteria** are the simplest of all organisms; they are classified as separate kingdoms by some. They also may be grouped together in the kingdom **Monera**, whose members consist of a single-cell that lacks a discrete nucleus and membrane-bound organelles. Cells of this type are called *prokaryotic* and such organisms are called *prokaryotes*. The nuclear material of prokaryotes is spread throughout the cell. Included in Monera are the cyanobacteria (formerly called the blue-green algae) and heterotrophic bacteria. Many of the Archaea are found in unusually hot, acid, or salty anaerobic environments, although they are also present in less extreme marine environments. Recent discoveries have shown bacteria and archaea to be much more important to marine ecology than previously believed. They are found throughout the breadth and depth of the oceans.

Kingdom **Protista** represents a later stage of evolutionary development. The cells of protists are *eukaryotic*—that is, they contain a membrane-bound nucleus and membranous organelles. Among the protists are single-celled algae that produce food for most marine animals, slime and water molds, and heterotrophic single-celled organisms called **protozoa**.

The fungi, members of the Kingdom **Mycota**, appear to be poorly represented in the oceans. Less than 1 percent of the known 50,000 species are sea-dwellers. Although fungi exist throughout the marine environment, they are most common in the intertidal zone where they live in a relationship with cyanobacteria or green algae. This relationship creates what we call *lichen*, in which the fungal cells provide a protective covering that retains water during periods of exposure, and the algal cells provide food for the fungus through photosynthesis. Other fungi function primarily as decomposers in the marine ecosystem—they re-mineralize organic matter.

Members of the kingdom **Plantae** are restricted to the shallow coastal margins of the ocean and are not a major component of the productive marine community. However, they are very important as producers within restricted communities such as mangrove swamps and salt marshes.

Members of the kingdom **Animalia** range in complexity from the simple sponges to huge and complex vertebrates, such as whales.

Each kingdom is divided into increasingly specific groupings. The system of classification still used today was introduced by Carl von Linné (Linnaeus) in 1758. It includes the following major categories:

Kingdom
    Phylum
        Class
            Order
                Family
                    Genus
                        Species

Every organism has a scientific name that consists of its genus and species—for example, *Delphinus delphis* (common dolphin). Most organisms also have a common name, which is generally used by people who do not need the precision of the scientific name.

Before examining productivity and energy transfer among organisms in the ocean, let's first take a quick tour of the algae and plants of the sea.

Macroscopic Algae and Plants

We will start with a group of plants and algae with which you are most likely to be familiar. These are the attached forms of large algae and plants found in shallow waters along the ocean margins. The classification of algae is based partly on the pigments they contain (Figure 15-3).

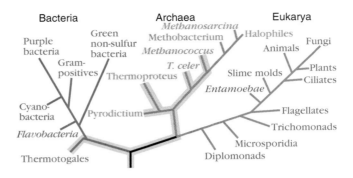

Figure 15-1 **The Three Domains of Life**

(a) Carl Woese discovered the Archaea in 1976. (b) The portions of the tree shaded in orange are hyperthermophobes (thrive in hot places). *(Part (a) Photo by Bill Wiegand/ University of Illinois News Bureau.)*

Brown Algae (Phaeophyta)

The brown algae of the phylum **Phaeophyta** include the largest members of the marine algal community that are attached to the bottom. Their color ranges from very light brown to black. Brown algae occur primarily in temperate and cold-water areas. Their sizes range widely. Smallest is the small black encrusting patch of *Ralfsia* of the upper and middle intertidal zones, where the algae may become crisp and dry in the sun without dying. Largest is the bull kelp (*Pelagophycus*), which may grow in water deeper than 30 m (100 ft) (Figure 15-3a, c).

Green Algae (Chlorophyta)

Green algae, members of the phylum **Chlorophyta**, are common in fresh water, but they are not well represented in the ocean. Most marine species are intertidal or grow in shallow bay waters. Green algae contain the pigment chlorophyll, which makes most of them grass green in color. They grow only to moderate size, seldom exceeding 30 cm (12 in) in the largest dimension. Forms range from finely branched filaments to flat, thin sheets.

Various species of sea lettuce (genus *Ulva*), a thin membranous sheet only two cell layers thick, are widely scattered throughout cold-water areas. Sponge weed (genus *Codium*), a two-branched form more common in warm waters, can exceed 6 m (20 ft) in length (Figure 15-3a, d).

Red Algae (Rhodophyta)

Red algae, members of the phylum **Rhodophyta**, are the most abundant and widespread of marine macroscopic algae. Over 4000 species of red algae occur from the very highest intertidal levels to the outer edge of the inner sublittoral zone. Many are attached to the bottom. They are very rare in fresh water. Red algae range from just visible to the unaided eye to lengths up to 3 m (10 ft). While found in both warm and cold-water areas, the warm-water varieties are relatively small.

The color of red algae varies considerably, depending on their depth in the intertidal or inner sublittoral zones. In the upper, well-lighted areas they may be green to black or purplish, changing through a brown to a pinkish red in deeper-water zones, where less light is available (Figure 15-3a, b).

The bulk of marine photosynthetic productivity is believed to occur above water depths where the amount of light is reduced to 1 percent of that available at the surface. This depth is about 100 m, or 330 ft. However, a red alga has been observed growing at a depth of 268 m (880 ft) on a seamount near San Salvadore, Bahamas. Available light at this sighting was thought to be only 0.05 percent of the light available at the ocean's surface.

Plants

The only plants found in a marine environment are seed plants, a group that includes the most familiar land plants—gymnosperms (pines, firs) and angiosperms (flowering plants). As noted above, marine plants are generally confined to coastal areas. Two seed plants found in the marine environment are eelgrass (*Zostera*) and surfgrass (*Phyllospadix*) (Figure 15-3e). Eelgrass, a grasslike plant with true roots, exists primarily in the quiet waters of bays and estuaries from the low-tide zone to a depth of about 6 m (20 ft). Surfgrass grows in the high-energy environment of an exposed rocky coast and can be found from the intertidal zones down to a depth of 15 m (50 ft). Both of these plants are important sources of detrital (fragmented organic matter) food for the marine animals

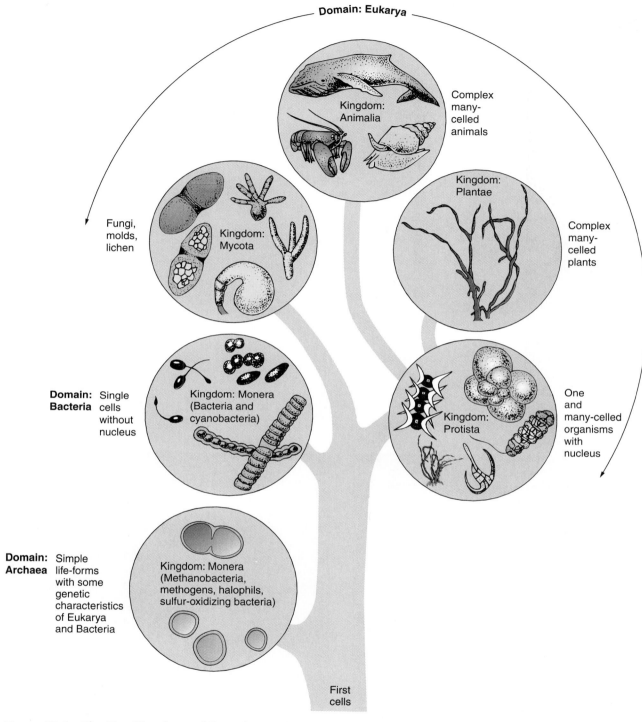

Domain: Eukarya

Kingdom: Animalia — Complex many-celled animals

Kingdom: Plantae — Complex many-celled plants

Kingdom: Mycota — Fungi, molds, lichen

Domain: Bacteria Single cells without nucleus

Kingdom: Monera (Bacteria and cyanobacteria)

Kingdom: Protista — One and many-celled organisms with nucleus

Domain: Archaea Simple life-forms with some genetic characteristics of Eukarya and Bacteria

Kingdom: Monera (Methanobacteria, methogens, halophils, sulfur-oxidizing bacteria)

First cells

Figure 15-2   The Five Kingdoms of Organisms

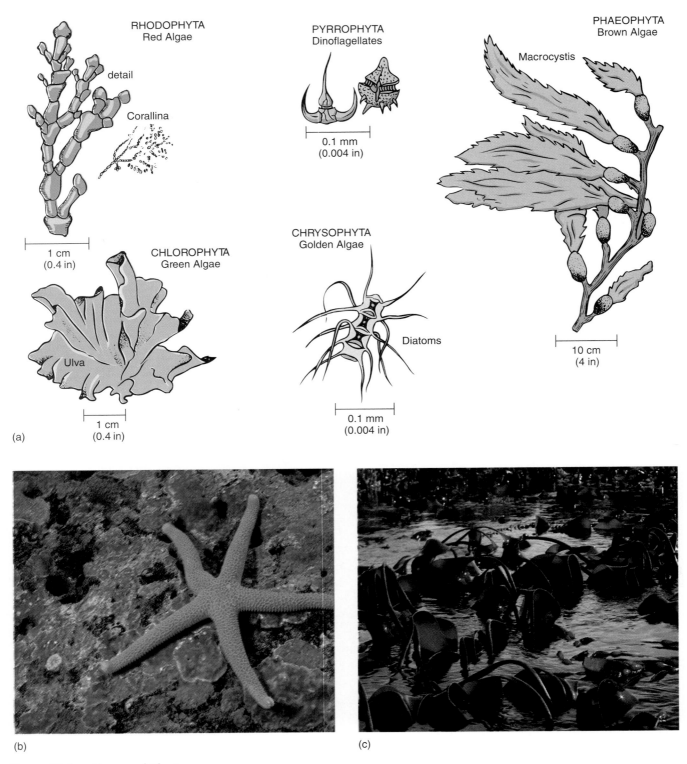

Figure 15-3   Algae and Plants

(a) Several types of algae. (b) Encrusting red alga, *Lithothamnion*, on sea floor beneath a sea star, *Henricia*. (c) Brown alga, *Laminaria*, a type of kelp; oarweed at low tide. **Opposite:** (d) Green alga, *Codium fragile*. This seaweed is also called *spongeweed*. (e) The spermatophyte, *Phyllospadix*, is also called surfgrass. (f) Two varieties of diatoms, *Rhizosolenia* (the slender form) and a larger unidentified form. (g) Coccoliths, disk-shaped calcium carbonate plates, cover a coccolithophore cell (diameter, 20 μm). (h) Photomicrograph of living *Gonyaulax polyedra*, (x1665). *Gonyaulax* is a large genus of phosphorescent marine dinoflagellates. In great abundance, they cause red tides along the shoreline.   *(Part (a) Line drawing by Phil David Weatherly; (b) Shane Anderson; (c) D. J. Wroebel, Monterrey Bay Aquarium/BPS; (e) Maria Redinger: (f) David Caron; (g) Courtesy of Deep Sea Drilling Project, Scripps Institution of Oceanography, University of California, San Diego.; (h) Biophoto Associates/Science Source/Photo Researchers, Inc.)*

(d)

(e)

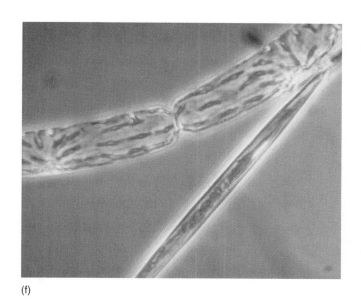

(f)

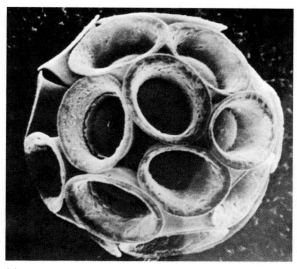

(g)

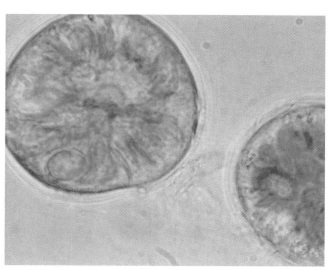

(h)

inhabiting their environment. Grasses, mostly of the genus *Spartina*, grow in salt marshes. Mangrove swamps contain primarily mangrove trees and shrubs (genera *Rhizophora* and *Avicennia*).

Microscopic Algae

Now we will introduce the important members of phytoplankton, which produce over 99 percent of the food supply for marine animals. As the plankton part of their name indicates, they are primarily floating forms, although some live on the bottom in the nearshore environment.

Golden Algae (Phylum Chrysophyta)

The microscopic algae in the phylum **Chrysophyta** contain the yellow pigment carotene and store food in the form of carbohydrates and oils. Diatoms are the most important marine algae of this group.

Diatoms are cells contained in a shell, or frustule, composed of opaline silica (SiO_2-nH_2O). These silica housings are important geologically because they accumulate on the ocean bottom and produce a siliceous sediment called diatomite. Some deposits of diatomite on land are mined and used primarily in the manufacture of filtering devices. The frustule of the diatom is similar in structure to a microscopic pill box. The top and the bottom of the box are called valves; the larger of the valves is the epitheca and the smaller the hypotheca. The cell contained within this housing exchanges nutrients and waste products with the surrounding water through slits or pores in the valves (Figure 15-3f).

The reproduction of diatoms is by simple cell division. To achieve this division, the valves of the frustule must separate. Each valve then serves as one-half the housing for each newly formed cell. A new valve must form to complete the enclosure of each daughter cell, and each of these newly formed halves forms as the smaller valve—the hypotheca. As can be seen in Figure 15-4, this process leads to the formation of smaller and smaller cells. Eventually, the newly formed daughter cell is so small that if it were decreased further, abnormalities would result. At this point, a change in the process occurs, namely, the formation of an auxospore. The auxospore forms between the two separating valves within a membrane that allows the auxospore to grow into a full-sized diatom so that the process of reproduction by the splitting of the frustule can recommence.

Coccolithophores are, for the most part, flagellum-bearing organisms covered with small calcareous plates ($CaCO_3$) called coccoliths. (A flagellum is a whiplike structure that extends from the cell and is used in locomotion.) The name of the group means *bearers of coccoliths*. The individual plates are about the size of a bacterium, and the entire organism is too small to be captured in plankton

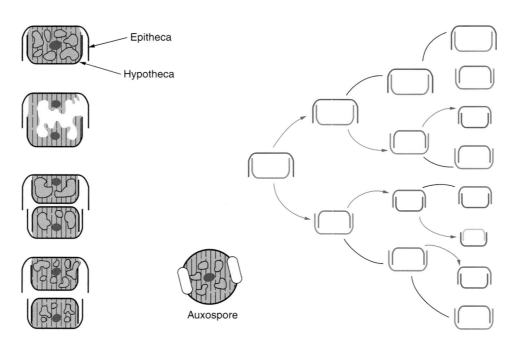

Figure 15-4　Diatom Reproduction

In diatom cell division, as the epitheca and hypotheca separate, each becomes the epitheca of a new cell. The new frustule is completed by the generation of a new hypotheca by each new diatom. Following the arrow through the formation of three generations of diatoms, it can be seen that some cells will be crowded into ever smaller frustules. When the size of the frustule becomes critically small, an auxospore forms by sexual reproduction to allow growth of a full-sized diatom.

Sea Lion Mortality and a Toxic Diatom Bloom

During May and June of 1998, 400 California sea lions died along the central California coast. This event has been tied to the occurrence of a harmful algal bloom (HAB). Figure 15A is a SeaWiFS satellite image showing the estimated chlorophyll a distribution scaled in $\mu g\ 1^{-1}$. The study of this event revealed that the phytoplankton species responsible was the diatom, *Pseudo-nitzschia australis* that secretes the toxin domoic acid (DA). This is shown in Figure 15B, which also shows that the concentration of DA in the tissue of sardines and anchovies increased along with the increased abundance of *P. australis*.

In addition to the sea lion deaths, many bird carcasses were found along the beaches of Monterey Bay. The symptoms of sick animals recovered were neurological and included seizures, head weaving, poor muscle coordination, and abnormal scratching. Examination of the tissues of 48 dead sea lions showed lesions of the brain and heart similar to those of mice, macaques and humans exposed to DA, as is shown in Figure 15C. DA was also found in the urine and feces of two sea lions tested and represented the first time it had been identified in body fluids of mammals associated with HAB mortality events. DA was also found in anchovies that are known to be prey of sea lions, indicating the transfer of the toxin from the algal source to the sea lions through the anchovies. Although this marks the first conclusive evidence linking the death of sea lions to HAB, similar events are known to have occurred in 1978, 1986, 1988 and 1992.

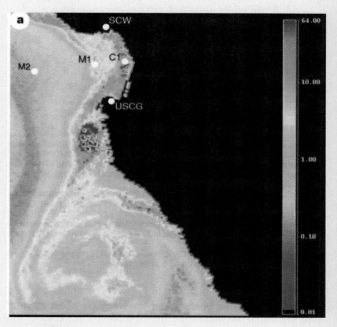

Figure 15A

U.S.C.G.–Coast Guard Pier
S.C.W.–Santa Cruz wharf
C1–mooring sample site
M1, M2–Ship sample sites.

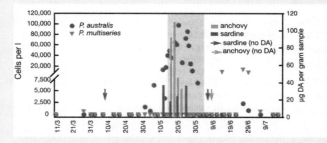

Figure 15B   Event Record

Shaded area is period of sea lion mortality. It coincides with high levels of DA in anchovy and sardine tissue and increased abundance of *P. australis*.

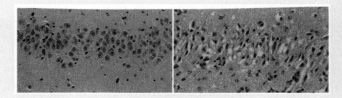

Figure 15C   Hippocampus Lesions

Left: normal sea lion brain tissue. Right: Lesions (green) in brain tissue from sea lion that died from DA Poisoning. *(15 A, B, and C, Courtesy of C.A. Scholin, M.B.A.R.I. Reprinted by permission from Nature 403:6765, 81 & 83. ©2000. www.nature.com)*

nets. Coccolithophores are included in the nanoplankton, which have dimensions of less than 0.06 mm (0.002 in). The coccolithophores contribute significantly to calcareous deposits in all of the temperate and warmer oceans (Figure 15-3g).

Dinoflagellates (Phylum Pyrrophyta)

A second group of phytoplankton important in marine productivity is the **dinoflagellates**, which make up the phylum **Pyrrophyta**. Possessing flagella for locomotion, they have a slight capacity to move into areas that are more favorable for photosynthesis. The dinoflagellates are not important geologically because many have no protective covering. Those that do, have coverings made of cellulose, which is easily decomposed by bacterial action when the organism dies. Many of the 1100 species are luminescent.

Sometimes up to 2 million dinoflagellates are found in 1 liter (1 quart) of water, causing a **red tide**. Two genera of dinoflagellates, *Ptychodiscus* and *Gonyaulax,* are mainly responsible for red tides (Figure 15-3h); both produce water-soluble toxins. *Gonyaulax* toxin is not poisonous to shellfish, but it concentrates in their tissue and is poisonous to humans who eat the shellfish. *Ptychodiscus* toxin kills both fish and shellfish. April through September are particularly dangerous months for red tides in the Northern Hemisphere, and in most areas, there is a quarantine against harvesting shellfish during those months. The shellfish feed on the dynoflagellates, and concentrate the poisons they secrete to levels dangerous to humans.

A potentially tragic epidemic of paralytic shellfish poisoning from a *Gonyaulax* red tide occurred in the coastal waters of Massachusetts in the fall of 1972. Fortunately, no deaths occurred, although 30 cases of poisoning were reported. Symptoms of paralytic shellfish poisoning are similar to those of drunkenness, including incoherent speech, uncoordinated movement, dizziness, and nausea. Documented cases throughout the world include 300 deaths and 1750 nonfatal cases. There is no known antidote for the toxin, which attacks the human nervous system; the critical period usually passes in 24 hours.

A new type of poisoning caused by domoic acid, a toxin produced by a diatom, occurred along the eastern coast of Canada in 1987. One hundred people were poisoned from eating mussels; of these, three people died and ten still suffer from memory loss. Domoic acid poisoning has been called amnesic shellfish poisoning because of the memory loss suffered by a number of the victims.

In September 1991, the first known occurrence of marine bird poisoning by domoic acid occurred along the northern California coast. Mass mortality occurred in Brown Pelicans and Brandt's Cormorants. The diatoms were ingested by anchovies, which were not affected by the toxin. The anchovies were eaten by the birds, which were killed. See "Sea Lion Mortality and a Toxic Diatom Bloom" on page 399 and "*Pfiesteria,* the Morphing Peril to Fish" on page 380.

Red tides are occurring more frequently, more toxic species are showing up, and larger areas are being affected. More marine species, including dolphins and humpback whales, are succumbing to the toxins. Human activities may be contributing to this increasing problem. For example, sewage and fertilizer that make their way into coastal waters can harm organisms by causing an overabundance of nutrients. This results in **eutrophic** conditions in which overproduction of organic matter causes anoxic conditions as microbes consume oxygen while decomposing the organic matter.

Another sickness caused by eating fish poisoned by dinoflagellates is ciguatera fish poisoning. The dinoflagellates live attached to seaweeds eaten by tropical fish. Ciguatera poisoning affects more humans than any other form of poisoning from seafood. It affects from 10,000 to 15,000 people annually in tropical regions.

Primary Productivity

Primary production is the amount of carbon fixed by organisms through the synthesis of organic matter using energy derived from solar radiation or chemical reactions. Most primary production is from photosynthesis and uses sunlight for energy.

Carbon fixation using reactions involving inorganic chemicals is called chemosynthesis. We have new knowledge of the role of chemosynthesis in supporting hydrothermal vent communities along oceanic spreading centers. However, chemosynthesis is much less significant in worldwide marine primary production than is photosynthesis. We will look at both methods, but study photosynthesis in more detail because of its far greater importance.

Photosynthetic Productivity

Photosynthesis is a chemical reaction in which energy from sunlight is converted into chemical bond energy and stored in organic molecules, in this manner:

$$6H_2O + 6CO_2 + \text{light energy} = C_6H_{12}O_6 + 6O_2$$

water + carbon dioxide   glucose + oxygen

The total amount of organic matter produced by photosynthesis per unit of time is the **gross primary production** of the oceans. Algae use some of this organic matter for their own maintenance through the process of respiration, which is the reverse of the reaction shown above. What remains is **net primary production**, which is manifested as growth and reproduction products. Net primary production supports the rest of the heterotrophic marine population—animals, protozoa, bacteria, and archaea.

Gross primary production has two components, new production and regenerated production. **New production** is that part supported by nutrients brought in from outside the local ecosystem, by processes such as upwelling. The higher the ratio of new production to gross primary production in a marine ecosystem, the greater its ability to support animal populations that we depend on for fisheries, such as pelagic fishes and benthic scallops. **Regenerated production** results from nutrients being recycled within the ecosystem. If we heavily fish a population primarily supported by recycled nutrients, we remove the organic matter needed to support future generations, thereby seriously decreasing the populations capacity to recover.

How is primary productivity measured? The Gran method, developed in the 1920s, is based on photosynthesis releasing oxygen in proportion to the amount of organic carbon synthesized. Equal quantities of phytoplankton and seawater are placed in series of bottles, all containing the same amount of dissolved oxygen. The bottles are then arranged in pairs, one being transparent and the other fully opaque. These bottle pairs are suspended on a line through the euphotic zone where photosynthesis occurs. Here they are left for a specific period of time. After the bottles are brought to the surface, the oxygen concentration is determined for each bottle.

Respiration consumes oxygen in both the transparent and opaque bottles. However, photosynthesis occurs only in the transparent bottles, where it releases oxygen to the water. Increased oxygen concentration in the transparent bottles represents the net production of the algae within. Decreased oxygen content within the opaque bottles simply corresponds to the respiration rate. For any depth, gross production can be estimated by adding the oxygen gain in the clear bottles to the oxygen loss in the opaque bottles. The depth at which the oxygen production and oxygen consumption are equal is called the **oxygen compensation depth**. This represents the level of light intensity below which algae do not survive.

Respiration goes on 24 hours a day. However, it is during the daylight hours that algae must produce biomass through photosynthesis. Biomass must be produced in excess of that which is consumed by respiration in any 24-hour period if the total biomass of the community is to increase (Figure 15-5). An analogy to a paycheck can be made—gross photosynthesis (gross pay earned) = oxygen change in clear bottle (take home pay) + oxygen loss in dark bottle (income tax and other withholding.)

A second method for measuring primary production using radioactive carbon (^{14}C) was developed in the 1950s. It has since been refined and is currently the most often used method for determining marine primary production. The procedure is similar to the Gran method in that it uses a series of paired clear and opaque bottles. Each bottle contains identical phytoplankton samples, equal amounts of CO_2 containing ^{14}C, and equal amounts of CO_2 containing stable carbon.

A third method used in studies of marine primary productivity measures chlorophyll in living phytoplankton samples from surface waters. This method is less precise than using radioactive carbon, but there is a direct relation between the amount of chlorophyll and primary productivity. The satellite-borne Coastal Zone Color Scanner (CZCS) senses the effect of phytoplankton pigment on the color of ocean water. CZCS data were used to prepare the pigment concentration map in Figure 15-6.

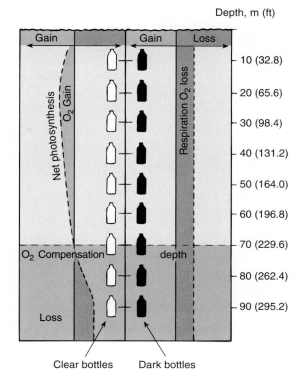

Figure 15-5   Oxygen Compensation Depth

A comparison of phytoplankton production and consumption of oxygen (O_2) with and without light. The clear bottles let in light needed for photosynthesis. The dark bottles, which prevent light from entering, form a control group. The comparison reveals how deeply light can penetrate water to sustain enough photosynthesis to increase phytoplankton mass. Phytoplankton growth can occur only above the *oxygen compensation depth*, the depth at which the amount of oxygen produced by photosynthesis equals the amount of oxygen consumed by respiration. For the clear bottles, the green area represents net photosynthesis (gross photosynthesis less the oxygen consumed by respiration). In the dark bottles, the oxygen loss gives a value for the amount of respiration that occurred. This value also applies to the clear bottles (violet). Oxygen loss, added to the net photosynthesis in the clear bottles, gives a value for the gross photosynthesis in the clear bottles.

Patterns of Biological Production and Temperature

Primary photosynthetic production in the oceans varies from about 0.1 gram of carbon per square meter of ocean surface per day ($gC/m^2/d$) in the open ocean, to more than 10 $gC/m^2/d$ in highly productive coastal areas. This variability is the result of the uneven distribution of nutrients throughout the photosynthetic zone, and seasonal changes in the availability of solar energy.

About 90 percent of the biomass generated in the euphotic zone of the open ocean is decomposed into inorganic nutrients before descending below this zone. Approximately 10 percent of this organic matter sinks into deeper water, where most of it is decomposed. About

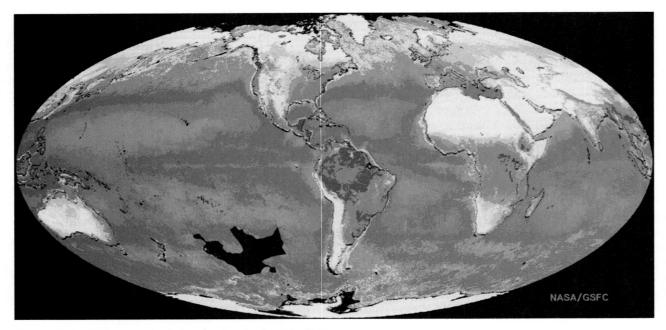

Figure 15-6   Photosynthetic Production in the World Ocean

Data from the Coastal Zone Color Scanner (CZCS) aboard the *Nimbus 7* satellite between November 1978 and June 1981 were used to produce this false-color image. The CZCS senses changes in seawater color caused by changing concentrations of photosynthetic pigment. Increasing photosynthetic pigment concentration is correlated with increasing photosynthetic productivity. It shows low values (magenta: below 0.1 milligrams per cubic meter (mg/m^3) in the oligotrophic open ocean and high values (red: exceeding 10.0 mg/m^3) along eutrophic continental margins. Productivity of 1 mg/m^3 is represented by the boundary between yellow and green. Productivity values may be off as much as 50 percent of their magnitude. Black indicates insufficient data.   *(Image is courtesy of Jane A. Elrod and Gene Feldman, NASA/Goddard Space Flight Center.)*

1 percent of the euphotic zone production reaches the deep-ocean floor. We call this process the biological pump because it removes (pumps) carbon dioxide and nutrients from the upper ocean and concentrates them in deep-sea waters and sediments.

The permanent thermocline prevents the return of these nutrients to the euphotic layer throughout much of the subtropical gyres. The thermocline and resulting pycnocline develop only during the summer season in temperate latitudes and are absent in polar regions. In the following sections, you will see the degree to which thermal stratification affects the patterns of biological production at different latitudes in the world ocean.

Productivity in Polar Oceans   As an example of seasonal productivity in a polar sea, consider the Barents Sea, which is north of the Arctic Circle off the northern coast of Europe. Diatom productivity peaks there in May and tapers off through July (Figure 15-7a). This feeds a zooplankton development, mostly small crustaceans (Figure 15-7d). The zooplankton biomass peaks in June and remains at a relatively high level until winter darkness begins in October. In this region above 70°N latitude, there is continuous darkness for about 3 months of winter and continuous illumination for about 3 months of summer.

In the Antarctic region, particularly at the southern end of the Atlantic Ocean, productivity is somewhat greater, most likely because of the continual upwelling of water from the North Atlantic. Moving southward as a deepwater mass, this North Atlantic Deep Water surfaces hundreds of years later, carrying high concentrations of nutrients (Figure 15-7b).

To illustrate the very great productivity that occurs during the short summer season in polar oceans, consider the growth rate of blue whales. The largest of all animals, the blue whale (see Figure 16-13) migrates through temperate and polar oceans at times of maximum zooplankton productivity. This excellent timing enables the whales to develop and support large calves (following a gestation of 11 months, the calves can exceed 7 m (23 ft) in length at birth). The mother blue whale suckles the calf for 6 months with a teat that actually pumps the youngster full of rich milk. By the time the calf is weaned, it is over 16 m (50 ft) long. In 2 years, it will attain a length of 23 m (75 ft). In 3.5 years, a 60-ton blue whale has developed. This phenomenal growth rate gives some indication of the enormous biomass of copepods and krill that these large mammals feed upon. See Lifestyles of the Large and Blubbery on page 420 for more details on this phenomenon.

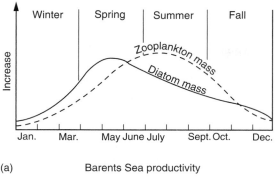

(a)   Barents Sea productivity

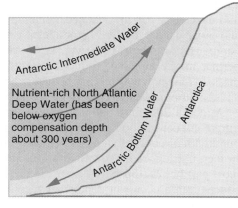

(b)   Antarctic upwelling

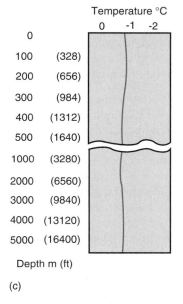

(c)

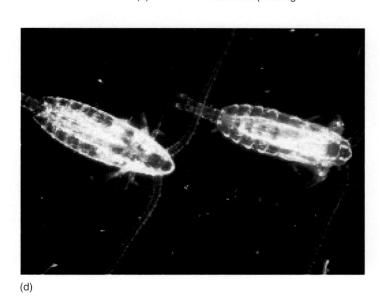

(d)

Figure 15-7   Productivity in Polar Oceans

(a) Diatom mass increases rapidly in the spring when the sun is high enough in the sky to cause deep penetration of sunlight. As soon as this diatom food supply develops, the zooplankton population begins feeding on it, and the zooplankton biomass peaks early in the summer. (b) The continuous upwelling of North Atlantic Deep Water keeps Antarctic waters rich in nutrients. When the summer sun provides sufficient radiation, there is an explosion of biological productivity. (c) Polar water shows nearly uniform temperature. (d) Copepods, *Calanus*, each about 8 mm (0.3 in) in length.   *(Courtesy of Scripps Institution of Oceanography, University of California, San Diego.)*

Polar waters have little vertical change in density or temperature (Figure 15-7c). Thus, in most polar areas, surface waters freely mix with deeper, nutrient-rich waters. However, some density stratification of water masses occurs when summer ice melts. The melting ice lays down a thin, low-salinity layer that does not readily mix with deeper waters. Such stratification is crucial to summer production, because it helps prevent phytoplankton from being carried into deeper, darker waters. The result is that they are concentrated in the sunlit surface waters. Without this density barrier, the summer bloom would not develop. It is becoming increasingly clear that high levels of biological productivity occur only when periods of deep mixing, which create

high nutrient levels in the sunlit surface waters, are followed by periods of density stratification.

Nutrient concentrations (phosphates and nitrates) usually are adequate in surface waters. Thus, photosynthetic productivity in the high latitudes is more commonly limited by solar energy availability than by nutrients. The productive season in these waters is relatively short but outstandingly productive.

Productivity in Tropical Oceans   In direct contrast with high productivity during the summer season in the polar seas, low productivity is the rule in tropical regions of the open ocean. This may sound contradictory, for warm, sunny tropical waters sound very productive—but there is a key limiting factor. True, light penetrates much deeper into the clear, open tropical ocean than into the temperate and polar waters, and this produces a very deep oxygen compensation depth. However, in the tropical ocean, a permanent thermocline produces a stratification of water masses. This prevents mixing between the surface waters and the nutrient-rich deeper waters (Figure 15-8).

At about 20° north and south latitude, phosphate and nitrate concentrations are commonly less than one one-hundredth as concentrated as in winter temperate oceans. In fact, nutrient-rich waters within the tropics lie for the most part below 150 m (500 ft), and the highest nutrient concentration occurs at depths between 500 and 1000 m (1640 and 3300 ft).

Primary productivity in tropical oceans generally occurs at a steady, low rate. However, when we compare the total annual productivity of tropical oceans with that of the more productive temperate oceans, we find that tropical productivity is generally at least one-half of that in temperate regions. Within tropical regions, three environments have unusually high productivity—regions of equatorial and coastal upwelling, and coral reefs.

- Equatorial upwelling—Where trade winds drive westerly equatorial currents on either side of the equator, surface water diverges as a result of the Ekman spiral. The surface water that moves off toward higher latitudes is replaced by nutrient-rich water that surfaces from depths of up to 200 m (660 ft). This condition of equatorial upwelling is probably best developed in the eastern Pacific Ocean.

- Coastal upwelling—Where the prevailing winds blow toward the equator and along western conti-

nental margins, surface waters are driven away from the coast. They are replaced by nutrient-rich waters from depths of 200 to 900 m (660 to 2950 ft). This nutrient-rich upwelling promotes high primary productivity in these areas, which support large fisheries. In the Pacific, such conditions exist along the southern coast of California and the southwestern coast of Peru; in the Atlantic, such conditions are found along the northwestern coast of Morocco and the southwestern coast of Africa.

- Coral reefs—The relatively high productivity of coral reef environment is not related to the upwelling process. It is discussed in Chapter 17.

Productivity in Temperate Oceans   We have discussed general productivity in the polar regions, where it is limited by available sunlight, and in the tropical low-latitude areas, where the limiting factor is nutrients. Now let us consider the temperate regions, where an alteration of these factors controls productivity in a more complex pattern. We will look at production season-by-season.

- Winter—Productivity in temperate oceans is very low during winter, despite high nutrient concentrations in the surface layers. Ironically, nutrient concentration is highest during winter. The limiting factor is solar energy. In Figure 15-9 (winter), you can see that the sun is at its lowest elevation above the horizon during this season. A high percentage of solar energy is reflected, leaving a smaller percentage to be absorbed into surface waters. The oxygen compensation depth for basic producers, such as diatoms, is so shallow that it does not allow growth of the diatom population.

- Spring—In Figure 15-9 (spring), as the sun rises higher in the sky, the oxygen compensation depth deepens as solar radiation is transmitted deeper into the surface water. Eventually, the sunlight warms the surface water, producing the seasonal thermocline, which helps to confine the phytoplankton to the euphotic zone and produce an exponential growth of the diatom population, a spring bloom. This expanding population puts a tremendous demand on the nutrient supply in the euphotic zone. In most Northern Hemisphere areas, decreases in the diatom population occur by May as a result of insufficient nutrient supply.

- Summer—In Figure 15-9 (summer), as the sun rises higher, surface waters in temperate parts of the ocean are warmed. The water becomes separated from deeper water masses by a strengthening seasonal thermocline that may develop at a depth of around 15 m (50 ft). As a result, little or no exchange of water occurs across this discontinuity. The nutrient supply of surface waters cannot be replenished from deep waters. Throughout the summer, the phytoplankton population remains at a relatively low level, but it increases again in some temperate areas during autumn.

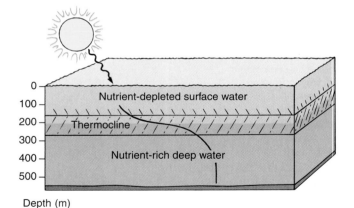

Figure 15-8   Productivity in Tropical Oceans

In typical tropical regions, deep penetration of sunlight produces a deep oxygen compensation depth, with a good supply of solar radiation available for photosynthesis. A permanent thermocline serves as an effective barrier to the mixing of surface and deep water. As algae consume nutrients in the surface layer, productivity is retarded because the thermocline prevents replenishment of nutrients from deeper water.

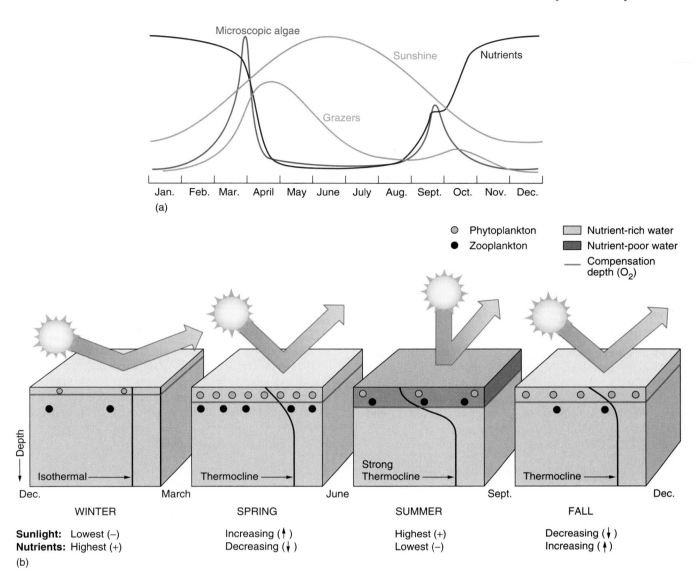

Figure 15-9   Productivity in Temperate Oceans

(a) Relationships among algae, grazers, amount of sunshine, and nutrient levels in surface waters in temperate latitudes. (b) In **winter**, the sun is low in the sky. Much solar energy is reflected, and little is absorbed into the ocean. The water column is isothermal. Nutrients are present throughout the water column. Algae and animal populations are at low levels due to lack of solar energy for photosynthesis (shallow oxygen compensation depth). In **spring**, the sun is higher in the sky. More solar energy is absorbed by the ocean. A thermocline begins to develop. A spring bloom of phytoplankton occurs as a result of the availability of both solar energy and nutrients. The oxygen compensation depth increases. In **summer**, the sun is high in the sky. Surface water warms and strong thermal stratification prevents mixing of surface and deep water. The thermocline reaches maximum development. When nutrients in the surface water are used up, the supply cannot be replenished from deep water. Even though the oxygen compensation depth is at its maximum, phytoplankton become scarce in late summer. The increase in animal population that followed spring bloom is followed by a decrease. In **fall**, the sun is lower in the sky. Surface water cools. Fall winds aid in mixing surface and deep water. The thermocline begins to disappear. A small fall bloom of phytoplankton, initiated by fresh nutrients from mixing, is terminated because too much solar energy is lost by reflection as the sun is lower in the sky.

- Fall—The autumn increase is much less spectacular than that of the spring. In Figure 15-9 (fall), solar radiation diminishes as the sun drops lower in the sky. This lowers surface temperatures, and the summer thermocline breaks down. A return of nutrients to the surface layer occurs as increased wind strength mixes the surface layer with the deeper water mass in which the nutrients have been trapped throughout the summer months. This bloom is very short-lived. The phytoplankton population declines rapidly. The limiting factor in the fall is the opposite of that which reduced the phytoplankton population in the spring. In the case of the spring bloom, solar radiation was readily available, and the decrease in nutrient supply was the limiting factor. When the fall bloom is terminated, nutrients are available, but available sunlight is limited.

How Accurate Are Productivity Measurements?

Coastal waters with high nutrient levels are highly productive, or eutrophic. Most of the open ocean has a low level of nutrients and productivity. Areas of low productivity are referred to as **oligotrophic**. Figure 15-6 shows the general patterns of ocean productivity based on photosynthetic pigment concentrations as observed by the Coastal Zone Color Scanner (CZCS) aboard the Nimbus 7 satellite.

Measurements of photosynthetic productivity of oligotrophic waters in the North Atlantic and North Pacific oceans indicate they may be from 2 to 7 (or even more) times as productive as ^{14}C data indicate. Instead of using the small bottles used in the ^{14}C measurements and suspending them for a short time in the ocean, some physical oceanographers have used a much larger test sample. In their studies of the circulation within the subtropical gyres, physical oceanographers use water samples that are capped by the pycnoclines produced by the thermoclines beneath the warm layers of surface water within the subtropical gyres of the open oceans. These samples contain (depending on the design of the study) from tens to thousands of cubic kilometers of water that record the average results of photosynthetic activity over periods ranging from a few months to decades.

The studies analyze the effect of photosynthesis on the oxygen concentrations (1) in the euphotic zone and (2) beneath the euphotic zone. In the North Pacific Ocean, oxygen saturations in **subsurface oxygen maximums** (SOMs) at depths between 50 and 100 m (164 and 328 ft) were from 110 to 120 percent at latitudes of from 30° to 40°N during summer months based on data obtained from 1962 through 1979. The excess oxygen (anything over 100 percent) may be due to trapped oxygen produced by photosynthesis. The North Atlantic study determined the rate at which oxygen is used up beneath the euphotic zone (below 100 m) by decomposing organic matter falling toward the ocean floor. **Oxygen utilization rates** (OURs)

of twice the accepted mean oligotrophic rate indicate there may be much more organic matter produced in the oligotrophic euphotic zone than is indicated by ^{14}C data.

The problem with the ^{14}C method may be that by using such a small sample and exposing it for such a short time, periods of unusually high productivity may be missed. An interesting find may help explain how some of the biological productivity of oligotrophic water may have been missed. Mats of diatoms belonging to the genus *Rhizosolenia* have been found floating in the North Pacific Gyre and the Sargasso Sea. Averaging 7 cm (3 in) in length, these mats possess symbiotic bacteria that fix molecular nitrogen (N_2) into nitrate ion (NO_3^-), which is a usable nutrient for the diatoms (Figure 15-10). Bundles of *Oscillatoria*, nitrogen-fixing cyanobacteria, have also been verified in the Sargasso Sea. Such mats are readily broken up by net tows and missed by standard bottle casts. Studies in the Pacific and Atlantic oceans indicate that a large percentage (60 percent in one study) of oligotrophic photosynthesis is achieved by picoplankton (0.2 to 2.0 μm; 0.000008 to 0.00008 in) that pass through many filters used in productivity studies. To add to the confusion, recent studies in the tropical North Atlantic indicate that nitrogen-fixing cyanobacteria account for more than 60 percent of the production, and picoplankton account for less than 10 percent. Continued investigation of this problem may reveal the overall productivity of the oligotrophic ocean waters has been greatly underestimated.

Chemosynthetic Productivity

Another source of potentially significant biological productivity in the oceans occurs in the rift valleys of oceanic spreading centers at depths of over 2500 m (8200 ft), where there is no light for photosynthesis. In some places, ocean water seeps down fractures into the ocean crust deep enough to become heated by underlying magma chambers. This creates hydrothermal springs that support remarkable benthic communities.

As the heated water rises to the ocean floor, it dissolves minerals from the crustal rocks it passes through. These minerals are deposited on the sea floor. Thus, potentially significant mineral deposits are associated with biotic communities along hydrothermal vents. Tube worms, clams, and mussels are significant members of these communities (Figure 15-11). The worms grow to over a meter in length; clams are typically 25 cm (10 in) long. They live in association with chemosynthetic bacteria that produce food using a chemical reaction rather than sunlight as an energy source. Hydrogen sulfide (H_2S) gas dissolved in hydrothermal waters is oxidized by the bacteria, releasing chemical energy used to synthesize organic matter.

Because these bacteria depend on the release of chemical energy, their food-making is called chemosynthesis. The true significance of bacterial productivity on the deep-ocean floor cannot be fully understood until much more research is conducted. Such research has the potential to increase our estimates of the ocean's biological productivity.

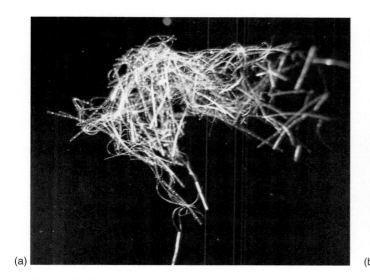
(a)

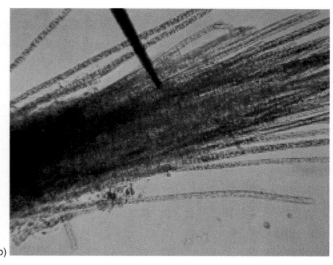
(b)

Figure 15-10 Nitrogen Fixation by Aggregates of Diatoms and Cyanobacteria in Oligotrophic Ocean Waters

(a) Typical mat of *Rhizosolenia* about 5 cm long. It is composed of intertwining chains of *R. castracanei* (wider cells) and *R. imbricata* (narrower cells). Within these cells, symbiotic bacteria fix nitrogen for uptake by the algal cells. (b) An aggregation of the nitrogen-fixing cyanobacteria *Oscillatoria* spp. with an O_2 probe inserted into it. The probe has a maximum diameter of about 5 μm.   *(Part (a) James M. King. (b) Courtesy of Hans W. Paerl, University of North Carolina.)*

(a)

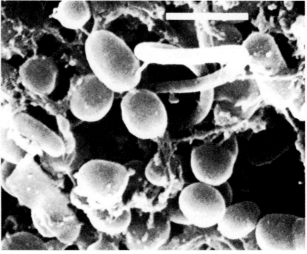
(b)

Figure 15-11 Chemosynthetic Life of the Galápagos Rift

(a) Tube worms up to 1 m long (3.3 ft), found at Galápagos Rift and other deep-sea hydrothermal vents. These worms contain sulfur-oxidizing bacteria, which produce food by chemosynthesis, a process that uses energy obtained from the oxidation of inorganic nutrients dissolved in the deep-ocean water. Similar communities have been discovered near cold water seeps at the base of the Florida Escarpment in the Gulf of Mexico and are discussed in Chapter 17. (b) Enlarged 20,000 times, these are the sulfur-oxidizing bacteria that live symbiotically within the tissues of tube worms, clams, and mussels found at hydrothermal vents. (White bar at top is 1 μm.)   *(Part (a) Fred Grassle. (b) Courtesy of Woods Hole Oceanographic Institution.)*

Biochemists recently discovered that bacteria can obtain chemical energy for the synthesis of their food through oxidation of various compounds containing the metals iron, manganese, copper, nickel, and cobalt. These microorganisms may be important in the formation of ore-quality deposits of oxides of these metals on the ocean floor in the form of manganese nodules.

Ecosystems and Energy Transfer ....

We have looked at nutrient availability. Now, let us turn our attention to the cycling of important classes of nutrients and the flow of energy. The term **biotic community** refers to an assemblage of organisms that live together within a definable area. An **ecosystem** includes the biotic community plus the environment with which it interacts in the exchange of energy and chemical substances. Within an ecosystem, there are three basic categories of organisms—producers, consumers, and decomposers. Plants, algae, and some bacteria are the **producers** in an ecosystem. These autotrophic organisms have the capacity to nourish themselves through chemosynthesis or photosynthesis. The consumers and the decomposers are heterotrophic organisms that depend, directly or indirectly, on the organic compounds produced by the autotrophs for their food supply.

Consumers can be divided into three categories: **herbivores**, which feed directly on plants or algae (herb = plant); **carnivores**, which feed only on other animals (carni- = meat); and **omnivores**, which feed on both (omni- = all). As the role of bacteria in the marine ecosystem becomes better understood, a fourth category of consumers, the **bacteriovores**, which feed on bacteria, may be identified as an important component of marine ecosystems.

The **decomposers**, the bacteria and fungi, break down the organic compounds in animal wastes and in the remains of dead plants and animals for their own energy requirements. They characteristically release simple inorganic salts that are re-used by plants and algae as nutrients.

Symbiosis is a relationship in which two or more organisms are closely associated in a way that benefits at least one of the participants, and sometimes both. Three types of symbiotic relationships are commensalism, mutualism, and parasitism.

- **Commensalism**—A smaller or less dominant participant benefits without harming its host, which affords subsistence or protection to the other. An example is the remora, a fish that attaches to a shark or other fish to obtain food and transportation, without harming its host (Figure 15-12a).
- **Mutualism**—Both participants benefit. Such a relationship exists between the large reef fish and the *cleaner fish* that eat their parasites (Figure 15-12b).
- **Parasitism**—One participant, the parasite, benefits at the expense of the host. Many fish are hosts to isopods, which attach to and derive their nutrition from the body fluids of the fish, thereby robbing the host of its energy supply (Figure 15-12c).

Energy Flow

It is important to understand that energy flow in an ecosystem is not cyclic, but unidirectional, originating from the sun and continually dissipating as organisms within the system utilize it for life processes. Solar energy is converted into chemical energy when organic compounds are synthesized by photosynthesis or chemosynthesis; energy is released from mass when organic compounds are broken down by respiration, and this energy is lost from the system as waste heat.

Figure 15-13 depicts the flow of energy through an algae-supported biotic community. Energy enters the system as solar radiation. Algae and plants, by photosynthesis, convert this electromagnetic energy into chemical energy in the form of organic compounds. These compounds may be incorporated into the structure of the organism or broken down by respiration for energy. Plant and algal mass consumed by feeding animals is converted into energy for metabolic processes or stored as animal mass. When these animals are eaten, again, mass is either converted to energy or stored as new animal mass. With each step, energy released by respiration is lost to the system, and progressively smaller amounts are left to be stored as biomass.

Composition of Organisms Having discussed the noncyclic, unidirectional flow of energy through the biotic community, let us now consider the flow of nutrients, which is cyclic. These are the biogeochemical cycles, so-called because they involve chemicals that are transferred between the biosphere (life system) and geosphere (physical earth systems). In these cycles, mass does not dissipate, but is cycled from one chemical form to another.

Living organisms are made up of millions of complex organic molecules such as carbohydrates, fats, and proteins. These molecules are composed mostly of 20 elements that can be divided into these three groups according to their abundances:

1. Primary constituents: Major elements in organic materials, each of which make up 1 percent or more (dry weight) of organic material; in order of abundance, carbon, oxygen, hydrogen, nitrogen, and phosphorus.
2. Secondary constituents: Elements in concentrations from 500 to 10,000 parts per million (ppm) (dry weight); including sulfur, chlorine, potassium, sodium, calcium, magnesium, iron, and copper.
3. Tertiary constituents: Elements in concentrations of less than 500 ppm (dry weight); including boron, manganese, zinc, silicon, cobalt, iodine, and fluorine.

The primary constituents make up the bulk of organisms—the visible tissue, bone, and body fluids. If not found in sufficient quantity, algal and plant growth or productivity may be limited. Carbon is the basic component of all organic compounds (including carbohydrates, protein, and fats). There is no scarcity of carbon for photosynthetic productivity; only about 1 percent of the total

(a)

(b)

Figure 15-12   Symbiotic Relationships

(a) Commensalism. A remora swims below a Caribbean jewfish in hopes of sharing the food of its host. The remora uses a sucking organ on the top of its head to attach itself to larger fish. (b) Mutualism. Juvenile bluehead wrasses clean parasites from a stoplight parrot fish, *Sparisoms viride*, at a reef at Bonaire Island, Netherlands Antilles. (c) Parasitism. A parasitic isopod on the head of a blackbar soldierfish.   *(Part (a) © Marty Snyderman; (b) © Fred Bavendam/Peter Arnold, Inc. (c) All rights reserved by Coral Reef Research Foundation, Palau.)*

carbon in the sea is involved in photosynthetic productivity. The remainder is in seawater as bicarbonate ions, or is bound into calcium carbonate shells. Comparatively, nitrogen compounds involved in photosynthetic productivity may be ten times the total nitrogen compound concentration that can be measured as a yearly average. This level implies that soluble nitrogen compounds must be cycled

completely up to ten times per year. Available phosphates may be turned over up to four times per year.

The secondary constituents usually are sufficiently concentrated in natural waters that they do not limit productivity. However, studies done in the equatorial Pacific and in the South Pacific Ocean have shown that where high-nutrient (nitrogen and phosphorus), low-chlorophyll

Figure 15-13 Energy Flow through Photosynthetic Ecosystem

Energy enters the system as radiant solar energy. It is converted to and stored as plant and algal biomass through photosynthesis. Metabolism releases the stored energy.

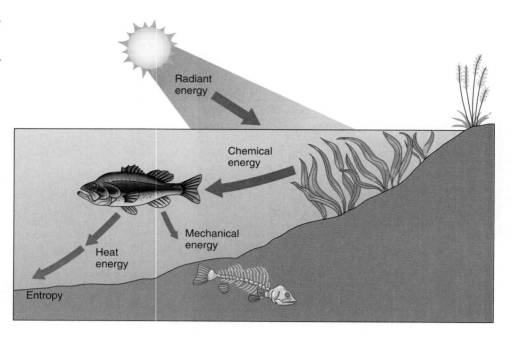

conditions exist, primary production is limited by the availability of iron. Tertiary constituents occur in very low concentrations in organisms, but are also found in very low concentrations in the marine environment. Their absence may limit productivity.

Microbes in the Marine Environment

Although it has been widely accepted that zooplankton are the primary grazers of phytoplankton, it is becoming clear that free-living bacteria and archaea my consume up to 50 percent of the production of phytoplankton. The bacteria are thought to consume the dissolved organic matter that is lost from the conventional food web by these three processes:

1. *Phytoplankton exudate.* As phytoplankton age, they lose some of their cytoplasm directly into the ocean.

2. *Phytoplankton munchate.* As phytoplankton are eaten by zooplankton, cytoplasm is spilled into the ocean.

3. *Zooplankton excretions.* The liquid excretions of zooplankton are dissolved into ocean water.

Free-living heterotrophic microbes (archaea and bacteria) absorb this dissolved organic matter and re-enter the conventional food web primarily when they are consumed by microscopic flagellates (Figure 15-14).

Free-living microbes may compete very effectively with zooplankton for the photosynthetic products of phytoplankton. Although some investigators disagree, evidence suggests that organic matter consumed by bacteria may not reenter the basic food web that supports the ocean's fish populations. It appears that much of the organic matter consumed by free-living microbes may be confined to a **microbial loop** and lost to the world fish populations because larger zooplankton may not graze efficiently on the protozoans thought to be the main bacteriorvores. (See Figure 7-15)

Viruses are also being discovered to play an important role in marine biological processes. Viruses are small particles about 20-200 nm long. They consist of genetic material surrounded by a protein or lipid coat. They are parasites that must depend on a host cell for their metabolism. The abundance of viruses is about ten billion per liter of seawater—about 5 to 25 times that of microbes. Viruses are most abundant in coastal waters as is true for most groups of organisms. They have also been identified as playing an important role in limiting the abundance of microbes and may be responsible for 8 to 34 percent of microbe mortality. Through infection of microbes, phytoplankton, and zooplankton, viruses may play an important role in biogeochemical cycling, as well as biodiversity and species distribution of microbes, algae and zooplankton (Figure 15-15).

Food Chains and Food Webs

In biogeochemical cycles, inorganic elements are taken in by autotrophic organisms and used to synthesize organic molecules. These molecules are passed as food through a sequence of organisms forming food chains and food webs. These sequences end in the bacterial and fungal decomposition of the organic molecules, returning the elements to inorganic forms that can be used again by other autotrophs.

A **food chain** is a sequence of organisms through which energy is transferred, starting with the primary producers, through herbivores and then through successive levels of carnivores to the highest-level carnivore, which does not

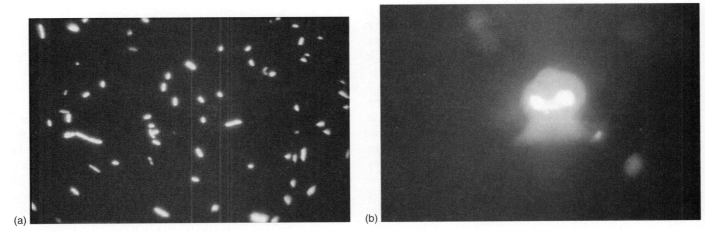

Figure 15-14   Free-Living Bacteria and Bacteriovore

(a) Bacteria made visible by treatment that makes them fluorescent. They are about $1\,\mu m \times 0.5\,\mu m$ in size. (b) This 25-μm ciliate is a voracious bacteriovore from a Georgia tidal creek.   *(Courtesy of Evelyn and Barry Sherr.)*

Figure 15-15   Fluorescent Image of Marine Viruses

The small dots in this image are viruses while the larger ones are either archaea or bacteria which can not be differentiated by this technique.   *(Courtesy of Jed Fuhrman.)*

have any predators. More commonly, however, the top carnivores in a food chain feed on a number of animals, each of which has its own simple or complex feeding relationships. This constitutes a **food web** (Figure 15-16).

Animals that are part of food webs, such as the North Sea Herring in Figure 15-16b, are less likely to starve should one of their food sources diminish in quantity, or even become extinct. Newfoundland Herring eat only copepods, so extinction of copepods would have a catastrophic effect on that herring population (Figure 15-16a). The Newfoundland Herring population does, however, have an advantage over its relatives who feed through a broader-based food web. The Newfoundland Herring is likely to have a larger biomass to feed on because these Herring are only two steps removed from the producers, whereas the North Sea Herring is at the fourth level in some of the food chains within its web.

Chemical energy stored in the ocean's algae is transferred to the animal community through feeding. Herbivores eat diatoms and other microscopic marine algae. Larger herbivores feed on larger macroscopic algae and grasses that grow attached to the ocean bottom. In turn, the herbivores are eaten by larger animals, the carnivores. They, in turn, are eaten by even larger carnivores, and so on. Each of these feeding levels is called a **trophic level**. Generally, the individuals of a trophic level are larger—but not too much larger—than the individuals in the trophic level they eat, although there are outstanding exceptions to this condition. The blue whale, the largest animal ever to exist on Earth, feeds upon krill that attain maximum lengths of only 6 cm (2.4 in).

Transfer Efficiency and Biomass Pyramid   Only about 2 percent, on average, of light energy absorbed by algae to synthesize food is made available to herbivores. The gross **ecological efficiency** at any trophic level is the ratio of energy passed on to the next higher trophic level, divided by the energy received from the trophic level below. An example is the energy consumed by a carnivore population (tuna) divided by the energy received from the herbivore population (anchovy). Figure 15-17 shows that some of the energy taken in as food by herbivores is passed as feces and the rest is assimilated. Of the assimilated chemical energy, much is quickly converted via respiration to kinetic energy for maintaining life; what remains is available for growth and reproduction. Of the food energy taken in by an

organism, only a small percentage is passed on to the next trophic level through feeding. Figure 15-18 shows the passage of energy between trophic levels through an entire ecosystem, from the solar energy assimilated by phytoplankton to the mass of the highest-level carnivore, which, in this example, is us. (Think of this diagram the next time you enjoy a fish dinner.)

The efficiency of energy transfer between trophic levels has been intensively studied. Many variables must be considered. For example, young animals often process available food more efficiently than older animals. Also, when food is plentiful, animals expend more energy in digestion and assimilation than when food is not readily available. Most efficiencies range between 6 and 15 percent, and ecological efficiencies in natural ecosystems average approximately 10 percent. There is some evidence that, in populations important to our present fisheries, this efficiency may run as high as 20 percent. The true value of this efficiency is of practical importance to us because it largely determines the fish harvest we can anticipate from the oceans. Because there is a decrease in available energy at each higher feeding level, there are also decreases in biomass and in the number of organisms that can be supported.

Biomass Pyramid   At any point in a food chain, only a small percentage of the biomass is passed on to higher feeding levels, because some mass is converted to energy and some is lost as waste. As a result, the total mass of producers required to support the marine populations is many times greater than the total mass of the ultimate consumers, such as sharks or killer whales.

Because of the 90-percent energy loss that occurs at each higher level of a food chain, there are a limited number of

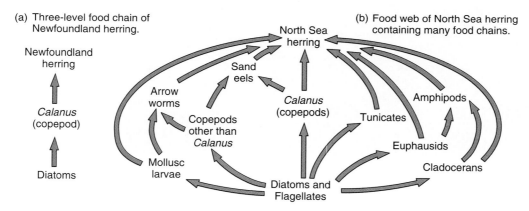

(a) Three-level food chain of Newfoundland herring.

(b) Food web of North Sea herring containing many food chains.

Figure 15-16   A Food Chain and a Food Web

(a) A food chain is a linear sequence of organisms through which energy passes, such as from diatoms to copepods to Newfoundland herring. (b) The food sources of the North Sea herring involve multiple paths or food chains, forming a food web—where this fish may be at the third or fourth trophic (feeding) level. Because the Newfoundland herring is always at the third trophic level, it has a potentially larger biomass available to it than the North Sea herring. However, a disruption in the food chain of Newfoundland herring would cause serious trouble because it has no alternative food sources. In contrast, if one chain of the North Sea herring's food web were disrupted, it would not pose as great a threat, because of all the other food chains in the web.

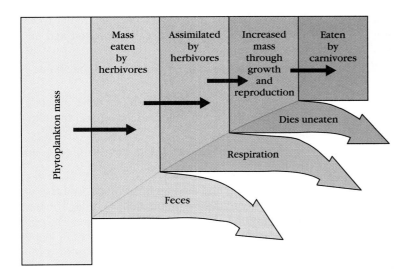

Figure 15-17   Passage of Energy through a Trophic Level

The mass of phytoplankton taken in by an herbivore is processed by the herbivore and converted into chemical compounds usable by the herbivore. Material that cannot be used is lost as feces. Some of the material assimilated is broken down by respiration and the energy released used for the herbivore's life activities. The remaining material becomes part of the herbivores body structure for growth or increased body weight, or goes toward producing young. This increased mass is available for a carnivore to eat, but not all herbivores are consumed, and some die uneaten. Thus, a small percentage (about 10 percent) of the food mass consumed by the herbivore is available to the carnivore that feeds upon it.

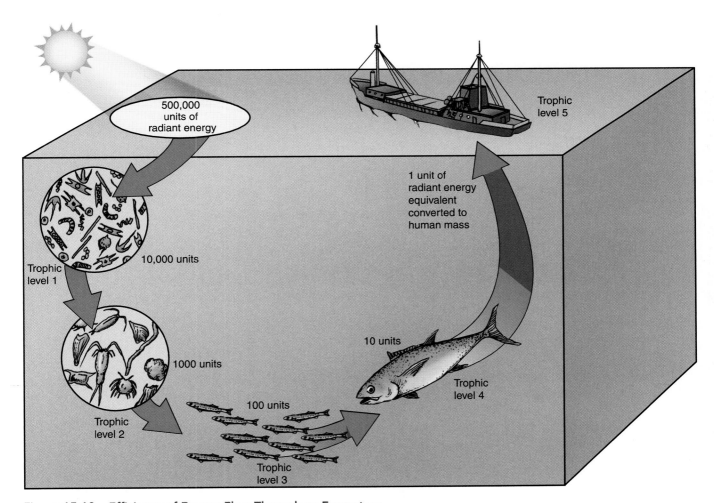

Figure 15-18   Efficiency of Energy Flow Through an Ecosystem

This diagram shows probable effects of energy transfer within an ecosystem. One unit of mass is added to the fifth trophic level (humans) for every 500,000 units of equivalent radiant energy input available to the producers (algae). This value is based on a 2-percent efficiency of transfer of radiant energy by algae and 10-percent efficiency at all feeding levels.

feeding levels, and each feeding level has about 90 percent less biomass than the level below it—the trophic level it feeds on. Individuals of a trophic level generally are larger than their prey, but the total biomass at that trophic level is approximately 10 percent of the total biomass in the trophic level below it.

Because of the inefficiency of energy transfer between trophic levels, it is beneficial for fishermen to choose a trophic level that feeds as close to the primary-producing trophic level as possible. This increases the biomass available for food and the number of individuals available to the fishery.

The ultimate effect of energy transfer between trophic levels can be seen in Figure 15-19. It depicts the progressive decrease in numbers of individuals and total biomass at successive trophic levels resulting from decreased amounts of available energy.

A study of food webs in 55 marine ecosystems indicated an average of 4.45 trophic levels.

- Only two trophic levels were encountered at rocky shores in New England, and Washington and at a Georgia salt marsh.
- Seven links were found in the longest food chains in Antarctic seas.
- Two tropical plankton communities were observed in the Pacific Ocean, one had seven and the other had ten trophic levels.
- Tropical epipelagic seas and Pacific Ocean upwelling zones had eight trophic levels.

The study found that the amount of primary productivity and the variability of the environment have little influence on the length of food chains within an ecosystem. The only variable that showed a clear relation to food chain length was whether or not the environment was two or three-dimensional. Three-dimensional (pelagic) marine ecosystems averaged 5.6 trophic levels, compared to an average of 3.5 for two-dimensional (benthic) ecosystems.

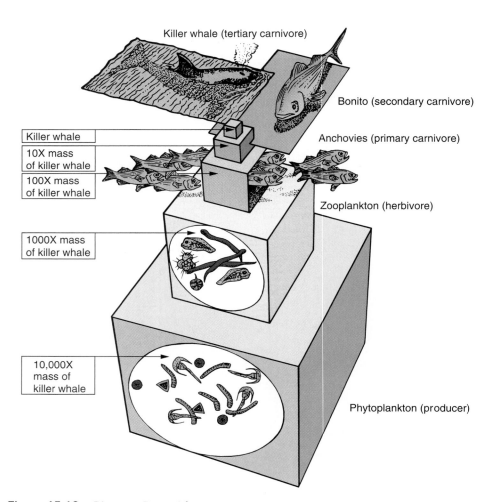

Killer whale (tertiary carnivore)

Bonito (secondary carnivore)

Anchovies (primary carnivore)

Killer whale

10X mass of killer whale

100X mass of killer whale

Zooplankton (herbivore)

1000X mass of killer whale

10,000X mass of killer whale

Phytoplankton (producer)

Figure 15-19 Biomass Pyramid
The higher on the food chain an organism lives, the physically larger it is as an individual. However, the total biomass represented by a population high on the food chain is less than that for a population at a lower feeding level.

Ozone Depletion and Phytoplankton Production

In 1985, researchers established that ozone (O_3) in the Antarctic stratosphere (part of the upper atmosphere) was being significantly depleted by the actions of human-made chemicals called chlorofluorocarbons (CFCs). Stratospheric ice particles and aerosols catalyze a CFC chemical reaction that produces chlorine monoxide (ClO). Cold temperatures also seem to accelerate the reaction. It is the ClO that combines with the ozone and destroys it. Key ingredients for this destruction are low temperature ($-78°$) and sunshine. During the Antarctic spring, the combination of these factors can reduce O_3 by as much as 50 percent. Over about a 3-month period each year, this creates a hole in the stratospheric ozone layer about the size of Antarctica (Figure 15-20).

In the Arctic, low temperatures do not last as long but in 1999, Arctic O_3 depletion peaked at about 60 percent. Ironically, warming of the lower atmosphere helps cool the stratosphere. If global lower atmosphere warming continues, this could lead to a colder stratosphere and an increased threat to polar stratospheric O_3 during the next decades.

The Total Ozone Mapping Spectrometer (TOMS) on the Nimbus 7 satellite began monitoring atmospheric ozone in 1978. It recorded an average decrease of 3 percent between the latitudes 65°N and 65°S over an 11.5-yr period ending in May 1990 (Figure 15-21). Between latitudes 40° and 50°N, the decrease exceeded 5 percent over the same time period. Ozone is presently monitored by a TOMS unit on NASA's Earth Probe satellite. According to models that have been established, this results in an increase of 10 percent in the ultraviolet B (UVB) radiation—the wavelength of radiation that most damages DNA (280 to 320 nm).

Clearly, ozone depletion has implications for land-dwelling organisms. Exposure to increased UVB radiation is predicted to cause an increase in human skin cancer. How will it affect life in the oceans, where the water absorbs UV radiation and provide some degree of protection? A study conducted throughout the spring and summer of 1990 investigated this question in the marginal ice zone (MIZ). This is the zone at the edge of the sea-ice development where summer melting reduces the surface salinity and stabilizes the surface layer. This confines phytoplankton to the euphotic zone and helps produce the large summer algal bloom. Parts of the MIZ are exposed to increasing intensity of UVB radiation as parts of the Antarctic ozone hole move across it (Figure 15-20). During this period, UVB radiation can be detected to a depth of 70 m (230 ft), and phytoplankton production may

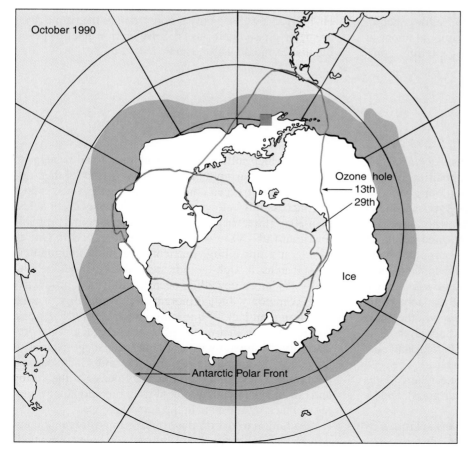

October 1990

Ozone hole
— 13th
— 29th

Ice

Antarctic Polar Front

Figure 15-20   Antarctic Ozone Hole Positions, October 13 and 29, 1990

The ozone hole changes position over the Antarctic. The green lines delineating its position are determined by the 200-Dobson-unit concentration. Ozone concentration is less than 200 Dobson units inside the hole. The pack ice edge is the seaward edge of the white zone, and the Antarctic Polar Front is at the outer edge of the dark blue region. The area investigated, indicated by the red rectangle, is the MIZ, a narrow zone at the seaward edge of the pack ice. *(After Smith et al., 1992.)*

Figure 15-21   Ozone Depletion, 1978–1990

The average annual percentage depletion of total ozone as measured by the Total Ozone Mapping Spectrometer (TOMS) for the months shown over an 11.5-yr period. The maximum depletion of more than 3 percent occurred in the Antarctic during the months of September through October. In the Arctic, ozone decreased a maximum of 1.5 percent during the months of February and March. A significant reduction of more than 0.8 percent occurred at midlatitudes in the Northern Hemisphere during the months of January through April. Little reduction occurred in equatorial and tropical latitudes. *(After Stolarski et al., 1991.)*

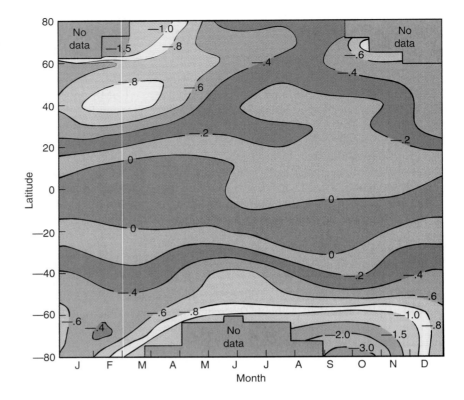

be reduced by as much as 12 percent. This represents a reduction in MIZ carbon production of about 7×10^{12} g, or about 2 percent, per year. Although 2 percent may seem a small reduction, it is a very significant cause for concern.

Considering the future prospects for improvement in stratospheric ozone depletion, ozone-depleting compounds dropped 3 percent below their 1993-94 peak in 1997, and the 1999 Antarctic ozone hole was the smallest it has been in this decade. Very short-lived gases such as HCFC-123 that pose only a minor risk to the ozone layer are being phased in for use in refrigeration and it is possible that ozone depletion can be halted by 2010.

Summary

Organisms are divided into three domains, *Archaea, Bacteria* and *Eukarya*. Based on the earlier classification system, it is still convenient to use five kingdoms: Monera, single-celled organisms without a nucleus; Protista, single-celled and simple many-celled organisms with a nucleus; Mycota, fungi; Plantae, many-celled plants; and Animalia, many-celled animals. Taxonomic classification of organisms involves dividing the kingdoms into the increasingly specific groupings; phylum, class, order, family, genus, and species. Photosynthetic protists include the macroscopic brown algae, Phaeophyta (kelp and *Sargassum*); the green algae, Chlorophyta; and the red algae, Rhodophyta. The microscopic algae include Chrysophyta (diatoms and coccolithophores) and Pyrrophyta (dinoflagellates). The seed plants are represented by a few genera of nearshore plants such as *Zostera* (eelgrass), *Phyllospadix* (surfgrass), *Spartina* (marshgrass), and mangrove trees.

Photosynthetic productivity of the oceans is limited by the availability of solar radiation and/or nutrients. The depth to which sufficient light penetrates to allow algae to produce only that amount of oxygen required for their respiration is the oxygen compensation depth. Algae cannot live successfully below this depth, which may occur at less than 20 m (66 ft) in turbid coastal waters, or at a probable maximum of 150 m (492 ft) in the open ocean. Due to runoff and upwelling, nutrients are most abundant in coastal areas. In high-latitude areas, thermoclines are generally absent, so upwelling occurs readily, and productivity is commonly limited more by the availability of solar radiation than lack of nutrients. In low-latitude regions, where a strong thermocline may exist year-round, productivity is limited by lack of nutrients except in areas of upwelling. In temperate regions, where distinct seasonal patterns are developed, productivity peaks in the spring and fall and is limited by lack of solar radiation in the winter and lack of nutrients in the summer.

In addition to the primary production of organic compounds through algal photosynthesis, organic compounds are produced by bacterial chemosynthesis. Chemosynthesis, observed on oceanic spreading centers in association with hydrothermal springs, is based on chemical energy released by the oxidation of hydrogen sulfide.

As biomass is transferred from phytoplankton to herbivore and the various carnivore trophic levels, only about 10 percent of the mass taken in at one level is passed on to the next. The ultimate effect of this decrease is a decrease in the number of individuals and total biomass of populations higher in the food chain. Upon the death of organisms, the biomass is decomposed to inorganic forms ready again for use as nutrients for phytoplankton.

Human-made CFCs are reducing the concentration of stratospheric ozone. The increased level of UVB radiation is having a negative impact on phytoplankton in the Antarctic marginal ice zone.

Key Terms

Animalia	Dinoflagellates	Mycota	Primary production
Archaea	Ecological efficiency	Net primary production	Producers
Bacteria	Ecosystem	New production	Protista
Bacteriovores	Eukarya	Nitrogen-fixing bacteria	Protozoa
Biotic community	Eutrophic	Oligotrophic	Pyrrophyta
Carnivores	Food chain	Omnivores	Red tides
Chlorophyta	Food web	Oxygen compensation depth	Regenerated production
Chrysophyta	Gross primary production	Oxygen utilization rate	Rhodophyta
Coccolithophores	Herbivores	Ozone	Spermatophyta
Commensalism	Microbial loop	Parasitism	Subsurface oxygen maximum
Consumers	Monera	Phaeophyta	Symbiosis
Decomposers	Mutualism	Plantae	Trophic level

Questions and Exercises

1. Compare the various kinds of macroscopic algae in terms of the maximum depth in which they grow, common species, and size.

2. Discuss and compare the contributions of the Pyrrophyta genera *Ptychodiscus* and *Gonyaulax* to red tide development.

3. Define oxygen compensation depth, and explain the use of the opaque and transparent bottle technique for its determination. Discuss how the quantity of oxygen produced by photosynthesis in each clear bottle is determined.

4. Compare the biological productivity of polar, temperate, and tropical regions of the oceans. Include a discussion of seasonal variables, thermal stratification of the water column, and the availability of nutrients and solar radiation.

5. Discuss chemosynthesis as a method of primary production. How does it differ from photosynthesis?

6. Explain why nitrogen is much more likely than phosphorus to be a limiting factor in marine productivity.

7. How is the energy taken in by individuals of a specific trophic level lost so that only a small percentage is made available to the next trophic level? What is the average efficiency of energy transfer between trophic levels?

8. If a killer whale is a third-level carnivore, how much phytoplankton mass is required to add each gram of new mass to the whale? Assume 10 percent efficiency of energy transfer between trophic levels. Include a diagram.

9. Describe the probable advantage to being the top carnivore in a food web over being part of a single food chain.

10. Discuss the effects of primary production, environmental variability, and environmental dimensionality on the length of food chains.

11. Describe the effect of CFCs in terms of stratospheric ozone depletion and the resulting increased intensity of UVB radiation penetrating Antarctic MIZ euphotic waters.

References

Burkholder, J. A. M., et al. 1992. New "phantom" dinoflagellate is the causative agent of major estuarine fish kills. *Nature* 358:6385, 404–10.

Carpenter, E. J., and K. Romans. 1991. Major role of the cyanobacterium, *Trichodesmium*, in nutrient cycling in the North Atlantic Ocean. *Science* 254:5036, 1356–58.

de Baar, H. J. W., *et al.* 1995. Importance of iron for plankton bloom and carbon dioxide drawdown in the Southern Ocean. *Nature* 373:6513, 412–15.

Ducklow, H. W. 1983. Production and fate of bacteria in the oceans. *Bioscience* 33:8, 494–501.

Dugdale, R. C., and F. P. Wilkerson. 1998. Silicate regulation of new production in the equatorial Pacific upwelling. *Nature* 391: 6664, 270–73.

Fraser, P. J and M. J. Prather. 1999. Uncertain road to ozone recovery. *Nature* 398:6729, 663–64.

Fuhrman, J. A. 1999. Marine viruses and their biogeochemical and ecological effects. *Nature* 399:6736, 541–48.

George, D., and J. George. 1979. *Marine life: An Illustrated Encyclopedia of Invertebrates in the Sea.* New York: Wiley-Interscience.

Grassle, J. F., *et al.* 1979. Galápagos '79: Initial findings of a deep-sea biological quest. *Oceanus* 22:2, 2–10.

Jenkins, W. J. 1982. Oxygen utilization rates in North Atlantic subtropical gyre and primary production in oligotrophic systems. *Nature* 300, 246–48.

Littler, M. M., et al. 1985. Deepest known plant life discovered on an uncharted seamount. *Science* 227:4683, 57–59.

Macdonald, R. W., and E. C. Carmack. 1991. Age of Canada Basin deep waters: A way to estimate primary production for the Arctic Ocean. *Science* 254:5036, 1348–50.

Martinez, L., et al. 1983. Nitrogen fixation by floating diatom mats: A source of new nitrogen to oligotrophic ocean waters. *Science* 221:4606, 152–54.

Paerl, H. W., and B. M. Bebout. 1988. Direct measurement of O_2-depleted microzones in marine *Oscillatoria*: Relation to N_2 fixation. *Science* 241:4864, 442–45.

Parsons, T. R., M. Takahashi and B. Hargrave. 1984. *Biological Oeanographic Processes*, 3rd Ed. New York: Pergamon Press.

Pimm, S. L., J. H. Lawton, and J. E. Cohen. 1991. Food web patterns and their consequences. *Nature* 350:6320, 669–74.

Platt, T., and S. Sathyendranath. 1988. Oceanic primary production: Estimation by remote sensing at local and regional scales. *Science* 241:4873, 1613–19.

Platt, T., D. V. Subba Rao, and B. Irwin. 1983. Photosynthesis of picoplankton in the oligotrophic ocean. *Nature* 301:5902, 702–4.

Russell-Hunter, W. D. 1970. *Aquatic Productivity*. New York: Macmillan.

Sathyendranath, S., et al. 1991. Estimation of new production in the ocean by compound remote sensing. *Nature* 353:6340, 129–33.

Scholin, C. A., et al. 2000. Mortality of sea lions along the central California coast linked to a toxic diatom bloom. *Nature* 403:6765, 80–84.

Sherr, B. F., E. B. Sherr, and C. S. Hopkinson. 1988. Trophic interactions within pelagic microbial communities: Indications of feedback regulation of carbon flow. *Hydrobiologia* 159:1, 19–26.

Shulenberger, E., and J. L. Reid. 1981. The Pacific shallow oxygen maximum, deep chlorophyll maximum, and primary productivity reconsidered. *Deep Sea Research* 28A:9, 901–19.

Smith, R. C., *et al.* 1992. Ozone depletion: Ultraviolet radiation and phytoplankton biology in Antarctic waters. *Science* 255:5047, 952–55.

Stolarski, R. S., et al. 1991. Scientific assessment of ozone depletion: 1991. *Journal of Geophysical Research Letters* 18, 1015–18.

Sullivan, C. W., *et al.* 1993. Distributions of phytoplankton blooms in the Southern Ocean. *Science* 262:5141, 1832–36.

Suggested Reading

Sea Frontiers

Arehart, J. L. 1972. Diatoms and silicon. 18:2, 89–94. Readable article describes the important role of silicon and other elements in the ecology of diatoms, including microphotographs showing the varied forms of diatoms.

Coleman, B. A., R. N. Doetsch, and R. D. Sjblad. 1986. Red tide: A recurrent marine phenomenon. 32:3, 184–91. Discusses the problem of periodic red tides along the Florida, New England, and California coasts.

Idyll, C. P. 1971. The harvest of plankton. 17:5, 258–67. The potential of zooplankton as a major fishery is discussed.

Jensen, A. C. 1973. Warning—red tide. 19:3, 164–75. An informative discussion of what is known of the cause, nature, and effect of red tides.

Johnson, S. 1981. Crustacean symbiosis. 27:6, 351–60. Describes the various symbiotic relationships entered into by tropical shrimps and crabs.

McFadden, G. 1987. Not-so-naked ancestors. 33:1, 46–51. The nature of the coverings of marine phytoplankton cells is revealed by the electron microscope.

Mistry, R. 1992. The lilliputian world of plankton. 38:1, 42–47. High-technology methods of studying the distribution of plankton are discussed.

Oremland, R. S. 1976. Microorganisms and marine ecology. 22:5, 305–10. Discusses the role of phytoplankton and bacteria in cycling matter in the oceans.

Philips, E. 1982. Biological sources of energy from the sea. 28:1, 36–46. The potential for converting marine biomass to energy sources useful to society is discussed.

Scientific American

Anderson, D. M. 1994. Red tides. 271:2, 62–69. Investigators believe red tides are occurring more frequently and in more regions of the world as a result of increased nutrient influx into the coastal oceans.

Benson, A. A. 1975. Role of wax in oceanic food chains. 232:3, 76–89. Reports findings from observations and implications of the wax content in the bodies of many marine animals from copepods to small deep-water fishes.

Burkholder, J. M. 1999. The lurking perils of *Pfiesteria*. 281:2, 42–49. This article reviews the history of *Pfiesteria* outbreaks that harm marine life and humans in coastal waters of Maryland and North Carolina.

Childress, J. J., H. Feldback, and G. N. Somero. 1987. Symbiosis in the deep sea. 256:5, 114–21. Deep-sea hydrothermal vent animals have a symbiotic relationship with sulfur-oxidizing bacteria that allows them to live in the darkness of the deep ocean.

Doolittle, W. F. 2000. Uprooting the Tree of Life. 282:2, 90–95. It appears that the interrelationships of the three domains of life, bacteria, archaea and eukaryota, are more complex than they were believed to be ten years ago.

Govindjee, and W. J. Coleman. 1990. How plants make oxygen. 262:2, 50–67. Explains the production of oxygen by plants.

Levine, R. P. 1969. The mechanisms of photosynthesis. 221:6, 58–71. This article reveals what was known in 1969 of the process by which energy is captured by plants and converted to useful forms of chemical energy while freeing oxygen to the atmosphere.

Pettit, J., S. Ducker, and B. Knox. 1981. Submarine pollination. 244:3, 134–44. Describes pollination of sea grasses by wave action.

Toon, O. B., and R. P. Turco. 1991. Polar stratospheric clouds and ozone depletion. 264:6, 68–75. Discusses the role of ice and nitric acid clouds in creating the reduction of ozone in the Antarctic ozone hole.

Oceanography on the Web

Visit the Introductory Oceanography home page for on-line resources for this chapter. There you will find an on-line study guide with review exercises and links to ocean- ography sites to further your exploration of the topics in this chapter. Introductory Oceanography is at http://prenhall.com/thurman (click on the Table of Contents menu and select this chapter).

Lifestyles of the Large and Blubbery: How to Grow a Blue Whale[1]

- What enables the blue whale to grow so large?
- Why is the Antarctic marine ecosystem so productive?
- What are krill and why are they considered a 'keystone species'?
- What are the threats to the Antarctic marine ecosystem?

Introduction

What is the largest living organism on Earth?

If you answered "blue whale" you'd be wrong. The designation actually belongs either to the redwood tree *Sequoia sempervirens* or to the soil fungus *Armillaria bulbosa,* which lives in northern Michigan.

But the blue whale (*Balaenoptera musculus;* Figure 1) is the largest animal that ever lived, reaching a length of about 30 m (98 ft) and weighing around 100 tons (91,000 kg).

Have you ever wondered why this is the only living animal that grows to this size? Exactly what permits this species to achieve its dimensions but prohibits others from doing so?

An Introduction to Antarctic Marine Ecology

The blue whale spends its summers in the waters surrounding Antarctica, the planet's fifth largest continent. Antarctica is completely surrounded by the southernmost extensions of the Pacific, Atlantic, and Indian Oceans, a physically distinct body of water collectively referred to as the Southern Ocean. The closest continent, South America is 970 km (600 mi) away.

Antarctica is, of course, a land mass, but you wouldn't know it from looking at it: nearly 98% of its 14 million km^2 (5.4 million mi^2) is covered with ice up to 4.5 km (nearly 3 mi) thick. Including sea ice, Antarctica holds 90% of the world's ice and 70% of the world's fresh water. Ironically, despite this much fresh water, because there is no rain in the interior (less than 4 cm of fine crystals known as 'diamond dust' per year!), the continent is essentially a barren desert.

The Southern Ocean

In contrast to the Antarctic land mass, the Southern Ocean supports a richly productive assemblage of organisms.

What makes these cold waters so productive is the southerly flow of deep, nutrient-rich water known as Circumpolar Deep Water. This water rises to the surface (upwells) near the continent from depths of 3000 m (9800 ft) and makes nutrients available to photosynthetic organisms (principally diatoms), which must live in the sunlit surface waters. It also transfers heat to the Antarctic atmosphere. In addition, during the Antarctic summer there is six months of daylight. Thus, photosynthesis can occur 24 hours a day during the summer.

Figure 1

Blue whale, the largest animal that ever lived, and a major consumer of Antarctic krill. *(Courtesy of M. Carwardine/Still Picutres)* **Figure 2** (Insert) Antarctic krill, a keystone species in the Southern Ocean. *(Photo by Frank T. Awbrey/Visuals Unlimited)*

The northern limit of this productivity is a narrow belt of water from 20 to 30 miles (32–48 km) wide known as the Antarctic Convergence, or Polar Front. Here, cold Antarctic surface waters sink below warmer waters flowing south. This is also the northern boundary of the Southern Ocean.

Krill

The key indicator of this Antarctic productivity, in terms of biomass, is not the diatoms or other tiny photosynthetic organisms that actually use the nutrients in the first place. It is a 5 to 6 cm (2–2.5 in) long crustacean called Antarctic krill (*Euphausia superba,* Figure 2) which can top the scales at 2 g (0.1 ou). Krill, meaning "young fish" in Norwegian, is a term applied to a group of about 85 species of shrimplike organisms collectively called *euphausiids* (Phylum Arthropoda) that inhabit waters from the poles to the tropics. Krill are filter-feeding organisms which use their setae-covered appendages to trap the tiny diatoms. They are heavier than water and must swim continuously to avoid sinking. While the life span of krill is not known with certainty, scientific estimates range from 5–11 years. Krill are eaten by whales, seals, birds, squid, and fish, and to a lesser but growing extent by humans.

It is quite possible that krill are the most abundant animal species on earth. Dense aggregates weighing an estimated 2 million tons (1.8 billion kg) and covering as much as 450 km^2 (170 mi^2) have been observed. Density of krill schools reportedly reach 30,000 animals per cubic meter of seawater! The total standing stock (the biomass at any time) has been estimated to be between 200 and 700 million tons (200–700 billion kg). In contrast, the 1997 total world fish catch was less than 100 million tons (100 billion kg).

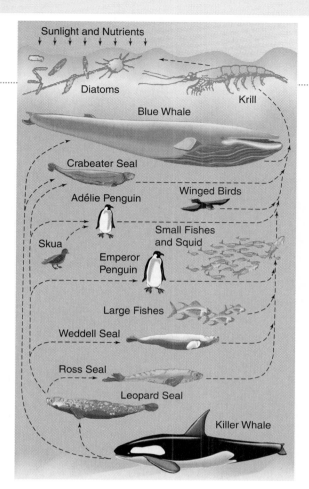

Sunlight and Nutrients

Diatoms

Krill

Blue Whale

Crabeater Seal

Winged Birds

Adélie Penguin

Small Fishes
and Squid

Skua

Emperor
Penguin

Large Fishes

Weddell Seal

Ross Seal

Leopard Seal

Killer Whale

Figure 3

Simplified Antarctic marine food web.

Thus, the numerical abundance and biomass of krill, as well as its important position in the food web (see Figure 3), make it a keystone species in its ecosystem. A keystone species is one whose impact on its ecosystem is disproportionately large and whose loss would severely disrupt the system.

Krill Fisheries

Before the advent of European whaling in the 19th century, baleen whales consumed huge volumes of krill. After the near-extinction of these whales by the 1960's, some viewed krill to be present in "surplus" amounts (estimated at 150 million tons!) which could be harvested without impacting either the species of the marine environment.

Large-scale fisheries for krill (Antarctic and North Pacific krill) currently occur only in Antarctic waters and off the coast of Japan. In the mid-70s the Soviets were the first to operate full-scale fisheries for Antarctic krill. Japan, Poland, Ukraine, and Poland currently conduct Antarctic krill fisheries.

As conventional world fisheries decline from overfishing, there will be greater emphasis on increasing commercial catches of species from more distant waters. This increase in demand

Figure 4

Icebergs being towed from shipping lanes in Antarctica. Will this costly procedure be more commonplace as temperatures increase due to global warming?   *(Photo by Steve Berkowitz)*

will lead to a greater pressure on Antarctic krill, which are the largest known krill stock.

Currently, most krill taken are used as bait and aquaculture feed, although a significant proportion is used for human consumption. Krill are fed to farmed fish because of their nutritional value and also because krill contain high concentrations of the red pigment group carotenoids, which heightens the red color of fish. Japanese consumers consider red as a sign of good luck and as an appetite stimulant.

The Antarctic krill fishery is not increasing currently due to the immense expense of conducting a fishery in Antarctic waters. However, it is likely that in the future this will change because of the growing pressure to catch krill commercially for use as feed in aquaculture. Krill fisheries may also respond to demand for krill concentrate, a health food because of its omega-3 fatty acids content. What this portends for krill is not known because fishery managers have little experience in dealing with an organism such as the krill, which occupies a low but vital position on the food web. Thus, while krill fisheries may increase the world's ocean harvest, they also could cause harm, perhaps catastrophic harm, to this species and to marine ecosystems.

Threats to the Antarctic Marine Environment

Even though it is far removed from the kinds of environmental problems most of the world is facing (fisheries notwithstanding), the Southern Ocean ecosystem is facing some major threats. Temperature increases due to global warming will likely result in accelerated melting of ice sheets (Figure 4) and the addition of massive volumes of freshwater to the marine environment.

Now, go to our web site to complete this analysis.

See the complete issue at:
www.prenhall.com/thurman

[1] Written with Steve Berkowitz of the Department of Marine Science of Coastal Carolina University

16 Animals of the Pelagic Environment

Reef Shark.   (Photo by Bert Yates)

The physical and chemical properties of the oceans have resulted in organisms with a wonderful variety of adaptations and lifestyles that allow them to live successfully in the oceans. In addition to physical adaptations that make it possible for individual animals to function, other adaptations, such as migration and schooling, involve all members of the species and may be related to successful feeding and breeding.

Phytoplankton depend primarily on their small size to provide a high degree of frictional resistance to sinking below the sunlit surface waters. Large animals possess bodies that are more dense than ocean water, have less surface area per unit of body mass, and, therefore, tend to sink more rapidly into the ocean. To remain in surface waters where the food supply is greatest, they must increase their buoyancy or swim. Various animals apply one or both of these strategies to remain in the upper pelagic environment. The ones we will discuss range from ctenophores (1 cm or 0.4 in) to the blue whales (30 m or 100 ft). Most marine animals obtain their oxygen from seawater and are cold-blooded, meaning that their body temperature equals that of the water in which they live. However, some fish, and all mammals, are warm blooded and maintain a constant, characteristic body temperature. Of course, mammals must obtain oxygen by breathing air.

Staying Above the Ocean Floor

Some animals depend on increased buoyancy to maintain themselves in near-surface waters. Their buoyancy results either from the presence of gas containers that significantly reduce their average density, or soft bodies void of dense hard parts. As swimmers, larger animals with bodies denser than seawater must exert more energy to propel themselves through the water. In this section, we will elaborate on these strategies.

Gas Containers

Air is approximately 0.001 the density of water at sea level, so just a small amount inside an organism increases its buoyancy. Some cephalopods have rigid gas containers in their bodies. Members of the genus *Nautilus* have an external shell, whereas the cuttlefish *Sepia* and deep-water squid *Spirula* have internal chambered structures (Figure 16-1). Because the pressure in their air chambers is always at 1 atm, these animals have a limited range of depths. *Nautilus* must stay above a depth of approximately 500 m (1640 ft) to prevent collapse of its chambered shell as the external pressure approaches 50 atm; it is rarely observed below a depth of about 250 m (800 ft).

In some slow-moving bony fish, neutral buoyancy is achieved by the presence of a gas bladder, or **swim bladder** (Figure 16-2). The swim bladder is normally not present in very active swimmers, such as the tuna, or in fish that live on the bottom, because neither need it. Some fish have a pneumatic duct that connects the swim bladder to the esophagus (Figure 16-2). These fish can add or remove air through this duct. In other fish, the gases of the swim bladder must be added or removed more slowly by an exchange with the blood.

Because change in depth causes the gas in the swim bladder to expand or contract, the fish removes or adds gas to the bladder to maintain a constant volume. Those fish without a pneumatic duct are limited in the rate at which they can make these adjustments and, therefore, cannot withstand rapid changes in depth.

The composition of gases in the swim bladders of shallow-water fish is similar to that of the atmosphere. With increasing depth (fish with swim bladders have been captured from depths of 7000 m, or 23,000 ft), the concentration of oxygen in the swim bladder gas increases from the 20 percent common near the surface to more than 90 percent. At the 700-atm pressure that exists at a depth of 7000 m, gas is compressed to a density of 0.7 g/cm³. This is about the density of fat, so many deep-water fish have swim bladders that are filled with fat instead of compressed gas.

Floating Forms

Some relatively large plankton, like the familiar jellyfish, are found floating at the surface. These animals all have soft, gelatinous bodies with little, if any, hard tissue.

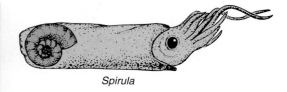

Spirula

Figure 16-1   Gas Containers of the Cephalopods *Nautilus*, *Sepia*, and *Spirula*

The nautilus (left) has an external chambered shell, and the spiny cuttlefish (*Sepia*) (middle) and *Spirula* (right) have internal chambered structures that can be filled with gas to provide buoyancy.   *(Photos courtesy of Bruce Carlson, Waikiki Aquarium.)*

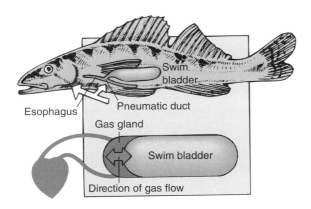

Figure 16-2   Swim Bladder of Some Bony Fishes

Example of a swim bladder connected to the esophagus by the pneumatic duct, allowing air to be added or removed rapidly. In fish with no pneumatic duct, all gas must be added or removed through the blood. This requires more time and is achieved by a network of capillaries associated with the gas gland.

Coelenterates   Coelenterates have soft bodies that are more than 95 percent water. There are two basic *pelagic* types, the **siphonophores** and the **scyphozoans**.

Siphonophore coelenterates are represented in all oceans by the Portuguese Man-of-War (genus *Physalia*) and By-the-Wind Sailor (genus *Velella*). Their gas floats, called pneumatophores, serve as floats and sails that allow the wind to push these colonies of creatures across the ocean surface (Figure 16-3). A colony of tiny individuals is suspended beneath the float. Portuguese man-of-war tentacles may be many meters long and possess **nematocysts**, structures with threadlike poisonous stingers, that can penetrate the skin of humans; they have been known to inflict painful and, for some people, life-threatening neurotoxin poisoning. The colonies grow from the initial polyp through asexual budding.

Jellyfish (scyphozoan coelenterates) have a bell-shaped body with a fringe of tentacles; there is a mouth at the end of a clapperlike extension that hangs beneath the bell-shaped float. Most jellyfish are less than 0.5 m (1.6 ft) in diameter, but species range in size from nearly microscopic to 2 m (6.6 ft) in diameter, and some have tentacles extending to 60 m (200 ft).

Jellyfish move by muscular contraction (see Figure 16-3a). Water enters the cavity under the bell and is forced out by contractions of muscles that circle the bell, jetting the animals ahead in short spurts. To allow the animal to swim in a generally upward direction, sensory organs are spaced around the outer edge of the bell. These may be light-sensitive or gravity-sensitive. This orientation ability is important because the jellyfish feed by swimming to the surface and sinking slowly back down, trapping organisms under the umbrella spread of their bells and with their tentacles.

Tunicates   Tunicates are barrel-shaped animals with openings on each end for current to flow in (incurrent opening) and out (excurrent opening) (Figure 16-4a). They move by a feeble form of jet propulsion created by contraction of bands of muscles that force water into one opening and out the other. Reproduction of solitary tunicates is complicated, involving alternating sexual and asexual reproduction. Salps (genus *Salpa*) include solitary forms reaching a length of 20 cm (8 in) and smaller aggregate forms that produce new individuals by budding. Individual members of an aggregate chain may be 7 cm (2.8 in) long, and the chain of newly budded members may reach great lengths (Figure 16-4b).

The genus *Pyrosoma* is luminescent and colonial. Individual members have their incurrent openings facing the outside surface of a tube-shaped colony that may be a few meters long. One end of the tube is closed, and the excurrent openings of the thousands of individuals all empty

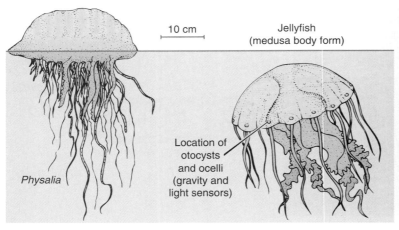

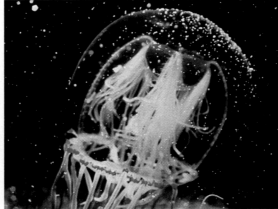

(a)

(b)

Figure 16-3   Planktonic Cnidarians

(a) *Physalia* and jellyfish. (b) Medusa.   (*Photo by Larry Ford.*)

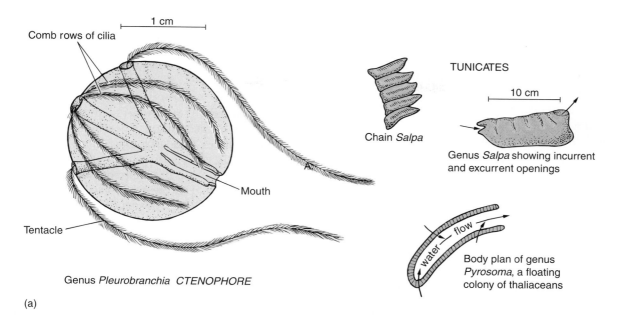

Comb rows of cilia

1 cm

TUNICATES

10 cm

Chain *Salpa*

Genus *Salpa* showing incurrent and excurrent openings

Mouth

Tentacle

water flow

Genus *Pleurobranchia* CTENOPHORE

Body plan of genus *Pyrosoma*, a floating colony of thaliaceans

(a)

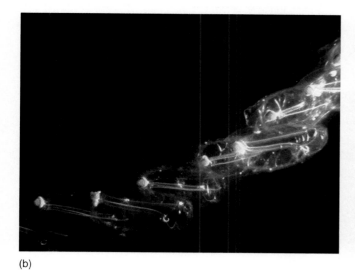

(b)

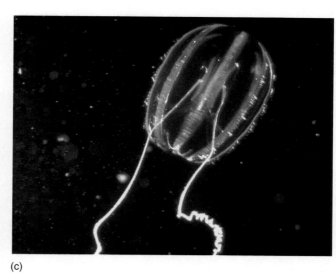

(c)

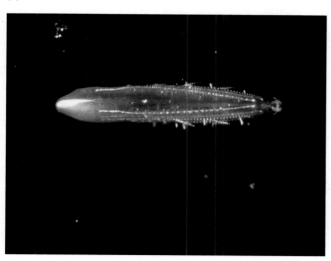

(d)

Figure 16-4   Pelagic Tunicates and Ctenophores

(a) Body structure. (b) Chain of salps. (c) Ctenophore (*Pleurobranchia*). (d) Ctenophore (*Beroe*).
(*Parts (b), (c), (d) James M. King, Graphic Impressions.*)

into the tube. Muscular contraction forces water out the open end to provide propulsion.

Ctenophores   **Ctenophores**, called comb-bearing animals or comb jellies, are closely related to the coelenterates. The body form of most ctenophores is basically spherical, with eight rows of cilia spaced evenly around the sphere. It is from these structures that the name was derived; when magnified, they look like miniature combs. Ctenophores are entirely pelagic and confined to the marine environment.

Some ctenophores have a pair of tentacles that contain adhesive organs instead of stinging cells to capture prey. Sea gooseberries (genera *Pleurobranchia*) are gooseberry-sized (Figure 16-4c), and the pink, elongated *Beroe* are over 15 cm (6 in.) (Figure 16-4d).

Chaetognaths   **Chaetognaths**, or arrowworms, are transparent and difficult to see, although they may grow to more than 2.5 cm (1 in) in length (Figure 16-5). The name *chaetognath* means bristle-jawed and refers to the hairlike attachments around their mouths, which they use to grasp prey while they devour it. Arrowworms are voracious feeders, primarily eating small zooplankton. In turn, they are eaten by fishes and larger planktonic animals such as

jellyfish. These exclusively marine, hermaphroditic animals are usually more abundant in the surface waters some distance from shore.

Swimming Forms

Larger pelagic animals, the nekton (swimmers), have substantial powers of locomotion. They include rapid-swimming invertebrates such as squids, as well as fish and marine mammals.

Swimming squid include the common squid (genus *Loligo*), flying squid (*Ommastrephes*), and giant squid (*Architeuthis*). Active predators of small fish, the smaller squid varieties have long, slender bodies with paired fins (Figure 16-6). Unlike the less-active *Sepia* and *Spirula* shown in Figure 16-1, squid have no hollow chambers in their bodies and therefore require more energy to remain in the upper water of the oceans without sinking. These invertebrates can swim about as fast as any fish their size, and do so by trapping water in a cavity between their soft body and penlike shell and forcing it out through a siphon. To capture prey, they use two long arms with pads containing suction cups at the ends. Eight shorter arms with

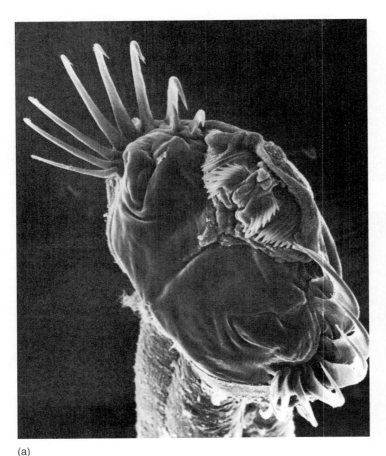

(a)

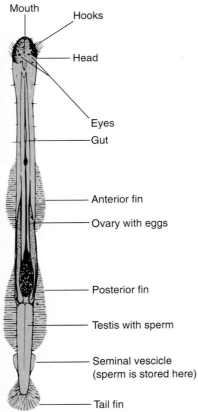

(b) General Anatomy of an Arrowworm

Figure 16-5   Arrowworm

(a) Head of chaetognath (*Sagitta tenuis*) from Gulf of Mexico magnified 161 times.   *(Howard J. Spero)* (b) Adult chaetognaths reach lengths ranging from 1 to 5 cm (0.4 to 2 in).

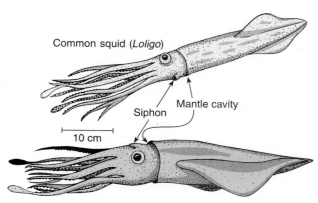

Figure 16-6   Squid

Squid move by trapping water in their mantle cavity between the soft body and penlike shell. They then force the water out through the siphon for rapid propulsion.

suckers convey the prey to the mouth, where it is crushed by a beaklike mouthpiece.

Locomotion in fish is more complex than in the squid because of body motion and the role of fins. The basic movement of a swimming fish is the passage of a wave of lateral body curvature from the front to the back of the fish. This is achieved by the alternate contraction and relaxation of muscle segments along the sides of the body. These muscle segments are called **myomeres**. The backward pressure of the fish's body and fins produced by this wave provides the forward thrust (Figure 16-7).

Most active swimming fish have two sets of paired fins, pelvic and pectoral, used in maneuvers such as turning, braking, and balancing. When they are not in use, the fins can be folded against the body. Vertically oriented fins, dorsal and anal, serve primarily as stabilizers. The fin most important in propelling a high-speed fish is the tail, or cau-

dal, fin. Caudal fins flare dorsally and ventrally to increase the surface area available to develop thrust against the water. This is equivalent to humans donning dive flippers on their feet to swim more efficiently. Increased surface area also increases frictional drag. The efficiency of the design of a caudal fin depends on its shape and can be expressed mathematically as an **aspect ratio**, calculated as

$$(\text{fin height})^2/\text{fin area}$$

There are five basic shapes of caudal fins, keyed to the following descriptions (Figure 16-8):

(a) The rounded fin (aspect ratio 1) is flexible and useful in accelerating and maneuvering at slow speeds.

(b) and (c) The somewhat flexible truncate tail (aspect ratio 3) and forked tail (aspect ratio 5) are found on faster fish and may also be used for maneuvering.

(d) The lunate caudal fin (aspect ratio 7–10) is found on fast-cruising fishes such as tuna, marlin, and swordfish; it is very rigid and useless in maneuverability, but very efficient in propelling.

(e) The heterocercal fin (ratio 7–10) is asymmetrical (hetero = uneven, cercal = tail), with most of its mass and surface area in the upper lobe.

The heterocercal fin produces a significant lift to sharks as it is moved from side to side. This lift is important because sharks have no swim bladder and tend to sink when they stop moving. To aid this lifting, the pectoral (chest) fins are large . Positioned on the shark's body like airplane wings, they in fact function like wings, or a hydrofoil, to lift the front of the shark's body to balance the rear lift supplied by the caudal fin. The shark gains tremendous lift but loses a lot in maneuverability as a result of this adaptation of the pectoral fins. This is why sharks tend to swim in broad circles, like a circling airplane. See Sharks on page 144 for more information on sharks.

Figure 16-7   Fish—Swimming Motions and Fins

Alternate contraction and relaxation of the myomeres sends a wave of body curvature back along the body to produce a forward thrust.

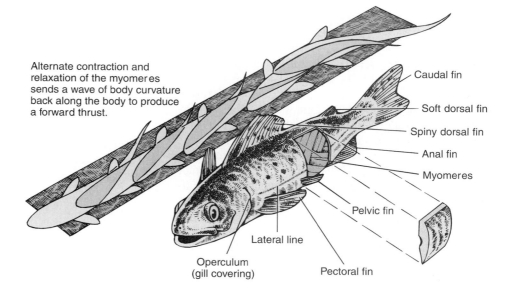

Figure 16-8 Caudal Fin Shapes and Aspect Ratios

(a) Rounded fin (aspect ratio 1) of peacock flounder (*Bothus lunatus*); sculpin and angel also have rounded fins. *(Photo © Charles Seaborn/Odyssey Productions.)* (b) Truncate fin (aspect ratio 3) of Nassau grouper (*Epinephelus striatus*) near North Caicos, British West Indies (Bahamas). *(Photo © Fred Bavendam/Peter Arnold, Inc.)* (c) Forked fin (aspect ratio 5) of yellowtail on a Caribbean reef; herring and goatfish also have forked fins. *(Photo © Greg Ochocki/The Stock Market.)* (d) Lunate fin (aspect ratio 7–10) of blue marlin; bluefish and tuna also have lunate fins. *(Photo © Bob Gomel/The Stock Market.)* (e) Heterocercal fin (aspect ratio 7–10) of blue shark (*Prionace glauca*) in the Pacific Ocean off Santa Barbara, California. *(Photo © Charles Seaborn/Odyssey Productions.)*

In many fish, pectoral fins do more than just help in maneuvering (Figure 16-9). Wrass and sculpins use them to row through the water in jerky motions while the rest of the body remains motionless. Skates and rays swim by sending undulating motions across the greatly modified pectoral fins, and manta rays flap them like wings. The flying fish, *Exocoetus*, uses greatly enlarged pectoral fins to glide up to 400 m (1312 ft) across the ocean surface after propelling itself with the enlarged ventral lobe of its caudal fin into the air to escape dolphins and other predators. Other fish have pectoral fins modified into fingerlike structures used to walk on the ocean floor.

Many fish in no great hurry propel themselves by undulation of the dorsal fin. Examples are triggerfish, sea horses, ocean sunfish, and trunkfish. The sunfish also uses its anal fin and stubby caudal fin to aid in the process, still without impressive results.

Adaptations for Seeking Prey

Several factors affect the ability of species to capture food. These include mobility (lunging versus cruising), speed, body length, body temperature, and circulatory system.

Mobility

Some fish spend most of their time waiting patiently for prey and exert themselves only in short bursts as they lunge at the prey. Others cruise relentlessly through the water seeking prey. There is a marked difference in the muscula-

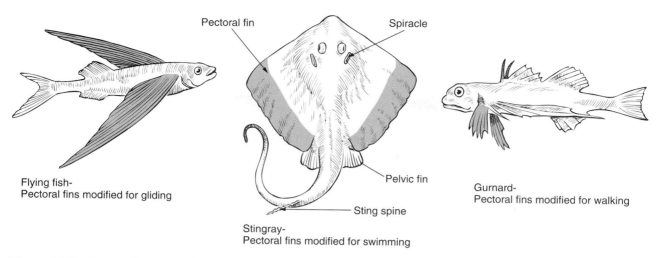

Pectoral fin

Spiracle

Flying fish-
Pectoral fins modified for gliding

Pelvic fin

Sting spine

Stingray-
Pectoral fins modified for swimming

Gurnard-
Pectoral fins modified for walking

Figure 16-9   Pectoral Fin Modifications

ture of fishes that use these different styles to obtain food. **Lungers**, like the grouper, sit and wait (Figure 16-10a). Groupers have truncate caudal fins for speed and maneuverability, and almost all their muscle tissue is white. **Cruisers**, like the tuna, on the other hand, actively seek game (Figure 16-10b). Less than half of a cruiser's muscle tissue is white; most is red.

What is the significance of red vs. white muscle tissue? Red muscle fibers are much smaller in diameter (25 to 50 μm, or 0.01 to 0.02 in) than white muscle fibers (135 μm, or 0.05 in). White muscles fibers contain lower concentrations of **myoglobin**, a red pigment with an affinity for oxygen. Their small size and the presence of myoglobin allow the red fibers to obtain a much greater oxygen supply than is possible for white fibers. This sup-

ports a metabolic rate six times that of white fibers; this high rate is needed by some animals for endurance. Red muscle tissue is abundant in cruisers that constantly swim. Lungers can get along quite well with little red tissue because they need not constantly move. White tissue, which fatigues much more rapidly than red tissue, is used by the tuna for short periods of acceleration while attacking. It is also quite adequate for propelling the grouper and other lungers during their quick passes at prey.

Speed

The speed of fish is thought to be closely related to body length. For tuna, which are well-adapted for sustained cruising and short bursts of high-speed swimming, cruising

(a)

(b)

Figure 16-10   Feeding Styles—Lungers and Cruisers

(a) Lungers like this tiger grouper sit patiently on the bottom and capture prey with quick, short lunges. Their muscle tissue is predominantly white.   *(Photo © Fred Bavendam/Peter Arnold Inc.)* (b) Cruisers like these yellowfin tuna swim constantly in search of prey and capture it with short periods of high-speed swimming. Their muscle tissue is predominantly red.   *(Photo courtesy of National Marine Fisheries Service.)*

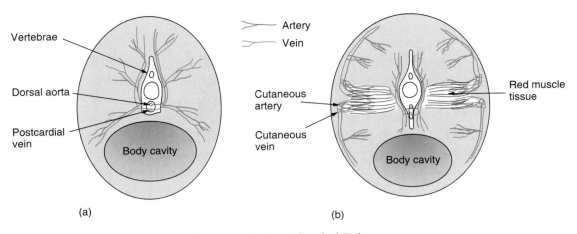

Figure 16-11 Circulatory Modifications in Warm-Blooded Fishes

(a) Most cold-blooded fishes have major blood vessels arranged in this pattern. All blood flows to muscle tissue from the dorsal aorta and returns to the postcardinal vein beneath the vertebral column. (b) Warm-blooded fishes like the tuna have cutaneous arteries and veins that help maintain high blood temperature by using heat energy generated by contracting muscle tissue.

speed averages about 3 body lengths/s. They can maintain maximum speed of about 10 body lengths/s only for 1 s. A yellowfin tuna, *Thunnus albacares*, has been clocked at 74.6 km/h (46 mi/h), which is a rate of more than 20 body lengths/s, but only for about 0.2 s.

Theoretically, a 4-m (13-ft) bluefin tuna, *Thunnus thynnus*, swimming at a rate of 10 body lengths/s could reach speeds of up to about 144 km/h (90 mi/h). Many of the toothed whales are known to be capable of high speeds. The porpoise *Stenella* has been clocked at 40 km/h (25 mi/h), and it is believed that the top speed of killer whales may exceed 55 km/h (34 mi/h).

Body Temperature

Metabolic processes proceed at greater rates at higher temperatures. Although fish are generally considered to be cold blooded, there are some fast swimmers whose body temperatures are higher than that of the surrounding water. Some are only slightly above the temperature of the water. For example, the mackerel (*Scomber*), yellowtail (*Seriola*), and bonito (*Sarda*) have body temperature elevations of from 1.3 to 1.8°C (2.3 to 3.2°F). Members of the mackerel shark genera, *Lamna* and *Isurus*, as well as the tuna, *Thunnus*, have much higher body-temperature elevations. Warm-blooded bluefin tuna maintain a body temperature of 30° to 32°C (86° to 90°F), regardless of the temperature of the water in which they are swimming. Although bluefin tuna commonly are found in warm waters where the temperature difference is no more than 5°C (9°F), they apparently maintain body temperatures of 30°C (86°F) even when swimming in 7°C (45°F) water.

Why do these fish exert so much energy to maintain their body temperatures at high levels when other fish do quite well with ambient body temperatures? Their mode of behavior is that of a cruiser, and any adaptation (high temperature and metabolic rate) that can increase the power output of their muscle tissue helps them search out and capture prey.

Circulatory System

Mackerel sharks and tuna are aided in maintaining their high body heats by modified circulatory systems (Figure 16-11). While most fish have a dorsal aorta just beneath the vertebral column that provides blood to the swimming muscles, mackerel sharks and tuna have additional cutaneous arteries just beneath the skin on either side of the body. As cool blood flows into red muscle tissue, its temperature is increased by heat generated by muscle metabolism (muscle contractions). A fine network of tiny blood vessels within the muscle tissue is designed to minimize heat loss. The vessels that return the blood to the cutaneous vein, parallel to the cutaneous artery along the side of the fish, are all paired with small vessels carrying blood into the muscle tissue. In this way, the warm blood leaving the tissue helps to heat the cooler blood entering from the cutaneous artery.

Marine Mammals

Coastal ocean mammals include herbivores, the **sirenia** (dugongs and manatees), and a variety of carnivorous forms (Figure 16-12). Sea otters and **pinnipeds** (sea lions, seals, and walruses) spend all or most of their lives in coastal waters and come ashore to breed. The truly oceanic mammals, spending all their time in the oceans, are the **cetaceans**—whales, porpoises, and dolphins (Figure 16-13).

We know that mammals evolved from reptiles on land some 200 million years ago. No marine mammals are known earlier than 60 million years ago. Therefore, it is believed that all marine mammals evolved from land-dwelling forms.

Figure 16-12   Marine Mammals—Manatee, Pinnipeds and Sea Otter

(a) Manatee, (b) Northern elephant seals, (c) Harbor seals, (d) Sea Otter.   *(Part (a) © Fred Bavendam/Peter Arnold, Inc. (b) © Tim Ragen. (c) © Allan Morgan/Peter Arnold, Inc. (d) © Jeff Foott/Bruce Coleman, Inc.)*

Figure 16-13   Baleen and Toothed Whales

No creature in the sea captures the imagination like whales, huge mammals that returned to the sea about 60 million years ago. Two suborders exist today, toothed whales *(Odontoceti)* and baleen whales *(Mysticeti)*. Represented here are members of each group drawn to scale. The toothed whales—bottle-nosed dolphin, narwhal, killer whale, and sperm whale—are active predators and may track down their prey by use of echolocation. The baleen whales— humpback whale, right whale, and blue whale—fill their mouths with water and force it out between long baleen slats that hang down from the upper jaw. Small fish, krill, and other plankton are trapped behind the baleen. The Pacific gray whale has short baleen slats and feeds by sucking in sediment from the shallow bottom of its North Pacific feeding grounds. Thus, it strains out benthic amphipods and other invertebrates. Although many great whale populations were pushed to near extinction by nineteenth- and twentieth-century whaling efforts, it appears all will survive this threat. The major threat to their survival may now be the disruption of their feeding and breeding grounds.

Atlantic Right Whale (*Balaena glacialis*)
This cold-water whale was the first recorded target of whaling by Basque seamen during the middle ages. Similar species are found near Japan and in high southern latitudes. Length is to 18 m (60 ft).

SPERM WHALE (*Physeter catodon*)
Found mostly in tropical waters, this deep diver has a huge snout that contains a large amount of oil. Length of male is to 19 m (63 ft). Length of female is to 10.5 m (35 ft).

BOTTLE-NOSED DOLPHIN (*Tursiops truncatus*)
Found in the North Atlantic, this is one of the many species of beaked dolphins. Length is to 3 m (10 ft).

Killer Whale (*Orcinus orca*)
Cosmopolitan in distribution, this whale is unique in that it not only feeds on fish but also on seals, birds, and other whales. The male grows to twice the female's size. Length is to 9 m (31 ft).

NARWHAL (*Monodon monoceros*)
The scientific name, which means one tooth, one horn, is accurate for the male of this Arctic whale. Length is to 6 m (20 ft).

PACIFIC GRAY WHALE (*Eschrichtius robustus*)
This primitive baleen whale with a small head and very reduced dorsal fin stays close to shore. It even enters the surf zone. Length is to 14 m (45 ft).

HUMPBACK WHALE (*Megoptera novaeangliae*)
The great communicators of the Cetacea have a cosmopolitan distribution. The genus name is derived from their extraordinary wing-like flippers. Length is to 16 m (52 ft).

BLUE WHALE (*Balaenoptera musculus*)
The largest animal to inhabit the Earth will consume over five tons of krill per day. Length is to 30 m (100 ft).

Modifications to Allow Deep Diving

People can free-dive to a maximum recorded depth of about 100 m (330 ft) and hold their breath in rare instances for up to 6 min. In contrast, the sperm whale (*Physeter*) is known to dive deeper than 2200 m (7200 ft); the North Atlantic bottle-nosed dolphin (*Hyperoodon ampullatus*) can stay submerged for up to 2 h. What adaptations do marine mammals have to permit such long and deep dives?

Breathing Adaptations Whales and other marine mammals can alternate between periods of normal breathing and cessation of breathing. The periods of cessation occur while the animal is submerged. To understand how some cetaceans are able to go for long periods without breathing requires a general knowledge of their lungs and associated structures.

In Figure 16-14, you can see that inhaled air finds its way to tiny terminal chambers, the alveoli. The alveoli are lined by a thin membrane that is in contact with a dense bed of capillaries; the exchange of gases between the inhaled air and the blood (oxygen in, carbon dioxide out) occurs across the alveolar membrane. Some cetaceans have an exceptionally large concentration of capillaries surrounding the alveoli (Figure 16-14b). The contraction and relaxation of muscle fibers within the lungs moves air against the membrane.

Cetaceans take from 1 to 3 breaths per minute while resting, compared to about 15 in humans. Because they hold the inhaled breath much longer, and because of the large capillary mass in contact with the alveolar membrane and the circulation of the air by muscular action, many cetaceans can extract almost 90 percent of the oxygen in each breath, compared to only 4 to 20 percent extracted by terrestrial mammals.

To use this large amount of oxygen, which can be taken into the blood efficiently during long periods underwater, cetaceans may apply two strategies: (1) storing the oxygen and (2) reducing oxygen use. The storage of so much oxygen is possible because prolonged divers have a much greater blood volume per unit of body mass than those that dive for only short periods.

Compared to terrestrial animals, some cetaceans have twice as many red blood cells per unit of blood volume and up to 9 times as much myoglobin, an oxygen storing substance, in the muscle tissue. Thus large supplies of oxygen can be stored chemically bound to the **hemoglobin** that stores oxygen in red blood cells and the myoglobin of the muscles. In addition, muscles that start a dive with a significant oxygen supply can continue to function through anaerobic respiration when the oxygen is used up. The muscle tissue is relatively insensitive to high levels of carbon dioxide and lactic acid—a product of anaerobic respiration.

Because the swimming muscles can function without oxygen during a dive, they and other organs, such as the digestive tract and kidneys, may be sealed off from the circulatory system by constriction of key arteries. The circulatory system then serves primarily the heart and brain. Because of the decreased circulatory requirements, the heart rate can be reduced by 20 to 50 percent of its normal rate. Although pinnipeds are known to use this method, recent research has shown that no such reduction in heart rate occurs during dives by the common dolphin (*Delphinus delphis*), the white whale (*Delphinapterus leucas*), or the bottle-nosed dolphin (*Tursiops truncatus*).

Gas Volume Changes with Pressure Another problem with deep and prolonged dives results from the change in pressure from 1 atm at the surface to one additional atm for every 10 m (33 ft) of depth. *Barotrauma* refers to tissue injury when gas-filled body organs such as the lungs, middle ear, and sinuses, fail to equalize internal pressures with external pressures. Gases occupy smaller volumes and dissolve to a greater extent in fluids with increasing pressure.

When a diver returns to the surface, gases that were forced into the body tissues at the higher pressures encountered underwater expand and come out of solution. If humans do not return to surface pressures gradually after prolonged dives, they are likely to develop **decompression sickness** (also known as the bends) and **nitrogen narcosis**. Both of these are due to the increased quantities

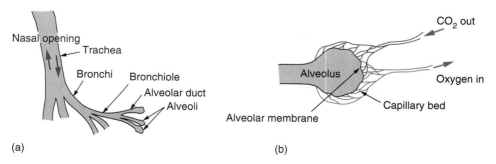

(a)   (b)

Figure 16-14   Cetacean Modifications to Allow Prolonged Submergence

(a) Basic lung design. Air enters the lung through the trachea, and oxygen is absorbed into the blood through the walls of the alveoli. (b) Oxygen exchange in the alveolus. A dense mat of capillaries receives oxygen through the alveolar membrane. Because air is normally held in the lungs of cetaceans for up to 1 min before exhaling, as much as 90 percent of the oxygen can be extracted.

of nitrogen that become dissolved in body tissues at higher pressures. Nitrogen narcosis, which occurs when divers go too deep or stay too long at depths greater than 30 m (100 ft), refers to the narcotic effect brought on by too much nitrogen in nervous system tissue. Symptoms are similar to drunkenness, but can include hallucinations and unconsciousness, and may lead to death. Decompression sickness results when a diver surfaces and the reduced pressure causes small bubbles of gas to form in the blood and other tissues. The bubbles interfere with blood circulation, causing excruciating pain, severe physical debilitation, embolism, and sometimes death.

Cetaceans and other marine mammals do not suffer from these difficulties. One main defense against absorbing too much nitrogen seems to be a more flexible rib cage. By the time a cetacean has reached a depth of 70 m (230 ft), the rib cage has collapsed under the 8 atm of pressure. The lungs within the rib cage also collapse, removing all air from the alveoli. Because most absorption of gases by the blood occurs across the alveolar membrane, the blood cannot absorb additional gases, and the problem of nitrogen narcosis is avoided. It is, however, possible that the collapsible rib cage is not the main defense against the bends. An experiment put enough nitrogen into the tissue of a dolphin to give a human a severe case of the bends, but the dolphin suffered no ill effects.

Attaching video cameras to diving whales and seals has revealed that once the lungs are fully collapsed below 80 m (262 ft), they spend over 78 percent of the time in prolonged glides between short periods of active swimming. With air removed from the lungs, the bodies are dense enough to sink under their own weight. This behavior is believed to increase aerobic capabilities up to 59 percent over what would be expected with continuous swimming.

Cetaceans (Whales, Porpoises, and Dolphins)

There are over 75 species of cetaceans of two basic types. The toothed whales, the **odontoceti**, include Sperm Whales, Killer Whales, porpoises, and dolphins; the Baleen Whales, the **mysticeti**, include the Blue, Humpback, Gray, and Right whales

Baleen Whales probably evolved from the toothed whales some 30 million years ago. In place of teeth, they have *baleen*, plates of horny material that hang from the upper jaw and operate as a sieve. Toothed whales are predators that feed mostly on smaller fish and squid, although the Killer Whale is known to feed on a variety of larger animals, including other whales. Baleen Whales feed primarily by filtering crustaceans from the surface down to and including the sediment of shallow ocean basins.

The cetacean body is more or less cigar-shaped, nearly hairless, and insulated with a thick layer of blubber. Cetacean forelimbs are modified into flippers that move only at the shoulder joint. The hind limbs are vestigial, not attached to the rest of the skeleton, and not externally visible. The skull is highly modified, with one nasal opening (in toothed whales) or two openings (in baleen whales)

near the top. Cetaceans propel themselves by vertical movements of a horizontal tail fin called a *fluke*.

Baleen Whales include these three families:

1. Gray Whales have short, coarse baleen, no dorsal fin, and only two to five ventral grooves beneath the lower jaw.
2. Rorqual Whales have short baleen, many ventral grooves, and are divided into these two subfamilies: (1) The *balaenopterids* have long, slender bodies; small, sickle-shaped dorsal fins; and flukes with smooth edges (Minke, Baird's, Bryde's, Sei, Fin, and Blue whales). (2) The *megapterids*, or Humpback whales, have more robust bodies; long flippers; flukes with uneven trailing edges; tiny dorsal fins; and tubercles on the head.
3. Right Whales have long, fine baleen; broad triangular flukes; no dorsal fin; and no ventral grooves. The Northern Right Whale is the baleen whale most threatened with extinction. The Southern Right Whale and the Bowhead Whale that remains near the Arctic pack-ice edge are the other members of this family.

Modifications to Increase Swimming Speed Cetaceans' muscles are not vastly more powerful than those of other mammals, and it is believed that their ability to swim at high speed must result from modifications that reduce frictional drag by creating streamlined (laminar) rather than turbulent flow near their bodies. A small dolphin would require muscles five times more powerful than it has in order to swim at 40 km/h (25 mi/h) if it had no streamlining.

In addition to a streamlined body, cetaceans are believed to actually modify the flow of water around their bodies to a smooth flow with the aid of a specialized skin structure. The skin is composed of two layers: a soft outer layer that is 80 percent water and has narrow canals filled with spongy material and a stiffer inner layer composed mostly of tough connective tissue. The soft layer tends to reduce the pressure differences at the skin–water interface by compressing under regions of higher pressure and expanding in regions of low pressure.

Use of Sound It has long been known that cetaceans make a variety of sounds, despite their lack of vocal cords. Speculation about the sounds' purpose range from echolocation (clearly true) to a highly developed language (doubtful). In fact, what we know about cetacean use of sound is limited.

All marine mammals have good vision, but conditions often limit its effectiveness. In coastal waters, where suspended sediment and dense plankton blooms make the water turbid, and in the deeper waters, where light is limited or absent, echolocation surely would assist pursuit of prey or location of objects.

Using lower-frequency clicks at great distance and higher frequency at closer range, the bottle-nosed dolphin (*Tursiops truncatus*) can detect a school of fish at distances exceeding 100 m (330 ft). It can pick out an individual fish 13.5 cm (5.3 in) long at a distance of 9 m (30 ft). It has been estimated that sperm whales can detect their main prey, squid, from a distance up to 400 m (1300 ft) by use of their low-frequency scanning clicks.

To locate something with these sounds, the animal's brain processes the sound it hears to automatically determine the distance. The distance to an object is obtained by multiplying the velocity at which sound travels to and returns from the object by the time required for travel, and divides this product by two (Figure 16-15).

Baleen Whales produce sounds at frequencies generally below 5000 Hz, and they are not known to use echolocation. Gray Whales produce pulses, possibly for echolocation, and moans that may be a means of communication with other gray whales. Rorqual Whales produce moans that last from one to many seconds. These sounds are extremely low in frequency, in the 10 to 20 Hz range, and are probably used to communicate over distances up to 50 km (31 mi). Moans of Fin Whales may be related to reproductive behavior, but much more observation is required to ascertain the message content of these sounds.

Youthful human ears are sensitive to frequencies from about 16 to 20,000 Hz. Some toothed whales respond to frequencies as high as 150,000 Hz. The clicks of the bottle-nosed dolphin are of a frequency range that is partly audible to humans. However, some clicks are at ultrasonic frequencies (above human hearing) and are repeated up to 800 times/s. How the pulses of clicks are produced and how the returning sound is received are not fully understood. Two hypotheses are presented in Figure 16-16. In most mammals, the bony housing of the inner ear structure is fused to the skull. When submerged, sounds transmitted through the water are picked up by the skull and travel to the hearing structure from many directions. This makes it impossible for such mammals to locate accurately the source of the sounds. Obviously, such a hearing structure would not work for an animal that depends on echolocation to find objects in water.

All cetaceans have evolved structures that insulate the inner ear housing from the rest of the skull. In toothed whales, the inner ear is separated from the rest of the skull and surrounded by an extensive system of air sinuses (cavities). The sinuses are filled with an insulating emulsion of oil, mucus, and air and are surrounded by fibrous connective tissue and venous networks. In many toothed whales, it is believed that sound is picked up by the thin, flaring jawbone and passed to the inner ear via the connecting oil-filled body.

Group Behavior

Depending on the size, feeding behavior, reproductive style, and other requirements for maintaining the species, many animals inhabiting the open ocean have developed patterns of group behavior that allow them to exploit their environment efficiently. Two of these patterns are schooling and migration.

Schooling

Obtaining food occupies most of the time of many inhabitants of the open ocean. Some animals are fast and agile, and obtain food through active predation. Other animals move at a more leisurely rate as they filter small food particles from the water. Examples of predators and filter feeders can be found in populations of pelagic animals, from the tiny zooplankton to the massive whales. Patches of zooplankton may occur simply because the nutrients that support their food, the phytoplankton, may be highly concentrated in certain coastal waters. We usually do not refer to these patches as schools. The term **schooling** is usually reserved for well-defined social organizations of fish, squid, and crustaceans.

The number of individuals in a school can vary from a few larger predaceous fish to hundreds of thousands of small filter feeders. Within the school, individuals of the same size move in the same direction with equal spacing between them. This spacing probably is maintained through visual contact, and in the case of fish, by use of the lateral line system that detects vibrations of swimming neighbors (see Figure 16-7). The school can turn abruptly or reverse direction, as individuals at the head or rear of the school assume leadership positions (Figure 16-17).

Why schooling? The advantage of schooling seems obvious from the reproductive point of view. During spawning, it assures that there will be males to release sperm to fertilize the eggs shed into the water or deposited on the bottom by females. However, most investigators believe the most important function of schooling in small fish is protection from predators.

At first, it may seem illogical that schooling would be protective. Any predator lunging into a school would surely catch something, just as land predators chase a herd of grazing animals until one weakens and becomes dinner. So, aren't the smaller fish making it easier for the predators by forming a large target? Based more on conjecture than research, the consensus of scientists is no.

Figure 16-15 Echolocation

Clicking sound signals are generated by cetaceans; the signals bounce off objects in the ocean to determine their size, shape, and distance.

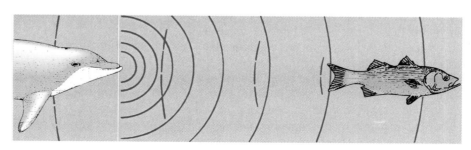

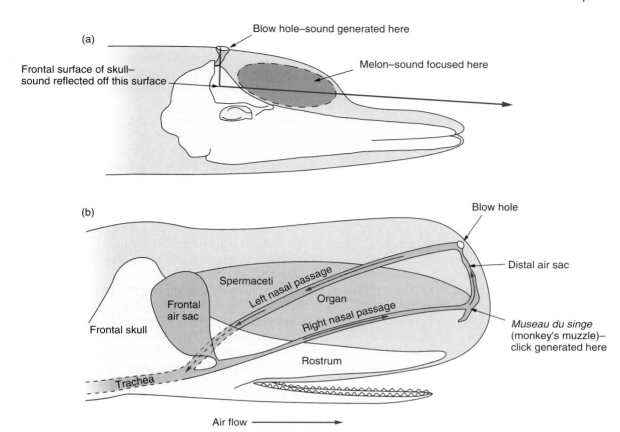

Figure 16-16   Generation of Echolocation Clicks in Small-Toothed and Sperm Whales

(a) In small-toothed whales, clicks may be generated within the blowhole mechanism, reflected off the frontal surface of the skull, and focused by the fatty melon into a forward-directed beam. (b) Structures that may be related to the generation of clicks by sperm whales. Air may pass from the trachea through the right nasal passage across the museau du singe, where clicks are generated. It passes along the distal air sac past the closed blowhole and returns to the lungs along the left nasal passage. The initial click is followed by 8 progressively weaker clicks that may result from sound energy bouncing back and forth between the frontal air sac and the distal air sac, being transmitted through the oil-filled spermaceti organ.

Figure 16-17   Schooling

Schooling grunts near Haemulon, Florida Keys.   *(© Larry Lipsky/Bruce Coleman, Inc.)*

Over 2000 fish species are known to form schools. This fact alone indicates that this behavior has evolved. Schooling is so pervasive among fish populations because, for fish with no other means of defense, it somehow provides a better chance of survival than swimming alone.

How schooling is protective may grow out of the following considerations:

1. If members of a species form schools, they reduce the percentage of ocean volume in which a cruising predator might find one of their kind.
2. Should a predator encounter a large school, it is less likely to consume the entire unit than if it encounters a small school or an individual.
3. The school may appear as a single large and dangerous opponent to the potential predator and prevent some attacks.
4. Predators may find the continually changing position and direction of movement of fish within the school confusing, making attack particularly

difficult for predators, who can attack only one fish at a time.

There may be other more subtle reasons for schooling. However, this gregarious behavior must enhance species survival because it is so widely practiced among pelagic animals.

Migration

Many oceanic animals undertake **migrations** of varying magnitudes. This behavior is observed among sea turtles, fish, and mammals. Some animals, such as sea turtles, return to dry beaches to lay their eggs. Other populations, including many of the baleen whales, may migrate because the physical environment of the feeding grounds does not meet the needs of young baleens.

Migratory routes of commercially important baleen whales have been well known since the mid-1800s. The paths of these and other air-breathing mammals are easy to observe, because the animals must surface periodically for air. The migratory paths of many fish have been more difficult to identify, despite their high commercial value. Studies of tagged fish have helped us understand their movement patterns. Sampling the distribution patterns of eggs, larvae, young, and adult populations and radio-tracking individuals have also helped to identify migratory routes.

Orientation During Migration How do migratory species orient themselves in time and space? To put the problem another way: How do they know where they are in relationship to where they want to go? How do they know when to leave so as to arrive at their destination on time?

Answers are still being sought, but researchers believe that all migratory species have an innate sense of time, referred to as a biological clock. Evidence of a species' biological clock are the physiological changes, for example, in respiratory rate and body temperature, that occur independent of changes in the environment. These variations may tell animals when to begin migration. However, some changes in the external environment can alter *circadian rhythms.* Thus, it is possible that changes in food availability, water temperature, and duration of daylight may trigger seasonal migrations.

Orientation in space is another complex problem. Animals, such as mammals and turtles, that migrate at the surface could use their sight for orientation. Gray whales, which migrate close to land over a large part of their migratory route, might identify landmarks along the shore. When out of sight of land, mammals and turtles could use the relative positions of the sun, moon, and stars to guide them on their way. Fish that migrate beneath the surface are thought to use smell, relative movement within ocean

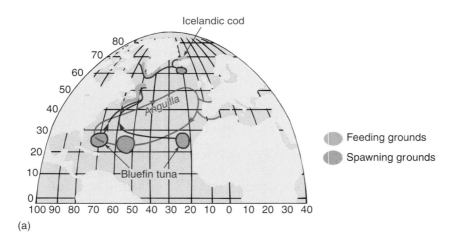

(a)

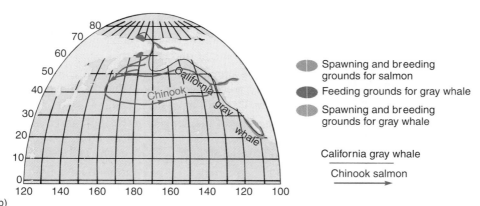

(b)

Figure 16-18 Migration Routes

(a) Migration routes of the Icelandic cod, bluefin tuna, and Atlantic eel (*Anguilla*). (b) Migration routes of the Chinook salmon and the California gray whale.

Feeding grounds

Spawning grounds

Spawning and breeding grounds for salmon

Feeding grounds for gray whale

Spawning and breeding grounds for gray whale

California gray whale

Chinook salmon

currents, and other small currents related to Earth's magnetic field to orient themselves in ocean space.

Migration Routes Many important food fish of high northern latitudes deposit pelagic egg masses that are transported by currents. These fish actively migrate upcurrent during spring or summer and spawn where the current will carry the eggs to a nursery ground with an appropriate food supply for newly hatched fry. Migrations of this type are usually from ten to a few hundred kilometers.

This behavior is observed in the population of Atlantic Cod (*Gadus morhua*). The cod spawn along the southern shore of Iceland (Figure 16-18a, green area). Adults that occupy feeding grounds along the northern and eastern coasts of Iceland and around the southern tip of Greenland migrate to the spawning grounds in the late winter and early spring. This migration involves swimming against the East Greenland Current and Irminger Current, and includes only mature adults at least 8 years of age. These spawning migrations occur annually until the adults die at 18 to 20 years of age.

The spawning grounds are bathed in the relatively warm water of the Irminger Current, a northward-flowing branch of the Gulf Stream. Each female releases up to 15 million eggs, which float in the surface current. The drifting eggs, carried by the East Greenland and Irminger currents, hatch in about 2 weeks. The larvae feed at midwater depths while the currents are carrying them to the adult feeding grounds (brown area in figure). They remain there, undertaking only short onshore-offshore migrations, until they mature and make their first spawning migration.

Longer migrations are undertaken by two Atlantic bluefin tuna populations that spawn in tropical waters near the Azores in the eastern Atlantic and the Bahamas in the western Atlantic (green areas in Figure 16-18a). Little is known about where the newly hatched tuna feed, but the adults are known to move northward along the seaward edge of the Gulf Stream in May and June and feed on the herring and mackerel populations off the coasts of Newfoundland and Nova Scotia (brown area). They may move even further northward, but their return route to the spawning grounds is unknown.

Catadromous Fish The Atlantic eel (*Anguilla*) undertakes what appears to be a single round trip migration during its lifetime. Its behavior is **catadromous**; After being spawned in the ocean, it enters freshwater streams for its adult life and returns near the end of its life to the ocean spawning site.

Anguilla spawning is thought to occur in the Sargasso Sea southeast of Bermuda (Figure 16-18a, green area). The larvae, 5 mm (0.2 in) long, are carried into the Gulf Stream and northward along the North American coast. After 1 yr in the ocean, some metamorphose into young eels, called elvers, and move into North American fresh water streams. Others migrate to the European coast before metamorphosing and move into European freshwater streams.

After spending up to 10 yr in the freshwater environment, the mature eels undergo yet another change. They develop a silvery color pattern and enlarged eyes that are typical of fish inhabiting the mesopelagic zone. They swim downriver to the ocean and presumably back to the spawning ground, where they are thought to die.

Anadromous Fish Salmon of the north Atlantic and Pacific oceans show **anadromous** behavior, which includes spawning in freshwater streams and spending most of their adult lives in the ocean. Pacific species die after spawning; Atlantic salmon return to the ocean.

Six species of Pacific salmon (*Oncorhynchus*) follow similar paths in their migrations (Figure 16-18b). Spawning occurs during the late summer and early fall. Eggs are deposited and fertilized in gravel beds far upstream. The chinook salmon (Figure 16-19) spawns from central California to Alaska. The young chinook hatching in the spring may head downstream as a silvery, filter-feeding smolt (a young salmon ready to go to sea) during its first or second year. By the time it reaches the ocean, the young chinook has become a predator, feeding on smaller fish.

After spending about 4 yr in the North Pacific, mature salmon of more than 10 kg (22 lb) in weight and 1 m (3.3 ft) in length return to their home streams to spawn. Although it is not known how salmon accurately return home, most investigators think the salmon's best guides are odors and currents.

California Gray Whale Migration Some of the longest migrations known in the open ocean are the seasonal migrations of baleen whales. Many baleen species feed in colder, high-latitude waters and breed and calve in warm tropical waters. Feeding occurs during summer when the long hours of sunlight iradiate the nutrient-rich waters to

Figure 16-19 Varieties of Salmon

Top, chum (*Oncorhynchus keta*); center, coho (*O. kisutch*); bottom, chinook (*O. tschawyscha*). (*Courtesy U.S. Fish and Wildlife Service.*)

produce a vast feast of crustaceans. Only this bountiful food enables the whales to sustain themselves during the long migrations and mating and calving season, when feeding is minimal.

The long migrations baleen whales now undertake may have resulted from productive feeding grounds, once near the tropical calving grounds, moving progressively pole-ward because of climate change. If this is the case, the breeding and calving grounds must be very important, for they would surely have been abandoned for locations clos-er to the present feeding grounds.

An alternate explanation for leaving the colder waters to calve is to avoid killer whales, which are more numer-ous in colder waters and a major threat to young whales. Because of the energy demands faced by female baleen whales in producing large offspring (the gestation period is up to 1 yr) and providing them with fat-rich milk for

several months, it is not uncommon for them to mate only once every 2 or 3 yr.

The migration route of the California gray whale demonstrates how the conditions just discussed are met by whale migration patterns. Gray whales are moderate-ly large, reaching lengths of 15 m (50 ft) and weighing over 30 metric tn. They feed during summer in the Okhot-sk, Bering, and Chukchi Seas (see purple areas in Fig-ure 16-18b). They are unique among baleen whales in that they do not feed by straining pelagic crustaceans and small fish from the water; instead, they stir up bottom sed-iment with their snouts and feed on bottom-dwelling am-phipods (Figure 16-20).

The western populations winter along the Korean coast, but the population known as the California Gray Whale migrates from the Chukchi and Bering Seas to winter in lagoons of the Pacific coast of Baja California

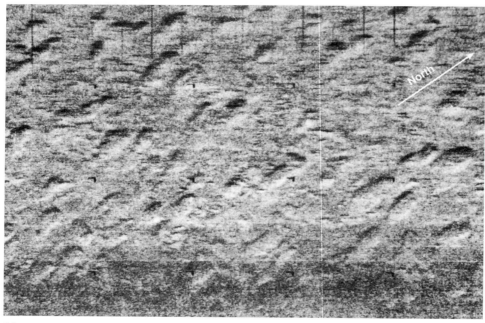

(a)

(b)

Figure 16-20   California Gray Whale Feeding

(a) Side-scan sonograph showing pits on the floor of the northern Bering Sea created by California gray whales feeding on bottom-dwelling amphipods and other invertebrates. The smaller 2- to 3-m (6.5- to 10-ft)-long pits produced by the initial feeding do not show up well at this scale. All of the readily visible pits have been enlarged by currents that flow in a northerly direction. The pits show this north–south alignment. *(Sonograph courtesy of William K. Sacco, Yale University.)*
(b) Amphipod mat. A mat of mucus-lined burrows is created by amphipods that are the preferred prey of California gray whales.   *(Photo courtesy of Kirk R. Johnson, Yale University.)*

and the Mexican mainland coast near the southern end of the Gulf of California (Figure 16-18b). The migration usually begins in September when pack ice begins to form over the continental shelf areas that are their feeding grounds. This migration, which is the longest known migration undertaken by any mammal, may involve a round trip distance of 22,000 km (13,700 mi). First to leave are the pregnant females. They are followed by a procession of mature females that are not pregnant, immature females, mature males, and immature males. After cutting through the Aleutian Islands, they follow the coast throughout their southern journey. Traveling at an average rate of about 200 km/day (125 mi/day), most reach the lagoons of Baja California by the end of January.

In these warm-water lagoons, the pregnant females give birth to 2-tn calves. The calves nurse and put on weight quickly during the next 2 months. While the calves are nursing, the mature males breed with the mature females that did not bear calves. Late in March, they return to the feeding grounds, with the procession order reversed. Most of the whales are back in the feeding grounds by the end of June, feeding on prodigious quantities of amphipods to replenish their depleted store of fat and blubber before the next trip south.

Reproduction

Reproduction of pelagic animals involves bringing the males and females of the species together on some periodic basis. These gatherings usually occur during the warmer spring or summer months, when water temperatures are higher and primary productivity is at its peak.

Oviparous Reproduction   Most of the invertebrates and fish that inhabit the open ocean lay eggs that hatch into larval forms in the open water or on the bottom. The larvae become meroplankton (temporary floating plankton that will become nekton or benthos when mature). These creatures are **oviparous**, which simply means they reproduce by laying eggs.

Animals that reproduce in this way usually produce enormous numbers of eggs—15 million eggs per season in the migrating Icelandic cod—because most will be consumed by predators before they hatch, or the larvae will be eaten by other zooplankton. However, some oviparous fish may produce only one or a few well-protected eggs per season; the cartilaginous sharks, skates, and rays are examples of such species (Figure 16-21).

Ovoviviparous Reproduction   The females of other fish species keep their fertilized eggs in their reproductive tract until they hatch. The young come into the world alive, making them **viviparous** (giving birth to living young). However, the process by which the embryos develop is the same as for those that develop from eggs laid in the open water. Thus, this reproduction scheme is called **ovoviviparous**, reproduction in which eggs are incubated internally.

Figure 16-21   Egg Case
Swell shark with yolk sac.   (© Alex Kerstitch/Bruce Coleman, Inc.)

Figure 16-22   Ovoviviparous Behavior in Sea Horses
The female deposits the eggs in the ventral pouch of the male, where they are protected during incubation.   (Photo by Larry Ford.)

The sea horse and pipefish exemplify a special twist to internal incubation where the female deposits the eggs in a pouch on the male (Figure 16-22). The male carries the eggs until they hatch, and the young fish enter the ocean directly from this pouch. The additional protection against loss of eggs to predators results in the reduction of the number of eggs that must be produced. For example, the dogfish shark, *Squalus*, usually produces fewer than a dozen eggs at a time.

Viviparous Reproduction   Viviparous animals, like mammals, give birth to live young. However this method also requires the young to be given more than incubation space in the mother's reproductive tract. Part of the process includes feeding the embryo, in addition to nourishment in the egg yolk. Some sharks and rays secrete a rich milk in the uterus. With this additional nutrition, the stingray, *Pteroplatea*, produces young that enter the ocean with a mass up to 50 times that of the mass of the egg yolk available to them.

Recruitment of Reef Fish on the Great Barrier Reef

It is generally believed that fish larvae are dispersed by currents from where they hatch to become juveniles at a location some distance downcurrent. This assumption was first tested in 1999 by Geoffrey Jones and his colleagues at Lizard Island on the northern Great Barrier Reef of Australia (Figure 16A).

At six sites around Lizard Island, they marked the eggs in nearly 2,000 nests of the damselfish, *Pomacentrus amboinensis* with tetracycline, a fluorescent dye that made the otoliths (ear bones) glow under ultraviolet light. The eggs hatched, and at the end of the three-week larval period, 7,327 larval damselfish larvae (Figure 16B) were captured at three locations within the marking sites.

Using standard methods, they estimated that from 15 to 60 percent of the juveniles present in the waters around Lizard Island were fish hatched in these local waters rather than immigrants carried in by currents from distant locations. This result challenges the assumption that long-distance dispersal is the rule—at least in the reef-island environment. It has implications for the management of fisheries and the preservation of species diversity throughout the sea.

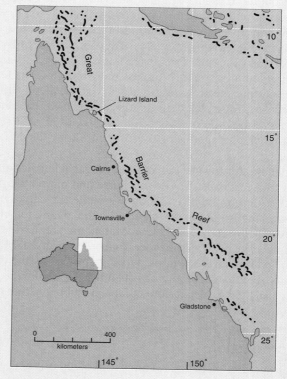

Figure 16A

Great Barrier Reef—Lizard Island.

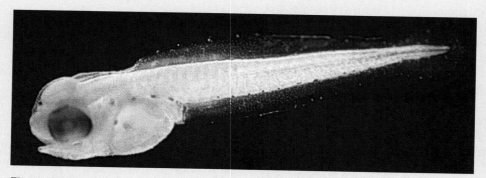

Figure 16B

Three-week old larvae of the damselfish, Pomacentrus amboinensis, captured at Lizard Island. (Courtesy of Geoff Jones)

Mammals exhibit the highest level of viviparous behavior in that the embryo is encased in the placental sac. The mother's blood flows through the placenta, and the embryo receives nutrition from the mother through an umbilical cord that attached it to the placenta. Essentially all nutrition is provided by the mother. This provides strong protection for the developing young, but puts a high energy demand on the mother. Thus mammalian births typically involve only one or a few young, which remain dependent on their mothers for protection and food for some time after their birth.

Other Adaptive Behaviors

As is true for land animals, marine animals also exhibit a variety of specialized defensive mechanisms to ward off predators or to enable them to be better predators. These include secreting poison, mimicry of other poisonous or distasteful species, camouflage, and symbiotic relationships. One extraordinary example is abduction, summarized below.

Abduction

In 1990, pelagic amphipods were observed to capture pteropods (sea butterflies) and carry them around on their backs beneath the Antarctic ice. Some investigation revealed that the 1.25 cm (0.5 in)-amphipod, *Hyperiella dilatata*, uses the 0.65 cm (0.25 in)-pteropod, *Clione limacina*, as a chemical defense against being eaten by predator fish. The pteropod contains an awful tasting chemical, and while the amphipod holds it captive, fish will not eat it (Figure 16-23).

The fish that prey on the amphipods depend on sight to find their food. Therefore, it is not surprising that up to 75 percent of the amphipods observed at depths of less than 9 m (28 ft) were carrying pteropods on their backs. In

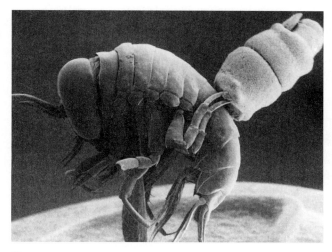

Figure 16-23   Captive Pteropod Protects Pelagic Amphipod from Predators

Scanning electron micrograph of an amphipod (*Hyperiella dilatata*) carrying its captive pteropod (*Clione limacina*), which provides chemical protection against predatory fish.   *(Photo courtesy of James B. McClintock with permission from Nature, Vol. 346, 462–64, 1990. Photo by Phil Oshal.)*

the darker waters at 50 m (160 ft), only 6 percent of the amphipods observed were protected by their foul-tasting captives.

This pattern of capturing another organism for the purpose of using its chemical defense against predators does not fit into any of the categories of symbiotic relationships previously observed between marine species. It is believed that the pteropods are eventually released unharmed, although they may experience only a short reprieve until they once again become hostages.

Summary

Frictional resistance to sinking, which helps tiny plankton stay near the surface, is not a major factor in keeping larger nekton from sinking; these animals depend primarily on buoyancy or swimming. The rigid gas containers of some cephalopods and the expandable swim bladders of many bony fishes are adaptations that increase buoyancy. Many invertebrates, such as the jellyfish, tunicates, and arrow worms, have soft, gelatinous bodies of relatively low density. The Portuguese Man-of-War has a gas-filled float that supports this pelagic colony. Many invertebrate forms are weak swimmers and depend primarily on buoyancy to maintain their positions near the surface.

Squid, fish, and marine mammals are strong swimmers. Marine mammals have developed specialized ways of adapting land-evolved bodies to life in the oceans. These include streamlining to reduce drag and altered pulmonary and circulatory functions that allow them to dive to great depths without suffering ill effects. Cetaceans also use sound to communicate and to locate prey.

Group behavior includes schooling, migration, and reproduction. Schooling behavior in active swimmers such as fish, squid, and crustaceans, is not fully understood, but likely serves a protective function. Migrations are observed among sea turtles, fish, and mammals and are related to reproductive needs and finding food. It is believed that orientation is maintained during migrations through use of visible landmarks, smell, and Earth's magnetic field. Baleen whales may migrate from their cold-water summer feeding grounds to warm, low-latitude lagoons in winter so their young can be born in warm water. Fishes such as the Atlantic cod swim upcurrent to deposit their eggs so that when they hatch, the cod fry will be in water where suitable food is available.

Migrating species may be catadromous, spawning in the ocean and living in freshwater, or anadromous, spawning in fresh water streams and spending their adult lives in the open ocean. Most fish are oviparous, depositing their eggs in the ocean. Some sharks and rays maintain their eggs in a body cavity until they hatch; this ovoviviparous strategy provides a greater protection for the eggs. Where oviparous fish may produce millions of eggs, ovoviviparous fish may produce fewer than a dozen. The stingray, White-tip Shark, and mammals are viviparous (giving birth to live young). They not only provide space in their bodies for the eggs to develop but also provide nutrition in addition to the egg yolk. In mammals, the young remain dependent on their mothers for nutrition for some time after birth.

Key Terms

Anadromous	Ctenophore	Mysticeti	Schooling
Aspect ratio	Decompression sickness	Nematocyst	Scyphozoan
Catadromous	Hemoglobin	Nitrogen narcosis	Siphonophore
Cetacean	Lunger	Odonticeti	Sirenia
Chaetognaths	Migration	Oviparous	Swim bladder
Coelenterates	Myoglobin	Ovoviviparous	Tunicate
Cruiser	Myomere	Pinniped	Viviparous

Questions and Exercises

1. Discuss how the rigid gas chambers in cephalopods may be more effective than the flexible swim bladders of bony fish in limiting the depth to which they can descend.

2. What are the major structural and physiological differences between the fast-swimming cruisers and the lungers that patiently lay in wait for their prey?

3. List the modifications that are thought to allow some cetaceans to (1) dive to great depths without suffering the bends and (2) stay submerged for long periods of time.

4. Describe the process by which the sperm whale may produce echolocation clicks.

5. Although there is disagreement about how it is achieved, discuss what most investigators believe to be the method by which sound reaches the inner ear of toothed whales.

6. Summarize the reasons some investigators believe schooling increases the safety of fishes from predators.

7. What are the methods believed to be used by migrating animals to maintain their orientation?

8. How are the migrations of the North Atlantic eels and Pacific salmon fundamentally different?

9. Compare the reproductive behavior of oviparous, ovoviviparous, and viviparous animals.

10. Explain how the amphipod, *Hyperiella dilatata*, is protected from predators by abducting the pteropod, *Clione limacina*. Why does the incidence of these abductions decrease with increasing depth?

References

Carey, F. G. 1973. Fishes with warm bodies. *Scientific American* 228:2, 36–44.

George, D., and J. George. 1979. *Marine Life: An Illustrated Encyclopedia of Invertebrates in the Sea.* New York: Wiley-Interscience.

Johnsen, S. 2000. Transparent animals. 282:2, 80–89. The ultimate form of defense for helpless animals living in the open sea is transparency. Researchers are learning how they achieve transparency and how predators are still able to find them.

Jones, G. P., et al. 1999. Self-recruitment in a coral reef fish population. *Nature* 402:6763, 802–804.

Kanwisher, J. W., and S. H. Ridgway. 1983. The physiological ecology of whales and porpoises. *Scientific American* 248:6, 110–21.

Lecomte-Finiger, R. 1992. The early life of the european eel. *Research in Marine Biology* 114, 205–210.

MacGinitie, G. E., and N. MacGinitie. 1968. *Natural History of Marine Animals.* 2nd ed. New York: McGraw-Hill.

McClintock, J. B., and J. Janssen. 1990. Pteropod Abduction as a Chemical Defense in a Pelagic Antarctic Amphipod. *Nature* 346:6283, 462–64.

Melamed, Y., A. Shupak, and H. Bitterman. 1996, Medical problems associated with underwater diving. In Pirie, R.G., ed. 1996. *Oceanography: Contemporary Readings in Ocean Sciences*, 3rd Ed. Oxford University Press, New York: 23–31.

Norris, K. S., and G. W. Harvey. 1972. A theory for the function of the spermaceti organ of the sperm whale (*Physeter catodon*). *Animal Orientation and Navigation,* pp. 397–417. Washington, D.C.: National Aeronautics and Space Administration.

Royce, W., L. S. Smith, and A. C. Hartt. 1968. Models of oceanic migrations of pacific salmon and comments on guidance mechanisms. *Fishery Bulletin* 66:441–62.

Thorson, G. 1971. *Life in the Sea.* New York: McGraw-Hill.

Vaughan, T. A. 1972. *Mammalogy.* Philadelphia: W. B. Saunders Company.

Williams, T. M., 2000. Sink or Swim: Strategies for cost-efficient diving by marine mammals. Science 288:5463, 133–36.

Würsig, B. 1989. Cetaceans. *Science* 244:4912, 1550–57.

Suggested Reading

Sea Frontiers

Alper, J., 1993. How fishes see may shed light on how they behave. 39:5, 20–24. Learning that fishes such as salmon can see polarized ultraviolet light, may help explain some of their perplexing behavior.

Bachand, R. G., 1985. Vision in marine animals. 31:2, 68–74. An overview of the types of eyes in marine animals.

Bleecker, S. E., 1975. Fishes with electric know-how. 21:3, 142–48. A survey of fishes that use electrical fields to navigate, to capture prey, and to defend themselves.

Bushnell, P. G., and K. N. Holland, 1989. Tunas: Athletes in a can. 35:1, 42–48. The physiology of various tuna species, specifically that related to high-speed swimming.

Cranston, C., Logging Time with Right Whales. 38:3, 39–43. Filmmakers spend a week filming and learning about right whales in Argentine waters.

Hersh, S. L., 1988. Death of the dolphins: Investigating the east coast die-off. 34:4, 200–207. The 1987–1988 east coast die-off of dolphins is discussed by a marine mammalogist.

Kleen, S., 1989. The diving seal: A medical marvel. 35:6, 370–74. The physiology of seals that allows them to dive deep and stay submerged for long periods of time is covered.

Klimley, A. P., 1976. The White Shark: A matter of size. 22:1, 2–8. Describes procedure used by Dr. John E. Randall for determining the size of sharks from the perimeter of the upper jaw and height of teeth.

Lineaweaver, T. H., III., 1971. The Hotbloods. 17:2, 66–71. Discusses the physiology of the bluefin tuna and sharks that have high body temperatures.

Maranto, G., 1988. The Pacific Walrus. 34:3, 152–59. A summary of the natural history of walruses of the Bering Sea and the role of humans as their predators.

McAuliffe, K., 1994. When whales had feet. 40:1, 20–33. Details the work of paleontologists in the Egyptian desert where they found the remains of a 45-million-year-old whale with feet.

Netboy, A., 1976. The mysterious eels. 22:3, 172–82. Describes what is known of the migrations of the catadromous eels and their importance as a fishery.

O'Feldman, R., 1980. The dolphin project. 26:2, 114–18. A description of the response of Atlantic spotted dolphin to musical sounds.

Reeve, M., 1971. The deadly arrow worm. 17:3, 175–83. This important group of plankton is described in terms of body structure and life style.

Volger, G., and S. Volger, 1988. Northern Elephant Seals. 34:6, 342–47. The history of the elephant seal populations off the coast of California and Baja California.

Scientific American

Denton, E., 1960. The buoyancy of marine animals. 303:11, 118–28. How various marine animals use buoyancy to maintain their position in the water column.

Donaldson, L. R., and T. Joyner, 1983. The salmonid fishes as a natural livestock. 249:1, 50–69. The genetic adaptability of the salmonid fishes may help them adapt to ranching operations.

Gosline, J. M., and M. E. DeMont, 1985. Jet propelled swimming in squids. 252:1, 96–103. Squids can move as fast as the speediest fishes by expelling water through their siphons, a form of jet propulsion.

Gray, J., 1957. How Fishes Swim. 197:2, 48–54. The roles of musculature and fins in the swimming of fishes.

Horn, M. H., and R. N. Gibson, 1988. Intertidal fishes. 258:1, 64–71. A discussion of a wide variety of fishes that inhabit tide pools.

Kooyman, G. L., 1969. The Weddell Seal. 221:2, 100–107. The life style and problems related to this mammal's living in water permanently covered with ice.

Leggett, W. C., 1973. Migration of shad. 228:3, 92–100. Describeshe routes of the anadromous shad in the rivers and Atlantic waters are described, along with factors that may control the migrations.

Moon, R. E., R. D. Vann, and P. B. Bennett, 1995. The physiology of decompression illness. 273:2, 70–77. Research has provided new knowledge of the source of bubbles that cause decompression illness and it may make for safer diving in the near future.

O'Shea, T. J., 1994. Manatees. 271:1, 66–73. Manatees evolved from the same ancestors as elephants and aardvarks. The threats to their survival are discussed.

Rudd, J. T., 1956. The Blue Whale. 195:6, 46–65. A description of the ecology of the largest animal that ever lived.

Sanderson, S. L., and R. Wassersug, 1990. Suspension-feeding vertebrates. 263:3, 96–102. The ecology of filter-feeding fishes and cetaceans is covered.

Shaw, E., 1962. The Schooling of Fishes. 206:6, 128–36. Theories seeking to explain the schooling behavior of fishes.

Triantafyllou, M. S., and G. S. Triantafyllou, 1995. An Efficient swimming machine. 272:3, 64–71. Engineers attempt to design a swimming machine as maneuverable as a dolphin.

Whitehead, H., 1985. Why whales leap. 252:3, 84–93. Whales seem to communicate with their spectacular lunges above the ocean's surface.

Würsig, B., 1988. The behavior of baleen whales. 258:4, 102–107. A summary of the facts concerning the behavior of baleen whales.

Zapol, W. M., 1987. Diving adaptations of the Weddell Seal. 256:6, 100–107. The physiological adaptations that enable the Weddell seal to make deep, long dives.

Oceanography on the Web

Visit the Introductory Oceanography home page for on-line resources for this chapter. There you will find an on-line study guide with review exercises and links to ocean-ography sites to further your exploration of the topics in this chapter. Introductory Oceanography is at http://prenhall.com/thurman (click on the Table of Contents menu and select this chapter).

17 Animals of the Benthic Environment

Brachiopods photographed in New Zealand. (All rights reserved. Coral Reef Reseach Foundation, Palau.).

There are more than 235,000 animal species known to live in the ocean. More than 98 percent of them are found in or on the ocean floor. Ranging from the rocky, sandy, and muddy environments of the intertidal zone to the muds of deep-ocean trenches more than 11 km (6.8 mi) below the surface, the ocean floor provides varied environments for a diverse benthic community. Living at or near the interface of the ocean floor and seawater, an organism's success is closely tied to its ability to cope with the physical conditions of the water, the nature of the substrate (rock or soft sediment), and other members of the biological community. The vast majority of known benthic species live on the continental shelf; only about 400 benthic species are found in the hadal zone of the deep-ocean trenches.

Distribution of Benthic Animals

The benthos includes protists, plants, and animals that live on, in, or attached to the ocean bottom. Recalling from Chapter 14, there are two terms used to describe benthic organisms: *infauna* are organisms that live buried in sand or mud, like clams; *epifauna* are organisms that live on the bottom, either attached to it like rockweed, or moving over it, like crabs. Thus, soft-bodied burrowing worms, hard massive staghorn corals, and delicate, long-armed brittle stars are all members of the benthic community. The most prominent variables affecting species diversity are temperature, currents, and wave energy. Animals are also dependent on primary producers, and thus their distributions are ultimately tied to the distribution of their food source.

We have previously discussed temperature's effect on species diversity with latitude. However, even at the same latitude, a significant difference in the number of benthic species is found on opposite sides of an ocean basin because of the effect of ocean currents on coastal water temperature. For example, over three times the benthic species exist along the European coast, where the Gulf Stream warms the water from the northern tip of Norway to the Spanish coast, than along the similar latitudinal range of the Atlantic coast of North America, where the Labrador Current cools water as far south as Cape Cod.

Figure 17-1 shows that the distribution of benthic biomass is patterned after the distribution of photosynthetic productivity in the surface waters (shown in Figure 14–8). This tells us that life on the ocean floor is very much dependent upon the primary photosynthetic productivity of the ocean-surface waters. A notable exception are the benthic communities that depend on deep-sea vent and seep chemosynthesis. We will also explore some unique communities developed around hydrocarbon seeps.

Animals of Rocky Shores

Figure 17-2 shows a typical rocky shore. The **spray zone**, above the spring high tide line, is covered by water only during storms. The **intertidal zone** lies between the high and low tidal extremes.

Along most shores, the intertidal zone can be divided into three subzones:

1. The **high-tide zone**, mostly dry, covered by the highest high tide but not by the lowest high tides.
2. The **middle-tide zone**, exposed and covered equally—covered by all high tides and exposed during all low tides.
3. The **low-tide zone**, mostly wet, covered during the highest low tides and exposed during the lowest low tides.

Along rocky shores, these divisions of the intertidal zone are often obvious because of the sharp boundaries between the communities of organisms that attach themselves to the surface. Because each centimeter of the littoral rocky shore has a significantly different character from the centimeter above and below it, evolution has been able to produce organisms with the abilities to withstand very specific degrees of exposure to the atmosphere. This results in the most finely defined biozones known in the marine environment.

Rocky intertidal ecosystems have a moderate diversity of species, with the greatest animal diversity being at lower tropical latitudes. Interestingly, the diversity of algae is greater in temperate latitudes.

Spray (Supralittoral) Zone

Throughout the world, the most obvious inhabitant of the rocky supralittoral zone is the periwinkle snail. The spray zone can easily be identified if the middle of the periwinkle belt is considered to be the boundary between the intertidal zone and the spray zone. The periwinkle genus *Littorina* includes species able to breathe air, like land snails (Figure 17-2b).

Hiding among the cobbles and boulders covering the floors of sea caves well above the high-tide line are isopods of the genus *Ligia*—often called rock lice or sea roaches. Neither name is particularly flattering to these little scavengers that reach lengths of 3 cm (1.2 in) and scurry about at night feeding on organic debris (Figure 17-2c). Another indicator of the spray zone is a distant relative of the periwinkle snails, a limpet (genus *Acmaea*) with a flattened conical shell that feeds in a manner similar to periwinkles (Figure 17-2d).

High-Tide Zone

As noted above, periwinkles can venture above high tide into the spray zone, but many other inhabitants of the high-tide zone must stay within its limits. Buckshot barnacles (Figure 17-2e), for example, are limited to the high-tide shoreline. They are confined for two reasons; they filter-feed from seawater and their larval forms are planktonic.

The most conspicuous algae in the high-tide zone are rockweeds—members of the genus *Fucus* in colder latitudes and *Pelvetia* in warmer latitudes (Figure 17-2f). Both have thick cell walls to reduce water loss during periods of low tide. On a clean, rocky shore, rockweeds establish

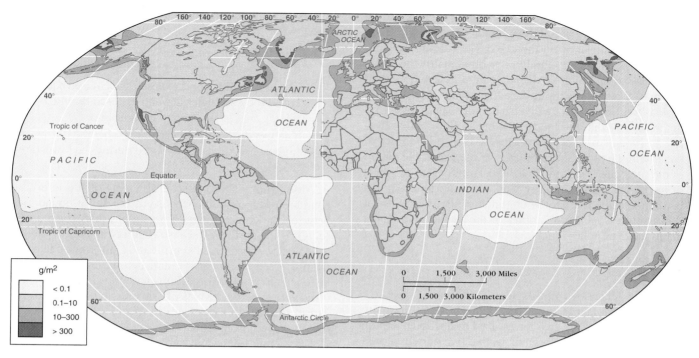

Figure 17-1   Distribution of Benthic Biomass of the World Ocean (g/m², wet weight)

Although the distribution pattern shown here is defensible in concept, it is based on too few samples to be regarded as accurate. The pattern is similar to that of photosynthetic pigment distribution shown in Figure 15-6. This suggests that most of the benthic community is directly dependent on the photosynthetic production in the surface waters of the oceans for its nutritional support. High concentrations of biomass associated with hydrothermal vents and cold seeps are not reflected in this distribution. The concentrations beneath the centers of the subtropical gyre waters are 20,000 times less than those of the continental-shelf benthos. Up to 60 percent of the organic matter reaching the deep-ocean floor may enter the food chain through bacteria, and deep-sea bacteria may metabolize organic matter at a rate at least 10 times slower than bacteria on the continental shelf. This could reduce the rates of biological productivity on the deep-ocean floor to levels of 1/200,000 that occurring on the continental shelves.   *(After Zenkevitch et al., 1971.)*

themselves before sessile animal forms. However, once barnacles or mussels move in, rockweeds seem doomed.

Middle-Tide Zone

Rockweeds flourish into the middle-tide zone, where the variety of life forms is much greater than in the high-tide zone. Not only does variety increase, but total biomass is much greater. There is, therefore, a greater competition for rock space by forms that attach themselves to the surface. Such forms are called **sessile**.

The barnacle most characteristic of this zone is the goose barnacle (*Pollicipes*) (Figure 17-2), which attaches itself to the rock surface by a long muscular neck. In the competition for space in the middle-tide zone, various mussels (genera *Mytilus* and *Modiolus*) are even more successful than barnacles. They attach to bare rock, algae, or barnacles by tough threads, settling on these surfaces as larval forms. Mussels are fed upon by varioius predators, including sea stars and carnivorous snails.

Two common genera of sea stars are the *Pisaster* and *Asterias*. To pry apart the two halves of the calcium carbonate mussel shell to reach the edible tissue inside, sea stars exert a continuous pull on the shells using their tube feet to apply suction. The mussel eventually becomes fatigued and can no longer hold its shell halves closed. When the shell opens ever so slightly, the sea star turns its stomach inside out and slips the stomach through the opening of the two shell halves; it then digests the mussel without having to take it out of its shell (Figure 17-3).

The dominant feature of the middle-tidal zone along most rocky coasts is a mussel bed that thickens toward the bottom until it reaches an abrupt bottom limit. This may be so pronounced that it appears that an invisible horizontal plane has prevented the mussels from growing below this depth. Protruding from the mussel bed are numerous goose barnacles, and concentrated in the lower levels of the bed are sea stars browsing on the mussels. Less conspicuous forms common to the mussel beds are algae, worms, clams, and crustaceans.

Where the rock surface flattens out within the middle tidal zone, tide pools trap water as the tide ebbs. These pools support interesting micro-ecosystems containing a wide variety of organisms. The largest member of this community will often be the sedentary relative of the jellyfish, the sea anemone (Figure 17-4). Shaped like a sack, anemones have a flat foot disk that provides suction for attachment to the rock surface. The only opening to the gut cavity, the mouth, is surrounded by rows of tentacles covered with cells that contain a stinging threadlike nematocyst. The nematocyst is automatically fired when an organism brushes against the tentacles.

Swimming in the tide pools are a variety of small fish. The opaleye reaches lengths of 5 cm (2 in) in the pools and grows larger in deeper water. This fish is easily identified by the tiny white spots on its back on either side of

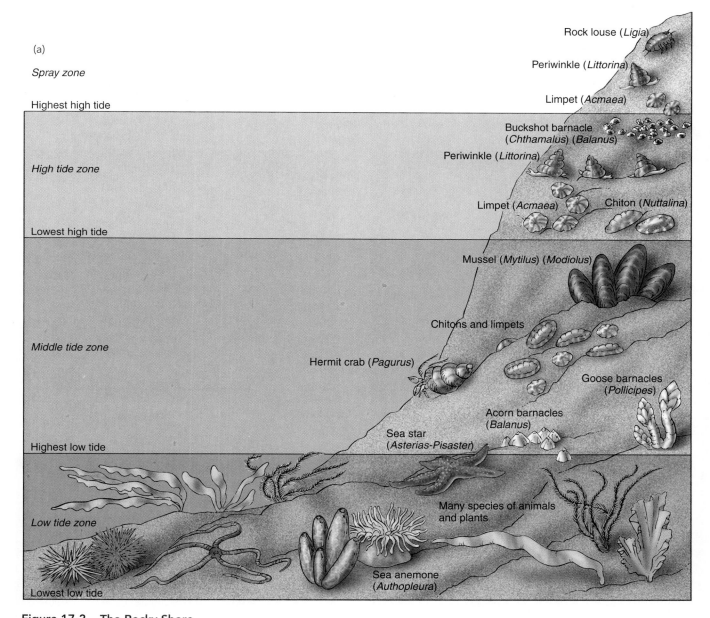

(a)

Spray zone

Rock louse (*Ligia*)

Periwinkle (*Littorina*)

Highest high tide

Limpet (*Acmaea*)

High tide zone

Buckshot barnacle (*Chthamalus*) (*Balanus*)

Periwinkle (*Littorina*)

Limpet (*Acmaea*)

Chiton (*Nuttalina*)

Lowest high tide

Mussel (*Mytilus*) (*Modiolus*)

Chitons and limpets

Middle tide zone

Hermit crab (*Pagurus*)

Goose barnacles (*Pollicipes*)

Acorn barnacles (*Balanus*)

Sea star (*Asterias-Pisaster*)

Highest low tide

Low tide zone

Many species of animals and plants

Sea anemone (*Authopleura*)

Lowest low tide

Figure 17-2　The Rocky Shore

(a) A typical rocky intertidal zone and some organisms found in its subzones (not to scale). (b) Periwinkles (*Littorina*) nestled in a depression near the upper limit of the high-tide zone. This behavior helps reduce exposure to direct sunlight. (c) Rock louse (*Ligia*). (d) Rough keyhole limpet (*Diodora aspera*) with encrusting red algae (*Lithothamnion*).　(*Photo © Norbert Wu/Peter Arnold, Inc.*) (e) Buckshot barnacles (*Chthamalus*). (f) *Fucus filiformes*. (*Photo © Breck P. Kent.*) Sharing the middle-tide zone with these rockweeds are (g) acorn barnacles (*Balanus*), and (h) *Pelvetia fastigata*. (i) Goose barnacles (*Pollicipes*) and mussels. (*Photo © Eda Rogers.*)

(b)

(c)

(d)

(e)

(f)

(g)

(h)

(i)

Figure 17-2   (continued)

the dorsal fin. The woolly sculpin is a permanent resident of tide pools and grows longer than 15 cm (6 in). It is covered with hairlike *cirri* and is usually resting on the bottom or walking on its large pectoral fins. Readily identified by its blunt head and continuous dorsal fin is the 15-cm (6-in) rockpool blenny (Figure 17-5).

The most interesting inhabitant of tide pools is the hermit crab, *Pagurus*. These animals have a well-armored pair of claws and upper body, but they also have a soft, unprotected abdomen. Hermit crabs protect their abdomen by moving into abandoned snail shells. Their abdomens have even evolved a right-hand curl to make them fit properly into the shells. Once in the snail shell, the crab can close off the opening with its large claws (Figure 17-6a).

In tide pools near the lower limit of the middle-tide zone, sea urchins may be found feeding on algae (Figure 17-6b). They have a five-toothed structure centered on the bottom side of a hard spherical covering that supports many spines. The hard protective covering is called a *test* and is composed of fused calcium carbonate plates that are perforated to allow tube feet and gills to pass through. They have a structure called Archimedes lantern at the

Figure 17-3   Sea Star Feeding on a Mussel

(Photo © Joy Sparr/Bruce Coleman, Inc.)

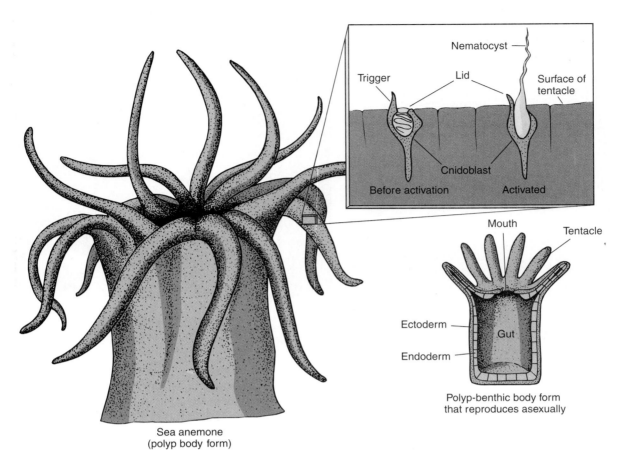

Sea anemone
(polyp body form)

Polyp-benthic body form
that reproduces asexually

Figure 17-4   Sea Anemone

Basic body plan of an anemone and detail of the stinging mechanism.

Figure 17-5   Rockpool Blenny

(© Larry Lipsky/Bruce Coleman, Inc.)

(a)

(b)

**Figure 17-6   Hermit Crab and
Sea Urchin**
....
(a) Hermit crab (*Pagurus*) in the shell
of a snail (*Maxwellia gemma*).
(*Photo by James McCullagh*). (b) Sea
urchins burrowed into the bottom of
a lower middle-tide zone tide pool.

center of the ventral surface that has five pointed teeth (one for each of the five segments of symmetry) that scrape algae off surfaces.

Low-Tide Zone

Unlike the upper and middle-tide zones, the low-tide zone is dominated by plants and algae rather than animals. A diverse community of animals exists, but they are less obvious because they are hidden by the great variety of seaweeds and surf grass (*Phylospadix*) (Figure 17-7). The encrusting red alga, *Lithothamnion*, also seen in middle zone tide pools, becomes very abundant in the lower tide pools (Figure 17-2d). In temperate latitudes, moderate-sized red and brown algae provide a drooping canopy beneath which much of the animal life is found.

Scampering from crevice to crevice and in and out of tide pools across the full range of the intertidal zone are various species of shore crabs (Figure 17-8). These scavengers help keep the shore clean. Shore crabs spend most

of the daylight hours hiding in cracks or beneath overhangs. They do most of their eating at night, shoveling in algae as rapidly as they can tear it from the rock surface with their large front claws (chelae). Shore crabs can spend long periods out of the water.

Animals of Sediment-Covered Shores

The sediment-covered shore ranges from steep boulder beaches, where wave energy is high, to the mud flats of quiet, protected embayments. However, the sediment-covered shore we know best is the sand beach typical of areas where wave energy is usually moderate.

The sediment-covered shore includes what are commonly called beaches, salt marshes, and mud flats, all of which represent lower-energy environments. As the energy level diminishes (meaning as the strength of the longshore current diminishes), particle size becomes smaller

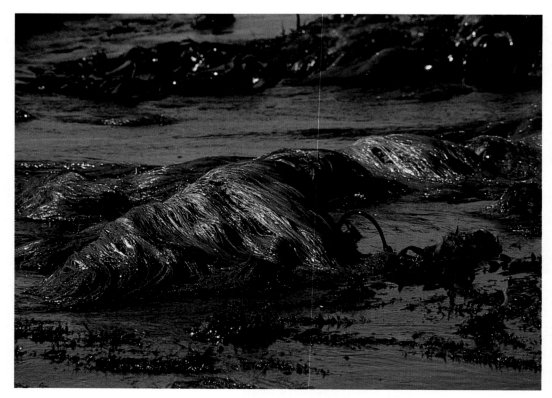

Figure 17-7   Algal Colony and Surf grass of the Low-Tide Zone Exposed during Extremely Low Tide

The dark-colored sea palms (brown algae) and green surf grass are common inhabitants of the California low-tide zone.

(a)

(b)

Figure 17-8   Shore Crabs

(a) Coral crab, or queen crab, Bonaire Island, Netherlands Antilles.   (© Fred Bavendam/Peter Arnold, Inc.) (b) Shore crab, *Pachygrapis crassipes*, female with eggs.   (Photo © Eda Rogers.)

and the sediment slope is reduced. Consequently, sediment stability increases. The water from breaking waves rapidly percolates down through coarse sands carrying oxygen that replaces the oxygen consumed by animals that live in the sediment. The readily available oxygen supply also enhances bacterial decomposition of dead tissue.

Life on and in the sediment requires very different adaptations than life on the rocky coast. The sandy beach supports fewer species than the rocky shore, and mud flats support fewer still; however, the total number of individuals may be nearly equal. In the low-tide zone of some beaches and on mud flats, as many as 5000 to 8000 burrowing clams have been counted in only 1 m² (10.8 ft²).

Burrowing is the most successful adaptation for life in the sediment-covered shore, so life here is less visible. By burrowing a few centimeters beneath the surface, organisms find a stable environment where they are not bothered by fluctuations in temperature and salinity, and the threat of drying out is greatly reduced. Burrowing in the sediment does not prevent animals from suspension feeding—straining plankton from the clear water above the sediment. With various techniques, the water above the sediment surface can be stripped of its plankton by buried animals. Two of these methods (suspension feeding and deposit feeding) are shown in Figure 17-9.

The Sandy Beach

When you go to a sandy beach, you do not see animals exposed at the surface as you do along a rocky shore. There is no stable, fixed surface that animals can attach to, so most of them burrow into the sand, safely hidden from view.

A bivalve is an animal having a two-part hinged shell, like a clam. Bivalve mollusks are well adapted for life in the sediment. The greatest variety of clams is found burrowed into the low-tide region of sandy beaches; their numbers decrease where the sands become muddier. Bi-

valve mollusks possess a soft body, a portion of which secretes the calcium carbonate valves (shells) that hinge together. A muscular foot digs into the sediment to pull the creature down into the sand. Siphons protrude vertically for feeding (Figure 17-9a). The procedure used by a bivalve to bury itself is shown in Figure 17-10. How deeply a bivalve can bury itself depends on the length of its siphons; they must reach above the sediment surface to pull in water from which plankton will be filtered. Oxygen is also extracted in the gill chamber before the water is expelled. Periodically, undigestible matter is forced back out the siphon by quick muscular contractions.

Some segmented worms, or annelids, are also well adapted for life in the sediment. Most common of the sand worms is the lugworm (*Arenicola* species). It lives in a U-shaped burrow with walls reinforced with mucus (Figure 17-9b). The worm moves forward to feed and extends its proboscis (snout) up into the head shaft of the burrow, loosening sand with quick pulsing movements. As sand continually slides into the burrow and is ingested by the worm, a cone-shaped depression forms at the surface over

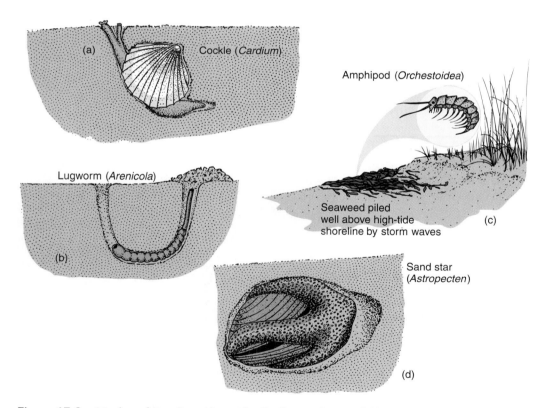

Figure 17-9   Modes of Feeding Along the Sediment-Covered Shore

(a) Suspension feeding. This method is used by clams that bury themselves in sediment and extend siphons to the surface. Water is pumped in through the siphons, and the clams feed by filtering plankton and other organic matter from the water. (b), (c) Deposit feeding. Some deposit feeders, like (b) the segmented worm *Arenicola*, feed by ingesting sediment and extracting organic matter from it. Others, like (c), the amphipod *Orchestoidea*, feed on more concentrated deposits of organic matter (detritus) on the sediment surface. (d) Carnivorous feeding. The sand star, *Astropecten*, cannot climb rocks like its sea star relatives, but it can burrow rapidly into the sand, where it feeds voraciously on crustaceans, mollusks, worms, and other echinoderms.

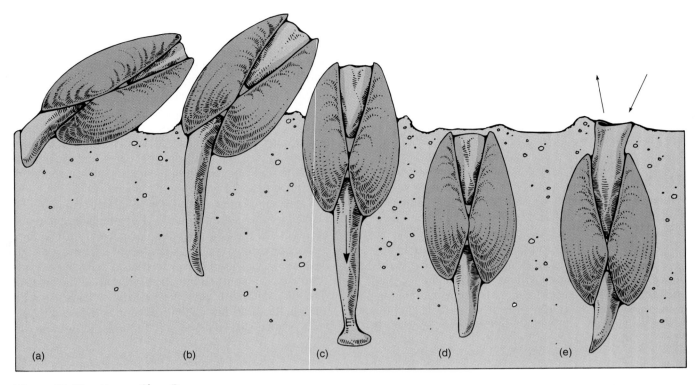

(a) (b) (c) (d) (e)

Figure 17-10 How a Clam Burrows

Clams exposed at the sediment surface can quickly burrow into the sediment by
(a) extending the pointed foot into the sediment, and (b) forcing the foot deeper into the
sediment and using this increasing leverage to bring the exposed, shell-clad body toward
vertical. (c) When the foot has penetrated deep enough, it changes shape, forming a bulbous
anchor at the tip. (d) A quick muscular contraction pulls the entire animal into the sediment.
(e) The siphons are then pushed up above the sediment to pump in water from which the
clam will extract food and oxygen.

the head end of the burrow. As sand passes through the worms digestive tract, the organic content is digested, and the processed sand is deposited.

Staying high on the beach and feeding on kelp cast up by storm or high tide waves are numerous crustaceans called beach hoppers. They are known to jump more than 2 m (6.6 ft). A common genus is *Orchestoidea*, which usually range from 2 to 3 cm (0.8 to 1.2 in) in length. Laterally flattened, beach hoppers usually spend the day buried in the sand or hidden in the kelp on which they feed. They become active at night and so many hop at once that they may form large clouds above the masses of seaweed on which they are feeding (Figure 17-9c).

Figure 17-11 shows a sand crab, a larger crustacean that may be harder to find on sandy beaches (genera *Blepharipoda*, *Emerita*, and *Lepidopa*). Ranging in length from 2.5 to 8 cm (1 to 3 in), they move up and down the beach near the shoreline. They bury their bodies into the sand, leaving their long, curved, V-shaped antennae pointing up the beach slope. These little crabs filter food particles from the water.

Echinoderms are represented in beach deposits by sand stars (*Astropecten*) and heart urchins (*Echinocardium*). Sand stars are well adapted to prey on invertebrates

Figure 17-11 Sand Crab, *Emerita*

The head of a sand crab emerges from the beach surface. Sand crabs are usually buried just beneath the surface.

that burrow into the low-tide region of sandy beaches. The sand star is well designed for moving through sediment, with five tapered spiny legs and a smooth back (Figure 17-9d). More flattened and elongated than sea urchins of the rocky shore, heart urchins live buried in the sand near the low-tide shoreline. They gather sand grains into their mouths, where the coating of organic matter is scraped off and ingested (Figure 17-12).

Meiofauna (meio = lesser) live in the spaces between sediment particles. These animals are 0.1 to 2 mm (0.004 to 0.08 in) long (Figure 17-13) and feed primarily on bacteria removed from the surface of sediment particles. The meiofauna, composed primarily of polychaetes, mollusks, arthropods, and nematodes, are found in sediment from the intertidal zone to the deep-ocean trenches.

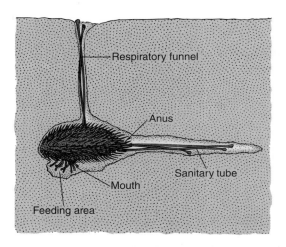

Figure 17-12 Heart Urchin, *Echinocardium*

Intertidal Zonation

The faunal distribution across the intertidal range of sediment-covered shore is similar to that observed on rocky shores. The animal species found in each of the corresponding intertidal zones are appropriately adapted to a soft substrate instead of hard one. However, life forms found in intertidal rocky shores and those found in sediment-covered shores have two common characteristics. The maximum number of species and the greatest biomass are found near the low-tide shoreline, and both decrease toward the high-tide shoreline. This zonation, shown in Figure 17-14, is best developed on steeply sloping, coarse-sand beaches and is less readily identified on gently sloping, fine-sand beaches. On mud flats, there is essentially no slope, which eliminates the possibility of much zonation within such protected, low-energy environments.

The Mud Flat

Two widely distributed plants associated with mud flats are eelgrass (*Zostera*) and turtlegrass (*Thalassia*). They occupy the low-tide zone and adjacent, shallow sublittoral regions bordering the flats. Numerous openings at the surface of mud flats attest to a large population of bivalve mollusks and other invertebrates.

Quite visible and interesting inhabitants of the mud flats are the fiddler crabs (*Uca*), living in burrows that may be more than 1 m (3 ft) deep. Relatives of the shore crabs, they usually measure no more than 2 cm (0.8 in) across the body. Fiddler crabs get their name because the males have one small claw and one outsized claw (up to 4 cm or 1.6 in long). This large claw is waved around in such a manner that the crab seems to be playing an imaginary fiddle. This large claw is used to court females and to fight competing males (Figure 17-15). The females have two normal-sized claws.

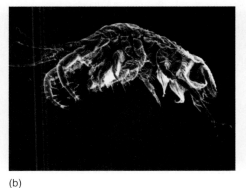

(a)   (b)   (c)

Figure 17-13 Meiofauna

Scanning electron micrographs of meiofauna. (a) Nematode head (x 804). The projections and pit on right side are possibly sensory structures. (b) Amphipod (x 20). This organism builds a burrow of cemented sand grains. (c) Polychaete with its proboscis extended (x 55). *(Courtesy of Howard J. Spero.)*

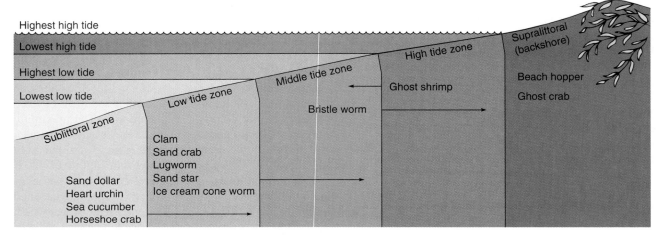

Figure 17-14   Intertidal Zonation on the Sediment-Covered Shore

Zonation is best displayed on coarse-sand beaches with steep slopes. As the sediment becomes finer and the beach slope decreases, zonation becomes less distinct. It disappears entirely on flat mud flats.

Figure 17-15   Life of the Mud Flat

Zostera, eelgrass, and *Uca*, fiddler crab, at Cape Hatteras, North Carolina.   *(Stephen J. Kraseman, © Peter Arnold, Inc.)*

Animals of the Shallow Offshore Ocean Floor

Extending from the spring low-tide shoreline to the seaward edge of the continental shelf is an environment that is mainly sediment-covered, although bare rock exposures may occur locally near shore. The sediment-covered shelf has a moderate-to-low species diversity. The diversity of benthos is lowest beneath upwelling regions, because upwelling carries nutrients to the surface. Here pelagic production is great, producing an excess of dead organic matter. When this material rains down on the bottom and decomposes, which consumes oxygen, the oxygen supply can become locally depleted.

The Rocky Bottom (Sublittoral)

A rocky bottom within the shallow inner sublittoral region is usually covered with algae. Along the North American Pacific coast, the giant bladder kelp (*Macrocystis*) attaches to rocks as deep as 30 m (100 ft) if the water is clear enough to allow sunlight to support algal growth at this depth. The giant bladder kelp and another fast-growing kelp (*Nereocystis*) often form bands of kelp forest along the Pacific coast. Smaller tufts of red and brown algae are found on the bottom, living on the kelp fronds. Commonly found growing along with the algae tufts on the fronds are smaller life forms that serve as food for many of the animals living within the kelp forest community. Surprisingly, very few animals feed directly on the living kelp plant. Among those that do are the large sea hare (*Aplysia*) and sea urchins (Figure 17-16b).

The large crustaceans that we call lobsters are common to rocky bottoms. They are a somewhat varied group, with robust external skeletons. Spiny lobsters are named for

(a)

(b)

Figure 17-16   Life of the Sublittoral Zone

(a) Giant bladder kelp (*Macrocystis*).   (*Mia Tegner*)   (b) Sea hares (*Aplysia californica*) and sea urchins in a kelp forest.   (*James McCullagh*)

(a)

(b)

Figure 17-17   Spiny and American Lobsters

(a) Spiny lobster.   *(B. Kiwala.)* (b) American lobster.
(Harold W. Pratt/Biological Photo Service.)

their spiny coverings; they have two very large, spiny antennae with noise-making devices near their base (Figure 17-17a). The genus *Palinurus* is a culinary delicacy, living deeper than 20 m (65 ft) along the European coast, and reaching lengths to 50 cm (20 in). The Caribbean species *Panulirus argus* sometimes migrates single-file for several kilometers, for unknown reasons.

Panulirus interruptus is the spiny lobster of the American west coast. All spiny lobsters are taken for food, but none is so highly regarded as the so-called true lobsters (genus *Homarus*), which include the American lobster *Homarus americanus*. Although they are scavengers like their spiny relatives, the true lobsters also feed on live animals, including mollusks, crustaceans, and members of their own species (Figure 17-17b).

Oysters are sessile (anchored) bivalve mollusks found in estuarine environments. They prefer a steady flow of clean water to provide plankton and oxygen. Because of their great commercial importance throughout the world, they have been closely studied.

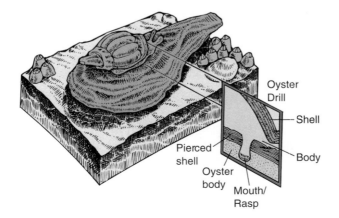

Figure 17-18   An Oyster Drill Feeding on an Oyster

Oyster beds are simply the empty shells of many oyster generations cemented to a rocky bottom or to one another, with the living generation on top. Each female produces many millions of eggs each year; when fertilized, the eggs become planktonic larvae. After a few weeks as plankton, the larvae attach themselves to the bottom. Attached oyster larvae are most abundant on live oyster shells, less abundant on dead oyster shells, and least abundant on rock. Oysters are food for a variety of sea stars, fishes, crabs, and boring snails that drill through the shell and rasp away the soft tissue of the oyster (Figure 17-18).

Coral Reefs

Coral reefs have a greater diversity of animal species than any other marine community. They contain 25 percent of all marine species. Corals are found throughout the ocean, but coral accumulations that might be classified as reefs are restricted to warmer-water regions where the average monthly temperature exceeds 18°C (64°F) throughout the year (Figure 17-19). Such temperature conditions are found primarily between the tropics. However, reefs also grow at latitudes approaching 35°N and 35°S on the western margins of ocean basins, where warm-water masses move into high-latitude areas and raise average temperatures.

Also shown in Figure 17-19 is the greater diversity of reef-building corals on the western side of ocean basins. More than 50 genera of corals thrive in a broad area of western Pacific Ocean and a narrow belt of the western Indian Ocean. Fewer than 30 genera occur in the Atlantic Ocean, with the greatest diversity occurring in the Caribbean Sea. This pattern of diversity may be related to past ocean current patterns, but such an explanation is controversial.

Because of changes in wave energy, salinity, water depth, temperature, and other less obvious factors, there

El Niño and Marine Iguanas of the Galapagos Islands

The iguanas (*Amblyrhynchus cristatus*) of the Galapagos Islands are the only known marine lizards. The equatorial islands they inhabit lie directly in the path of the significant physical changes in the marine environment associated with El Niño Southern Oscillation (ENSO) events discussed in Chapter 9. Two studies covering 8 and 18 years have shown that some iguanas shrink by as much as 20 percent (6.8 cm) in two years during ENSO events and regrow when conditions again become favorable (Figure 17A).

When water rolls in from the warm pool to the west, surface temperatures rise from an average 18°C to as high as 32°C as the upwelling of cool, nutrient-rich waters is disrupted. This results in the iguana's preferred food, green and red algae, being replaced by brown algae that are harder for the iguanas to digest. During the 1997–98 El Niño, larger animals shrank the most, and females shrank more than males of the same size. This was probably because of the energy females expended producing eggs the previous year.

Shrinking appears to be an adaptive response to low food availability and stress. About half of the shrinkage can be attributed to decreases in the mass of cartilage and connective tissue, while bone absorption may account for the remainder. As is experienced by astronauts that spend long periods in weightlessness, some of the shrinkage may result from the decrease in exercise associated with foraging for food. They get little exercise feeding during El Niño events. It may well be that lizards are the only vertebrates that can shrink and regrow repeatedly during their lifetime in response to changes in the environment.

Figure 17A Marine Iguana of the Galapogos Islands.

(Photo courtesy of Martin Wikelski, University of Illinois).

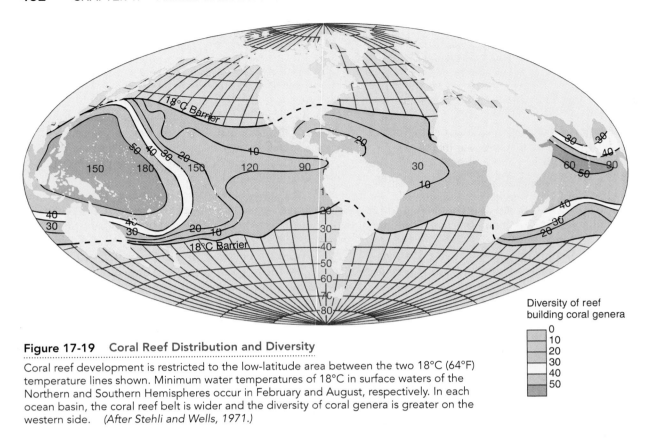

Figure 17-19 **Coral Reef Distribution and Diversity**

Coral reef development is restricted to the low-latitude area between the two 18°C (64°F) temperature lines shown. Minimum water temperatures of 18°C in surface waters of the Northern and Southern Hemispheres occur in February and August, respectively. In each ocean basin, the coral reef belt is wider and the diversity of coral genera is greater on the western side. *(After Stehli and Wells, 1971.)*

Diversity of reef building coral genera

0
10
20
30
40
50

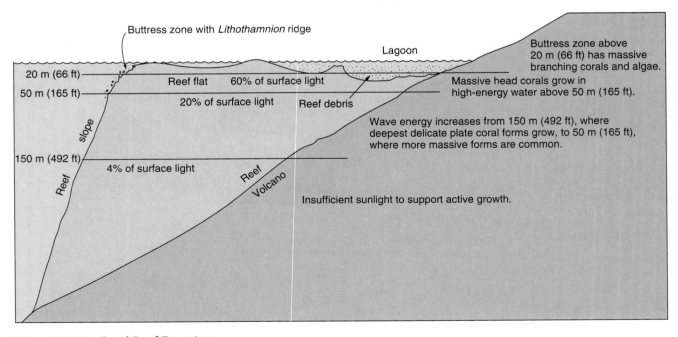

Figure 17-20 **Coral Reef Zonation**

As depth increases, wave energy decreases and light intensity decreases. Massive branching corals occur above 20 m (66 ft) where wave energy is great. Corals become more delicate with increasing depth until they die out around 150 m (500 ft). Below this depth, there is too little solar radiation to support the symbiotic algae that corals must have to survive.

is a well-developed vertical and horizontal zonation of the reef slope (Figure 17-20). These zones are readily identified from the assemblages of algae and animal life found in and near them. Reef growth also requires that the water have a relatively normal salinity and that it be free from particulate matter. Therefore, we see very little coral reef growth near the mouths of rivers, where the salinity is lower and the water carries large quantities of suspended material that would choke the reef colony.

Symbiosis of Coral and Algae

Coral reefs are more than just coral. Algae, mollusks, and foraminifera also make important contributions to the reef structure. Reef-building corals are **hermatypic**, which means that they have a mutualistic relationship with algae called **zooxanthellae**. These algae live within the tissue of the coral polyp. Not only reef-building corals but other reef animals have similar symbiotic relationships with algae. Those that derive part of their nutrition from their algae partners are called **mixotrophs**. This group includes coral, foraminifera, sponges, and mollusks (Figure 17-21).

Because light is essential for algal photosynthesis, reef-building corals are restricted to clear, shallow waters. The algae not only nourish the coral, but may contribute to their calcification capability by extracting carbon dioxide from the coral's body fluids. Corals contribute to the mutual relationship by providing nutrients to the zooxanthellae. This exchange between algae and corals, as well as other mixotrophs, supports high levels of biological productivity within the reef community despite the low nutrient levels in the surrounding water.

Coral reefs actually contain up to three times more algal biomass than animal biomass. The zooxanthellae account for less than 5 percent of the reef's overall algal mass; most of the rest is filamentous green algae. However, zooxanthellae account for up to 75 percent of the biomass of the reef-building corals, and they provide the corals with up to 90 percent of their nutrition.

Because the algae require sunlight for photosynthesis, the greatest depth to which active coral growth extends is 150 m (500 ft), below which there is not enough sunlight. Water motion is decreased at these depths, so relatively delicate plate corals can live on the outer slopes of

(a)

(c)

Figure 17-21   Coral Reef Inhabitants That Depend on Algal Symbionts

(a) Staghorn corals (*Acropora sp.*) spawn on the Great Barrier Reef. As do most reef corals, they depend primarily on symbiont algae for most of their nutrition. (b) The blue-gray sponge on the left is *Niphates digitalis*, a totally heterotrophic sponge. On the right is *Angelas* sp., which contains some cyanobacterial symbionts—it is a mixotroph. Photo was taken in 20 m (60 ft) of water at Carrie Bow Cay, Beliz. (c) A giant clam, *Tridacna gigas*. These suspension feeders also depend primarily on symbiotic algae living in the mantle tissue that account for its green color.   *(Part (a) Terry P. Hughes, James Cook University. (b) C. R. Wilkerson, Australian Institute of Marine Science; (c) All rights reserved. Coral Reef Research Foundation, Palau.)*

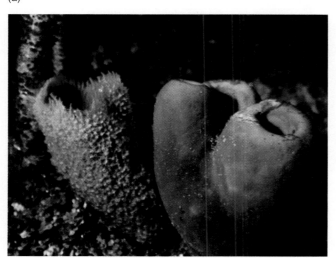

(b)

Ruth D. Turner—Investigator of Deep-Sea Benthos

Although her initial interest was in birds, Ruth Dixon Turner has gained international eminence as a benthic marine biologist specializing in the molluscan anatomy and systematics. Dr. Turner authored definitive monographs on the clam families Teredinidae (shipworms) and Pholadidae (includes deep-sea wood borers), and she is the first woman to dive in the DSRV *Alvin* (Figure 17B).

At an age of 70 plus, Ruth Turner still talks exictedly as she details anecdotes from her experiences—ones usually including the contributions of her colleagues who were involved directly or indirectly in her investigations. One of her major interests is the comparison of the ecosystems that develop around wood islands (artificially deposited masses of wood on the deep-ocean floor) with those of deep-sea hydrothermal vents. Dr. Turner is particularly interested in securing data related to feeding types, life histories, and growth rates. Both ecosystems are located in the deep ocean, where the physical environment has long-term stability, and each is supported by a high level of productivity (wood and chemosynthetic bacteria). Long-term observations of the wood islands and hydrothermal vents should help test the hypotheses that stable environments with adequate primary productivity will produce complex, diverse communities, with the greatest species diversity occurring at the predator level.

Ruth Turner's view of science is that it should be fun, it can't be fully scheduled, and it is wonderful. It requires the cooperative effort of many individuals and, above all, "Nothing replaces knowing your animals."

Figure 17B

Dr. Ruth Turner is shown preparing to enter the deep-sea research vessel, *Alvin*. Dr. Turner was the first woman to dive in this marvelous tool, which has allowed so many deep-sea researchers to see firsthand the realm they wish to learn more about. Many experiments that helped Dr. Turner research her deep-sea wood borers were made possible only because of the availability of *Alvin*. In which she has made approximately 50 dives. *(Photo courtesy of the Museum of Comparative Zoology, Harvard University.)*

the reef from 150 m (500 ft) up to about 50 m (165 ft), where light intensity exceeds 4 percent of the surface intensity (Figure 17-20).

From 50 m to about 20 m (66 ft), the strength of water motion from breaking waves increases on the side of the reef facing the prevailing current flow. Correspondingly, the mass of coral growth and the strength of the coral structure supporting it increase toward the top of this zone, where light intensity exceeds 60 percent of surface value. Above 20 m, the **buttress zone** is inhabited by massive, branching corals such as the staghorn corals shown in Figure 17-21a. The reef flat may be under a few centimeters to a few meters of water at low tide. A variety of beautiful reef fish, as well as sea cucumbers, worms, and a variety of mollusks inhabit this shallow water. In the protected water of the reef lagoon, there are gorgonian coral, anemones, crustaceans, mollusks, and a great variety of echinoderms (Figure 17-21).

An example of commensalism can be seen in the relationship of shrimpfish and sea urchins. The shrimpfish swim head down among the long slender spines of sea urchins on the reef, the spines are a significant deterrent to any predator. The sea urchin is neither hindered nor aided by the presence of the little fish. The anemone fish receives similar protection by swimming among the tentacles of two species of sea anemones. This relationship is believed to be mutual in that the anemones benefit by the anemone fish serving as bait to draw other fish within reach of anemone tentacles (Figure 17-22).

Threats to Coral Reefs

When human populations increase on lands adjacent to coral reefs, the reefs deteriorate. Many aspects of human behavior can damage a reef—fishing, trampling, boat collisions with the reef, sediment increase due to development, and collection of reef inhabitants by visitors all take their toll.

Some of the most harmful and difficult factors to measure in terms of reef damage are the inevitable increase in reef water nutrient levels caused by sewage discharge and fertilizer runoff, and increased water turbidity from soil erosion caused by coastal development. Pollution and shell collecting may also have caused population imbalances between coral and coral predators like the crown-of-thorns seastar. Global warming may also be implicated in coral bleaching.

Nutrient Increases   As nutrient levels increase in reef waters, the dominant benthic community changes. At low nutrient levels, hermatypic corals and other reef animals that contain algal symbiotic partners thrive. Moderate nutrient levels favor the development of fleshy benthic algae, and high nutrient levels favor suspension feeders like clams. At high nutrient levels, the phytoplankton mass exceeds the benthic algal mass, so benthic populations that are tied to the phytoplankton food web dominate.

The clarity of water is reduced by increased phytoplankton biomass. The fast-growing members of the phytoplankton-based ecosystem destroy the reef structure by

Figure 17-22   Mutualism

The anemone fish, which lives unharmed among the stinging tentacles of the sea anemone, brings food to the anemone. This symbiotic relationship, which benefits both participants, is called mutualism.   *(Scott Johnson)*

overgrowing the slow-growing coral and through **bioerosion**, which is erosion of the reef by organisms. Sea urchins and sponges particularly, damage the reef by bioerosion.

The Crown-of-Thorns Phenomenon   Since 1962, the crown-of-thorns sea star (*Acanthaster planci*) has destroyed living coral on many reefs throughout the western Pacific Ocean (Figure 17-23). Some investigators believe this is a modern phenomenon brought about by the activities of

Figure 17-23   Crown-of-Thorns Sea Star

This crown-of-thorns sea star is being attacked by one of its few predators, the Pacific triton.   *(Scott Johnson)*

humans. However, there is little evidence to point to such a cause. A 1989 study of the Great Barrier Reef indicated that, during the past 80,000 years, the crown-of-thorns sea star has periodically been even more abundant on the reefs than it is today. If this is true, the sea star may be an integral part of the reef ecology in this region, rather than a destructive upstart taking advantage of human actions that have modified the reef in some way favorable to its proliferation.

Coral Bleaching   Loss of color in coral reef organisms is called **coral bleaching** (Figure 17-24). The cause of the bleaching is expulsion of the coral's symbiotic partner, the zooxanthellae algae. Recall that essentially all reef-building corals and some other reef mixotrophs are nourished by these algae, which live within their tissues. The loss of this nourishment can kill the coral; if it does not regain its zooxanthellae algae, it dies.

Coral bleaching has occurred locally numerous times in the past. However, a mass mortality of at least 70 percent of the corals along the Pacific Central American coast occurred as a result of a bleaching episode associated with the severe El Niño of 1982–1983. This eastern Pacific bleaching is believed to have been caused by warmer water temperatures associated with the El Niño. Two species of Panamanian coral became extinct during this event. Whatever the cause, it will take many years for these reefs to recover.

In addition to bleaching, numerous diseases, some of which can be attributed to fungal or bacterial attack, have been identified in the Florida Keys since 1995. Increased nutrient concentrations and increased temperatures may be contributing to the increase in disease incidence. Figure 17-25 shows examples of four diseases that are on the increase.

Reef corals are either spawners, which release sperm and eggs into the water (Figure 17-21a), or brooders, which release larvae. Studies in the Caribbean Sea and on the Great Barrier Reef show that the annual recruitment of new corals by spawning may vary by as much as 25 times. Because recruitment on the Great Barrier Reef is up to 100 times greater, on average, than that of Caribbean Sea, Caribbean reefs are expected to recover much more slowly from natural or human-induced disturbances.

Animals of the Deep-Ocean Floor

We know much less of life in the deep ocean than of life within any of the shallow nearshore environments because of the great difficulty of investigating the deep sea. However, advances in technology are making it possible to observe and sample even the deepest reaches of the ocean. During the next decade, we should learn more about life in the deep ocean than we have learned throughout all of history.

The deep-ocean floor includes the bathyal, abyssal, and hadal zones, as described in Chapter 14. Light penetrates down to only 1000 m (3300 ft), and is absent below this depth. The temperature is low, rarely exceeding 3°C (37°F), and falling as low as −1.8°C (28.8°F) at high latitudes. Pressure exceeds 200 atm on the oceanic ridges; it ranges between 300 and 500 atm on the deep-ocean abyssal plains, and it exceeds 1000 atm in the deepest trenches.

Much of the deep-ocean floor is covered by at least a thin layer of sediment. Sediments range from the muddy clay deposits of the abyssal plains and deep trenches, to carbonate oozes of the oceanic ridges and rises, to coarser sediments deposited by rivers or glaciers on the continental rise (Figure 17-26). Near the crests of the oceanic ridges and rises, and down the slopes of seamounts and oceanic islands, sediment is absent and basaltic ocean crust forms the bottom.

A 1989 study of sediment-dwelling animals in the North Atlantic revealed an unexpectedly large diversity of species.

Figure 17-24   Bleached Coral

Figure 17-26   Deep-Sea Sediment

A sablefish and fine sediment with occasional pebbles were photographed from *Triest* at 1280 m (4200 ft) off San Diego, California.   *(U. S. Navy)*

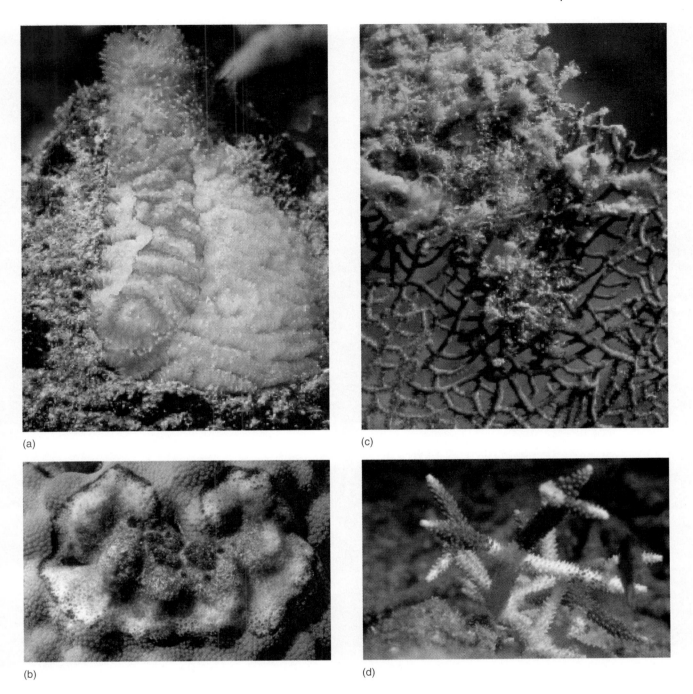

(a)

(b)

(c)

(d)

Figure 17-25   Diseased Coral

Dendrogyra cylindrus suffering from white plague disease. (b) *Montastrea annularis* suffering from black band disease. (c) Fungus *Aspergillus* attacks *Gorgonia flabellum*. (d) *Aacropora cervicornis* suffering from white band disease.   *(Courtesy of John Porter, University of Georgia).*

An area of 21 m² (225 ft²) contained 898 species. Of these, 460 were new to science. The rate that new species were discovered just from analysis of the 200 samples in this study suggested millions of undiscovered deep-sea species. However, deep-sea sediment-dwelling benthos have been insufficiently studied to determine accurately their distribution patterns and diversity.

Deep-Sea Hydrothermal Vent Biocommunities

In 1977, the first active hydrothermal vent field was discovered below 2500 m (8200 ft) in the Galápagos Rift, near the equator in the eastern Pacific Ocean (Figures 17-27 and 17-28). Water temperature immediately around the vents

Figure 17-27 *Alvin* Approaches a Hydrothermal Vent Community Typical of That Found on the East Pacific Rise

In the lower left corner are a sea anemone and three octacorals, which are close relatives of the anemone. They may feed primarily on sulfur-oxidizing bacteria suspended in the water near the vents. Grenadier or rattail fish are common in the deep ocean and are usually the first to arrive at bait placed on the deep-ocean floor. Vestimentiferan tube worms (*Riftia*) and giant clams (*Calypotogena*) do not possess guts: they are nourished by chemosynthetic bacteria that live in their tissues. White brachyuran crabs swarm over the lava pillows, and a black smoker spews hot (350°C) sulfide-rich water from its metallic sulfide chimney.

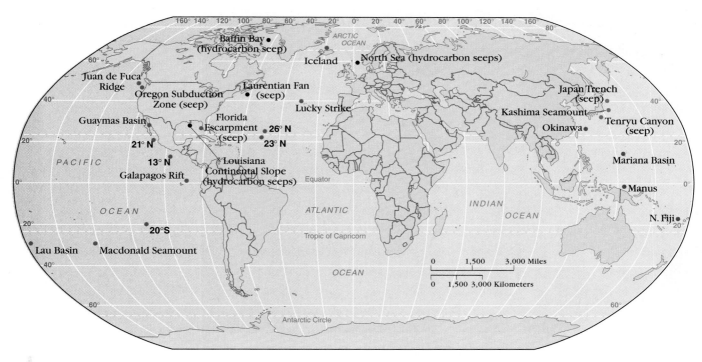

Figure 17-28   **Vents and Seeps Known to Support Communities**
Hydrothermal vents (red), cold seeps (blue), and hydrocarbon seeps (black).

was 8° to 12°C (46° to 54°F), whereas normal bottom-water temperature at these depths is about 2°C (36°F). These vents were found to support the first known **hydrothermal vent communities**, consisting of unusually large organisms for these depths. The more prominent members are tube worms over 1 m long, giant clams up to 25 cm (10 in) in length, large mussels, and two varieties of white crabs.

At 21°N latitude on the East Pacific Rise, south of the tip of Baja California (location, Figure 17-28), tall underwater chimneys were found to belch hot vent water (350°C or 662°F) so rich in metal sulfides that it was black. These chimney vents, first observed in 1979, were composed primarily of sulfides of copper, zinc, and silver and came to be called black smokers. A new species of vent fish was first observed here.

The most important members of these hydrothermal vent communities are the **sulfur-oxidizing bacteria**. Through chemosynthesis, they manufacture organic molecules that feed the community; these bacteria are the base of the food web around the vents. The bacteria use the chemical energy released by sulfur oxidation much as marine algae use the Sun's energy to carry on photosynthesis. Although some animals feed on the bacteria and larger prey, many of them depend primarily on a symbiotic relationship with the bacteria. For instance, the tube worms and giant clams depend entirely on sulfur-oxidizing bacteria that live symbiotically within their tissues.

An interesting problem relating to the study of hydrothermal vent communities is the determination of the temperature range within which the various species can live. It is generally believed that 50°C (122°F) is the highest temperature in which invertebrates can survive. However, in April 1992, a polychaete worm, *Alvinella pompejana*, was observed living at a much higher temperature on the side of a smoker chimney. The worm was coiled around a temperature probe that read 105°C (221°F). A video camera captured the event (Figure 17-29).

In 1981, the Juan de Fuca Ridge community was observed (Figure 17-30a; see location in Figure 17-28). Although vent fauna at this site are less robust than at the Galápagos Rift and on the East Pacific Rise, the metallic sulfide deposits from the vents aroused much interest because they are so near the U.S. coast. In 1982, the Guaymas Basin, in the Gulf of California (location in Figure 17-28), was the site of the first observation of hydrothermal vents beneath a thick layer of sediment. Sediment and sulfide samples recovered were saturated with hydrocarbons, which may have entered the food chain through bacterial uptake (Figure 17-30b). The abundance and diversity of life may exceed that of the rocky bottom vents, such as those on the East Pacific Rise. (There is no sediment on the Rise, where the vents were found.)

A phenomenon similar to that of the Guaymas Basin was reported in 1987 at the Mariana Basin, a small spreading-center system beneath a sediment-filled back-arc basin (Figure 17-30c; location, Figure 17-28). Subsequent exploration has revealed numerous hydrothermal vent communities in the Pacific Ocean, including one in

(a)

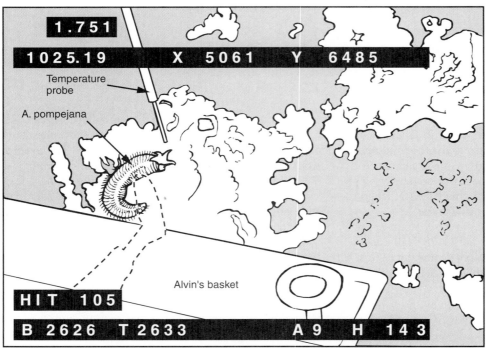

(b)

Figure 17-29 Alvinellid Polychaete Worm in Hot Water

(a) Image from a video taken from *DSRV Alvin* at hydrothermal vent located at 13°N on the East Pacific Rise. The worm *Alvinella pompejana* is living in 105°C-water, as shown on the temperature probe on which the worm rests. (b) Drawing clarifies the position of the worm and the temperature probe. *(Courtesy of Pierre Chevaldonne, Daniel Desbruyeres and James J. Childress.)*

the Southern Hemisphere on the East Pacific Rise at 20°S (see Figure 17-28).

The first active hydrothermal vents with associated communities in the Atlantic Ocean were discovered in 1985 at depths below 3600 m (11,800 ft) near the axis of the Mid-Atlantic Ridge at 23°N and 26°N. The predominant fauna of these vents consists of shrimp that have no eye lens but can detect light emitted by the black smoker chimneys that is invisible to the human eye (Figure 17-31). In 1993, a hydrothermal vent community was discovered on a flat-topped volcano rising to 1525 m

(5000 ft) above the ocean floor—well above the walls of the Mid-Atlantic Ridge rift valley. Called the Lucky Strike vent field, it is about 1000 m (3300 ft) shallower than most other sites. It is the only Mid-Atlantic Ridge site to possess the mussels common at many Pacific Ocean vent sites, and it is the only location where a new species of pink sea urchin was found.

An ancient vent area found on the Galápagos spreading center indicates that vent communities may have relatively short life spans. This inactive vent was identified by an accumulation of dead clams. It appears that the vent

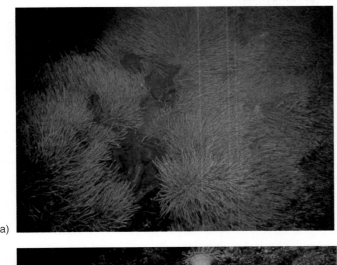

(a)

(b)

(c)

Figure 17-30   Hydrothermal Vent Biocommunities of the Juan de Fuca Ridge, Guaymas Basin, and Mariana Back-Arc Basin

(a) Vestimentiferan tube worms, *Ridgeia phaeophiale*, found at Juan de Fuca vents.   *(Photo by Robert W. Embly, courtesy of William Chadwick, Jr., Oregon State University, Hatfield Marine Science Center.)* (b) At the Guaymas Basin in the Gulf of California, the tube worms, *Riftia pachyptila*, intergrown with a mat of bacteria. The mats are composed of very large bacteria about 150 μm in diameter.   *(Photo by Robert Hessler, courtesy of Scripps Institution of Oceanography, University of California, San Diego.)* (c) This view of the community of the Mariana Back-Arc Basin includes a new genus and species of sea anemone, *Marianactis bythios*; a new family, genus, and species of gastropod, *Alviniconcha hessleri* (the first known snail to contain chemosynthetic bacterial symbionts); and the galatheid crab, *Munidopsis marianica*.   *(Photo courtesy of Robert Hessler.)*

Figure 17-31   Atlantic Ocean Hydrothermal Vent Organisms

Swarm of particulate-feeding shrimp, the predominant animals observed at hydrothermal vents near 23°N and 26°N on the Mid-Atlantic Ridge. This swarm was photographed at 26°N. *(Courtesy of Peter A. Rona, NOAA.)*

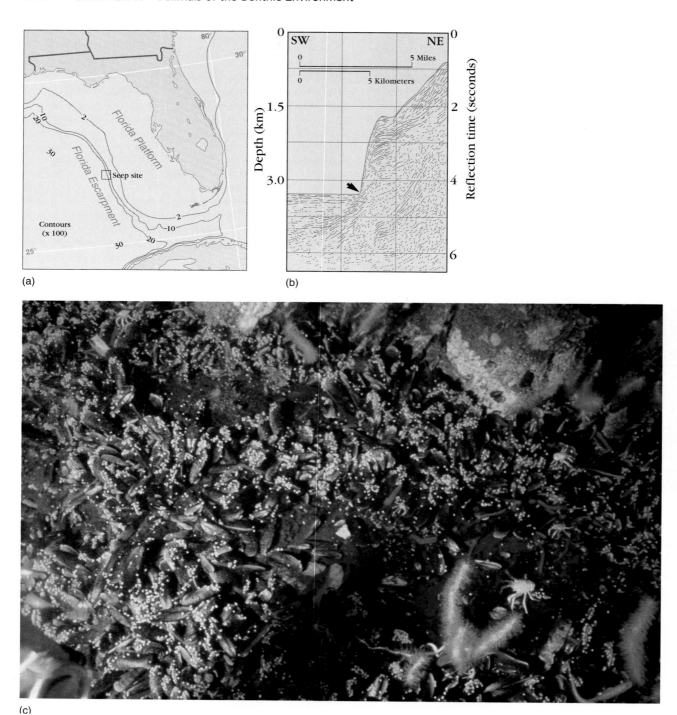

(a)

(b)

(c)

Figure 17-32 Hypersaline Seep Community at Base of Florida Escarpment

(a) Rectangle is the location of the seep and community. (b) Seismic reflection profile of limestone Florida Escarpment and abyssal bedded sediments at its base. Arrow marks location of seep. (c) Florida Escarpment seep community of dense mussel beds covers much of the image. White dots are small gastropods on mussel shells. At lower right are tube worms covered with hydrozoans and galatheid crabs. Fractures in the escarpment limestone are visible along top. *(Courtesy of C. K. Paull, Scripps Institution of Oceanography, University of California, San Diego.)*

became inactive after a lifespan of only a few years. At that point, the hydrogen sulfide that served as the source of energy for the community was no longer available, and the community died. Present evidence indicates that newly formed vents are populated by larval forms that travel hundreds of kilometers along the ridge or rise axis from existing vent sites. Low species diversities have been found to date. There are just over 375 known animal species from this very unstable environment. Most of these species exist only in hydrothermal vent communities.

Low-Temperature Seep Communities

Three additional submarine spring environments also have been found to support communities that carry on chemosynthesis. In 1984, a **hypersaline seep** (46.2‰ salinity) of ambient temperature was studied at the base of the Florida Escarpment in the Gulf of Mexico (Figure 17-32a). Researchers discovered a community similar in many respects to the hydrothermal vent communities. The seeping water appears to flow from joints at the base of the limestone escarpment (Figure 17-32b) and move out across the clay deposits of the abyssal plain at a depth of about 3200 m (10,500 ft). The hydrogen sulfide-rich waters support a number of white bacterial mats that carry on chemosynthesis in a fashion similar to the bacteria at hydrothermal vents. These and other chemosynthetic bacteria may provide most of the support for a diverse community of animals. The community includes sea stars, shrimp, snails, limpets, brittle stars, anemones, tubeworms, crabs, clams, mussels, and some fish (Figure 17-32c).

Also observed in 1984 were dense biological communities associated with oil and gas seeps on the Gulf of Mexico continental slope (Figure 17-33). Trawls at depths of

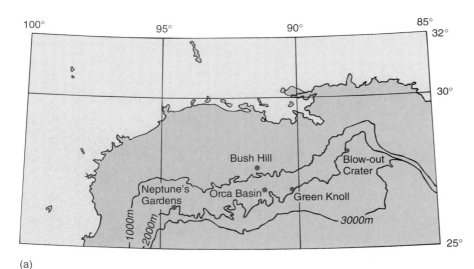

(a)

Figure 17-33 Hydrocarbon Seeps on the Continental Slope of the Gulf of Mexico

(a) locations of known hydrocarbon seeps with associated communities. (b) chemosynthetic mussels and tube worms from the Bush Hill seep. The mussel has demonstrated that it can grow with methane as its sole carbon and energy source. The tube worm is dependent on hydrogen sulfide. *(Courtesy of Charles R. Fisher, Penn State University.)* (c) Alaminos Canyon site (Neptune's Garden) discovered in 1990 is a significant oil seep with a new species of mussel that probably depends on methane and two species of vestimentiferan tube worms that depend on hydrogen sulfide. One of the tube worms may be a new species. *(Courtesy of Charles R. Fisher, Penn State University.)*

(b)

(c)

Low Oxygen Symbiosis in the Santa Barbara Basin

The Santa Barbara Basin is a 600 meter deep basin between the California mainland and the Channel Islands (Figure 17C). A sill depth of 475 meters provides circulation restriction that produces an Oxygen Minimum Zone at its bottom. Surface sediments are very oxygen-depleted ($O_2 < 1 \ \mu M$ in the top 2-3 mm) and sulphide concentrations are greater than $0.1 \ \mu M$ in the top centimeter. These conditions support a mat of the sulphide-oxidizing bacterium *Beggiatoa* but are toxic to all macrofauna except one surface-dwelling snail. The sediments contain meiofauna that include 28 protistans and five metazoans. Euglenozoan flagellates are the most numerous, while foraminifera represent the greatest biovolume. Most of the eukaryotic meiofauna possessed symbiont bacteria.

Scanning (SEM) and transmission (TEM) electron micrographs document the prokaryotic-eukaryotic symbiosis. Figure 17D (part a) shows the symbionts in the cytoplasm of a foraminifer. The b are dividing bacteria and the m is a mitochondrum. Part (b) is a euglenoid veiled in bacterial rods that extend beyond its posterior as a "tail." Another euglenoid is shown in part (c) that is in a condition similar to that of part (b). The ciliate in part (d) has rod-like bacteria attached to its surface. Macronuclei within the ciliate are labeled n. The insert shows the full body of the ciliate with stained bands of the bacteria running across the body. Part (e) shows the rod-shaped bacteria aligned end to end between bands of cuticle covering the surface on a nematode. Rod-shaped bacteria cover the surface of a polychaete worm in part (f).

Although the relationships are still being studied, it is suspected that some of the symbionts will be sulphide-oxidizing forms. Ciliate symbionts that are sulphate reducers have previously been described, and one of the ciliates is known to harbor methanogens (methane producers) at other locations. More instances of bacteria-eukaryotic symbiotic communities will surely be found in the future. Environments that support them such as silled basins, hydrocarbon seeps, hydrothermal vents and Oxygen Minimum Zones are widespread in todays oceans. Human activities are reapidly increasing the number of oxygen-depleted environments, so they will be even more widespread in the future.

Figure 17C Santa Barbara Basin

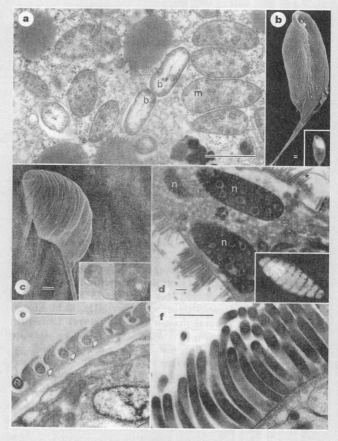

Figure 17D Basterial symbionts all scale bars = 1 μm.

between 600 and 700 m (2000 and 2300 ft) recovered fauna similar to those observed at the hydrothermal vents and the hypersaline seep at the base of the Florida Escarpment. Subsequent investigations identified seeps with associated communities to depths of 2200 m (7300 ft) on the continental slope. Carbon-isotope analysis indicates that **hydrocarbon seep communities** are based on a chemosynthetic productivity that derives its energy from hydrogen sulfide and/or methane. Bacterial oxidation of methane results in the production of calcium carbonate slabs found here and at other hydrocarbon seeps shown in Figure 17-28.

Finally, a third environment of **subduction zone seep communities** was observed from Alvin during 1984. It is located at a subduction zone of the Juan de Fuca Plate at the base of the continental slope off the coast of Oregon (Figure 17-34a). Here, the trench is filled with sediments. At the seaward edge of the slope, clastic sediments are folded into a ridge. At the crest of the ridge, pore water escapes from the 2 million-year-old folded sedimentary rocks into a thin overlying layer of soft sediment. At a depth of 2036 m (6678 ft), the seeps produce water that is only slightly warmer (about 0.3C°, or 0.5F°) than ambient conditions. The vent water contains methane and sulfide that is probably produced by decomposition of organic material in the sedimentary rocks. These gases are sources of energy for bacteria that oxidize them and produce food by chemosynthesis for themselves and the rest of the community, which contains many of the same genera found at other vent and seep sites (Figure 17-34b).

During 1985, similar communities were located in subduction zones of the Japan Trench and Peru-Chile Trench. All the seeps are located on the landward side of the trenches at depths of from 1300 to 5640 m (4265 to 18,500 ft).

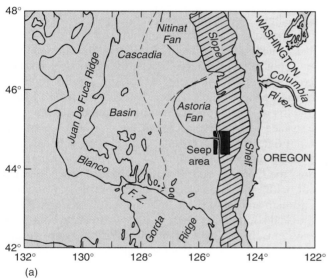

(a)

Figure 17-34 Subduction Zone Seep Communities

(a) A detailed map of the location of the vent communities off the coast of Oregon. They are associated with the subduction zone of the Juan de Fuca Plate. Sediment filling the trench is folded into a ridge with vents at its crest. (b) A male giant white clam (*Calyptogena soyoae*) spawns at a depth below 1,100 m (3,608 ft) in Sagami Bay, Japan. The clams associated with the Japan Tranch host sulfide-oxidizing microorganisms, from which they derive their nutrition. *(Photo courtesy of JAMSTEC.)*

(b)

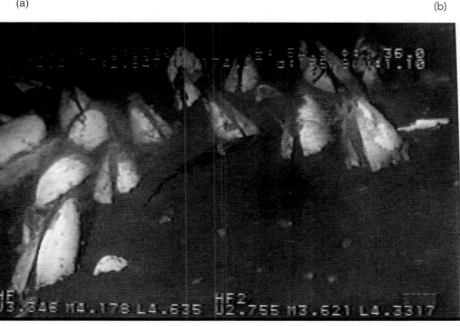

Summary

Over 98 percent of the more than 235,000 species of marine animals live in or on the ocean floor. Most of these are found on the continental shelves, with only about 400 species found in deep-ocean trenches. Temperature, currents, wave energy, and primary productivity are all important controls on species distribution. Because of tidal motions, the intertidal zone can be divided into the high-tide zone (mostly dry), middle-tide zone (equally wet and dry), and the low-tide zone (mostly wet). The intertidal zone is bounded by the supralittoral (covered only by storm waves) and the sublittoral, which extends below the low-tide shoreline. The supralittoral zone along rocky coasts is characterized by the presence of the periwinkle snail, rock louse, and limpet. The barnacles most characteristic of the high-tide zone are the tiny buckshot barnacles. The most conspicuous algae of the high-tide zone are *Fucus* along colder shores and *Pelvetia* in warmer latitudes. These algae become more abundant in the middle-tide zone, and in general, the diversity and abundance of the flora and fauna increase toward the lower intertidal zone.

Larger species of acorn barnacles are found in the middle-tide zone. An assemblage common to rocky middle-tide zones includes goose barnacles, mussels, and sea stars. Joining the sea stars as predators of the barnacles and mussels are carnivorous snails. Tide pools within the middle-tide zone commonly house sea anemones, fishes, and hermit crabs. At the lower limit of the middle-tide zone, sea urchins become numerous. The low-tide zone in temperate latitudes is characterized by a variety of moderately sized red and brown algae.

Moving into more protected segments of the shore, lower levels of wave energy allow deposition of sand and mud. Sand deposits are usually well oxygenated, and mud deposits are anaerobic below a thin surface layer. Compared to the rocky shore, species diversity is reduced on and in beach deposits and is quite restricted in mud deposits, but populations can still be high. Although life is less visible, up to 8000 burrowing clams have been recovered from 1 m^2 (10.8 ft^2) of mud flat. Suspension feeders (filter feeders) characteristic of the rocky shore are still found in the sediment, but there is a great increase in the relative abundance of deposit feeders that ingest sediment and detritus.

Bivalve (two-shelled) mollusks are well suited for sediment-covered shores. Beach hoppers feed on kelp deposited high on the beach by storm waves. Sand crabs filter food from the water with their long, curved antennae while buried in the sand near the shoreline. Echinoderms are represented in the sandy beach by the sand star, which feeds on buried invertebrates, and the heart urchin, which scrapes organic matter from sand grains. As is true for the rocky shore, the diversity of species and abundance of life on the sediment-covered shore increases toward the low-tide shoreline. Zonation is best developed on steep, coarse-sand beaches and is probably missing on mud flats with little or no slope.

Attached to the rocky sublittoral bottom just beyond the shoreline is a band of algae including large kelp. Growing on the large fronds of the kelp are small varieties of algae and fauna; sea hares and sea urchins feed on the kelp and other algae. Spiny lobsters are common to rocky bottoms. Oyster beds found in estuarine environments consist of individuals that attach themselves to the bottom or to the empty shells of previous generations.

Living coral reef can be found off the shores of islands and continents above a depth of 150 m (500 ft) in clear, nutrient-poor tropical waters. Reef-building corals and other mixotrophs are hermatypic, containing symbiotic algae (zooxanthellae) in their tissues. Delicate corals are found at depths of 150 m; more massive corals are found near the surface where wave energy is higher. Above a depth of 20 m (66 ft), the buttress zone, is reinforced with calcium carbonate deposited by algae. Waves cut surge channels across the reef, and debris channels extend down the reef front below the surge channels. Many instances of commensalism and mutualism are found within the coral reef biological community. There are many threats to coral reef ecosystems, including population explosions of crown-of-thorns seastar, pollution from sewage and soil erosion, and bleaching of coral reefs under stress from elevated water temperatures.

Although little is known of the deep-ocean benthos, it is clear that it is much more varied than previously thought. The discovery of hydrothermal vent communities in 1977 on the Galápagos Rift has shown that, at least locally, chemosynthesis is an important means of primary productivity at these vents and at cold-water and hydrocarbon seeps.

Key Terms

Bioerosion	Hydrocarbon seep community	Meiofauna	Subduction zone seep community
Buttress zone	Hydrothermal vent community	Middle-tide zone	
Coral bleaching	Hypersaline seep	Mixotroph	Sulfur-oxidizing bacteria
Hermatypic	Intertidal zone	Sessile	Zooxanthellae
High-tide zone	Low-tide zone	Spray zone	

Questions And Exercises

1. Discuss the general distribution of life in the ocean. Include distribution of benthic biomass, as well as species diversity differences between pelagic and benthic environments and within the benthic environment.

2. Diagram the intertidal zones of the rocky shore, and list characteristic organisms of each zone.

3. Describe the mussel bed, the dominant feature of the middle-tide zone along rocky coasts. Include a discussion of other organisms associated with the mussels.

4. Discuss how sandy and muddy shores differ in sediment stability and oxygenation.

5. Other than predation, discuss the two types of feeding styles characteristic of animals of rocky, sandy, and muddy shores. One of these feeding styles is rather well represented in all these environments; name it and give an example of an organism that uses it in each environment.

6. How does the diversity of species on sediment-covered shores compare with that of rocky shores? Can you think of any reasons why this should be so? If you can, discuss them.

7. Discuss the dominant species of kelp, their epifauna, and animals that feed on kelp in the Pacific coast kelp forest.

8. Describe the environment suited to development of coral reefs.

9. Describe the zones of the reef slope, the characteristic coral types, and the physical factors related to zonation.

10. As one moves from the shoreline to the deep-ocean floor, what changes in the physical environment can be expected?

References

Bernhard, J. M., et al. 2000. The Santa Barbara Basin is a symbiosis oasis. *Nature* 403:6765, 77–80.

Chadwick, W. W., R. W. Embley, and C. G. Fox. 1991. Evidence for volcanic eruption on the Southern Juan de Fuca Ridge between 1981 and 1987. *Nature* 350, 416–18.

Childress, J. J., et al. 1986. A methanotrophic marine molluscan (*Bivalvia, Mytilidae*) symbiosis: Mussels fueled by gas. *Science* 233:4770, 1306–8.

Fisher, C. R. 1990. Chemoautrophic and methanotrophic symbioses in marine invertebrates. *Reviews in Aquatic Sciences,* 2:3 and 4, 399–436.

Fujiwara, Y., et al. 1998. In situ spawning of a deep-sea vesicomyid clam: Evidence for an environmental cue. *Deep-sea Research I* 45, 1881–1889.

George, D., and J. George. 1979. *Marine Life: An Illustrated Encyclopedia of Invertebrates in the Sea.* New York: Wiley-Interscience.

Grassle, F. J., and N. J. Maciolek. 1992. Deep-sea species richness: Regional and local diversity estimates from qantitative bottom samples. *American Naturalist* 139:2, 313–41.

Hallock, P., and W. Schlager. 1986. Nutrient excess and the demise of coral reefs and carbonate platforms. *Palaios* 1:389–98.

Hessler, R. R., and P. F. Lonsdale. 1991. Biogeography of Mariana trough hydrothermal vent communities. *Deep-Sea Research* 38:2, 185–99.

Hughes, T. P., et al. 1999. Patterns of recruitment and abundance of corals along the Great Barrier Reef. *Nature* 397: 6714, 59–63.

Kennicutt, M. C., II, et al. 1985. Vent-type Taxa in a Hydrocarbon Seep Region on the Louisiana Slope. *Nature* 317:6035, 351–53.

Klum, L. D.,et al. 1986. Oregon subduction zone: Venting fauna and carbonates. *Science* 231:4738, 561–66.

Marshall, A. T. 1996. Calcification in hermatypic and ahermatypic corals. *Science* 271:5249, 637–42.

Ricketts, E. F., J. Calvin, and J. Hedgpeth. 1968. *Between Pacific Tides.* Stanford, CA: Stanford University Press.

Rona, P. A., et al. 1986. Black smokers, massive sulphides and vent biota at the Mid-Atlantic Ridge. *Nature* 321:6065, 33–37.

Takami, H., et al. 1998. Microbial diversity in the deep-sea environment. *American Geophysical Union,* 1998 Spring Meeting S171.

Thorne-Miller, B., and J. Catena. 1991. *The Living Ocean: Understanding and Protecting Marine Biodiversity.* Washington, D.C.: Island Press.

Tunnicliffe, V., and C. M. R. Fowler. 1996. Influence of sea-floor spreading on the global hydrothermal vent fauna. *Nature* 379:6565, 531–33.

Walbran, P. D., et al. 1989. Evidence from sediments of long-term *Acanthaster planci* predation on corals of the Great Barrier Reef. *Science* 245:4920, 847–50.

Wikelski, M., and C. Thom. 2000 Marine iguanas shrink to survive El Niño. *Nature* 403:6765,37–38.

Williams, E. H., Jr., C. Goenaga, and V. Vicente. 1987. Mass bleaching on Atlantic coral reefs. *Science* 238:4830, 877–78.

Yonge, C. M. 1963. *The Sea Shore.* New York: Atheneum.

Zenkevitch, L. A., et al. 1971. Quantitative distribution of zoobenthos in the world ocean. *Bulletin der Moskauer Gen der Naturforscher, Abt. Biol.* 76, 27–33.

Suggested Reading

Sea Frontiers

Alper, J. 1990. The Methane Eaters. 36:6, 22–29. An account of the exploration of hydrocarbon seep communities in the Gulf of Mexico.

Coleman, N. 1974. Shell-less molluscs. 20:6, 338–42. Nudibranchs, gastropods without shells, are described.

Fellman, B. 1993. Going with the flow. 39:4, 20–24. Adaptations in algae, plants, and animals that live in high- energy rocky coastal waters.

George, J. D. 1970. The curious bristle-worms. 16:5, 291–300. Describes locomotion, feeding, and reproductive habits of polychaetes.

Gibson, M. E. 1981. The plight of Allopora. 27:4, 211–18. Explains why the author believes the unusual California hydrocoral is headed for the endangered species list.

Humann, P. 1991. Loving the reef to death. 37:2, 14–21. How divers and others degrade the coral reef environment.

McClintock, J. 1994. Out of the oyster. 40:3, 18–23. The history of pearls as jewelry and the development of the cultured pearl industry.

Ruggiero, G. 1985. The giant clam: Friend or foe? 31:1, 4–9. The ecology and behavior of the giant clam (*Tridacna gigas*) as possible danger to divers.

Shinn, E. A. 1981. Time capsules in the sea. 27:6, 364–74. How geologists determine past climatic and environmental conditions by studying coral reefs.

Viola, F. J. 1989. Looking for exotic marine life? Don't leave the dock. 35:6, 336–341. Marine life on and near dock pilings. Good color photos.

Winston, J. E. 1990. Intertidal space wars. 36:1, 47–51. Florida intertidal life.

Scientific American

Caldwell, R. L., and H. Dingle. 1976. Stomatopods. 234:1, 80–89. Presents the ecology of these interesting crustaceans that have appendages specialized for spearing and smashing prey.

Feder, H. A. 1972. Escape responses in marine invertebrates. 227:1, 92–100. The escape responses of limpets, snails, clams, scallops, sea urchins, and sea anemones. Some interesting photographs.

Wicksten, M. K. 1980. Decorator crabs. 242:2, 146–57. Species of spider crabs use materials from their environment to camouflage themselves.

Yonge, C. M. 1975. Giant clams. 232:4, 96–105. The distribution and general ecology of the tridacnid clams.

Oceanography on the Web

Visit the Introductory Oceanography home page for on-line resources for this chapter. There you will find an on-line study guide with review exercises and links to ocean-ography sites to further your exploration of the topics in this chapter. Introductory Oceanography is at http://prenhall.com/thurman (click on the Table of Contents menu and select this chapter).

18 Exploitation and Pollution of Marine Resources

- Laws and Regulations
- Ecosystem Health and Fisheries
- Mariculture
- Petroleum Resources
- Other Pollutants
- Mineral Resources
- Summary

Lobsterman checks and baits traps on a commercial lobster boat in Quahog Bay, Maine (Photo courtesy of Researchers Inc.)

As the human population of Earth grows, there is an ever-increasing demand on ocean resources and ever-increasing stress on marine environments. It is only recently that we have stopped viewing the oceans as being so large as to be unaffected by resource exploitation and pollution. At the beginning of this new millenium, the effects of cumulative stresses on the oceans have become large enough for us to finally acknowledge the finiteness and fragility of the world environment. Because our ability to reap resources from the ocean is so closely related to how much we damage ocean ecosystems by pollution, we have chosen to discuss these two issues together.

Laws and Regulations

One of the most urgent global problems in managing ocean resources is international agreement on who governs the seafloor and the resources contained above and below it. This has been a problem in coastal areas for most of human history, typically solved by bloody conflicts that determined who ruled the high seas. As we develop technology that allows us access to mineral and energy resources beyond the continental shelves, and space on land for waste disposal becomes more limited, the problem grows in magnitude.

In 1609 Hugo Grotius, a Dutch jurist and statesman whose writings helped formulate much of current international law, established the doctrine that the ocean is free to all nations. Although controversy continued over whether nations could control portions of the oceans, Cornelius van Bynkershonk produced an accepted solution in 1702 by extending national domain over the sea to the distance that could be protected by cannons on the shore. The first official dimension of this zone was established in 1772 when the British decided to exercise control over a distance of 1 league (3 naut mi) from the shores of the Commonwealth, thus defining the **territorial sea.**

The Law of the Sea

The rapid development of technology for drilling beneath the ocean prompted the first United Nations Conference on the Law of the Sea, held in Geneva in 1958. This conference produced a treaty stating that the prospecting and mining of minerals on the continental shelf were controlled by the country owning the nearest land. Unfortunately, the seaward limit of the continental shelf were not well defined in the treaty. A second, less productive United Nations Convention on the Law of the Sea (UNCLOS) was held in Geneva in 1960; the third such convention had its first meeting in New York in 1973. Subsequent meetings were held at least once per year through 1982.

In April of 1982, U.N. member nations, by a vote of 130 to 4, with 17 abstentions, adopted a new Law of the Sea (LOS) Convention. Most of the developing nations who could benefit significantly from the provisions of the treaty voted for its adoption. The opposition was led by

the United States, who along with Turkey, Israel, and Venezuela, voted against it. These countries support private companies planning seabed mining operations and believe certain provisions of the treaty would make the mining unprofitable. Among the abstainers were the former Soviet Union, Britain, Belgium, the Netherlands, Italy, and West Germany, all of which have expressed interest in seabed mining.

By the end of 1993, 60 nations had ratified the United Nations Convention on the Law of the Sea and cleared the way for it to become part of international law. Negotiations removed the objections of nations interested in seabed mining. The United States signed on July 19, 1994 and the LOS Convention entered into force on November 16, 1994.

The primary features of the treaty provide for the following:

- Coastal Nations Jurisdiction: A uniform 12-mi (19.2-km) territorial sea and a 200-naut-mi (370-km) **Exclusive Economic Zone (EEZ)** were established. The coastal nation has jurisdiction over mineral resources, fishing, and pollution within the EEZ and beyond it, up to 350 mi (564 km) from shore, if the continental shelf (defined geologically) extends beyond the EEZ (Figure 18-1).

- Ship-Passage: The right of free passage on the high seas is maintained; it is also provided for within territorial seas through straits used for international navigation.

- Deep-Ocean Mineral Resources: Private exploitation of seabed resources may proceed under the regulation of the **International Seabed Authority** (ISA), which includes a mining company, called Enterprise, chartered by the United Nations. Private mining companies are required to fund both their own operations and those of Enterprise, a provision that created most of the contention during the convention.

- Arbitration of Disputes: A U.N. Law of the Sea tribunal will arbitrate any treaty or ownership disputes.

The Law of the Sea places 42 percent of the world ocean area under the jurisdiction of coastal nations. Benefits include making those nations responsible for their marine resources and preventing a free-for-all exploitation of deep-sea bed resources by those with the most advanced technology.

Marine Pollution Control in the United States

The United States has long been concerned with legal protection of the marine environment. The first **Rivers and Harbors Act** of 1899 provided a straightforward prohibition against dumping refuse into the navigable waters of the United States. Most current regulations on waste disposal within the 200-naut-mi (370-km) Exclusive Economic Zone (Figure 18-2) are contained in two major statutes. The **Federal Water Pollution Control Act**

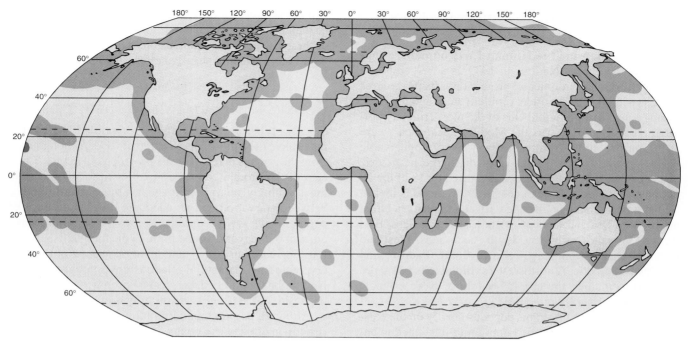

Figure 18-1   The Exclusive Economic Zone (EEZ) of the Law of the Sea Treaty

The Law of the Sea Treaty places 42 percent of the world's oceans under the control of coastal nations.   *(Source: U.S. State Department.)*

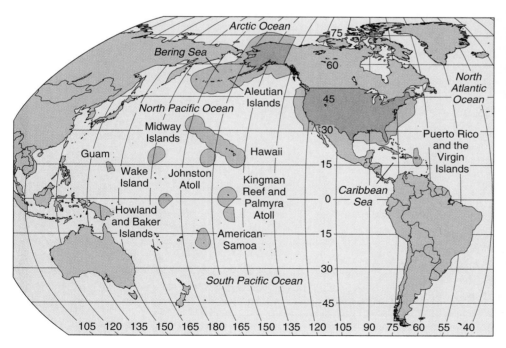

Figure 18-2   Exclusive Economic Zone of the United States

The boundaries of this region were established by the March 1, 1977 declaration of U.S. control of fisheries within 200 naut mi of its coastal waters.   *(Map courtesy of National Marine Fisheries Service.)*

(Clean Water Act) of 1948 and its amendments address point sources of municipal and industrial waste and spills of oil and hazardous materials. The **Marine Protection, Research, and Sanctuaries Act** (Ocean Dumping Act) of 1972 controls dumping of wastes at sea, research, and the establishment of marine sanctuaries.

Although these regulations are quite complex, these laws basically (1) prohibit the dumping of materials known to be harmful, and (2) specify the criteria under which other materials may be dumped. The basic premise of each seems to be that land disposal is preferable to marine disposal. If it cannot be proved to the satisfaction of the regulating agencies that dumping of a material will not adversely affect human health, it will not be allowed in the ocean. In 1991, the **Ocean Dumping Ban Act** banned all sewage and industrial-waste dumping offshore.

International Efforts to Protect the Marine Environment

The **Action Plan for the Human Environment**, established by the United Nations in 1972 as a first step in a global effort at environmental protection, sets procedures for assessing the most critical problem areas and for monitoring sources of pollution. One of the most critically polluted areas identified was the Mediterranean Sea where pollution has come from essentially every conceivable source—domestic sewage, industrial discharge, pesticides, and petroleum. An action plan was ratified by 13 of the 18 nations with Mediterranean coasts. Other regions identified as environmental units and with action plans in effect are the Red Sea and Gulf of Aden, East Asian Seas, East Africa, Kuwait, Caribbean, South Asian Seas, West and Central Africa, Southeast Pacific, South Pacific, and Southwest Atlantic. A northwest Pacific plan is being developed to include the coastal waters of Korea, Japan, China, and Russia (Figure 18-3).

Success of these programs depends greatly on the existence of a regional scientific community interested in environment improvement work and government officials that respect the scientific community and the importance of their work. These preconditions are met in only a few of these regions. Political tensions and frequent wars also have had a dramatic negative effect in some areas.

Ecosystem Health and Fisheries ....

Since well before the beginning of recorded history, humans have used the sea as a source of food. Although food from the sea has in recent years provided only about 5 percent of the protein consumed by the world's population, this seemingly small percentage has meant the difference between starvation and an adequate diet for millions of people. If we are to continue to provide even this small amount of the world's food through the marine fishery, we must increase our understanding of all aspects of the fishery, including the natural relationships among the organisms in the food web, the critical environmental factors for the health of the fishery, and the effects of our exploitation.

Although environmentalists often target pollution and fishermen often blame competition from natural predators, as the greatest threats to fish populations, most investiga-

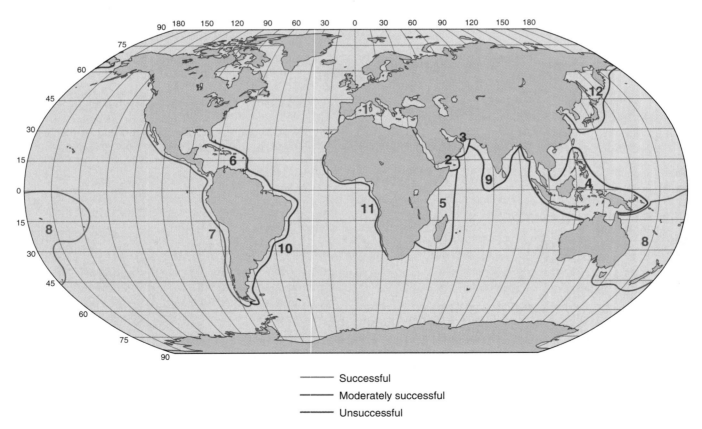

Successful
Moderately successful
Unsuccessful

Figure 18-3   Areas Covered by UNEP's Regional Seas Program

According to the United Nations Environment Program, definition of boundaries is the responsibility of the governments concerned, and those shown here are only illustrative. Environmental units with action plans in effect or in preparation include (1) Mediterranean Sea, (2) Red Sea and Gulf of Aden, (3) Kuwait region, (4) East Asian Seas, (5) East Africa, (6) Caribbean Sea, (7) Southeast Pacific, (8) South Pacific, (9) South Asian Seas, (10) Southwest Atlantic, and (11) West and Central Africa, (12) Northwest Pacific.

tors agree that overfishing has had the greatest negative impact on fish populations. *Overfishing* occurs when adult fish are harvested faster than their natural rate of reproduction such that their populations begin a decline that can only be reversed by curtailing fish harvests. The population soon falls below the size necessary to produce a **maximum sustainable yield**. These already stressed populations are even more susceptible to decline if pollution levels increase in their environments.

Predicting the Effects of Pollution on Marine Organisms

Laboratory **bioassays** are most widely used to determine at what concentration a pollutant begins to negatively affect living organisms. Regulatory agencies such as the Environmental Protection Agency use a bioassay that determines the concentration of a pollutant that causes 50 percent mortality among the test organisms to set allowable concentration limits of pollutants in coastal water effluents. One shortcoming of the bioassay is that it does not predict the long-term, or chronic, effects of sublethal doses.

It is desirable to measure a pollutant's effects at these four **levels of biological response** that require increasingly long periods of observation:

1. Biochemical–cellular: minutes to hours. Exposure to organic pollutants such as hydrocarbons and PCBs produce immediate dysfunction in metabolic processes, causing death.
2. Organism: hours to months. Pollution stress may create changes in physiological processes such as metabolic rate or digestive efficiency that severely reduce the energy available for growth and reproduction.
3. Population dynamics: months to decades. These responses manifest as changes in the abundance and distribution of a species, population structure on an age class basis, growth rates within age classes, fecundity, and incidence of disease.
4. Community dynamics and structure: years to decades. Communities, as assemblages of populations, become less diverse, and more-resistant populations increase in numbers.

Much more research is needed to identify early warning signs of stress at each level of biological response. Because the degree of system complexity and the time needed to measure a response increase exponentially from the biochemical–cellular level to the community level, the predictive difficulties also increase at each level.

Fish Survival

Over the years, huge fluctuations in fish stocks that were not clearly related to fishing pressures convinced many fisheries biologists that detailed study of fisheries ecology was needed. One object of the studies was to produce dependable data on the survival rates of larval and juvenile fishes to determine how many young adult fish were added to the fishery each year. Studies in progress have produced some preliminary findings:

- ***Recruitment***: The addition of young adult fish to the fishery, or **recruitment**, was found to depend on survival at critical stages in the life of a fish. The mortality rate is very high for eggs and larvae. Mortality decreases in juvenile fish and reaches its lowest level in adults. Much less than one percent of larvae survive to become breeding adults.
- ***Larval survival***: After the larva uses up its yolk reserves, it must find appropriate food or die. For instance, anchovy larvae require at least 20 phytoplankton cells/ml of water within 2.5 days, and the minimum cell size must be at least 40 µm in diameter.
- ***Juvenile survival***: Once a fish survives the larval stage, it still must survive a juvenile existence before it can be recruited into the fishery. Juvenile fish may die from predation, disease, parasitism, genetic defects, and the effects of pollution. Availability of adequate food is still important because rapid growth is the best insurance against predation; if juvenile fish grow faster than their adult predators, they become too large to be suitable prey. The food source of juvenile fish may change as they grow, so that they are typically feeding on larger plankton as they increase in size.

Primary Productivity Effect on Fisheries

In most ecosystems, nitrogen is the limiting nutrient that curtails primary productivity. The cycling of nitrogen within and the influx of new nitrogen into the ecosystem thus can be used to estimate the maximum amount of protein that can be removed from the ecosystem by fishing while still maintaining a healthy fishery. The amount of fish removed must be less than or equal to the proportional influx of new nitrogen if the fishery is to increase or maintain its size.

Influx of new nitrogen accounts for less than 10 percent of the primary productivity in oligotrophic waters, but about 50 percent in areas of upwelling, such as coastal Peruvian waters (Figure 18-4). If, on a worldwide basis, we assume an average of about 20 percent of primary production results from the influx of new nitrogen, we can estimate the theoretical mass of fish that can be removed from the oceans per year, or the potential world fishery. Such estimates must be considered tentative, but if 0.1 percent of this primary production sustained by new nitrogen found its way into the world fishery, the **potential world fishery** would be about 100 million MT. The average fishery take for 1994–1997 was close to 86.12 million MT, plus an additional discarded incidental catch of 24.5 million MT, suggesting that the amount of biomass removed by fishing exceeds that produced by the influx of new nutrients (Table 18-1).

The latest assessment of annual primary production for the world ocean is 37,318.6 million MT of carbon. Based

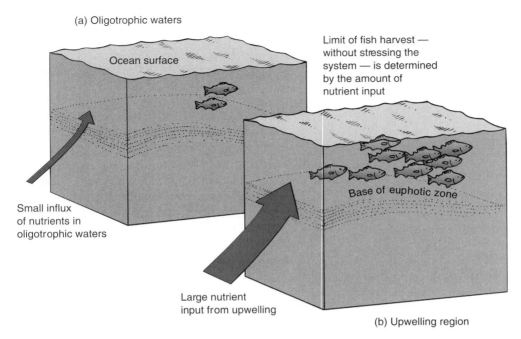

Figure 18-4 Potential Fishery and Influx of Nitrogen

All marine ecosystems are supported by nutrients recycled within the euphotic zone containing algal populations and by nutrients added from surrounding and deeper waters. The biomass of fish taken from the ecosystem cannot exceed that which is supported by the influx of nutrients or ecosystem populations will decline. The arrow in a represents a small amount of nutrient flux into an oligotrophic open-ocean ecosystem. The much larger arrow in b represents a large influx of nutrients into the euphotic zone of an upwelling region. This greater influx of nutrients makes possible a much greater fishery in the upwelling region than in the oligotrophic ecosystem.

on the annual average catch levels given above and the trophic levels for target fish species, we can estimate that the world marine fishery consumes 7.3 percent of annual primary production. However, this percentage varies greatly among ecosystems.

The world fishery is drawn from five ecosystems— (1) open ocean, (2) upwellings, (3) tropical shelves, (4) nontropical shelves, and (5) coastal and coral systems. The open ocean produces 78 percent of the world ocean primary production, but only 1.8 percent of it is used by the cruisers targeted by fisheries—tunas, bonitos, billfishes, and krill (Table 18-2). The most important members of this fishery are the yellowfin and skipjack tuna that cruise the breadth of the desert-like subtropical gyres searching for scattered patches of food.

A large variety of small fishes, bivalves, seaweeds, turtles, shrimps and other crustaceans, and other invertebrates are taken from coastal and coral ecosystems. They consume 8.3 percent of primary production. This value is relatively low because of the high level of primary production, large catches of seaweeds and invertebrates at low trophic levels, and overfishing, which has so reduced the fish population that it cannot make use of the available primary production.

The percentage of the primary production consumed by target fishes in upwellings, tropical shelves, and non-

tropical shelves is much higher than for open ocean and coastal and coral systems. The shelf fisheries consume from 24.2 (tropical) to 35.3 percent (nontropical) of primary production. The upwelling and shelf fisheries provide most of the fish taken commercially. Given the high percentage of primary production they require, it seems unlikely that the take from these areas can increase much without negatively impacting these fish populations.

Other methods of assessing fisheries include using statistics on fish harvesting and scientific sampling of fish populations. However, the wide range of numbers and difficulties in obtaining accurate data resulted in large margins of error. It is much easier and more accurate to sample phytoplankton and zooplankton populations and to determine the productivity of these populations. However, there are still many problems in converting from phytoplankton standing stock to fishery standing stock. The **standing stock** of a population is the mass present at a given time. Methods of estimating the potential fishery require knowledge of size grades, standing stocks, and size-doubling times of each component of the food chain from the phytoplankton up to the fishery species. When such methods are applied to fisheries, the usual conclusion is that the fishermen are taking an amount equal to the capacity of the fishery. Among the major fish stocks for which information is available, about 16 percent are

TABLE 18-1 World Commercial Fish Catch, 1951–1997 (Million Mt).

Year	Fresh Water	Peruvian Anchovy	Other Marine	Total Marine	Total	Year	Fresh Water	Peruvian Anchovy	Other Marine	Total Marine	Total
1953	3.0	NA*	22.9	22.9	25.9	1976	5.9	4.3	59.7	63.9	69.8
1954	3.2	NA	24.4	24.4	27.6	1977	6.1	0.8	62.3	62.8	68.9
1955	3.4	NA	25.5	24.5	28.9	1978	5.8	1.4	63.3	64.6	70.4
1956	3.5	0.1	27.2	27.3	30.8	1979	5.9	1.4	63.8	65.2	71.1
1957	3.9	0.3	27.5	27.8	31.7	1980	6.2	0.8	65.4	65.8	72.0
1958	4.5	0.8	28.0	28.8	33.3	1981	6.6	1.5	66.7	68.2	74.8
1959	5.1	2.0	29.8	31.8	36.9	1982	6.8	1.8	67.9	69.7	76.5
1960	5.6	3.5	31.1	34.6	40.2	1983	7.2	0.1	69.2	69.3	76.5
1961	5.7	5.3	32.6	37.9	43.6	1984	9.7	0.1	73.0	73.1	82.8
1962	5.8	7.1	31.9	39.0	44.8	1985	8.0	0.1	76.8	76.9	84.9
1963	5.9	7.2	33.5	40.7	46.6	1986	11.3	0.1	80.8	80.9	92.3
1964	6.2	9.8	35.9	45.7	51.9	1987	12.2	0.1	80.4	80.5	92.7
1965	7.0	7.7	38.5	46.2	53.2	1988	NA	NA	NA	85.4	NA
1966	7.3	9.6	40.4	50.0	57.3	1989	NA	NA	NA	85.8	NA
1967	7.2	10.5	42.7	53.2	60.4	1990	6.6	NA	NA	82.8	89.4
1968	7.4	11.3	45.2	56.2	63.9	1991	NA	NA	NA	81.7	NA
1969	7.6	9.7	45.4	55.1	62.7	1992	6.3	NA	NA	80.0	86.3
1970	8.4	13.1	46.6	59.7	68.1	1993	NA	NA	NA	NA	NA
1971	9.0	11.2	48.3	59.5	68.5	1994	6.9	NA	NA	85.8	92.7
1972	5.7	4.8	53.7	58.5	64.2	1995	7.4	NA	NA	85.6	93.0
1973	5.8	1.7	55.3	57.0	62.8	1996	7.6	NA	NA	87.1	94.7
1974	5.8	4.0	56.8	60.7	66.5	1997	7.7	NA	NA	86.0	93.7
1975	6.2	3.3	57.0	60.2	66.4						

*NA—Not available
Source: Ocean Yearbook 10, 1993; FAO internet site.

TABLE 18-2 World Fishery Ecosystems, Primary Production (PP) And Percentage of Primary Production Required (PPR) By Fisheries.

Ecosystem	Primary Production 10^6 MT C/yr	PPR (%) 10^6 MT C/yr
Open ocean (3.75)*	29,138.9	524.5 (01.8)
Coastal and coral (18.68)	1,355.5	112.5 (08.3)
Upwellings (20.88)	675.3	169.5 (25.1)
Tropical shelves (21.08)	2,162.8	523.4 (24.2)
Nontropical shelves (35.61)	3,986.1	1,407.1 (35.3)
World ocean	37,318.6	2,737.0 (7.3)

After Pauly and Christensen, 1995.
*The number in parentheses following each ecosystem indicates the percentage of the world fishery that comes from that ecosystem.

overfished, 44 percent are fully exploited. Another 6 percent are depleted, while 3 percent are recovering slowly from near depletion.

Upwelling and Fisheries

Because so much of the productivity of upwelling areas is supported by the influx of nutrients, these areas, which represent about 0.1 percent of the ocean surface area, account for around 21 percent of the world fishery. Duration and rate of upwelling have the most important effects on the fisheries (Figure 18-5). Generally, upwelling of long duration (more than 250 days) produces ideal conditions for fishery development. However, even with shorter times, high levels of productivity may result at moderate rates of upwelling. Regardless of duration, extremely low or high rates of upwelling produce a decreased fishery. Waters that are too stratified or too well mixed—where upwelling rates are too low or high, respectively—tend to develop smaller-sized phytoplankton species that cannot be grazed efficiently by fishes such as sardine and anchovy. This decrease in phytoplankton size is thought to result from the decreased concentration of nutrients that results from either condition. If upwelling rates are low, nutrients are simply not brought into the system fast enough to support large phytoplankton; where rates are too high, the turbulence produced carries nutrients (and phytoplankton) back down and out of the photic zone as rapidly as upflowing waters bring them in. When phytoplankton are small, the food chain is longer and less efficient because herbivorous zooplankton must first feed on the tiny algal cells and the zooplankton are subsequently grazed by fish. This has the effect of reducing the standing stock of a fishery such as anchovy to 10 percent of what it might be as a result of feeding directly on large diatoms.

Where upwelling rates are moderate, the water column is less turbulent, and higher concentrations of nutrients remain in the photic zone to support large species of diatoms that are grazed directly by fish. The elimination of herbivorous zooplankton greatly increases the energy available to the fish and thus the size of the potential fishery. Therefore, it appears that areas of upwelling of long duration and moderate rates provide the optimum environment to produce large anchovy and sardine fisheries.

Incidental Catch

Marine organisms that are caught incidentally by fishers going after commercial species are called *incidental catch*, or sometimes, *bykill*. Close to one-fourth of the catch, on average, is discarded, although for some fisheries, such as shrimp, the incidental catch may be up to eight times larger than the catch that is kept. In most cases, these animals die before they are thrown back overboard. Incidental catch includes birds, turtles, sharks, and dolphins, as well as many species of non-commercial fish.

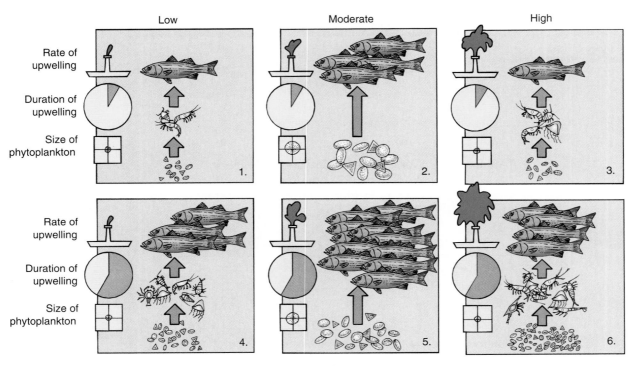

Figure 18-5 **Classification of Upwelling Systems in Terms of Duration and Rate of Upwelling**

Duration determines the absolute amount of production, and rate of upwelling determines the algal cell size. *(After Wyatt, 1980.)*

Some of these species are protected by United States and international law.

The problem of dolphin deaths caused by tuna fishing was graphically presented to the world in March 1988 through video footage taken by biologist Samuel F. La Budde. Just over 2 years later, smarting from a successful boycott of tuna products initiated by the environmental group Earth Island Institute, the United States tuna canning industry declared it would not buy or sell tuna caught by methods that kill or injure dolphins, and a special adjunct was added to the **Marine Mammal Protection Act** in 1992.

The policy of the U.S. tuna canning industry has greatly reduced the practice of using dolphin sightings to locate yellowfin tuna in the eastern Pacific Ocean. For unknown reasons, schools of the tuna are commonly found swimming beneath spotted and spinner dolphins. The dolphins are caught in the fisher's purse seine, a large net drawn around a school and closed when a line is passed through its bottom (Figure 18-6). In 1991, Mexico challenged the U.S. ban on purchasing dolphin-caught tuna to the General Agreement on Tariffs and Trade (GATT) panel and was successful. The panel concluded that the GATT agreement only covers products and not processing standards. It also concluded that the U.S. claim that the ban was the only way it could protect dolphin life and health outside its jurisdiction was not tenable. Instead of taking the decision to a full GATT council, Mexico and the U.S agreed to solve the controversy through compromise during negotiation of the North American Free Trade Agreement (NAFTA).

Another means of netting tuna is by use of **driftnets** or **gill nets**, which are made of monofilament line that is virtually invisible and cannot be detected by dolphin echolocation. Up until 1993, Japan, Korea, and Taiwan had the largest driftnet fleets, deploying as many as 1500 fishing vessels into the North Pacific and setting over 48,000 km (30,000 mi) of nets in one day. Although driftnetting was supposed to be restricted to specific fisheries, some fishers, claiming to be fishing for squid, were illegally taking large quantities of salmon and steelhead trout by this method. Driftnetters also were targeting immature tuna in the South Pacific, and there is considerable concern that this has depleted the South Pacific tuna fishery. In addition to dolphin, tens of thousands of birds and turtles were killed annually in these nets. A prohibition against the importation of fish caught in driftnets was included in the International Convention on Pacific Long Driftnet Fishing, signed in 1989. Driftnets are banned by international law, but pirate fishers still use them in the Pacific and Atlantic.

Figure 18-6 Spotted Dolphin (*Stenella attenuata*)

(Photo by W. High, Courtesy of National Marine Fisheries Service.)

Fisheries Management

Fisheries management historically has been more concerned with maintaining human employment than with preserving fisheries ecosystems. The ecosystems, however, make employment of fishers possible, and so it was surely short-sighted to relegate the welfare of the ecosystem to any position other than number one in fisheries management.

Primitive societies that rely on fish as their main food source have always understood that fish must be permitted to reproduce if the fishery is to be maintained. However, the governments of the industrialized nations have not had the will to regulate their modern, high-technology fishing industries in a manner that would assure sufficient reproduction to maintain the fishery. Fisheries such as the anchovy, cod, flounder, haddock, herring, and sardine are suffering from overfishing.

A major regulatory failure has been the absence of restrictions on the number of fishing vessels. From 1970 to 1989, the size of the world fishing fleet nearly doubled. The number of vessels continues to grow, but more slowly (Table 18-3). Fishing vessels account for 41 percent of ocean-going vessels today—there are more than 3 million of them, including over 1.6 million small nondecked vessels primarily found in Asia and Africa. Many of the larger vessels use nets that can haul in 27,000 kg (60,000 lb) of fish after each lowering. This increase in vessels began with the extension of coastal countries' right to control fishing to a distance of 325 km (200 mi) from their shores.

Governments made loans available to build fishing vessels and then subsidized their nation's fishing fleet. The world fishing fleet spent $92 billion to catch $70 billion worth of fish in 1989. When the catches began to drop off after 1989, the number of vessels did not decrease. The government subsidies just increased.

The effects of inadequate fisheries management are illustrated by the history of fisheries in the northwest Atlantic. Under management by the International Commission for the Northwest Atlantic Fisheries, the fishing capacity of the international fleet increased 500 percent from 1966 to 1976; the total catch, however, rose by only 15 percent. This was a significant decrease in the catch per unit of effort, a good indication that the fishing stocks of the Newfoundland–Grand Banks area were being overexploited. The biologists who recommended the total allowable fishing quotas for the major species within this region complained that enforcement was lacking and that the quotas per country were being bartered in a game of international politics such that they exceeded the total allowable catch set by the commission.

The difficulty of enforcing regulations by the international commission was largely responsible for Canada's unilateral decision to extend its right to control fish stocks for a distance of 200 naut mi (370 km) from its shores beginning January 1, 1977. The United States followed with a similar action on March 1, 1977.

This, however, was not the solution. After essentially all coastal nations assumed regulatory control over their coastal waters, things continued to deteriorate. Overfishing became even more of a problem. The Canadian government has had to shut down the Grand Banks fishery off Newfoundland at a cost of about 40,000 jobs, resulting in government outlays of more than $3 billion in welfare. This cost far exceeds the value of the fishery, which generated no more than $125 million during one of its best years. A similar situation has developed in Canadian and U. S. waters of Georges Bank.

U.S. policy is in part to blame for a continuing fisheries management crisis in the EEZ off Alaska, including the Bering Sea and the Gulf of Alaska. With the government providing guaranteed loans totaling $300 million, the U.S. fishing fleet in this area grew to more than 65 factory trawlers by 1992. These vessels can catch up to 240 tons of fish in a single trawl. It does not require much imagination to see how such a fleet, working 16-hour days, could inflict great damage to a fishery in just a few weeks. The Alaskan king crab fishery, developed in the late 1970s, was depleted by 1985. The pollack fishery appears to be destined for the same fate; during 1979–1980, only 100,000 tons of pollack and other bottom fish were taken; by 1990, the catch totaled 2 million MT, with a value of $2 billion.

Another continuing problem in the Bering Sea is the unregulated waters that are not part of the EEZs of Russia or the U. S. (Figure 18–7). Foreign fleets wait in this area and then make quick forays into Russian and U.S. waters to poach pollack. It is assumed that while waiting, they also must have fished the unregulated waters very heavily.

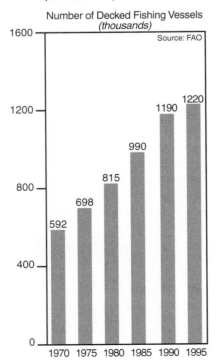

TABLE 18-3 Number of Decked Fishing Vessels (Thousands).

Number of Decked Fishing Vessels (thousands)

Source: FAO

Year	Vessels
1970	592
1975	698
1980	815
1985	990
1990	1190
1995	1220

Source: FAO

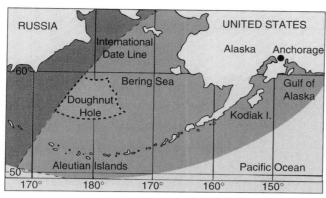

- U.S. EEZ
- Russian EEZ

Figure 18-7   Bering Sea and Gulf of Alaska

Because they contain the most prolific U.S. fishing waters, the Bering Sea and Gulf of Alaska have attracted a large fishery effort. One difficult aspect of regulating fishing activities in the Bering Sea is the existence of the unregulated "doughnut hole" over the deep Aleutian Basin. Foreign fishing fleets congregate in these waters, where fishing is not so good, and make short, illegal forays into the surrounding waters regulated by Russia and the United States.

Russia, the United States, Japan, China, South Korea, and Poland have agreed on a moratorium on fishing in unregulated waters for fear the pollack stock is being depleted.

The Food and Agriculture Organization (FAO) of the United Nations has estimated, on the basis of the yields of traditional fisheries, that the maximum catch that might be expected from the oceans is about 120 million MT. This estimate may be considered conservative and could well be exceeded, particularly with the addition of new species to the significant fisheries. A candidate for addition to the list of major fisheries might be the Antarctic krill. Grenadiers and lantern fishes, found in deeper waters, and the pelagic red crab of the Eastern Pacific and subantarctic regions may also be added in the near future. Another significant addition would be cephalopods such as squid, cuttlefish, and octopus, which yield about 80 percent edible flesh compared to approximately 20 to 50 percent for the most common representative of our present fishery. For further information on the state of the world fishery and a solution to the problems of creating a major fisheries out of "trash fish"—fish taken as a bycatch that have no market value, see Catch of the Day: Chilean Sea Bass on page 276.

When Fishing And Industrial Pollution Collide

The stage was set for the first tragic occurrence of mercury poisoning with the establishment of a chemical factory on Minamata Bay, Japan, in 1938. One of the products of this plant was acetaldehyde, which requires a mercury catalyst in its production. The first ecological changes in Minamata Bay were reported in 1950, human effects were

noted in 1953, and the disease (**Minamata disease**) became epidemic in 1956. It was not until 1968 that the Japanese government declared mercury the cause of the disease, which involves a breakdown of the nervous system. The plant was immediately shut down, but by 1969 more than 100 people were known to suffer from the disease. Almost half of these victims died. A second occurrence of mercury poisoning resulted from pollution by an acetaldehyde factory, shut down in 1965, that was located in Niigata, Japan. Between 1965 and 1970, 47 fishing families contracted the disease.

During the 1960s and 1970s, much attention was given to the problem of mercury contamination in seafood. Studies done on the amount of seafood consumed by various populations of humans have led to the establishment of safe levels of mercury content in fish to be marketed. To establish these levels for a given population, these three variables must be considered:

1. The fish consumption rate of the human population under consideration

2. The mercury concentration in the fish being consumed by that population

3. The minimum ingestion rate of mercury that induces symptoms of disease

If dependable data on these variables are available and a safety factor of 10 is applied to the determination, a maximum concentration level allowable for the contaminant (in this case mercury) can be established with a high degree of confidence that it will protect the health of the human population. Figure 18-8 shows how such data can be used to arrive at a safe level of mercury concentration in fish to be consumed by a population. For the general populations of Japan, Sweden, and the United States, the average individual fish consumption rates are 84, 56, and 17 g/day, respectively. By comparison, the members of the Minamata fishing community averaged 286 g/day during winter and 410 g/day during summer. Based on the minimum level of mercury consumption known to bring on symptoms of poisoning (0.3 mg/day over a 200-day period), determined by Swedish scientists using Japanese data, the three populations shown in Figure 18-8 would begin to show symptoms if they ate fish with the following mercury concentrations: Japan, 4 ppm, Sweden, 6 ppm, and United States, 20 ppm. If a safety factor of 10 is applied, the maximum mercury concentrations in fish that could be safely consumed by these populations would be, for Japan, 0.4 ppm; for Sweden, 0.6 ppm; and for the United States, 2.0 ppm.

Although the U.S. Food and Drug Administration (FDA) initially established an extremely cautious limit of 0.5 mg/kg (ppm), the present limit of 1 mg/kg adequately protects the health of U.S. citizens. Essentially all tuna falls below this concentration, and most swordfish is acceptable. The present limit amounts to the use of a safety factor of 20 instead of 10, so unless a person eats an unusually large amount of tuna or swordfish, one should have little concern about consuming these fishes.

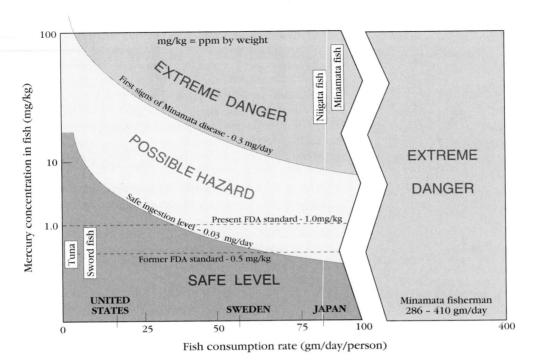

Figure 18-8   Mercury Concentrations in Fish vs. Consumption Rates for Various Populations

Plotted curves show the safe ingestion level of mercury until the first signs of Minamata disease can be expected (0.3 mg/day) and the safe ingestion level of mercury after the safety factor of 10 is included (0.03 mg/day). For the United States, it appears that tuna and swordfish are safe to consume, although some swordfish is banned because its mercury concentrations are above the present 1.0 mg/kg (ppm) FDA limit.

Mariculture

Marine aquaculture, or **mariculture**, has been conducted for years throughout the Far East, making major contributions to the available food supply in that part of the world. Worldwide in 1997, marine (11.14) and freshwater (17.13) aquaculture account for about 28.27 million MT, or 23 percent of the world fishery. Of the aquaculture total, mariculture presently accounts for about 39.4 percent, and is growing rapidly (Table 18-4).

A major force driving the increase in mariculture is the declining harvests of natural fisheries. In 1995, the fishing industry spent $124 billion to catch $70 billion worth of fish; the deficit was made up by government subsidies. In spite of increases in technology, harvests haven't improved much. Although it too is fraught with problems, mariculture may offer a better way to serve consumer demand for marine products.

Organisms chosen for mariculture are popular marine products that command a high price, are easy and inexpensive to grow, and reach marketable size within a year or less. Candidates also should be hardy and resistant to disease and parasites. Severe economic problems can result if the chosen organism is not able to reproduce or cannot be brought to sexual maturity in captivity.

Algae

About 17 percent of mariculture production is seaweed algae, grown and marketed as luxury food. The life cycle of most of these algae is very complicated, and most of the successful operations are conducted in Japan, where spore-producing plants are cultivated in laboratories. When the plants release their spores into the water, local growers come to the laboratories to dip devices into the water, typically poles or nets, that allow the spores to attach themselves to the surfaces. These are submerged in estuarine waters, where the spores grow into mature plants using naturally available nutrients.

Bivalves

Cultivation of oysters and mussels is probably the most successful form of mariculture. Commercially, no bivalves have yet been reared from the larval stage to marketing size in captivity because sufficient phytoplankton to feed them cannot be supplied economically. Instead, hatchery-produced juveniles are allowed to mature in natural environments such as bays, estuaries, and other protected coastal waters where the mollusks feed on phytoplankton carried to them by the tides and other water movements (Figure 18-9). This simple procedure results in yields

TABLE 18-4 World fisheries production and utilization.

	1990	1992	1994	1995	1996	1997
(million tonnes)						
PRODUCTION						
INLAND						
Aquaculture	8.17	9.39	12.11	13.86	15.61	17.13
Capture	6.59	6.25	6.91	7.38	7.55	7.70
Total inland	**14.76**	**15.64**	**19.02**	**21.24**	**23.16**	**24.83**
MARINE						
Aquaculture	4.96	6.13	8.67	10.42	10.78	11.14
Capture	79.29	79.95	85.77	85.62	87.07	86.03
Total marine	**84.25**	**86.08**	**94.44**	**96.04**	**97.85**	**97.17**
Total aquaculture	13.13	15.52	20.77	24.28	26.38	28.27
Total capture	85.88	86.21	92.68	93.00	94.63	93.73
Total world fisheries	**99.01**	**101.73**	**113.46**	**117.28**	**121.01**	**122.00**
UTILIZATION						
Human consumption	70.82	72.43	79.99	86.49	90.62	92.50
Reduction	28.19	29.29	33.47	30.78	30.39	29.50

Source: FAO

Figure 18-9   Bamboo Raft Used in Culturing Oysters

Oysters are suspended from bamboo rafts in nutrient-rich coastal waters off Hirado Island, Japan. This method of culturing off the bottom helps protect oysters from predators such as boring snails and sea stars.   *(Photo from M. Grant Gross.)*

ranging from 10 to more than 1000 MT/acre of edible meat per year.

An interesting symbiosis has developed between the oil industry and mariculture in California. The substructure of offshore oil platforms in the coastal ocean off the Santa Barbara, California are colonized by mussels so rapidly that it was becoming very expensive for the oil industry to remove them. Ecomar Marine Consulting saw the opportunity to offer the oil companies free-cleaning for their platforms by harvesting 3500 pounds of mussels per day, and selling them to the restaurant industry. Ecomar also cultures up to 50,000 scallops and oysters annually on trays suspended from oil platforms.

Crustaceans

Shrimp and prawn are, on a worldwide basis, the most valuable commodity produced by mariculture. The total value of prawn and shrimp mariculture production in 1996 was over $4 billion. Asian nations lead the way, with China responsible for 22 percent of world production. Ecuador, Taiwan, Indonesia, Thailand, and the Philippines also are major producers. Worldwide, shrimp mariculture exceeds 450,000 MT per year, which is about 22 percent of the world supply. The brightest spot in shrimp mariculture in the Western Hemisphere is Ecuador, where a very profitable operation is producing 45,000 MT per year. The success in Ecuador, where the cost of labor and dealing with government regulation is low, contrasts with high costs in the United States, where a similar operation might have to cope with more than 35 regulating agencies and high labor costs. However, it is clear that shrimp can be produced in mariculture operations at much lower cost than they can be provided by fishing. This is being felt by the world shrimp fishing fleet. The largest problem the shrimp mariculture industry faces is the production of larvae, which are susceptible to bacterial and viral infections.

American lobsters (*Homarus americanus*) can easily live through their complete life cycle in captivity, but they have a cannibalistic nature. To be successfully grown, these lobsters would have to be compartmentalized and maintained at an optimum temperature of about 20°C (68°F) for 2 years. Artificial feeding would also be necessary. Encouraging results have been achieved in rearing American lobster in the warm effluent from an electrical generating plant in Bodega Bay, California. Similar projects are being tried in the coastal waters of Maine. The spiny lobster is not so easily reared through its long, complicated larval development, and less effort has been exerted toward developing commercial mariculture with this animal.

Fish

Mariculture with fish is difficult, and only in Japan and Norway have significant marine programs been developed. Salmon, yellowtail, tuna, puffers, and a few other fishes are being raised in Japan on an experimental basis. Norway has pioneered the rearing of Atlantic salmon in suspended pens with success far greater than that achieved throughout the rest of the world in salmon-ranching efforts (Figure 18-10). Norwegian salmon production reached 100,000 MT in 1990, and this technology is now being used in the Pacific Northwest. In Washington, Oregon, and British Columbia, it has replaced the failed salmon ranching, which released young salmon into the ocean in hope of their return. The returns were very poor. This new activity is controversial in the Pacific Northwest because of the introduction of Atlantic salmon into Pacific waters. Although the technology does not involve releasing the salmon from the cages, some have escaped and created a worrisome situation. There is concern about the effect the escaped Atlantic salmon might have on indigenous populations.

Good returns of salmon released by the Alaska Fish and Game hatcheries are being reported. In Canada, the addition of nitrates and phosphates to lakes provided a larger biomass of plankton for young salmon to feed on before going to sea. This rather simple modification, which had no ill effects on the ecology of the lake, increased salmon production in the lake and salmon returns from the open ocean. It is hoped that when the management of commercial operations is refined, we will be hearing success stories from the Pacific Northwest.

Some estuary fish farming involving mullet and milkfish has been practiced successfully in the Far East for centuries. These fishes are hardy and tolerate salinity ranging

Figure 18-10 Salmon Pen Mariculture
(Photo by Mark C. Burnett, courtesy Photo Researchers, Inc.)

from that of fresh water to full seawater. They do not have to be artificially fed. However, because they cannot be spawned and raised to sexual maturity in an artificial environment, the fry must be collected from a natural nursery environment. The costs of operating the farms are relatively low, and yields of about 1 MT/acre are common.

Petroleum Resources

The rocks and sediments in coastal marine environments have been exploited for their mineral and oil deposits for many years. As we continue to deplete our land-based and nearshore resources, we seek to explore in deeper and deeper waters at a pace limited only by how rapidly the needed technology develops and by international law.

Petroleum and Natural Gas

In spite of the occasional glut of oil, the search for petroleum will be the major focus of mineral exploration beneath the oceans in the near future. In 1995, offshore petroleum and natural gas generated over $200 billion in revenues worldwide. In North America, most exploration for new reserves is focused in deepwater Gulf of Mexico and offshore eastern Canada, an area known as iceberg alley. There are immense technological challenges to be met in order to drill and produce oil from rocks beneath water that is over a mile deep, not to mention the high latitude hazards of icebergs ramming the platforms. Figure 18-11 shows some of the platform designs that are being used to extract oil from deep water and hazardous offshore locations.

Offshore oil and gas production and new exploration is taking place in many parts of the world including western Africa, eastern South America, western Australia, the Philippines and Indonesia, the North Sea, the China Sea, the Caspian Sea, the Caribbean, and the Persian Gulf. In spite of low prices, world production is accelerating as many developing countries seek to fund their development with money from oil. This increasing supply assures that prices remain low, and only encourages the world to increase its dependence on this finite resource.

Environmental concerns are one of the few factors that impede exploration and development of new oil and gas fields. For example, plans for further exploration off the shore of California have been abandoned by almost all companies because of pressure from environmental groups and prohibitively costly environmental regulations. Surprisingly, however, offshore platforms have rarely been sources of oil spills. Some exceptions are the 1969 Santa Barbara oil spill off the shore of California; the explosion of the Mexican IXTOC platform in 1979 (Figure 18-12); and the numerous spills created during the Gulf War when the Iraqi army sabotaged Kuwaiti wells in the Persian Gulf. Oil tankers and land discharges are the biggest part of the oil spill problem (Table 18-5).

Petroleum Pollution

Oil contains many different organic compounds, collectively termed *hydrocarbons*. Because hydrocarbons are organic substances that are at least partly biodegradable by microorganisms, they are less damaging than many other pollutants in the open ocean, but spills that reach shorelines can devastate intertidal flora and fauna, including the birds and mammals that live on or near beaches.

A number of natural processes act to destroy oil slicks once they form (Figure 18-13). After oil is spilled onto the ocean surface, the volatile, easily evaporated components of the crude disappear over the first few days into the atmosphere, leaving behind a more viscous oil that may begin to aggregate into tarry lumps that eventually sink to the bottom. The residual oil also coats suspended particles,

Figure 18-11   Offshore oil platforms

(Courtesy of John Stroder, Chevron Petroleum Technology Company)

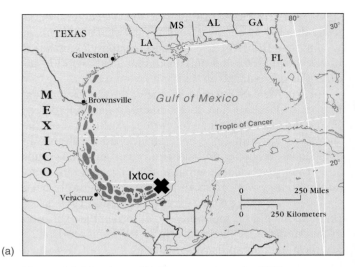

(a)

(b)

Figure 18-12 Ixtoc #1 Blowout, June 3, 1979

(a) Location of the Gulf of Campeche blowout on June 3, 1979.
(b) Firefighting equipment spraying water on the flaming oil directly above the wellhead. Oil can be seen spreading across the ocean surface. Before it was capped on March 24, 1980, it spewed 468,000 MT of oil into the Gulf of Mexico. *(Photo courtesy of NOAA.)*

TABLE 18-5	Sources of Oil to the Oceans (Data from National Academy of Sciences).			
Source	Milion Metric Tons Annually (1973)	Percent of total (1973)	Million Metric Tons Annually (1981)	Percent of total (1981)
Tanker and transportation operations	1.90	31.3	1.10	34.3
Municipal and industrial wastewater discharges and runoff	2.50	41.1	1.00	31.3
Tanker accidents	0.20	3.2	0.40	2.5
Offshore oil production	0.08	1.3	0.05	1.6
Natural seeps and erosion	0.60	9.9	0.25	7.8
Atmospheric fallout	0.60	9.9	0.30	9.4
Refinery wastewater	0.20	3.3	0.10	3.1
Total	**6.08**		**3.20**	

which eventually settle to the seafloor. Wave and wind action disperse the slick and mix the oil into the water to make a frothy emulsion. Bacteria and photo-oxidation act to break down the hydrocarbon molecules into compounds that dissolve in water. Oil cleanup efforts often focus on helping or enhancing these natural processes.

As long as the oil slick stays at sea, damage to marine ecosystems is minimal and short-lived. The *Argo Merchant* grounding on Fishing Rip Shoals off Nantucket on December 15, 1976, provides a well-studied example (Figure 18-14). Winds were such that no oil came ashore, and the surface slick moved eastward out to sea. It was gone by mid-January. Damage to plankton and planktonic larvae was not insignificant, however. Scientists from Woods Hole Oceanographic Institution, the National Marine Fisheries Service, and local institutions studied the embryon-

ic development of fish eggs sampled from plankton tows shortly after the spill. Of the 49 pollock eggs recovered, 94 percent were oil-fouled, as were 60 percent of the cod eggs. Twenty percent of the cod eggs and 46 percent of the pollock eggs were dead or dying (Figure 18-15). Unfortunately, there were no comparative data on the natural mortality rate of these fish eggs, but a sample of cod eggs spawned in a laboratory showed only a 4-percent mortality at the same development stage. Because pollock females spawn about 225,000 eggs and cod spawn about 1 million eggs each season and their populations range from New Jersey to Greenland, it is unlikely that this single spill had a major effect on those fisheries.

How long does it take for a shore to recover from an oil spill? The oldest well-studied spill in the U. S. occurred when an oil barge, the *Florida*, came ashore and ruptured

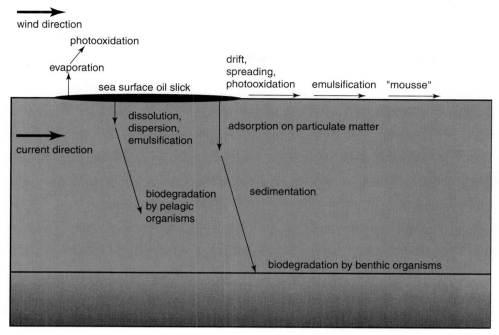

Figure 18-13   Processes Acting on Oil Spills

wind direction

photooxidation

evaporation

sea surface oil slick

drift, spreading, photooxidation

emulsification

"mousse"

current direction

dissolution, dispersion, emulsification

adsorption on particulate matter

biodegradation by pelagic organisms

sedimentation

biodegradation by benthic organisms

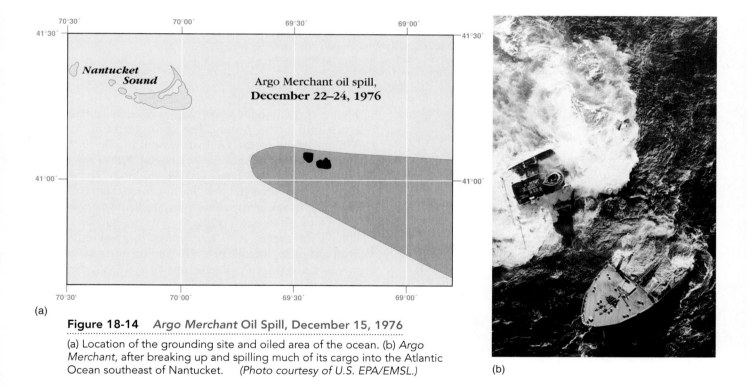

(a)

Nantucket Sound

Argo Merchant oil spill, **December 22–24, 1976**

Figure 18-14   *Argo Merchant* Oil Spill, December 15, 1976

(a) Location of the grounding site and oiled area of the ocean. (b) *Argo Merchant*, after breaking up and spilling much of its cargo into the Atlantic Ocean southeast of Nantucket.   *(Photo courtesy of U.S. EPA/EMSL.)*

(b)

on September 16, 1969, near West Falmouth Harbor in Buzzards Bay, Massachusetts. Currents carried fuel oil north into Wild Harbor, where the most severe damage occurred (Figure 18-16). The initial kill was almost total for intertidal and subtidal animals in the most severely oiled area, but was followed by rapid increases during the first year in the population of polychaete worms, resistant to the oil. Species diversity did not increase appreciably until well into the third year, and marsh grasses and animals had reentered the area by the fifth year. No damage was visible after 10 years. After 20 years, there was virtually no oil in the subtidal sediments, and intertidal marsh sediments were more than 99 percent free of *Florida* oil. At the most heavily oiled site in Wild Harbor, oil was still present at a depth of 15 cm (6 in). In this region, there was enough oil to kill animals that burrowed into the sediments. However, the history at Wild Harbor since 1969 shows clearly that even in quiet, protected marsh environments with long

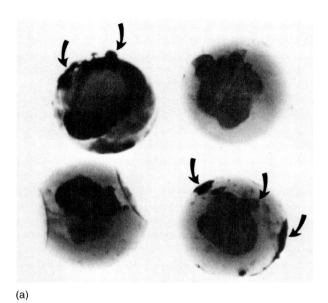

(a)

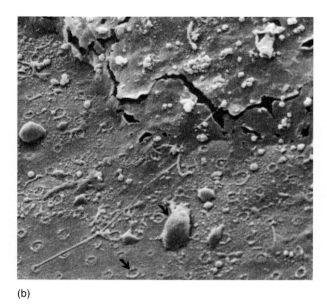

(b)

Figure 18-15 Pollock Eggs Affected by the *Argo Merchant* Spill

(a) These eggs, in the tail-bud and tail-free embryo stages, were taken from the edge of the oil slick. The outer membranes of the eggs at the upper left and lower right are contaminated with a tarlike oil; arrows point to some of the oil masses. The uncontaminated egg at the upper right has a malformed embryo; that at the lower left is collapsed and also has an abnormal embryo. The actual size of pollock eggs is about 1 mm (0.04 in); of cod eggs, about 1.5 mm (0.06 in). (b) A portion of the surface of an oil-contaminated egg. The upper arrow points to one of many oil droplets; the lower arrow to one of the membrane pores. Scanning electron microscope: about 5000 μ.   *(Courtesy of A. Crosby Longwell.)*

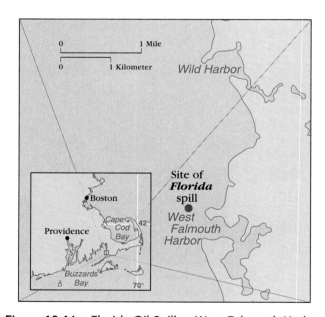

Figure 18-16   Florida Oil Spill at West Falmouth Harbor

residence times, recovery from an oil spill can occur much more rapidly than some researchers thought possible.

The first major oil spill resulting from the development of the North Slope of Alaska petroleum reserves occurred March 23, 1989 when the California-bound *Exxon Valdez* went aground on rocks 40 km (25 mi) out of Valdez, Alaska (Figure 18-17). About 32,000 MT of oil were spilled, the largest spill in U.S. history. The spill spread from Prince William Sound to the Gulf of Alaska. In all, 1775 km (1102 mi) of shoreline were damaged with a recorded death toll of 994 sea otters and 34,434 birds. The actual kill could have been 10 times that amount by some official estimates. It was predicted that the Alaskan waters would have a long, slow recovery, but the fisheries bounced back with record takes in 1990. Encouraged by good results on small test plots, Exxon spent $10 million to spread phosphorus- and nitrogen-rich fertilizers on Alaskan shorelines to boost the development of indigenous oil-eating bacteria. It resulted in a cleanup rate that achieved 7 years of predicted progress within 2 to 3 years.

In January 1991, during the Persian Gulf War, Iraqi forces began purposefully releasing oil into the Persian Gulf from the Kuwait Sea Island terminal. Later they opened wellheads on land. More than 500,000 MT of oil are believed to have been introduced into the Gulf—the largest single oil pollution event known. More than 770 km (478 mi) of Saudi coastline was oiled (Figure 18-18). Contamination was severe along about 400 km (248 mi) of Saudi Arabian coastline. Satellite surveys show that the original slick deteriorated rapidly and was not detectable by May 1991. Normal chronic pollution of the northern Gulf from production and transportation of oil is estimated at about 270,000 MT per year. Pollution was actually at a lower level during the 1991 surveys than during a 1983-1986 survey. The reduced pollution level surely resulted

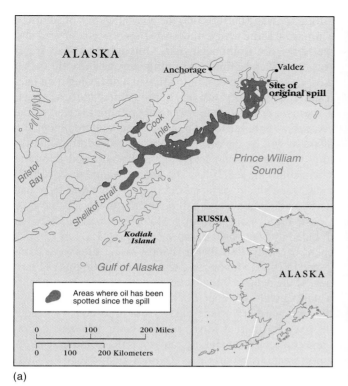

(a)

(b)

Figure 18-17 *Exxon Valdez* Oil Spill

(a) Location and extent of contamination resulting from the spill of heavy crude oil from the supertanker *Exxon Valdez*. (b) Oil from the spill damaged shorelines and killed sea otters and birds. *(Photo courtesy of U.S. Coast Guard.)*

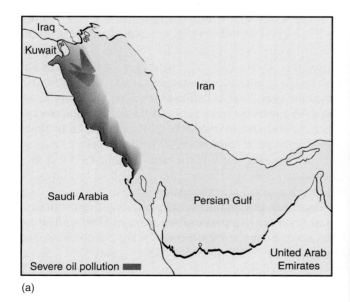

(a)

(b)

Figure 18-18 Persian Gulf War Oil Pollution, 1991

(a) The approximately 500,000 MT of oil spilled into the Persian Gulf as a result of Iraqi sabotage of Kuwait production facilities was confined primarily to the northwest coasts of the Gulf by currents and prevailing southeasterly winds. (b) Saudi Arabian government official examines a pool of oil collected on the beach from a Kuwaiti oil spill. *(Bob Jordan/AP/Wide World Photos)*

from the severe curtailment in the production and transportation of oil because of the war. Subtidal sediments were not severely polluted, and marine life there survived and is recovering well. However, surveys in May and December 1991 showed little improvement in the condition of the polluted intertidal zone. The once-common ghost crabs (*Ocypode*) and bryozoan (*Cleistostoma*) returned by January 1992. In the lower intertidal zone, which is regularly covered during high tides, mats dominated by cyanobacteria have developed over the oiled surfaces. These mats, which contain other photosynthetic organisms, such as diatoms, as well as oil-degrading bacteria, represented the only life in this environment. Never observed elsewhere in association with oil spills, these mats appear to be unique to the extreme climatic conditions of the Persian Gulf.

Other Pollutants

Plastics

During the last 30 years, the use of plastics has increased at a tremendous rate. The disposal of this very durable substance has already strained the capacity of our land-based disposal systems. Such waste is also an increasingly abundant component of oceanic flotsam. Small pellets used in the production of plastic products are transported in bulk aboard commercial vessels. Found throughout the oceans, the pellets probably find their way into the oceans as a result of spillage at loading terminals. In coastal waters, plastic products used in fishing and thrown overboard by recreational and commercial vessels are common. They also find their way into the open ocean waters from careless dumping by commercial vessels.

The best documentation of negative effects of plastics in the ocean on marine organisms is found in the strangulation of seals and birds caught in plastic netting and packing straps (Figure 18-19). Marine turtles are known to mistake plastic bags for jellyfish or other transparent plankton on which they typically feed.

All plastics that enter the ocean may eventually be removed by shorelines that filter out the particles. Beaches throughout the world are probably increasing their plastic pellet content as a result of the filtering process. A 1972 survey compared to a 1987 survey found that concentrations of plastic pellets had doubled on many beaches. Bermuda beaches show up to 10,000 pellets/m^2 and Menemsha Harbor on Martha's Vineyard yielded 16,000/m^2. Thin plastic film products are eventually broken into smaller particles by photochemical degradation, and other particles may sink as they become denser because of photochemical degradation and encrustation by epifauna such as bryozoans and hydroids. It is now illegal to dump plastic materials in U.S. coastal or any open ocean waters.

Sewage

More than 500,000 MT per year of sewage sludge containing a mixture of human waste, oil, zinc, copper, lead, silver, mercury, PCBs, and pesticides have been dumped through sewage outfalls into the coastal waters of southern California. More than 8 million MT have been dumped into the New York Bight each year. Although the Clean Water Act of 1972 prohibited dumping sewage into the ocean after 1981, the high cost of treating and disposing it on land resulted in extended waivers being granted to these areas.

It now appears that the political pressure to clean up the coastal water is sufficient to assure that land treatment and disposal will be the primary disposal procedure of the future. For the present, Los Angeles has stopped pumping sewage into the sea. If it is allowed to resume this practice, the sewage sludge pumped through the lines will have had its toxic chemicals and pathogens removed. Off the East Coast, more than 8 million MT of sewage sludge was dumped by barge over dump sites totaling 150 km^2 (58 m^2) each year until 1986. These shallow-water sites were abandoned in 1986 in favor of a deep-water site 106 mi (171 km) out. At the deep-water site, beyond the shelf break, there is usually a well-developed density gradient separating low-density, warmer surface water from high-density, colder deep water. Internal waves moving along this density gradient can retard the sinking rate of particles and may allow a horizontal transport rate 100 times greater than the sinking rate. Clearly, bottom conditions were not visibly altered at the disposal site in deeper water. However, fishers began reporting adverse effects on their fishery as soon as the dumping began. A 1990 survey detected some sewage sludge tracers in two surface deposit-feeding species, a sea urchin and a sea cucumber. Some species did become rarer as the ecology of the dump site gradually degraded. The deep-site dumping was stopped in 1993, and the rarer species have since shown an increase in numbers.

Some 48 different communities have been depositing their sewage into Boston Harbor, making it one of the most polluted water bodies in the nation. A system is now in place to stop this dumping by carrying sewage through a 15.3 km (9.5 mi) long tunnel to Massachusetts Bay (Figure 18-20). Since the year 2000, all effluent receives secondary treatment. To pay the $4 billion cost, the average annual sewage bill for a Boston-area household will be about $1200. There is presently much controversy about the project, caused by fear of degrading the environment in Cape Cod Bay and Stellwagen Bank, an important whale habitat recently designated a national marine sanctuary.

Beach advisories and closings resulting from high bacterial levels reached record numbers in 1998 indicating that sewage contamination is still a huge problem. As an example, El Niño produced heavy rains that caused sewage overflows and contaminated runoff requiring California to issue swimming advisories and closings that jumped from 1,141 in 1997 to 3,273 in 1998.

Radioactive Waste

Since 1944, artificial **radionuclides** (radioactive atoms) have been reaching the oceans as fallout from the production and testing of nuclear weapons. This pollution has been about four times greater in the Northern Hemisphere than in

Figure 18-19   Elephant Seal with Plastic Packing Strap Tight Around Its Neck

(a) This animal was found on a beach of San Clemente Island off the Southern California coast. It was saved from strangulation by the removal of the plastic packing strap by scientists from the National Marine Fisheries Service.   *(Photo by Wayne Perryman, NMFS.)* (b) Herring gull (*Larus argentatus*) with a plastic 6-pack ring around its neck near Cape May, New Jersey.   *(© Joe McDonald.)*

the Southern Hemisphere. Because of the nature of the fall-out process, no high concentration of any radionuclide has resulted. Table 18-6 shows the amount of radiation in thousands of curies that has reached the oceans as fallout. (A *curie* is a quantity of any radioactive nuclide in which there are exactly 3.7×10^{10} disintegrations per second.)

The second largest source of radionuclides is nuclear fuel. Nuclear power generating plants release little radiation to the oceans, but nuclear fuel reprocessing plants contribute significant quantities. Such plants are found at Sellafield and Dounreay in Great Britain and Cape de la Hague, France. The Irish Sea, which has been receiving reprocessing discharges since the early 1950s, is now the most radioactive area in the world ocean.

Another significant source of **radioactive waste** is an international dumping operation carried out over the past 30 years. Since 1967, the only significant dumping has occurred in the northeast Atlantic under coordination by the

Nuclear Energy Agency of the **Organization for Economic Cooperation and Development** (Table 18-6).

For nearly five decades, high-level nuclear waste has been accumulating from the production of nuclear weapons and commercial power generation. The United States alone has more than 75 million gal of waste from weapons production and 12,000 MT of spent reactor fuel. This material will be radioactive for more than a million years, but safe disposal would likely require a much shorter time of highly secure confinement. The wastes are composed of over 50 isotopes, each with a different chemistry and decay half-life. Scientists think that a reasonable criteria would be to allow no more release of radiation into the atmosphere than that emitted by the natural uranium ore used to generate these radioactive materials (Figure 18-21).

Given that the waste must be kept out of the way of human activities, safe from exposure by natural erosion, and away from seismically active regions, the centers of

TABLE 18-6 Artificial Radionuclides in the Oceans and Their Sources.					
	Plutonium-239, -240 (kCi)	Cesium-137 (kCi)	Strontium-90 (kCi)	Carbon-14 (kCi)	Tritium (kCi)
Total worldwide fallout by early 1970s	320	16,700	11,500	6,000	3,000,000
North Atlantic Ocean—early 1970s	63	3,300	2,300		650,000
Windscale, 1957–1978 discharge	14	830	130		370
	Total α-emitters	Total β/γ-emitters (other than tritium)			
NEA* dumpsite (1967–1979)	8.3	258			262

*Nuclear Energy Agency

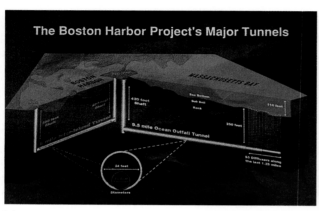

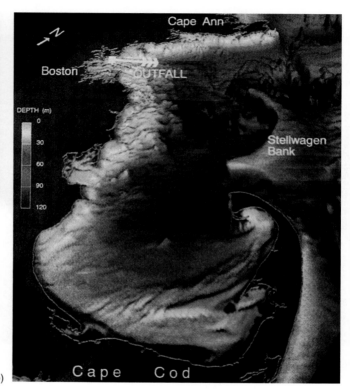

(a)

Figure 18-20   Boston Harbor Sewage Project

(a) Main components of the sewage treatment and disposal system. Sewage will be treated at the Deer Island facility. Effluent will be pumped into the ocean through diffuser lines rising from the main tunnel 76 meters (250 feet) beneath the ocean floor. *(Courtesy of Massachusetts Water Resources Authority.)* (b) Computer-generated bathymetric map of coastal ocean in Boston–Cape Cod area. There is concern that the discharge of sewage effluent in the outfall area will degrade the environment of Stellwagen Bank and Cape Cod Bay to the south. (Vertical exaggeration ×100.)   *(Courtesy of U.S. Geological Survey, Woods Hole, Massachusetts.)*

(b)

oceanic lithospheric plates might be ideal disposal sites. This possibility was studied by the U.S. and several other countries in the late 1970s and early 1980s. Such regions are beneath at least 5000 m (16,400 ft) of water and far from human activities. The lithospheric plates are covered with up to 1000 m (3280 ft) of fine sediment that has been accumulating for over 100 million years in some regions. This sediment accumulation indicates long periods of stability as the plates move across Earth's surface. The high clay content of these sediments also makes them an ideal medium in which to bury nuclear waste because clays have a high adsorption capacity for these compounds should they escape from the canisters (projected to fail within 1000 years). This would slow diffusion of radioac-

tivity away from the burial site. A number of midplate, midgyre sites (MPGs) away from ocean currents have been identified in the North Atlantic and North Pacific oceans (Figure 18-22a). The initial plan would be to place the canisters under 30 to 100 m (98 to 328 ft) of sediment at intervals of at least 100 m. (Figure 18-22b).

Models indicate that heat released from the waste would be transferred almost entirely by conduction, so upward convection of radiation carried by sediment pore water might be negligible. This is a crucial consideration. Although these sites appear promising, questions remain, such as how well the burial holes would reseal themselves and how escaped radiation would be transported through the water column once it reaches the sediment surface.

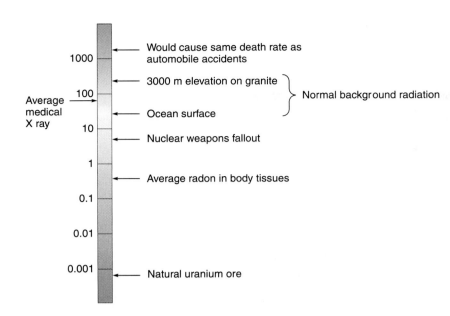

Figure 18-21   Radiation Doses from Various Sources in Millirems per Year

The Environmental Protection Agency is responsible for setting the minimum performance requirements (levels of radiation emitted from high-level disposal sites). It is believed that they will be no higher than that of natural uranium ore.

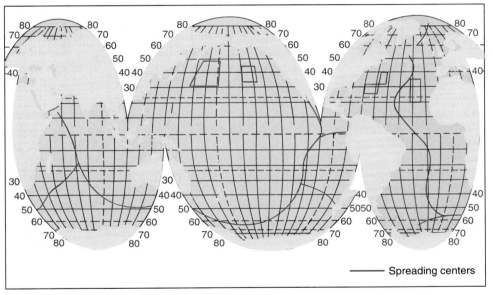

(a)

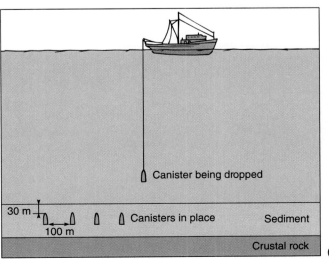

(b)

Figure 18-22   MPG Regions

(a) Mid-plate, mid-gyre (MPG) regions in northern oceans where the environment can be expected to be stable for millions of years. (b) Plan of canister emplacement shows them buried at least 30 m (98 ft) beneath ocean floor spaced 100 m (328 ft) apart. Because of the softness of the sediment, it may be possible to implant the canisters by simply dropping them from an appropriate distance above the sediment surface. (Figure not to scale.)

Peruvian Anchoveta Fishery

The magnificent anchoveta fishery at the north end of the Peru Current shows the importance of placing the ecology of a fishery above all other considerations in fisheries management. The unfortunate expansion of this fishery was based in large part on the consideration of the jobs that were provided. In spite of this intent, many of those jobs are gone forever.

Southeasterly winds blow along the coast of Peru and, because of Ekman transport, drive the surface waters away from the coast. The surface water is replaced by an upwelling of cold water from depths of 200 to 300 m (660 to 990 ft). The Peru Current (Figure 18A), the northward-flowing eastern boundary of the South Pacific current gyre, seldom exceeds 0.5 km/h (0.3 mi/h), and extends to a depth of 700 m (2300 ft) near its seaward edge, some 200 km (120 mi) from the coast. The Peru Countercurrent knifes its way between the coastal and oceanic regions of the Peru Current near the equator. This southbound surface flow of warm equatorial water seldom reaches more than 2° or 3° south of the equator, except during the summer when it reaches its maximum strength. Beneath the surface current is the southward-flowing Peru Undercurrent.

Biologically, the significance of the cold water and nutrients provided by upwelling is great. During the years 1964–1970, 20 percent of the world fishery was represented by anchoveta taken off the coast of Peru. The food web starts with diatoms and other phytoplankton that support a vast zooplankton population of copepods, arrow worms, fish larvae, and other small animals. Energy is passed through a series of predators, but most stops with the anchoveta (Figure 18B), which once attained a biomass of 20 million MT. Natural predators that feed on the anchoveta are marine birds—boobies, cormorants, and pelicans—and large fishes and squid.

Periodically during the summer, the southeasterly trade winds weaken. The northward-flowing current and up-welling slow down, and the Peru Countercurrent may move as far as 1000 km (621 mi) south of the equator, laying down a warm, low-salinity layer over the coastal water. As discussed in Chapter 9, this event is called El Niño. The increase in water temperature and decrease in nutrients brought about by the slowdown or complete halt of upwelling cause a severe decrease in the anchoveta population, the remnants of which seek deeper, cooler water. Marine birds that feed primarily on the anchoveta have their populations cut severely. In 1972, a severe El Niño occurred. The total biomass of anchoveta decreased from an estimated 20 million MT in 1971 to an estimated 2 million MT in 1973.

Before blaming this decrease solely on the natural event of El Niño, with which the anchoveta have contended for thousands of years, another factor must be evaluated. The Peruvian anchoveta fishery began to develop into a major industry in 1957. In 1960, the Instituto del Mar del Peru was established with the aid of the United Nations to study the fishery and recommend a management program to the government. It was hoped that early study would prevent the fishery from going the way of the Japan herring and California sardine fisheries that had collapsed as a result of overexploitation.

Biologists estimated that 10 million MT was the maximum sustainable yield of the anchoveta. Figure 18C shows that this limit was exceeded in 1968, 1970, and 1971. The 1970 catch of 12.3 million MT would probably reach 14 million MT if processing losses and spoilage were included. Dr. Paul Smith, who has studied the anchoveta fishery problem for the National Marine Fisheries Service in the Southwest Fisheries Center at La Jolla, California, believes, as do other biologists familiar with the problem, that the yearly anchoveta catches may have been much larger than the official figures indicate. Many of the catches contained large numbers of pelladilla, young an-

Figure 18A   Cross section of Peru Current

The Peru Countercurrent is a warm-water mass that spreads over the surface and separates the Peru Current into coastal and oceanic regions.

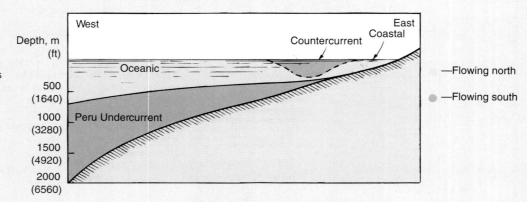

Figure 18B   Anchovies

These small fish represented one of the world's major fisheries until the collapse of 1972. An active fish meal industry based on anchoveta made Peru, for a while, the world's leading fishing nation.   *(Photo courtesy of National Marine Fisheries Service.)*

choveta. Under reporting would have resulted if buyers reduced the total tonnage of a catch by a percentage determined by the proportion of high water-content and low oil-content pelladilla in the catch. For example, a 100-MT catch may have been recorded as a 70-MT catch because it contained a large number of pelladilla.

Dr. Smith feels the crisis may have resulted from a combination of the effects of El Niño and the industry's lack of understanding of the anchoveta fishery, which is reflected in the monthly catch allotments it set. In January 1972, the allotment was set at 1.2 million MT. This figure was reached easily by the industry with about one-half the expected effort. Because the fishing vessels were able to meet their allotment so easily without going far from shore, the fishery was thought to have developed handsomely, and the February allotment was raised to 1.8 million MT. Much more effort was required to meet this allotment. In March the fishery collapsed. What had happened?

Apparently, El Niño, or the Peru Countercurrent, had come south in January but remained offshore. The anchovy were concentrated in the nearshore portion of the Peru Current that had not been affected (Figure 18D). In January the nets were filled easily near shore, and it was natural to think the high concentration of anchovy extended over the normal fishery area as well. This one event cannot explain the collapse of the fishery, but it is a symptom common to fisheries that are poorly understood. It may be that even with upwelling, ammonia (NH_3) from decomposition of anchoveta excrement is required to support enough phytoplankton to allow the anchoveta population to expand. Because anchoveta populations are now small and scattered, they may swim through great expanses of water without adding much nutrient enrichment. When the population was larger, the whole region received that enrichment. It may require a long time, with a gradual increase in the population, for this nutrient to be made available to the phytoplankton throughout the entire region.

The anchoveta are gradually being replaced by sardines, a more valuable variety of fish. The entire anchoveta catch was converted to fish meal and exported as poultry and hog feed, but the sardine fishery will be directed at producing canned and frozen products for direct human consumption. With an emphasis on species conservation in Peru's future fishing activity, the anchoveta

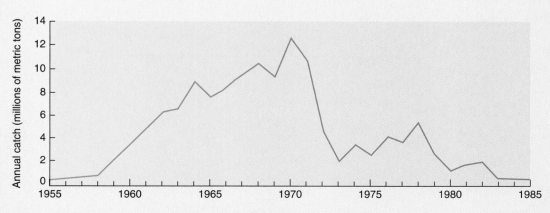

Figure 18C   Annual Catch of Peruvian Anchoveta, 1955–1985

After a peak catch of 12.3 million MT in 1970, the fishery declined.

fish-meal fishery may be viable in the future. However, it will never reach previous levels. Although it is much smaller in biomass, the sardine fishery may eventually develop into a major source of revenue because of the higher value of the products that can be produced from it. To improve the present state of affairs, the Ministry of Fisheries will have to conduct a prudent program of fishery assessment and management.

Figure 18D   Full Nets When the Peruvian Anchoveta Fishery was Booming

A catch of Peruvian anchoveta during the early 1960s when the fishery was undergoing rapid development.   *(Photo courtesy of the Food and Agriculture Organization.)*

Much additional testing will be conducted before such a form of disposal is put into operation.

Halogenated Hydrocarbons (DDT and PCBs)

DDT (dichlorodiphenyltrichlorethane) and **PCBs** (polychlorinated biphenyls) are found throughout the marine environment. They are persistent, biologically active chemicals that have been put into the oceans entirely as a result of human activities. At the time of a near-total ban on U.S. production of the pesticide DDT in 1971, 2.0×10^6 mt (4.4×10^9 lb) had been manufactured, most of it in the United States. Since 1972, its use in the Northern Hemisphere has virtually ceased. PCBs are industrial chemicals found in a variety of products from paint to plastics. They have been indicated as causes of spontaneous abortions in sea lions and the death of shrimp in Escambia Bay, Florida. Both of these chemicals bioaccumulate—they are stored in animal tissue and are passed in increasing concentrations up the food chain.

The danger of excessive use of DDT and similar pesticides was first manifested in the oceans as declining marine bird populations. During the 1960s, there was a serious decline in the brown pelican population of Anacapa Island off the coast of California; high concentrations of DDT in the fish eaten by the birds had caused them to lay eggs with excessively thin shells. A decline in the osprey population of Long Island Sound that began in the late 1950s and continued throughout the 1960s was also caused by thinning egg shells brought on by DDT contamination. Studies showed a 1 percent increase in brown pelican and osprey egg shell thickness from 1970 to 1976, while the concentration of pesticide residue decreased. Because there was an increased egg hatching rate associated with these changes, it was concluded that DDT was the cause of the decline in these and other bird populations during the 1960s.

The main route by which DDT and PCBs enter the ocean is the atmosphere. They are concentrated initially in the thin surface slick of organic chemicals at the ocean surface; then they gradually sink to the bottom attached to particles. A study off the coast of Scotland indicated that open-ocean concentrations of DDT and PCBs are 10 and 12 times less, respectively, than in coastal waters. Long-term studies have shown that DDT residue in mollusks along the U.S. coasts reached a peak in 1968. The pervasiveness of DDT and PCBs in the marine environment can best be demonstrated by the fact that Antarctic marine organisms contain measurable quantities. Obviously there has been no agriculture or industry on that continent to explain this presence; these substances have been transported from distant sources by winds and ocean currents.

Mineral Resouces

The rich deposits of copper, lead, zinc, and silver associated with the margins of lithospheric plates originate during the processes of plate construction and destruction (Figure 18-23). The Troodos Massif along the southern coast of Cyprus is an example of a **copper sulfide ore** associated with ophiolites (See Special Feature, Chapter 3) formed when part of the Mediterranean seafloor was obducted. The Troodos deposits have been mined since the beginnings of Mediterranean civilization

Metal-rich hydrothermal solutions associated with oceanic ridges are probably the source of metallic enrichment of ocean crust. The metals are predominantly iron and manganese, with significant amounts of copper, nickel, cobalt, zinc, and barium. Analysis of sediments in

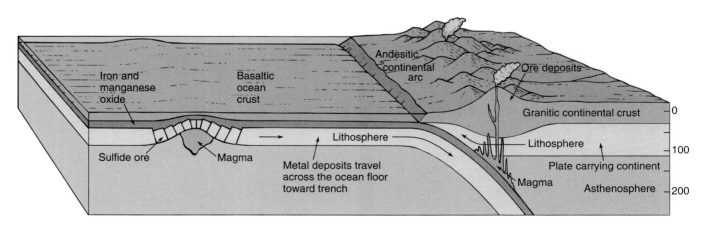

Figure 18-23   Theory of Metallic Ore Production

At constructive plate margins (left side of figure), ocean water enters fractures in basaltic crust and leaches metals from the rock. The metal-enriched water rises to the surface near the axis of the mid-ocean ridge, precipitating sulfides rich in copper, zinc, and lead. After the water returns to the ocean, iron and manganese oxides are precipitated. At destructive plate margins (right side of figure), descending metal-rich ocean crust melts and fluids rich in metals rise. Metals are precipitated as ore bodies in the overlying volcanic rocks.

a 200-km² area of the Atlantis Deep in the Red Sea have recorded concentrations that would result in an ore deposit with more than 3 million MT of zinc, 1 million MT of copper, 80,000 MT of lead, and 5000 MT of silver. Similar iron-and manganese-rich deposits were recovered by submersibles diving along the Mid-Atlantic Ridge during the French-American Mid-Ocean Undersea Study (FAMOUS), as well as on the East Pacific Rise and the Juan de Fuca Ridge.

Manganese nodules have been known to exist in the deep ocean since the voyage of the *H.M.S. Challenger* (1872–1876). Containing significant concentrations of manganese and iron and smaller concentrations of copper, nickel, and cobalt, these nodules caught the attention of mining companies, which began in the 1960s to assess the potential for mining them. Explorations found the richest metallic content in deposits in the eastern Pacific Ocean between Hawaii and Mexico (Figure 18-24). Recently, interest has faded in the face of a depressed metals market.

Major enrichments of cobalt in Earth's crust are confined to central southern Africa and deep-ocean nodules and crusts. Considering the unstable political situation in Africa, the United States has been looking to the ocean floor as a more dependable source. In 1981, **cobalt-rich manganese crusts** were found on the upper slopes of islands and seamounts that lie within the EEZs of the United States and its allies. The cobalt concentrations in these crusts are on average half again as rich as in the richest African ores and at least twice as rich as in deep-sea manganese nodules. The greatest concentrations seem to be associated with depths between 1000 and 2500 m (3280 and 8200 ft)

on the flanks of islands and seamounts in the central Pacific Ocean. There are at least 100 of these seamounts in the EEZs of the Line Islands and Hawaiian Islands, and each could yield up to 4 million metric tons of ore.

Deposits of phosphorite in nodules or bedded crusts may be found in water close to the edge of the continental shelf. These deposits are not likely to be mined in the near future because of the extensive deposits of phosphate on land. Closer to shore and much easier to mine are **placer deposits** on the continental shelf. Placer deposits are mechanically concentrated deposits of economically important minerals in surface sediments. Such deposits of gravel and sand are formed through erosion of minerals from nearby landmasses and by concentration in old river channels or along beaches that were subsequently submerged by the rising ocean. The offshore sand and gravel industry is second in value only to the offshore petroleum industry. This resource, which includes rock fragments and shells of marine organisms, is mined by technology similar to that used on land. It is used primarily for beach fill, landfill, and concrete aggregate. Offshore deposits are a major source of sand and gravel in New England, New York, and throughout the Gulf Coast (Figure 18-25). Many European countries, as well as Iceland, Israel, and Lebanon also depend heavily on such deposits.

Some offshore deposits also are rich in valuable minerals (Figure 18-26). Diamonds have been found in gravel deposits in South Africa and Australia. Sediments rich in tin have been mined for years from Thailand to Indonesia. Platinum and gold have been found in deposits in gold mining areas throughout the world. One of the major concerns of

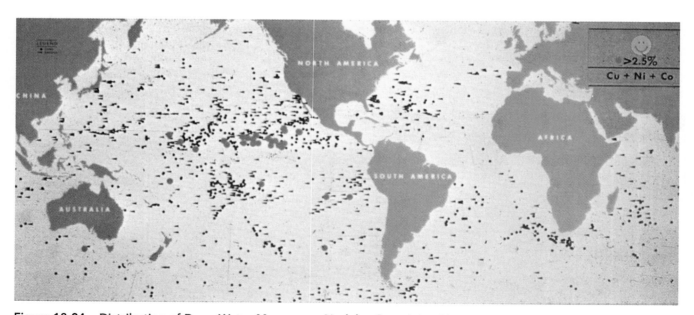

Figure 18-24   Distribution of Deep-Water Manganese Nodules Containing More Than 2.5 Percent Combined Copper, Nickel, and Cobalt

The manganese (ferromagnesian) nodule deposits with the greatest economic potential are those with the highest concentrations of Cu, Ni, and Co. They are found between the Clarion and Clipperton fracture zones in the eastern Pacific Ocean (see Figure 4-8).   *(Courtesy Scripps Institution of Oceanography, University of California, San Diego.)*

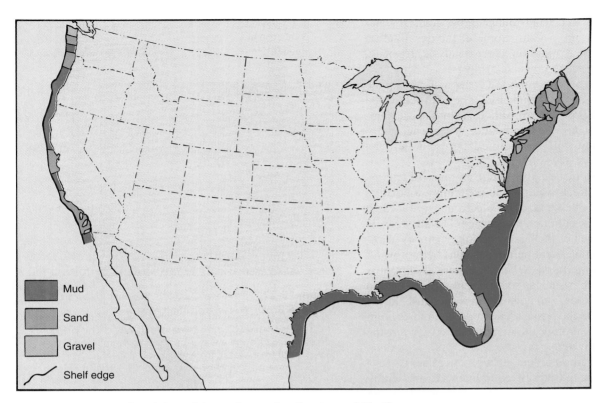

Figure 18-25   Sand and Gravel Deposits on the Continental Shelf
(From U.S. Department of Commerce Public Bulletin 188717.)

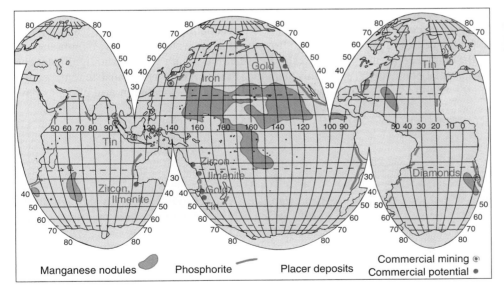

Figure 18-26   Manganese Nodule, Phosphorite, and Placer Deposits

mining sand and gravel deposits is that mining not be conducted too near shore. Nearshore mining operations create turbidity that is detrimental to marine life and aesthetics, and also may result in damage to beaches as beach material transported by longshore drift is trapped in offshore pits.

Water and Salt

With great expanses of the continents being arid and the available freshwater supplies throughout the world being used at greater rates each year, it is becoming urgent that practical methods of **desalination**, or salt removal, be developed so we can take advantage of the greatest water supply on Earth—the oceans. Some of the desalination methods in use are distillation, electrodialysis, freezing, solar humidification, and reverse osmosis.

Distillation involves boiling salt water and passing the resulting water vapor to a condenser, where it is collected as fresh water. This procedure, although simple, is expensive because it requires large amounts of heat energy. Increased efficiency will be required before it will be practical on a large-scale basis.

Electrodialysis also requires large amounts of energy. Two volumes of fresh water—one containing a positive electrode and the other a negative electrode—are located on either side of a volume of seawater. The seawater is separated from each of the freshwater reservoirs by semipermeable membranes that permit salt ions but not water molecules to pass through. When an electrical current is applied, positive ions such as sodium ions are attracted to the negative electrode, and negative ions such as chlorine ions are attracted to the positive electrode. In time, enough ions are removed through the membranes to convert the seawater to fresh water.

Freezing is a natural process that produces nearly salt-free sea ice. Again, the energy requirements to artificially freeze seawater make this approach cost-prohibitive, but imaginative thinkers have proposed towing icebergs from the polar regions to the coastal waters of arid lands. Here, fresh water produced as the icebergs melt could be captured and pumped ashore.

Solar humidification does not require supplemental heating and has been used successfully in large-scale agricultural experiments in Israel, West Africa, and Peru. It involves the evaporation of water in a covered container placed in direct sunlight. Salt water in the container is subjected to evaporation, and the water vapor that condenses on the cover runs into trays from which it is collected.

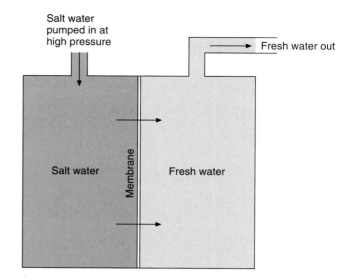

Figure 18-27 Reverse Osmosis Process Used to Produce Fresh Water from Salt Water

Pressure is applied to a reservoir of salt water. The pressure forces water through a water-permeable membrane into a reservoir of fresh water, from which it is collected. The salt, which cannot pass through the membrane, is left behind.

Reverse osmosis has perhaps the greatest potential for large-scale projects. Osmosis is the net flow of water molecules through a water-permeable membrane from a freshwater solution into a saltwater solution. In reverse osmosis, pressure is applied to the saltwater solution, forcing it to flow through the water-permeable membrane into the freshwater solution (Figure 18-27). The significant cost of this method is the membranes that must be replaced frequently. At least 30 nations located in arid climates have built reverse osmosis units.

Summary

International efforts at creating a *law of the sea* date back to a 1609 doctrine that the ocean was free to all nations. The present law of the sea is detailed in the United Nations Convention on the Law of the Sea. Pollution is the introduction of substances or energy into the marine environment that results in harm to the living resources of the oceans or humans that use these resources. Scientists are trying to determine the ocean's assimilative capacity—the amount of a substance it can absorb without becoming polluted. The U.S. government has been concerned about coastal pollution since it passed the Rivers and Harbors Act in 1899. More recent laws have been passed to prohibit the dumping of materials known to be harmful and to specify criteria under which other materials may be dumped. With its Action Plan for the Human Environment, initiated in 1972, the United Nations Environment Program established 12 environmental units throughout the world, within which assessment and monitoring have begun or are planned.

As a result of a rapidly expanding human population and a growing need for food and minerals, technology has developed to the point where exploitation of the ocean's resources is accelerating. Attempts at regulation of fisheries have not been successful to date, probably due to their social, rather than ecological, motivations. From the overfishing of the northwest Atlantic to the collapse of the Peruvian anchoveta fishery, we see many examples of fisheries mismanagement. Present studies of fishery ecology allow scientists to make better predictions of recruitment of young adults to the fishery and levels of maximum sustainable yield. About 21 percent of the world fishery comes from about 0.1 percent of the ocean surface area where upwelling occurs. Studies of duration and rate of upwelling in these regions reveal that moderate rates of upwelling over long periods of time produce larger fisheries than high or low rates of upwelling.

The first major human disaster resulting from fishing in polluted waters reached epidemic proportions in 1956 within Minamata Bay, Japan. Of more than 100 people affected by mercury poisoning, almost half died. A similar episode occurred during the late 1960s at Niigata, Japan. Mercury contamination levels have been set on the basis of fish consumption by human populations throughout the world, and they appear to have been effective in preventing further poisonings.

Mariculture projects around the world have, in some cases, indicated potential for increasing the food taken from the ocean. The most successful mariculture to date has been directed at algae, oysters, mussels, salmon, mullet, and milkfish.

Inorganic resources such as petroleum, metal ores, phosphorite, and placer deposits of diamonds, gold, and other minerals and aggregates are targets of exploration and exploitation. Petroleum is already being extensively exploited, and plans are well developed for the mining of manganese nodules.

Study of areas such as Wild Harbor, Massachusetts, and Chedabucto Bay, Nova Scotia, have shown that oil pollution reduces species diversity and persists longer on muddy bottoms than on sandy or rocky bottoms. Oil spills that do not come ashore, like that of the *Argo Merchant,* do considerably less environmental damage. The greatest damage resulting from the *Argo Merchant* spill was probably to plankton, especially fish eggs. The largest oil pollution event occurred during the Persian Gulf War in 1991 when 500,000 MT of oil contaminated the shores of Kuwait and Saudi Arabia.

Desalination of ocean water to provide fresh water for domestic, and agricultural use may be achieved by distillation, electrodialysis, freezing, solar humidification, and reverse osmosis of seawater. The last two methods show the greatest potential for practical applications.

The amount of plastic accumulating in the oceans has increased dramatically. Certain forms of plastic are known to be lethal to marine mammals and turtles, and national and international legislation is being considered to ban the disposal of plastic in the oceans.

Although 1972 legislation required an end to sewage dumping in the coastal ocean by 1981, exceptions continue to be made. Especially off the southern California coast and in New York Bight, large amounts of sewage continued to be dumped. Increased public concern resulted in new legislation that prohibits sewage dumping in the ocean. It has been stopped in the L.A. basin and in New York Bight. A newly built system now dumps treated Boston-area sewage offshore in Massachusetts Bay instead of Boston Harbor.

Most of the radioactive waste in the ocean comes from fallout resulting from the testing of nuclear explosives since 1944. However, areas with the greatest concentrations of radionuclides entering the ocean are near fuel-reprocessing plants. After nearly five decades of production of high-level nuclear waste, disposal of this material still represents an unsolved and complex problem. Although the presently favored option is land disposal, research has been done to identify sediments near the centers of the mid-ocean plates as the most suitable marine disposal sites.

DDT and PCBs are pollutants that bioaccumulate, thereby having the worst effects on predatory species. Birds are particularly susceptible to DDT, which was blamed for a decline in the Long Island osprey population in the 1950s and the brown pelican population of the California coast in the 1960s. Virtual cessation of DDT use in the Northern Hemisphere in 1972 allowed the recovery of both populations. DDT thinned the egg shells and reduced the number of successful hatchings.

Key Terms

Action Plan for the Human Environment	Exclusive Economic Zone (EEZ)	Marine Mammal Protection Act	Placer deposits
Bioassay	Federal Water Pollution Control Act	Marine Protection, Research, and Sanctuaries Act	Potential world fishery
Cobalt-rich manganese crusts		Maximum sustainable yield	Radioactive waste
Copper sulfide ores	Fisheries management	Minamata disease	Radionuclides
DDT	Freezing	Nuclear Energy Agency	Recruitment
Desalination	Gill nets	Ocean Dumping Ban Act	Reverse osmosis
Distillation	International Seabed Authority (ISA)	Organization for Economic Cooperation and Development	Rivers and Harbors Act
Driftnets			Solar humidification
Electrodialysis	Levels of biological response	PCBs	Standing stock
	Mariculture		Territorial sea

Questions And Exercises

1. List and discuss some positive and negative effects that you think may have resulted from the establishment of the 200-naut-mi (370-km) Exclusive Economic Zones.

2. Discuss the different levels of biological response to the effects of pollution. Discuss the symptoms by which these effects can be recognized and the time typically required for them to manifest themselves.

3. Do you think it is a better policy to (1) prohibit dumping into the ocean of any substance that cannot be proven to be harmless to the ocean, or (2) consider the effects of disposing of the substance on land and in the ocean before deciding where to dispose of it? Discuss the reasons for your choice.

4. What are the two critical survival stages that must occur before young adults can be recruited to a fishery, and what factor is most important in increasing the chance of survival at each stage?

5. Explain the relationship between the influx of nutrients (nitrogen) into an ecosystem and the amount of the fishery that can safely be removed each year.

6. Discuss the desirability of setting a mercury contamination level for fish at the 1.0 mg/kg level for the citizens of the United States, Sweden, and Japan.

7. What do algae and bivalve cultivation have in common that helps make those organisms good choices for mariculture operations?

8. Why are oil spills that come ashore more destructive to marine life than those that stay offshore?

9. Describe the relationship of metallic ore deposits in the lithosphere to the plate tectonics process.

10. What components of the nuclear energy generation industry introduce the most radioactive wastes into the oceans?

References

Beer, T., 1997. *Environmental Oceanography*, 2ⁿᵈ ed. Boca Raton, FL.:CRC Press.

Borgese, E. M., et al. Eds. 1998. *Ocean Yearbook*, Vol. 13. Chicago: University of Chicago Press.

Bragg J. R., et al. 1994. Effectiveness of bioremediation for the *Exxon Valdez* oil spill. *Nature* 368:6470, 413–18.

Clark, W. C., 1989. Managing planet Earth. *Scientific American* 261:3, 47–54.

International Atomic Energy Agency. 1980. International Symposium on the Impacts of Radionuclide Releases into the Marine Environment.

Cramer, D.,1995. Troubled waters. *The Atlantic Monthly* 275:6, 22–26.

Kerr, R., 1998. The next oil crisis looms large–and perhaps close. *Science* 281: 5380, 1128-31.

Kurlansky, M., 1997. *Cod: Biography of the Fish that Changed the World.* New York: Walker

Lawrence, W. W., 1979. *Of Acceptable Risk: Science and the Determination of Safety.* Los Altos, Calif.: William Kaufman.

Longhurst, A. R., and D. Pauly, 1987. *Ecology of Tropical Oceans.* San Diego: Academic Press.

Masters, C. D., D. H. Root, and E. D. Attanasi, 1991. Resource constraints in petroleum production potential. *Science* 253:146–52.

Mearns, A. J., and J. Q. Word, 1981. Forecasting effects of sewage solids on marine benthic communities. In *Ecological Effects of Environmental Stress*, Eds. J. O'Connor and G. Mayer. Estuarine Research Foundation. Special publication.

Myers, N. 1998. Lifting the veil on perverse subsidies. *Nature* 392:6674, 327–28.

Pauly, D., and V. Christensen, 1995. Primary production required to sustain global fisheries. *Nature* 374:6519, 255–57.

Pauly, D., et al., 1998. Fishing doen marine food webs. *Science* 279: 5352, 860-63.

Ramanathan, V., 1988. The greenhouse theory of climate change: A test by an inadvertent global experiment. *Science* 240:4850, 293–99.

Readman, J. W., et al. 1992. Oil and combustion-product contamination of the Gulf marine environment following the war. *Nature* 358:6388, 622–65.

Royce, W. F., 1987. *Fishery Development.* San Diego: Academic Press.

Van Dover, C. L., et al., 1992. Stable isotope evidence for entry of sewage-derived organic material into a deep-sea food web. *Nature* 360:6400, 153–56.

Wilber, R. J., 1987. Plastics in the North Atlantic. *Oceanus* 30, 61–68.

Wyatt, T., 1980. The growth season in the sea. *Journal of Plankton Research* 2:81–97.

Suggested Reading

Sea Frontiers

Alper, J. 1992. Seabird comeback. 38:2, 12–17, 62. Restoring habitats on the east and west coasts of the United States has aided the resurgence of puffin and tern populations.

Behnken, L. 1992. Strike in the shrinking grounds. 38:6, 28–31. Because of overfishing Alaskan fishermen have to adjust to an 11-day black cod fishing season.

Cox, V. 1993. A forum: The ocean's canaries. 39:1, 16–18. The die-offs of dolphins and pinnipeds from 1987 through 1991 may be a sign of polluted seas.

Cuyvers, L. 1984. Milkfish: Southeast Asia's protein machine. 30:3, 173–9. The importance of milkfish mariculture operations to the food supply in Southeast Asia.

Driessen, P. K. 1987. Oil rigs and sea life: A shotgun marriage that works. 33:5, 362–72. Despite resistance to offshore drilling because of aesthetic and environmental concerns, there is evidence that offshore drilling rigs enhance the ecology of many areas.

Hersch, S. L. 1988. Death of the dolphins: Investigating the East Coast die-off. 34:4, 200–207. A marine mammalogist investigating the cause of the deaths of dolphins along the East

Coast discusses the evidence developed since the deaths began in June 1987.

Hinman, K. 1992. The real cost of shrimp on your table. 38:1, 14–19. Discusses cost of wasted bycatch in shrimp nets in the Gulf of Mexico. Author supports use of the TED (turtle excluder device or trawling efficiency device).

Hull, E. W. S. 1978. Oil spills: The causes and the cures. 24:6, 360–69. General discussion of oil spills, their frequency, and methods of clean-up.

Idyll, C. P. 1971. Mercury and fish. 17:4, 230–40. An article written just after the mercury scare of 1970 discusses the effects of mercury on marine life.

Iverson, E. S., and D. E. Jory. 1986. Shrimp culture in Ecuador: Farmers without seed. 32:6, 442–53. The rapidly growing shrimp mariculture industry of Ecuador encounters some growing pains.

Loupe, D. 1991. The food factor. 37:2, 22–27. How some spin-off mariculture projects contribute to the profitability of ocean thermal energy conversion energy-producing projects.

McCredie, S. 1990. Controversy travels with driftnet fleets. 36:1, 13–21. Problems related to development of driftnet fisheries in the Pacific Ocean.

Nicol, S. 1987. Krill: Food of the future? 33:1, 12–17. Before krill can be considered as a viable fishery many questions about the cost of the fishery and processing as well as the fishery's effect on krill-eating mammal populations must be answered.

Parker, P. A. 1990. Clearing the oceans of plastics. 36:2, 18–27. The history and scope of the problems of plastics in the ocean.

Reynolds, J. E., and J. R. Wilcox. 1987. People, power plants, and manatees. 33:4, 263–69. The effects of water heating by power plants on the distribution of manatees.

Sletto, B. 1992. When fish break out. 38:4, 40–43. When pen-raised Atlantic salmon break out of their pens, natural populations are threatened with genetic and epidemic disasters.

Wacker, R. 1991. The bay killers. 37:6, 44–51. An overview of the problems facing coastal waters near large U.S. metropolitan areas.

Wacker, R. 1994. Politics. 40:3, 14–17. The health of the world fishery is troubled by poor management—politics!

Scientific American

Bascom, W. 1974. Disposal of waste in the ocean. 231:2, 16–25 Misconceptions about the disposal of waste in the oceans and methods that may allow safe use of the oceans for such a purpose.

Bongaarts, J. 1994. Can the growing human population feed itself? 270:3, 36–43. Focusing on the role of agriculture in feeding the 10 billion people projected to inhabit Earth by 2050, this article provides insight on the pressure fisheries will face in the future.

Brimhall, G. 1991. The genesis of ores. 126:5, 84–91. The origin of ore deposits by the transport of metals by magma, water, and air.

Butler, J. N. 1975. Pelagic tar. 232:6, 90–97. The origin and distribution of lumps of tar found floating at the ocean surface and their possible implications for marine biology.

Grove, R. H. 1992. Origin of western environmental action. 276:1, 42–47. Concern for the environment is traced back to the Age of Discovery.

Halloway. M. 1994. Nurturing nature. 270:4, 98-108, Focusing on the Florida Everglades, the potential to restore damaged wetlands is investigated.

_____1996. Sounding out science. 275:4, 106-112. An update on the recovery of Prince Williiam Sound after the *Exxon Valez* oil spill, with a focus on how science is used to analyze the recovery of affected areas.

Houghton, R. A., and G. M. Woodwell. 1989. Global climatic change. 260:4, 36–44. The increases in atmospheric carbon dioxide and methane caused by human activities will surely change Earth's climate. Drastic changes in human behavior are required to reduce the threat of change.

McDonald, I. R. 1998. Natural oil spills. 279:5, 56-61. A look at the effects of natural oil seeps in the Gulf of Mexico and the unique biologic comunities that consume these hydrocarbons.

Theme Issue. 1989. Managing planet earth. 261:3, 1–175. Articles covering the broad scope of environmental problems.

Oceanography on the Web

Visit the Introductory Oceanography home page for on-line resources for this chapter. There you will find an on-line study guide with review exercises and links to ocean-ography sites to further your exploration of the topics in this chapter. Introductory Oceanography is at http://prenhall.com/thurman (click on the Table of Contents menu and select this chapter).

Appendix I

Logarithms and Scientific Notation

A logarithm is the exponent of the power to which it is necessary to raise a fixed number (the base) to produce a given number (the antilogarithm). For instance, the base 10 must be raised to the power 2 to produce the number 100.

$$10(\text{base})^{2(\text{logarithm})} = 100(\text{antilogarithm})$$

As we commonly work with a number system to the base 10, in the common system of logarithms, the logarithm of 100 is 2. The logarithm of 1,000 is 3, of 10,000 is 4, etc.

Logarithms	0	1	2	3	4
Natural numbers	1	10	100	1,000	10,000

The logarithms of numbers between 1 and 10 consist of decimals; of numbers between 10 and 100 consist of the integer 1 and a decimal; of numbers between 100 and 1,000 consist of the integer 2 and a decimal; etc. The integer part of a logarithm is the *characteristic,* and is always less by 1 than the number of integers in the natural number. The decimal part of the logarithm is the *mantissa.* The use of logarithms simplifies the arithmetic operations, especially the multiplication and division of large numbers.

To simplify writing very large and very small numbers, scientists indicate the number of zeros by scientific notation. One integer is placed to the left of the decimal, and a multiplication times a power of 10 tells which direction and how far the decimal is moved to write the number out in its long form. For example:

$$2.13 \times 10^5 = 213{,}000$$

or

$$2.13 \times 10^{-5} = 0.0000213$$

Further examples showing numbers that are powers of 10 are:

$$1{,}000{,}000{,}000 = 1.0 \times 10^9, \text{ or } 10^9$$
$$1{,}000{,}000 = 1.0 \times 10^6, \text{ or } 10^6$$
$$1{,}000 = 1.0 \times 10^3, \text{ or } 10^3$$
$$100 = 1.0 \times 10^2, \text{ or } 10^2$$
$$10 = 1.0 \times 10^1, \text{ or } 10^1$$
$$1 = 1.0 \times 10^0, \text{ or } 10^0$$
$$0.1 = 1.0 \times 10^{-1}, \text{ or } 10^{-1}$$
$$0.01 = 1.0 \times 10^{-2}, \text{ or } 10^{-2}$$
$$0.001 = 1.0 \times 10^{-3}, \text{ or } 10^{-3}$$
$$0.000001 = 1.0 \times 10^{-6}, \text{ or } 10^{-6}$$
$$0.000000001 = 1.0 \times 10^{-9}, \text{ or } 10^{-9}$$

To add or subtract numbers written as powers of 10, they must be converted to the same power:

Addition

$$2.1 \times 10^3 \qquad 0.021 \times 10^5$$
$$\underline{+1.0 \times 10^5} = \underline{+1.000 \times 10^5}$$
$$1.021 \times 10^5$$

Subtraction

$$3.4 \times 10^4 \qquad 3.4 \times 10^4$$
$$\underline{-2.0 \times 10^3} = \underline{-0.2 \times 10^4}$$
$$3.2 \times 10^4$$

To multiply or divide numbers written as powers of 10, the exponents are added or subtracted:

Multiplication

$$6.04 \times 10^2$$
$$\underline{\times 2.1 \times 10^4}$$
$$12.684 \times 10^6$$

Division

$$3.0 \times 10^3$$
$$\underline{\div 1.5 \times 10^2}$$
$$2.0 \times 10^1$$

To raise a number expressed as a power by a power, the powers are multiplied:

$(10^5)^2 = 10^{10}$

Appendix II

Latitude and Longitude Determination

Lines of latitude and longitude provide a grid system that makes it possible to describe the position of any point on Earth's surface. It is illustrated in Figure 1.

Latitude is defined as the angular distance measured (in degrees of arc) north or south of the equatorial plane with a vertex at the center of the Earth. This process is illustrated in Figure 2. Connecting all points at a given latitude produces a circular line parallel to the equator. Thus lines of latitude are also called parallels.

Longitude is defined as the angular distance measured (in degrees of arc) east or west of an arbitrary chosen line that runs from pole to pole, crossing the equator at right angles. All lines of longitude meet at the poles and are most distant from one another at the equator. Longitude lines are also called meridians because in the early days of determining longitude, it was convenient to do it when the sun was at its highest point in the sky—local noon. This process is described in Figure 3. Figure 1 illustrates the nature of longitude and latitude lines on Earth's surface.

A British battle fleet commanded by Sir Cloudesley Shovell ran aground in the Scilly Islands in 1707 with the loss of four ships and 2000 men. This happened because they lost track of their longitude (location east or west) after weeks at sea. To determine longitude at any location in the ocean, it was necessary to know the time at a reference meridian (we now use the Greenwich Meridian, running through Greenwich, England). However, clocks in 1707 were driven by pendulums and would not work for long on a ship rocking at sea.

In 1714 the British government offered a £20,000 prize (about $2,000,000 today) for the development of a clock that would work well enough at sea to determine longitude within 0.5° after a voyage to the West Indies. A cabinetmaker in Lincolnshire named John Harrison began working on such a timepiece in 1728. His chronometer was driven by a helical balance spring and remained horizontal regardless of the attitude of the ship. It was complicated, costly, and delicate. In 1736 his first chronometer was successfully tested and he received £500. Eventually, his fourth version was tested in 1761. Upon reaching Jamaica, it was so accurate that the longitude error was only 1.25' (recall that a minute is only $1/60°$) and it had lost only 5 seconds of time.

This performance greatly exceeded the requirements of the government, and Harrison claimed the prize. The government was slow in paying because they wanted to be convinced that reliable versions of the chronometer could be produced in large numbers. Finally, after the intervention of King George III, Harrison received the balance of his prize in 1773, at age 80.

Figure 1   Earth's Grid System

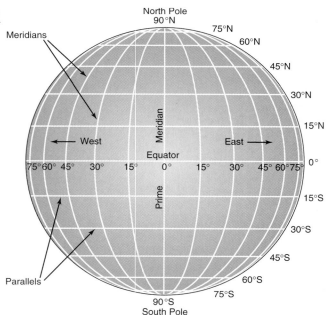

514

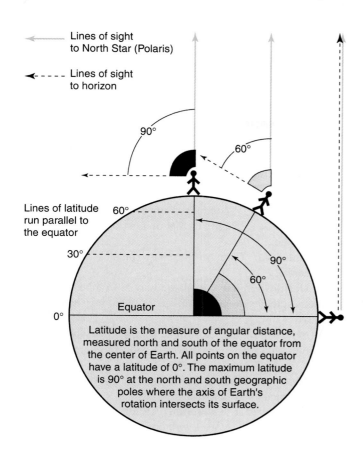

Lines of sight to North Star (Polaris)

Lines of sight to horizon

90°

60°

Lines of latitude run parallel to the equator

60°

30°

90°

60°

0°

Equator

Latitude is the measure of angular distance, measured north and south of the equator from the center of Earth. All points on the equator have a latitude of 0°. The maximum latitude is 90° at the north and south geographic poles where the axis of Earth's rotation intersects its surface.

Figure 2 Determining Latitude

The method of determining latitude used by Pytheas was to measure the angle between the horizon and the North Star (Polaris), which is directly above the North Pole. Latitude north of the equator is the angle between the two sightings. Similar determinations may be made in the Southern Hemisphere by using the Southern Cross, which is directly overhead at the South Pole.

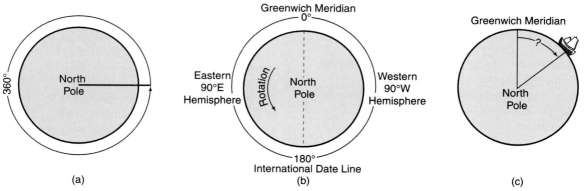

(a)

360°

North Pole

(b)

Greenwich Meridian
0°

Eastern
90°E
Hemisphere

Rotation

North Pole

Western
90°W
Hemisphere

180°
International Date Line

(c)

Greenwich Meridian

?

North Pole

Figure 3 Determining Longitude

View of Earth from outer space, looking down on North Pole. (a) As Earth turns on its rotational axis, it moves through 360° of angle every 24 hours. (b) The meridian that runs from pole to pole through Greenwich, England, was selected as the reference meridian, dividing Earth into a Western and Eastern hemisphere. After John Harrison's chronometer was developed, many ships carried it, showing the time on the Greenwich Meridian—Greenwich time. (c) Since Earth rotates through 15° of angle, or longitude, per hour (360° ÷ 24 hours = 15°/hour), a ship's captain could easily determine longitude each day at noon. For example, a ship sets sail west across the Atlantic Ocean, checking its longitude each day at noon (when the sun crosses the meridian of the observer). One day when the sun is at the noon position, the captain checks the chronometer. It reads 16:18 hours (0:00 is midnight and 12:00 is noon). What is the ship's longitude?

Longitude solution:

16:18 hours = 4:18 P.M.
Earth rotates through 1/4° (15')
of angle per minute of time.
(One degree of arc is divided into 60 minutes.)
4 hours × 15°/hour = 60° of longitude
18 minutes of time × 15 minutes of angle/minute
of time = 270 minutes of arc
270 minutes ÷ 60 minutes/degree (°) = 4.5°
of longitude
60° + 4.5° = 64.5°W longitude

Appendix III

Taxonomic Classification of Common Marine Organisms

Kingdom Monera

Organisms without nuclear membranes; nuclear material is spread throughout cell; predominantly unicellular.

Phylum Schizophyta Smallest known cells; bacteria (1500 species).

Phylum Cyanophyta Blue-green algae; chlorophyll *a*, carotene and phycobilin pigments (200 species).

Kingdom Protista

Organisms with nuclear material confined to nucleus by a membrane.

Phylum Chrysophyta Golden-brown algae; includes diatoms, coccolithophores, and silicoflagellates; chlorophyll *a* and *c*, xanthophyll, and carotene pigments (6000 + species).

Phylum Pyrrophyta Dinoflagellate algae; chlorophyll *a* and *c*, xanthophyll, and carotene pigments (1100 species).

Phylum Chlorophyta Green algae; chlorophyll *a* and *b* and carotene pigments (7000 species).

Phylum Phaeophyta Brown algae; chlorophyll *a* and *c*, xanthophyll, and carotene pigments (1500 species).

Phylum Rhodophyta Red algae; chlorophyll *a*, carotene, and phycobilin pigments (4000 species).

Phylum Protozoa Nonphotosynthetic, heterotrophic protists (27,400 species).

Class Mastigophora Flagellated; dinoflagellates (5200 species).

Class Sarcodina Ameboid; foraminiferans and radiolarians (11,500 species).

Class Ciliophora Ciliated (6000 species).

Kingdom Fungi

Phylum Mycophyta Fungi, lichens; most fungi are decomposers found on the ocean floor, whereas lichens inhabit the upper intertidal zones (3000 species of fungi, 160 species of lichen).

Kingdom Metaphyta

Multicellular, complex plants.

Phylum Tracheophyta Vascular plants with roots, stems, and leaves that are serviced by special cells that carry food and fluids (287,200 species).

Class Angiospermae Flowering plants with seeds contained in a closed vessel (275,000 species).

Kingdom Metazoa

Multicellular animals.

Phylum Porifera Sponges; spicules are the only hard parts in these sessile animals that do not possess tissue (10,000 species).

Class Calcarea Calcium carbonate spicules (50 species).

Class Desmospongiae Skeleton may be composed of siliceous spicules or spongin fibers or be nonexistent (9500 species).

Class Sclerospongiae Coralline sponges; massive skeleton composed of calcium carbonate, siliceous spicules, and organic fibers (7 species).

Class Hexactinellida Glass sponges; six-rayed with siliceous spicules (450 species).

Phylum Cnidaria (Coelenterata) Radially symmetrical, two-cell, layered body wall with one opening to gut cavity; polyp (asexual, sexual, benthic) and medusa (sexual, pelagic) body forms (10,000 species).

Class Hydrozoa Polypoid colonies such as pelagic Portuguese man-of-war and benthic Obelia common; medusa present in reproductive cycle but reduced in size (3000 species).

Class Scyphozoa Jellyfish; medusa up to 1 m in diameter is dominant form; polyp is small if present (250 species).

Class Anthozoa Corals and anemones possessing only polypoid body form and reproducing asexually and sexually (6500 species).

Phylum Ctenophora Predominantly planktonic comb jellies; basic eight-sided radial symmetry modified by secondary bilateral symmetry (80 species).

Phylum Platyhelminthes Flatworms; bilateral symmetry; hermaphroditic (25,000 species).

Phylum Nemertea Ribbon worms; as long as 30 m; benthic and pelagic (800 species).

Phylum Nematoda Roundworms; marine forms are primarily free-living and benthic; most 1 to 3 mm in length (5000 marine species).

Phylum Rotifera Ciliated, unsegmented forms less than 2 mm in length (1500 species, only a few marine).

Phylum Bryozoa (Ectoprocta) Moss animals; benthic, branching or encrusting colonies; lophophore feeding structure (4500 species).

Phylum Branchiopoda Lamp shells; lophophorate benthic bivalves (300 species).

Phylum Phoronida   Horseshoe worms; 24-cm-long lophorate tube worms that live in sediment of shallow and temperate shallow waters (15 species).

Phylum Sipuncula   Peanut worms; benthic (325 species).

Phylum Echiura   Spoon worms; sausage shaped with spoon-shaped proboscis; burrow in sediment or live under rocks (130 species).

Phylum Pogonophora   Tube-dwelling, gutless worms 5 to 80 cm long; absorb organic matter through body wall (100 species).

Phylum Tardigrada   Marine meiofauna that have the ability to survive long periods in a cryptobiotic state (diversity is poorly known).

Phylum Mollusca   Soft bodies possessing a muscular foot and mantle that usually secretes calcium carbonate shell (75,000 species).

Class Monoplacophora   Rare trench-dwelling forms with segmented bodies and limpetlike shells (10 species).

Class Polyplacophora   Chitons; oval, flattened body covered by eight overlapping plates (600 species).

Class Gastropoda   Large, diverse group of snails and their relatives; shell spiral if present (64,500 species).

Class Bivalvia   Bivalves; includes mostly filter-feeding clams, mussels, oysters, and scallops (7500 species).

Class Aplacophora   Tusk shells; sand-burrowing organisms that feed on small animals living in sand deposits (350 species).

Class Cephalopoda   Octopus, squid, and cuttlefish that possess no external shell except in the genus *Nautilus* (600 species).

Phylum Annelida   Segmented worms in which musculature, circulatory, nervous, excretory, and reproductive systems may be repeated in many segments; mostly benthic (10,000 marine species).

Phylum Arthropoda   Jointed-legged animals with segmented body covered by an exoskeleton (30,000 marine species).

Subphylum Crustacea   Calcareous exoskeleton, two pairs of antennae; cephlon, thorax, and abdomen body parts; includes copepods, ostracods, barnacles, shrimp, lobsters, and crabs (26,000 species).

Subphylum Chelicerata

Class Merostomata   Horseshoe crabs (4 species).

Class Pycnogonide   Sea spiders.

Subphylum Uniramia   Insects; genus *Halobites* is the only true marine insect.

Phylum Chaetognatha   Arrowworms; mostly planktonic, transparent and slender; up to 10 cm long (50 species).

Phylum Echinodermata   Spiny-skinned animals; benthic animals with secondary radial symmetry and water vascular system (6000 species).

Class Asteroidea   Starfishes; free-living, flattened body with five or more rays with tube feet used for locomotion; mouth down (1600 species).

Class Ophiuroidea   Brittle stars and basket stars; prominent central disc with slender rays; tube feet used for feeding; mouth down (200 species).

Class Echinoidea   Sea urchins, sand dollars, and heart urchins; free-living forms without rays; calcium carbonate test; mouth down or forward (860 species).

Class Holothuroidea   Sea cucumbers; soft bodies with radial symmetry obscured; mouth forward (900 species).

Class Crinoidea   Sea lilies and feather stars; cup-shaped body attached to bottom by a jointed stalk or appendages; mouth up (630 species).

Phylum Hemichordata   Acorn worms and pterobranchs; primitive nerve chord; gill slits; benthic (90 species).

Phylum Chordata   Notochord; dorsal nerve chord and gills or gill slits (55,000 species).

Subphylum Urochordata   Tunicates; chordate characteristics in larval stage only; benthic sea squirts and planktonic thaliaceans and larvaceans (1375 species).

Subphylum Cephalochordata   Amphioxus or lancelets; live in coarse temperate and tropical sediment (25 species).

Subphylum Vertebrata   Internal skeleton; spinal column of vertebrae; brain (52,000 species).

Class Agnatha   Lampreys and hagfishes; most primitive vertebrates with cartilaginous skeleton, no jaws, and no scales (50 species).

Class Chondrichthyes   Sharks, skates, and rays; cartilaginous skeleton; 5 to 7 gill openings; placoid scales (625 species).

Class Osteichthyes   Bony fishes; cycloid scales; covered gill opening; swim bladder common (30,000 species).

Class Amphibia   Frogs, toads, and salamanders; Asian mud flat frogs are the only amphibians that tolerate marine water (2600 species).

Class Reptilia   Snakes, turtles, lizards, and alligators; orders Squamata (snakes) and Chelonia (turtles) are major marine groups (6500 species).

Class Aves   Birds; many live on and in the ocean but all must return to land to breed (8600 species).

Class Mammalia   Warm-blooded; hair; mammary glands; bear live young; marine representatives found in the orders Sirenia (sea cows, dugong), Cetacea (whales), and Carnivora (sea otter, pinnipeds) (4100 species).

Appendix IV

Roots, Prefixes, and Suffixes*

a- not, without
ab-, abs- off, away, from
abysso- deep
acanth- spine
acro- the top
aero- air, atmosphere
aigial- beach, shore
albi- white
alga- seaweed
alti- high, tall
alve- cavity, pit
amoebe- change
an- without, not
annel- ring
annu- year
anomal- irregular, uneven
antho- flower
apex- tip
aplysio- sponge, filthiness
aqua- water
arachno- spider
arena- sand
arthor- joint
asthen- weak, feeble
astro- star
auto- self
avi- bird
bacterio- bacteria
balaeno- whale
balano- acorn
barnaco- goose
batho- deep
bentho- the deep sea
bio- living
blast- a germ
botryo- bunch of grapes
brachio- arm
branchio- gill
broncho- windpipe
bryo- moss
bysso- a fine thread
calci- limestone (CaCO$_3$)
calori- heat
capill- hair
cara- head
carno- flesh
cartilagi- gristle
caryo- nucleus
cat- down, downward
cen-, ceno- recent

cephalo- head
ceta- whale
chaeto- bristle
chiton- tunic
chlor- green
choano- funnel, collar
chondri- cartilage
cilio- small hair
circa- about
cirri- hair
clino- slope
cnido- nettle
cocco- berry
coelo- hollow
cope- oar
crusta- rind
cteno- comb
cyano- dark blue
cypri- Venus, lovely
-cyst bladder, bag
-cyte cell
deca- ten
delphi- dolphin
di- two, double
diem- day
dino- whirling
diplo- double, two
dolio- barrel
dors- back
-dram- run, a race
echino- spiny
eco- house, abode
ecto- outside, outer
edrio- a seat
en- in, into
endo- inner, within
entero- gut
epi- upon, above
estuar- the sea
eu- good, well
exo- out, without
fauna- animal
fec- dregs
fecund- fruitful
flacci- flake
flagell- whip
flora- flower
fluvi- river
fossili- dug up
fuco- red

geno- birth, race
geo- Earth
giga- very large
globo- ball, globe
gnatho- jaw
guano- dung
gymno- naked, bare
halo- salt
haplo- single
helio- the sun
helminth- worm
hemi- half
herbi- plant
herpeto- creeping
hetero- different
hexa- six
holo- whole
homo-, homeo- alike
hydro- water
hygro- wet
hyper- over, above, excess
hypo- under, beneath
ichthyo- fish
-idae members of the animal family of
infra- below, beneath
insecti- cut into
insula- an island
inter- between
involute- intricate
iso- equal
-ite rock
juven- young
juxta- near to
kera- horn
kilo- one thousand
lacto- milk
lamino- layer
larvi- ghost
latent hidden
latero- side
lati- broad, wide
lemni- water plant
limno- marshy lake
lipo- fat
litho- stone
litorial near the seashore
lopho- tuft
lorica- armor
luci- light

luna moon
lux light
macro- large
mala- jaw, cheek
mamilla- teat
mandibulo- jaw
mantle- cloak
mari- the sea
masti- chewing
mastigo- whip
masto- breast, nipple
maxillo- jaw
madi- middle
medus- a jellyfish
mega- great, large
meio- less
meridio- noon
meros- part
meso- middle
meta- after
meteor- in the sky
-meter measure
-metry science of measuring
mid- middle
milli- thousandth
mio- less
moll- soft
mono- one, single
-morph form
myo- muscle
myst- mustache
nano- dwarf
necto- swimming
nemato- thread
neo- new
neph- cloud
nerito- sea nymph
noct- night
-nomy the science of
nucleo- nucleus
nutri- nourishing
o-, oo- egg
ob- reversed
ocellus- little eye
octa- eight
oculo- eye
odonto- teeth
oiko- house, dwelling
oligo- few, scant
-ology science of

*Source: D.J. Borror, 1960, *Dictionary of Word Roots and Combining Forms*. Palo Alto, Calif.: National Press Books.

omni- all	**pisci-** fish	**pseudo-** false	**strati-** layer
ophi- a serpent	**plani-** flat, level	**ptero-** wing	**sub-** below
opto- the eye, vision	**plankto-** wandering	**pulmo-** lung	**supra-** above
orni- a bird	**pleisto-** most	**pycno-** dense	**symbio-** living together
-osis condition	**pleuro-** side	**quadra-** four	**taxo-** arrangement
oto- hear	**plio-** more	**quasi-** almost	**tecto-** covering
ovo- egg	**pluri-** several	**radi-** radial	**terra-** earth
pan- all	**pneuma-** air, breath	**rhizo-** root	**terti-** third
para- beside, near	**pod-** foot	**rhodo-** rose-colored	**thalasso-** the sea
pari- equal	**poikilo-** variegated	**sali-** salt	**trocho-** wheel
pecti- comb	**poly-** many	**schizo-** split, division	**tropho-** nourishment
pedi- foot	**polyp** many footed	**scyphi-** cup	**tunic-** cloak, covering
penta- five	**poro-** channel	**semi-** half	**turbi-** disturbed
peri- all around	**post-** behind, after	**septi-** partition	**un-** not
phaeo- dusky	**-pous** foot	**sessil-** sedentary	**vel-** veil
pholado- lurking in a hole	**pre-** before	**siphono-** tube	**ventro-** underside
-phore carrier of	**pro-** before, forward	**-sis** process	**xantho-** yellow
photo- light	**procto-** anus	**spiro-** spiral, coil	**xipho-** sword
phyto- plant	**proto-** first	**stoma-** mouth	**zoo-** animal
pinnati- feather			

Appendix V

A Chemical Background

The following discussion provides a definition of some chemical terms and an explanation of chemical bonds.

An **element** is a substance comprised entirely of like particles, called atoms, that cannot be broken into smaller particles by chemical means. The **atom** is the smallest particle of an element that can combine with similar particles of other elements to produce compounds. The periodic table of elements, shown inside this book's front cover, lists the elements and describes their atoms. A **compound** is a substance containing two or more elements combined in fixed proportions. A **molecule** is the smallest particle of an element or compound that, in the free state, retains the characteristics of the substance.

As an illustration of these terms, consider **Sir Humphrey Davey**'s use of electrical dissociation to break the compound water into its component elements, hydrogen and oxygen. Atoms of the elements hydrogen (H) and oxygen (O) combine in the proportion 2 to 1, respectively, to produce molecules of water (H_2O). As an electric current is passed through the water, the molecules dissociate into hydrogen atoms that collect near the cathode (negatively charged electrode) and oxygen atoms that collect near the anode (positively charged electrode). Here they combine to form the diatomic gaseous molecules of the elements hydrogen (H_2) and oxygen (O_2). Because there are twice as many hydrogen atoms as oxygen atoms in a given volume of water, twice as many molecules of hydrogen gas (H_2) as oxygen gas (O_2) are formed. Further, because under identical conditions of temperature and pressure the volume of gas is proportional to the number of gas particles (molecules) present, two volumes of hydrogen gas are produced for each volume of oxygen.

A Look at the Atom

Building upon earlier discoveries, the Danish physicist **Niels Bohr** (1884–1962) developed his theory of the atom as a small solar system in which a positively charged nucleus takes the place of the sun and the planets that orbit around it are represented by negatively charged electrons. Although this theory has since been diagrammatically altered, it is still commonly used to demonstrate the arrangement of electrons and nuclear particles in the atom.

Bohr's earliest concern was with the atom of hydrogen, which he considered to consist of a single positively charged **proton** in its nucleus orbited by a single negatively charged **electron**. Since the mass of the electron is only about 1/1840 the mass of the proton, it will be considered negligible in our coming discussion of atomic masses.

According to Bohr, the number of protons, or units of positive charge in the nucleus, coincides with the **atomic number** of the element. Thus hydrogen, with an atomic number of 1, has a **nucleus** containing a single proton. Helium, the next heavier element, having an atomic number of 2, contains two nuclear protons, and so forth. The atomic number also indicates the number of electrons in a normal atom of any element, because this number is equal to the number of protons in the nucleus (Figure 1).

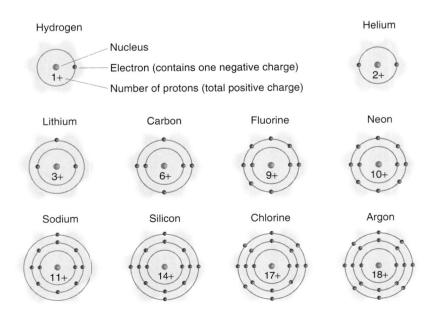

Figure 1 Bohr-Stoner Orbital Models for Atoms

Each atom is composed of a positively charged nucleus with negatively charged electrons around it. The nucleus, which occupies very little space, contains most of the mass of the atom. The atomic number is equal to the number of protons (positvely charged nuclear plarticles) in the nucleus of an atom of an element. Note that the first shell holds only two electrons. Other shells can hold no more than eight electrons when they are in the outer position.

Hydrogen
Nucleus
Electron (contains one negative charge)
1+
Number of protons (total positive charge)

Helium
2+

Lithium
3+

Carbon
6+

Fluorine
9+

Neon
10+

Sodium
11+

Silicon
14+

Chlorine
17+

Argon
18+

An **isotope** is an atom of an element that has a different **atomic mass** than other atoms of the same element. As we will see later in some detail, chemical properties of atoms are determined by the electron arrangement that surrounds the nucleus. This arrangement, in turn, is determined by the number of protons in the nucleus. Because some isotopes have different atomic masses but identical chemical characteristics, we suspect that the proton is not the only nuclear particle that can influence the atomic mass of an atom. There must be a nuclear particle that has no effect on the atom's chemical properties, in that it does not affect the electron structure surrounding the nucleus (Figure 2).

Nuclear research has discovered the additional particle, the **neutron**, which was postulated by **Ernest Rutherford** in 1920 and was first detected by his associate **James Chadwick** in 1932. It has a mass very similar to that of the proton but no electrical charge. This characteristic of being electrically neutral has made it one of the particles most utilized by physicists. These scientists continue to explore the atomic nucleus and to make new discoveries of nuclear particles. We will not consider these particles, as knowledge of them is not necessary for our understanding of the chemical nature of atoms.

We should summarize here that an atom of a given element can be changed to an ion of that element by adding or taking away one or more electrons. It is changed to an isotope of the element by adding or taking away one or more neutrons. Adding or taking away one or more protons will change the atom to an atom of a different element.

We can now divide the atom into two parts—the nucleus, which contains neutrons and protons, and the **electron** cloud surrounding the nucleus that is involved in chemical reactions. It is not possible to determine the precise location of electrons in this cloud at any instant, but it is possible to estimate the most probable position of an electron in this cloud. We will picture these regions in which we would more likely find particular electrons as concentric spheres, or shells, that surround the nucleus (see Figure 1).

Chemical Bonds

In considering the chemical reactions in which atoms are involved, we will be concerned primarily with the distribution of electrons in the outer shell. With some exceptions, when atoms combine to form compounds, they combine in one of two ways:

1. **Ionic bonding**—some of the atoms lose electrons from their outer shell, and others gain electrons in their outer shell
2. **Covalent bonding**—sharing electrons in the outer shells of the atoms

Ionic bonding produces an **ion**, an electrically charged atom that no longer has the properties of a neutral atom of the element it represents. A positively charged ion, a **cation**, is produced by the loss of electrons from the outer shell, the positive charge being equal to the number of electrons lost. A negatively charged ion, an **anion**, is produced by the gain of electrons in the outer shell of an atom, and its charge is equal to the number of electrons gained.

An example of a compound formed by ionic bonding is the metallic salt sodium chloride (Figure 3). In the formation of this compound, the sodium atom, which has one electron in its outer shell, this electron, forming a sodium

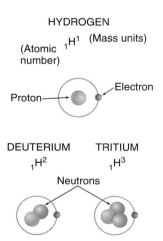

HYDROGEN

(Atomic number) $_1H^1$ (Mass units)

Proton → ○ ← Electron

DEUTERIUM $_1H^2$ TRITIUM $_1H^3$

Neutrons

Figure 2 Neutrons as Nuclear Particles. Hydrogen Isotopes

Isotopes are atoms of an element that have different atomic masses. The hydrogen atom ($_1H^1$) accounts for 99.98% of the hydrogen atoms on Earth. It contains one nuclear particle, a proton. Deuterium ($_1H^2$) contains one neutron and one proton in its nucleus and combines with oxygen to form "heavy water" with a molecular mass of 20. Tritium ($_1H^3$) is a very rare radioactive isotope of hyrogen that has a nucleus containing one proton and two neutrons.

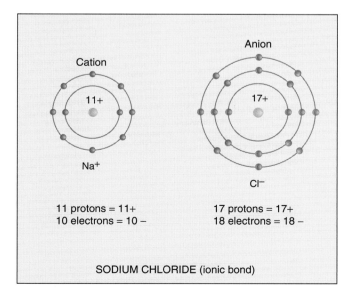

Cation

11+

Na⁺

Anion

17+

Cl⁻

11 protons = 11+
10 electrons = 10 −

17 protons = 17+
18 electrons = 18 −

SODIUM CHLORIDE (ionic bond)

Figure 3 Ionic Bond

Notice the contrast in the outer shells of the ions of sodium (Na⁺) and chlorine (Cl⁻) with the atoms of these elements in Figure 1.

ion with a positive electrical charge of one. The chlorine atom, which contains seven electrons in its outer shell, completes its shell by gaining an electron and becoming a chloride ion with a negative charge of one. These two ions are held in close proximity by an electrostatic attraction between the two ions of equal and opposite charge.

In a statement of great oversimplification, let us say it is the tendency of an individual atom to assume the outer-shell electron content of the inert gases such as helium (2), neon (8), and argon (8) that makes it chemically reactive (*see* Figure 1). Normally, if an atom can assume this configuration by either sharing one or two electrons with an atom of another element or by losing or gaining one or two electrons, the elements are highly reactive. By contrast, the elements are less reactive if three or four electrons must be shared, gained or lost to achieve the desired configuration.

We can describe the combining power of atoms of different elements by the concept of valence. **Valence** can be defined as the number of hydrogen atoms with which an atom of a given element can combine. Valence can be either ionic or covalent. Elements with lower valences of one or two combine chemically in a more highly reactive manner than do those with higher valences of three or four or, in certain instances, more than four. Although those elements with higher valences do not react as violently as low-valence elements, they have a greater combining power and can gather about them larger numbers of atoms of other elements than can the atoms with lower valence values.

An example of a covalent bond is the sharing of electrons by hydrogen and oxygen atoms in the water molecule. In the formation of this molecule, both hydrogen atoms and the oxygen atom assume the inert gas configuration they seek by sharing electrons (Figure 4).

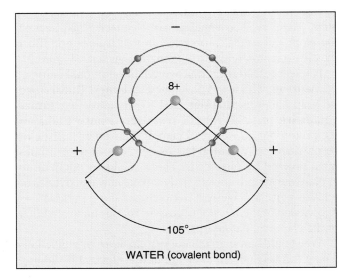

WATER (covalent bond)

Figure 4   Covalent Bond

Glossary

Abiogenic   Not of biological origin

Abiotic   Without life.

Absolute dating   The use of radioisotope half-lives to determine the age of rock units in years within 2 or 3%.

Abyssal clay   Deep-ocean (oceanic) deposits containing less than 30% biogenous sediment.

Abyssal hill   Volcanic peaks rising less than 1 km (0.621 mi) above the ocean floor.

Abyssal hill province   Deep-ocean regions, particularly in the Pacific Ocean, where oceanic sedimentation rates are so low that abyssal plains do not form and the ocean floor is covered with abyssal hills.

Abyssal plain   A flat depositional surface extending seaward from the continental rise or oceanic trenches.

Abyssal zone   The benthic environment between 4000 and 6000 m (13,120 and 19,680 ft).

Abyssopelagic zone   Ocean-ocean (oceanic) environment below 4000 m (13,120 ft) depth.

Acid   A solution in which the H^+ concentration is greater than the OH^- concentration. In a water solution, the hydrogen ion concentration is greater than one part per 10^7.

Acoelomate   Without a secondary body cavity (coelom).

Achitarch   An apparently unicellular microfossil of unknown affinity, possessing a central cavity enclosed by a wall of organic composition.

Active arc   An island arc that is volcanically active because of its position above a subducting lithospheric plate.

Active margin   A continental margin that is tectonically deformed as it collides with another tectonic plate. It is the leading edge of the continent as it moves away from an oceanic spreading center.

Adhesive force   An attractive force that exists between two objects composed of different materials (for example, water and glass).

Adiabatic   Pertaining to a change in the temperature of a mass resulting from compression or expansion. It requires no addition of heat to or loss of heat from the substance.

Agulhas Current   A warm current that carries Indian Ocean water around the southern tip of Africa and into the Atlantic Ocean.

Alaskan gyre   A small Pacific Ocean subpolar current gyre that rotates counterclockwise south of Alaska.

Algae   One-celled or many-celled plants that have no root, stem, or leaf systems; simple plants.

Alkaline   Pertaining to a solution in which the OH^- concentration is greater than the H^+ concentration. In a water solution, the hydrogen ion concentration is less than one part per 10^7.

Amino acid   One of more than 20 naturally occurring compounds that contain NH_2 and $COOH$ groups. They combine to form proteins.

Ammonia (NH_3)   A colorless, pungent gas composed of nitrogen and hydrogen.

Amnesic shellfish poisoning (ASP)   Poisoning caused by domoic acid secreted by a diatom. It has been known to kill birds and humans and obtains its name from the fact that a pervasive symptom in humans is amnesia.

Amphidromic point   A nodal or no-tide point in the ocean or sea around which the crest of the tide wave rotates during one tidal period.

Amphineura   Class of mollusks with eight dorsal calcareous plates; includes chitons.

Amphipoda   Crustacean order containing laterally compressed members such as the "sand hoppers."

Amplitude   Height of wave crest above or trough below still water.

Anadromous   Pertaining to a species of fish that spawns in fresh water, then migrates into the ocean to grow to maturity.

Anaerobic respiration   Respiration carried on in the absence of free oxygen (O_2). Some bacteria and protozoans carry on respiration this way.

Anchoveta   A small, herringlike fish found in the Peruvian upwelling region.

Andesite   A fine-grained igneous rock that is the mineralogic equivalent of diorite. Its name is derived from the Andes Mountains of South America where it is common in association with volcanoes produced by the subduction of the Nazca Plate.

Animalia   Kingdom of many-celled animals. They are members of the domain, eukarya.

Anion   A negatively charged ion.

Annelida   Phylum of elongated segmented worms.

Anomalistic month   The time required for the moon to go from perigee to perigee, 27.5 days.

Anoxic   Without oxygen.

Antarctic Bottom Water   A water mass that forms in the Weddell Sea, sinks to the ocean floor, and spreads across the bottom of all oceans.

Antarctic Circle   The latitude of 66.5° south.

Antarctic Circumpolar Current   The eastward-flowing current that encircles Antarctica and extends from the surface to the deep-ocean floor. The largest volume current in the oceans.

Antarctic Convergence   The zone of convergence along the northern boundary of the Antarctic Circumpolar Current where the southward-flowing boundary currents of the sub-tropical gyres converge on the cold Antarctic waters.

Antarctic Divergence   The zone of divergence separating the westward-flowing East Wind Drift and the easterly-flowing West Wind Drift.

Antarctic krill   A species of krill found in Antarctic waters and a major food for baleen whales. It is being considered as a possible fishery.

Antarctic Intermediate Water   Antarctic zone surface water that sinks at the Antarctic convergence and flows north at a depth of about 900 m (2952 ft) beneath the warmer upper-water mass of the South Atlantic subtropical gyre.

Antarctic Polar Front   Name applied to the Antarctic Convergence

Antilles Current   This warm current flows north seaward of the Lesser Antilles from the north equatorial current of the Atlantic Ocean to join the Florida Current.

Antinode   Zone of maximum vertical particle movement in standing waves where crest and trough formation alternate.

Aphelion   The point in the orbit of a planet or comet where it is farthest from the sun.

Aphotic zone   Without light. The ocean is generally in this state below 1000 meters (3280 ft).

Apogee   The point farthest from Earth in the orbit of the moon or a human-made satellite.

Apparent polar wandering   Assuming little movement of Earth's axis of rotation relative to Earth's surface, the path the poles appear to have followed as a result of motions of lithospheric plates.

Arabian Sea   A sea at the northern margin of the Indian Ocean that lies between India and the Arabian penninsula.

Arabs   Natives of Arabia or members of the Arabian branch of the Semitic peoples.

Aragonite   A form of $CaCO_3$ that is less common and less stable than calcite. Pteropod shells are usually composed of aragonite.

Archaea   A recently identified domain of prokaryotic organisms differentiated from bacteria by genetic characteristics.

Archipelagic apron   A gently sloping sedimentary feature surrounding an oceanic island or seamount.

Arctic Circle   The latitude of 66.5° north.

Arctic Convergence   Regions in the North Atlantic and Pacific oceans where cold water from the subpolar gyres meets warmer water of the subtropical gyres.

Arrowworm   A member of the phylum Chaetognatha. It averages about 1 cm in length and is an important member of the plankton.

Aschelminthes   Phylum of wormlike pseudocoelomates.

Aspect ratio   The index of propulsive efficiency obtained by dividing the square of fin height by the area of the fin.

Asthenosphere   A plastic layer in the upper mantle from depths as shallow as 10 km (6.2 mi) to as deep as 800 km (497 mi) which may allow lateral movement of lithospheric plates and isostatic adjustments.

Atlantic salmon   Any of a variety of fish belonging to the genus *Salmo* and living in the North Atlantic Ocean. Salmon have been moved to many locations in the world as objects of mariculture.

Atlantic-type margin   The passive trailing edge of a continent that is subsiding because of lithospheric cooling and increasing sediment load.

Atlantic II Deep   A deep basin in the Red Sea where hypersaline, hyperthermal water is emitted from vents associated with the spreading center. Metal-rich deposits are found there.

Atoll   A ring-shaped coral reef growing upward from a submerged volcanic peak. It may have low-lying islands composed of coral debris.

Atom   The smallest particle of an element that can combine with similar particles of other elements to produce compounds.

Atomic mass   A number equal to the sum of protons and neutrons in the nucleus of an atom of an element.

Atomic number   A number representing the relative position of an element in the periodic table of elements. It is equal to the number of positive charges in the atom's nucleus.

Augite   A dark mineral, usually black, rich in iron and magnesium. An important constituent of basalt, the rock that is characteristic of the ocean crust.

Authigenic mineral   A mineral that forms in place, such as phosphorite and those found in manganese nodules.

Autolytic decomposition   The breakdown of organic matter by enzymes activated when an organism dies.

Autotomy   Pertaining to the ability of some organisms to slough off certain body parts as a defensive mechanism.

Autotroph   Plant or bacterium that can synthesize organic compounds from inorganic nutrients.

Autumn bloom   The increase in temperate ocean photosynthetic production resulting from decreased autumnal thermal stratification and increased wind-driven mixing of nutrients into the sunlit surface waters.

Autumnal equinox   The passage of the sun across the equator as it moves from the Northern Hemisphere into the Southern Hemisphere, September 23.

Auxospore   A diatom cell that has shed its frustule to allow growth.

Back-arc spreading center   Spreading centers behind island arcs that result from the tensional stresses created by the seaward migration of the associated oceanic trench.

Backshore   The inner portion of the shore, lying landward of the mean spring-tide high-water line. Acted upon by the ocean only during exceptionally high tides and storms.

Bacteria   A domain of prokaryotic organisms.

Bacterioplankton   Bacteria that live as plankton.

Bacteriovore   An organism that feeds primarily on bacteria.

Bar-built estuary   A shallow estuary (lagoon) separated from the open ocean by a bar deposit such as a barrier island. The water in these estuaries usually exhibits vertical mixing.

Barnacle   *See* Cirripedia.

Barrier flat   Lying between the salt marsh and dunes of a barrier island, it is usually covered with grasses and even forests if protected from overwash for sufficient time.

Barrier island   A long, narrow, wave-built island separated from the mainland by a lagoon.

Barrier reef   A coral reef separated from the nearby landmass by open water.

Basalt   A dark-colored volcanic rock characteristic of the ocean crust. Contains minerals with relatively high iron and magnesium content.

Base   A compound that releases hydroxide ions in aqueous solution.

Bathyal zone   Benthic environment from 200 to 4000 m (656 to 13,120 ft).

Bathymetry   The study of ocean depth.

Bathypelagic   Open-ocean environment of approximately 1000 to 4000 m (3280 to 13,120 ft).

Bathysphere   A spherical chamber with windows that can be lowered into the sea so humans can observe marine life and the ocean floor.

Bay   A coastal body of water enclosed by land on three sides and open to the ocean on one side.

Bay barrier   A marine deposit attached to the mainland at both ends and extending entirely across the mouth of a bay, separating the bay from the open water.

Bay of Bengal   A sea at the northern margin of the Indian Ocean that is located between India and the Malay peninsula.

Beach   Sediment seaward of the coastline through the surf zone that is in transport along the shore and within the surf zone.

Benguela Current   The cold eastern boundary current of the South Atlantic subtropical gyre.

Benthic   Pertaining to the ocean bottom.

Benthic nepheloid layer (BNL)   A layer of turbid water adjacent to the ocean floor; created by bottom currents.

Benthos   The forms of marine life that live on the ocean bottom.

Bering Sea   A sea at the northern margin of the Pacific Ocean that is bounded on the south by the Aleutian Islands.

Bicarbonate ion   An ionic molecule (HCO_3^-) representing the form in which most CO_2 is stored in ocean water.

Big Bang   The theoretical explosion that set the universe into its expansive motion 12 to 15 billion years ago.

Biocommunity   A distinctive community of organisms associated with a specific physical environment such as hydrothermal vents

Bioerosion   Erosion of coral and rock surfaces on the ocean floor by organisms such as sponges and sea urchins.

Biogenous particles   Material produced by plants or animals, e.g., coral reefs, shell fragments, and housing of diatoms, radiolarians, foraminifera, and coccolithophores.

Biogeochemical cycling   The natural cycling of compounds among the living and nonliving components of an ecosystem.

Biological pump   The movement of CO_2 that enters the ocean from the atmosphere through the water column to the sediment on the ocean floor by biological processes—photosynthesis, secretion of shells, feeding, and dying.

Bioluminescence   Light produced by chemical reaction. Found in bacteria, phytoplankton, and metazoans.

Biomass   Total weight of the organisms in a particular habitat, species, or group of species.

Biotic community   All the organisms that live within some definable area.

Black smoker   Submarine hydrothermal vents that have vent water temperatures around 350°C. These waters are rich in black particles of metallic sulfides.

Bladder kelp   Species of kelp with air bladders that serve as floats to help the organism stand erect within the water column.

Body wave   A longitudinal or transverse wave that transmits energy through a body of matter.

Boiling point   The temperature at which a substance changes state from a liquid to a gas at a given pressure.

Bore   A steep-fronted tide crest that moves up a river in association with high tide.

Bosphorus   A narrow strait between the Black Sea and the Sea of Marmara through which Mediterranean and Black Sea water may mix.

Boundary current The northward- or southward-flowing currents that form the western and eastern boundaries, respectively, of the subtropical circulation gyres.

Brazil Current The warm western boundary current of the South Atlantic subtropical gyre.

Breaker A wave that obtains a steepness of 1/7 on approaching the shore, breaks, and releases its energy.

Breakwater Any artificial structure constructed to protect a coastal region from the force of ocean waves.

Bryozoa Phylum of colonial animals that often share one coelomic cavity. Encrusting and branching forms secrete a protective housing (zooecium) of calcium carbonate or chitinous material. Possess lophophore feeding structure.

Buffering Of chemical or other processes, any process that reduces impact.

Buoyancy The ability or tendency to float or rise in a liquid.

Buttress zone The reef slope above 20 meters depth exposed to maximum wave energy.

Calcareous Containing calcium carbonate.

Calcite The most common form of $CaCO_3$.

Calcium carbonate ($CaCO_3$) A chalklike substance secreted by many organisms in the form of coverings or skeletal structures.

California Current The slow moving, cold boundary current that forms the eastern side of the North Pacific gyre.

Calorie Unit of energy defined as the amount of heat required to raise the temperature of 1 g of water $1°C$ ($1.8°F$).

Canary Current The cold eastern boundary current of the North Atlantic subtropical gyre.

Capillarity The process by which fluid (such as water) is drawn up into small interstices or tubes as a result of surface tension.

Capillary wave Ocean wave whose wavelength is less than 1.74 cm (0.7 in). The dominant restoring force for such waves is surface tension.

Carapace Chitinous or calcareous shield that covers the cephalothorax of some crustaceans. Dorsal portion of a turtle shell.

Carbohydrate An organic compound consisting of carbon, hydrogen, and oxygen. Sugars and starches are examples.

Carbonate ion An ionic molecule (CO_3^{2-}) that combines with Ca^{2+} to produce calcium carbonate ($CaCO_3$).

Carbonate compensation depth (CCD) The depth at which carbonate particles falling from above are dissolved.

Carbon dioxide (CO_2) A gas containing one atom of carbon and two atoms of oxygen.

Carbonic acid An acid (H_2CO_3) that forms when carbon dioxide is dissolved in water and combines with it.

Caribbean Sea A sea at the western margin of the North Atlantic Ocean. It is bounded on the north and east by the Greater and Lesser Antilles Islands.

Caribbean Surface Water The surface water mass of the Caribbean Sea which generally has salinity in excess of 35.5‰ in open water away from coastal freshwater runoff.

Carnivore An animal that depends solely or chiefly on other animals for its food supply.

Carotene (carotin) A red to yellow pigment found in plants.

Catadromous Pertaining to a species of fish that spawns at sea, then migrates into a freshwater stream or lake to grow to maturity.

Cation A positively charged ion.

Celsius temperature scale Scale in which $0°C = 273.16$ K; $0°C$ = freezing point of water; $100°C$ = boiling point of water.

Center of gravity The point where the entire mass of a body may be considered to be concentrated.

Centrifugal force A force that seems to make an object move away from the center of a curved path it is following. It results from the application of a centripetal force acting against the inertia of the object.

Centripetal acceleration The acceleration of an object in motion that results from the application of centripetal force to the object.

Centripetal force A center-seeking force that tends to make rotating bodies move toward the center of rotation.

Cephalopoda A class of the phylum Mollusca with a well-developed pair of eyes and a ring of tentacles surrounding the mouth. The shell is absent or internal on most members. The class includes the squid, octopus, and *Nautilus*.

Cetacea An order of marine mammals that includes the whales.

Chaetognatha A phylum of elongate, transparent, wormlike pelagic animals commonly called *arrowworms*.

Characteristic period The period with which a seiche, a single-node free-standing wave, will possess within any given body of water.

Chela Arthropod appendage modified to form a pincer.

Chemical weathering The weathering of rock by chemical reactions. Carbonic acid is an important cause of chemical weathering.

Chemosynthesis The formation of organic compounds from inorganic substances using energy derived from oxidation.

Chesapeake Bay A bay formed by the flooding of the mouths of the Susquehanna River and its tributaries. It is bounded by the states of Virginia and Maryland.

Chiton Common name for any member of the Amphineura class of mollusks with eight dorsal plates.

Chloride ion (Cl^-) A chlorine atom that has become negatively charged by gaining one electron.

Chlorinity The chloride content of seawater expressed in grams per kilogram (g/kg) or parts per thousand (‰) by weight. It includes all the halide ions (F^-, Cl^-, Br^-, I^-, and At^-).

Chlorophyll A group of green pigments that make it possible for plants to carry on photosynthesis.

Chlorophyta Green algae; characterized by the presence of chlorophyll and other pigments.

Chronometer, John Harrison's The first practical marine timepiece that made possible the accurate determination of longitude. It was first put into service by the British admiralty in 1762.

Chrysophyta An important phylum of planktonic algae, including the diatoms. The presence of chlorophyll is masked by the pigment carotin, which gives the plants a golden color.

Cilia Short hairlike structures common on lower animals. Beating in unison, they may be used for locomotion or to create water currents that carry food toward the mouth of the animal.

Circadian rhythm Behavioral and physiological rhythms of organisms related to the 24-hour day. Sleeping and waking patterns are an example.

Circumpolar Current Eastward-flowing current that extends from the surface to the ocean floor and encircles Antarctica.

Cirripedia An order of crustaceans with up to six pairs of thoracic appendages that strain food from the water. They are barnacles that attach themselves to a substrate and secrete an external calcareous housing.

Clastic Pertaining to a rock or sediment composed of broken fragments of pre-existing rocks. Two common examples are beach deposits and sandstone.

Clay A term relating to particle size between silt and colloid. Clay minerals are hydrous aluminum silicates with plastic, expansive, and cation exchange properties.

Cnidaria Phylum that contains some 10,000 species of predominantly marine animals with a sacklike body and stinging cells on tentacles that surround the single opening to the gut cavity. There are two basic body forms. The medusa is a pelagic form represented by the jellyfish. The polyp is a predominantly benthic form found in sea anemones and corals. Also called *Coelenterata*.

Cnidoblast Stinging cell of the phylum Cnidaria; contains a stinging mechanism (nematocyst) used in defense and capturing prey.

Coal A black, hard fuel that results from the burial and partial decomposition of organic matter in the absence of oxygen and under elevated temperature.

Coast A strip of land that extends inland from the coastline as far as marine influence is evidenced in the landforms.

Coastal downwelling When winds blow water toward shore, the water piles up against the shore and sinks beneath the surface.

Coastal plain estuary An estuary formed by rising sea level flooding a coastal river valley.

Coastal upwelling The movement of deeper nutrient-rich water into the surface water mass as a result of windblown surface water moving offshore.

Coastline Landward limit of the effect of the highest storm waves on the shore.

Cobalt-rich manganese crust Hydrogenous deposits found on the flanks of volcanic islands and seamounts.

Coccolith Tiny calcareous discs averaging about 3 μm in diameter that form the cell wall of coccolithophores.

Coccolithophorid A microscopic planktonic form of algae, encased by a covering composed of calcareous discs (coccoliths).

Coelenterata Phylum of radially symmetrical animals that includes two basic body forms, the medusa and the polyp. Includes jellyfish (medusoid) and sea anemones (polypoid). Also called *Cnidaria*.

Coelom Secondary body cavity (gut cavity is primary cavity). Forms within the mesoderm in higher animals, is lined with peritoneum, and contains vital organs.

Cold boundary current Slow drifting currents on the eastern side of a subtropical gyre that carry cold water toward the equator.

Cold current A current carrying cold water into areas of warmer water. It generally travels toward the equator.

Cold front A weather front in which a cold air mass moves into and under a warm air mass. It creates a narrow band of intense precipitation.

Colloid Substance having particles of a size smaller than clay.

Colonial animal An animal that lives in groups of attached or separate individuals. Groups of individuals may serve special functions.

Columbia River estuary An estuary at the border between the states of Washington and Oregon that has been most adversely affected by the construction of hydroelectric dams.

Comb jelly Common name for members of the phylum Ctenophora. (*See* Ctenophora.)

Commensalism A symbiotic relationship in which one organism benefits at no expense to its host.

Compound A substance containing two or more elements combined in fixed proportions.

Condensation The conversion of water from the vapor to the liquid state. When it occurs, the energy required to vaporize the water is released. This is about 585 cal/g of water at 20°C (68°F).

Conduction The transmission of heat by the passage of energy from particle to particle.

Conjunction An apparent closeness of two or more heavenly bodies. During the new moon, the sun and moon are in conjunction on the same side of Earth.

Conservative property A property of surface ocean water that is changed only by mixing and diffusion after the water sinks below the surface.

Conservative tracer A chemical that has a very low rate of chemical reactivity in ocean water. It can be used to trace subsurface water motion over long periods of time.

Constancy of composition, Rule of The major constituents of ocean-water salinity are found in the same relative concentrations throughout the ocean-water volume.

Constructive interference A form of wave interference in which two waves come together in phase, e.g., crest to crest, to produce a greater displacement from the still-water line than that produced by either of the waves alone.

Consumer A heterotrophic organism that consumes an external supply of organic matter.

Continent About one-third of Earth's surface that rises above the deep-ocean floor to be exposed above sea level. Continents are composed primarily of granite, an igneous rock of lower density than the basaltic oceanic crust.

Continental arc A volcanic arc confined to the margin of a continent and associated with an ocean trench. The Andes Mountains and the Peru-Chile Trench constitute such a system.

Continental borderland A highly irregular portion of the continental margin that is submerged beneath the ocean and is characterized by depths greater than those characteristic of the continental shelf.

Continental drift A term applied to early theories supporting the possibility that the continents are in motion over Earth's surface.

Continental flood basalt (CFB) Large volume flows of lava that spread out over continents after a plume of hot mantle material breaks through the lithosphere. Typically, more than 1 million km^3 of lava will be deposited in about 1 million years.

Continental margin Extending from the shoreline to the deep-ocean basin, this feature includes the continental shelf, continental slope, and continental rise.

Continental rise A gently sloping depositional surface at the base of the continental slope.

Continental shelf A gently sloping depositional surface extending from the low-water line to a marked increase in slope around the margin of a continent.

Continental slope A relatively steeply sloping surface lying seaward of the continental shelf.

Convection Process by which, in a fluid being heated, the warmer part of the mass will rise and the cooler portions will sink. If the heat source is stationary, **convection cells** may develop as the rising warm fluid cools and sinks in regions on either side of the axis of rising.

Convergence Coming together of water masses in polar, tropical, and subtropical regions of the ocean. Along these lines of convergence, the denser mass will sink beneath the others.

Convergent plate boundary A lithospheric plate boundary where adjacent plates converge, producing ocean trench island arc systems, ocean trench continental volcanic arcs, or folded mountain ranges. Also called *destructive boundary*.

Copepoda An order of microscopic to nearly microscopic crustaceans that are important members of the zooplankton in temperate and subpolar waters.

Copper sulfide ore Ore containing sulfides of metals such as nickel and copper.

Coral A group of benthic coelenterates that exist as individuals or in colonies and may secrete external skeletons of calcium carbonate.

Coral bleaching A condition, caused by stresses such as high temperature, that results in hermatypic corals expelling their algal symbionts for their survival. This causes the coral to appear bleached as the algae provide color to coral tissue.

Coral reef A calcareous organic reef composed significantly of solid coral and coral sand. Algae may be responsible for more than half of the $CaCO_3$ reef material. Found in waters where the minimum average monthly temperature is 18°C.

Corange lines A circular line, circling an amphidromic point and crossing cotidal lines at right angles, that is composed of all points with a given value of tidal range.

Core The core of Earth is composed primarily of iron and nickel. It has a liquid outer portion 2270 km (1410 ft) thick and a solid inner core with a radius of 1216 km (755 ft).

Core-mantle boundary (CMB) The boundary between Earth's core and mantle at a depth of about 2885 km (1792 ft).

Coriolis effect An effect resulting from Earth's rotation that causes particles in motion to be deflected to the right in the Northern Hemisphere and to the left in the Southern Hemisphere.

Cosmogenous particles   All sediment derived from outer space.

Cotidal lines   Lines connecting points where high tides occur simultaneously.

Covalent bond   A chemical bond in which atoms combine to form compounds by sharing electrons. The water molecule is an example.

Crest (wave)   The portion of an ocean wave that is displaced above the still-water line.

Cromwell Current   A ribbonlike eastward-flowing current embedded in the South Equatorial Current that flows from Samoa to the Galápagos Islands.

Cruiser   Fish such as the bluefin tuna that constantly cruise the pelagic waters in search of food.

Crust   Unit of Earth's structure that is composed of basaltic ocean crust and granitic continental crust. The total thickness of the crustal units may range from 5 km (3.1 mi) beneath the ocean to 50 km (31 mi) beneath the continents.

Crustacea   A class of the phylum Arthropoda that includes barnacles, copepods, lobsters, crabs, and shrimp.

Crystalline rock   Igneous or metamorphic rocks. These rocks are made up of crystalline particles with orderly molecular structures.

Ctenophora   A phylum of gelatinous organisms that are more or less spheroidal with biradial symmetry. These exclusively marine animals have eight rows of ciliated combs for locomotion, and most have two tentacles for capturing prey.

Curie point   The temperature above which thermal agitation prevents spontaneous alignment of magnetic particles with Earth's magnetic field.

Current   A horizontal movement of water.

Cutaneous artery   The artery that runs down both sides of some cruiser-type fish to help maintain a constant elevated temperature in the myomere musculature used for swimming.

Cutaneous vein   The vein that runs down both sides of some cruiser-type fish to help maintain a constant elevated temperature in the myomere musculature used for swimming.

Cyanobacteria   Photosynthetic bacteria that may have been the first organisms to use aerobic respiration.

Cypris   Advanced free-swimming larval stage of barnacles. After attaching to substrate, it metamorphoses into an adult.

Davidson Current   A northward-flowing current along the Washington-Oregon coast that is driven by geostrophic effects on a large freshwater runoff.

Dayglow   The absorption and reradiation of solar energy by atomic oxygen 290 km (180 mi) above sea level creates this glow that can be imaged by satellites in space.

DDT   An insecticide that caused damage to marine bird populations in the 1950s and 1960s. Its use is now banned throughout most of the world.

Decapoda   1. An order of crustaceans with five pairs of thoracic "walking legs," including crabs, shrimp, and lobsters. 2. Suborder of cephalopod mollusks with 10 arms that includes squids and cuttlefish.

Declination   The angular distance of the sun or moon above or below the plane of Earth's equator.

Decomposer   An organism, primarily bacteria, that breaks down nonliving organic material, extracts some of the products of decomposition for its own needs, and make available the compounds needed for plant production.

Decompression sickness   Also known as the bends. It results from too rapid ascent from deep dives. Nitrogen bubbles form in the blood and tissue, causing great pain. It can be deadly.

Deep boundary current   Relatively strong deep current flowing across the continental rise along the western margin of an ocean basin.

Deep-ocean basin   The deep part of the ocean beyond the continental margin.

Deep scattering layer (DSL)   A layer of marine organisms in the open ocean that scatters signals from an echo sounder. The organisms migrate daily from near the surface at night to more than 800 m (2624 ft) during the day.

Deep-sea fan   A large fan-shaped deposit commonly found on the continental rise at the mouth of submarine canyons. They are particularly well developed seaward of such sediment-laden rivers as the Amazon, Indus, or Ganges-Brahmaputra.

Deep-sea system   System that includes all benthic environments beneath the littoral (sublittoral, bathyal, abyssal, and hadal).

Deep water mass   That portion of the water column from the base of the permanent thermocline or pycnocline to the ocean floor. Water temperatures are relatively uniform throughout.

Deep-water wave   Ocean wave traveling in water that has a depth greater than one-half the average wavelength. Its velocity is independent of water depth.

Delta   A low-lying deposit at the mouth of a river.

Denitrifying bacterium   Bacterium that reduces oxides of nitrogen to produce free nitrogen (N_2).

Density   Mass per unit volume of a substance. Usually expressed as grams per cubic centimeter. For ocean water with a salinity of 35‰ at 0°C, the density is 1.028 g/cm^3. (*See* sigma.)

Density (σ, or sigma)   Density of ocean water.

Density, In situ (σ_T, or sigma T)   Density of water in place.

Density, Potential (σ_Θ, or sigma theta)   Density of ocean water with the adiabatic effect removed. It is always less than *in situ* density except at the surface where the adiabatic effect is zero.

Depth of frictional influence   The depth to which surface currents driven by wind extend. At this depth, the energy put in the water by the wind has been consumed by friction.

Desalination   The removal of salt ions from ocean water to produce pure water.

Desiccation   Process of drying out.

Destructive interference   A form of wave interference in which two waves come together out of phase, e.g., crest to trough, and produce a wave with less displacement than the larger of the two waves would have produced alone.

Detritus   Any loose material produced directly from rock disintegration. (Organic: material resulting from the disintegration of dead organic remains.)

Diamond   A precious mineral that is the hardest natural substance on Earth, has great brilliance, and is a pure crystalline form of carbon.

Diatom   Member of the class Bacillariophyceae of algae; possesses a wall of overlapping silica valves.

Diatomite   A deposit composed primarily of the frustules of diatoms.

Diatom ooze   An oceanic deposit that is composed of at least 30% biogenous particles, most of which are diatom shells.

Diffraction   Any bending of a wave around an obstacle that cannot be interpreted as refraction or reflection.

Diffusion   The transfer of material or a property by random molecular movement. The movement is from a region in which the material or the property is high in concentration to regions of low concentration.

Dinoflagellate   Single-celled microscopic organism that may possess chlorophyll and belong to the plant phylum Pyrrophyta (autotrophic) or may ingest food and belong to the class Mastigophora of the animal phylum Protozoa (heterotrophic).

Diploblastic   Pertaining to body structure composed of two cell layers: ectoderm and endoderm. Diploblastic phyla are Porifera, Cnidaria, and Ctenophora.

Dipolar   Having two poles. The water molecule possesses a polarity of electrical charge with one pole being more positive and the other more negative in electrical charge.

Discoaster   A tiny star-shaped plate that may have formed on a coccolithophorelike algal cell which died out at the start of the Pleistocene epoch.

Discontinuity   An abrupt change in a property, such as temperature or salinity, at a line or surface.

Disphotic zone   The dimly lit zone corresponding approximately with the mesopelagic in which there is not enough light to carry on photosynthesis. Sometimes called the *twilight zone*.

Dissolved organic carbon (DOC)   The carbon contained in dissolved organic compounds.

Dissolved oxygen   Oxygen that is dissolved in ocean water.

Dissolved oxygen minimum   The depth at which dissolved oxygen concentration in the water column reaches its lowest value. It is usually between 800 and 1000 m (2624 and 3280 ft).

Distillation   A method of purifying liquids by heating them to their boiling point and condensing the vapor.

Distributary   A small stream flowing away from a main stream. Such streams are characteristic of deltas.

Distributary channel   One of a system of divergent channels leading away from the mouth of a submarine canyon and across the deep-sea fan deposited by turbidity flows.

Diurnal inequality   The difference in the heights of two successive high or low waters during a lunar (tidal) day.

Diurnal tide   A tide with one high water and one low water during a tidal day. Tidal period is 24 h 50 min.

Divergence   A horizontal flow of fluid from a central region, as occurs in upwelling.

Divergent plate boundary   A lithospheric plate boundary where adjacent plates diverge, producing an oceanic ridge or rise (spreading center). Also called *constructive boundary*.

Doldrums   A belt of light, variable winds 10° to 15° north and south of the equator, resulting from the vertical flow of low-density air within this equatorial belt. Doldrums is the common name for the *Intertropical Convergence Zone*.

Dolomite   A common rock-forming mineral; a calcium-magnesium carbonate [$CaMg(CO_3)_2$].

Dolphin   1. A brilliantly colored fish of the genus *Coryphaena*. 2. The name applied to the small, beaked members of the cetacean family Delphinidae.

Dorsal   Pertaining to the back or upper surface of most animals.

Dorsal aorta   For most fish, this is the only major artery that runs the length of the fish through openings in the vertebrae and supplies blood. Some pelagic cruisers also have a cutaneous artery.

Downwelling   In the open or coastal ocean where Ekman transport causes surface waters to converge or impinge on the coast, surface water will be carried down beneath the surface.

Drift net   Long net set in the sea to passively catch anything that swims into it.

Drift   Thick sediment deposit on the continental rise produced where the western boundary undercurrent slows and loses sediment as it changes direction to follow the base of the continental slope.

Drowned beach   An ancient beach now beneath the coastal ocean because of rising sea level.

Drowned river   River mouth that is inundated by rising sea level to produce a coastal plain estuary like Chesapeake Bay.

Dune   Coastal deposit of sand lying landward of the beach and deriving its sand from onshore winds that transport beach sand inland.

Dynamical tide theory   The theory of tidal behavior that takes into account friction between the ocean water and the ocean floor, the effects of changing depth of the ocean floor, and the interference of the continents on the passage of tidal waves.

Dynamic topography   A surface configuration resulting from the geopotential difference between a given surface and a reference surface of no motion. A contour map of this surface is useful in estimating the nature of geostrophic currents.

Earthquake   A sudden motion or trembling in the earth caused by the sudden release of slowly accumulated strain by faulting (movement along a fracture in Earth's crust) or volcanic activity.

East Australian Current   A warm boundary current flowing south along the east coast of Australia.

Eastern boundary current   A slow, drifting cold current that flows toward the equator on the eastern side of a subtropical gyre.

East Pacific Rise   A fast-spreading divergent plate boundary extending southward from the Gulf of California through the eastern South Pacific Ocean.

East Wind Drift   The coastal current driven in a westerly direction by the polar easterly winds blowing off of Antarctica.

Ebb current   Seaward-flowing current during a decrease in the height of the tide.

Echinodermata   Phylum of animals that have bilateral symmetry in larval forms and usually a five-sided radial symmetry as adults. Benthic and possessing rigid or articulating exoskeletons of calcium carbonate with spines, this phylum includes sea stars, brittle stars, sea urchins, sand dollars, sea cucumbers, and sea lilies.

Echosounder   A device that transmits sound from a ship's hull to the ocean floor where it is reflected back to receivers. Knowing the speed of sound in the water, the depth can be determined from the travel-time of the sound signal.

Ecliptic   The plane of the center of the Earth-moon system as it orbits around the sun.

Ecological efficiency   Efficiency with which energy is transferred from one trophic level to the next, or the ratio of the amount of protoplasm added to a trophic level to the amount of food required to produce it.

Ecosystem   All the organisms in a biotic community and the abiotic environmental factors with which they interact.

Ectoderm   Outermost layer of cells in an animal embryo. In vertebrates it gives rise to the skin, nervous system, sense organs, etc.

Eddy   A current of any fluid forming on the side of or within a main current. It usually moves in a circular path and develops where currents encounter obstacles or flow past one another.

Ekman layer   The surface layer of water within which occurs the Ekman spiral.

Ekman spiral   A theoretical consideration of the effect of a steady wind blowing over an ocean of unlimited depth and breadth and of uniform viscosity. The result is a surface flow at 45° to the right of the wind in the Northern Hemisphere. Water at increasing depth will drift in directions increasingly to the right until at about 100 m (328 ft) depth it is moving in a direction opposite to that of the wind. The net water transport is 90° to the wind, and velocity decreases with depth.

Ekman transport   The net transport of surface water set in motion by wind. Because of the Ekman spiral phenomenon, it is theoretically in a direction 90° to the right and 90° to the left of the wind direction in the Northern Hemisphere and Southern Hemisphere, respectively.

Electrodialysis   The removal of undesired ions from a solution (seawater) by the application of direct current to electrodes in dialysis system.

Electromagnetic energy   Energy that travels as waves or particles with the speed of light. Different kinds possess different properties based on wavelength. The longest wavelengths (up to 10 km, or 6.21 mi, in length) belong to radio waves. At the other end of the spectrum are cosmic rays with great penetrating power and wavelengths of less than 0.000001 μm.

Electromagnetic spectrum   The spectrum of radiant energy emitted from stars and ranging between cosmic rays with wavelengths less than 10^{-11} cm and very long waves with wavelengths in excess of 100 km (62 mi).

Electron   A negatively charged particle in orbit around the nucleus of an atom.

Electron cloud   The organized assemblage of electrons surrounding the nucleus of an atom.

Element   One of a number of substances, each of which is composed entirely of like atoms.

El Niño   A southerly-flowing warm current that generally develops off the coast of Ecuador shortly after Christmas. Occasionally it will move farther south into Peruvian coastal waters and cause the widespread death of plankton and fish.

El Niño–Southern Oscillation   The correlation of El Niño events with an oscillatory pattern of pressure change in a persistent high pressure cell in the southeastern Pacific Ocean and a persistent low pressure cell over the East Indies.

Emergent shoreline   A shoreline resulting from the emergence of the ocean floor relative to the ocean surface. It is usually rather straight and characterized by marine features usually found at a greater depth.

Endoderm   Innermost cell layer of an embryo. Develops into the digestive and excretory systems and forms the lining for the respiratory system, etc., in vertebrates.

Endothermic reaction   A chemical reaction that absorbs energy. For example, energy is stored in the organic products of the chemical reaction photosynthesis.

Enterprise   The mining entity established by the United Nations Law of the Sea convention to benefit the developing nations.

Entropy   A quantity reflecting the degree of uniform distribution of heat energy in a system. It increases with time and represents a state in which energy is unrecoverable for work.

Environment   The sum of all physical, chemical, and biological factors to which an organism or community is subjected.

Epicenter   The point on Earth's surface that is directly above the focus of an earthquake.

Epifauna   Animals that live on the ocean bottom, either attached or moving freely over it.

Epipelagic zone   The upper region of the oceanic province, extending to a depth of 200 m (656 ft).

Epitheca   The top valve of a diatom frustule.

Equatorial Countercurrent   Eastward-flowing currents found between the north and south equatorial currents in all oceans, but particularly well developed in the Pacific Ocean.

Equatorial Current   A warm current in the equatorial region driven in a westerly direction by the trade winds.

Equatorial tide   A semimonthly tide occurring when the moon is over the equator. It displays a minimal diurnal inequality.

Equatorial Undercurrent   A thin, ribbonlike east-flowing current embedded in the Pacific South Equatorial Current that flows over 6000 km from the western Pacific to the Galápagos Islands.

Equatorial upwelling   Water that rises from beneath the surface to replace surface water that moves away from the equator at the surface due to Ekman transport.

Equilibrium tide theory   A tidal hypothesis that considers the ocean to be of uniform nature and depth throughout Earth's surface. It is further assumed that this ocean will respond instantly to the gravitational forces of the sun and moon.

Equinox   The times when the sun is over the equator, making day and night of equal length throughout Earth. *Vernal equinox* occurs about March 21 as the sun is moving into the Northern Hemisphere. *Autumnal equinox* occurs about September 21 as the sun is moving into the Southern Hemisphere.

Estuarine circulation   Circulation characteristic of an estuary and other bodies of water having restricted circulation with the ocean; this results from an excess of runoff and precipitation as compared to evaporation. Surface flow is toward the ocean with a subsurface counter flow.

Estuary   The mouth of a river valley, or a bay or lagoon receiving fresh water, where marine influence is manifested as tidal effects and increased salinity of the fresh water.

Eukaryotic cells   Cells that possess a nucleus, other intracellular bodies such as mitochondria and plastids, and a cell wall more complex than that of prokaryotic cells.

Euphausiacea   An order of planktonic crustaceans ranging in length from 5 to 30 cm (1.97 to 11.82 in). Most possess luminous organs, and some are the principal food for baleen whales.

Euphotic zone   The surface layer of the ocean that receives enough light to support photosynthesis. The bottom of this zone, which is marked by the oxygen compensation depth, varies and reaches a maximum value of around 150 m (492 ft) in the very clearest open ocean water.

Euryhaline   Pertaining to the ability of a marine organism to tolerate a wide range of salinity.

Eurythermal   Pertaining to the ability of a marine organism to tolerate a wide range of temperature.

Eustatic   "Truly stationary." A term used in reference to world-wide changes in sea level resulting from increase or decrease in ocean water volume or the volume capacity of ocean basins. The term derives from the ideal consideration that the continents remain stationary during the process.

Eutrophic   Characterized by an abundance of nutrients.

Evaporation   The physical process of converting a liquid to a gas. Commonly considered to occur at a temperature below the boiling point of the liquid.

Excess volatile   Any of the volatile compounds found in the oceans, sediments, and atmosphere in quantities greater than the chemical weathering of crystalline rock could produce. They are considered to have been produced by volcanic action.

Exclusive Economic Zone (EEZ)   A coastal zone 200 naut mi wide over which the coastal nation has jurisdiction over mineral resources, fishing, and pollution. If the continental shelf extends beyond 200 miles, the EEZ may be up to 350 miles in width.

Exothermic reaction   A chemical reaction that liberates energy. For example, the energy stored in the products of photosynthesis is released by the chemical reaction respiration.

Extrusive rock   Igneous rock formed from lava that flows out onto Earth's surface and cools rapidly.

Fahrenheit temperature scale (°F)   Scale in which the freezing point of water is 32°, boiling point of water is 212°.

Falkland Current   A northward-flowing cold current found off the southeastern coast of South America.

Fallout   The radioactive atomic nuclei that were released into the atmosphere by nuclear bomb testing and fell out onto the continents and into the oceans.

Fan   A gently sloping, fan-shaped feature normally located at the lower end of a submarine canyon.

Fast ice   Sea ice that is attached to the shore and therefore remains stationary.

Fat   A colorless, odorless organic compound consisting of carbon, hydrogen, and oxygen; insoluble in water.

Fathom   A unit of ocean depth commonly used in countries using the English system of units. It is equal to 1.83 m, or 6 ft.

Fault   A fracture or fracture zone in Earth's crust along which displacement has occurred.

Fault block   A crustal block bounded on at least two sides by faults. Usually elongate; if it is down-dropped, it produces a graben; if uplifted, it is a horst.

Fauna   The animal life of any particular area or of any particular time.

Fecal pellet   Organic excrement found in marine sediment and produced primarily by invertebrates. Usually of an ovoid form that is less than 1 mm (0.04 in) in length.

Ferromagnesian mineral   Mineral rich in iron and magnesium.

Fetch   1. Area of the open ocean over which the wind blows with constant speed and direction, thereby creating a wave system. 2. The distance across the fetch (wave-generating area) measured in a direction parallel to the direction of the wind.

Fishery assessment   The conduct of research on the ecological and economic factors related to a fishery and the application of the knowledge gained to its regulation.

Fisheries biologist   A biologist who studies the abundance and/or ecology of important fishery species.

Fisheries management   The organized effort directed at regulating fishing activity with the goal of maintaining a long-term fishery.

Fjord   A long, narrow, deep, U-shaped inlet that usually represents the seaward end of a glacial valley that has become partially submerged after the melting of the glacier.

Flagellum   A whiplike living process used by some cells for locomotion.

Flocculent   A loose, open-sturctured deposit of organic debris composed of tiny particles.

Floe   A piece of floating ice other than fast ice or icebergs. May range in dimension from about 20 cm (7.9 in) across to more than a kilometer.

Flood basalt   A basalt plateau composed of from 1 to 2 million km^3 of basalt deposited in a period of from 1 to 2 million years.

Flood current   A tidal current associated with increasing height of the tide, generally moving toward the shore.

Flora   The plant life of any particular area or of any particular time.

Florida Current   A warm current flowing north along the coast of Florida. It becomes the Gulf Stream.

Flux   The amount of flow per unit of cross-sectional area per unit of time.

Fog   The condensation of water in relatively warmer air of high humidity when it moves over water that is much cooler.

Folded mountain range   Mountain ranges formed as a result of the convergence of lithospheric plates. They are characterized by masses of folded sedimentary rocks that formed from sediments deposited in the ocean basin that was destroyed by the convergence.

Food chain   The passage of energy materials from producers through a sequence of herbivores and carnivores.

Food web   A group of interrelated food chains.

Foraminifer   A member of an order of planktonic and benthic protozoans that possess protective coverings usually composed of calcium carbonate.

Foraminiferan ooze   An oceanic deposit composed of more than 30% by weight biogenous particles, most of which are the shells of foraminifers.

Forced wave   A wave that is generated and maintained by a continuous force such as the gravitational attraction of the moon.

Fore-arc   The ocean floor that lies between an island arc and its associated trench.

Foreshore   The portion of the shore lying between the normal high- and low-water marks—the intertidal zone.

Fortnight   Half a synodic month (29.5 days), or about 14.75 days. Normally used in reference to a period of time equal to 2 weeks. The time that elapses between the new moon and full moon.

Fossil   Any remains, trace, or imprint of an organism that has been preserved in rocks.

Fracture zone   An extensive linear zone of unusually irregular ocean floor topography, characterized by large seamounts, steep-sided or asymmetrical ridges, troughs, or long, steep slopes. Usually represents ancient, inactive transform fault zones.

Free wave   A wave created by a sudden rather than a continuous impulse that continues to exist after the generating force is gone.

Freezing   Cooling a substance below its freezing point. This is one of the methods used in the desalination of seawater.

Freezing point   The temperature at which a liquid becomes a solid under any given set of conditions. The freezing point of water is 0°C (32°F) under atmospheric pressure.

Frequency   *See* wave frequency (*f*).

Fringing reef   A reef that is directly attached to the shore of an island or continent. It may extend more than 1 km (0.62 mi) from shore. The outer margin is submerged and often consists of algal limestone, coral rock, and living coral.

Frustule   The siliceous covering of a diatom, consisting of two halves (epitheca and hypotheca).

Fucoxanthin   The reddish-brown pigment that gives brown algae its characteristic color.

Full moon   The phase of the moon that occurs when the sun and moon are in opposition, on opposite sides of Earth.

Fully developed sea   The maximum average size of waves that can be developed for a given wind speed when it has blown in the same direction for a minimum duration over a minimum fetch.

Furrow   Parallel, troughlike structure cut into mud waves by bottom currents. Furrows are aligned in the direction of current flow.

Gabbro   Dark-colored intrusive rocks that are the mineralogic equivalent of basalt.

Galápagos Rift   A divergent plate boundary extending eastward from the Galápagos Islands toward South America. The first deep-sea hydrothermal vent biocommunity was discovered here in 1977.

Galaxy   One of the billions of large systems of stars that make up the universe.

Gaseous state   A state of matter in which molecules move by translation and only interact through chance collisions.

Gastropoda   A class of mollusks, most of which possess an asymmetrical spiral one-piece shell and a well-developed flattened foot. A well-developed head will usually have two eyes and one or two pairs of tentacles. Includes snails, limpets, abalone, cowries, sea hares, and sea slugs.

Geologic history   The history of Earth recorded in the rocks of Earth's crust.

Geostrophic current   A current that develops out of Earth's rotation and is the result of a near balance between gravitational force and the Coriolis effect.

Gill   A thin-walled projection from some part of the external body or the digestive tract; used for respiration in a water environment.

Gill net   A net set in the water to passively catch fish by having them swim into the mesh. The mesh is sized to slide over the gills and trap the fish.

Glacial deposit   Deposit of rock fragments carried to the ocean by glaciers. The deposits form as the glacier ice melts and releases the rock fragments to fall to the ocean floor.

Glacial Epoch   The Pleistocene Epoch, the earlier of two divisions of the Quaterary Period of geologic time. During this time, high-latitude continental areas now free of ice were covered by continental glaciers.

Glacier   A large mass of ice formed on land by the recrystallization of old compacted snow. It flows from an area of accumulation to an area of wasting, where ice is removed from the glacier by melting or calving.

Glacioeustatic oscillation   Rising and falling of sea level resulting from the formation and melting of continental ice sheets.

Glauconite   A group of green hydrogenous minerals consisting of hydrous silicates of potassium and iron.

Global plate tectonics   The process by which lithospheric plates are moved across Earth's surface to collide, slide by one another, or diverge to produce the topographic configuration of Earth.

Globigerina ooze   An oceanic deposit composed of at least 30% by weight biogenous particles, most of which are the shells of foraminifers belonging to the genus *Globigerina*.

Gold   A soft, heavy, yellow mineral, the native metallic element, Au. It has been mined from placer deposits in the coastal ocean.

Gondwanaland   A hypothetical protocontinent of the Southern Hemisphere named for the Gondwana region of India. It included the present continental masses of Africa, Antarctica, Australia, India, and South America.

Graben   An elongate, downdropped crustal block that is bounded by faults. The rift valleys along the axes of spreading centers are grabens.

Graded bedding   Stratification in which each layer displays a decrease in grain size from bottom to top.

Gradient   The rate of increase or decrease of one quantity or characteristic relative to a unit change in another. For example, the slope of the ocean floor is a change in elevation (a vertical linear measurement) per unit of horizontal distance covered. Commonly measured in m/km.

Granite   A light-colored igneous rock characteristic of the continental crust. Rich in nonferromagnesian minerals such as feldspar and quartz.

Gran Method   A system involving the observation of the change in dissolved oxygen within paired transparent and opaque bottles containing phytoplankton and suspended in the ocean surface water to determine the base of the euphotic zone.

Gravitational force   The force of attraction that exists between any two bodies in the universe that is proportional to the product of their masses and inversely proportional to the distance between the centers of their masses.

Gravitational stability   The degree to which segments of the water column tend to remain stationary. The portion of the water column with the greatest degree of gravitational stability is called the pycnocline, where water at the top is much less dense than water at the bottom. This zone of density stratification normally coincides with the thermocline.

Gravity wave   A wave for which the dominant restoring force is gravity. Such waves have a wavelength of more than 1.74 cm (0.7 in), and their velocity of propagation is controlled mainly by gravity.

Gray whale   Pacific baleen whales of the species *Eschrichtius robustus* that feed in the Chukchi Sea and Bering Sea and breed and calve in the warm waters of lagoons in Baja California and Japan.

Greenhouse effect   The heating of Earth's atmosphere that results from the absorption by components of the atmosphere such as water vapor and carbon dioxide of infrared radiation from Earth's surface.

Grenadier   A rattailed demersal fish that is usually the first to arrive at bait deposited on the ocean floor.

Groin   A low, artificial structure projecting into the ocean from the shore to interfere with longshore transportation of sediment. It usually has the purpose of trapping sand to cause the buildup of a beach.

Gross ecological efficiency   The amount of energy passed on from a trophic level to the one above it divided by the amount it received from the one below it.

Gross primary production   The total amount of organic material produced by autotrophs.

Guano   A phosphate deposit formed by the leaching of bird droppings in arid climates.

Guiana Current   The current that diverts South Equatorial Current water north along the Guiana coast of South America into the Northern Hemisphere.

Gulf of California   A sea on the eastern margin of the Pacific Ocean between the Mexican mainland and the Baja California pennisula.

Gulf of Mexico   A sea on the western margin of the North Atlantic Ocean. It is bounded by the U.S., the Mexican mainland, and Cuba.

Gulf Stream   The high-intensity western boundary current of the North Atlantic Ocean subtropical gyre that flows north off the east coast of the United States.

Guyot   A tablemount; a conical volcanic feature on the ocean floor that has had the top truncated to a relatively flat surface.

Gyre   A circular motion. Used mainly in reference to the circular motion of water in each of the major ocean basins centered in subtropical high-pressure regions.

Habitat   A place where a particular plant or animal lives. Generally refers to a smaller area than environment.

Hadal zone   The deepest ocean environment, specifically ocean trenches deeper than 6 km (3.7 mi).

Half-life   The time required for half the atoms of a radioactive isotope sample to decay to atoms of another element.

Halocline   A layer of water which has a high rate of salinity change in the vertical dimension.

Harmonic analysis   Mathematical analysis of the partial tides to predict the tides in a given region.

Headland   A steep-faced irregularity of the coast that extends out into the ocean.

Heat   Energy moving from a high-temperature system to a lower-temperature system. The heat gained by the one system may be used to raise its temperature or to do work.

Heat budget (global)   The equilibrium that exists on the average between the amount of heat absorbed by Earth and its cipal movement of air masses at these latitudes is one of vertical descent.

Heat capacity   The amount of heat required to raise the temperature of 1 gram of a substance 1°C.

Heat flow   The measurement of heat moving from Earth's interior to the surface. It is measured in heat flow units (10^{-6} cal/cm^2/sec).

Hemoglobin   A red proteinaceous pigment containing iron and found in red blood cells of vertebrates. It combines with oxygen and carries oxygen from the lungs to tissues.

Herbivore   An organism that eats only algae and/or plants.

Hermatypic   Pertaining to corals and other reef animals that depend on algal symbionts for their metabolic energy.

Heterotrophs   Organisms that depend on an external food supply

Highly stratified estuary   A relatively deep estuary, with a high volume of fresh water flowing seaward at the surface and a large mass of marine water at depth, producing a well-developed halocline.

High-tide zone   Shore covered by the highest high tides but not by the lowest high tides.

Hot spot   The relatively stationary surface expression of a persistent jet of molten mantle material rising to the surface.

Hurricane   A tropical cyclone in which winds reach velocities above 120 km/h (73 mi/h). Generally applied to such storms in the North Atlantic Ocean, eastern North Pacific Ocean, Caribbean Sea, and Gulf of Mexico. Such storms in the western Pacific Ocean are called *typhoons*.

Hydrated   To be chemically combined with water or surrounded by water.

Hydration sphere   A sphere of water molecules that surround ions dissolved in water.

Hydrocarbon   An organic compound consisting solely of hydrogen and carbon. Petroleum is a mixture of many hydrocarbon compounds.

Hydrocarbon seep biocommunity   Deep bottom-dwelling community associated with a hydrocarbon seep from the ocean floor. The community depends on methane and sulfur-oxidizing bacteria as producers. The bacteria may live free in the water, on the bottom, or symbiotically in the tissues of some of the animals.

Hydrogen bond   An intermolecular bond that forms within water because of the dipolar nature of water molecules.

Hydrogen chloride (HCl)   This compound is a chemically active gas. Dissolved in water, it forms hydrochloric acid.

Hydrogenous sediment   Sediment that forms from ocean water precipitation or ion exchange between existing sediment and ocean water. Examples are manganese nodules, phosphorite, glauconite, phillipsite, and montmorillonite.

Hydrogen sulfide (H$_2$S)   The gas that smells like rotten eggs. Its molecule is composed of two atoms of hydrogen and one atom of sulfur.

Hydrological cycle   The cycle of water exchange among the atmosphere, land, and ocean through the processes of evaporation, precipitation, runoff, and subsurface percolation.

Hydrophilic   Pertaining to the property of attracting water.

Hydrophobic   Pertaining to the property of being impossible or difficult to wet with water.

Hydrothermal spring   Vents of hot water found primarily along the spreading axes of oceanic ridges and rises.

Hydrothermal vent   Ocean water that percolates down through fractures in recently formed ocean floor is heated by underlying magma and surfaces again through these vents. They are usually located near the axis of spreading on oceanic ridges and rises.

Hydrothermal vent biocommunity   Deep bottom-dwelling community associated with a hydrothermal vent. The hot water vent is usually associated with the axis of a spreading center, and the community is dependent on sulphur-oxidizing bacteria that may live free in the water, on the bottom, or symbiotically in the tissue of some of the animals of the community.

Hydrozoa A class of coelenterates that characteristically exhibit alternation of generations, with a sessile polypoid colony giving rise to a pelagic medusoid form by asexual budding.

Hypersaline lagoon Shallow lagoons such as Laguna Madre, which may become hypersaline due to little tidal flushing and seasonal variability in freshwater input. High evaporation rates and low freshwater input can result in very high salinities.

Hypersaline seep Seeps in which the vented water has a salinity much higher than ocean water.

Hypertonic Pertaining to the property of an aqueous solution having a higher osmotic pressure (salinity) than another aqueous solution separated by a semipermeable membrane allowing osmosis. The hypertonic fluid will gain water molecules through the membrane from the other fluid.

Hypotheca The lower valve of a diatom frustule.

Hypotonic Pertaining to the property of an aqueous solution having a lower osmotic pressure (salinity) than another aqueous solution separated by a semipermeable membrane allowing osmosis. The hypotonic fluid will lose water molecules through the membrane to the other fluid.

Hypsographic curve A cumulative frequency profile representing the statistical distribution of the areas of Earth's solid surfaces at elevations above or below sea level.

Iceberg A massive piece of glacier ice that has broken from the front of the glacier (calved) into a body of water. It floats with its tip at least 5 m (16 ft) above the water's surface and at least four-fifths of its mass submerged.

Ice floe *See* floe.

Ice shelf A thick layer of ice with a relatively flat surface that is attached to and nourished by a continental glacier from one side. The shelf, which is for the most part afloat, may extend above water level by more than 50 m (164 ft) along its seaward cliff formed by the break-off of large tabular chunks of ice that become icebergs.

Igneous rock One of the three main classes into which all rocks are divided, i.e., igneous, metamorphic, and sedimentary. Rock that forms from the solidification of molten or partly molten material (magma).

In situ In place, i.e., *in situ* density of a sample of water is its density at its original depth.

In situ **density** *See* Density, *In situ*.

Inertia Newton's first law of motion. It states that a body at rest will stay at rest and a body in motion will remain in a uniform motion in a straight line unless acted on by some external force.

Infauna Animals that live buried in the soft substrate (sand or mud).

Infrared radiation Electromagnetic radiation between the wavelengths of 0.8 μm (0.00032 in.) and about 1000 μm (0.394 in.). It is bounded on the shorter-wavelength side by the visible spectrum and on the long side by microwave radiation.

Inner sublittoral zone The section of ocean floor from the low-tide shoreline to the point where attached plants stop growing.

Insolation The rate at which solar radiation is received per unit of surface area at any point at or above Earth's surface.

Interface A surface separating two substances of different properties, i.e., density, salinity, or temperature. In oceanography, it usually refers to a separation of two layers of water with different densities caused by significant differences in temperature and/or salinity.

Interface wave An orbital wave that moves along an interface between fluids of different density. An example is ocean surface waves moving along the interface between the atmosphere and the ocean, which is 1000 times more dense.

Interference pattern The pattern of wave development that develops from the combined interference effects of more than one wave system as they pass in the open ocean.

Intermediate water Water masses that usually form at the Arctic or Antarctic convergences and sink to a depth of 800 to 1000 m (2624 to 3280 ft) before spreading out beneath the surface-water masses.

Intermolecular bond A relatively weak bond that forms between molecules of a given substance. The hydrogen bond and the van der Waals bonds are intermolecular bonds.

Internal wave A wave that develops below the surface of a fluid, the density of which changes with increased depth. This change may be gradual or occur abruptly at an interface.

Intertidal zone Littoral zone, the foreshore. The ocean floor covered by the highest normal tides and exposed by the lowest normal tides and the water environment of the tide pools within this region.

Intertropical Convergence Zone Zone where northeast trade winds and southeast trade winds converge. Averages about 5°N in the Pacific and Atlantic Oceans and 7°S in the Indian Ocean.

Intrusive rock Igneous rock such as granite that cools slowly beneath Earth's surface.

Invertebrate Animal without a backbone.

Ion An atom that becomes electrically charged by gaining or losing one or more electrons. The loss of electrons produces a positively charged cation, and the gain of electrons produces a negatively charged anion.

Ionic bond A chemical bond resulting from the electrical attraction that exists between cations and anions.

Irminger Current A warm current that branches off from the Gulf Stream and moves up along the west coast of Iceland.

Iron spherule Magnetic, iron-rich spherical particle about 30 μm (0.012 in) in diameter that rains down on Earth as a component of space dust. Iron spherules may be the products of collisions of asteroids.

Island arc system A linear arrangement of islands, many of which are volcanic, usually curved so the concave side faces a sea separating the islands from a continent. The convex side faces the open ocean and is bounded by a deep-ocean trench.

Island mass effect As surface current flows past an island, surface water is carried away from the island on the downcurrent side. This water is replaced in part by upwelling of water on the downcurrent side of the island.

Isobar Lines connecting values of equal pressure on a map or graph.

Isohaline Of the same salinity.

Isopoda An order of dorsoventrally flattened crustaceans that are mostly scavengers or parasites on other crustaceans or fish.

Isostasy A condition of equilibrium, comparable to buoyancy, in which the rigid crustal units float on the underlying mantle.

Isostatic rebound Isostatic uplift of continental regions resulting from the removal of glacial mass.

Isotherm A line connecting points of equal temperature.

Isothermal Of the same temperature.

Isotonic Pertaining to the property of having equal osmotic pressure. If two such fluids were separated by a semipermeable membrane that allows osmosis to occur, there would be no net transfer of water molecules across the membrane.

Isotope One of several atoms of an element that has a different number of neutrons, and therefore a different atomic mass, than the other atoms, or isotopes, of the element.

Jellyfish 1. A free swimming, umbrella-shaped medusoid member of the coelenterate class, Schyphozoa. 2. Also frequently applied to the medusoid forms of other coelenterates.

Jet stream An easterly moving air mass at an elevation of about 10 km (6.2 mi). Moving at speeds that can exceed 300 km/h (186 mi/h), the jet stream follows a wavy path in the midlatitudes and influences how far polar air masses may extend into the lower latitudes.

Jetty A structure built from the shore into a body of water to protect a harbor or a navigable passage from being shoaled by deposition of longshore (littoral) drift material.

Juan de Fuca Ridge   A divergent plate boundary off the Oregon-Washington coast.

Juvenile water   Water that is derived directly from magma, being released for the first time at Earth's surface as the magma crystallizes to igneous rock.

Kelp   Large varieties of Phaeophyta (brown algae).

Kelp forest community   All of the organisms found in the kelp forests of shallow coastal waters where water temperatures are relatively cool.

Kelvin temperature scale (K)   Scale in which 0 K = −273.16°C. One degree on the Kelvin scale equals the same temperature range as one degree on the Celsius scale. 0 K is the lowest temperature possible.

Kelvin wave   Wave that results when a progressive tide wave moves from the open ocean into and out of a relatively narrow body of water during a tidal cycle. The tidal range will be greater on the right side of the narrow body of water during flood tide. This results from the fact that the channel rotates in a counterclockwise direction as Earth rotates, while the wave tends to move in a straight line.

Key   A low, flat island composed of sand or coral debris that accumulates on a reef flat.

Killer whale   The toothed whale, *Orcinus orca*, which grows to a length of 9 m (28 ft). It is cosmopolitan and feeds on the widest variety of animals of any whale—from fish to other whales.

Kinetic energy   Energy of motion. It increases as the mass or velocity of the object in motion increases.

Knot (kn or kt)   Unit of speed equal to 1 naut mi per hour (approximately 51 cm/s).

Krill   A common name frequently applied to members of the crustacean order Euphausiacea (euphausids).

Kuroshio Current   The north-flowing western boundary current of the North Pacific Ocean subtropical gyre.

Kuroshio Extension   The eastern extension of the Kuroshio Current that becomes the North Pacific Current.

Labrador Current   A cold current flowing south along the coast of Labrador in the northeastern Atlantic Ocean.

Lagoon   A shallow stretch of water partly or completely separated from the open ocean by an elongate narrow strip of land such as a reef or barrier island.

Laguna Madre   A hypersaline lagoon behind Padre Island along the south Texas coast.

Lamina   A layer.

Laminar flow   Flow in which a fluid flows in parallel layers or sheets. The direction of flow at any point does not change with time; nonturbulent flow.

Langmuir circulation   A cellular circulation set up by winds that blow consistently in one direction with velocities above 12 km/h (7.5 mi/h). Helical spirals running parallel to the wind direction are alternately clockwise and counterclockwise.

La Niña   An event where the surface temperature in the waters of the eastern South Pacific falls below average. It usually occurs at the end of an El Niño–Southern Oscillation event.

Lantern fish   A Mesopelagic fish that measures about 13 cm (5 in) in length. The name is derived from the photophores the fishes possess.

Larva   An embryo that is on its own before it assumes the characteristics of the adults of the species.

Latent heat   The quantity of heat gained or lost per unit of mass as a substance undergoes a change of state (liquid to solid, etc.) at a given temperature and pressure.

Latent heat of evaporation   The heat energy that must be added to 1 g of a liquid substance to convert it to a vapor at a temperature below its boiling point. For water, it is 585 cal at 20°C.

Latent heat of melting   The heat energy that must be added to 1 g of a substance at its melting point to convert it to a liquid. For water, it is 80 cal.

Latent heat of vaporization   The heat energy that must be added to 1 g of a substance at its boiling point to convert it to a vapor. For water, it is 540 cal.

Lateral line system   A sensory system running down both sides of fishes to sense subsonic pressure waves transmitted through ocean water.

Lateritic soil   A red subsoil rich in secondary iron and aluminum oxides. It is characteristics of intertropical soils, may contain much quartz and kaolinite, and can harden to a bricklike substance.

Latitude   Location on Earth's surface based on angular distance north or south of the equator. Equator, 0°, North Pole, 90°N; South Pole, 90°S.

Laurasia   A hypothetical protocontinent of the Northern Hemisphere. The name is derived from Laurentia, pertaining to the Canadian Shield of North America, and Eurasia, of which it was composed.

Lava   Fluid magma coming from an opening in Earth's surface, or the same material after it solidifies.

Law of gravitation   *See* gravitational force.

Leeuwin Current   A warm current that flows south out of the East Indies along the western coast of Australia.

Leeward   Direction toward which the wind is blowing or waves are moving.

Lesser Antilles   The arc of West Indian Islands from the Virgin Islands south to the islands of coastal Venezuela.

Levee   Natural (resulting from deposition during flooding) or human-made low ridges on either side of a river channel.

Levels of biological response   The effects of pollution show up on these different levels within the biotic community: biochemical–cellular, organism, population dynamics, and community dynamics and structure.

Lichen   Organism involving a photosynthetic, mutualistic relationship between an alga and a fungus. The alga is protected by the fungus, which is dependent on the alga for photosynthetically produced food.

Light-year   The distance traveled by light during 1 year at a speed of 300,000 km/s (186,000 mi/s). It equals 9.8 trillion kilometers or 6.2 trillion miles.

Limestone   A class of sedimentary rock composed of at least 50% calcium or magnesium carbonate. Limestone may be either biogenous or hydrogenous.

Limpet   A mollusk of the class Gastropoda that possesses a low conical shell exhibiting no spiraling in the adult form.

Lipid   Fats which, along with proteins and carbohydrates, are the principal structural components of living cells.

Liquid state   A state of matter in which a substance has a fixed volume but no fixed shape.

Lithogenous particles   Mineral grains derived from the rock of continents and islands and transported to the ocean by wind or running water.

Lithosphere   The outer layer of Earth's structure, including the crust and the upper mantle to a maximum depth of about 200 km (124 mi). It is this layer that breaks into the plates that are the major elements of global plate tectonics.

Lithothamnion ridge   A feature common to the seaward edge of a reef structure, characterized by the presence of the red alga, *Lithothamnion*.

Littoral zone   The benthic zone between the highest and lowest normal water marks; the intertidal zone.

Lobster   Large marine crustacean used as food. *Homarus americanus* (American lobster) possesses two large chelae (pincers) and is found off the New England coast. *Panulirus* sp. (spiny lobsters or rock lobsters) have no chelae but possess long spiny antennae effective in warding off predators. *Panulirus argus* is found off the coast of Florida and in the West Indies, while *P. interruptus* is common along the coast of southern California.

Longitude   Location on Earth's surface based on angular distance east or west of the Greenwich Meridian (0° longitude). 180° longitude is the International Date Line.

Longitudinal wave   A wave in which particle vibration is parallel to the direction of energy propagation.

Longshore current A current located in the surf zone and running parallel to the shore as a result of breaking waves.

Longshore drift The load of sediment transported along the beach from the breaker zone to the top of the swash line in association with the longshore current.

Loop Current A clockwise-flowing current followed by water moving north from the Yucatán Strait through the Gulf of Mexico. It becomes the Florida Current when it enters the Florida Straits.

Lophophore Horseshoe-shaped feeding structure bearing ciliated tentacles characteristic of the phyla Bryozoa, Brachiopoda, and Phoronidea.

Lower continental rise hills Deposits also called *drifts* or *ridges* that are deposited where the western boundary undercurrents slow with change in direction.

Lower high water (LHW) The lower of two high waters occurring during a tidal day where tides are mixed.

Lower low water (LLW) The lower of two low waters occurring during a tidal day where tides are mixed.

Low marsh The marsh that is found between mean sea level and neap high-tide level.

Low slack water The zero velocity that occurs when the ebb current reaches low water before reversing as a flood current.

Low-tide zone Mostly covered with water. It lies between the lowest spring-tide levels and the highest neap-tide levels.

Low water (LW) The lowest level reached by the water surface at low tide before the rise toward high tide begins.

Lunar day The time interval between two successive transits of the moon over a meridian (approximately 24 h and 50 min of solar time).

Lunar hour One twenty-fourth of a lunar day (about 62.1 min).

Lunar tide The part of the tide caused solely by the tide-producing force of the moon.

Lunger Fish such as groupers that sit motionless on the ocean floor waiting for prey to appear. A quick burst of speed over a short distance suffices to capture the prey.

Macroplankton Plankton larger than 2 cm (0.8 in) in their smallest dimension.

Magma Fluid rock material from which igneous rock is derived through solidification.

Magma chamber A reservoir of molten rock that underlies the axes of spreading centers. Magma chambers are more extensively developed beneath fast-spreading rises like the East Pacific Rise where they may be found at a depth of about 1.5 km (0.9 mi) beneath the ocean floor and have a width of about 4 km (2.5 mi).

Magnetic anomaly Distortion of the regular pattern of Earth's magnetic field, resulting from the various magnetic properties of local concentrations of ferromagnetic minerals in Earth's crust.

Magnetic dip The dip of magnetite particles in rock units of Earth's crust relative to sea level. It is approximately equivalent to the latitude at which the rock formed.

Manganese nodule Concretionary lump containing oxides of iron, manganese, copper, or nickel found scattered in groups over the ocean floor.

Mangrove swamp A marshlike environment that is dominated by mangrove trees. They are restricted to latitudes below 30°.

Mantle The zone between the core and crust of Earth; rich in ferromagnesian minerals. In pelecypods, the portion of the body that secretes shell material.

Marginal ice zone (MIZ) The water marginal to the pack ice accumulation.

Marginal sea A semienclosed body of water adjacent to a continent.

Mariculture The application of the principles of agriculture to the production of marine organisms.

Marine terrace Wave-cut benches that have been exposed above sea level by a drop in sea level.

Marsh An area of soft, wet land. Flat land periodically flooded by salt water, common in portions of lagoons.

Maximum sustainable yield The maximum fishery biomass that can be removed yearly and be sustained by the fishery ecosystem.

Meander A sinuous curve, bend, or turn in the course of a current.

Mean high water (MHW) The average height of all the high waters occurring over a 19-year period.

Mean low water (MLW) The average height of the low waters occurring over a 19-year period.

Mean sea level (MSL) The mean surface water level determined by averaging all stages of the tide over a 19-year period, usually determined from hourly height observations along an open coast.

Mean tidal range The difference between mean high water and mean low water.

Mechanical energy Energy manifested as work being done; the movement of a mass some distance.

Mediterranean circulation Circulation characteristic of bodies of water with restricted circulation with the ocean that results from an excess of evaporation as compared to precipitation and runoff. Surface flow is into the restricted body of water with a subsurface counterflow as exists between the Mediterranean Sea and the Atlantic Ocean.

Mediterranean Intermediate Water Water that sinks below the surface in the eastern Mediterranean Sea to a depth between 200 and 600 m (656 to 1968 ft) and flows toward the Strait of Gibraltar. At the time of formation, it has a temperature of 15°C (59°F) and a salinity of 39.1%.

Mediterranean Sea A sea on the eastern margin of the North Atlantic Ocean. It lies east of the Strait of Gibraltar between Europe and Africa.

Mediterranean Water The Mediterranean Intermediate Water becomes the Mediterranean Water of the North Atlantic Ocean after it flows across the Gibraltar sill, sinks to a depth of about 900 m (2890 ft), and spreads out across the North Atlantic Ocean.

Medusa Free-swimming, bell-shaped coelenterate body form with a mouth at the end of a central projection and tentacles around the periphery. Reproduces sexually.

Meiofauna Small species of animals that live in the spaces among particles in a marine sediment.

Mercury The heavy, liquid, native metallic element, Hg.

Meridian of longitude Half a great circle terminating at the North and South poles.

Meroplankton Planktonic larval forms of organisms that are members of the benthos or nekton as adults.

Mesoderm A primitive cell layer in the embryo that develops between the endoderm and ectoderm. In vertebrates, it gives rise to the skeleton, muscles, circulatory and excretory organs, and most of the reproductive system.

Mesopelagic zone That portion of the oceanic province from about 200 to 1000 m (656 to 3280 ft). Corresponds approximately with the disphotic (twilight) zone.

Metamorphic rock Rock has undergone recrystallization while in the solid state in response to changes of temperature, pressure, and chemical environment.

Metaphyta Kingdom of many-celled plants.

Metazoa Kingdom of many-celled animals.

Meteorite The remains of a meteor that enters the Earth's atmosphere and falls as a solid to the surface of the Earth. They probably originate as fragments of asteroids.

Methane (CH₄) The simplest hydrocarbon.

Microbial loop Cycling of energy and matter among phytoplankton, heterotrophic bacteria, and protozoans in the pelagic ecosystem without this energy and matter being passed on to larger animals.

Microcontinent   A submarine plateau that is an isolated fragment of the continental crust. It usually has linear features; found primarily in the Indian Ocean.

Microplankton   Net plankton. Plankton not easily seen by the unaided eye, but easily recovered from the ocean with the aid of a fine-mesh plankton net.

Mid-Atlantic Ridge   A slow-spreading divergent plate boundary running north-south and bisecting the Atlantic Ocean.

Middle-tide zone   That portion of the intertidal zone that lies between the highest low-tide shoreline and the lowest high-tide shoreline.

Mid-ocean ridge basalt (MORB)   Basalt typical of oceanic ridges and rises. It is depleted in potassium, rubidium, cesium, uranium, and thorium compared to basalt produced at hot spots. This may be evidence that hot spot basalts arise from near the core-mantle boundary while typical spreading center basalts are formed in the upper mantle.

Migration   Long journeys undertaken by many marine species for the purpose of successful feeding and reproduction. The stimuli that cause the animals to initiate their migrations are for the most part still unknown.

Milky Way   The galaxy to which the solar system belongs.

Milkfish   A fish capable of withstanding a wide range of salinity that is an important object of mariculture in the Far East.

Minamata Bay, Japan   The site of the 1953 occurrence of human poisoning by mercury contained in marine food victims consumed.

Minamata Disease   A neurological disease caused by mercury contaminated seafood consumed by fishing families living near Minamata Bay, Japan.

Mineral   An inorganic substance occurring naturally in the earth and having distinctive physical properties and a chemical composition that can be expressed by a chemical formula.

Mixed interference   A pattern of wave interference in which there is a combination of constructive and destructive interference.

Mixed layer   The surface layer of the ocean water mixed by wave and tide motions to produce relatively isothermal and isohaline conditions.

Mixed tide   A tide having two high and two low waters per tidal day with a marked diurnal inequality. Such a tide may also show alternating periods of diurnal and semidiurnal components.

Mixotroph   An organism that depends on a combination of autotrophic and heterotrophic behavior to meet its energy requirements. Many coral reef species exhibit such behavior.

Mohorovicic discontinuity   A sharp seismic discontinuity between the crust and mantle of Earth. It may be as shallow as 5 km (3.1 mi) below the ocean floor or as deep as 60 km (37 mi) beneath some continental mountain ranges. Also known as the *Moho*.

Mole   The weight of a substance in grams numerically equal to its molecular weight (gram molecule). One mole of water (H_2O) is 18 g.

Molecular motion   Molecules move in three ways: vibration, rotation, and translation.

Molecule   The smallest particle of an element or compound that, in the free state, retains the characteristics of the substance.

Mollusca   Phylum of soft unsegmented animals usually protected by a calcareous shell and having a muscular foot for locomotion. Includes snails, clams, chitons, and octopuses.

Molt   Periodic shedding of exoskeleton by arthropods to permit growth.

Monera   Kingdom of organisms that do not have nuclear material confined within a sheath but spread throughout the cell. Bacteria and blue-green algae.

Mononodal   Pertaining to a standing wave with only one nodal point or nodal line.

Monsoon   A name for seasonal winds derived from the Arabic word for season, *mausim*. The term was originally applied to winds over the Arabian Sea that blow from the southwest during summer and the northeast during winter.

Montmorillonite   A clay mineral rich in calcium and sodium ions. It forms by the alteration of ferromagnesian minerals, calcium-rich feldspars, and volcanic glass.

Moraine   Unsorted material deposited at the margins of glaciers. Many such deposits have become economically important as fishing banks after being submerged by the rising level of the ocean.

Mud   Sediment consisting of silt- and clay-sized particles smaller than 0.06 mm (0.002 in). Actually, small amounts of larger particles will also be present.

Mud waves   Wave feature with lengths of 2 to 3 km (1.2 to 1.9 mi) that bottom currents produce on the surfaces of drifts or ridges.

Mullet   A group of fishes that are able to live in fresh and salty water.

Mutualism   A symbiotic relationship in which both participants benefit.

Mycota   The kingdom of fungi. In the marine environment they are found living symbiotically with algae as lichen in the intertidal zone and as decomposers of dead organic matter in the open sea.

Myoglobin   A red, oxygen-storing pigment found in the muscle tissue.

Myomere   A muscle fiber.

Mysticeti   The baleen whales.

Nadir   The point on the celestial sphere directly opposite the zenith and directly beneath the observer.

Nanoplankton   Plankton less than 50 µm in length that cannot be captured in a plankton net and must be removed from the water by centrifuge or special microfilters.

Nansen bottle   A device once used by oceanographers to obtain samples of ocean water from beneath the surface.

Natural selection   The process described by Charles Darwin by which the forces of nature select for survival those most fit to live in a given environment.

Nauplius   A microscopic free-swimming larval stage of crustaceans such as copepods, ostracoids, and decapods. Typically has three pairs of appendages.

Neap tide   Tides of minimal range occurring when the moon is in quadrature, first and third quarters.

Nearshore   That zone from the shoreline seaward to the line of breakers.

Nektobenthos   Those members of the benthos that can actively swim and spend much time off the bottom.

Nekton   Pelagic animals such as adult squids, fish, and mammals that are active swimmers to the extent they can determine their position in the ocean by swimming.

Nematocyst   The stinging mechanism found within the cnidoblast of members of the phylum Cnidaria (Coelenterata).

Nepheloid layer   Well-mixed, turbid layer of water at the base of the oceanic water column. It is particularly well developed in the western boundary undercurrent.

Neritic province   That portion of the pelagic environment from the shoreline to a depth of 200 m (656 ft).

Neritic sediment deposit   That sediment composed primarily of lithogenous particles and deposited relatively rapidly on the continental shelf, continental slope, and continental rise.

Net primary production   The remaining amount of organic material produced by autotrophs after they have met their respiration needs.

Neutron   An electrically neutral particle found in the nucleus of most atoms. It has a mass approximately equal to that of a proton.

New ice   Newly formed sea ice.

New moon   The phase of the moon that occurs when the sun and the moon are in conjunction, on the same side of Earth.

New production   Photosynthetic production supported by nutrients supplied from outside the immediate ecosystem by upwelling or other physical transport.

Niche   The ecological role of an organism and its position in the ecosystem.

Niigata, Japan   In the 1960s, the site of mercury poisoning of humans by mercury-contaminated seafood.

Nitrate   A chemical radical (NO_3) that is an important component of nutrients required for biological production.

Nitrogen fixation   Conversion by bacteria of atmospheric nitrogen (N_2) to oxides of nitrogen (NO_2 NO_3) usable by plants in primary production.

Nitrogen-fixing bacterium   Any of the bacteria that convert atmospheric nitrogen (N_2) to oxides of nitrogen (NO_2, NO_3) usable by algae in primary production.

Nitrogen narcosis   A sickness that effects divers. It results from too much N_2 being dissolved in the blood and reducing the flow of O_2 to tissues. The threat of this problem increases with increasing pressure (depth).

Node   The point on a standing wave where vertical motion is lacking or minimal. If this condition extends across the surface of an oscillating body of water, the line of no vertical motion is a nodal line.

Nonconservative property   A property of ocean water attained at the surface and changed by processes other than mixing and diffusion after the water sinks below the surface. For example, dissolved oxygen content will be altered by biological activity.

Nonconservative tracer   Chemicals such as oxygen and carbon dioxide that have a high level of chemical reactivity in the ocean. Their concentrations are significantly altered by photosynthesis and respiration, and they can thus be used to trace biological processes.

Nonferromagnesian mineral   Any of a group of common igneous rock-forming minerals that do not contain iron and magnesium.

North Atlantic Deep Water   A deep-water mass that forms primarily at the surface of the Norwegian Sea and moves south along the floor of the North Atlantic Ocean.

Northeast Monsoon   A northeast wind that blows off the Asian mainland onto the Indian Ocean during the winter season.

North Equatorial Current   The westward-flowing equatorial segments of North Pacific and North Atlantic Oceans' subtropical gyres.

North Pacific Current   The eastward-flowing northern segment of the subtropical gyre in the North Pacific Ocean.

Norwegian Current   A warm current that branches off from the Gulf Stream and flows into the Norwegian Sea between Iceland and the British Isles.

Nuclear fuel cycle   The recovery of usable radionuclides from spent fuel rods is achieved at plants that specialize in this process. These plants release radioactive waste into the ocean.

Nucleic base   One of the four basic units that combine to form nucleotides that are arranged in a single strand in RNA and two strands coiled as a double helix in DNA.

Nucleus   1. A central, membrane-bound mass in eukaryotic cells; containing chromosomes. 2. The central, positively charged part of an atom; containing protons and neutrons.

Nudibranch   Sea slug. A member of the mollusk class Gastropoda that has no protective covering as an adult. Respiration is carried on by gills or other projections on the dorsal surface.

Nutrient   Any organic or inorganic compound used by plants in primary production. Nitrogen and phosphorus compounds are important examples.

Obduction   The reverse of subduction. In the case of ophiolites, the rock is pushed up on the continent instead of subducting beneath it.

Ocean Acoustical Tomography   A method by which changes in water temperature may be determined by changes in the speed of transmission of sound. It has the potential to help map ocean circulation patterns over large ocean areas.

Ocean beach   Beach on the open-ocean side of a barrier island.

Ocean Drilling Program (ODP)   In 1983, this program replaced the Deep Sea Drilling Project. It focuses more on drilling the continental margins using the drill vessel *JOIDES Resolution.*

Ocean Thermal Energy Conversion (OTEC)   A technology by which the temperature difference between surface waters and deep waters in low latitude regions is used to generate electricity.

Oceanic Common Water   The deep-water mass that enters the Indian Ocean and Pacific Ocean from the Antarctic Circumpolar Current. It forms from the mixing of North Atlantic Deep Water and Antarctic Bottom Water in the South Atlantic.

Oceanic crust   A mass of rock with basaltic composition that is about 5 km (3 mi) thick and underlies the ocean basin.

Oceanic province   That division of the pelagic environment where the water depth is greater than 200 m.

Oceanic ridge   A linear, seismic mountain range that extends through all the major oceans, rising 1–3 km above the deep-ocean basins. Averaging 1500 km in width, rift valleys are common along the central axis. Source of new oceanic crustal material.

Oceanic rise   Oceanic spreading centers where high spreading rates produce a gently sloping mountain range such as the East Pacific Rise.

Oceanic sediment deposit   The inorganic abyssal clays and the organic oozes that accumulate particle by particle on the deep-ocean floor.

Oceanic sediment   Deep sea sediment in which at least 50 percent of the mass is composed of particles less then 5 μm in size and in which less than 25 percent of the mass of particles larger than 5 μm in size can be lithogenous particles. It includes organic oozes and abyssal clay.

Oceanic spreading center   The axes of oceanic ridges and rises that are the locations at which new lithosphere is added to lithospheric plates. The plates move away from these axes in the process of sea-floor spreading.

Ocelli   Light-sensitive organ around the base of many medusoid bells.

Odontoceti   Toothed whales.

Offshore   The comparatively flat submerged zone of variable width extending from the breaker line to the edge of the continental shelf.

Old ice   Sea ice more than one year of age.

Oligotrophic   Areas such as the mid-subtropical gyres where there are low levels of biological production.

Omnivore   An animal that feeds on both plants and animals.

Oolite   A deposit formed of small spheres from 0.25 to 2 mm in diameter. They are usually composed of concentric layers of calcite.

Ooze   A pelagic sediment containing at least 30% skeletal remains of pelagic organisms, the balance being clay minerals. Oozes are further defined by the chemical composition of the organic remains (siliceous or calcareous) and by their characteristic organisms (diatom ooze, foraminifera ooze, radiolarian ooze, pteropod ooze).

Opal   An amorphous form of silica ($SiO_2 \cdot nH_2O$) that usually contains from 3 to 9% water. It forms the shells of radiolarians and diatoms.

Opposition   The separation of two heavenly bodies by 180° relative to Earth. The sun and moon are in opposition during the full moon phase.

Orbital wave   A wave phenomenon in which energy is moved along the interface between fluids of different density. The wave form is propagated by the movement of fluid particles in orbital paths.

Organic chemistry   The branch of chemistry dealing with carbon compounds.

Organic molecule   Molecule of a compound that is naturally produced by organisms. ATP, DNA, carbohydrates, and lipids are examples.

Orthogonal lines   Lines drawn perpendicular to wave fronts and spaced uniformly so equal amounts of energy are contained by the segments of the wave front lying between any two orthogonal lines in a series. The areas where energy is concentrated as the waves break on the shore can be identified by the convergence of the orthogonal lines.

Orthophosphate   Phosphoric oxide (P_2O_5) can combine with water to produce orthophosphates ($3H_2O \cdot P_2O_5$ or H_3PO_4) that may be used by plants as nutrients.

Osmosis   Passage of water molecules through a semipermeable membrane separating two aqueous solutions of different solute concentration. The water molecules pass from the solution of lower solute concentration into the other.

Osmotic pressure   A measure of the tendency for osmosis to occur. It is the pressure that must be applied to a solution to prevent the passage of water molecules into it from a reservoir of pure water.

Osmotic regulation   Physical and biological processes used by organisms to counteract the osmotic effects of differences in osmotic pressures of their body fluids and the water in which they live.

Ostracoda   An order of crustaceans that are minute and compressed within a bivalve shell.

Otocyst   Gravity sensitive organs around the bell of a medusa.

Outer sublittoral zone   The section of ocean floor from the seaward edge of the inner sublittoral zone to a depth of 200 m (656 ft). No attached plants grow here.

Outgassing   A process, resulting from heating, by which gases and water vapor are released from molten rocks. It has produced Earth's atmosphere and oceans.

Overlapping spreading center   Spreading center more common on fast spreading rises. The segments range in length from 30 km (20 mi) where spreading rates are high to 140 km (90 mi) where they are lower. The ends of these spreading centers are offset from 2 to 10 km (1.2 to 6.2 mi) and overlap instead of being truncated as they are at greater transform fault offsets.

Oviparous   Pertaining to an animal that releases eggs which develop and hatch outside its body.

Ovoviviparous   Pertaining to an animal that incubates eggs inside the mother until they hatch.

Oxygen (O_2) compensation depth   The depth at which marine plants photosynthesize at a rate which exactly meets their respiration needs (the base of the euphotic zone).

Oxygen utilization rate (OUR)   The rate of use of dissolved oxygen caused by the respiration of animals and bacterial decomposition of dead organic matter descending to the bottom. The OUR is highest just beneath the euphotic zone (from about 100 to 1000 m, or 328 to 3280 ft) and gradually decreases with depth as the mass of dead organic matter and living animals decreases. Beneath 1000 m, oxygen values may increase owing to the inflow of oxygen carried by deep- and bottom-water masses that descend from the high-latitude ocean surface.

Oyashio Current   A cold current flowing south along the coast of Japan to converge with the warm Kuroshio Current.

Ozone (O_3)   A triatomic form of oxygen. It is formed and destroyed by ultraviolet radiation in the stratosphere, thus reducing the level of ultraviolet radiation that reaches Earth's surface.

Pacific-type margin   Leading edge of a continent that undergoes tectonic uplift as a result of lithospheric plate convergence.

Pack ice   Any area of sea ice other than fast ice. Less than 3 m (9.8 ft) thick, it covers the ocean sufficiently to make navigation possible only by icebreakers.

Paleoceanography   Branch of oceanography pertaining to the biological and physical character of ancient oceans.

Paleogeography   Branch of geography pertaining to the shapes and positions of ancient continents and oceans.

Paleomagnetism   The record of the past polarization of Earth's magnetic field. Needlelike iron oxide crystals frozen into Earth's crustal rocks indicate the polarity of the magnetic field at the time of the rocks' formation.

Pancake ice   Circular pieces of newly formed sea ice from 30 cm to 3 m (0.98 to 9.8 ft) in diameter that form in early fall in polar regions.

Pangaea   A hypothetical supercontinent of the geologic past that contained all the continental crust of Earth.

Panthalassa   A hypothetical proto-ocean surrounding Pangaea.

Paralytic shellfish poisoning (PSP)   A form of poisoning that results from humans ingesting shellfish contaminated with toxins secreted by dinoflagellate algae.

Parapodia   Flat protuberances on each side of most segments of polychaete worms. Most possess cirri and setae (bristlelike projections); may be modified for special functions such as feeding, locomotion, and respiration.

Parasite   An organism that takes its nutrients from the tissues of another organism and benefits at the host's expense.

Parasitism   A symbiotic relationship in which the parasite harms the host from which it takes its nutrition.

Partial tide   One of the harmonic components comprising the tide at any location. The periods of the partial tides are derived from the various combinations of the angular velocities of Earth, sun, and moon relative to one another.

Particulate Organic Carbon (POC)   The carbon contained in organic particles found in the ocean. It includes all particles larger than 0.2 μm in diameter.

Passive margin   The margin of a continent that is not significantly deformed by tectonic processes because it is the trailing edge of the continent. It does not directly collide with other lithospheric plates. The Atlantic coast of North America is an example.

PCBs (Polychlorinated biphenyls)   A group of industrial chemicals used in a variety of products; responsible for several episodes of ecological damage in coastal waters.

Pedicellariae   Minute stalked or unstalked pincerlike structures around the base of spines and dermal branchiae in certain echinoderms, especially Asteroidea, Echinoidea, and Ophiuroidea. They snap shut on debris and small organisms to keep the surface of the echinoderm clean.

Pelagic environment   The open-ocean environment, divided into the neritic province (water depth 0 to 200 m, or 0 to 656 ft) and the oceanic province (water depth greater than 200 m, or 656 ft).

Pelagic red crab   A species of pelagic crab characteristic of tropical waters of the eastern Pacific Ocean.

Pelecypoda   A class of mollusks characterized by two fairly symmetrical lateral valves with a dorsal hinge. These filter feeders pump water through the filter system and over gills through posterior siphons. Many possess a hatchet-shaped foot used for locomotion and burrowing. Includes clams, oysters, mussels, and scallops.

Pelladilla   One-year-old Peruvian anchoveta. They are of low value compared to older fish because they have a high water content and lower lipid content than older fish.

Peridotite   An ultramafic mantle rock composed primarily of olivine.

Perigee   The point on the orbit of an Earth satellite (moon) that is nearest Earth.

Perihelion   That point on the orbit of a planet or comet around the sun that is closest to the sun.

Period   *See* wave period (*T*).

Peritoneum   Thin membrane lining the coelom and covering all organs in the coelom.

Periwinkle snail   Snails of the genus, *Littorina,* characteristically found in the high-tide zone and spray zone.

Permeability   A condition that allows the passage of liquids through a substance.

Peru Current   The eastern boundary current of the South Pacific subtropical gyre.

Petroleum   A naturally occurring liquid hydrocarbon.

pH   The negative of the logarithm of the hydrogen ion concentration in an aqueous solution.

pH scale   A scale indicating the hydrogen ion content or acidity of a solution. Ranging from 1 to 13, a lower value indicates a higher hydrogen ion concentration.

Phaeophyta   Brown algae characterized by the carotenoid pigment fucoxanthin. Contains the largest marine algae.

Phase   A state of matter; solid, liquid, gas.

Phosphate   A compound containing the radical PO_4. It is an important component of nutrients required by algae for primary production.

Phosphorite   A sedimentary rock composed primarily of phosphate minerals.

Photic zone   The upper ocean in which the presence of solar radiation is detectable. It includes the euphotic and disphotic zones.

Photophore   One of several types of light-producing organs found primarily on fishes and squids inhabiting the mesopelagic and upper bathypelagic zones.

Photosynthesis   The process by which plants produce carbohydrate from carbon dioxide and water in the presence of chlorophyll, using light energy and releasing oxygen.

Phycoerythrin   A red pigment characteristic of the Rhodophyta (red algae).

Phytoplankton   Plant plankton. The most important community of primary producers in the ocean.

Picoplankton   Small plankton within the size range of 0.2 to 2.0 µm in size. Composed primarily of bacteria.

Pinniped   A suborder of marine mammals that includes the sea lions, seals, and walruses.

Placer deposit   Concentration of economically important minerals in a deposit that has been worked by ocean waves. Placers of gold, diamonds, and tin have been mined from coastal waters.

Plankton   Passively drifting or weakly swimming organisms that are dependent on currents. Includes mostly microscopic algae, protozoans, and larval forms of higher animals.

Plankton bloom   A very high concentration of phytoplankton, resulting from a rapid rate of reproduction as conditions become optimum during the spring in high-latitude areas. Less obvious causes produce blooms that may be destructive in other areas.

Plankton net   Plankton-extracting device that is cone-shaped and typically of synthetic material. It is towed through the water or lifted vertically to extract plankton down to a size of 50 µm (0.02 in).

Plastic   Any of variety of nonmetallic compounds synthesized by polymerization of organic compounds.

Pleistocene Epoch   The time in Earth history from 1.6 million years ago to 10,000 years ago during which pronounced glacial advances occurred.

Plume   Rising jets of molten mantle material that create hot spots when they penetrate the crust of Earth.

Plunging breaker   Impressive curling breakers that form on moderately sloping beaches.

Pneumatic duct   An opening into the swim bladder of some fishes that allows rapid release of air into the esophagus.

Pneumatophore   An air-containing float found on siphonophores.

Pogonophora   A phylum of entirely marine tube worms that have no gut and are found only in water deeper than 20 m (65.6 ft).

Poikilotherm   An organism whose body temperature varies with and is largely controlled by the temperature of its environment.

Polar   Pertaining to the polar regions.

Polar easterly winds   Cold air masses that move away from the polar regions toward lower latitudes.

Polar emergence   The emergence of low- and midlatitude temperature-sensitive deep-ocean benthos onto the shallow shelves of the polar regions where temperatures similar to that of their deep-ocean habitat exist.

Polar ice   The accumulation of sea ice in the Arctic polar region.

Pollution (marine)   The introduction of substances or energy into the marine environment that results in harm to the living resources of the ocean or humans that use these resources.

Polychaeta   Class of annelid worms that includes most of the marine segmented worms.

Polynya   A nonlinear opening in sea ice.

Polyp   A single individual of a colony or a solitary attached coelenterate.

Population   A group of individuals of one species living in an area.

Porifera   Phylum of sponges. Supporting structure composed of $CaCO_3$ or SiO_2 spicules or fibrous spongin. Water currents created by flagella-waving choanocytes enter tiny pores, pass through canals, and exit through a larger osculum.

Potential density   *See* Density, Potential.

Potential world fishery   The mass of fish that are supported annually by nutrients that originate outside of an ecosystem (world fishery ecosystems) and flow into it through upwelling or other processes. This amount of fish can be removed on an annual basis without degrading the ecosystem.

Precession   Regarding the moon's orbit around Earth, the axis of this orbit slowly changes its direction and describes a complete cone every 18.6 years. This is accompanied by a clockwise rotation of the plane of the moon's orbit that is completed in the same time interval.

Precipitation   In a meteorological sense, the discharge of water in the form of rain, snow, hail, or sleet from the atmosphere onto Earth's surface.

Primary coast   Coast that has been formed recently by nonmarine processes and little modified by marine processes.

Primary crystalline rock   Igneous rock.

Primary production   The amount of organic matter synthesized by organisms from inorganic substances within a given volume of water or habitat in a unit of time.

Prime meridian   The meridian of longitude 0 ° used as a reference for measuring longitude; the Greenwich Meridian.

Prokaryotic cells   Cells with no central nucleus. The first cells and present day bacteria and archaea are examples of prokaryotic organisms.

Producer   The autotrophic component of an ecosystem that produces the food that supports the biocommunity.

Progressive wave   A wave in which the waveform progressively moves.

Propagation   The transmission of energy through a medium.

Protein   A very complex organic compound made up of large numbers of amino acids. Proteins make up a large percentage of the dry weight of all living organisms.

Protista   A kingdom of organisms that includes all one-celled forms with nuclear material confined to a nuclear sheath. Includes the animal phylum Protozoa and the phyla of algal plants.

Protoearth   Earth early in its development. It may have had a diameter 1000 times greater and a mass 500 times greater than at present.

Proton   A fundamental particle of the nucleus of all atoms with a mass number of 1 and a unit positive electrical charge.

Protoplanet   The form taken by any planet early in its development.

Protoplasm   The complicated self-perpetuating living material making up all organisms. The elements carbon, hydrogen, and oxygen constitute more than 95%; water and dissolved salts make up from 50 to 97% of most plants and animals, with carbohydrates, lipids (fats), and proteins constituting the remainder.

Protozoa   Phylum of one-celled animals with nuclear material confined within a nuclear sheath.

Pseudopodia   Extensions of protoplasm in broad, flat, or long needlelike projections used for locomotion or feeding. Typical of amoeboid forms such as foraminifers and radiolarians.

Pteropoda   An order of pelagic gastropods in which the foot is modified for swimming and the shell may be present or absent.

Pteropod ooze   An oceanic deposit composed of more than 30% biogenous particles by weight. Pteropod shells are the dominant biogenous component.

Purse seine   A curtainlike net that can be used to encircle a school of fish. The bottom is then pulled tight much the way a purse string is used to close a baglike purse.

Pycnocline   A layer of water in which a high rate of change in density in the vertical dimension is present.

Pycnogonid   A spiderlike arthropod found on the ocean bottom at all depths. The more commonly observed nearshore varieties are usually less than 1 cm across, while deeper water varieties may reach spreads of over 1 m.

Pyrrophyta   A phylum of microscopic algae that possesses flagella for lo-comotion—the dinoflagellates.

Quadrature   The first and third quarter moon phases occur when the sun and moon are in quadrature, at right angles to one another relative to Earth.

Quarter moon   First and third quarter moon phases which occur when the sun and moon are in quadrature one week after the new moon and full moon phases, respectively.

Radiata   A grouping of phyla with primary radial symmetry—phyla Cnidaria and Ctenophora.

Radioactive waste   Radioactive materials that are no longer useful for the purpose they were designed. They must be stored safely away from contact with the biosphere.

Radioactivity   The spontaneous breakdown of the nucleus of an atom resulting in the emission of radiant energy in the form of particles or waves.

Radiolaria   An order of planktonic and benthic protozoans that possess protective coverings usually made of silica.

Radiolarian ooze   An oceanic deposit containing more than 30% by weight biogenous particles. The biogenous content is predominantly the remains of radiolarians.

Radiometric dating   *See* Absolute dating.

Radionuclide   Nucleus of a radioactive atom.

Radula   Filelike calcium carbonate rasp used by snails to scrape algae off surfaces, drill through shells, and rasp away the tissue of their prey. Plural *radulae*.

Ray   A cartilaginous fish in which the body is dorsoventrally flattened, eyes and spiracles are on the upper surface, and gill slits are on the bottom. The tail is reduced to a whiplike appendage. Includes electric rays, manta rays, and stingrays.

Recruitment (fishery)   The year-class (number of fish or mass of fish) of young adults added to a fishery following each spawning season.

Red muscle fiber   Fine muscle fibers rich in myoglobin that are abundant in cruiser-type fishes.

Red Sea   A sea at the northern margin of the Indian Ocean. It is between Africa and the Arabian peninsula.

Red Sea Water   A water mass that spills into the Arabian Sea from the Red Sea. It spreads as a thin layer at a depth of about 900 m (2950 ft) into the northern Indian Ocean.

Red tide   A reddish-brown discoloration of surface water, usually in coastal areas, caused by high concentrations of microscopic organisms, usually dinoflagellates. It probably results from increased availability of certain nutrients for various reasons. Toxins produced by the dinoflagellates may kill fish directly, or large populations of animal forms that spring up to feed on the plants, along with decaying plant and animal remains, may use up the oxygen in the surface water to cause asphyxiation of many animals.

Reef   A consolidated rock (a hazard to navigation) with a depth of 20 m (65.6 ft) or less.

Reef flat   A platform of coral fragments and sand that is relatively exposed at low tide.

Reef front   The upper seaward face of a reef from the reef edge (seaward margin of reef flat) to the depth at which living coral and coralline algae become rare (16 to 30 m, or 52 to 98 ft).

Reflection   The process in which a wave has part of its energy returned seaward by a reflecting surface.

Refraction   The process by which the part of a wave in shallow water is slowed down to cause the wave to bend and tend to align itself with the underwater contours.

Refractory organic matter   Organic matter that resists bacterial decomposition.

Regenerated production   The portion of gross primary production that is supported by nutrients recycled within an ecosystem.

Relative dating   The determination of whether certain rock units are older or younger than others by the use of fossil assemblages. It was not possible to tell the actual age of rocks by use of fossils until radiometric dating was developed.

Relict beach   A beach deposit laid down and submerged by a rise in sea level. It is still identifiable on the continental shelf, indicating no deposition is presently taking place at that location on the shelf.

Relict sediment   A sediment deposited under a set of environmental conditions that still remains unchanged although the environment has changed and it remains unburied by later sediment. An example is a beach deposited near the edge of the continental shelf when sea level was lower.

Remnant arc   An inactive volcanic arc that has been split away from an active arc by back-arc spreading.

Renewable energy   Energy sources that can not be depleted because they are renewed by radiant energy from the sun.

Reservoir   A container in which a substance is stored; e.g., in the hydrologic cycle, water is contained at least temporarily, in reservoirs such as streams, atmosphere, ground water, glaciers, lakes and the ocean.

Residence time   The average length of time a particle of any substance spends in the ocean. It is calculated by dividing the total amount of the substance in the ocean by the rate of its introduction into the ocean or the rate at which it leaves the ocean.

Resonance tide   A condition in which the natural free period of oscillation and the forced tidal period of oscillation are identical or multiples of one another and in phase. Thus constructive interference produces an increased tidal range.

Respiration   The process by which organisms use organic materials (food) as a source of energy. As the energy is released, oxygen is used and carbon dioxide and water are produced.

Restoring force   A force such as surface tension or gravity that tends to restore the ocean surface displaced by a wave to that of a still-water level.

Reverse osmosis   A method of desalinating ocean water that involves forcing water molecules through a water-permeable membrane under pressure.

Reversing current   The tide current as it occurs at the margins of landmasses. The water flows in and out for approximately equal periods of time separated by slack water where the water is still at high and low tidal extremes.

Rhodophyta   Phylum of algae composed primarily of small encrusting, branching, or filamentous plants that receive their characteristic red color from the presence of the pigment phycoerythrin. With a worldwide distribution, they are found at greater depths than other algae.

Rhyolite   A volcanic rock equivalent to granite in its mineral composition.

Ridge   *See* Drifts; oceanic ridge.

Right whale   A whale of the genus *Balaena* that was the favorite target of early whalers.

Rip current   A strong narrow surface or near-surface current of short duration (up to 2 h) and high speed (up to 4 km/h) flowing seaward through the breaker zone at nearly right angles to the shore. It represents the return to the ocean of water that has been piled up on the shore by incoming waves.

Ripple   Capillary wave 10–15 cm (4–6 in) long. Ripples are found on the sides of furrows cut into mud waves by current action.

Rise   A long, broad elevation that rises gently and rather smoothly from the deep-ocean floor.

Rock lice   Isopods belonging to the genus, *Ligia,* that are common in the supralittoral zone.

Rogue wave   A rare wave that may exceed a height of 30 m (100 ft) that forms as a result of constructive wave interference in seas where the average wave height is less than 5 m.

Rorqual whale   Any of several baleen whales with many ventral grooves; the minke, Baird's, Bryde's, sei, fin, blue, and humpback whales.

Rotary current   Tidal current as observed in the open ocean. The tidal crest makes one complete rotation during a tidal period.

Sabellid   A member of the annelid family Sabellidae that lives in a tube composed of shell fragments, sand, and agglutinous material. Featherlike gills and feeding structures filter food from the water above the tube opening.

Salinity   A measure of the quantity of dissolved solids in ocean water. Formally, it is the total amount of dissolved solids in ocean water in parts per thousand by weight after all carbonate has been converted to oxide, the bromide and iodide to chloride, and all the organic matter oxidized. It is normally computed from conductivity, refractive index, or chlorinity.

Salinometer   A conductance-measuring device that measures the salinity of ocean water to a precision of 0.003‰.

Salpa   Genus of pelagic tunicates that are cylindrical, transparent, and found in all oceans.

Salt   Any substance that yields ions other than hydrogen or hydroxyl. Salts are produced from acids by replacing the hydrogen with a metal.

Salt marsh   A relatively flat area of the shore where fine sediment is deposited and salt-tolerant grasses grow. One of the most biologically productive regions of Earth.

Salt wedge estuary   A very deep river mouth with a very large volume of freshwater flow beneath which a wedge of salt water from the ocean invades. The Mississippi River is an example.

San Andreas Fault   A transform fault that cuts across the state of California from the northern end of the Gulf of California to Pt. Arena north of San Francisco.

Sand   Particle size of 0.0625 to 2 mm. It pertains to particles that lie between silt and granules on the Wentworth scale of grain size.

Sardine   Any of the small fishes with a maximum length of 40 cm (16 in) that are common in the coastal upwellings of the major subtropical gyres.

Sargasso Sea   A region of convergence in the North Atlantic lying south and east of Bermuda where the water is a very clear deep blue in color and contains large quantities of floating *Sargassum.*

Sargassum   A brown alga characterized by a bushy form, substantial holdfast when attached, and a yellow-brown, green-yellow, or orange color. Two species, *S. fluitans* and *S. natans,* make up most of the macroscopic vegetation in the Sargasso Sea.

Scaphopoda   A class of mollusk commonly called *tusk shells.* The shell is an elongate cone open at both ends. The conical foot surrounded by threadlike tentacles extends from the larger end to aid the animal in burrowing.

Scarp   A linear, steep slope on the ocean floor separating gently sloping or flat surfaces.

Scavenger   An animal that feeds on dead organisms.

Schooling   Well-defined social organizations of fish, squid, and crustaceans that aid them in survival. Often the precise benefit to the schooling species eludes the human observer.

Scyphozoa   A class of cnidarians that includes the true jellyfish, in which the medusoid body form predominates and the polyp is reduced or absent.

Sea   1. A subdivision of an ocean. Two types of seas are identifiable and defined. They are the *mediterranean seas,* where a number of seas are grouped together collectively as one sea, and *marginal seas* that are connected individually to the ocean. 2. A portion of the ocean where waves are being generated by wind.

Sea anemone   A member of the class Anthozoa whose bright color, tentacles, and general appearance resemble a flower.

Sea arch   An opening through a headland caused by wave erosion. Usually develops as sea caves are extended from one or both sides of the headland.

Sea cave   A cavity at the base of a sea cliff; formed by wave erosion.

Sea cow   An aquatic, herbivorous mammal of the order Sirenia that includes the dugong and manatee.

Sea cucumber   A common name given to members of the echinoderm class Holotheuroidea.

Sea-floor spreading   A process producing the lithosphere when convective upwelling of magma along the oceanic ridges moves away at rates of from 1 to 10 cm (0.4 to 4 in) per year.

Sea hare   A shelless snail common in the low-tide zones of temperate rocky shores.

Sea ice   Any form of ice originating from the freezing of ocean water.

Seamount   An individual peak extending more than 1000 m (3280 ft) above the ocean floor.

Sea otter   A seagoing otter that has recovered from near extinction along the North Pacific coasts. It feeds primarily on abalone and sea urchins.

Sea roach   An isopod of the genus, *Ligia,* that is common in the rocky supralittoral zone.

Sea smoke   When very cold air moves over warmer water, the bottom air is heated and rises. As it rises it carries evaporated water into the colder upper air, where it condenses to produce the "smoke."

Sea snake   A reptile belonging to the family Hydrophiidae with venom similar to that of cobras. Found primarily in the coastal waters of the Indian Ocean and the western Pacific Ocean.

Sea state   A description of the ocean surface that includes the average height of the highest one-third of the waves observed in a wave train; referred to a numerical code.

Sea turtle   Any turtle of the reptilian order Testudinata; widely found in warm water.

Sea urchin   An echinoderm belonging to the class Echinoidea; possessing a fused test (external covering) and well-developed spines.

Seasonal thermocline   A thermocline that develops due to surface heating of the oceans in mid- to high latitudes. The base of the seasonal thermocline is usually above 200 m (656 ft).

Seawall   A wall built parallel to the shore to protect coastal property from the waves.

Secondary coast   Coast that has aged enough that its nonmarine origin has been destroyed or hidden by marine biological or physical processes.

Sediment   Particles of organic or inorganic origin that accumulate in loose form.

Sediment maturity   A condition in which the roundness and degree of sorting increase and clay content decreases within a sedimentary deposit.

Sedimentary rock   A rock resulting from the consolidation of loose sediment, or a rock resulting from chemical precipitation, i.e., sandstone and limestone.

Seiche   A standing wave of an enclosed or semienclosed body of water that may have a period ranging from a few minutes to a few hours, depending on the dimensions of the basin. The wave motion continues after the initiating force has ceased.

Seismic   Pertaining to an earthquake or earth vibration, including those that are artificially induced.

Seismic sea wave   *See* Tsunami.

Seismic surveying   The use of sound-generating techniques to identify features on or beneath the ocean floor.

Semidiurnal tide   Tide having two high and two low waters per tidal day with small inequalities between successive highs and successive lows. Tidal period is about 12 h and 25 min solar time. Semidaily tide.

Serpulid   A polychaete worm belonging to the family Serpulidae that builds a calcareous or leathery tube on a submerged surface.

Sessile   Permanently attached to the substrate and not free to move about.

Seta   Hairlike or needlelike projections on the exoskeleton of arthropods. Similar structures exist on some annelids.

Sewage   Human waste that is usually treated before disposal into the marine environment. This practice is essentially stopped now except for occasional accidental releases.

Shallow-water wave A wave on the surface of the water whose wavelength is at least 20 times water depth. The bottom affects the orbit of water particles, and speed (S) is determined by water depth.

$$S\ (m/s) = 3.1\ \sqrt{\text{water depth (m)}}$$

Shelf break The depth at which the gentle slope of the continental shelf steepens appreciably. It marks the boundary between the continental shelf and continental rise.

Shelf ice Thick shelves of glacial ice that push out into Antarctic seas from Antarctica. Large tabular icebergs calve at the edge of these vast shelves.

Shoal Shallow.

Shore Seaward of the coast, extends from highest level of wave action during storms to the low-water line.

Shoreline The line marking the intersection of water surface with the shore. Migrates up and down as the tide rises and falls.

Shoreline of emergence Shorelines that indicate a lowering of sea level by the presence of stranded beach deposits and marine terraces above it.

Shoreline of submergence Shorelines that indicate a rise in sea level by the presence of drowned beaches or submerged dune topography.

Sialic Pertaining to the composition of the granitic continental crust, which is rich in silica and aluminum.

Side-scan sonar A method of mapping the topography of the ocean floor along a strip up to 60 km (37 mi) wide using computers and sonar signals that are directed away from both sides of the survey ship.

Sigma (σ) Symbol for *in situ* density.

Sigma-T (σ_T) Symbol for in situ density.

Sigma theta (σ_Θ) Symbol for potential density.

Silica Silicon dioxide (SiO_2).

Silicate chrondrule Cosmic particle that is grouped according to two size ranges, 30 μm or 125 μm (0.012 or 0.05 in). Silicate chondrules are believed to form from the collision of asteroids.

Silicate tetrahedron The basic pattern of silicate structure—a silicon atom surrounded by four oxygen atoms that form a tetrahedron.

Siliceous A condition of containing abundant silica (SiO_2).

Sill A busmarine ridge partially separating bodies of water such as fjords and seas from one another or from the open ocean.

Silt A particle size of 1/128–1/16 mm. It is intermediate between sand and clay.

Simatic Pertaining to the portion of Earth's crust underlying the oceans and the sialic continents; rich in magnesium and iron and therefore more dense than the sialic continental crust.

Siphonophora An order of hydrozoan coelenterates that form pelagic colonies containing both polyps and medusae. Examples are *Physalia* and *Velella*.

Sirenia An order of vegetarian marine mammals that includes the dugong and manatee.

Slack water Occurs when a reversing tidal current changes direction at high or low water. Current speed is zero.

Slick A smooth patch on an otherwise rippled surface caused by a monomolecular film of organic material that reduces surface tension.

Slightly stratified estuary An estuary of moderate depth in which marine water invades beneath the freshwater runoff. The two water masses mix so that the bottom water is slightly saltier than the surface water at most places in the estuary.

Slope current Another name for the western boundary undercurrent; derived from the fact that these currents flow along the base of the continental slope.

Sofar Channel Sound fixing and ranging channel. This is a low velocity sound travel zone that coincides with the permanent thermocline in low and midlatitudes.

Solar humidification A process by which ocean water can be desalinated by evaporation and the condensation of the vapor on the cover of a con-

tainer. The condensate then runs into a separate container and is collected as fresh water.

Solar system The sun and the celestial bodies, asteroids, planets, and comets that orbit around it.

Solar tide The partial tide caused by the tide-producing forces of the sun.

Solar wind The motion of interplanetary plasma or ionized particles away from the sun.

Solid state A state of matter in which the substance has a fixed volume and shape. A crystalline state of matter.

Solstice The time during which the sun is directly over one of the tropics. In the Northern Hemisphere the summer solstice occurs on June 21 or 22 as the sun is over the Tropic of Cancer, and the winter solstice occurs on December 21 or 22 when the sun is over the Tropic of Capricorn.

Solute A substance dissolved in a solution. Salts are the solute in salt water.

Solution A state in which a solute is homogeneously mixed with a liquid solvent. Water is the solvent for the solution that is ocean water.

Solvent A liquid that has one or more solutes dissolved in it.

Somali Current This current flows north along the Somali coast of Africa during the southwest monsoon season.

Sonar An acronym for sound navigation and ranging. A method by which objects may be located in the ocean.

Sounding Measuring the depth of water beneath a ship.

South Equatorial Current The equatorial segment of the subtropical gyres in the three major ocean basins.

Southwest Monsoon A southwest wind that develops during the summer season. It blows off the Indian Ocean onto the Asian mainland.

Southwest Monsoon Current During the southwest monsoon season, this eastward-flowing current replaces the west-flowing North Equatorial Current in the Indian Ocean.

Species diversity The number or variety of species found in a subdivision of the marine environment.

Specific gravity The ratio of density of a given substance to that of pure water at 4°C and at atmospheric pressure.

Specific heat The quantity of heat required to raise the temperature of 1 g of a given substance 1°C. For water it is 1 cal.

Spermatophyta Seed-bearing plants.

Sperm whale The largest toothed whale. It has a worldwide distribution.

Spicule A minute needlelike calcareous or siliceous form found in sponges, radiolarians, chitons, and echinoderms that acts to support the tissue or provide a protective covering.

Spilling breaker A type of breaking wave that forms on a gently sloping beach which gradually extracts the energy from the wave to produce a turbulent mass of air and water that runs down the front slope of the wave.

Spit A small point, low tongue, or narrow embankment of land commonly consisting of sand deposited by longshore currents and having one end attached to the mainland and the other terminating in open water.

Sponge *See* Porifera.

Spray zone The shore zone lying between the high-tide shoreline and the coastline. It is covered by water only during storms.

Spreading center A divergent plate boundary.

Spreading rate The rate of divergence of plates at a spreading center.

Spring bloom The rapid increase in biological production that occurs in the spring season in temperate and polar regions.

Spring tide Tide of maximum range occurring every fortnight when the moon is new or full.

Stack An isolated mass of rock projecting from the ocean off the end of a headland from which it has been detached by wave erosion.

Standard laboratory bioassay Tests conducted in a laboratory setting to determine the percentage of dying caused within a population by a given concentration of foreign material.

Standard Seawater   Ampules of ocean water for which the chlorinity has been determined by the Institute of Oceanographic Services in Wormly, England. The ampules are sent to laboratories all over the world so equipment and reagents used to determine the salinity of ocean water samples can be calibrated by adjustment until they give the same chlorinity as is shown on the ampule label.

Standing stock (crop)   The biomass of a population present at any given time.

Standing wave   A wave, the form of which oscillates vertically without progressive movement. The region of maximum vertical motion is an *antinode*. Between antinodes are *nodes* where there is no vertical motion but maximum horizontal motion.

State of matter   Solid, liquid, or gas.

Steepness   *See* Wave steepness.

Stenohaline   Pertaining to organisms that can withstand only a small range of salinity change.

Stenothermal   Pertaining to organisms that can withstand only a small range of temperature change.

Storm surge   A rise above normal water level resulting from wind stress and reduced atmospheric pressure during storms. Consequences can be more severe if it occurs in association with high tide.

Strait of Gibraltar   The narrow opening between Europe and Africa through which the waters of the Atlantic Ocean and Mediterranean Sea mix.

Stranded beach   An ancient beach deposit found above present sea level because of a lowering of sea level.

Streamlining   The shaping of an object so it produces the minimum of turbulence while moving through a fluid medium. The teardrop shape displays a high degree of streamlining.

Stromatolite   A calcium carbonate sedimentary structure in which algal assemblages trap sediment and bind it into forms that are often dome-shaped. They are known to form only in shallow-water environments.

Subaerial   Beneath the atmosphere.

Subduction   A process by which one lithospheric plate descends beneath another. The surface expression of such a process may be an island arc-trench system or a folded mountain range.

Subduction zone seep biocommunity   Seeps of pore water squeezed out of sediments in subduction zone support a distinctive fauna that depend on sulfur-oxidizing carbon fixing bacteria.

Subduction zone seep   Seeps of pore water squeezed out of sediments in subduction zones. A distinctive fauna that depends on sulfur-oxidizing carbon fixing bacteria lives in association with the seeps.

Sublimation   The transformation of the solid state of a substance to a vapor without going through the liquid phase, and vice versa.

Sublittoral zone   That portion of the benthic environment extending from low tide to a depth of 200 m (656 ft). Some consider it to be the surface of the continental shelf.

Submarine canyon   A steep V-shaped canyon cut into the continental shelf or slope.

Submerged dune topography   Ancient coastal dune deposits found submerged beneath the present shoreline because of a rise in sea level.

Submergent shoreline   Shoreline formed by the relative submergence of a landmass in which the shoreline is on landforms developed under subaerial processes. It is characterized by bays and promontories and is more irregular than a shoreline of emergence.

Subneritic province   The benthic environment extending from the shoreline across the continental shelf to the shelf break. It underlies the neritic province of the pelagic environment.

Suboceanic province   Benthic environments seaward of the continental shelf.

Subpolar   Pertaining to the oceanic region that is covered by sea ice in winter. The ice melts away in summer.

Subpolar gyre   Oceanic gyres of ocean water centered at about 60° latitude. They rotate counterclockwise in the Northern Hemisphere and clockwise in the Southern Hemisphere.

Substrate   The base on which an organism lives and grows.

Subsurface chlorophyll maximum (SCM)   The concentration of chlorophyll in many places in the ocean is highest immediately below the oxygen compensation depth.

Subsurface current   A current usually flowing below the pycnocline, generally at slower speed and in a different direction from the surface current.

Subsurface oxygen maximum (SOM)   Generally observed within the lower euphotic zone, the SOM may represent supersaturations of oxygen of up to 120% as a result of photosynthetically produced oxygen.

Subtropical   Pertaining to the oceanic region poleward of the tropics (about 30° latitude).

Subtropical convergence   The zone of convergence that occurs within all subtropical gyres as a result of Ekman transport driving water toward the interior of the gyres.

Subtropical gyre   The trade winds and westerly winds initiated in the subtropical regions of all ocean basins, with the influence of the Coriolis effect, set large regions of ocean water in motion. They rotate clockwise in the Northern Hemisphere and counterclockwise in the Southern Hemisphere, and they are centered in the subtropics.

Subtropical Underwater   A high-salinity water mass centered at a depth of about 50 m (164 ft) in the southeastern Caribbean Sea. It slopes to a depth of 200 m (650 ft) as it flows north through the Yucatán Strait.

Sulfur   A yellow mineral composed of the element sulfur. It is commonly found in association with hydrocarbons and salt deposits.

Sulfur-oxidizing bacterium   Any of the bacteria that support many deep-sea hydrothermal vent and cold water seep biocommunities by using energy released by oxidation to synthesize organic matter chemosynthetically.

Summer solstice   In the Northern Hemisphere, it is the instant when the sun moves north to the Tropic of Cancer before changing direction and moving southward toward the equator, June 21.

Superswell   This enigmatic feature is located in the French Polynesian region of the Pacific Ocean. It is 1000 m (3280 ft) too shallow for the age of the ocean floor (60–80 million years). The lithosphere is 50 km too thin and releases 30% of all hot spot heat release from 3% of Earth's surface area.

Supralittoral zone   The splash or spray zone above the spring high-tide shoreline.

Surface tension   The tendency for the surface of a liquid to contract owing to intermolecular bond attraction.

Surf zone   The region between the shoreline and the line of breakers where most wave energy is released.

Suspended cage   A form of mariculture where caged species such as salmon are suspended into a natural coastal environment to feed.

Sverdrup (sv)   A unit of volume transport equal to 1 million m^3/s.

Swash   A thin layer of water that washes up over exposed beach as waves break at the shore.

Swell   A free ocean wave by which energy put into ocean waves by wind in the sea is transported with little energy loss across great stretches of ocean to the margins of continents where the energy is released in the surf zone.

Swim bladder   A gas-containing, flexible, cigar-shaped organ that aids many fishes in attaining neutral buoyancy.

Symbiosis   A relationship between two species in which one or both benefit or neither or one is harmed. Examples are commensalism, mutualism, and parasitism.

Tablemount   A flat-topped seamount; a guyot.

Tectonic estuary   An estuary, the origin of which is related to tectonic deformation of the coastal region.

Tectonics   Deformation of Earth's surface by forces generated by heat flow from Earth's interior.

Temperate   Pertaining to the oceanic region where pronounced seasonal change occurs (about 40° to 60° latitude).

Temperature   A direct measure of the average kinetic energy of the molecules of a substance.

Temperature gradient   The rate of temperature change within the water column. It is steep within the thermocline, where water temperature changes rapidly with changing depth, but less steep in the deep water, where distribution patterns are more uniform.

Temperature of maximum density   The temperature at which a substance reaches its highest density. For water, it is 4°C.

Territorial sea   A 12-naut-mi-wide strip of ocean adjacent to land over which the coastal nation has control over the passage of ships.

Tethered-float breakwater   Floating hollow spheres tethered to the bottom and acting as an upside-down pendulum to extract energy from waves near the shore without interfering with the longshore drift of sediment.

Tethys Sea   An ancient body of water that separated Laurasia to the north and Gondwanaland to the south. Its location was approximately that of the present Alpine-Himalayan mountain system.

Thermocline   A layer of water beneath the mixed layer in which a rapid change in temperature can be measured in the vertical dimension.

Thermohaline circulation   The vertical movement of ocean water driven by density differences resulting from the combined effects of variations in temperature and salinity.

Tidal bore   A steep-fronted wave that moves up some rivers when the tide rises in the coastal ocean.

Tidal day   *See* Lunar day.

Tidal period   Elapsed time between successive high or low waters.

Tidal range   The difference in height between consecutive high and low waters. The time frame of comparison may also be a day, month, or year.

Tide   Periodic rise and fall of the ocean surface and connected bodies of water resulting from the unequal gravitational attraction of the moon and sun on different parts of Earth.

Tide datum   The tide level from which the heights of high and low tidal extremes are measured. It is usually mean low tide (MLT) or mean lower low tide (MLLT).

Tide generating force   The magnitude of the centripetal force required to keep all particles of Earth with identical mass moving in identical circular paths required by the movements of the Earth–moon system is identical. This force is provided by the gravitational attraction between the particles and the moon. This gravitational force is identical to the required centripetal force only at the center of Earth. For ocean tides, the horizontal component of the small force that results from the difference between the required and provided forces is the tide-generating force on that individual particle. These forces are such that they tend to push the ocean water into bulges toward the tide-generating body on one side of Earth and away from the tide-generating body on the opposite side of Earth.

Tide wave   The long-period gravity wave generated by tide-generating forces and manifested in the rise and fall of the tide.

Tin   The bluish white native metallic element, Sn.

Tintinnid   A ciliate protozoan of the family Tintinnidae with a tubular to vase-shaped outer shell.

Tissue   An aggregate of cells and their products developed by organisms for the performance of a particular function.

Tombolo   A sand or gravel bar that connects an island with another island or the mainland.

Topography   The configuration of a surface. In oceanography, it refers to the ocean bottom or the surface of a mass of water.

Total allowable catch   The permissible annual catch of a given species of fish that will not degrade the fishery given the knowledge of the ecosystem from which the fishery is being removed.

Trade winds   The air masses moving from subtropical high pressure belts toward the equator. They are northeasterly in the Northern Hemisphere and southeasterly in the Southern Hemisphere.

Transform fault   A fault characteristic of oceanic ridges along which they are offset.

Transform plate boundary   The shear boundary between two lithospheric plates formed by a transform fault. Also called *shear boundary*.

Transitional crust   Thinned section of continental crust at the trailing edge of a continent created by the breaking apart of an ancient continent over a newly formed spreading center.

Transitional wave   A wave moving from deep water to shallow water that has a wavelength more than twice the water depth but less than 20 times the water depth. Particle orbits are beginning to be influenced by the bottom.

Transverse ridge   A ridge within a fracture zone that contains large concentrations of mantle rock, periodotite that has been tectonically squeezed up.

Transverse wave   A wave in which particle motion is at right angles to energy propagation.

Trawl   A sturdy bag or net that can be dragged along the ocean bottom or at various depths above the bottom, to catch fish.

Trench   A long, narrow, and deep depression on the ocean floor with relatively steep sides.

Triple point   The one combination of temperature and pressure at which a substance can exist simultaneously as a solid, liquid and gas. For water, it is 0.006 atmospheres and 0.0098°C.

Troodos Massif   An ophiolite ore deposit located in western Cyprus.

Trophic level   A nourishment level in a food chain. Algal producers constitute the lowest level, followed by herbivores and a series of carnivores at the higher levels.

Tropical   Pertaining to the regions of the tropics (about 23.5° latitude).

Tropical tide   A tide occurring twice monthly when the moon is at its maximum declination north and south of the equator. It is in the tropical regions where tides display their greatest diurnal inequalities.

Tropic of Cancer   The latitude of 23.5° north.

Tropic of Capricorn   The latitude of 23.5° south.

Trough   The part of an ocean wave that is displaced below the still-water line.

Tsunami   Seismic sea wave. A long-period gravity wave generated by a submarine earthquake or volcanic event. Not noticeable on the open ocean but builds up to great heights in shallow water.

Tube worm   *See* Sabellidae and Serpulidae.

Tunicate   A member of the chordate subphylum Urochordata, which includes sacklike animals. Some are sessile (sea squirts) while others are pelagic (salps).

Turbidite   A sediment or rock formed from sediment deposited by turbidity currents characterized by both horizontally and vertically graded bedding.

Turbidity   A state of reduced clarity in a fluid caused by the presence of suspended matter.

Turbidity current   A gravity current resulting from a density increase brought about by increased water turbidity. Possibly initiated by some sudden force such as an earthquake, the turbid mass continues under the force of gravity down a submarine slope.

Turbulence   A disorderly flow of ocean water.

Turbulent flow   Flow in which the flow lines are confused owing to random velocity fluctuations.

Typhoon   A severe tropical storm in the western Pacific.

Ultraplankton   Plankton smaller than 5 μm (0.002 in). Very difficult to separate from the water.

Ultrasonic   Pertaining to sound frequencies above human range (above 20,000 cycles per second).

Ultraviolet radiation   Electromagnetic ratiation shorter than visible radiation and longer than X rays. The approximate range is from 1 to 400 nanometers (nm).

Upper water mass   That portion of the water column from the ocean surface to the base of the permanent thermocline or pycnocline. Surface temperatures are relatively high and uniform to the base of the mixed layer (100 to 200 m, or 328 to 656 ft) and decrease rapidly to the base of the thermocline (about 1000 m, or 3280 ft).

Upwelling   The process by which deep, cold, nutrient-laden water is brought to the surface, usually by the wind divergence of equatorial currents or coastal winds pushing water away from the coast.

Valence   The combining capacity of an element measured by the number of hydrogen atoms with which it will combine.

Van der Waals force   Weak attractive force between molecules; a result of the interaction between the nuclear particles of one molecule and the electrons of another.

Vapor pressure   A measure of the tendency for molecules of a liquid to escape into the gaseous phase. It increases with increased temperature.

Vector   A physical quantity that has magnitude and direction. Examples are force, acceleration, and velocity.

Veliger   A planktonic larval stage of many gastropods with two ciliated lobes.

Ventral   Pertaining to the lower or under surface.

Vernal equinox   The passage of the sun across the equator as it moves from the Southern Hemisphere into the Northern Hemisphere, March 21.

Vertebrata   Subphylum of chordates that includes those animals with a well-developed brain and a skeleton of bone or cartilage; includes fish, amphibians, reptiles, birds, and animals.

Vertically mixed estuary   Very shallow estuaries such as lagoons in which fresh water and marine water are totally mixed from top to bottom so that the salinity at the surface and the bottom is the same at most places within the estuary.

Viscosity   The property of a substance to offer resistance to flow; internal friction.

Viviparous   Pertaining to an animal that gives birth to living young.

Volcanogenic particles   Sediment particles produced by volcanic eruptions.

Vadati-Benioff seismic zone   A plane descending from beneath oceanic trenches, at an angle of   approximately 45°, in which earthquake foci are located. It is thought to represent the core of subducting lithosphere.

Walker Circulation   The pattern of atmospheric circulation that involves the rising of warm air over the East Indies low pressure cell and its descent over the high pressure cell in the southeastern Pacific Ocean off the coast of Chile. It is the weakening of this circulation that accompanies an El Niño event which has led to the development of the term El Niño–Southern Oscillation event.

Warm boundary current   An intense current flowing along the western margin or boundary of a subtropical gyre. It flows away from the equator. The Gulf Stream is an example.

Warm current   A current carrying warm water into areas of colder water. It generally moves poleward.

Warm front   A weather front in which a warm air mass moves into and over a cold air mass producing a broad band of gentle precipitation.

Warm-water vent   An opening in Earth's crust along the axes of spreading centers where water with temperatures ranging between 10°C and 20°C (50°F and 68°F) is vented.

Water (H_2O)   The oxide of hydrogen that makes life possible on the Earth.

Water mass   A body of water identifiable by its temperature, salinity, or chemical content.

Water vapor   The gaseous state of H_2O. At sea level it forms at the boiling point of 100°C (212°F).

Wave   A disturbance that moves over or through a medium with a speed determined by the properties of the medium.

Wave-cut bench   A gently sloping surface produced by wave erosion and extending from the base of the wave-cut cliff out under the offshore region.

Wave-cut cliff   A cliff produced by wave erosion cutting landward.

Wave dispersion   The separation of waves as they leave the sea area by wave size. Larger waves travel faster than smaller waves and thus leave the sea area first to be followed by progressively smaller waves.

Wave frequency (f)   The number of waves that pass a fixed point in a unit of time, usually one second.

Wave height (H)   Vertical distance between a crest and the preceding trough.

Wavelength (L)   Horizontal distance between two corresponding points on successive waves, such as from crest to crest.

Wave period (T)   The elapsed time between the passage of two successive wave crests past a fixed point.

Wave steepness   Ratio of wave height to wavelength.

Wave train   A series of waves from the same direction.

Weathering   A process by which rocks are broken down by chemical and mechanical means.

Wentworth scale   The scale of sediment particle size developed by C. K. Wentworth in 1922.

West Australian Current   This cold current forms the eastern boundary current of the Indian Ocean subtropical gyre. It is separated from the coast by the warm Leeuwin Current except during El Niño-Southern Oscillation events when the Leeuwin Current weakens.

Westerlies   The air masses moving away from the subtropical high pressure belts toward higher latitudes. They are southwesterly in the Northern Hemisphere and northwesterly in the Southern Hemisphere.

Western boundary current   Poleward-flowing warm currents on the western side of all subtropical gyres.

Western boundary undercurrent (WBUC)   A bottom current that flows along the base of the continental slope eroding sediment from it and redepositing the sediment on the continental rise. It is confined to the western boundary of deep-ocean basins.

Westward intensification   Pertaining to the intensification of the warm western limb of the subtropical gyre currents that is manifested in higher velocity and deeper flow compared to the cold eastern boundary currents that drift leisurely toward the equator.

West Wind Drift   The surface portion of the Antarctic Circumpolar Current driven in an easterly direction around Antarctica by the strong westerly winds.

Wetlands   Biologically productive regions bordering estuaries and other protected coastal areas. They are usually salt marshes at latitudes greater than 30° and mangrove swamps at lower latitudes.

Whitecap   A wind-generated wave that breaks in the open ocean.

White muscle fiber   Thick muscle fibers with relatively low concentrations of myoglobin that make up a large percentage of the muscle fiber in lunger-type fishes.

White smoker   A hot-water vent found along the axes of oceanic spreading centers that emits water with temperatures ranging between 30°C and 330°C (86°F and 364°F).

Wind-driven circulation   Any movement of ocean water that is driven by winds. This includes most horizontal movements in the surface waters of the world's oceans.

Windrow   Row of floating debris aligned parallel to the direction of the wind that results from Langmuir circulation.

Windward   The direction from which the wind is blowing.

Winter solstice   The instant the southward-moving sun reaches the Tropic of Cancer before changing direction and moving north back toward the equator (December 22).

Zenith   That point on the celestial sphere directly over the observer.

Zeolite   A group of aluminosilicate minerals that easily lose or gain water of hydration and have great capacity for ion exchange.

Zooplankton   Animal plankton.

Zooxanthellae   A form of algae that lives as a symbiont in the tissue of corals and other coral reef animals and provides varying amounts of their required food supply.

Index

The Phylogenetic Tree of Animals

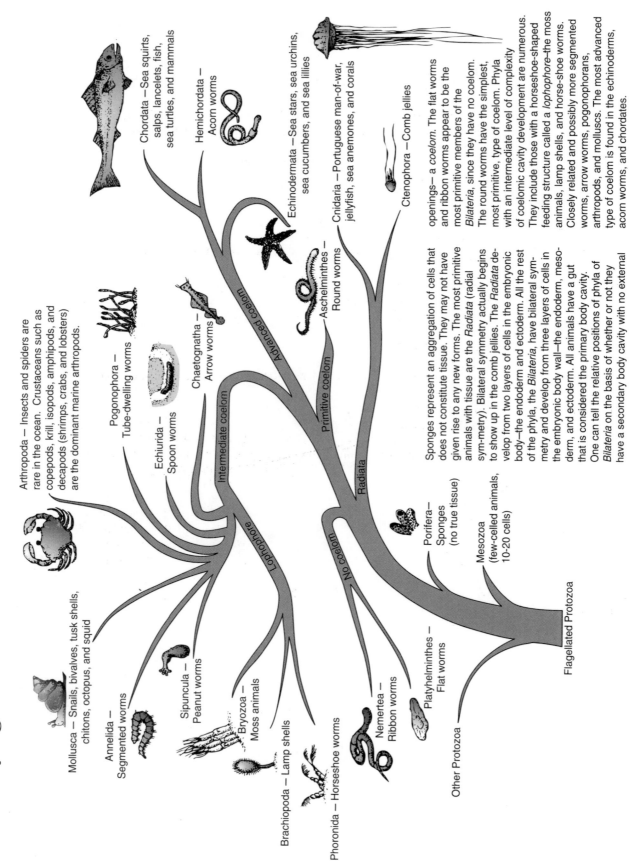

Arthropoda — Insects and spiders are rare in the ocean. Crustaceans such as copepods, krill, isopods, amphipods, and decapods (shrimps, crabs, and lobsters) are the dominant marine arthropods.

Chordata — Sea squirts, salps, lancelets, fish, sea turtles, and mammals

Hemichordata — Acorn worms

Pogonophora — Tube-dwelling worms

Echinodermata — Sea stars, sea urchins, sea cucumbers, and sea lillies

Echiurida — Spoon worms

Chaetognatha — Arrow worms

Cnidaria — Portuguese man-of-war, jellyfish, sea anemones, and corals

Aschelminthes — Round worms

Ctenophora — Comb jellies

Mollusca — Snails, bivalves, tusk shells, chitons, octopus, and squid

Annelida — Segmented worms

Sipuncula — Peanut worms

Bryozoa — Moss animals

Brachiopoda — Lamp shells

Phoronida — Horseshoe worms

Nemertea — Ribbon worms

Platyhelminthes — Flat worms

Other Protozoa

Porifera — Sponges (no true tissue)

Mesozoa (few-celled animals, 10-20 cells)

Flagellated Protozoa

Lophophore

Intermediate coelom

Advanced coelom

Primitive coelom

No coelom

Radiata

Sponges represent an aggregation of cells that does not constitute tissue. They may not have given rise to any new forms. The most primitive animals with tissue are the *Radiata* (radial sym-metry). Bilateral symmetry actually begins to show up in the comb jellies. The *Radiata* de-velop from two layers of cells in the embryonic body—the endoderm and ectoderm. All the rest of the phyla, the *Bilateria*, have bilateral sym-metry and develop from three layers of cells in the embryonic body wall—the endoderm, meso-derm, and ectoderm. All animals have a gut that is considered the primary body cavity. One can tell the relative positions of phyla of *Bilateria* on the basis of whether or not they have a secondary body cavity with no external openings— a *coelom*. The flat worms and ribbon worms appear to be the most primitive members of the *Bilateria*, since they have no coelom. The round worms have the simplest, most primitive, type of coelom. Phyla with an intermediate level of complexity of coelomic cavity development are numerous. They include those with a horseshoe-shaped feeding structure called a *lophophore*—the moss animals, lamp shells, and horse-shoe worms. Closely related and possibly more segmented worms, arrow worms, pogonophorans, arthropods, and molluscs. The most advanced type of coelom is found in the echinoderms, acorn worms, and chordates.